McGraw-Hill's

NATIONAL ELECTRICAL SAFETY CODE® (NESC®) 2017 HANDBOOK

About the Author

David J. Marne, P.E., B.S.E.E., is a registered professional electrical engineer. Mr. Marne is a nationally recognized speaker on the National Electrical Safety Code® (NESC®) and he serves on various NESC® Subcommittees. He is company president and senior electrical engineer for Marne and Associates, Inc. in Missoula, Montana, where he specializes in NESC® training and expert witness services. Mr. Marne has over 34 years of experience in the utility industry and is a senior member of the IEEE.

On a personal note, Dave was born and raised in Mount Pleasant, Pennsylvania. He graduated from Montana State University in Bozeman, Montana. His first engineering job was in Vallejo, CA and he now lives with his family in Missoula, MT. Dave is of Italian descent. His grandfather's name was Marmo but the family name was "Americanized" to Marne.

McGraw-Hill's

NATIONAL ELECTRICAL SAFETY CODE® (NESC®) 2017 HANDBOOK

Based on the Current 2017
National Electrical Safety Code® (NESC®)

David J. Marne

National Electrical Safety Code® and NESC®
are both registered trademarks and service marks
of the Institute of Electrical and Electronics Engineers, Inc. (IEEE)

Mc
Graw
Hill
Education

New York   Chicago   San Francisco   Athens   London
Madrid   Mexico City   Milan   New Delhi
Singapore   Sydney   Toronto

McGraw-Hill books are available at special quantity discounts to use as premiums and sales promotions, or for use in corporate training programs. To contact a representative please e-mail us at bulksales@mcgraw-hill.com.

McGraw-Hill's National Electrical Safety Code® (NESC®) 2017 Handbook

3 4 5 6 7 8 9 LCR 21 20 19 18

ISBN 978-1-25-958415-2
MHID 1-25-958415-1

The pages within this book were printed on acid-free paper.

Sponsoring Editor
Michael McCabe

Editorial Supervisor
Donna M. Martone

Acquisitions Coordinator
Lauren Rogers

Project Manager
Anju Joshi,
Cenveo® Publisher Services

Proofreader
Chris Anderson, Irina Burns

Indexer
Cenveo Publisher Services

Production Supervisor
Pamela A. Pelton

Composition
Cenveo Publisher Services

Art Director, Cover
Jeff Weeks

Contents

Part 3.   Safety Rules for the Installation and
          Maintenance of Underground Electric
          Supply and Communication Lines

Part 4.   Work Rules for the Operation of Electric Supply and Communications Lines and Equipment

Preface

McGraw-Hill's National Electrical Safety Code® (NESC®) 2017 Handbook is written for the engineer, staking technician, power lineman, communications lineman, inspector, and safety administrator of an electric power or communication utility company, contracting company, or consulting firm. Employees involved with substation, transmission, and distribution, design, construction, operation, and maintenance should have an understanding of the **NESC** to keep the public, utility workers, and themselves safe.

McGraw-Hill's NESC 2017 Handbook is designed to be used in conjunction with the **National Electrical Safety Code (NESC)** published by the Institute of Electrical and Electronic Engineers (IEEE). This *Handbook* presents hundreds of figures and photos, plus examples and discussions to explain and clarify the **NESC** rules. The straightforward and practical information in this *Handbook* is intended to aid the understanding of the sometimes confusing and complicated text in the **NESC**. This *Handbook* follows the order of parts, sections, and rules as presented in the **NESC** including the general rules, substation rules (Part 1), overhead line rules (Part 2), underground line rules (Part 3), and work rules (Part 4). This format ensures a quick and easy correlation between the **NESC** rules and the discussions and explanations in this *Handbook*. This *Handbook* is most effectively used by referring to the **NESC** for the precise wording of a rule and then referring to the corresponding rule number in this *Handbook* for a practical understanding of the rule. Appendix A of this *Handbook* includes numerous photos of **NESC** applications and violations. Appendix B of this *Handbook* contains Occupational Safety and Health Administration (OSHA) standards related to the **NESC** work rules.

The National Electrical Safety Code (NESC) is the "bible" for developing power and communication utility standards. All utility construction has some relevance to the NESC including items such as line design, substation design, standard drawings, material purchases, pole placement, work practices, signage, and others. The NESC should not be confused with the National Electrical Code (NEC) published by the National Fire Protection Agency (NFPA). The NEC is used for residential, commercial, and industrial building wiring. It does not apply to utility systems although there is some overlap of the NESC and the NEC at a building service. The NESC is written as a voluntary standard. It can be adopted as law by individual states or other governmental authorities. To determine the legal status of the NESC, the state public service commission or public utility commission, or other governmental authority should be contacted. The Code is written by various NESC committees. The organizations represented, subcommittees, and committee members are listed in the front of the Code book. The procedure and time schedule for revising the NESC are also described in the front of the Code book. The NESC has an interpretation committee that issues formal interpretations. The process for obtaining a formal interpretation is outlined in the front of the Code book. The NESC is currently published on a 5-year cycle. Urgent safety matters that require a change in between Code editions are handled through a Tentative Interim Amendment (TIA) process. Original work on the NESC began in 1913. This *Handbook* is based on the 2017 edition of the NESC.

McGraw-Hill's NESC Handbook is the only NESC *Handbook* that focuses on the practical application of the current edition of the NESC. The numerous figures, photos, and examples make *McGraw-Hill's NESC 2017 Handbook* the most complete and useful handbook available.

Additional reference material for the interested reader includes, *The Lineman's and Cableman's Handbook* by Thomas M. Shoemaker and James E. Mack, and the *Standard Handbook for Electrical Engineers*, by Donald G. Fink and H. Wayne Beaty, both published by McGraw-Hill. The transmission line, distribution line, and substation standards published by the Rural Utility Service (RUS), formerly known as the Rural Electric Administration (REA), are also invaluable references.

This *Handbook* should be used as a reference for understanding and applying the rules in the NESC. This *Handbook* is not an official Code document and does not contain official NESC committee interpretations. The information, figures, and photos in this *Handbook* are intended to be used as aids to the reader of the Code and are not intended to be a replacement for the comprehensive nature of the Code as it is written.

David J. Marne, P.E.

Acknowledgments

I wish to express my sincere appreciation to the many individuals who helped me bring the 2017 edition of this *Handbook* to print.

The following peer reviewers provided a technical review of my manuscript: Nelson Bingel, Grant Glaus, Mickey Gunter, Keith Reese, Lawrence Slavin, and Greg Wolven. The following people helped bring the text, figures, and photographs to life: Maria Melvin and Larry Coles. The following people at McGraw-Hill and their affiliates supported and guided my efforts: Michael McCabe, Anju Joshi, and their hardworking staff.

Finally, I wish to thank my wife, Patty, my daughter, Kirstin, and my son, David, for their constant help and support.

David J. Marne, P.E.

General Sections

GENERAL SECTIONS

01 INTRODUCTION   02 DEFINITIONS

03 REFERENCES   09 GROUNDING METHODS

GENERAL SECTIONS 01, 02, 03, 09

PART 1

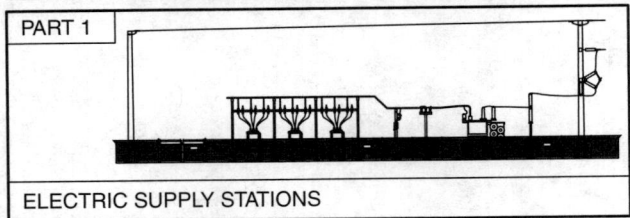

ELECTRIC SUPPLY STATIONS

PART 2

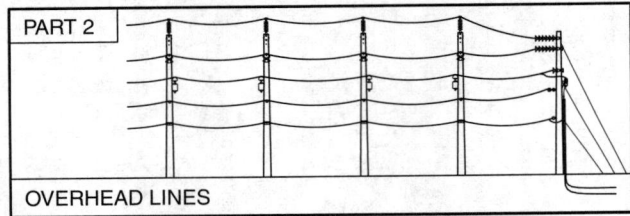

OVERHEAD LINES

PART 3

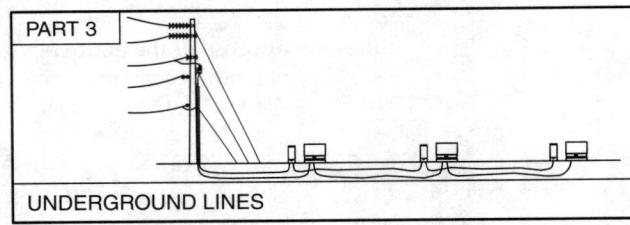

UNDERGROUND LINES

PART 4

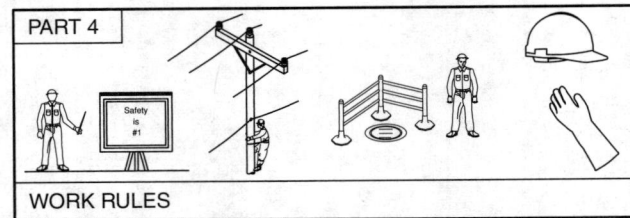

WORK RULES

Introduction to the National Electrical Safety Code

010. PURPOSE

The very first rule in the **NESC** outlines the purpose of the entire book. The rules contained in the **NESC** are provided for the "practical safeguarding of persons and utility facilities during the installation, operation, and maintenance of electric supply and communication facilities, under specified conditions." The persons the **Code** is referring to are both the public and utility workers (employees and contractors). The utility facilities the **Code** is referring to are electric supply stations (covered in Part 1), overhead supply and communication lines (covered in Part 2), and underground supply and communication lines (covered in Part 3). Utility facilities also include electric supply and communication equipment connected to utility facilities, for example, a pole-top transformer connected to an overhead supply line, a pad-mounted transformer connected to an underground supply line, or a communications amplifier connected to an overhead or underground communications line.

The **Code** uses the term electric supply for electric power. The **Code** also covers communications lines and equipment. Communications utilities include, but are not limited to, telephone, cable TV, and fiber utilities.

The **NESC** is a standard published by the Institute of Electrical and Electronic Engineers (IEEE). It is a recognized standard for utility company safety. It is adopted by utilities or by an authority having jurisdiction over utilities (i.e., a state public service commission or public utility commission) or by some other authority. To determine the specific legal status of the **NESC**, the authority having jurisdiction must be contacted. In general, utilities in 49 of the 50 United States use the NESC. The exception is the State of California which writes its own codes titled, General Order 95 (GO95) (Overhead Lines), General Order 128 (GO128) (Underground Lines), and General Order 165 (GO165) (Inspection).

The **Code** contains the basic provisions necessary for safety under specified conditions. The **Code** is not intended to be a design specification or instruction manual. The **Code** specifies what needs to be accomplished for safety, not how to accomplish it. Values for clearance and structure strength must be not less than the values indicated in the **Code** for safety purposes. Clearance and strength values can certainly be greater than the **Code** specifies, but greater values are not required for safety.

For example, if the **Code** clearance of a 12.47/7.2-kV phase conductor is 18.5 ft over the ground, the **Code** does not specify how high the poles have to be or where the wires are attached on the pole to obtain the 18.5-ft clearance. Selecting the proper pole heights and attachment heights is a design function. The **Code** does require that the phase wires be not less than 18.5 ft high (they can certainly be higher). Some utilities establish clearance values by using the **Code** clearance plus an adder. The adder can be thought of as a design or construction tolerance adder to maintain the required **Code** clearance. A line designed for exactly 18.5 ft of clearance over the ground could end up having problems meeting **Code**. There are several factors that could jeopardize the 18.5-ft clearance. One factor could be that the pole hole was dug 6 in deeper than it should have been dug. Another factor might be a slight rise in elevation at midspan that was not detected due to the fact that the ground line was not profiled or surveyed. Other factors that can jeopardize clearance include wire stringing issues, leaning poles, leaning cross-arms, etc. Using a clearance adder (e.g., 3 ft) would require designing the line to 21.5 ft instead of 18.5 ft. The clearance adder can also help meet and maintain the required **Code** clearance over time. See Rule 230I for an additional discussion of maintaining clearances over the life of an installation.

There are many design standards available for utility companies to reference. The Rural Utilities Service (RUS) publishes transmission, distribution, and sub-station design manuals. Many larger utilities develop their own design manuals. Each manual written must be in compliance with the **Code**, as the **Code** is the "bible" for all utility work.

The **Code** states that rules apply to specified conditions. One example of the specified conditions is overhead line clearance for open supply conductors, 750 V–22 kV, over a road subject to truck traffic (**NESC** Table 232-1, Row 2). In **NESC** Appendix A (**NESC** Fig. A-1), the **Code** shows the reference component (of a truck) to be 14 ft. The mechanical and electrical clearance component for a conductor shown in this figure is 4.5 ft, for a total of 18.5 ft of clearance when the conductor is at the maximum sag condition. The specified conditions for the maximum sag of the conductor can be found in **NESC** Rules 230B and 232A. These are the specified conditions associated with the 18.5 ft of clearance. If an overhead line is being designed for a mining installation and the conditions are that 22-ft-high trucks are used, then this condition requires that the same line be designed with a 26.5-ft clearance (22' + 4.5' = 26.5') at the maximum sag of the conductor.

The first rule in the Code is a good place to discuss the format and numbering system used in the Code. The following are the main parts of the Code:

- General Sections 01, 02, 03, and 09 (Introduction, Definitions, References, and Grounding Methods)
- Part 1–Electric Supply Stations (Commonly referred to as substations)
- Part 2–Overhead Lines (Power and communications)
- Part 3–Underground Lines (Power and communications)
- Part 4–Work Rules (Power and communications, similar OSHA Standards apply)

An example of rule numbering is shown in Fig. 010-1.

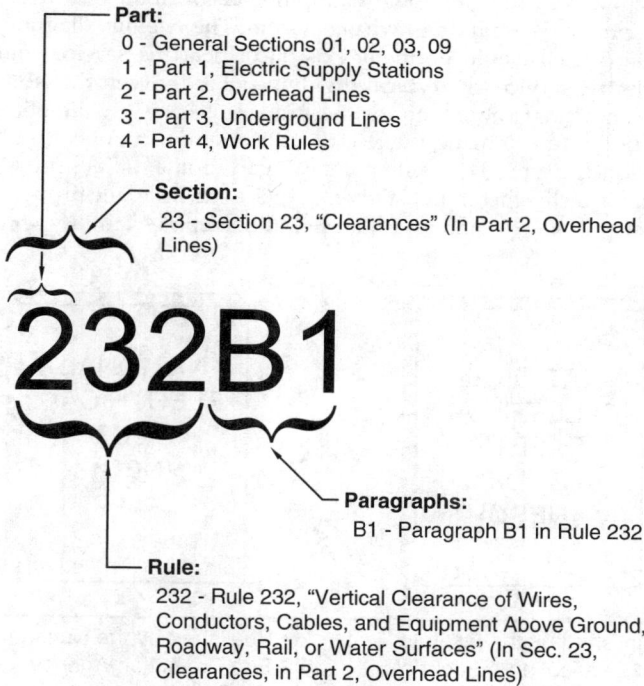

Part:
0 - General Sections 01, 02, 03, 09
1 - Part 1, Electric Supply Stations
2 - Part 2, Overhead Lines
3 - Part 3, Underground Lines
4 - Part 4, Work Rules

Section:
23 - Section 23, "Clearances" (In Part 2, Overhead Lines)

232B1

Paragraphs:
B1 - Paragraph B1 in Rule 232

Rule:
232 - Rule 232, "Vertical Clearance of Wires, Conductors, Cables, and Equipment Above Ground, Roadway, Rail, or Water Surfaces" (In Sec. 23, Clearances, in Part 2, Overhead Lines)

Fig. 010-1. Example of rule numbering (Rule 010).

011. SCOPE

This rule defines what is covered in the National Electrical Safety Code and what is not covered. The NESC covers supply and communication facilities and associated work practices carried out by a supply or communications utility company or an entity functioning as a utility. In general, the NESC applies to electric supply (power) and communications utilities. Communications utilities include, but are not limited to, telephone, cable TV, and fiber utilities. An example of an organization not normally thought of as a utility but functioning as a utility could be a university campus system or a large industrial complex that owns a utility-voltage distribution system. The NESC covers functions of utilities including generation of energy or communication signals and transmission and distribution to the service point.

The NESC does not cover utilization wiring in buildings. The standard that does cover building wiring is the National Electrical Code (NEC). The NEC is the "bible" for the electrical building industry and is used primarily by engineers and electricians. The NESC is the "bible" used primarily by utility engineers and utility linemen. See Fig. 011-1.

In places the NESC and the NEC overlap or come close to overlapping. One location is at the service to a building. The NESC contains NESC Fig. 011-1 which is a one-line diagram showing the dividing line between an electric supply utility and a premises (i.e., building) wiring system. The one-line diagram does not show a meter which would normally exist at or near the service point. For an overhead electric service, the typical dividing point between the NESC and the NEC is the conductor splice at the weatherhead. This is the dividing line between the electric utility function and the electric utilization function. For underground electric services the typical dividing point can vary. Some utilities provide service to the terminals on the meter base. Others provide service to the property line and the customer provides wiring after that point. Sometimes a utility will have the

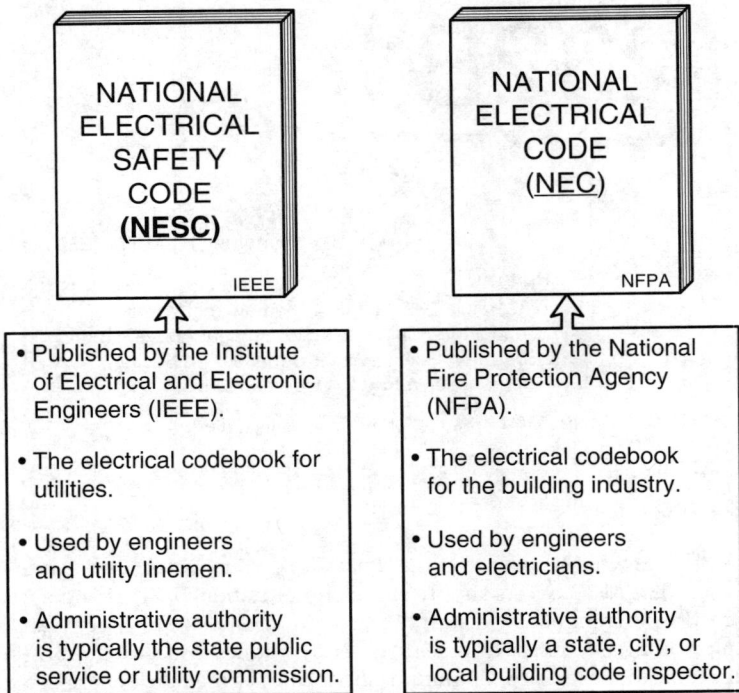

NATIONAL
ELECTRICAL
SAFETY
CODE
(NESC)

IEEE

- Published by the Institute of Electrical and Electronic Engineers (IEEE).

- The electrical codebook for utilities.

- Used by engineers and utility linemen.

- Administrative authority is typically the state public service or utility commission.

NATIONAL
ELECTRICAL
CODE
(NEC)

NFPA

- Published by the National Fire Protection Agency (NFPA).

- The electrical codebook for the building industry.

- Used by engineers and electricians.

- Administrative authority is typically a state, city, or local building code inspector.

Fig. 011-1. Differences between the NESC and the NEC (Rule 011).

customer install secondary wiring from the pad mount transformer to the meter but then the utility will take ownership of this wiring. To determine whether the circuit is covered under the **NESC** or the NEC, the ownership of the circuit and who maintains and controls the circuit are important factors to consider. Some utilities use a direct buried splice or a junction box under the meter base to provide a clear transition point between the two codes. The tariff that the utility has on file with the State Public Service Commission or Public Utility Commission may also help define the service point of the utility. See Fig. 011-2.

The dividing line between the **NESC** and the NEC for an overhead or underground communications (e.g., phone, cable TV, or fiber utility) service is typically the network interface device or demarcation box installed on the building being served. The demarcation box typically has a compartment for the communications company service wires and a separate compartment for the building communications cabling. This separates the utility communications function from the building communications function.

Another item that is covered by both the **NESC** and NEC is street and area lighting. Control of the street and area lighting is the important factor to consider when determining whether street and area lighting is covered under the **NESC** or the NEC. Street and area lighting that is metered usually falls under NEC requirements. Street and area lighting that is not metered and is owned, operated, and maintained by the

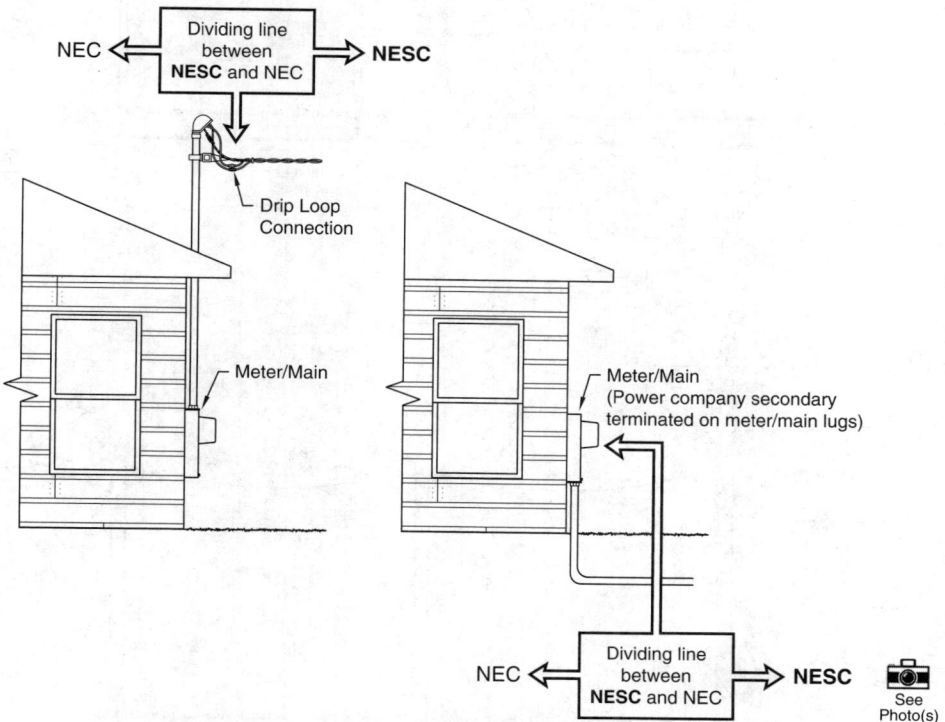

Fig. 011-2. Typical dividing lines between the **NESC** and the NEC (Rule 011).

utility typically is covered under the **NESC**. The main differences between street and area lighting covered under the **NESC** and street and area lighting covered under the NEC are grounding methods and overcurrent protection requirements. **NESC** street and area lights are typically operated and maintained by power linemen and NEC street and area lights are typically operated and maintained by electricians. Examples of street lighting systems covered by the **NESC** and NEC are shown in Fig. 011-3.

The **NESC** does not cover installations in mines (underground), ships, railway rolling equipment, aircraft, or automotive equipment. These industries have their own standards. See Rules 101 and 301 for a discussion of utilization wiring in Parts 1 and 3.

Section 02, Definitions of Special Terms, provides definitions of the terms "delivery point" for one utility delivering energy or signals to another utility and

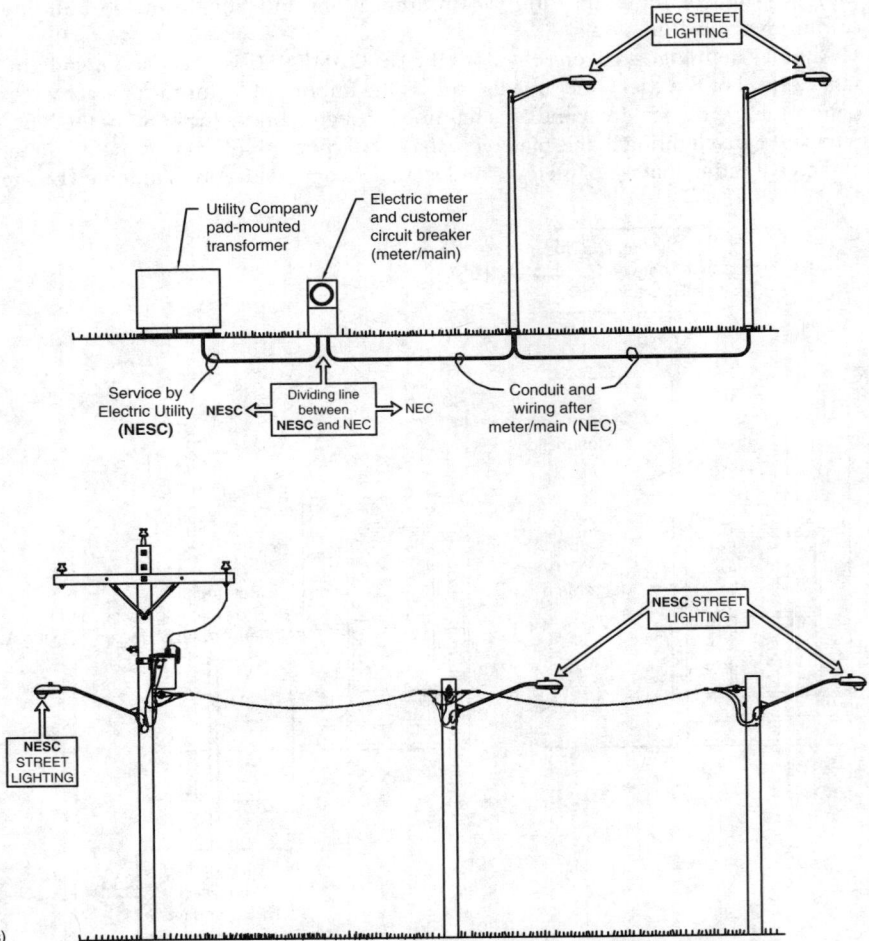

Fig. 011-3. Examples of **NESC** and NEC street lighting systems (Rule 011).

"service point" for determining the dividing line between the serving utility (supply or communication) and the premises wiring. A clear separation between the NESC and the NEC can be determined by applying NESC Rule 011 and using the definitions in NESC Sec. 02 and by referencing the corresponding scope rule in the front of the NEC.

Although not specifically addressed in Rule 011, the NESC does not cover easement conditions, environmental protection, raptor (bird) protection, FCC regulations, FAA regulations, NERC standards, electromagnetic fields (emf), settings for protective device coordination, or construction, operation, and maintenance cost issues. These items are either covered by other standards or they do not present a safety concern or they are simply addressed in the NESC and accepted good practice (Rule 012C) must be applied.

The NESC includes rules for physical loads (e.g., wind, ice, weight, wire tension, etc.) and strength of materials (e.g., wood, steel, concrete fiberglass, etc.), rules for clearances and spacings, rules for grounding, and rules for safe work practices.

012. GENERAL RULES

This rule provides three general rules for applying the Code. Rule 012A requires that the design, construction, operation, and maintenance of electric supply and communication lines equipment must be in accordance with the NESC. Rule 012B requires that utilities or other organizations performing work for the utility, such as a contractor, are the responsible parties for meeting NESC requirements for design, construction, operation, and maintenance. Rule 012C acknowledges that the Code cannot cover every conceivable situation. Where the scope of the NESC applies and the Code does not specify a rule to cover a particular installation, construction and maintenance should be done in accordance with "accepted good practice" for the local conditions known at the time. This does not allow the Code to be ignored for a condition that is covered in the Code. If "accepted good practice" is needed because a specific Code rule does not exist, a comparable NESC rule or the National Electrical Code (NEC) can be referenced to find an "accepted good practice." Other standards can also be referenced or "accepted good practice" may be determined by reviewing utility operating records to find out what safe practices are practical for the local conditions. Rule 012 applies to both supply and communication utilities during the design, construction, operation, and maintenance of lines and equipment. The NESC applies during the initial design and construction and during the life of the installation (i.e., during operation and maintenance).

013. APPLICATION

New installations and extensions are covered under Rule 013A. This rule clearly states that all new installations and extensions must adhere to the provisions of the NESC. Rule 013A1 does allow the administrative authority (i.e., the public service or public utility commission) to waive or modify the NESC rules if safety is provided in other ways. An example is listed in the Code to clarify this statement. Rule 013A2 recognizes that new types of construction and methods may be used experimentally to obtain information even if they are not covered in the Code. This can be done if three conditions are met. Qualified supervision must be provided. One example of qualified supervision is supervision under the direction of a registered professional engineer who is using engineering judgment and collecting data on the new construction. Equivalent safety must be provided. On joint-use (power and communications) facilities, all affected users must be notified in a timely manner. Since the Code is on a 5-year revision cycle, many times new construction methods arise during the 5-year period and then are included in a future edition of the Code.

Rule 013B discusses how the Code is applied to existing installations. Many people use the term "grandfather clause" when discussing this rule. The "grandfather clause" concept appears simple; however, caution must be used when applying this rule. Rule 013B1 states that if an existing installation meets or is altered to meet the current NESC, the installation is considered to be in compliance with the current edition and is not required to comply with any previous edition. Rule 013B2 states that existing installations, including maintenance replacements, that comply with a prior edition of the Code, need not be modified to comply with the current edition. Rule 013B2 can be applied to changes in the Code. Changes from the previous edition are indicated by a vertical black bar in the left margin of the NESC Codebook. Applying Rule 013B2 eliminates the need to alter existing installations that do not comply with a new rule in the current edition (assuming the installation complied with the Code at the time the line was built). Two exceptions apply: modifications may be required for safety reasons by the administrative authority (i.e., public service or public utility commission) and use of the current edition of Rule 238C is required when a structure (e.g., pole) is replaced. See Rules 202 and 238C for additional information.

Rule 013B3 specifically discusses conductors and equipment that are added, altered, or replaced on an existing structure. Rule 013B3 states that the structure or the facilities on the structure need not be modified or replaced if the resulting installation will be in compliance with either:
- The Code rules that were in effect at the time of the original installation,
- The Code rules in effect in a subsequent edition to which the installation has been previously brought into compliance, or
- The rules of the current edition.

Choosing the third bullet in the list above (the rules of the current edition) results in using the best Code information available as the NESC has evolved and improved over time. Choosing the first or second bullet in the list above is acceptable as the Code provides a choice of any one of the three options. However, if the first or second bullet is chosen, caution needs to be taken to determine if the existing installation complies with a prior Code. If the second bullet in the list above applies (bringing an installation into compliance with a subsequent edition), then the first bullet (the rules in effect at the time of the original installation) no longer applies.

The last paragraph of Rule 013B3 addresses how to work on a structure (e.g., pole) that has a violation. This paragraph points to the inspection rules in Part 2 (Overhead Lines), not Part 1 (Electric Supply Stations) or Part 3 (Underground Lines). Rule 214 requires inspections. When a condition or defect is found that affects compliance with the Code (i.e., a violation), Rules 214A4 and 214A5 address the timeline to correct the violation. The timeline is not addressed with hard numbers like a day, a week, a month, a year, etc., the rules only provide general wording. If a violation is not corrected, it must be recorded until the correction is done.

The last paragraph of Rule 013B3 addresses how to work on a structure (e.g., pole) that has a violation but the violation has not yet been corrected. If the work to be done on the structure in itself does not create a structural, clearance, or grounding violation or worsen an existing violation, then the work on the structure can be done without fixing the existing recorded violation. In other words, the existing recorded violation can be done at a later date. If these conditions are not met, the existing recorded violation must be corrected at the same time the other work is done or the other work must be postponed until the recorded violation is corrected. One example is adding a new service drop (power or communications) to a pole that has some other recorded violation. If adding the service drop meets the requirements in Rule 013B3 (in itself, adding the service drop does not create a structural, clearance, or grounding violation or worsen an existing violation), then the service drop can be added without fixing the other recorded violation at the same time.

Normally electric supply and communication engineers have the most current codebook sitting upon their desks and field workers do not have prior codebooks lying around in their trucks. Even if something as simple as adding a transformer to an existing installation or replacing an existing pole is being done, the utility performing the work must be careful not to blindly assume that the existing situation complies with a prior Code. Utilities may want to consult the advice of a legal professional when applying Rule 013B as interpreting this rule from a legal viewpoint can be as critical as interpreting it from a technical viewpoint. Although not a Code requirement, a simple solution to eliminate the complexities of applying the "grandfather clause" is to use the current edition of the NESC for existing installations that require maintenance replacements or additions, alterations, or replacements.

The application of a "grandfather clause" should not be confused with maintaining clearances or a change in land use under a supply or communications line. See Rule 230I for a discussion.

Rule 013C requires that inspections of new and existing facilities meet the inspection rules in the current edition of the NESC, not prior editions. Inspections of electric supply stations are covered in Rule 121, inspections of overhead lines are covered in Rule 214, and inspections of underground lines are covered in Rule 313. Rule 013C also requires that work performed on new and existing facilities meet the work rules in Part 4 of the current edition of the NESC, not prior editions.

014. WAIVER FOR EMERGENCY AND TEMPORARY INSTALLATIONS

In this rule the Code recognizes the need to waive or modify rules in cases of emergency or temporary installations. The Code grants the person responsible for the installation the ability to modify or waive rules with specific requirements. Rule 014A for emergency installations applies to both overhead and underground lines. Rule 014B for temporary installations applies only to overhead lines. The Code does not specify a time length for temporary installations or define an emergency installation. The Code makes it very clear that temporary installations may not have reduced clearance. Strength reductions for temporary service are also addressed in NESC Table 261-1, Footnote 4, which does not use the same wording as Rule 014. See Fig. 014-1 for a summary of the requirements of Rule 014.

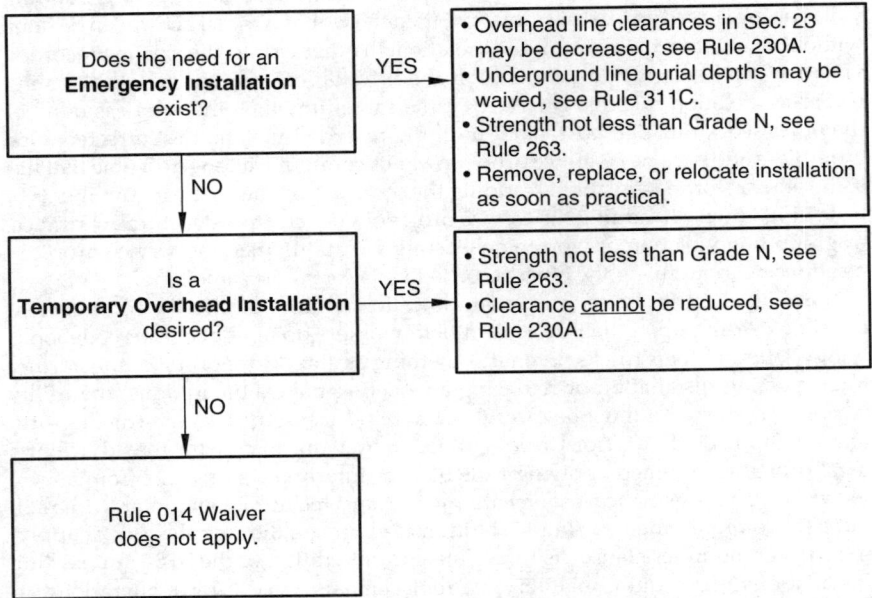

```
┌─────────────────────────┐        ┌───────────────────────────────────┐
│ Does the need for an     │  YES   │ • Overhead line clearances in Sec. 23│
│ Emergency Installation   │ ─────▶ │   may be decreased, see Rule 230A.   │
│ exist?                   │        │ • Underground line burial depths may be│
└─────────────────────────┘        │   waived, see Rule 311C.             │
            │                       │ • Strength not less than Grade N, see │
           NO                       │   Rule 263.                          │
            │                       │ • Remove, replace, or relocate installation│
            ▼                       │   as soon as practical.              │
┌─────────────────────────┐        └───────────────────────────────────┘
│ Is a                     │  YES   ┌───────────────────────────────────┐
│ Temporary Overhead Installation│ ──▶ │ • Strength not less than Grade N, see │
│ desired?                 │        │   Rule 263.                          │
└─────────────────────────┘        │ • Clearance cannot be reduced, see   │
            │                       │   Rule 230A.                         │
           NO                       └───────────────────────────────────┘
            │
            ▼
┌─────────────────────────┐
│ Rule 014 Waiver          │
│ does not apply.          │
└─────────────────────────┘
```

Fig. 014-1. Summary of the requirements for emergency and temporary installations (Rule 014).

015. INTENT

This rule defines three key **NESC** words: "shall," "should," and "RECOMMEN-DATION." **Code** rules containing the word "shall" indicate that the rule is man-datory and the rule must be met. Not complying with a rule that contains the word "shall" is a direct **Code** violation. **Code** rules containing the word "should" indicate requirements that are normally and generally practical for the specified conditions. The **Code** does recognize that under certain circumstances some rules may not be practical. Where this is the case, the word "should" is used. Under the definition of "should," the **Code** references Rule 012 that requires "accepted good practice" to be used. If at all possible the words "should" and "shall" should be considered the same. If it is not possible to treat "should" the same as "shall," it is prudent to carefully review the specific conditions that prohibit a utility from applying the **Code** rule that contains the word "should" and it is prudent to apply an "accepted good practice" that provides an equivalent degree of safety for the specific conditions. The **Code** uses the word "RECOMMENDATION" for provisions that are considered desirable but not intended to be mandatory. "RECOMMENDATION" is the least stringent of the three terms "shall," "should," and "RECOMMENDATION." Since the **Code** does consider a "RECOMMENDA-TION" desirable, it seems prudent that the utility make some effort to comply with a "RECOMMENDATION" even though it is not mandatory.

Rule 015 also contains clarifications of the words "NOTE," "EXAMPLE," and "footnote." "NOTES" and "EXAMPLES" are not mandatory, and they are provided for information and illustrative purposes but are not considered part of the Code requirements. "Footnotes," however, are used for tables throughout the Code and they carry the full force and effect of the table or rule with which they are associated.

An "EXCEPTION" to a rule has the same force as the rule itself. Exceptions are not reduced safety measures. For example, if a clearance value is reduced by an exception, some condition associated with the exception may be provided to maintain safety. Typically, the Code provides a larger value in the rule and smaller value in the exception.

The physical location of the words "RECOMMENDATION," "EXCEPTION," and "NOTE," and how these words are indented with respect to other text, signifies to what rule the "RECOMMENDATION," "EXCEPTION," or "NOTE" applies.

016. EFFECTIVE DATE

This rule states that the 2012 edition of the NESC may be used at any time on or after publication date. In addition, this edition shall become effective no later than the first day of the month after 180 days following the publication date. The effective date applies to the design and approval process, not just the construction date. If the design or approval for a new installation or extension was started before the effective date, the prior Code may be used. The example in this rule indicates the 2012 NESC publication date of August 1, 2011 and establishes February 1, 2012 as the effective date. The note in this rule explains that the 180-day (6-month) grace period allows utilities and other agencies to acquire copies of the Code and revise regulations, standards, and procedures. The note also clarifies that this edition is not required to be used before the 180-day period; however, it is not prohibited to use it during this period.

The NESC is a standard published by the Institute of Electrical and Electronic Engineers (IEEE). For the NESC to become a legal requirement, it is typically adopted by a state authority having jurisdiction over utilities (i.e., a public service commission or public utility commission) or by some other authority. To determine the specific legal status of the NESC, the authority having jurisdiction must be contacted.

017. UNITS OF MEASURE

The Code uses the metric system as the primary unit for numerical values. The customary (English or inch-foot-pound) system is the secondary system. In the text of the Code, metric values are shown first with the customary inch-foot-pound system shown second and inside parentheses. Some tables in the Code have the metric and English system in the same table. Other more complex tables have separate tables. When separate tables are used, the first table in the Code will be the metric table and the second table will be the English table. Metric values are based on the current version of the metric system titled, "The International System of Units" (or SI). The values in each system are rounded to convenient numbers. An exact conversion is not used so that the values appear functional for safety purposes (e.g., rounded to the nearest half foot or tenth of a meter).

IEEE (the publishers of the **Code**) have chosen the metric system as the primary unit of measure as IEEE publishes around the globe. Units of measure discussed in this Handbook are based on the customary (English or inch-foot-pound) system.

Rule 017B states that physical items referenced in the **Code** are in "nominal values" unless specific dimensions are provided. Other standards may set tolerances for manufacturing. An example of nominal values is shown in Fig. 017-1.

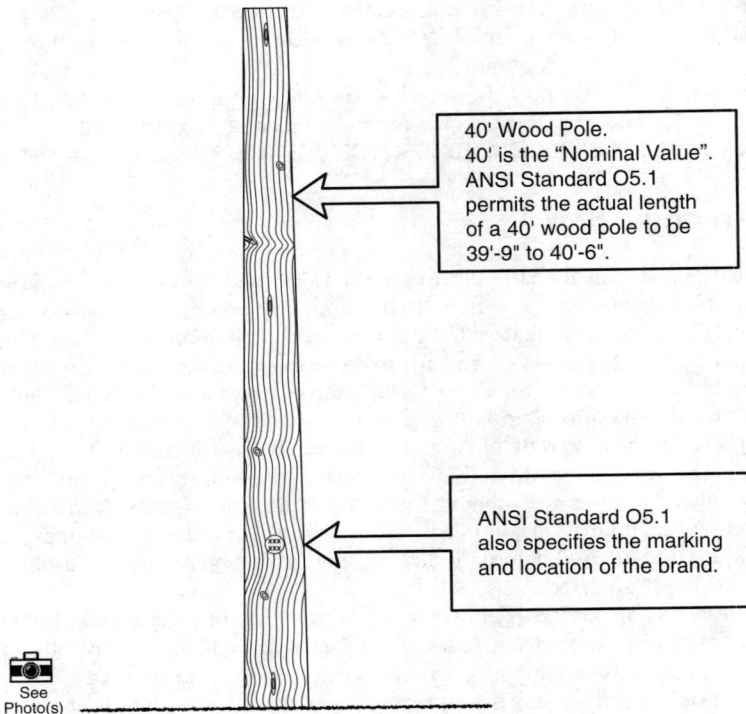

40' Wood Pole.
40' is the "Nominal Value".
ANSI Standard O5.1 permits the actual length of a 40' wood pole to be 39'-9" to 40'-6".

ANSI Standard O5.1 also specifies the marking and location of the brand.

See Photo(s)

Fig. 017-1. Example of nominal values (Rule 017B).

018. METHOD OF CALCULATION

Rule 018 provides rounding requirements. In general, rounding "off" to the nearest significant digit is required unless otherwise specified in applicable rules. One rule that requires a different rounding method is Rule 230A4, which requires rounding "up" for clearance calculations as Sec. 23 deals with various overhead line clearances which are typically specified as "not less than" clearances. Rounding "off" follows the rules of traditional rounding learned in math class. An example of rounding "off" is rounding 20.02 down to 20.0 or rounding 20.66 up to 20.7. An example of rounding "up" for "not less than" clearance is rounding 20.02 to 20.1 because rounding "off" to 20.0 would not meet a clearance required to be "not less than" 20.02. See Rule 230A4 for additional information.

Section 02

Definitions of Special Terms

The **NESC** provides several terms and their definitions for use in the codebook. Section 02 is the first place to look for definitions of special terms. The **Code** text under the title of Sec. 02 references *The IEEE Standards Definition Database* for definitions not contained in this section. The IEEE Standards Definition Database should be used as a second step if the definition is not provided in Sec. 02 of the **NESC**. The third and final step for definitions not provided in Sec. 02 of the **NESC** or in the IEEE Standards Definition Database is to look the word up in an ordinary dictionary. These three steps are outlined in Fig. 02-1.

Occasionally, terms are defined in individual rules instead of Sec. 02. One example of this is the words "shall," "should," and "RECOMMENDATION" that are defined in Rule 015. Another example is how the word "equipment" is defined relative to a specific application. Rule 238A defines equipment relative to clearance between communication and supply facilities located on the same overhead line structure. Rule 380A provides examples of equipment relative to underground construction.

Clarifications and drawings of key terms are provided in this handbook. They are provided in the individual rules in which the terms apply instead of this section with the exception of two terms, voltage and effectively grounded, which are discussed below.

The term voltage has six definitions in Sec. 02 of the **NESC**. In some code rules and tables, voltage is specifically stated as phase to phase or phase to ground. In some locations, a voltage is stated without a phase-to-phase or phase to ground reference. If a voltage is stated without a phase-to-phase or phase-to-ground reference, the voltage is dependent on the type of grounding system. For example, if the code states that a clearance adder is required for lines over 50 kV (without a phase-to-phase or phase-to-ground reference) and the line is fed from a wye-connected single grounded system/unigrounded system with ground fault relaying

```
┌─────────────────────────────┐
│ ┌───────┐                   │
│ │ Step 1│                   │
│ └───────┘                   │
│  Term definition in NESC    │
│  Sec. 02. (Some definitions are │
│  provided in individual rules.) │
└─────────────────────────────┘
              │
              ▼
┌─────────────────────────────┐
│ ┌───────┐                   │
│ │ Step 2│                   │
│ └───────┘                   │
│  If term is not defined in Step 1, │
│  find term definition in The IEEE  │
│  Standards Definition Database.    │
└─────────────────────────────┘
              │
              ▼
┌─────────────────────────────┐
│ ┌───────┐                   │
│ │ Step 3│                   │
│ └───────┘                   │
│  If term is not defined in Step 1 │
│  or 2, find term definition in an │
│  ordinary dictionary.             │
└─────────────────────────────┘
```

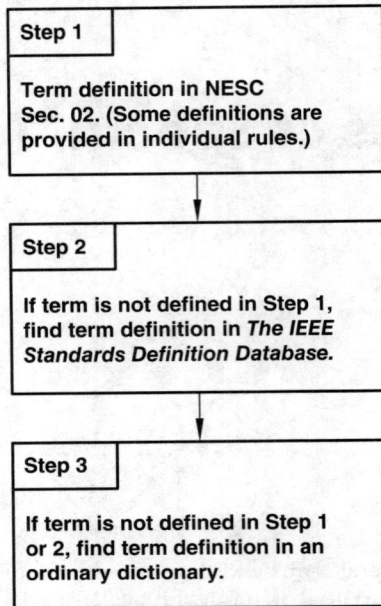

Fig. 02-1. Steps for finding definitions of terms (Sec. 02).

that is connected to an effective ground, the clearance adder may be applied to lines over 50 kV to ground (or 86.6 kV phase to phase). If the line is ungrounded, the clearance adder must be applied to lines over 50 kV phase to phase. Voltage as defined in Sec. 02 is the effective rms voltage. See Rule 230G for a discussion and figures related to AC rms voltage, DC voltage, phase-to-phase voltage, and phase-to-ground voltage.

The terms effective ground/effectively grounded, effectively grounded neutral conductor, multigrounded/multiple grounded system, and single grounded system/ungrounded system are all defined in Sec. 02 of the **NESC**. The important part of the definition of effective ground/effectively grounded is what the definition does not say. The definition of effective ground/effectively grounded does not provide a value in ohms (e.g., the **Code** does not say that an effective ground/effectively grounded system is 5 Ω or less). Multigrounded systems discussed in Rule 096C typically are considered effectively grounded, but in special cases (e.g., a very rocky 1-mile stretch of line), more than four grounds in each mile may be needed to make a multigrounded system an effectively grounded system. Rule 096D does specify a 25-Ω limit for single-grounded systems.

Section 03

References

Section 03 provides a list of standards that are referenced in the **Code**. The standards listed form a part of the **Code** to the extent that they are referenced in the **Code** rules. If a standard is cited for information purposes only, it appears in the Bibliography in Appendix E of the **NESC**. For example, ANSI O5.1, *American National Standard Specifications and Dimensions for Wood Poles,* is listed in Sec. 03 as it is referenced in **Code** Rule 261A2b(1). However, IEEE Standard 80, *IEEE Guide for Safety in AC Substation Grounding,* is listed in Appendix E of the **NESC** (Bibliography) as it is cited for informational purposes in a "NOTE" in Rule 092E. Rule 015 states that a "NOTE" is provided for information purposes only. Some standards are referenced in a **Code** rule in one location and cited in a "NOTE" in other locations. These standards with a dual reference are listed in Sec. 03. The **Code** recognizes that the standards listed in Sec. 03 provide information that does not need to be repeated in the **Code**. This helps keep the codebook from getting too wordy and utilizes standards that another agency has documented. The relationship between Sec. 03—References and Appendix E—Bibliography is shown in Fig. 03-1.

The **Code** acknowledges, with a note in Sec. 03, that current standards may be newer than ones listed, as the **Code** is updated on a 5-year revision cycle and some standards may be updated during the middle of this process. The standards listed in Sec. 03 are an important part of a technical library of reference material. Additional notes in Sec. 03 of the **NESC** provide mailing and website addresses to order the various standards.

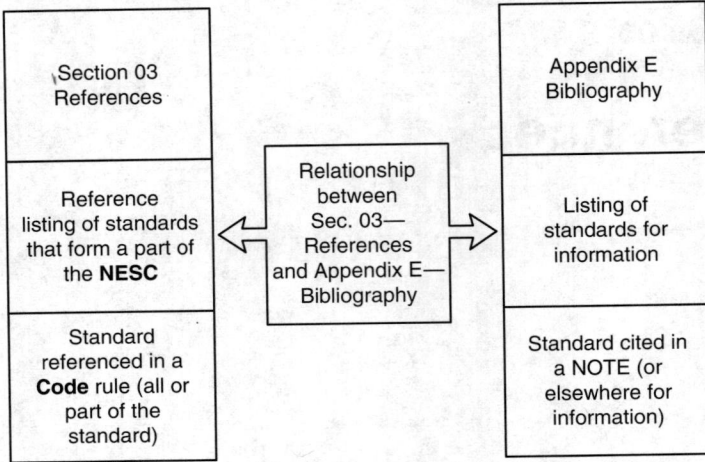

Fig. 03-1. Relationship between Sec. 03—References and Appendix E—Bibliography (Sec. 03).

Section 09

Grounding Methods for Electric Supply and Communications Facilities

090. PURPOSE

The purpose of Sec. 09 is to provide practical methods of grounding. Grounding is one of the ways to protect people from hazardous voltages. Grounding also allows protective devices to operate during a fault condition. The basic theory behind grounding is to keep the voltage of a grounded part (e.g., equipment case, neutral conductor, communications messenger, etc.) as close as possible to the potential of the earth so that a voltage difference does not exist between a person and a grounded metal object. The Code states in this rule that grounding is used as one of the means of safeguarding employees and the public from injury. Other means include, but are not limited to, guarding, adequate clearance above ground, proper burial depth, etc.

091. SCOPE

The scope of Sec. 09 is to provide the *methods* of protective grounding for supply and communication conductors and equipment. The *requirements* for grounding are listed in the other parts of the Code. The requirements for grounding electric supply stations are predominantly in Rule 123. The requirements for grounding overhead lines are predominantly in Rule 215. The requirements for grounding underground lines are predominantly in Rule 314. Rules 123, 215, and 314 all use the term effectively grounded. Rule 091 provides methods for effective grounding and points to the definition of effectively grounded in Sec. 02, "Definitions of special terms." The scope of Sec. 09 does not include the grounded return of electric railways or lightning protection not associated with supply and communication wires, for example, lightning protection wires connected to a lightning rod on top of a barn.

092. POINT OF CONNECTION OF GROUNDING CONDUCTOR

092A. Direct Current Systems That Are Required to Be Grounded. This rule has basic connection requirements for direct-current (DC) systems. For 750 V and less, the grounding conductor connection must be made only at the supply station. For three-wire DC circuits, the connection must be made to the neutral. For DC systems over 750 V, the grounding conductor connection must be made at both the supply and load points. The connection must be made to the neutral of the system. The ground or grounding electrode can be external or remote from each of the stations. This permits separating the electrode from areas with ground currents that can cause electrolytic damage. The Code permits one of the two stations to have its grounding connection made through a surge arrester as long as the other station has the neutral effectively grounded. An exception is provided for the 750-V and greater category for back-to-back DC converter stations that are adjacent to each other. For this condition the neutral of the system should be connected to ground at one point only.

092B. Alternating-Current Systems That Are Required to Be Grounded
092B1. 750 V and Below. The point of the grounding connection on alternating-current (AC), wye-connected, three-phase, four-wire and single-phase, three-wire systems operated at 750 V and below is shown in Fig. 092-1.

On other one-, two-, or three-phase systems feeding lighting circuits, a grounding connection must be made to a common circuit conductor. Common examples include a 120/240-V, three-phase, four-wire center tap delta service, a 120/208-V, single-phase, three-wire service fed from a 120/208-V, three-phase, four-wire service or a 120-V, single-phase, two-wire service fed from a 120/240-V, single-phase, three-wire service.

Wye and delta circuits that are not grounded or do not use a common (neutral) conductor for grounding cannot be used to serve lighting loads. See Fig. 092-2.

Grounding connections must be made at the source and line side of a service as shown in Fig. 092-3.
092B2. Over 750 V. Nonshielded conductors (e.g., bare neutral conductors) must be grounded as shown in Fig. 092-4.

The wording in Rule 092B2a requires unigrounding at the source (substation transformer) and permits, but does not require, multigrounding along the line. However, various rules in Part 2, Overhead Lines, and Part 3, Underground

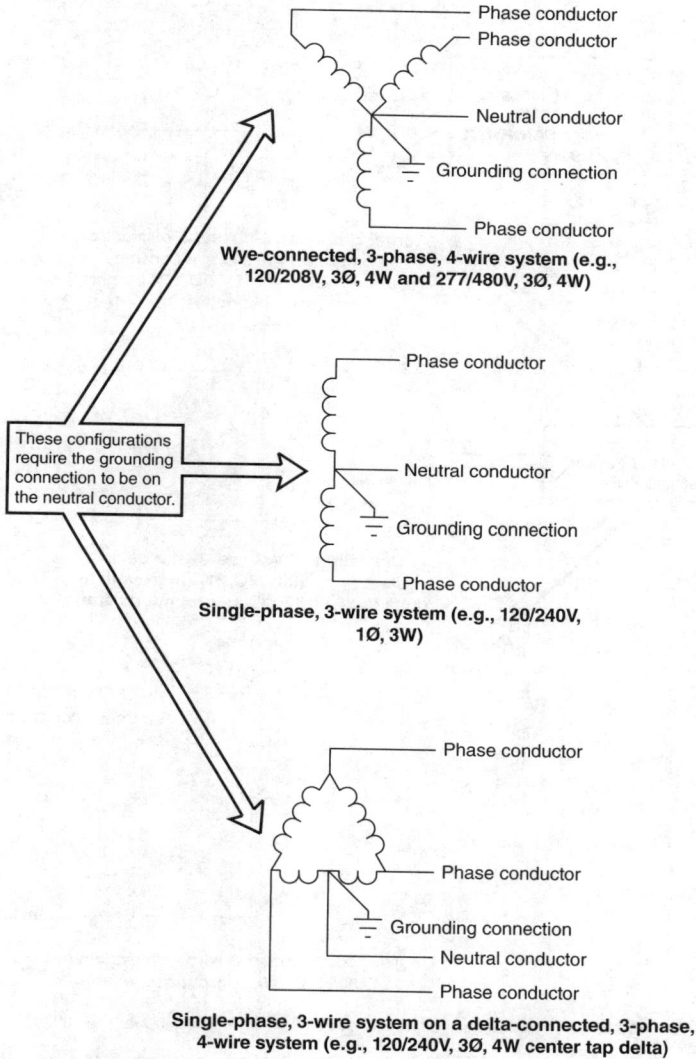

Fig. 092-1. Grounding connection on wye-connected, three-phase, four-wire and single-phase, three-wire systems (Rule 092B1).

Lines, will require systems to be effectively grounded. Effectively grounded systems typically need to be multigrounded to provide sufficiently low ground impedance. Multigrounded systems are discussed in Rule 096C.

Shielded conductors on riser poles must be grounded as shown in Fig. 092-5.

Shielded cables without an insulating jacket must be grounded as shown in Fig. 092-6.

A circuit conductor may be grounded (corner grounded) but this configuration **CANNOT** be used for lighting loads.

Phase conductor

Phase conductor
Grounded
phase conductor

Grounding
connection

Corner grounded 3-phase, 3-wire delta system (e.g., 480V, 3Ø, 3W corner grounded delta or 240V, 3Ø, 3W corner grounded delta)

These configurations **CANNOT** be used for lighting circuits.

Phase conductor

Phase conductor
Phase conductor

Ungrounded 3-phase, 3-wire delta system (e.g., 480V, 3Ø, 3W ungrounded delta or 240V, 3Ø, 3W ungrounded delta)

Phase conductor
Phase conductor

Phase conductor

Ungrounded 3-phase, 3-wire wye system (e.g., 480V, 3Ø, 3W ungrounded wye)

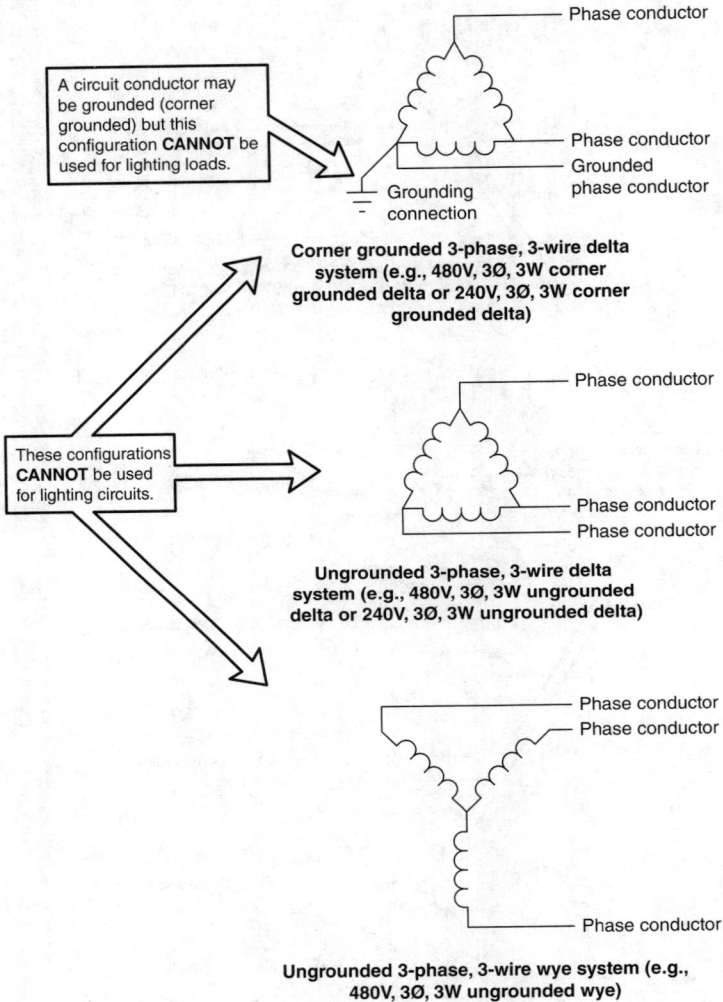

Fig. 092-2. Wye and delta systems not to be used for lighting loads (Rule 092B1).

Shielded cables with an insulating jacket must be grounded as shown in Fig. 092-7.

Shielded cable without an insulating jacket that is buried in direct contact with the earth has an advantage of being grounded all along its length. However, direct-buried shielded cable without an insulating jacket is susceptible to corrosion. The insulating jacket can prevent corrosion of the shield or concentric neutral. See Rule 096C for multigrounding requirements and special exceptions.

Neutral

Grounding connections must be made to the source (transformer) and the line side of the service equipment (meter/main).

Grounding connection

Grounding connections must be made to the source (transformer) and the line side of the service equipment (meter/main).

Grounding connection

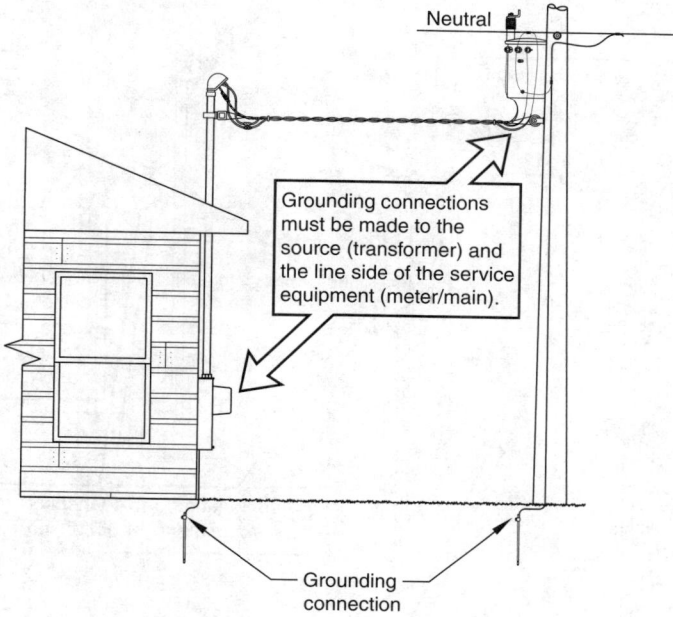

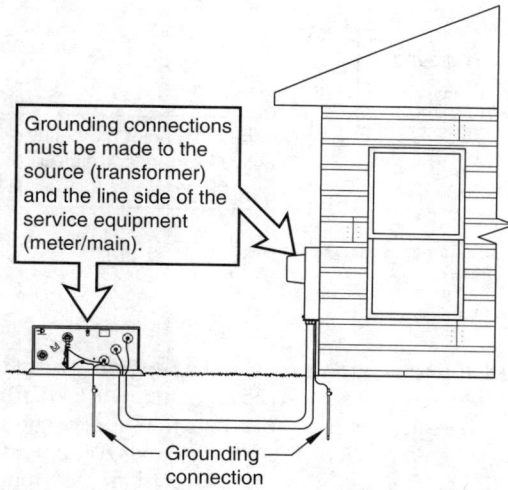

Fig. 092-3. Grounding connections at source and line side of a service (Rule 092B1).

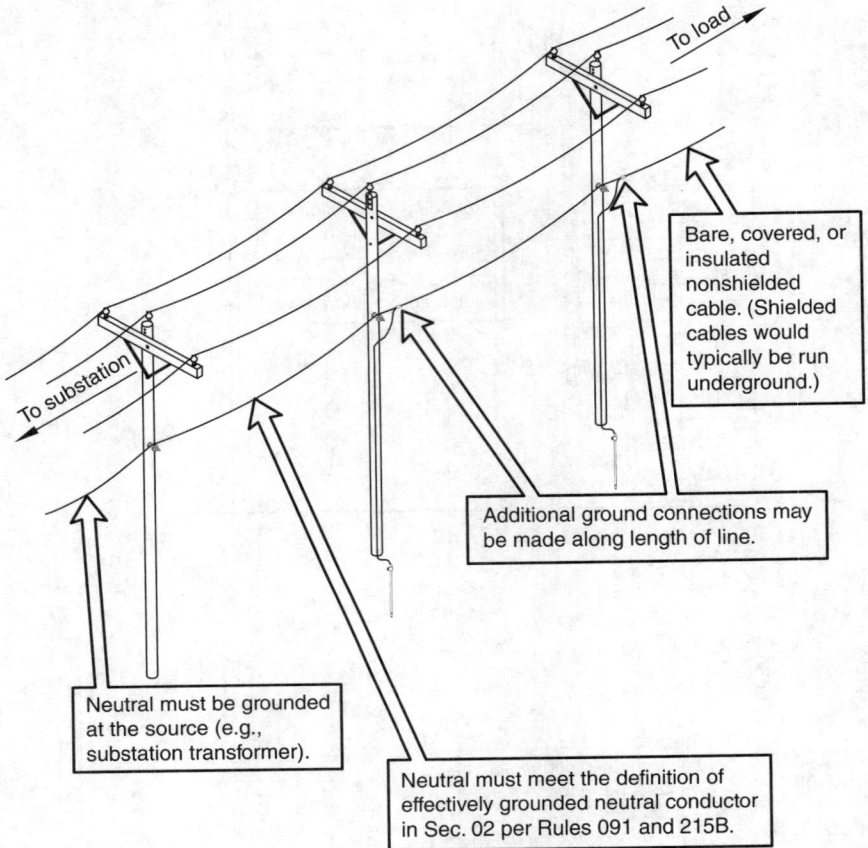

To load

Bare, covered, or insulated nonshielded cable. (Shielded cables would typically be run underground.)

To substation

Additional ground connections may be made along length of line.

Neutral must be grounded at the source (e.g., substation transformer).

Neutral must meet the definition of effectively grounded neutral conductor in Sec. 02 per Rules 091 and 215B.

See Photo(s)

Fig. 092-4. Grounding connections for nonshielded cables over 750 V (Rule 092B2a).

092B3. Separate Grounding Conductor. If a separate grounding conductor is used on an AC system to be grounded as an adjunct (joined addition) to a cable run underground, there are several conditions that apply. The separate grounding conductor must be connected directly or through the neutral to items that must be grounded. The conductor must be located as shown in Fig. 092-8.

Adjunct (joined addition) grounding conductors are typically used with shielded supply cables. If the shield on the supply cable is not a sufficient size to carry neutral current or fault current, an adjunct grounding cable can be used. An adjunct grounding conductor should not be used to replace a corroded concentric neutral conductor in a direct-buried cable. Rule 350B requires that a direct-buried cable operating above 600 V have a continuous metallic shield,

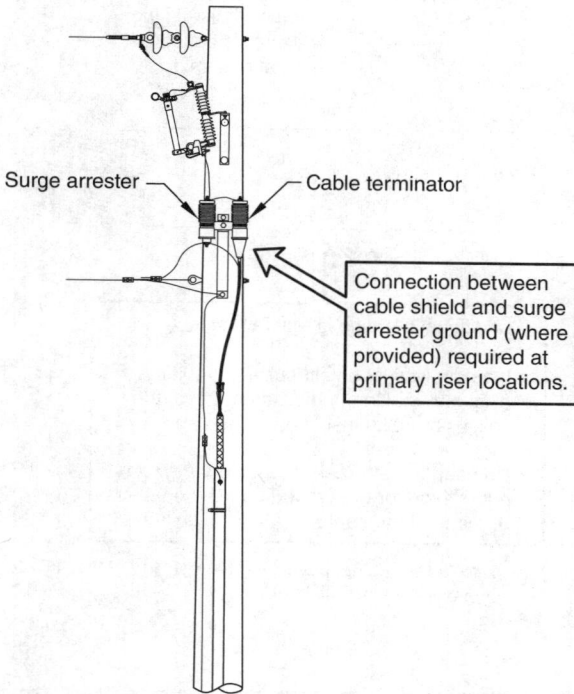

Surge arrester —

— Cable terminator

Connection between cable shield and surge arrester ground (where provided) required at primary riser locations.

See Photo(s)

Fig. 092-5. Surge arrester cable—shielding interconnection (Rule 092B2b(1)).

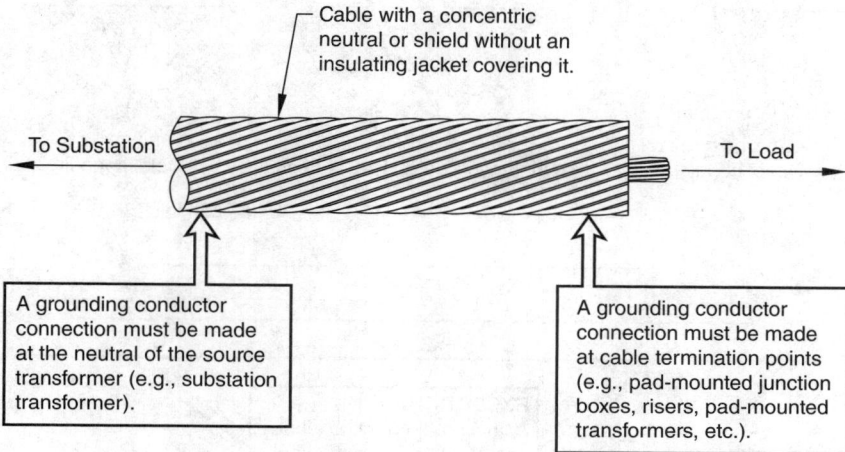

Cable with a concentric neutral or shield without an insulating jacket covering it.

To Substation

To Load

A grounding conductor connection must be made at the neutral of the source transformer (e.g., substation transformer).

A grounding conductor connection must be made at cable termination points (e.g., pad-mounted junction boxes, risers, pad-mounted transformers, etc.).

See Photo(s)

Fig. 092-6. Grounding points for a shielded cable without an insulating jacket (Rule 092B2b(2)).

Cable with insulating
jacket over concentric
neutral or shield.

- Additional bonding is recommended.

- If electrolysis or shield-current problems
  exist and multigrounding is not used, the
  splices must be insulated.

- Bonding transformers or reactors may be
  substituted for the ground connection at
  one end of the cable.

See
Photo(s)

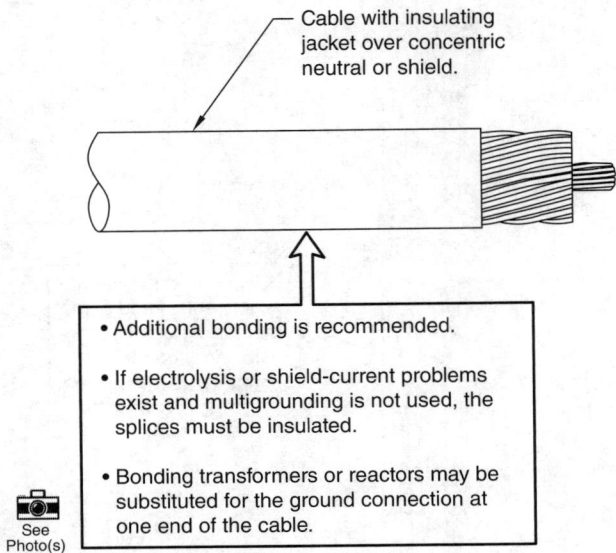

Fig. 092-7. Grounding points for a shielded cable with an insulating jacket (Rule 092B2b(3)).

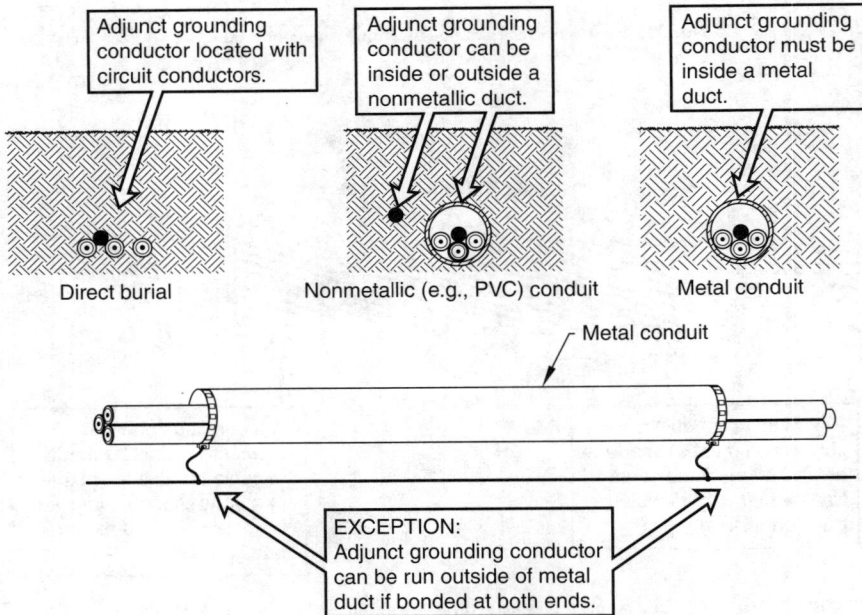

Adjunct grounding
conductor located with
circuit conductors.

Adjunct grounding
conductor can be
inside or outside a
nonmetallic duct.

Adjunct grounding
conductor must be
inside a metal
duct.

Direct burial

Nonmetallic (e.g., PVC) conduit

Metal conduit

Metal conduit

EXCEPTION:
Adjunct grounding conductor
can be run outside of metal
duct if bonded at both ends.

Fig. 092-8. Separate (adjunct) grounding conductor (Rule 092B3).

sheath, or concentric neutral. The adjunct grounding conductor can be used to supplement the concentric neutral but not replace it if it has corroded away.

092C. Messenger Wires and Guys

092C1. Messenger Wires.   The point of connection of the grounding conductor to messenger wires that are required to be grounded by other parts of the Code is shown in Fig. 092-9.

Communications messenger wires are required to be grounded in Part 2, Overhead Lines, Rule 215C. Communications messenger wires on joint-use (power and communication) poles are required to be grounded in Secs. 23 and 24 to meet certain clearance and grade of construction requirements. The messenger must meet certain ampacity and strength criteria defined

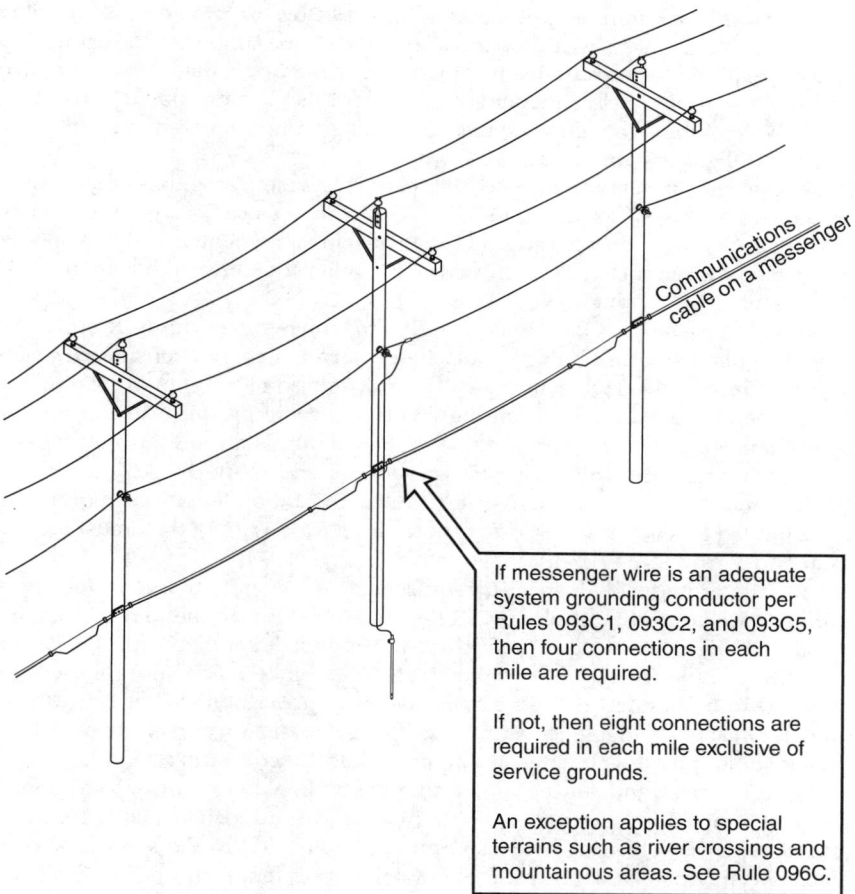

Communications cable on a messenger

If messenger wire is an adequate system grounding conductor per Rules 093C1, 093C2, and 093C5, then four connections in each mile are required.

If not, then eight connections are required in each mile exclusive of service grounds.

An exception applies to special terrains such as river crossings and mountainous areas. See Rule 096C.

See Photo(s)

Fig. 092-9. Grounding of messenger wires (Rule 092C1).

in Rules 093C1, 093C2, and 093C5. For the messenger (on a joint-use power and communication structure) to meet the ampacity requirement in Rule 093C2, the messenger wire ampacity must be rated not less than one-fifth of the neutral wire ampacity. It is sometimes difficult to find the ampacity rating of messenger wires as many communications messenger wires are actually guy wires. Manufacturers of guy wires typically provide mechanical strength ratings, not electrical ampacity ratings. The four grounds in each mile rule appears here for the first time in the **Code**. It is discussed in detail in Rule 096C.

092C2. Guys.   The point of connection of the grounding conductor to guys that are required to be grounded by other parts of the **Code** is shown in Fig. 092-10.

Guys (both supply and communication) must be either grounded (per Rule 215C2) or insulated (per the exception to Rule 215C2). If guys are grounded, they must be grounded using the methods in this rule.

092C3. Common Grounding of Messengers and Guys on the Same Supporting Structure.   When messengers and guys are on the same supporting structure and they are required to be grounded by other parts of the **Code**, they must be bonded together and grounded by the connection methods listed in this rule. The methods listed are a combination of the messenger and guy connection requirements provided in Rules 092C1 and 092C2.

092D. Current in Grounding Conductor.   This rule recognizes that multigrounded systems, for example, a 12.47/7.2-kV, three-phase, four-wire circuit that has four or more grounds in each mile, may develop objectionable current flow on the grounding conductor (pole ground). This rule provides methods to alleviate the objectionable current flow.

Objectionable current flow may exist due to stray earth currents or other reasons. Fault currents and lightning discharge currents are not considered objectionable current flows when applying this rule and some amount of current will always be present on the grounding conductor (pole ground) during normal operation. Separating primary and secondary neutrals on multigrounded systems to address stray voltage concerns is addressed in Rule 097D2.

092E. Fences.   When conductive electric supply station fences are required to be grounded by Part 1 of the **Code** (primarily in Rule 110A1), they must be connected to a grounding conductor as shown in Fig. 092-11.

This rule provides both specific requirements for conductive electric supply station fence grounding (Rules 092E1 through 092E6) and general requirements by noting IEEE Standard 80, which is the industry standard for substation grounding. Rule 093C6 also applies to fences. Fence mesh strands are only required to be bonded if the fence posts are nonconducting. For conducting (metal) fence posts, the fence mesh must be under tension and electrically connected to the post for the mesh to be grounded. A grounding conductor feed up to barbwire strands at the top of a fence can be woven through the chainlink mesh for added grounding continuity. A ground grid which is typically buried under an electric supply station and connected to the station fence is discussed in Rule 096B. An example of substation fence grounding is shown in Fig. 092-12.

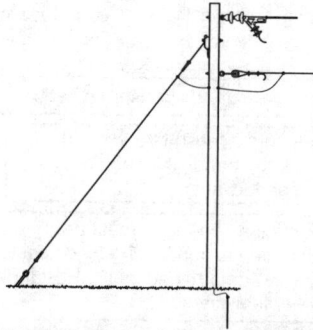

Metallic supporting structure (steel pole).
Guy connection to effectively grounded metal pole.

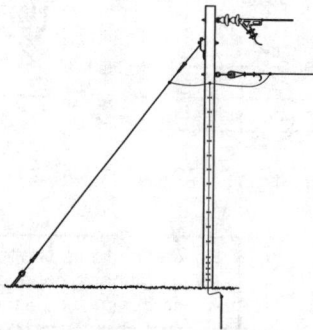

Guys required to be effectively grounded must be connected to one or more of these.

Nonmetallic supporting structure (wood pole).
Guy connection to effective ground (pole ground).

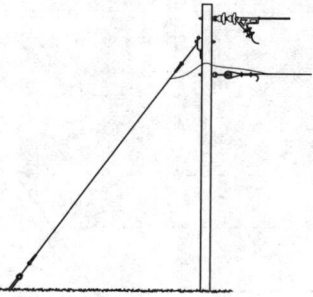

Guy connection to an effectively grounded neutral conductor.

Fig. 092-10. Grounding of guys (Rule 092C2).

Conductive supply station fence
(e.g., a chain-link fence)

Barbwire strands (if used) must be bonded.

Bond fence mesh strands and
barbwire strands if fence posts are
nonconducting.

Grounding conductor must be
connected to conducting fence posts
(spacing determined by IEEE Standard
80 or other method).

Gates must be bonded
(typically done with a flexible braid).

Ground on each side of gate.

A buried bonding jumper
must be used across the
gate.

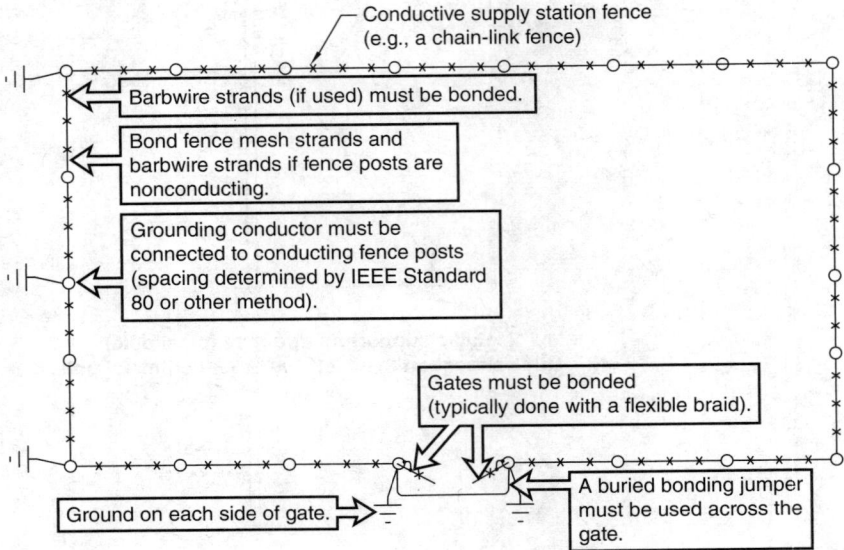

Fig. 092-11. Conductive supply station fence grounding (Rule 092E).

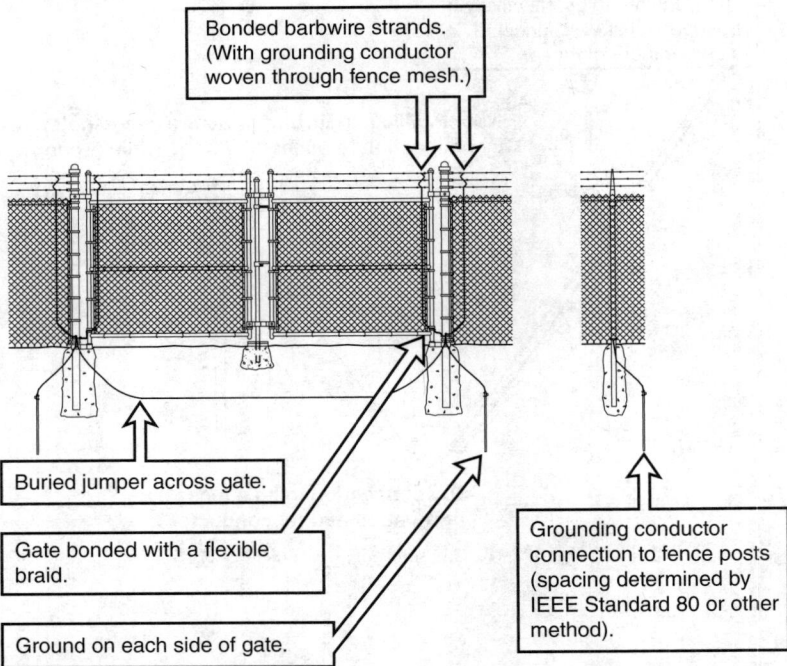

Bonded barbwire strands.
(With grounding conductor
woven through fence mesh.)

Buried jumper across gate.

Gate bonded with a flexible
braid.

See
Photo(s)

Ground on each side of gate.

Grounding conductor
connection to fence posts
(spacing determined by
IEEE Standard 80 or other
method).

Fig. 092-12. Example of conductive electric supply station fence grounding (Rule 092E).

093. GROUNDING CONDUCTOR
AND MEANS OF CONNECTION

093A. Composition of Grounding Conductors. Grounding conductors can be copper or other metals or combinations of metals that will not corrode during their expected service life under the existing conditions. Surge arrester connections must be short, straight, and free from sharp bends. Metallic electrical equipment cases or the structural metal frame of a building or structure can also be used as a grounding conductor. An example of a copper grounding conductor (pole ground) and a structural metal grounding conductor (steel pole) is shown in Fig. 093-1.

Many utilities use copper for the entire length of the grounding conductor (pole ground). The size of the grounding conductor (pole ground) is covered in Rule 093C. Some utilities use aluminum or ACSR. If aluminum or ACSR is used above grade, it is typically spliced to copper, which then runs below grade (see Rule 093E5). Some utilities utilize copper-coated steel. Copper substitutes have become popular due to copper theft.

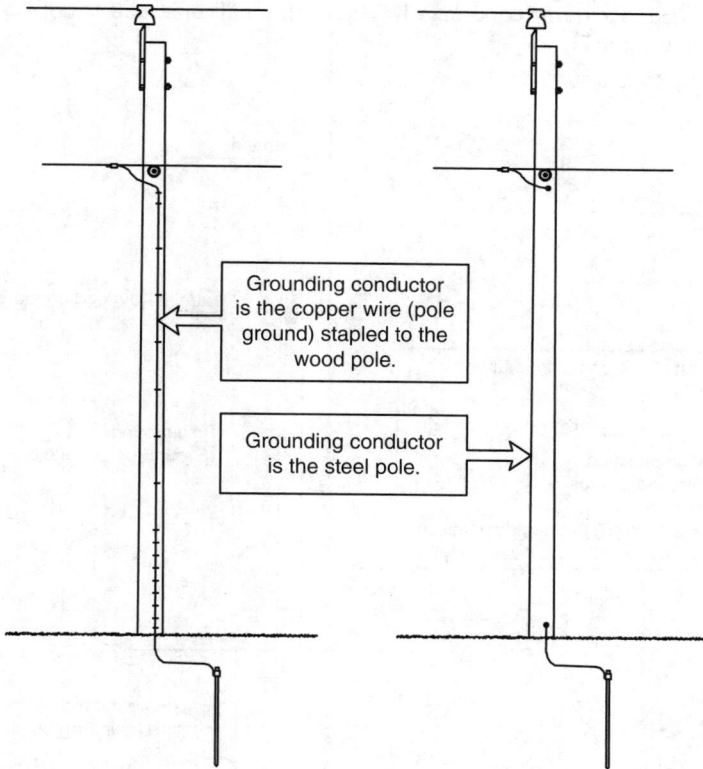

Grounding conductor is the copper wire (pole ground) stapled to the wood pole.

Grounding conductor is the steel pole.

See Photo(s)

Fig. 093-1. Example of a copper grounding conductor (pole ground) and a structural metal grounding conductor (steel pole) (Rule 093A).

The grounding conductor must not have a switching device connected to it. Some exceptions apply, including high-voltage DC systems, testing under competent supervision, and surge arrester operation. This rule provides an important note stating that the normally grounded base of the surge arrester may be at line potential (fully energized) following the operation of the disconnector.

093B. Connection of Grounding Conductors. The connection between the grounding conductor (pole ground) and grounded conductor (neutral) must be made considering the metals involved and exposure to the environment. The connector must not corrode and must be rated for the type of metals it is connecting. Dissimilar metals connected together with an improper connector will set up a battery action which will accelerate corrosion. Soldering is not acceptable, except on lead sheath cable, as fault currents will produce enough heat to melt the solder. Suitable connection methods and clarification of the terms grounded and grounding are shown in Fig. 093-2.

093C. Ampacity and Strength. This rule defines short-time ampacity requirements for bare and insulated grounding conductors. A bare conductor can carry a larger fault current than an insulated conductor of the same size because the bare conductor is only limited by melting or damaging the conductor material. The insulated grounding conductor has the additional constraint of not damaging the insulation. See Fig. 093-3.

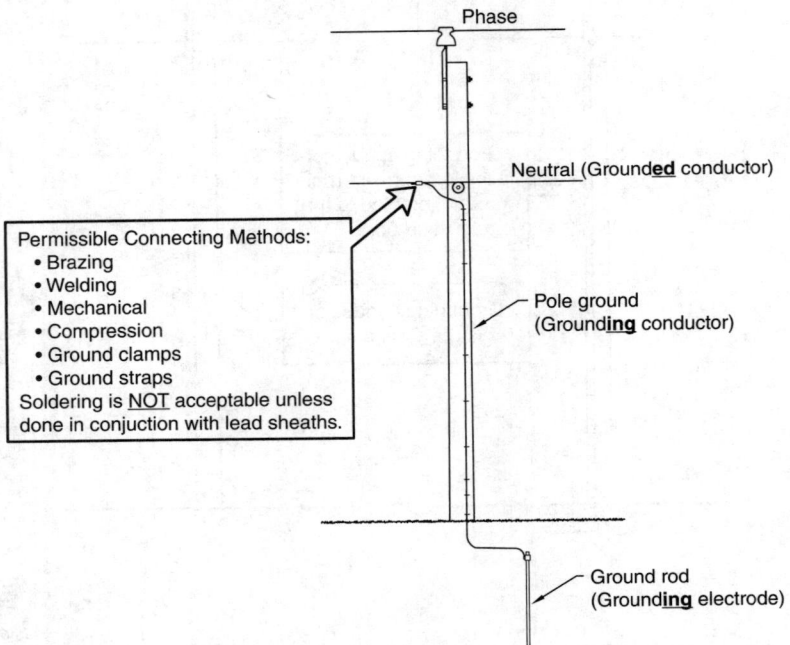

Fig. 093-2. Connection of grounding conductor to grounded conductor (Rule 093B).

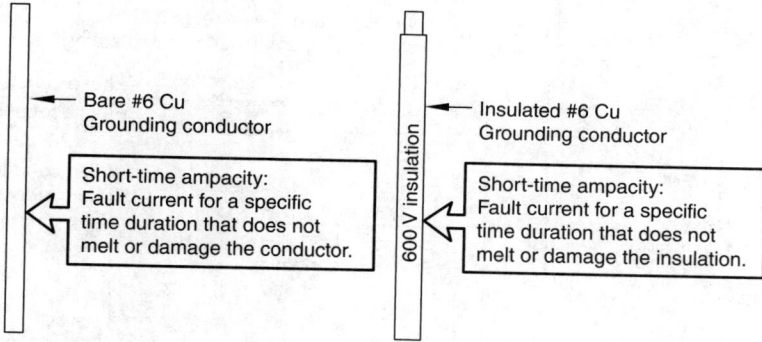

Fig. 093-3. Short-time ampacity of bare and insulated grounding conductors (Rule 093C).

Short-time ampacity of both bare and insulated conductors can be obtained from conductor manufacturers. This information is typically referred to as a conductor short-circuit withstand chart or a conductor damage curve.

Short-time ampacity for a single-grounded system is shown in Fig. 093-4.

Short-time ampacity for a multigrounded AC system is shown in Fig. 093-5.

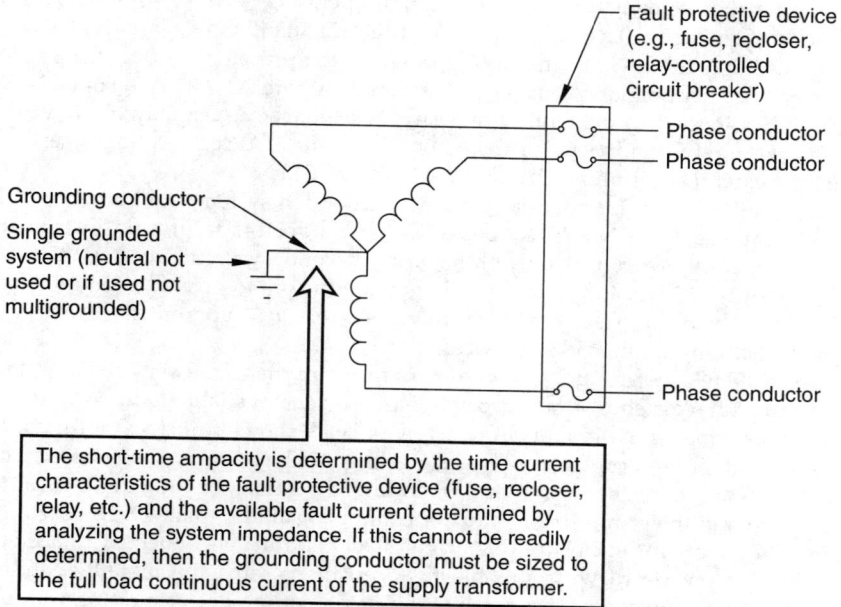

Fig. 093-4. System grounding conductor for single-grounded systems (Rule 093C1).

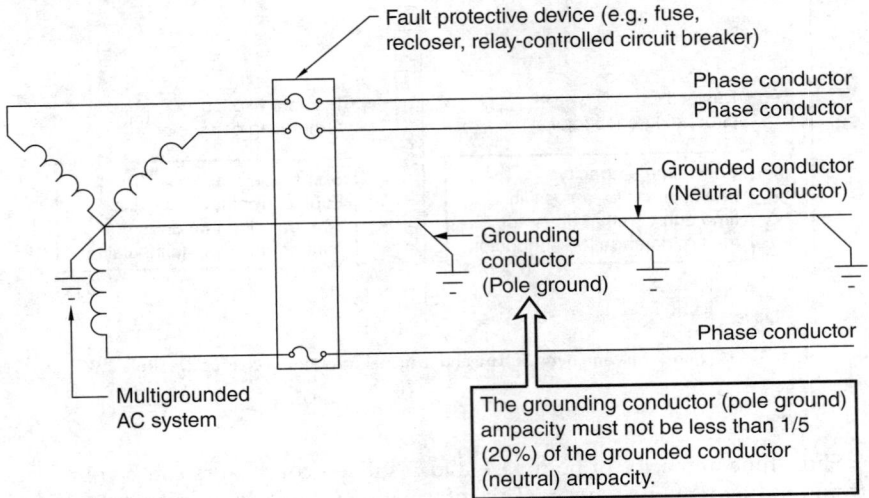

Fig. 093-5. System grounding conductors for multigrounded AC systems (Rule 093C2).

Rule 093C2 references Rule 093C8, which also specifies ampacity limits based on the ampacity of phase conductors and grounding electrode resistance. The one-fifth ampacity requirement applies to the normal operating current, not to the short-time fault ampacity. An example of pole ground sizing is shown in Fig. 093-6.

In addition to checking the pole ground to the primary neutral, the service transformer neutral should also be considered. A bare AWG No. 6 copper pole ground connected to the neutral of a large secondary service may not have the required one-fifth ampacity of the secondary neutral. Large secondary services require careful application of Rules 093C2 and 093C8.

In addition to single-grounded and multigrounded system requirements, Rule 093C requires AWG No. 12 copper or larger conductors to ground instrument transformer cases and instrument transformer secondary circuits. and AWG No. 6 copper or AWG No. 4 aluminum or larger conductors to ground primary surge arresters. The primary surge arrester rule has an exception permitting use of copper-clad or aluminum-clad steel wires.

Per Rule 093C5, grounding conductors for equipment, messenger wires, and guys must have a short-time ampacity based on the available fault current and operating time of the circuit protective device. If the circuit does not have an overcurrent or fault protection device (e.g., fuse, recloser, relay-controlled circuit breaker, etc.), then the design and operating conditions of the circuit must be analyzed and the grounding conductor cannot be smaller than AWG No. 8 copper. If a conductor enclosure (e.g., rigid steel conduit) is connected to a metal equipment enclosure with suitable lugs, bushings, etc., the metallic conduit and metallic equipment path can be used as an equipment-grounding conductor. Grounding conductors for equipment, messenger wires, and guys must be

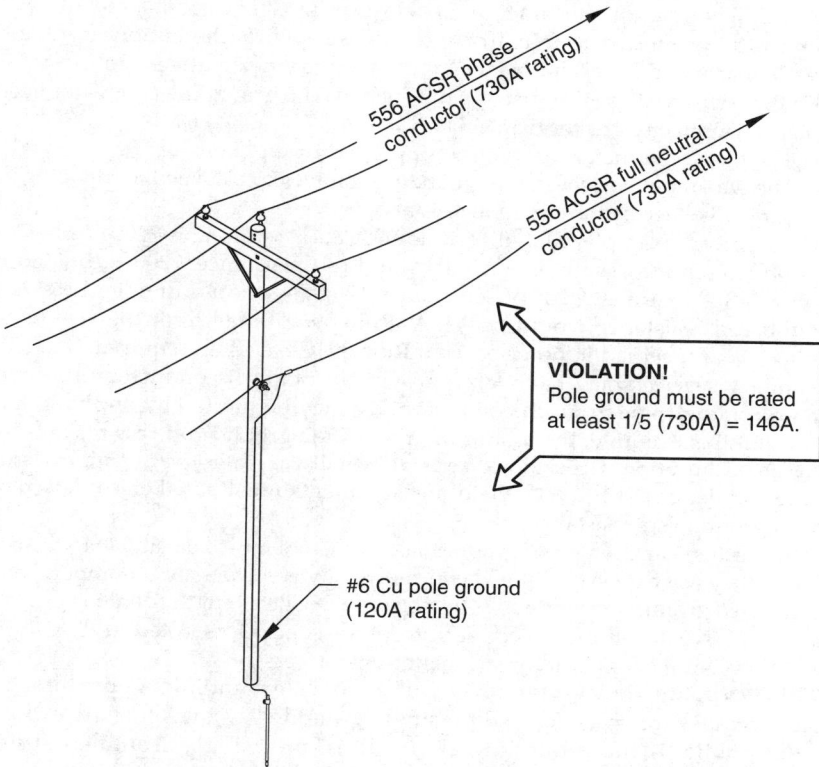

556 ACSR phase conductor (730A rating)

556 ACSR full neutral conductor (730A rating)

VIOLATION!
Pole ground must be rated
at least 1/5 (730A) = 146A.

#6 Cu pole ground
(120A rating)

See
Photo(s)

Fig. 093-6. Example of pole ground ampacity (Rule 093C2).

connected to a suitable lug, terminal, or other device without disruption during normal inspection, maintenance, or operation. The **Code** does not specifically address using the messenger or guy hardware as the bond between the messenger or guy and the grounding conductor. For example, a transmission tower static wire may be bonded to the steel tower through the static wire hardware or a separate bonding jumper may be used between the static wire and the transmission tower. Another example is an anchor guy may be bonded to a grounding conductor through the guy wire hardware or a separate bonding jumper may be used between the grounding conductor and the guy wire. Rule 012C, which requires accepted good practice, must be used for these examples. Use of a separate bonding jumper provides a connection that meets this rule, meets the definition of "bonding" in Sec. 09, and does not rely on hardware that can become loose or slack during the inspection, maintenance, or operation of the line.

The ampacity and strength of the grounding conductor used for grounding fences must also have adequate short-time ampacities or must be Stl WG No. 5 or larger.

Bonding of equipment frames and enclosures must consist of a metallic path back to the grounded terminal of the local supply. If the supply is remote, metallic parts within reach must be bonded and connected to ground.

Rule 093C8 specifies an ampacity limit such that no grounding conductor needs to have an ampacity greater than either:

- The phase conductor that would supply the ground fault, or
- The maximum current in the grounding conductor calculated by dividing the supply voltage by the electrode resistance

Consider an example related to Rule 093C8b. Assuming a 7200-V phase to ground circuit and assuming a 25-Ω ground rod resistance, 7200 V divided by 25 Ω = 288 A. For a 120/240-V secondary, 120 V to ground divided by a 25-Ω ground rod resistance would be 4.8 A. Rule 093C8 may limit the size of the ground wire specified in other parts of Rule 093C based on required ampacity. Secondary services may have large grounded (neutral) conductors; however, the grounding (pole ground) conductor size may be limited by applying Rule 093C8. In this example, the assumption of a 25-Ω ground rod resistance is just that, an assumption. Ground rod resistance will vary by type of soil, moisture in the soil, length of rod, etc. Field measurements must be taken to determine actual ground rod resistance.

The mechanical strength of grounding conductors must be suitable to the conditions they are exposed to (i.e., lawn mowers, weed eaters, car bumpers, etc.). Unguarded grounding conductors must have a tensile strength equal to or greater than AWG No. 8 soft-drawn copper except for conductors noted in Rule 093C3 (i.e., AWG No. 12 copper for instrument transformers).

093D. Guarding and Protection. Guards over grounding conductors (i.e., pole grounds) are only required for single-grounded systems that are exposed to the public. If the grounding electrode is on a single-grounded system that is not exposed to the public (e.g., in a fenced substation), it does not have to be guarded. Grounding conductors on multigrounded systems are not required to be guarded even if they are exposed to mechanical damages. A multigrounded system requires at least four grounds in each mile per Rule 096C, and Rule 214 requires inspection of overhead lines. These two requirements provide a method to assure safe grounding on multigrounded systems; therefore, guards on multigrounded systems are not required. If guards are not required but they are installed, they should be installed in a manner as if they were required.

Rules 239D and 360A provide additional guarding requirements for various types of conductors. If guarding of the grounding conductor is required, guards must be suitable for the damage to which they will be exposed and be at least 8 ft above the ground or other surface. If guarding of the grounding conductor is not required, a typical installation method is stapling the grounding conductor to a wood pole. If the grounding conductor that is not required to be guarded is exposed to mechanical damage, in addition to stapling, where practical, the grounding conductor is to be located on the side of the pole with least exposure to damage (e.g., away from car bumpers in a parking lot). The requirements for grounding conductors with or without guards are outlined in Fig. 093-7.

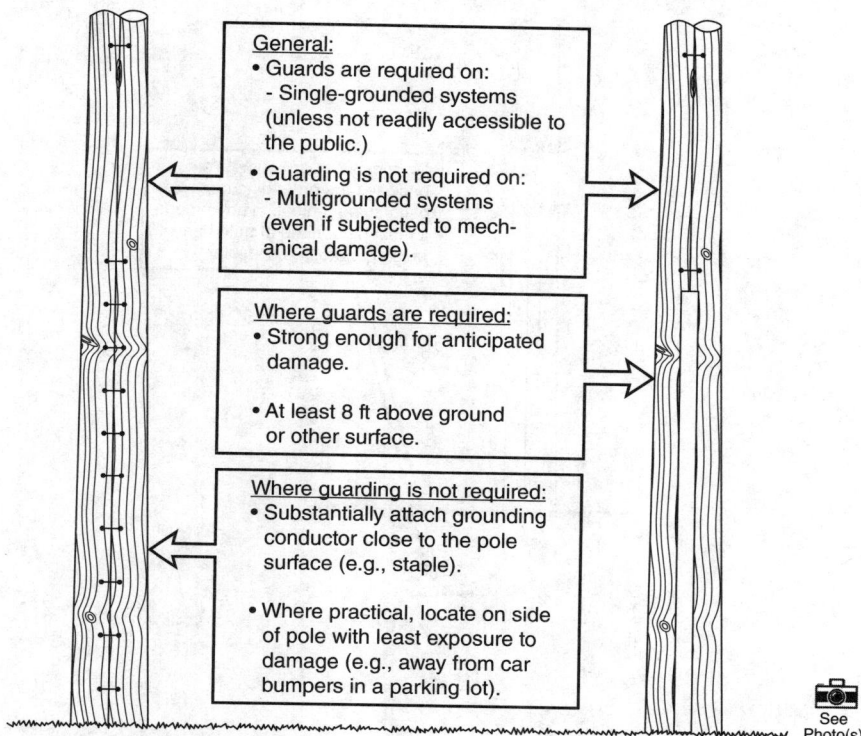

General:
• Guards are required on:
 - Single-grounded systems
 (unless not readily accessible to
 the public.)
• Guarding is not required on:
 - Multigrounded systems
 (even if subjected to mechanical damage).

Where guards are required:
• Strong enough for anticipated
 damage.

• At least 8 ft above ground
 or other surface.

Where guarding is not required:
• Substantially attach grounding
 conductor close to the pole
 surface (e.g., staple).

• Where practical, locate on side
 of pole with least exposure to
 damage (e.g., away from car
 bumpers in a parking lot).

See
Photo(s)

Fig. 093-7. Requirements for grounding conductors with or without guards (Rules 093D1, 093D2, and 093D3).

Rule 093D4 recognizes that an inductive choke is created when a conductor is run through a metallic raceway. This can create a hazardous voltage during a lightning strike (or even during a fault condition). The **Code** requires a nonmetallic guard to avoid this condition. The strength of nonmetallic materials (i.e., plastics) has increased to the point where they can be used for protection without cracking or breaking. A U-shaped metallic raceway is acceptable, as it does not completely enclose the grounding conductor. If a metallic guard similar to a steel pipe or rigid metal conduit is used, it must be bonded to the grounding conductor at both ends, as shown in Fig. 093-8.

093E. Underground. Grounding conductors laid underground require slack due to the settling of the earth. Direct-buried joints or splices must be made with corrosion resistance in mind. Splices should be kept to a practical minimum. A cable insulation shield (e.g., concentric neutral, metallic foil, braid, etc.) must be connected to other grounded supply equipment in manholes, handholes, and vaults. Exceptions exist for cathodic protection and cross bonding. An example of grounding interconnection in a manhole is shown in Fig. 093-9.

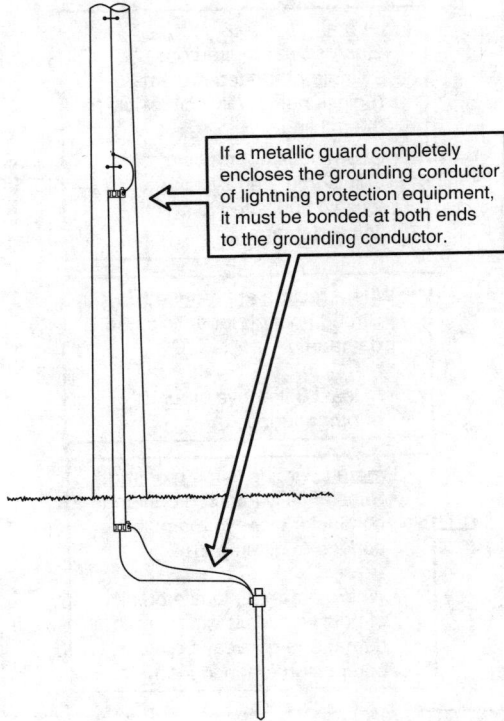

If a metallic guard completely encloses the grounding conductor of lightning protection equipment, it must be bonded at both ends to the grounding conductor.

Fig. 093-8. Requirements for a metallic guard that completely encloses the grounding conductor of lightning protection equipment (Rule 093D4).

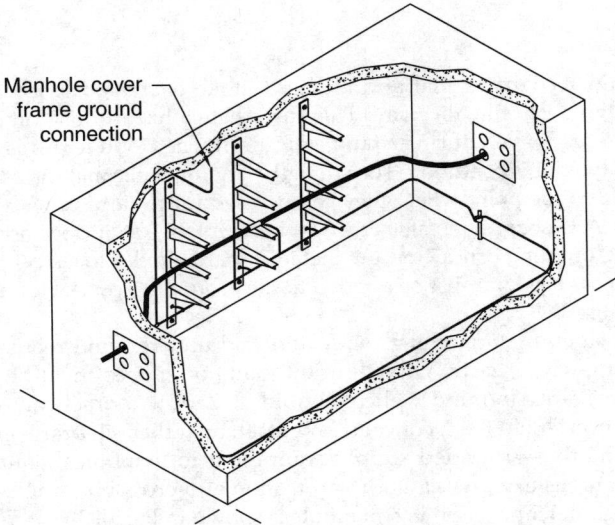

Manhole cover frame ground connection

Fig. 093-9. Example of grounding interconnection in a manhole (Rule 093E3).

Looped magnetic elements must not be positioned between the grounding conductor and the phase conductors. The metals used for grounding in earth, concrete, or masonry must not corrode. This rule specifically notes that aluminum is not generally acceptable when used underground. An example of an aluminum ground wire that transitions to copper for underground burial is shown in Fig. 093-10.

Sheath transposition connections, also termed cross bonding, are sometimes used to neutralize induced voltages and therefore eliminate or minimize circulating currents. Cross bonding of cable shields or sheaths involves insulating the cable shields or sheaths from ground at sectionalized points along the cable route. The insulation level for cross bonding the cable shields or sheaths must be 600 V or greater if required. The cross bonding jumpers must be sized to carry the available fault current. See Fig. 093-11.

093F. Common Grounding Conductor for Circuits, Metal Raceways, and Equipment. This rule allows one common grounding conductor for both the

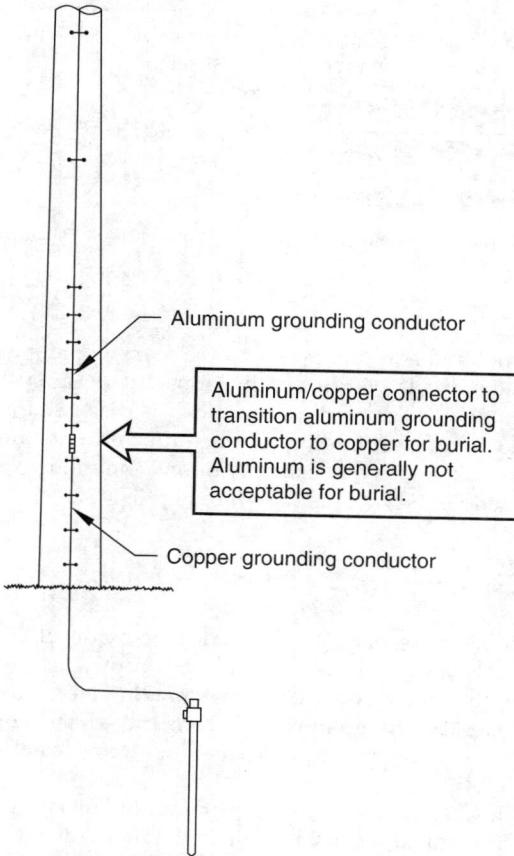

Aluminum grounding conductor

Aluminum/copper connector to transition aluminum grounding conductor to copper for burial. Aluminum is generally not acceptable for burial.

Copper grounding conductor

See Photo(s)

Fig. 093-10. Example of aluminum grounding conductor transitioning to copper for burial (Rule 093E5).

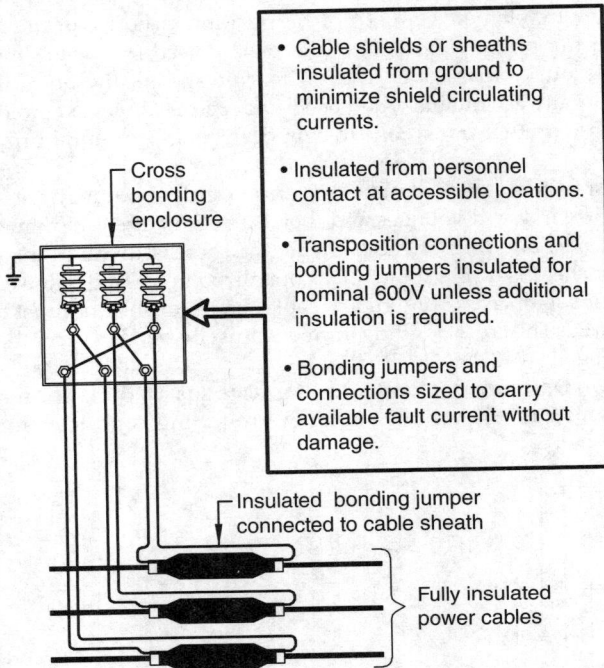

Fig. 093-11. Sheath transposition connections (cross bonding) (Rule 093E6).

supply system (neutral) and equipment (e.g., a recloser) where the ampacity of the grounding conductor is adequate for both. Ampacity for the system grounding conductor and equipment grounding conductor is discussed in Rule 093C. Rule 097 addresses a common grounding conductor for primary and secondary neutrals at transformer locations. An example of one common grounding conductor for the circuit and equipment is shown in Fig. 093-12.

094. GROUNDING ELECTRODES

Grounding electrodes can be existing electrodes or made electrodes. Existing electrodes are existing conductive items buried in the earth for a purpose other than grounding but can also serve as a grounding electrode. Most utilities use made electrodes, which are purposely constructed and buried to serve as grounding electrodes. Requirements for existing electrodes are outlined in Figs. 094-1 through 094-3.

Made electrodes must, as far as practical, penetrate the moisture level and be below the frost line. They must be metal or combined metals that do not corrode and they must not be painted, enameled, or covered in any way with an insulating

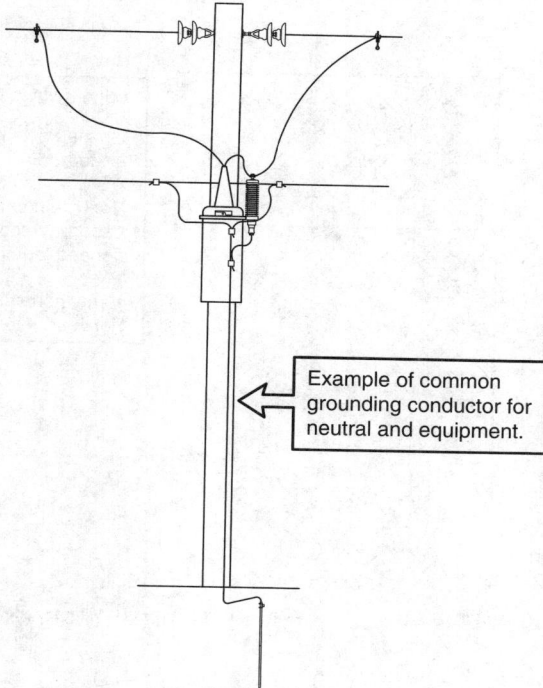

Example of common
grounding conductor for
neutral and equipment.

See
Photo(s)

Fig. 093-12. Example of common grounding conductor for neutral and equipment (Rule 093F).

material. For the purposes of this rule (primarily for strips, plates or sheets, and pole butt plates), stainless steel with appropriate non-corrosive properties is considered to be nonferrous metal. The driven ground rod is the most commonly used made electrode. However, buried wire, strips, and plates are considered equivalent if they meet the Code requirements. Other made

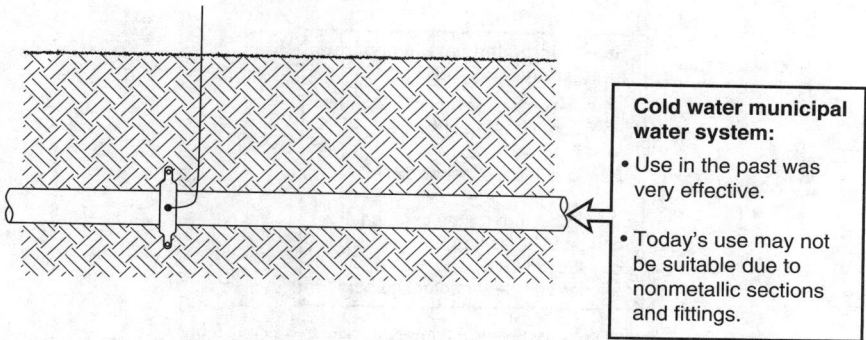

Cold water municipal water system:

• Use in the past was very effective.

• Today's use may not be suitable due to nonmetallic sections and fittings.

Fig. 094-1. Existing electrode—metallic water piping system (Rule 094A1).

Local (water piping) system:

- Isolated cold water piping system.

- Verify that parts that can become disconnected are bonded together.

- Verify nonmetallic piping is not used.

Local well

Submersible pump in well casing

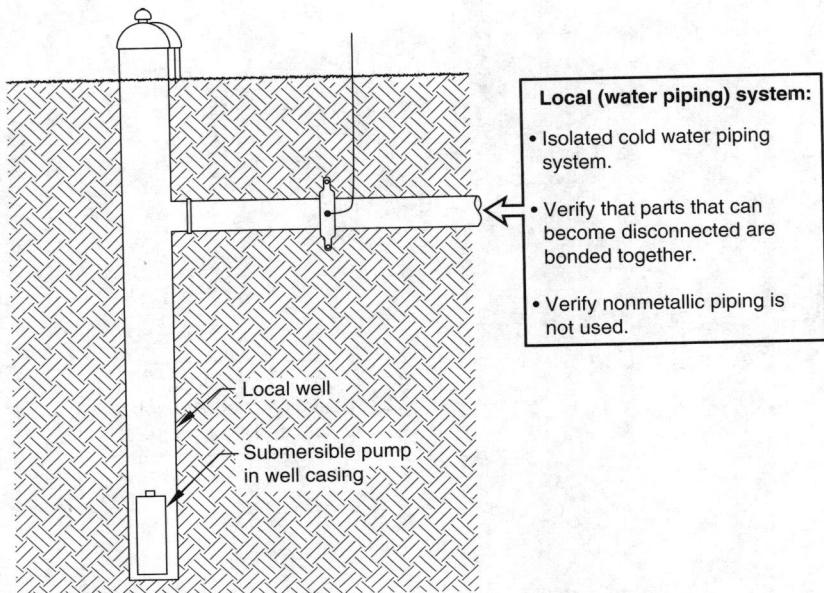

Fig. 094-2. Existing electrode—local (water piping) system (Rule 094A2).

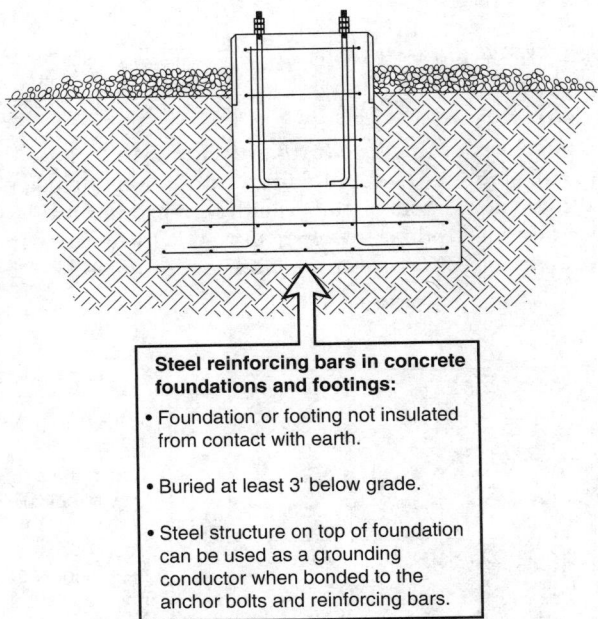

Steel reinforcing bars in concrete foundations and footings:

- Foundation or footing not insulated from contact with earth.

- Buried at least 3' below grade.

- Steel structure on top of foundation can be used as a grounding conductor when bonded to the anchor bolts and reinforcing bars.

Fig. 094-3. Existing electrode—steel reinforcing bars in concrete foundations and footings (Rule 094A3).

electrodes may be used if supported by a qualified engineering study. Many utilities require the ground rod to be located in undisturbed earth a fixed distance away from the pole hole, although no such requirement is provided in the **Code**. The rules for ground rods use the term "driven rods," which implies driven into the earth, not dropped in the pole hole or laid in a trench. Throughout Rule 094 the terms resistance and resistivity are used. Ground resistance of an electrode such as a ground rod is measured in ohms (Ω). Soil resistivity, which is a measure of how much the soil resists the flow of electricity, is measured in ohm-centimeters ($\Omega \cdot$ cm). The requirements for various types of made electrodes listed in the **Code** are outlined in Figs. 094-4 through 094-12.

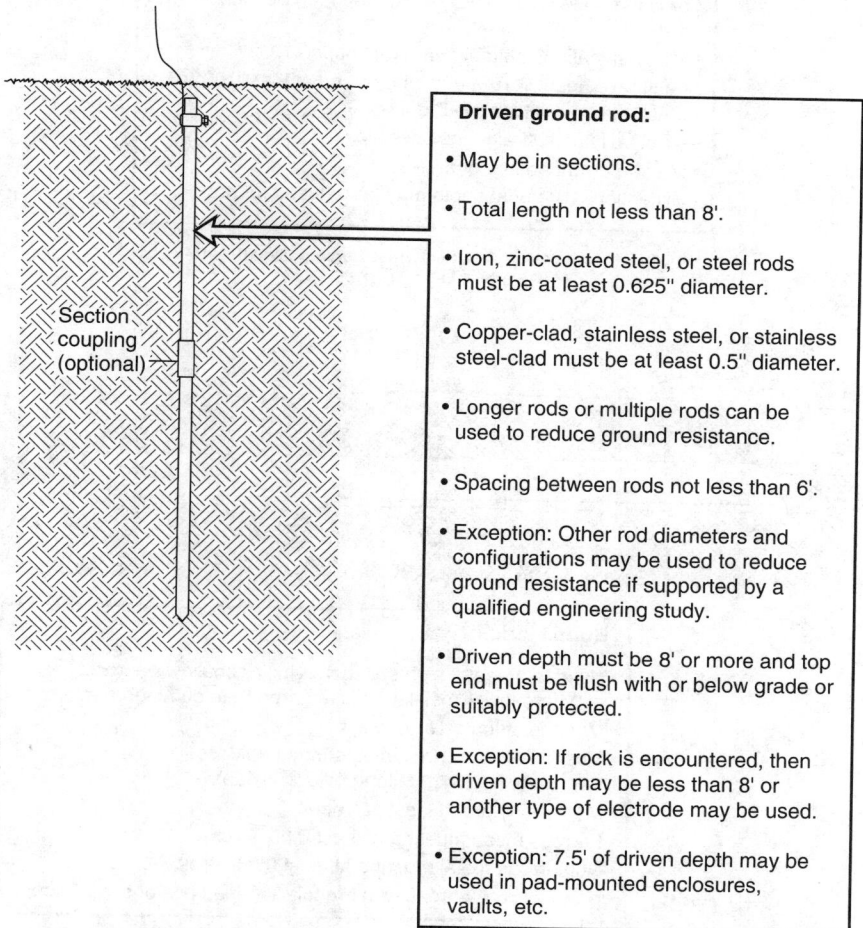

Driven ground rod:

• May be in sections.

• Total length not less than 8'.

• Iron, zinc-coated steel, or steel rods must be at least 0.625" diameter.

• Copper-clad, stainless steel, or stainless steel-clad must be at least 0.5" diameter.

• Longer rods or multiple rods can be used to reduce ground resistance.

• Spacing between rods not less than 6'.

• Exception: Other rod diameters and configurations may be used to reduce ground resistance if supported by a qualified engineering study.

• Driven depth must be 8' or more and top end must be flush with or below grade or suitably protected.

• Exception: If rock is encountered, then driven depth may be less than 8' or another type of electrode may be used.

• Exception: 7.5' of driven depth may be used in pad-mounted enclosures, vaults, etc.

Fig. 094-4. Made electrodes—driven ground rods (Rule 094B2a).

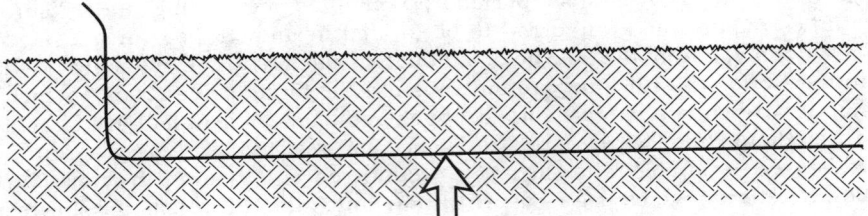

Buried wire (counterpoise):

- Used in areas of high soil resistivity, or shallow bedrock, or where lower resistance is required than obtainable with rods.
- Material must be suitable for direct burial.
- Must be at least 0.162" in diameter and at least 100' long.
- Must be buried at least 18" deep.
- Must be laid as straight as possible.
- May be arranged in a grid.
- Exception: 18" burial depth may be reduced for rock.

Fig. 094-5. Made electrodes—buried wire (counterpoise) (Rule 094B2b1).

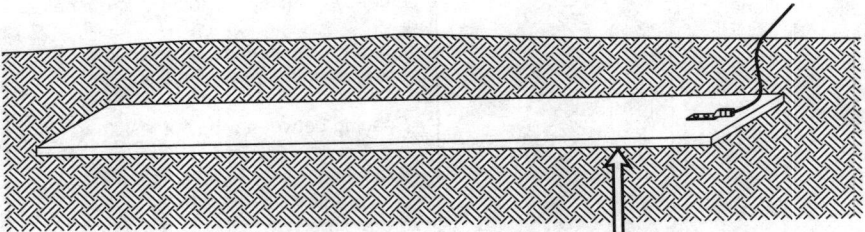

Buried strips:

- Used in areas of high soil resistivity, or shallow bedrock, or where lower resistance is required than obtainable with rods.
- Must be at least 10' in length.
- Must have total (two sides) surface not less than 5 sq. ft. (e.g., 10' long by 0.25' wide).
- Must be buried at least 18" deep.
- Ferrous metal must be at least 0.25" thick.
- Nonferrous metal must be at least 0.06" thick.
- Used for rocky areas with irregular-shaped pits of excavation.

Fig. 094-6. Made electrodes—buried strips (Rule 094B2b2).

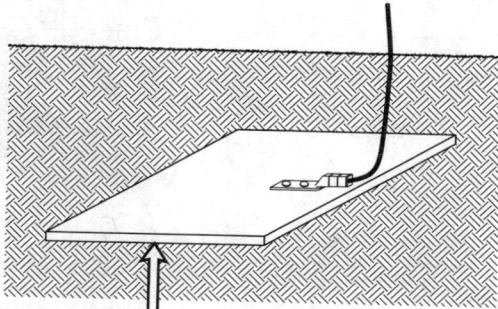

Buried plates or sheets:

• Used in areas of high soil resistivity, or shallow bedrock, or where lower resistance is required than obtainable with rods.
• Must have at least 2 sq. ft. of surface exposed to soil. Therefore, 1' × 1' if both top and bottom are exposed to soil. If the top was exposed to soil and the bottom was exposed to rock or if the bottom was exposed to soil and the top was exposed to the bottom and sides of a pole, then 1' × 2' would be required.
• Must be buried at least 5' deep.
• Ferrous metal must be at least 0.25" thick.
• Nonferrous metal must be at least 0.06" thick.

Fig. 094-7. Made electrodes—buried plates or sheets (Rule 094B2b3).

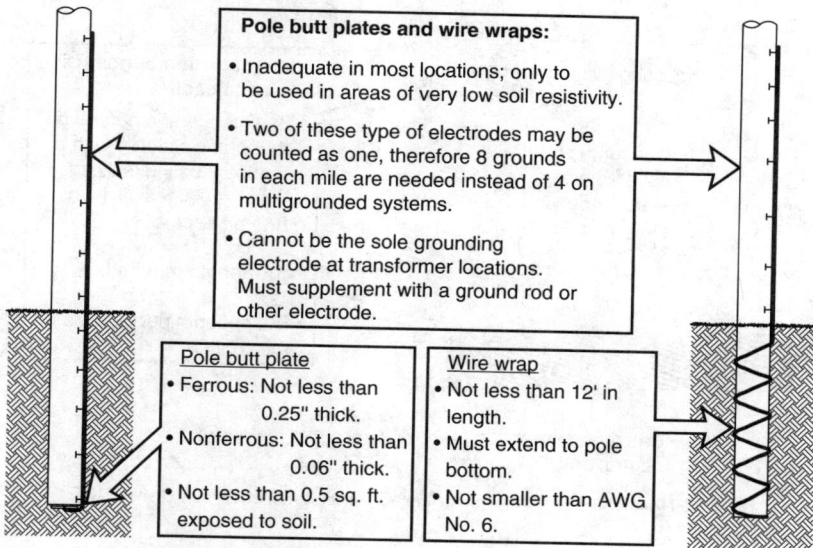

Pole butt plates and wire wraps:

• Inadequate in most locations; only to be used in areas of very low soil resistivity.

• Two of these type of electrodes may be counted as one, therefore 8 grounds in each mile are needed instead of 4 on multigrounded systems.

• Cannot be the sole grounding electrode at transformer locations. Must supplement with a ground rod or other electrode.

Pole butt plate
• Ferrous: Not less than 0.25" thick.
• Nonferrous: Not less than 0.06" thick.
• Not less than 0.5 sq. ft. exposed to soil.

Wire wrap
• Not less than 12' in length.
• Must extend to pole bottom.
• Not smaller than AWG No. 6.

Fig. 094-8. Made electrodes—butt plates and wire wraps (Rule 094B3).

VIOLATION!

Pole butt plates and wire wraps cannot be the sole grounding electrodes at transformer locations. A ground rod (or other electrode) must be added to these poles.

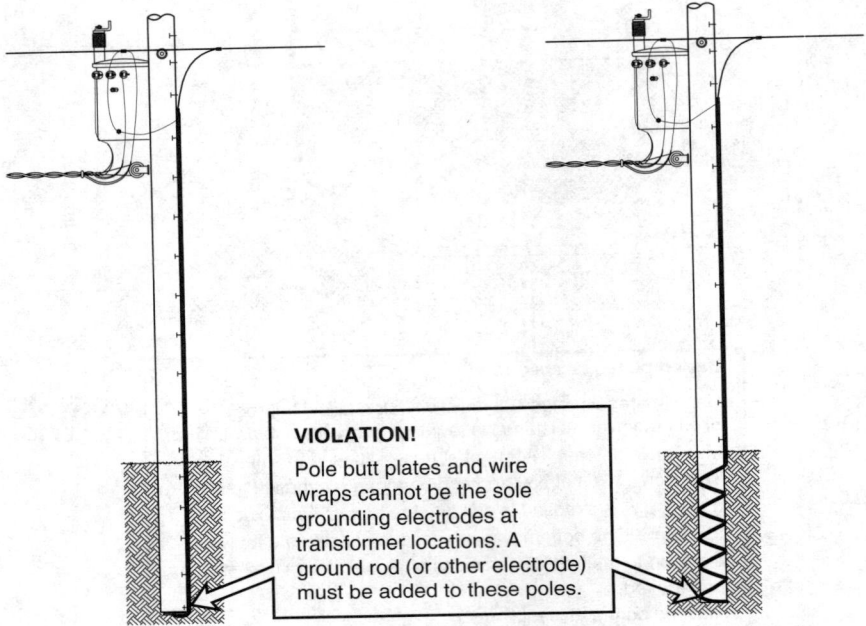

Fig. 094-9. Made electrodes—butt plates and wire wraps at transformer locations (Rule 094B3a).

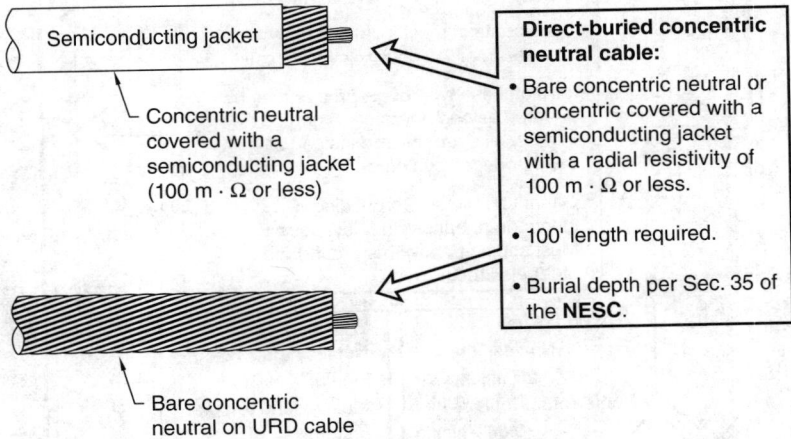

Semiconducting jacket

Concentric neutral covered with a semiconducting jacket (100 m · Ω or less)

Bare concentric neutral on URD cable

Direct-buried concentric neutral cable:
- Bare concentric neutral or concentric covered with a semiconducting jacket with a radial resistivity of 100 m · Ω or less.
- 100' length required.
- Burial depth per Sec. 35 of the **NESC**.

Fig. 094-10. Made electrodes—direct-buried concentric neutral cable (Rule 094B4).

Wire or rod
in concrete

Concrete-encased electrodes:

- Wire or rod which is suitable for burial and concrete encasement and not insulated from contact with the earth.
- Top of concrete not less than 1' below grade, 2.5' is recommended.
- Wire not smaller than AWG No. 4 if copper, or 3/8" diameter or AWG No. 1/0 if steel.
- Not less than 20' long inside the concrete except for the external connection.
- Run as straight as practical.
- Shorter pieces arrayed similar to a structural footing is acceptable.
- May be more practical or effective than driven rods, strips, or plates.
- Exception: Other lengths or configurations may be used per a qualified engineering study.

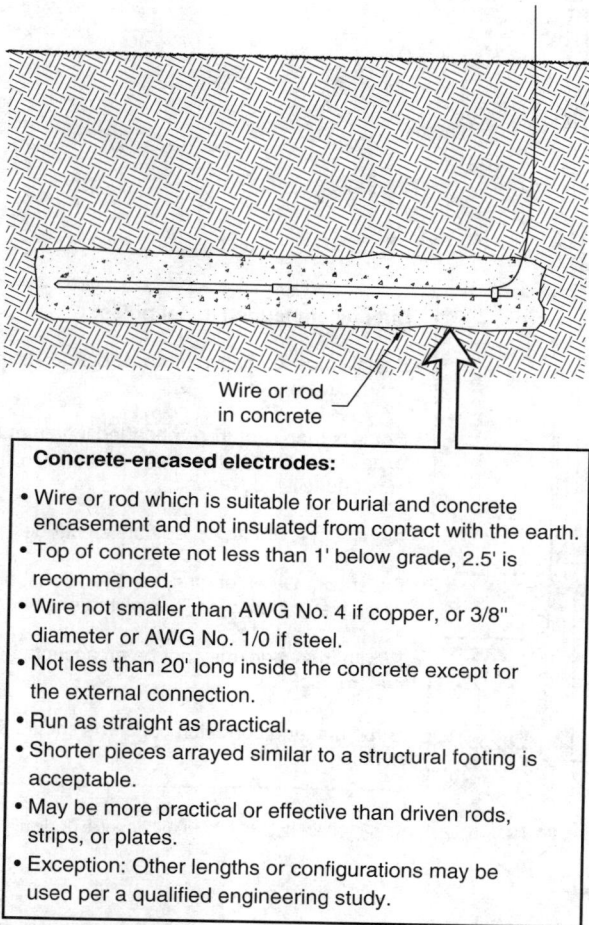

Fig. 094-11. Made electrodes—concrete-encased electrodes (Rule 094B5).

095. METHOD OF CONNECTION TO ELECTRODE

The connection to the grounding electrode must be permanent (except for removal due to inspection or maintenance) and be mechanically sound, corrosion-resistant, and have the required ampacity for the fault current to which it will be subjected. Suitable connection methods are shown in Fig. 095-1.

The **Code** also has specific rules for connecting to steel framed and non-steel-framed structures. The connection to water piping systems is also outlined. When

Directly embedded metal poles:

• Backfill must be native earth, concrete, or other conductive material.

• Not less than 5' of the embedded length must be exposed directly to the earth, without a nonconductive covering.

• Aluminum is not an acceptable electrode.

• Exception: Other lengths, configurations, or metal types may be used per a qualified engineering study.

• Weathering steel may not be an acceptable electrode.

• Structural and corrosion concerns should be investigated.

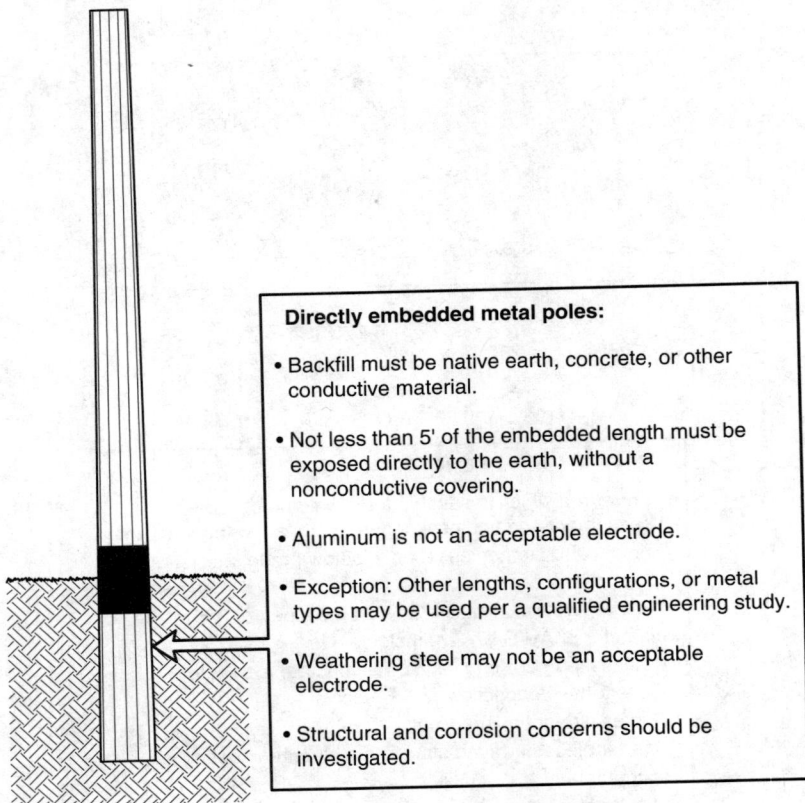

Fig. 094-12. Made electrodes—directly embedded metal poles (Rule 094B6).

water piping is used as the grounding electrode, bonds must be made around meters or other removable fittings.

The **Code** (in Sec. 094, "Grounding Electrodes") does not list gas piping as an acceptable electrode. Made electrodes or grounded structures should be separated from high-pressure (150 lb/in^2 or greater) pipelines containing flammable liquids or gases by a distance of 10 ft or more. No distances are specified for separating grounding electrodes from low-pressure gas lines. High-pressure pipelines are used as transmission facilities. Low-pressure gas lines are most commonly used to supply natural gas to homes. The requirements for separating grounding electrodes from high-pressure pipelines are shown in Fig. 095-2.

Rule 095C requires that the connection to the grounding electrode be free from rust, enamel, or scale. This may be done by cleaning or using fittings that penetrate such coatings.

Phase

Neutral (Ground**ed** conductor)

Pole ground
(Ground**ing** conductor)

Permissible Connecting Methods:
• Clamp
• Fitting
• Braze
• Weld
• Bronze plug tightly screwed

Ground rod
(Ground**ing** electrode)

See
Photo(s)

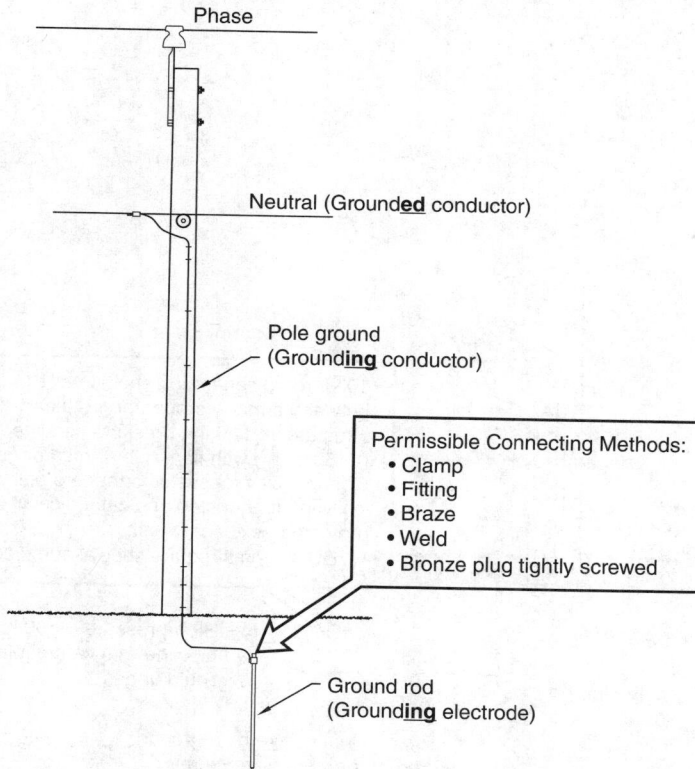

Fig. 095-1. Connection of grounding conductor to grounding electrode (Rule 095A).

096. GROUND RESISTANCE REQUIREMENTS

096A. General.   The main intent of Rule 096 is to assure a grounding resistance low enough to permit prompt operation of circuit protective devices (e.g., fuses, reclosers, relay-controlled circuit breakers, etc.).

096B. Supply Stations.   Supply stations typically require extensive grounding systems consisting of a ground grid or mat combined with grounding electrodes. They are designed to limit touch, step, mesh, and transferred potentials. The Code notes IEEE Standard 80 as a reference for substation grounding. Typically, the design of a substation ground grid starts with taking earth (soil) resistivity measurements. Earth resistivity is a measure of how much the soil resists the flow of electricity and is commonly expressed in ohm-centimeters ($\Omega \cdot$ cm). The final ground resistance (not resistivity) of the substation ground grid is measured in ohms (Ω) or sometimes a fraction of one ohm (Ω). Rules 092E and 093C6

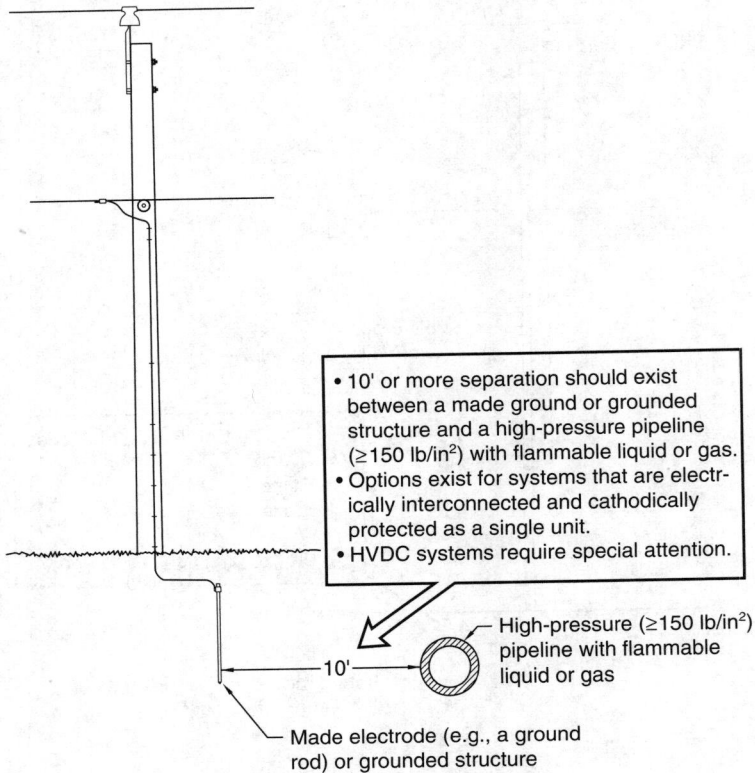

Fig. 095-2. Grounding electrode separation from high-pressure pipelines (Rule 095B2).

apply to grounding the fence enclosing the electric supply station. The requirements of Rule 096B are outlined in Fig. 096-1.

096C. Multigrounded Systems. Multigrounded systems are the most common type of distribution system. A typical 12.47/7.2-kV, three-phase, four-wire grounded-wye distribution system is multigrounded. For a system to be multigrounded, the following must occur:

- The circuit must have a neutral of sufficient size and ampacity.
- The neutral must be connected to a grounding electrode at each transformer location.
- The neutral must be connected to a grounding electrode not less than four times in each mile of the entire line. The grounds at transformers can be counted in the four grounds in each mile, but the grounds at individual services (i.e., meters) cannot be counted.

The intent of a multigrounded system is to always carry a neutral and to have not less than four grounds in each mile of the entire line. This results in grounds being placed approximately 1/4 mile or shorter apart, although some intervals may

> • Supply stations may require extensive grounding systems consisting of buried conductors, grounding electrodes, or interconnected combinations of both.
>
> • The grounding system must be designed to limit touch, step, mesh, and transferred potentials. (No specifics are provided for conductor sizes, conductor spacing, or number of ground rods. IEEE Standard 80 is noted as a reference.)
>
> • The fence grounding requirements of Rules 092E and 093C6 also apply.

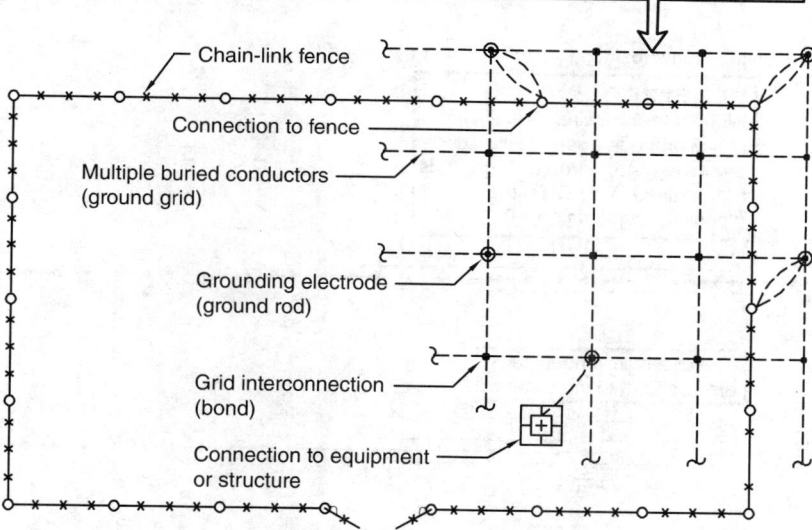

Fig. 096-1. Ground resistance requirements for supply stations (Rule 096B).

be shorter or longer. Typically overhead lines in urban areas or underground lines in subdivision areas have lots of service transformers and therefore lots of ground connections. Lines in rural areas or express feeders without many services typically need a review for not less than four grounds in each mile. To check the four grounds in each mile requirement, a "one-mile window" can be used. Examples are shown in Fig. 096-2.

The **Code** does not specify a ground resistance for multigrounded systems. The **Code** notes that multigrounded systems are dependent on the multiplicity of grounding electrodes, not the ground resistance of any individual electrode. See Sec. 02, Definitions, for additional information on multigrounded, effectively grounded, and effectively grounded neutral conductor.

For underground installations where the supply cable has an insulating jacket over the concentric neutral or the supply cable is in conduit, the cable must be terminated and grounded so that there are not less than four grounds in each mile along the line. If an express direct-buried underground feeder is constructed with an insulating jacket but without frequent termination points, the cable jacket

Secondary service (meter)
(Ground CANNOT be
counted.)

— Pole on distribution line
  12.47/7.2 kV, 3Ø, 4W

One-mile window has
4 grounds.

Transformer pole
(Ground CAN be
counted.) —

Secondary service
conductor 120/240V,
1Ø, 3W

— Distribution substation   └ Pole ground

VIOLATION!
When one-mile window is moved,
only 3 grounds exist. Additional
grounding is required as 4 grounds
are required IN EACH MILE
(special exceptions apply).

The one-mile window can
also be bent to check taps.

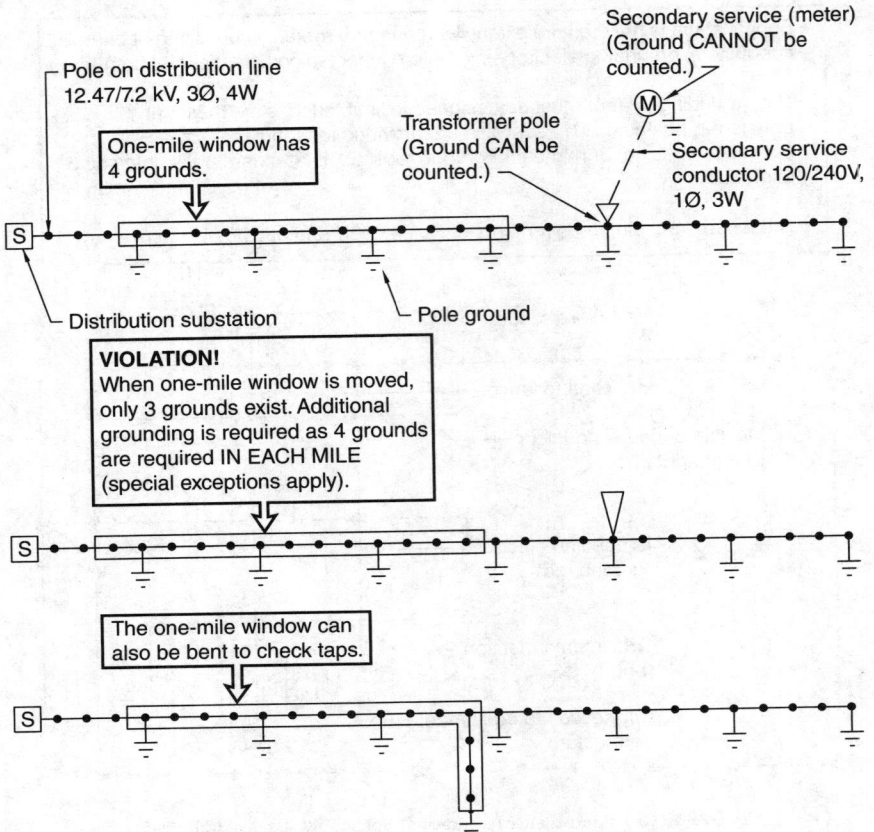

Fig. 096-2. Example of checking "four grounds in each mile" (Rule 096C).

must be stripped back and a suitable grounding electrode must be connected not less than four times in each mile along the line. If a supply cable has a semiconducting jacket, the cable can be treated similar to a bare concentric neutral cable. The semiconducting jacket must not exceed 100 m · Ω radial resistivity per Rule 094B5. Use of semiconducting jacketed cable is not very common due to the fact that these cables are higher in cost than insulated jacketed cable.

Rule 096C provides three exceptions to the four grounds in every mile requirement. The exceptions are outlined below:

- Underwater crossings.
- Underground where the cable is not accessible and would require removing the protective jacket only to install a ground.
- Overhead for special terrain areas such as river crossings or mountainous areas.

For these exceptions, the neutral conductor must be of sufficient size and ampacity. For the underwater and underground exceptions, the neutral conductor must be effectively grounded at locations that are accessible to personnel. For the overhead special terrain exception, all available structures should be grounded. Grounding on each side of the exceptions should be given special attention to make up for any lack of grounding in the area of the exceptions.

Rule 096C provides a note that discusses using this rule for shield wires (also referred to as overhead ground wires, static wires, and surge-protection wires) which are typically located at the top of transmission lines. See Fig. 096-3.

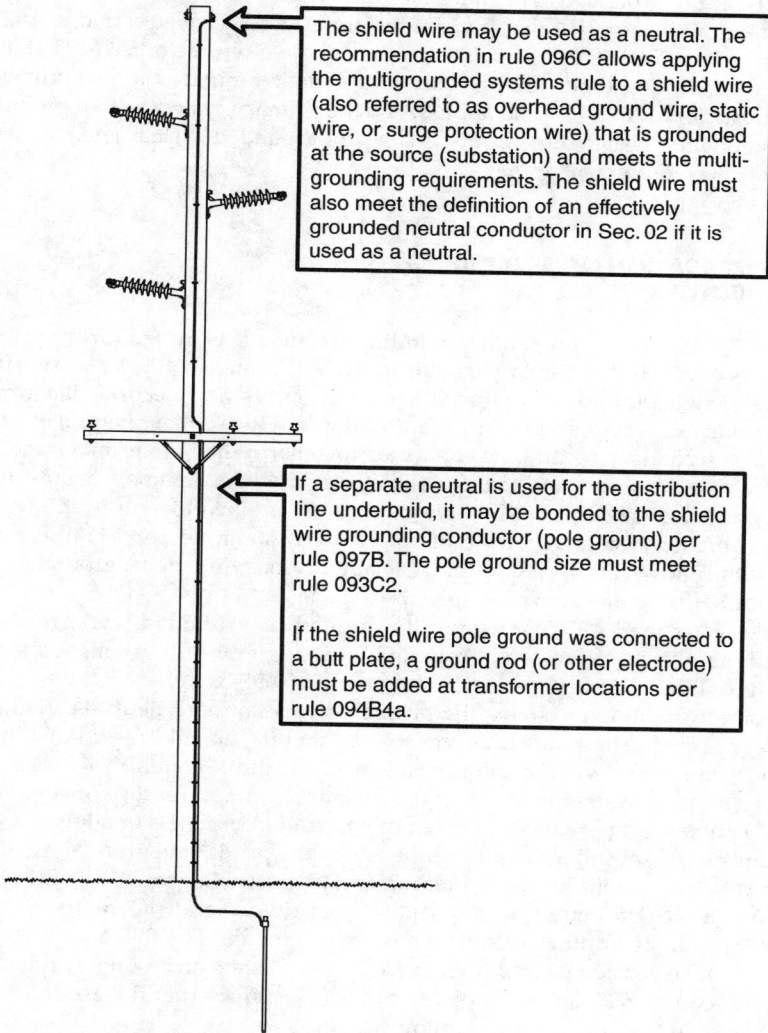

The shield wire may be used as a neutral. The recommendation in rule 096C allows applying the multigrounded systems rule to a shield wire (also referred to as overhead ground wire, static wire, or surge protection wire) that is grounded at the source (substation) and meets the multi-grounding requirements. The shield wire must also meet the definition of an effectively grounded neutral conductor in Sec. 02 if it is used as a neutral.

If a separate neutral is used for the distribution line underbuild, it may be bonded to the shield wire grounding conductor (pole ground) per rule 097B. The pole ground size must meet rule 093C2.

If the shield wire pole ground was connected to a butt plate, a ground rod (or other electrode) must be added at transformer locations per rule 094B4a.

Fig. 096-3. Recommendation permitting applying the multigrounded systems rule to shield wires (Rule 096C).

There are instances in the Code where eight grounds in each mile of line are required instead of four grounds in each mile. Rule 092C1 requires eight grounds in each mile for a communications messenger that is not an adequate grounding conductor. Rule 094B4 requires eight grounds in each mile for pole butt plates and wire wraps. Rule 354D3 requires eight grounds in each mile for direct-buried power and communications conductors in random separation (less than 12 in apart).

096D. Single-Grounded (Unigrounded or Delta) Systems. Single-grounded systems, typically grounded wye transmission systems, that do not carry a neutral and are grounded only at the source transformer must have a ground resistance not exceeding 25 Ω. This rule states that if a single electrode exceeds 25 Ω, then other grounding methods must be used. If the single-grounded system originates in a substation, Rule 096B also applies. A delta primary system with a grounded wye secondary system also results in a single grounded system and the not to exceed 25 Ω requirement applies.

097. SEPARATION OF GROUNDING CONDUCTORS

Rule 097A requires that separate grounding conductors be run for primary surge arresters over 750 V, secondary circuits under 750 V, and shield wires. Rule 097B allows a single grounding conductor and single grounding electrode if a ground connection exists at each surge arrester location and the primary neutral or shield wire and secondary neutral are connected together. When the primary and secondary neutrals are connected, Rule 097C requires the common neutral to be multigrounded (see Rule 096C). Rule 097A can be applied to a sufficiently heavy ground bus or system ground cable or Rule 097A can be applied in conjunction with Rule 097D1. Examples of these applications for a delta–grounded-wye transformer bank are shown in Figs. 097-1 and 097-2.

Rules 097B and 097C are typically applied to grounded-wye–grounded-wye three-phase systems and grounded-wye single-phase systems fed from a multigrounded primary system as shown in Fig. 097-3.

On multigrounded systems the primary and secondary neutrals should be interconnected. The NESC uses the word "should" in this case, not "shall," as there are times when separation of primary and secondary neutrals on a multigrounded system is applicable. The most common reason for separating primary and secondary neutrals on a multigrounded system is to minimize stray voltage on the secondary neutral imposed by the primary neutral. Normal and objectionable current in the grounding conductor (pole ground) is addressed in Rule 092D. The requirements separating primary and secondary neutrals for stray voltage or other valid reasons are outlined in Fig. 097-4.

If a made electrode is used to ground surge arresters on an ungrounded system exceeding 15 kV phase to phase, the NESC requires that the ground rod(s) be at least 20 ft from buried communication cables.

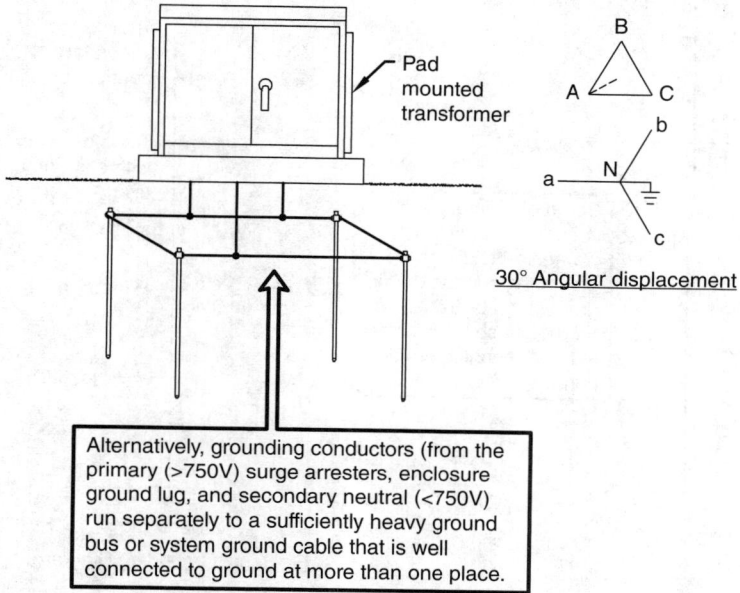

Pad
mounted
transformer

30° Angular displacement

Alternatively, grounding conductors (from the
primary (>750V) surge arresters, enclosure
ground lug, and secondary neutral (<750V)
run separately to a sufficiently heavy ground
bus or system ground cable that is well
connected to ground at more than one place.

Fig. 097-1. Example of grounding conductors from different voltage classes connected to a sufficiently heavy ground bus (Rule 097A).

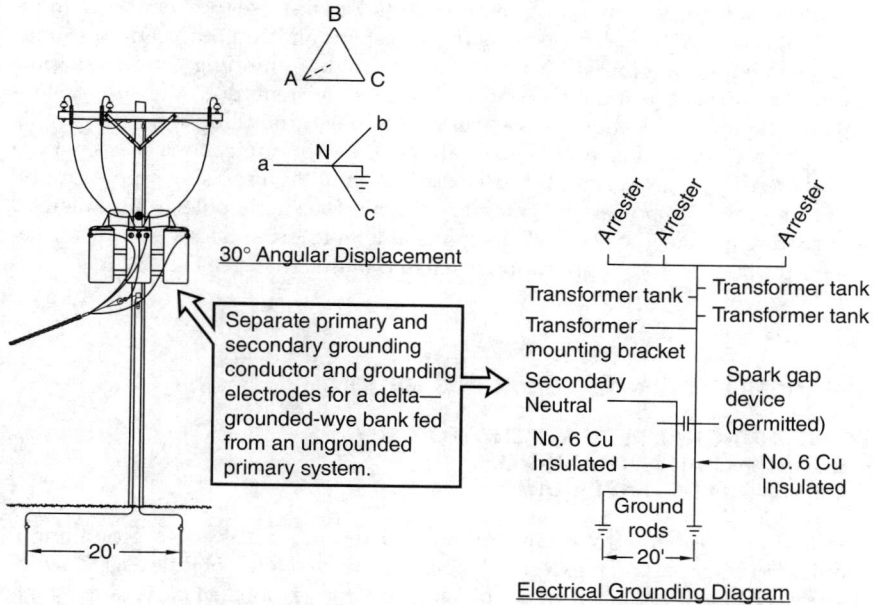

30° Angular Displacement

Separate primary and
secondary grounding
conductor and grounding
electrodes for a delta—
grounded-wye bank fed
from an ungrounded
primary system.

20'

Arrester   Arrester   Arrester

Transformer tank — Transformer tank
Transformer — Transformer tank
mounting bracket
Secondary   Spark gap
Neutral   device
No. 6 Cu   — (permitted)
Insulated —   — No. 6 Cu
                            Insulated
Ground
rods
20'

Electrical Grounding Diagram

Fig. 097-2. Example of separate primary and secondary grounding (Rules 097A and 097D1).

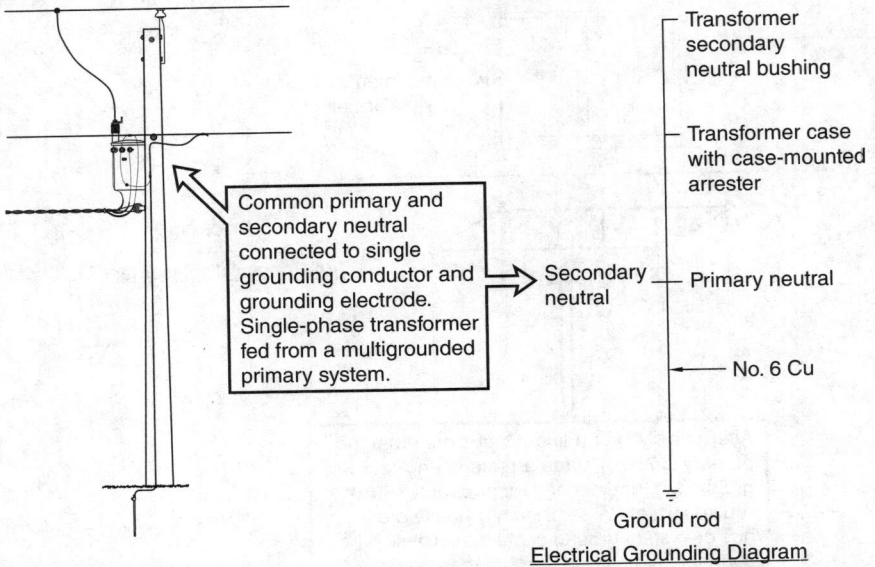

Transformer
secondary
neutral bushing

Transformer case
with case-mounted
arrester

Common primary and
secondary neutral
connected to single
grounding conductor and
grounding electrode.
Single-phase transformer
fed from a multigrounded
primary system.

Secondary
neutral

Primary neutral

No. 6 Cu

See
Photo(s)

Ground rod

Electrical Grounding Diagram

Fig. 097-3. Example of a common neutral with single grounding (Rules 097B and 097C).

Rule 097G focuses on bonding requirements for joint-use (power and communication) poles. Where both electric supply systems and communication systems are grounded on a joint-use structure and a single grounding conductor (pole ground) is present, it must be connected to both systems (i.e., the supply neutral and the communications messenger). If separate grounding conductors (pole grounds) are run to the supply neutral and the communications messenger, a bond between the pole grounds must exist. Most utilities use a single-pole ground for grounding both power and communications. The single-pole ground method will require a review for special cases like a delta to grounded-wye transformation or the stray voltage application discussed in this rule. See Fig. 097-5.

098. NUMBER 098 NOT USED IN THIS EDITION.

099. ADDITIONAL REQUIREMENTS FOR GROUNDING AND BONDING OF COMMUNICATION APPARATUS

This rule outlines how to ground communication apparatus when grounding is required in other parts of the Code. This rule references Note 2 of Rule 097D2, which discusses cooperation between supply and communications employees to isolate primary and secondary neutrals (typically for resolving stray-voltage problems).

Separating primary and secondary neutrals for stray voltage:

- Separate primary and secondary grounding conductors and grounding electrodes are required.

- Spark gap device with 60 Hz breakdown voltage of 3 kV or less must be used.

- Grounding electrodes must be 6' apart or more.

- Primary or secondary grounding conductors or both must be insulated for 600 V.

- The secondary grounding conductor must be guarded.

- Coordinate with phone, CATV, or other utilities to verify a communications ground or sheath is not maintaining primary and secondary bonding at the transformer or at a remote location (e.g., the customer's service).

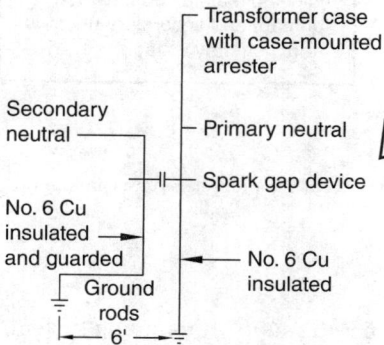

Transformer case with case-mounted arrester

Secondary neutral

Primary neutral

Spark gap device

No. 6 Cu insulated and guarded

No. 6 Cu insulated

Ground rods

6'

Electrical Grounding Diagram

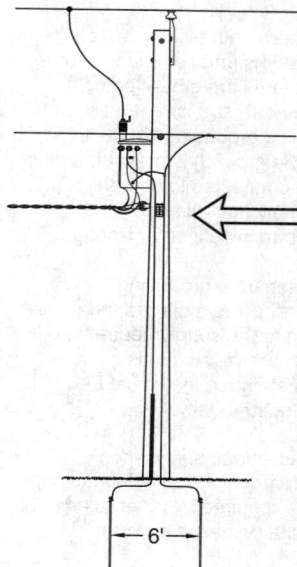

Fig. 097-4. Separating primary and secondary neutrals for stray voltage (Rule 097D2).

A communications grounding conductor shall preferably be made of copper or other material that will not corrode and shall not be less than AWG No. 6. The communications grounding conductor must be connected as shown in Fig. 099-1.

A separate communications ground rod is not required per Rule 099A. If a communications ground rod is used because a supply service does not exist, the communications ground rod may be smaller in diameter and length per the exception to Rule 099A3. However, if a supply service does exist and a com-

- Where both electric supply systems and communication systems are grounded on a joint-use structure, and a single grounding conductor (pole ground) is present, it must be connected to both the supply neutral and the communications messenger.

- If separate grounding conductors (pole grounds) are run to the supply neutral and the communications messenger, they must be bonded together.

- Exceptions apply when separate grounding conductors are required by other rules (e.g., delta primary systems).

- Exceptions apply when isolation is being maintained between primary and secondary neutrals (e.g., for stray voltage).

Supply Neutral

Communications Messenger

See Photo(s)

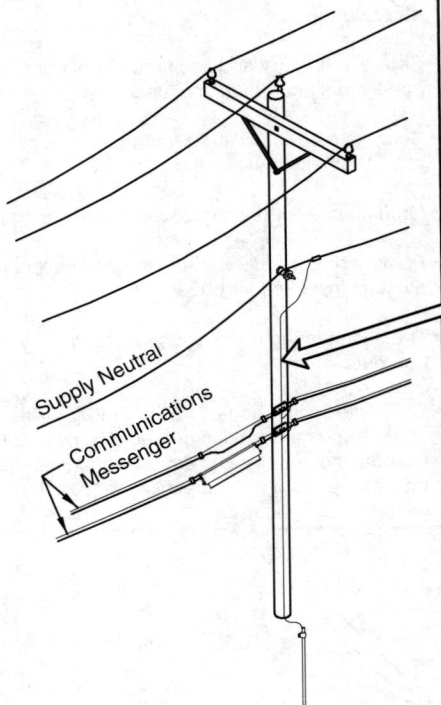

Fig. 097-5. Bonding of communication systems to electric supply systems on a joint-use (power and communication) structure (Rule 097G).

munications ground rod is used to supplement the supply grounding system, the exception to Rule 099A3 permitting smaller rods does not apply. Rule 099A does not prohibit a supplemental communications ground rod, but only if the supply service does not exist can the smaller communications-size ground rod be used. If a standard-size ground rod (per Rule 094B2) is used for communications grounding to supplement the supply ground rod, an AWG No. 6 copper or equivalent jumper must bond the two ground rods together. The No. 6 copper jumper is commonly used when the supply and communications service are adjacent to each other on a building. An equivalent jumper is typically used on a large building that has the supply service and communications service on opposite ends of the building. The requirements in this rule overlap the requirements in the National Electrical Code (NEC). The NEC should be reviewed to resolve any service entrance issues with the local building inspection authority. See Fig. 099-2.

Communications protector

Communications service conductor in conduit

Communications grounding conductor:
- Preferably copper or other noncorrosive material.
- AWG #6 minimum size.
- Attach to grounding electrode with bolted clamp or other suitable method.

Connect to:

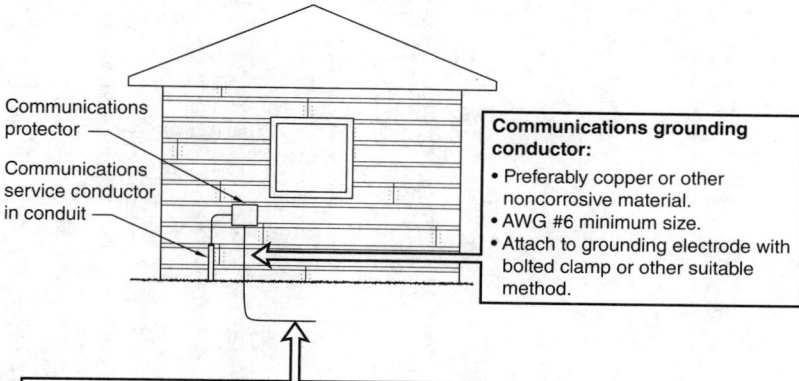

1. Grounded supply service electrode,
   Grounded supply service metallic conduit,
   Grounded supply service equipment enclosure,
   Grounded supply service grounding electrode conductor, or
   Grounded supply service grounding electrode metal enclosure.

2. If items in 1 are not available, then use existing grounding electrodes described in Rule 094A.

3. If items in 1 and 2 are not available, then use made electrodes described in Rule 094B. Exceptions apply to ground rod diameters and lengths.

Fig. 099-1. Additional requirements for communications grounding (Rules 099A and 099B).

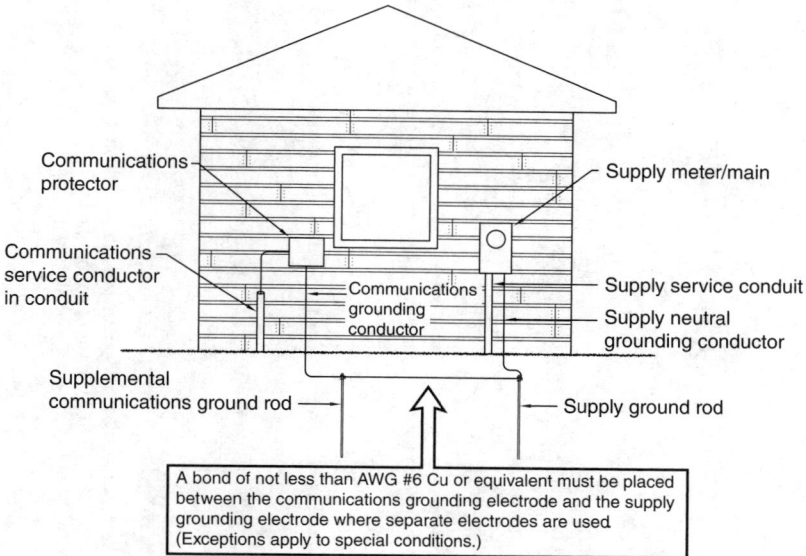

Communications protector

Supply meter/main

Communications service conductor in conduit

Communications grounding conductor

Supply service conduit

Supply neutral grounding conductor

Supplemental communications ground rod

Supply ground rod

A bond of not less than AWG #6 Cu or equivalent must be placed between the communications grounding electrode and the supply grounding electrode where separate electrodes are used (Exceptions apply to special conditions.)

Fig. 099-2. Bonding of communications and supply electrodes (Rule 099C).

Part 1

Safety Rules for the Installation and Maintenance of Electric Supply Stations and Equipment

GENERAL SECTIONS

01 INTRODUCTION   02 DEFINITIONS

03 REFERENCES   09 GROUNDING METHODS

GENERAL SECTIONS 01, 02, 03, 09

PART 1

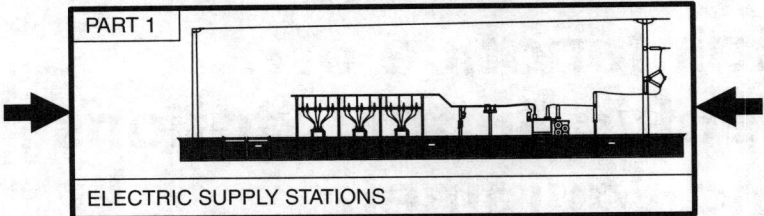

ELECTRIC SUPPLY STATIONS

PART 2

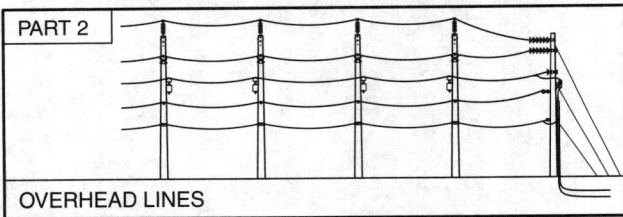

OVERHEAD LINES

PART 3

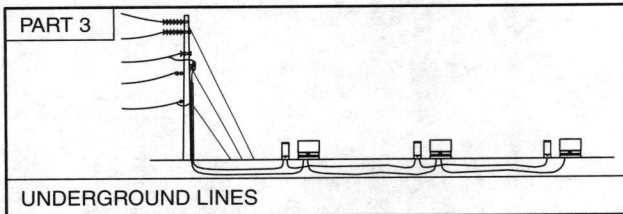

UNDERGROUND LINES

PART 4

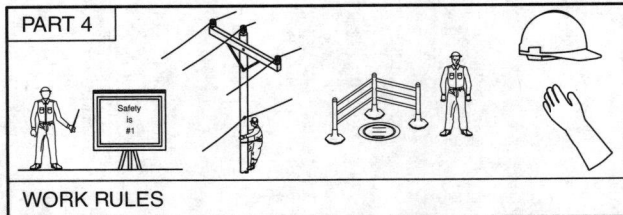

WORK RULES

Section 10

Purpose and Scope of Rules

100. PURPOSE

The purpose of Part 1, Electric Supply Stations, is similar to the purpose of the entire NESC outlined in Rule 010, except Rule 100 is specific to electric supply stations and equipment. Part 1 of the NESC focuses on the practical safeguarding of persons during the installation, operation, and maintenance of electric supply stations and equipment.

101. SCOPE

The scope of Part 1, Electric Supply Stations, includes electric supply conductors and equipment (in electric supply stations), and associated structural arrangements (in electric supply stations). Electric supply stations can consist of generating stations, substations, and switching stations. The term arrangements is important, as Part 1, Electric Supply Stations, provides rules for arranging items in electric supply stations for clearance purposes, but Part 1 does not provide strength and loading requirements in the form of ice and wind loads for the structural components inside the substation.

A key phrase in this rule is, "accessible only to qualified personnel." The rules of Part 1, Electric Supply Stations, assume that the general public is not exposed to the conductors and equipment located in the electric supply stations. For example, in Part 2, Overhead Lines, **NESC** Table 232-2 specifies a vertical clearance of 18.0 ft for a 12.47/7.2-kV, three-phase, four-wire rigid live part above a roadway, driveway, parking lot, or alley, and 14.0 ft when it is located in a pedestrian-only area. When this same 12.47/7.2-kV rigid live part is located inside a substation fence (accessible only to qualified personnel), **NESC** Table 124-1 specifies a vertical clearance of 9.0 ft for a 15-kV phase to phase, 110-kV BIL rigid live part above the ground or other accessible surface. This example shows that Part 1, Electric Supply Stations, is applicable when the supply facilities are accessible to qualified personnel only, via a fence, locked room, or other method (see Rule 110A), and that the clearance values are lower in electric supply stations than in Part 2, Overhead Lines, which are accessible to the general public.

The last sentence of Rule 101 clarifies the application of the **NESC** versus the National Electrical Code (NEC) to supply substations. Rule 011 discusses the scope of the **NESC** and the NEC. The **NESC** covers conductors and equipment in an electric supply station when they are serving a utility function (not an office building wiring function). The **NESC** electric supply station rules cover utility functions. Generation stations in particular and even the control buildings of substations and switching stations involve utilization wiring for lighting, ventilation, and controls. The **NESC** does not provide specific rules for utilization wiring. Rule 012C, which requires accepted good practice, must be applied when specific conditions are not covered. The NEC is an excellent reference for accepted good practice in this case.

As evidenced by reading all the rules in Part 1, Electric Supply Stations, the scope of Part 1 applies to both indoor and outdoor substations.

Communications utility personnel can skip over Part 1, Electric Supply Stations, as it does not apply to them. The definition of electric supply station in Sec. 02 includes generating stations and substations, but not communications central offices. A communications utility may provide service to a substation (e.g., a phone line, cable TV line, or fiber line) but the Part 1, Electric Supply Stations, rules do not address this installation. Rule 012C, which requires accepted good practice, must be applied. The communications protective requirements of Rules 223 and 315 provide guidance for overhead and underground lines near substations.

The definition of Electric Supply Station in Sec. 02 of the **NESC** includes generating stations, substations, and switching stations. Examples of various types of electric supply stations including generating stations and substations are shown in Fig. 101-1.

Typical nuclear, coal, or natural gas electric supply (generating) station.

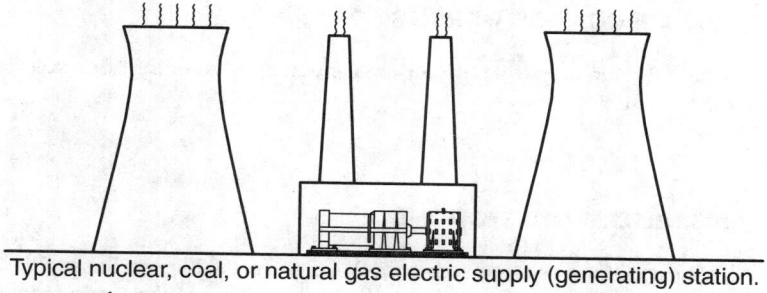

The scope of Part 1, Electric Supply Stations, includes generating stations and substations. See the definition of Electric Supply Station in **NESC** Sec. 02, definitions.

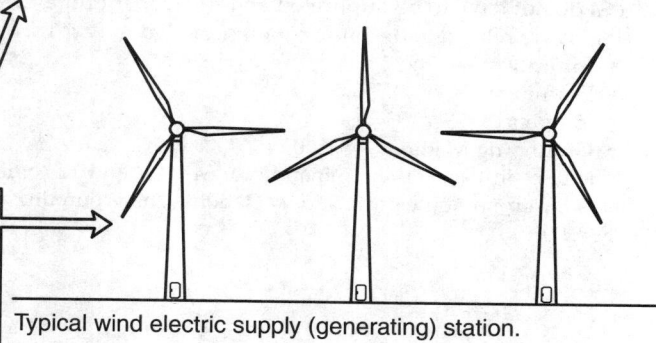

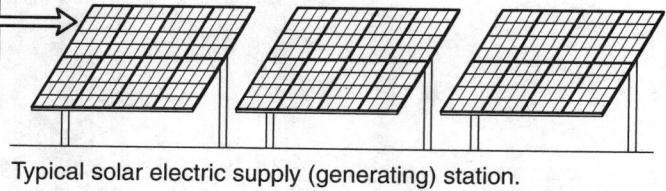

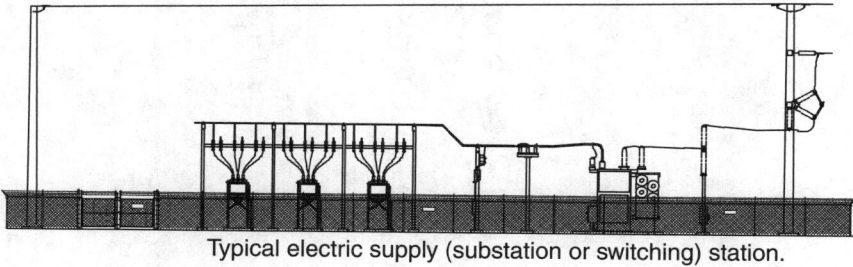

Typical wind electric supply (generating) station.

Typical solar electric supply (generating) station.

Typical electric supply (substation or switching) station.

Fig. 101-1. Examples of electric supply stations (Rule 101).

102. APPLICATION OF RULES

Rule 102 references Rule 013 for the general application of Code rules; see Rule 013 for a discussion.

103. REFERENCED SECTIONS

This rule references four sections related to Part 1, Electric Supply Stations, so that rules do not have to be duplicated and the reader of the Code realizes that other sections are related to the information provided in Part 1. The related sections are:

- Introduction—Sec. 01
- Definitions—Sec. 02
- References—Sec. 03
- Grounding Methods—Sec. 09

The rules in Part 1, predominantly Rules 110A and 123, will provide the requirements for grounding electric supply stations. The grounding methods are provided in Sec. 09.

Section 11

Protective Arrangements in Electric Supply Stations

110. GENERAL REQUIREMENTS

110A. Enclosure of Equipment.   Rule 110A is the defining rule of Part 1, Electric Supply Stations. If Rule 110A is met, the rules in Part 1, Electric Supply Stations, can be used. If Rule 110A is not met, then the rules in Part 2, Overhead Lines, or Part 3, Underground Lines, apply instead of Part 1, Electric Supply Stations. An example of how Rule 110A applies to Part 1, Electric Supply Stations, or Part 2, Overhead Lines, is shown in Fig. 110-1.

An example of how Rule 110A applies to Part 1, Electric Supply Stations, or Part 3, Underground Lines, is shown in Fig. 110-2.

The **NESC** discusses enclosures of rooms (for indoor applications) and spaces (for outdoor applications). The following barriers are required to enclose the room or space:

- Fences,
- Screens,
- Partitions, or
- Walls

The enclosure formed is required to "limit the likelihood" of entrance by unauthorized people (i.e., the general public) or unauthorized workers. Even the best prison system cannot avoid an escape, therefore, "limit the likelihood" is used rather than "prevent entry." The entrance to the room or space must be locked or under observation by an authorized attendant. This wording can become critical

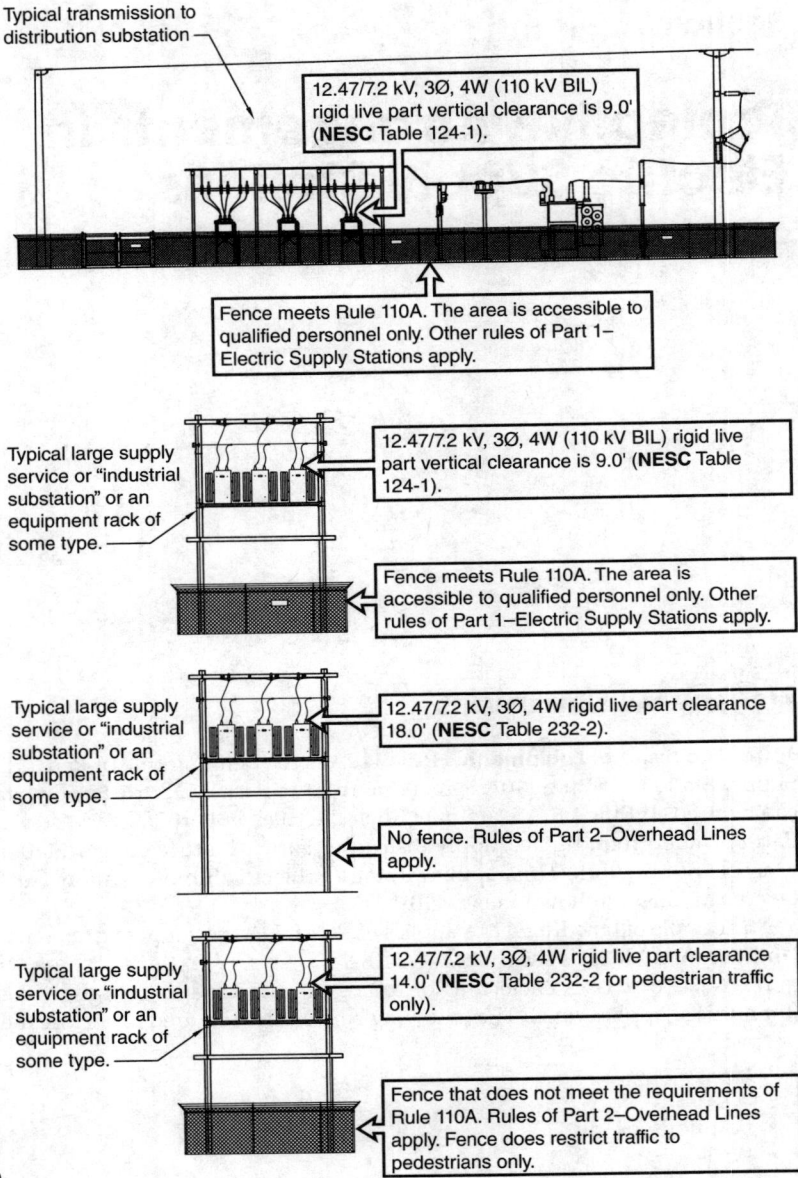

Typical transmission to
distribution substation

12.47/7.2 kV, 3Ø, 4W (110 kV BIL)
rigid live part vertical clearance is 9.0'
(**NESC** Table 124-1).

Fence meets Rule 110A. The area is accessible to
qualified personnel only. Other rules of Part 1–
Electric Supply Stations apply.

Typical large supply
service or "industrial
substation" or an
equipment rack of
some type.

12.47/7.2 kV, 3Ø, 4W (110 kV BIL) rigid live
part vertical clearance is 9.0' (**NESC** Table
124-1).

Fence meets Rule 110A. The area is
accessible to qualified personnel only. Other
rules of Part 1–Electric Supply Stations apply.

Typical large supply
service or "industrial
substation" or an
equipment rack of
some type.

12.47/7.2 kV, 3Ø, 4W rigid live part clearance
18.0' (**NESC** Table 232-2).

No fence. Rules of Part 2–Overhead Lines
apply.

Typical large supply
service or "industrial
substation" or an
equipment rack of
some type.

12.47/7.2 kV, 3Ø, 4W rigid live part clearance
14.0' (**NESC** Table 232-2 for pedestrian traffic
only).

Fence that does not meet the requirements of
Rule 110A. Rules of Part 2–Overhead Lines
apply. Fence does restrict traffic to
pedestrians only.

See
Photo(s)

Fig. 110-1. Example of how Rule 110A applies to Part 1, Electric Supply Stations, or Part 2,
Overhead Lines (Rule 110A).

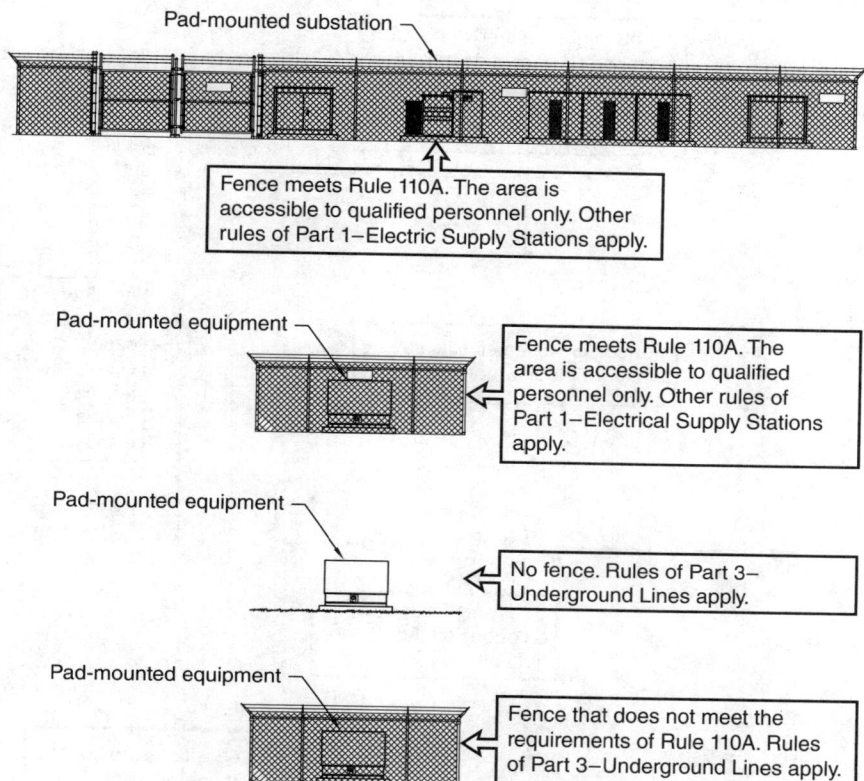

Pad-mounted substation

Fence meets Rule 110A. The area is accessible to qualified personnel only. Other rules of Part 1–Electric Supply Stations apply.

Pad-mounted equipment

Fence meets Rule 110A. The area is accessible to qualified personnel only. Other rules of Part 1–Electrical Supply Stations apply.

Pad-mounted equipment

No fence. Rules of Part 3–Underground Lines apply.

Pad-mounted equipment

Fence that does not meet the requirements of Rule 110A. Rules of Part 3–Underground Lines apply.

See Photo(s)

Fig. 110-2. Example of how Rule 110A applies to Part 1, Electric Supply Stations, or Part 3, Underground Lines (Rule 110A).

when utility employees are working inside a substation with the gate open. At least one employee must observe the unlocked gate, or the gate must be locked after the employees enter the substation.

The **Code** requires a safety sign on or beside the gate or door at each entrance of the electric supply station. Fenced or walled electric supply stations without roofs must have a safety sign located on each side of the fenced or walled enclosure. If the electric supply station is entirely enclosed by walls and a roof, a safety sign is only required at ground level entrances. Where entrance is gained through sequential doors, the safety sign should be located at the inner door position. Nothing in the **Code** prevents an additional sign at the outer door position if desired. The electric supply station may be a generating station, substation, or switching station. All of these installations may be enclosed in a building or by a fence. The requirements for safety signs for electric supply stations with and without roofs are shown in Fig. 110-3.

Although not required in the **Code**, two or more signs may be used on long sides so that at least one sign is in an approaching person's field of view.

Chain-link fenced substation

Fenced or walled electric supply station without a roof.

A safety sign is required on each exterior side of the fenced or walled enclosure.

Safety sign is required on or beside the door or gate at each entrance.

A safety sign is required only at ground level entrances.

If entrance is gained through sequential doors, the safety sign should be located on the inner door position.

Although not required in the code, nothing prevents an additional safety sign on the outer door position if desired.

Electric supply station entirely enclosed by walls and a roof.

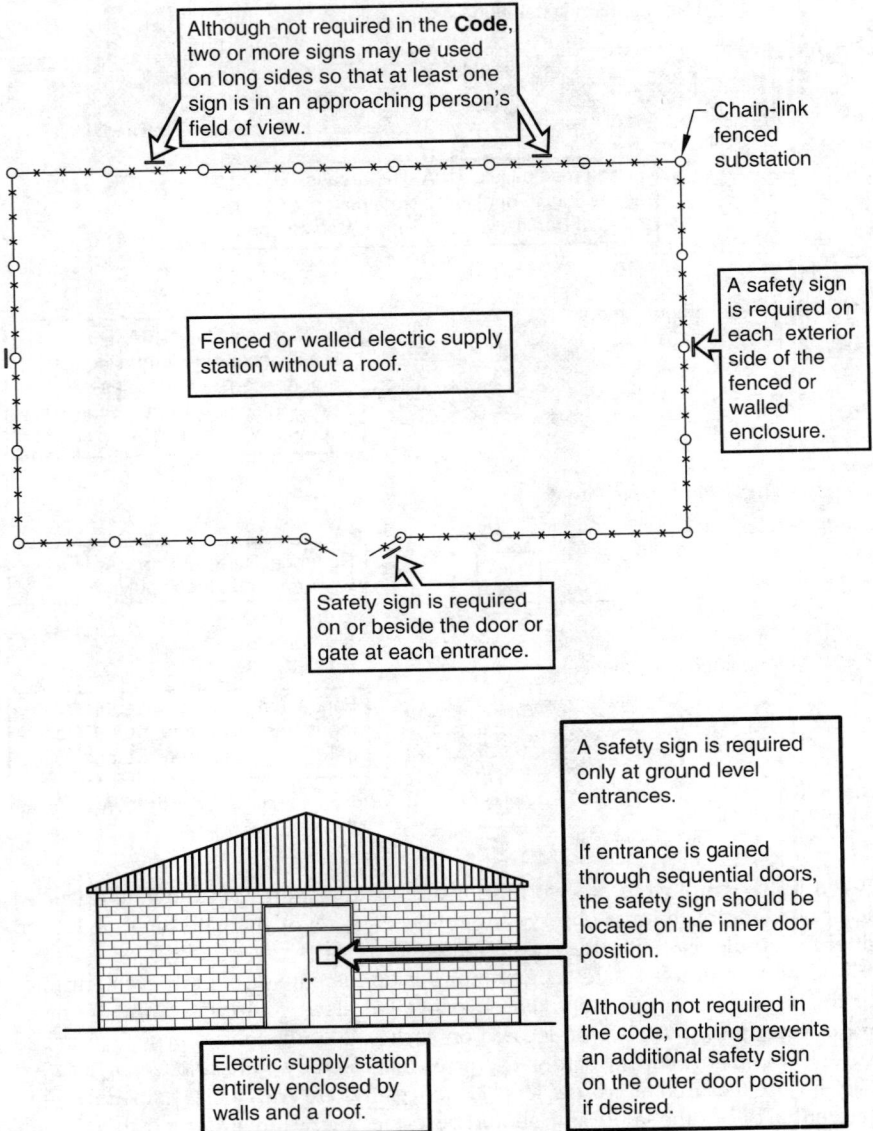

Fig. 110-3. Safety sign locations on an electric supply station with and without a roof (Rule 110A).

The **NESC** notes ANSI Z535 series documents for sign applications. Substation fences and pad-mounted transformers and enclosures are two of the most common signage applications for electric supply utilities. The ANSI Z535 approach to signage uses the philosophy that a "warning" sign is appropriate on the outer barrier (i.e., fence or enclosure), and if that barrier is breached, a "danger" sign is then appropriate. Traditionally, the "danger" sign was the most common choice for substation fence applications with little or no signage used inside the substation. Using the ANSI Z535 signage philosophy, "warning" signs would be placed on the substation fence and "danger" signs would be placed inside the substation on structures that support energized parts. This same philosophy can be applied to a pad-mounted transformer. A "warning" sign is placed on the outside of the enclosure and a "danger" sign on the inside. Utilities should consult the ANSI Z535 documents, federal or state regulatory agencies, and the utility's insurance company for signage applications. Neither the **NESC** nor the ANSI Z535 signing documents are specific as to what words or pictorials are required for individual signing applications. Rule 012C, which requires accepted good practice, must be used. It is important to note that **NESC** Rule 110A does not require safety signs inside the substation. Furthermore, **NESC** Rule 381G does not require a safety sign on the outside of pad-mounted equipment. Examples of safety signs are shown in Fig. 110-4.

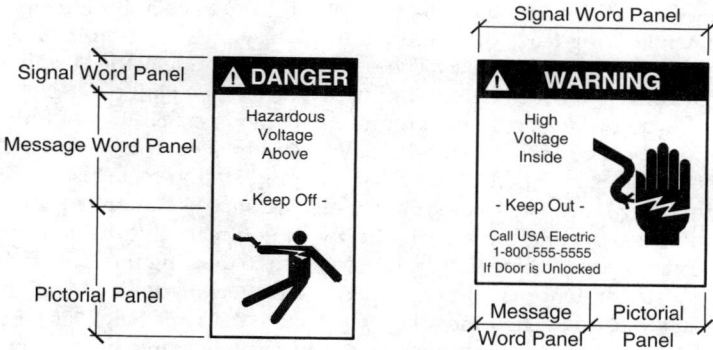

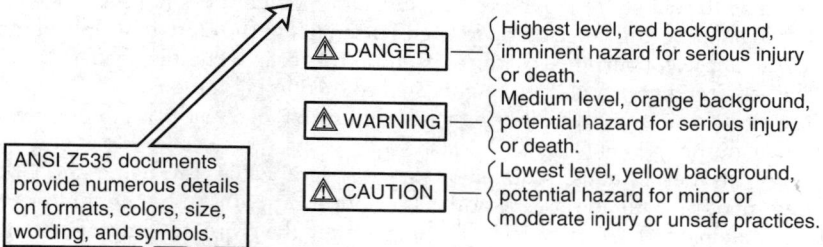

Fig. 110-4. Examples of ANSI Z535 safety signs (Rule 110A).

When a chain-link fence is used to enclose a substation, the **NESC** details very specific height requirements. The construction requirements discuss the fence fabric and barbed wire strands but do not provide details on the gauge of the fence mesh or diameter of the fence posts or rails. These details are left to the designer. Rule 110A1 starts out with the general statement that the enclosure must limit the likelihood of entrance of unauthorized persons. Other types of construction must present an equivalent barrier to climbing and unauthorized entry as the chain-link fence. To provide an equivalent barrier to climbing, fences should not have handholes or footholes more predominant than the mesh on a chain-link fence. Although not stated in the **Code**, pad-mounted equipment, park benches, parked vehicles, etc., should not be placed near a substation fence, as they can create "steps" for climbing the fence.

The **Code** does address neighboring fences or similar structures. Neighboring fences or similar structures must not be connected to or located within 6 ft of an electric supply station fence without concurrence of the substation owner. This requirement is addressing two primary issues related to neighboring fences. The first is a neighboring fence of a different height can create a "step" for climbing the electric supply station fence. The second is that the electric supply station fence that consists of metal chain-link fence or other metallic barrier must be grounded in accordance with the grounding methods in Sec. 09 (see Rule 092E). The question becomes what grounding methods are needed for the neighboring metal fence that is attached to the electric supply station metal fence. Keeping the neighboring fence 6 ft away from the electric supply station fence mitigates these concerns. Locating the electric supply station fence 6 ft inside the electric supply station property line may be one solution to meeting this rule. If the electric supply station fence is on the property line, working with the neighboring fence owner to use a 6 ft non-metallic fence section the same height as the substation fence may be another solution. This is just one example of applying some accepted good practice (Rule 012C) by the substation owner before concurrence is given by the substation owner. If the substation owner does not see any issues with the neighboring fence being located within 6 ft of the electric supply station fence, a simple concurrence by the substation owner is all that is needed.

The requirements for barrier heights are outlined in Fig. 110-5.

The requirements for neighboring fences within 6 ft of an electric supply station fence are outlined in Fig. 110-6.

Many utilities establish substation fence height values by using the **Code** requirement plus an adder. The adder (1 ft, for example) can be thought of as a design or construction tolerance adder to maintain the required fence height over time. There are several factors that can jeopardize the fence height. Factors could include the addition of gravel or some other type of fill outside the substation. Installing a substation fence with an overall height of 8 ft can help maintain the **Code**-required 7-ft height over the life of the installation. For additional information on the concept of **Code** plus an adder, see Rule 010.

In addition to the fence height requirements in Rule 110A1, Rule 110A2 specifies a safety clearance zone from the substation fence to exposed live (energized) parts inside the substation. The method used depends on the type of barrier around the supply station that keeps the public out of the substation.

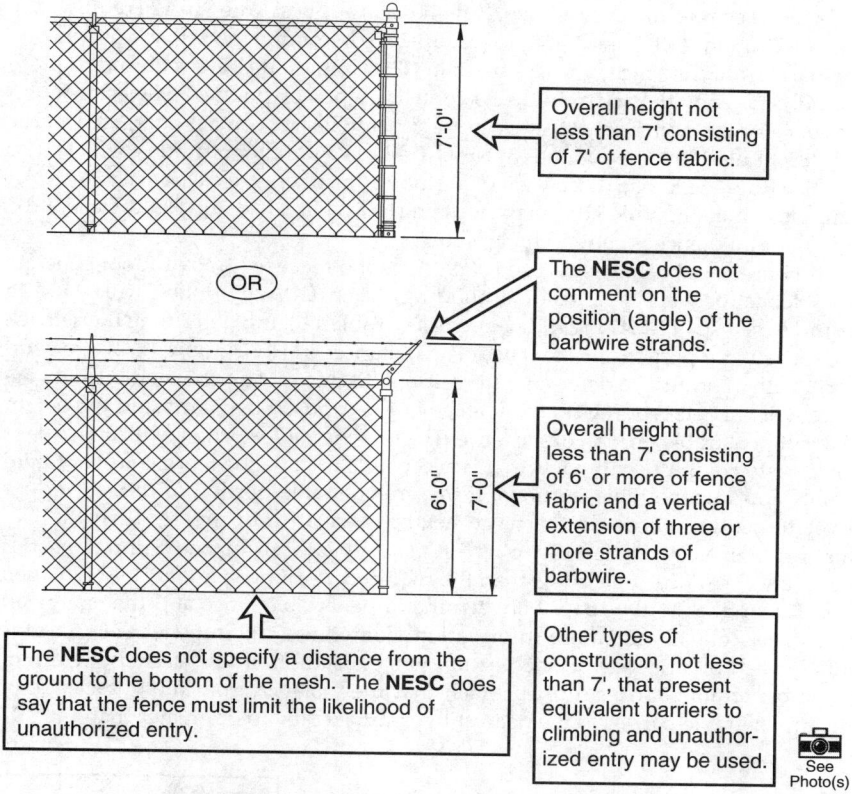

7'-0"

Overall height not less than 7' consisting of 7' of fence fabric.

OR

The **NESC** does not comment on the position (angle) of the barbwire strands.

Overall height not less than 7' consisting of 6' or more of fence fabric and a vertical extension of three or more strands of barbwire.

6'-0"   7'-0"

The **NESC** does not specify a distance from the ground to the bottom of the mesh. The **NESC** does say that the fence must limit the likelihood of unauthorized entry.

Other types of construction, not less than 7', that present equivalent barriers to climbing and unauthorized entry may be used.

See Photo(s)

Fig. 110-5. Barrier height requirements (Rule 110A).

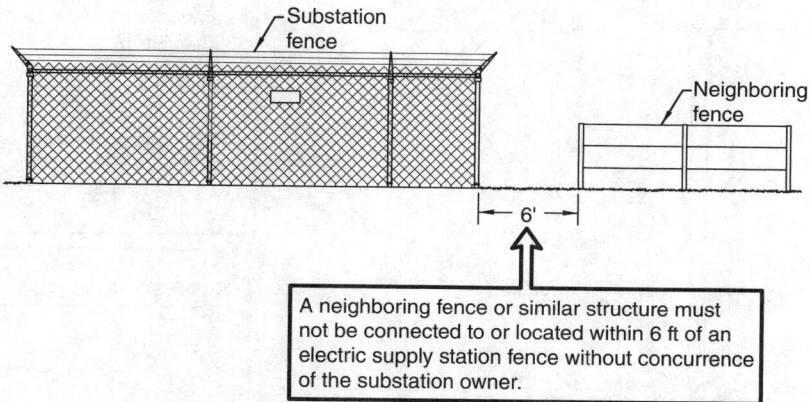

Substation fence

Neighboring fence

6'

A neighboring fence or similar structure must not be connected to or located within 6 ft of an electric supply station fence without concurrence of the substation owner.

Fig. 110-6. Neighboring fence requirements (Rule 110A).

If a metal chain-link fence is used, Rule 110A2a applies. Rule 110A2a requires the use of **NESC** Fig. 110-1 and **NESC** Table 110-1 to determine the setback of exposed live parts from the fence. The distances in **NESC** Table 110-1 are required to place live parts far enough back from the fence so that a person poking an object through the fence or swinging an object over the fence will not contact energized parts. Per **NESC** Rule 124A1, the values in **NESC** Table 110-1 are for altitudes of 3300 ft or less. See Rule 124A for a discussion of altitude adjustments to clearance values. An example of meeting Rule 110A2a by applying **NESC** Fig. 110-1 and **NESC** Table 110-1 to a chain-link fence is shown in Fig. 110-7.

If an impenetrable barrier is used such as a solid fence (e.g., an outdoor concrete block fence) or wall (e.g., inside a building), Rule 110A2b applies. Rule 110A2b requires the use of **NESC** Fig. 110-2 and **NESC** Table 110-1 to determine the setback of exposed live parts from the fence. The distances in **NESC** Table 110-1 are combined with a formula to determine the setback distance. The calculated setback distance will vary depending on the voltage of the exposed live parts and the height of the impenetrable portion of the fence. The impenetrable portion of the fence does not have to cover the entire fence. It is acceptable to have the fence consist of penetrable and impenetrable portions. If the impenetrable portion of the fence does cover the entire fence and the fence height is equal to or greater than the "R" dimension in **NESC** Table 110-1 plus 5 ft, a safety clearance zone will not exist. If the safety clearance zone is less than the required working space in Rule 125, then Rule 125 applies to the space between the impenetrable fence and the energized parts. See Rule 125 for more information. Sometimes the impenetrable fence is only located on one side or one portion of a substation, in this case the width of the impenetrable barrier must be such that the distance from the outer edge of the impenetrable barrier to the nearest live part is equal to or greater than the "R"

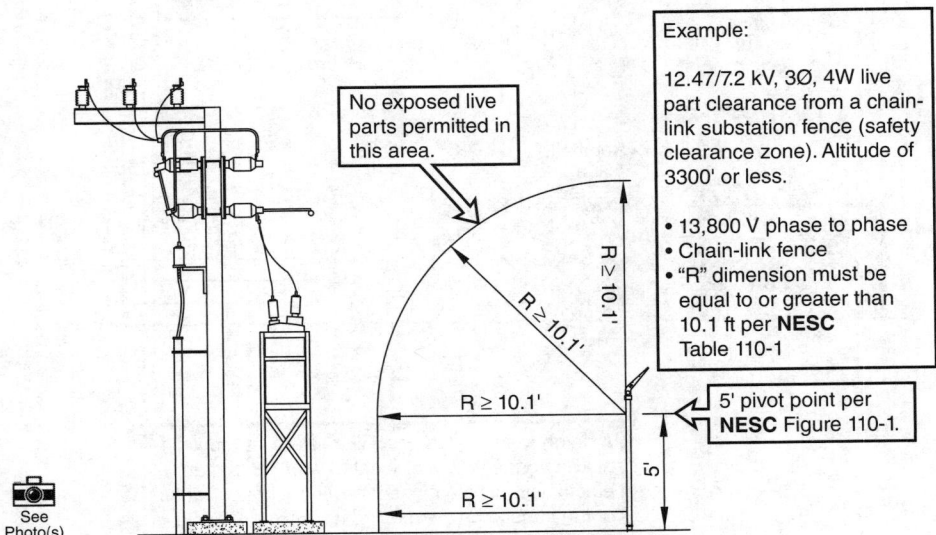

Fig. 110-7. Example of how to apply the safety clearance zone to a chain-link fence per NESC Table 110-1 and NESC Fig. 110-1 (Rule 110A2a).

dimension in **NESC** Table 110-1 for the voltage involved. If there are openings below the impenetrable portion of the fence, the "R" dimension of the **NESC** Table 110-1 for the voltage involved must be applied from the lowest impenetrable point to the closest energized part. An example of this application is using plywood sheets bolted onto a chain-link fence to shorten the safety clearance zone between the fence and exposed live (energized) parts inside the substation. There are times when a concern arises that the plywood sheets will greatly increase the wind load on the chain-link fence. Some utilities will not cover the fence near the bottom of the fence to minimize the wind load on the fence and in this case the "R" dimension of the **NESC** Table 110-1 must be applied from the lowest impenetrable point to the closest energized part. Examples of meeting Rule 110A2b by applying **NESC** Fig. 110-2 and **NESC** Table 110-1 are shown in Figs. 110-8 and 110-9.

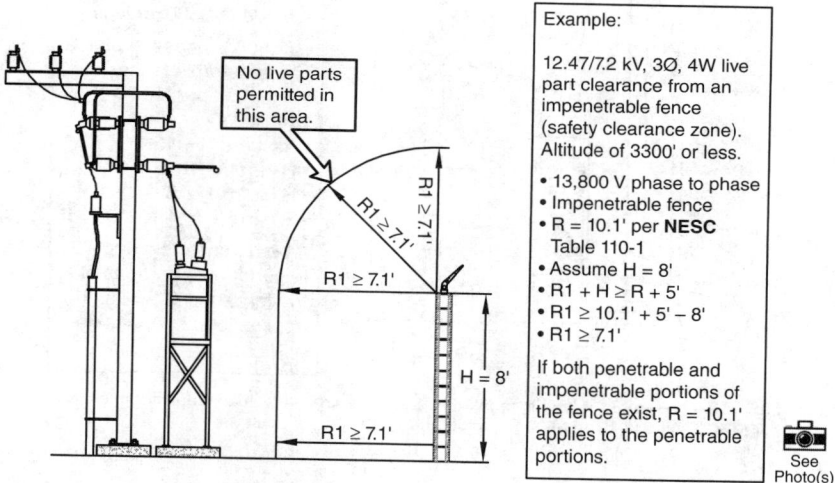

No live parts permitted in this area.

R1 ≥ 7.1'

R1 ≥ 7.1'

R1 ≥ 7.1'

R1 ≥ 7.1'

H = 8'

Example:

12.47/7.2 kV, 3Ø, 4W live part clearance from an impenetrable fence (safety clearance zone). Altitude of 3300' or less.

- 13,800 V phase to phase
- Impenetrable fence
- R = 10.1' per **NESC** Table 110-1
- Assume H = 8'
- R1 + H ≥ R + 5'
- R1 ≥ 10.1' + 5' – 8'
- R1 ≥ 7.1'

If both penetrable and impenetrable portions of the fence exist, R = 10.1' applies to the penetrable portions.

See Photo(s)

Fig. 110-8. Example of how to apply the safety clearance zone to an impenetrable fence per NESC Table 110-1 and NESC Fig. 110-2 (Rule 110A2b).

The exception to the safety clearance zone involves internal substation fences. An internal fence does not need to comply with the safety clearance zone as an interior fence inside the substation fence is accessible to qualified employees only. The exception is outlined in Fig. 110-10.

The North American Electric Reliability Corporation (NERC) standards apply to protecting transmission substations from a physical attack. NERC Standard CIP-014 provides information for transmission substation owners. The NERC physical security standards were developed to address cascading outages.

110B. Rooms and Spaces. The rooms (i.e., interior) or spaces (i.e., exterior) that comprise an electric supply station must be noncombustible. The **Code** uses the phrase, "as much as practical noncombustible." This wording recognizes the oil-filled equipment may be combustible; however, the substation structure should not be. Rules 152 and 172 provide additional requirements related to oil-filled equipment located in electric supply stations. See Rule 152B for additional information

For an impenetrable fence height (H) of 15.1 ft, the safety clearance zone (R1) is equal to or greater than 0 ft. Working space requirements of Rule 125 must still be met. Per Rule 125, using **NESC** Table 124-1 the horizontal working space for 15 kV (110 kV BIL) is 3 ft-6 in. If the safety clearance zone (R1) is equal to or greater than the 3 ft-6 in working space, then the height (H) of the impenetrable fence can be reduced from 15.1 ft to 11.6 ft. The minimum approach distances in **NESC** Rule 441 should be considered. Additional working room may be needed for tools (e.g., hot sticks).

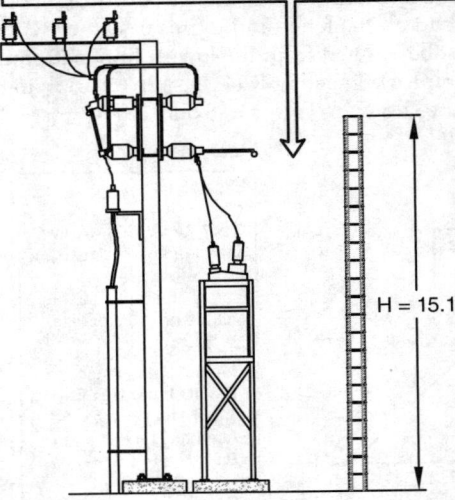

H = 15.1'

Example:

12.47/7.2 kV, 3Ø, 4W live part clearance from an impenetrable fence (safety clearance zone). Altitude of 3300' or less.

- 13,800 V phase to phase
- Impenetrable fence
- R = 10.1' per **NESC** Table 110-1
- Assume H = 15.1'
- $R1 + H \geq R + 5'$
- $R1 \geq 10.1' + 5' - 15.1'$
- $R1 \geq 0'$

If both penetrable and impenetrable portions of the fence exist, R = 10.1' applies to the penetrable portions.

Fig. 110-9. Example of how to apply the safety clearance zone to an impenetrable fence per NESC Table 110-1 and NESC Fig. 110-2 (Rule 110A2b).

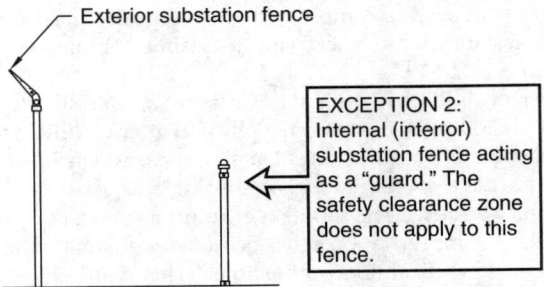

Exterior substation fence

EXCEPTION 2: Internal (interior) substation fence acting as a "guard." The safety clearance zone does not apply to this fence.

Fig. 110-10. Exception to the safety clearance zone requirements (Rule 110A2).

related to locating an electric supply station indoors. Steel is the most common choice for modern outdoor substation construction but the **Code** recognizes that wood poles are still commonly located within the substation fence. Dry grass or weeds should be removed from an outdoor substation to maintain the noncombustible requirement.

The substation room or space must not contain combustible materials or fumes and must not be used for manufacturing or storage. Three exceptions to Rule 110B2 apply to storage of materials in an electric supply station.

The first exception permits storage of material, equipment, and vehicles that are essential for maintenance of the electric supply station, for example, spare fuses, a spare substation transformer, or a bucket truck used for maintenance of the supply station equipment. The material, equipment, or vehicle must be guarded (e.g., stored in a shed) or separated from live parts per Rule 124. See Fig. 110-11.

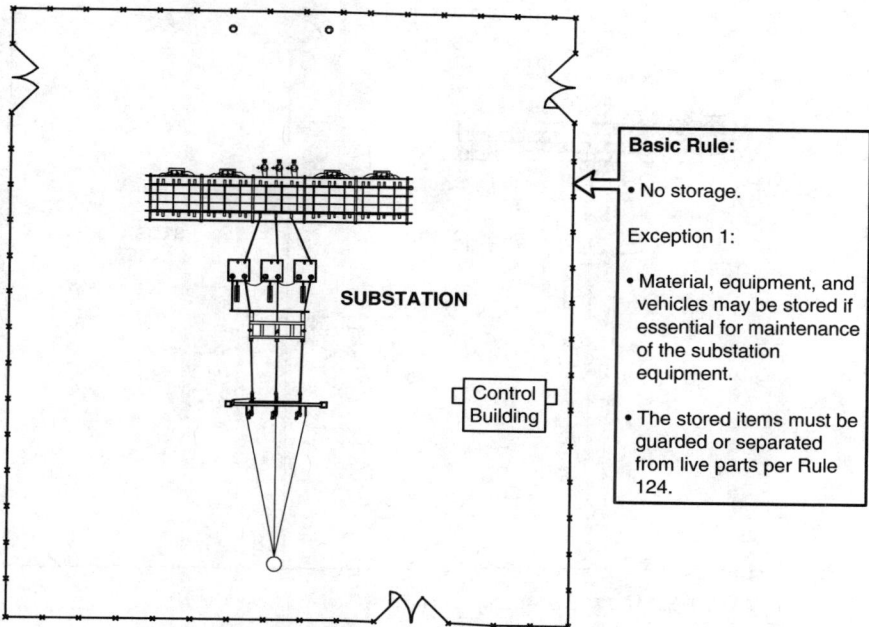

Fig. 110-11. Electric supply station use (storage) (Rule 110B2).

The second exception permits storage of material, equipment, and vehicles for construction, operations, or maintenance work of station, transmission, and distribution facilities (not just supply station facilities). This exception requires the stored items to be fenced separately from the electric supply substation equipment. The fence separating the electric supply equipment and the storage materials must meet the requirements of Rule 110A. If this exception is applied, the storage ends up not being in the electric supply station per se, but in a separate space adjacent to it. See Fig. 110-12.

Basic Rule:

• No storage.

Exception 2:

• Material, equipment, and vehicles may be stored for construction, operations, or maintenance work of station, transmission, and distribution facilities (not just supply station facilities).

• The stored items must be separated from the electric supply station equipment by a fence meeting Rule 110A.

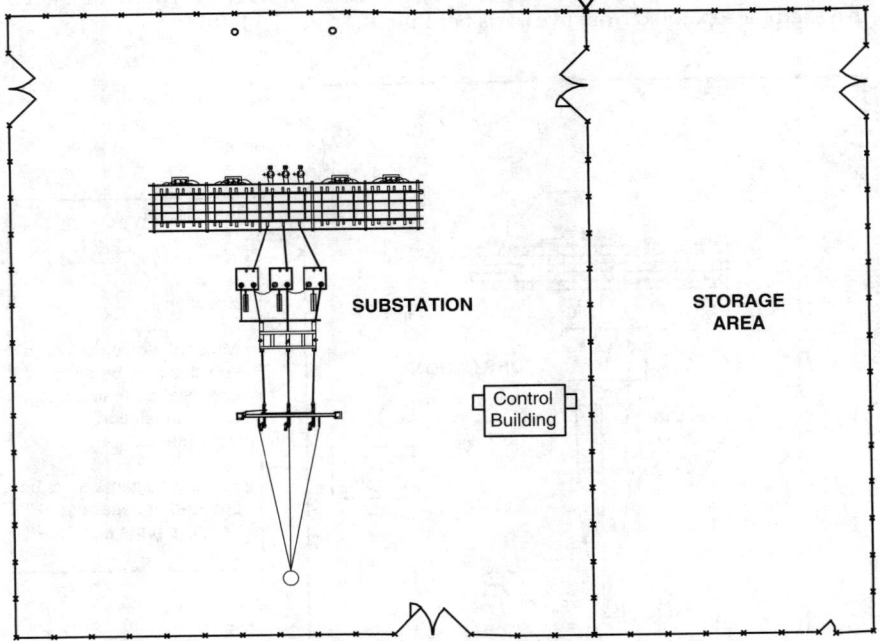

Fig. 110-12. Electric supply station use (storage) (Rule 110B2).

The third exception permits storage of material, equipment, and vehicles on a temporary basis (no time period is specified) for material, equipment, and vehicles for construction, operations, or maintenance of station, transmission, and distribution facilities (not just supply station facilities) work in progress. For example, if a new distribution line is being built near a substation, the substation may temporarily be used to store items for the project. This exception requires the stored items to be associated with work in progress. In other words, the substation site cannot be used as a storage yard for permanent warehousing. To apply exception three, the **Code** lists five conditions that must be met to maintain a safe working area. See Fig. 110-13.

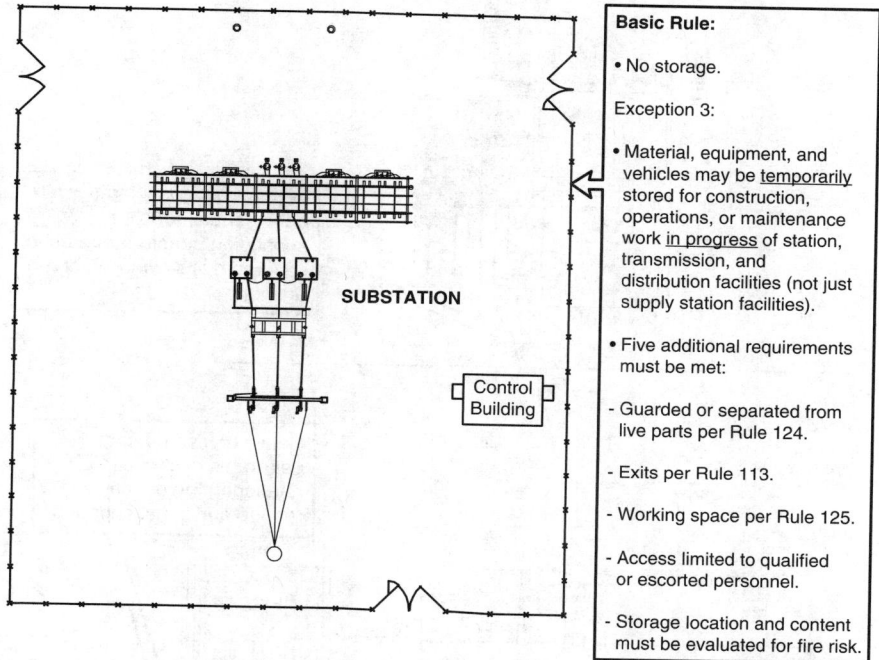

Basic Rule:

• No storage.

Exception 3:

• Material, equipment, and vehicles may be <u>temporarily</u> stored for construction, operations, or maintenance work <u>in progress</u> of station, transmission, and distribution facilities (not just supply station facilities).

• Five additional requirements must be met:

- Guarded or separated from live parts per Rule 124.

- Exits per Rule 113.

- Working space per Rule 125.

- Access limited to qualified or escorted personnel.

- Storage location and content must be evaluated for fire risk.

Fig. 110-13. Electric supply station use (storage) (Rule 110B2).

The ventilation in the room or space must be adequate and the room or space, if indoors, should be dry. If the electric supply station space is located outdoors, the equipment in the space must be designed for the atmospheric conditions.

110C. Electric Equipment. Electric equipment in the supply station must be supported and secured. The Code does not specifically address any seismic (earthquake) construction requirements. Rule 012C, which requires accepted good practice, must be applied in this case. IEEE Std. C57.114, *IEEE Seismic Guide for Power Transformers and Reactors,* is one reference for accepted good practice. With proper consideration, heavy equipment such as a substation transformer can be secured in place by its own weight. Heavy equipment such as an electric generator that has dynamic (rotating movement) forces will require supporting measures in addition to its own weight. See Fig. 110-14.

111. ILLUMINATION

111A. Under Normal Conditions. This rule provides illumination (lighting) levels for electric supply station rooms and spaces. NESC Table 111-1 provides illumination values in lux (metric) and foot-candles (English) for generating station areas, both interior and exterior. NESC Table 111-1 also contains illumination values for specific substation areas. The specific areas include control building

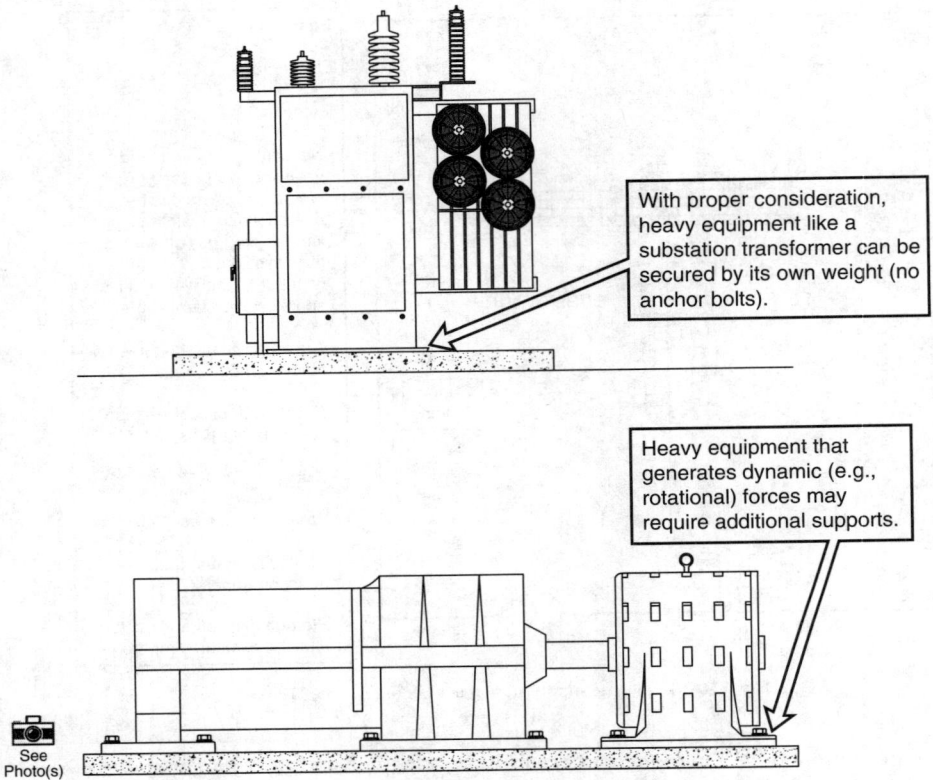

With proper consideration, heavy equipment like a substation transformer can be secured by its own weight (no anchor bolts).

Heavy equipment that generates dynamic (e.g., rotational) forces may require additional supports.

See Photo(s)

Fig. 110-14. Supporting and securing heavy equipment (Rule 110C).

interior, general exterior horizontal (e.g., the substation gravel) and equipment vertical (e.g., a row of disconnect switches), and remote areas. Multiple footnotes to the table provide additional information. Various light fixture manufacturers publish calculation aids for determining lighting levels. The Illuminating Engineering Society (IES) publishes books on lighting design applications. Rule 111A does not require that the lighting be permanently installed; therefore, portable lighting can be used to meet the rule. Rules 111A, 111B, and 111C discuss receptacles and portable cords for cord and plug light fixtures. The rules for illumination under normal conditions are outlined in Fig. 111-1.

111B. Emergency Lighting.   Attended electric supply stations must have automatically initiated emergency lighting for power failure. The exit paths in attended stations must have 1 foot-candle of lighting at all times. This can be done using an emergency generator or storage batteries. The duration of backup lighting should be evaluated, but in no case should the duration be less than 90 min (1.5 h). It is recommended that the wiring for the emergency lighting fixtures be kept independent from the normal wiring.

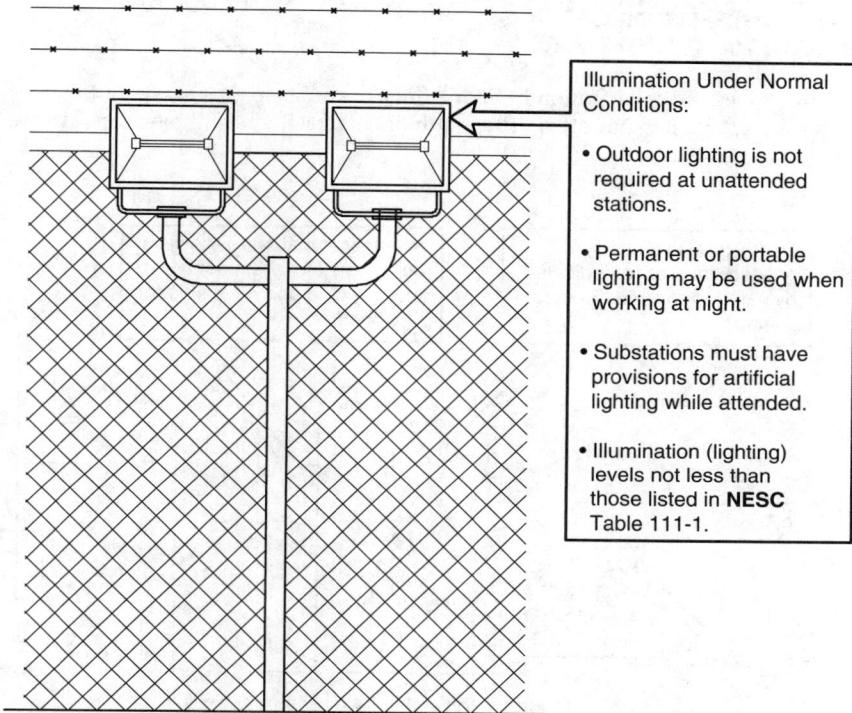

Illumination Under Normal Conditions:

• Outdoor lighting is not required at unattended stations.

• Permanent or portable lighting may be used when working at night.

• Substations must have provisions for artificial lighting while attended.

• Illumination (lighting) levels not less than those listed in **NESC** Table 111-1.

See Photo(s)

Fig. 111-1. Illumination under normal conditions (Rule 111A).

111C. Fixtures.   Portable cords must not be brought dangerously close to live parts (e.g., in a substation) or moving parts (e.g., in a generating station). Consideration needs to be given to the location of permanent fixtures and plug receptacles to avoid this hazard. Switches for lighting must be in a safely accessible location.

111D. Attachment Plugs and Receptacles for General Use.   Plugs and receptacles used in electric supply stations must disconnect all poles by one operation and must be of the grounding type. Special voltages, amperages, or frequencies must have plugs and receptacles that are not interchangeable. Manufacturers of wiring devices (e.g., receptacles, plugs, switches, etc.) use National Electrical Manufacturers Association (NEMA) standard configurations for various voltage, phase, and current ratings.

111E. Receptacles in Damp or Wet Locations.   If the receptacle is in a damp or wet location, it must have ground-fault interruption (GFI) as part of either the receptacle or the circuit breaker feeding the receptacle. As an alternative to using GFI protection, the **NESC** allows testing of a grounded circuit (e.g., using a Megger to verify adequate insulation resistance) as often as experience has shown necessary.

112. FLOORS, FLOOR OPENINGS, PASSAGEWAYS, AND STAIRS

This rule gives special attention to floors, floor openings, passageways, and stairs, as they are accident-prone areas. The requirements for floors and passageways are outlined in Fig. 112-1.

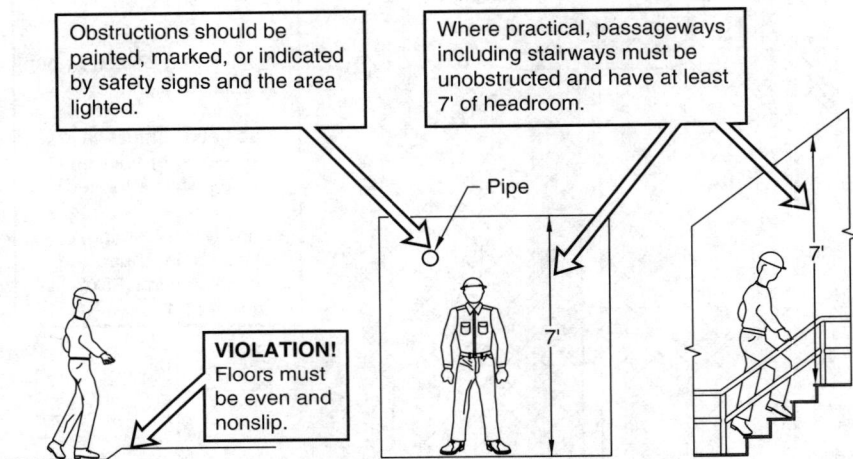

Fig. 112-1. Requirements for floors and passageways (Rule 112).

Railings are required for floor openings and raised platforms or walkways in excess of 1 ft in height. Handrails are required for stairways with four or more risers. A 3-in unobstructed clearance is required around handrails to assure an adequate grip. The rule requiring handrails for stairways consisting of four or more risers is outlined in Fig. 112-2.

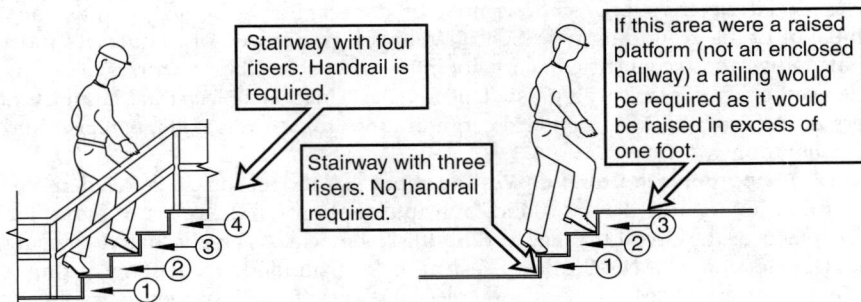

Fig. 112-2. Requirements for stair handrails (Rule 112).

113. EXITS

Exits in spaces and rooms must be kept clear of obstructions. Double exits must be provided if the arrangement of equipment and an accident can make a single exit inaccessible. The exit doors must swing out and have some type of panic hardware (e.g., a push bar, not a door knob) except for fence gates in outdoor substations and doors in rooms containing only low-voltage nonexplosive equipment. See Fig. 113-1.

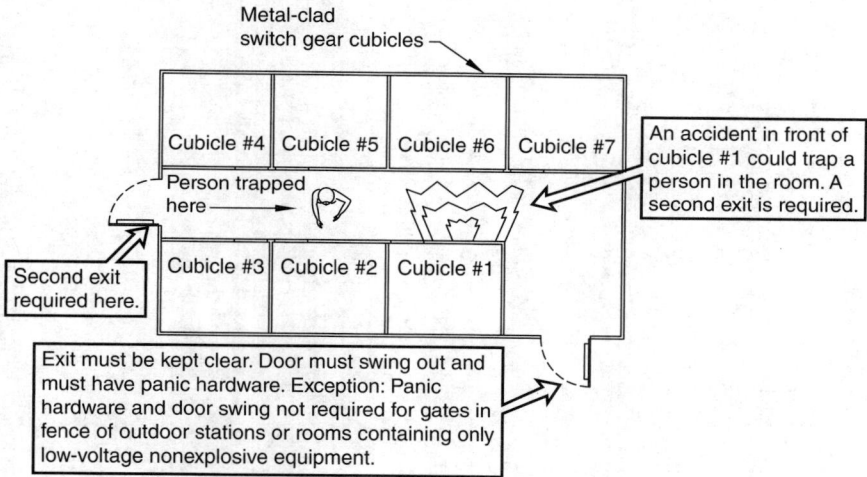

Fig. 113-1. Electric supply station exit requirements (Rule 113).

Section 12

Installation and Maintenance of Equipment

120. GENERAL REQUIREMENTS

This rule discusses installation and maintenance of electric supply station equipment. Safeguarding personnel during installation, construction, and maintenance is the primary concern of Sec. 12. The rules of Sec. 12 apply to both alternating-current (AC) and direct-current (DC) electric supply stations.

121. INSPECTIONS

The inspections discussed in this rule and the inspections discussed in Rule 214, Part 2, Overhead Lines, and in Rule 313, Part 3, Underground Lines, form the basic requirements for inspecting electric supply stations and supply and communication lines. The inspections required in this rule for electric supply stations are outlined in Fig. 121-1.

Any equipment or wiring found defective must be permanently disconnected or promptly corrected. New equipment must be tested in accordance with industry practice. The **Code** does not specify the details of the inspection program; it simply states that it be regular and scheduled. Accepted good practice, which is discussed in Rule 012C, is commonly used to help develop a substation inspection program.

STATUS OF EQUIPMENT   TYPE OF INSPECTION

In-service equipment

Idle equipment - Not energized

Idle equipment - Energized but not connected to load

Emergency equipment

New equipment

Inspect and **Maintain** at intervals as experience has shown to be necessary.

Inspect and **Test** before use.

Inspect and **Test** at intervals as experience has shown to be necessary.

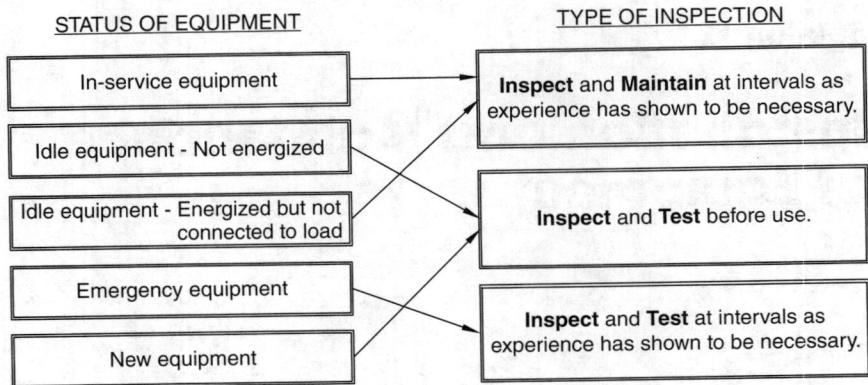

Fig. 121-1. Inspection requirements (Rule 121).

122. GUARDING SHAFT ENDS, PULLEYS, BELTS, AND SUDDENLY MOVING PARTS

Mechanical parts located in an electric supply station must be safeguarded. Mechanical transmission machinery is abundant in generating stations. Substations typically do not have much mechanical machinery. This rule requires the use of ANSI/ASME B15.1. Many times the Code makes note of a standard, but in this case, the standard is required, as it is part of the Code text. Suddenly moving parts must also be guarded or isolated.

123. PROTECTIVE GROUNDING

This rule states the requirements for electric supply station grounding. The methods of protective grounding are found in Sec. 09, "Grounding Methods for Electric Supply and Communications Facilities."

Non-current-carrying metal parts in the electric supply station must be grounded or isolated. Metallic fences must be grounded. IEEE Standard 80 is noted in this rule and in Sec. 09, as it is the standard for electric supply station grounding.

Provisions must also exist for grounding during maintenance. When a conductor, bus section, or piece of equipment is disconnected for maintenance, it must be grounded. The grounding can be done with permanent grounding switches or a readily accessible means for connecting portable grounding jumpers. The Part 4 Work Rules are referenced for proper procedures.

Direct-current (DC) systems have unique rules for grounding, which are discussed in Sec. 09.

124. GUARDING LIVE PARTS

124A. Where Required. Live parts in an electric supply station over 300 V phase to phase must be guarded, isolated by location (using vertical and horizontal clearance), or insulated to avoid inadvertent contact by qualified utility personnel in the substation. The basic intent of this rule is that utility personnel who are qualified to be in the substation (but not necessarily working on the substation) can walk around without accidentally contacting energized parts. The clearances in this section do not apply to the general public. Rule 110A, "General Requirements, Enclosure of Equipment," must be met before the clearances of Rule 124 apply. Otherwise Part 2, Overhead Lines, is applicable. See Rule 110A for an example.

Rule 124 constantly uses the word "guard," which implies a physical barrier between the utility employee and the live part. A common guard in a substation is the grounded enclosure of metal-clad switchgear. For substations consisting of open bus construction, physical guards are not nearly as common as the alternative to providing a guard, which is providing adequate clearance.

To determine the adequate clearance of live (i.e., energized) parts in the electric supply station, Rule 124A1 requires the use of **NESC** Table 124-1 and **NESC** Fig. 124-1 and the live parts must meet the safety clearance zone to the fence or wall which is covered in Rule 110A2. The vertical clearances in **NESC** Table 124-1 can use the taut-string method described in Rule 124D. The footnotes to **NESC** Table 124-1 provide information on Basic Impulse Insulation Levels (BIL) selection methods. Appendix D in the **NESC** provides additional information on per-unit overvoltage factors. **NESC** Rule 124A1 states that the clearance values in **NESC** Table 124-1 and **NESC** Table 110-1 are for altitudes of 3300 ft or less and appropriate atmospheric correction factors must be applied for higher altitudes. Applying the appropriate atmospheric correction factor may result in an increased clearance. The **NESC** does not provide a formula for increasing the substation clearance values due to higher altitudes. Altitude correction formulas are commonly used in the **NESC** in Part 2, Overhead Lines but no such formulas appear in Part 1, Electric Supply Stations. A note in Rule 124A1 references two IEEE Standards for additional information. Examples of altitude correction methods for increasing substation clearances can also be found in the Rural Utilities Service (RUS) Bulletin 1724E-300, *Design Guide for Rural Substations*.

In addition to clearance to live parts, Rule 124A3 specifies an 8-ft, 6-in (8.5-ft) vertical clearance to parts on indeterminate potential. A bushing or insulator has a surface along it of indeterminate potential. The top of the bushing has a known voltage, for example, 7.2 kV to ground. The bottom of the bushing has a known voltage, 0 V if it is grounded. The surface between the top of the bushing and the bottom of the bushing is of unknown voltage, or as the **Code** calls it, indeterminate potential. It is somewhere between 7.2 kV and 0 V. Per Rule 124A3, the 8-ft, 6-in (8.5-ft) vertical clearance must exist from the bottom of the part of indeterminate potential (bottom of the bushing) to the surface below. The vertical clearance in Rule 124A3 can use the taut-string method described in Rule 124D.

If the substation equipment purchased is not tall enough to meet the required vertical clearances to the top of the equipment bushings per **NESC** Table 124-1 and **NESC**

Fig. 124-1 and the 8-ft, 6-in (8.5-ft) clearance to the bottom of a part of indeterminate potential (bottom of the bushing), then a concrete pad and/or equipment stands must be used to provide the required height for both the live part clearance (top of the bushing) and the indeterminate voltage clearance (bottom of the bushing). Both clearances (top and bottom of the bushing) must be met or exceeded to avoid a **Code** violation. An example of how to apply the vertical clearances of **NESC** Table 124-1, **NESC** Fig. 124-1, and **NESC** Rule 124A3 is shown in Fig. 124-1.

Both Rule 124A1 and Rule 124A3 require clearance to any permanent supporting surface for workers inside the substation. If a concrete pad under substation equipment is large enough to stand on, then the measurement needs to be taken from the top of the concrete pad. The interpretation of what size concrete pad is a permanent supporting surface for workers is not well defined in the **Code**. If the concrete pad is oversized for easy maneuverability of a worker, the clearance measurement must be made from the top of the concrete pad. If the pad is used for additional vertical clearance of the equipment but a worker has to

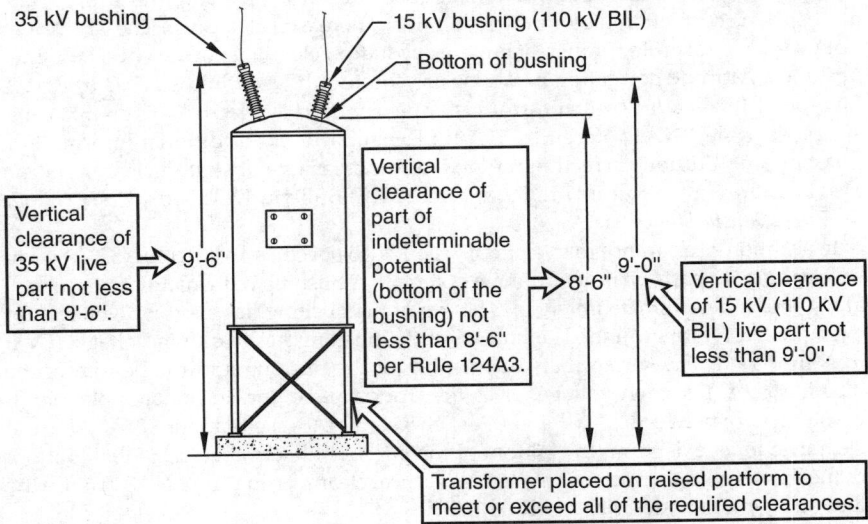

Fig. 124-1. Example of how to apply vertical clearances per **NESC** Table 124-1, **NESC** Fig. 124-1, and Rule 124A3 (Rules 124A1 and 124A3).

"hug the equipment" to stay standing on the concrete pad, the clearance measurement can be made from the substation surface (e.g., gravel) instead of the top of the concrete pad. This rule is outlined in Fig. 124-2.

Rule 124A2 recognizes that additional clearances or guarding may be needed where material may be carried such as passageways, corridors, storage areas, etc. (primarily indoor areas). Additional clearance values are not specified. If physical guards are used for these areas, they must be removed with tools or keys.

The **Code** does not specify energized conductor or bus clearances to vehicles in a substation. Many substations are designed for bucket truck access. The **NESC** Part 4 Work Rules and OSHA Standard 1910.269 applies. See Fig. 124-3.

The **Code** does not specify bus to bus clearances, conductor to bus clearances, conductor to conductor clearances, or energized part to grounded part clearances in Part 1, Electric Supply Stations. Rule 012C, which requires accepted good practice, must be applied. The conductor to conductor clearances given in Part 2, Overhead Lines, Sec. 23, are not required to be used in Part 1, Electric Supply Stations, but they are a reference for accepted good practice. Other common references for accepted good practice for bus to bus clearance in outdoor electrical substations are ANSI C37.32, NEMA SG6, and Rural Utilities Service (RUS) Bulletin 1724E-300, *Design Guide for Rural Substations*. The absence of a **Code** rule related to bus clearance is outlined in Fig. 124-4.

124B. Strength of Guards. Physical guards, when used instead of clearance, must be rigid and secure such that a person falling or slipping will not displace or deflect the guard. A common physical guard in a substation is the grounded enclosure of metal-clad switchgear.

124C. Types of Guards. The first sentence of this rule points out that meeting the safety clearance zone for the electric supply station fence in Rule 110A2

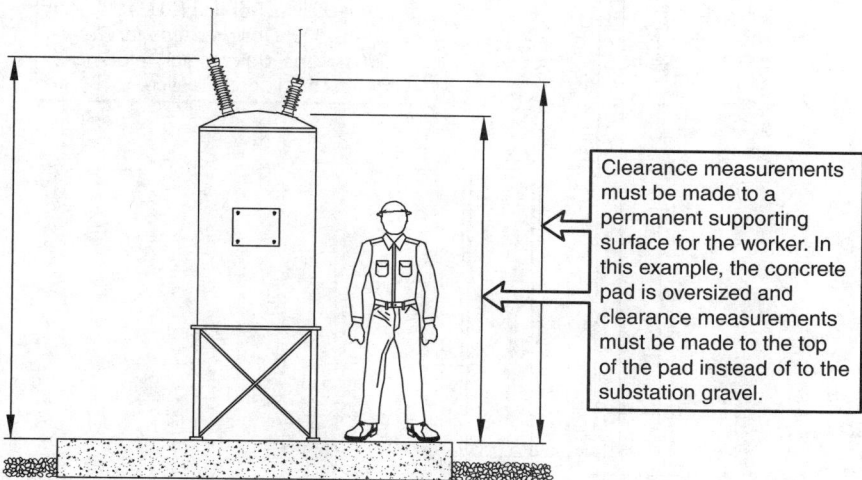

Clearance measurements must be made to a permanent supporting surface for the worker. In this example, the concrete pad is oversized and clearance measurements must be made to the top of the pad instead of to the substation gravel.

See Photo(s)

Fig. 124-2. Clearance measurements made to a permanent supporting surface (Rule 124A1).

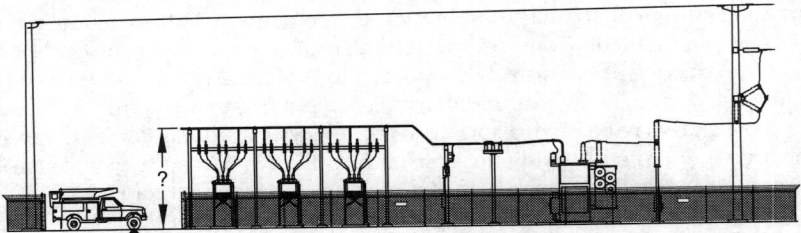

NO CODE RULE
The **NESC** does not specify a vertical
clearance for vehicles in a substation.
The **NESC** Part 4 Work Rules,
including minimum approach
distances, apply.

Fig. 124-3. Absence of a Code rule related to vertical clearance to vehicles in a substation (Rule N/A).

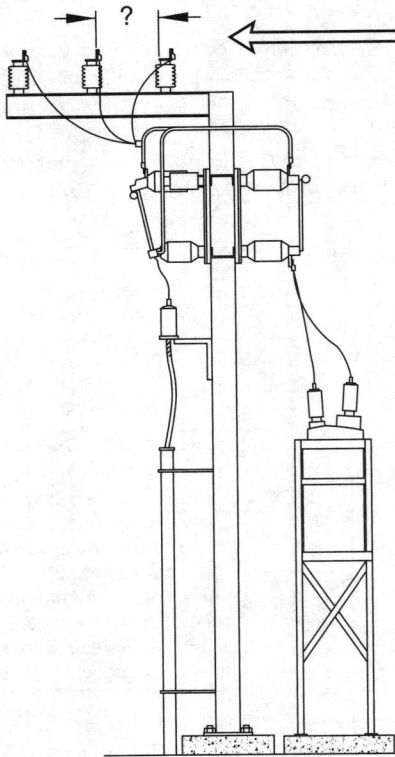

NO CODE RULE
The **NESC** does not specify bus
to bus clearances, conductor to
bus clearances, conductor to
conductor clearances, or energized
part to grounded part clearances
in electric supply stations. Rule 012C,
which requires accepted good
practice, must be applied.

ANSI C37.32, NEMA SG6, and
Rural Utilities Service (RUS) Bulletin
1724E-300, *Design Guide for Rural
Substations* are common references
for accepted good practice.

See
Photo(s)

Fig. 124-4. Absence of Code rule related to bus to bus clearances (Rule N/A).

and meeting the live part clearances in **NESC** Table 124-1 permits guarding by location.

Providing adequate clearance is the most common form of guarding by location. When guarding by isolation is used, entrances to the guarded space must be locked, barricaded, or roped off, and safety signs must be posted at entrances.

Rules 124C2 through 124C6 discuss various types of physical guards including shields, enclosures, barriers, mats, supporting surfaces for persons above live parts, and insulating covering. See Rule 163 for a discussion and a figure related to using insulation as a guard for a substation underground get-away. When railings or fences are used as guards, **NESC** Fig. 124-2 applies. The requirement in Rule 124C3 to locate the guard railing or fence "preferably not more than 4 ft" from the nearest point in the guard zone may not be practical in some cases. A NOTE to the rule indicates that additional working space may be required (more than 4 ft) when the working space in Rule 125 and the minimum approach distances in Rule 441 are considered for working with hot sticks. An example of a railing or fence used as a guard is shown in Fig. 124-5.

124D. Taut-String Distances. The taut-string clearance distance is composed of two components, the vertical clearance component which must be not less than 5 ft, and the shortest diagonal or horizontal clearance component. The shortest diagonal or horizontal clearance component allows the setback of a bushing at the top of a

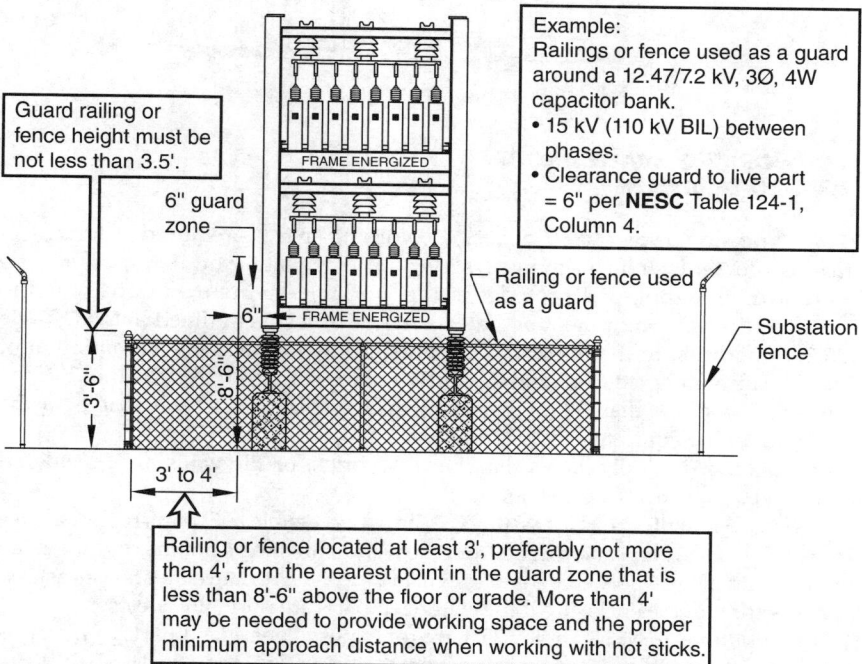

Fig. 124-5. Example of how to apply **NESC** Table 124-1 and **NESC** Fig. 124-2 (Rule 124C3).

piece of equipment to be factored into the overall vertical clearance distance. **NESC** Fig. 124-3 is provided to illustrate the taut-string measurement. The taut-string distance can be applied to the energized (live) part clearances in Rule 124A1 and to the part of indeterminate potential clearance in Rule 124A3. The taut-string vertical and diagonal components for an energized part (top of bushing) and the taut-string vertical and horizontal components for a part of indeterminate potential (bottom of the bushing) are shown in Fig. 124-6.

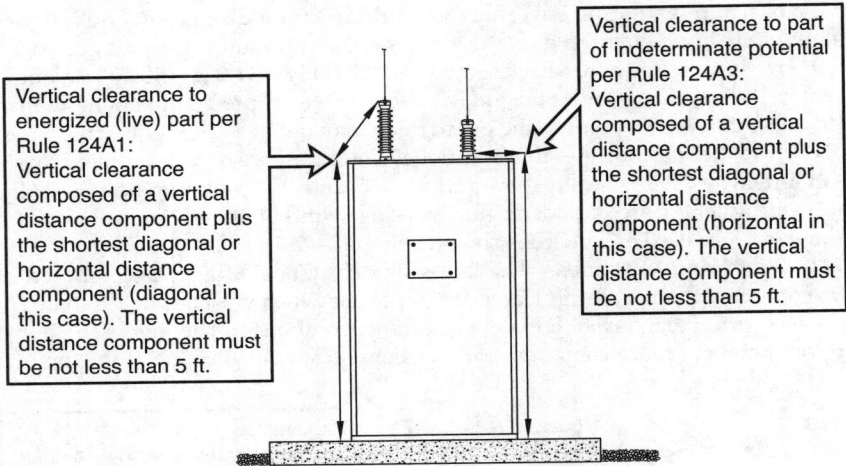

Vertical clearance to energized (live) part per Rule 124A1:
Vertical clearance composed of a vertical distance component plus the shortest diagonal or horizontal distance component (diagonal in this case). The vertical distance component must be not less than 5 ft.

Vertical clearance to part of indeterminate potential per Rule 124A3:
Vertical clearance composed of a vertical distance component plus the shortest diagonal or horizontal distance component (horizontal in this case). The vertical distance component must be not less than 5 ft.

Fig. 124-6. Taut-string distances (Rule 124D).

125. WORKING SPACE ABOUT ELECTRIC EQUIPMENT

125A. Working Space (600 V or Less).
Working space is required around electrical equipment for inspection or servicing. Adequate working space avoids equipment crowding and provides a safe working environment. The working space required for equipment operated at 600 V or less is outlined in **NESC** Table 125-1. In addition to the horizontal distances shown in the table, a minimum of 7 ft of headroom is required and a width of not less than 30 in is required. If the equipment is wider than 30 in, then the working space must be available for the full width of the equipment.

The **Code** specifically states that concrete, brick, or tile walls are considered grounded. A sheetrock wall is not referenced.

The back of a switchboard is assumed to be nonaccessible if all parts replacements, wire connections, etc., can be done from the front. If this were not the case, working space would also be needed behind a switchboard. The distance must be measured from the front of the enclosure if the energized parts are normally enclosed.

The conditions in **NESC** Table 125-1 are for exposed energized parts. If the parts are always de-energized during inspection, servicing, etc., then the **NESC** does not specify a workspace dimension. See Figs. 125-1 and 125-2.

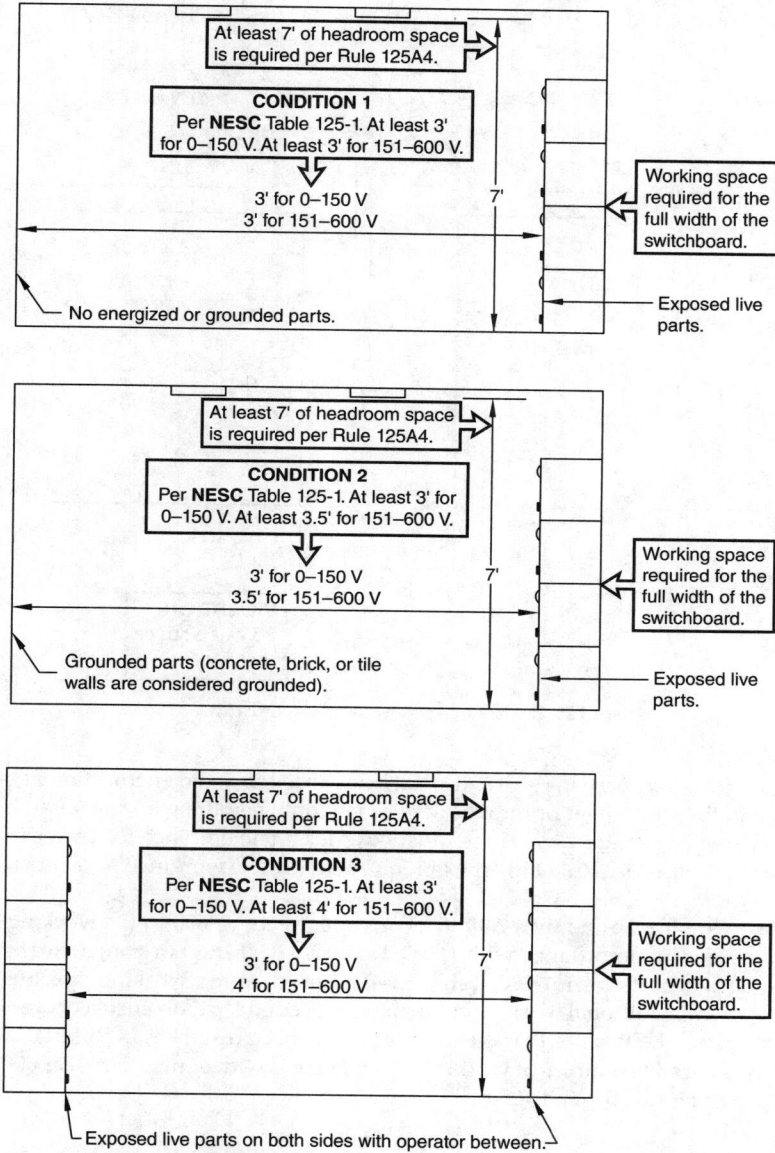

Fig. 125-1. Working space for 600 V or less (Rule 125A).

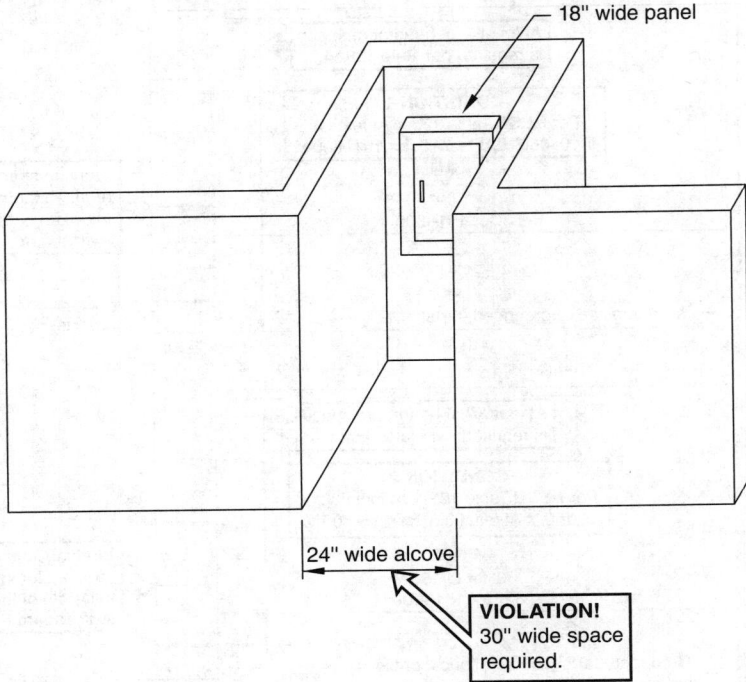

Fig. 125-2. Minimum width of working space (Rule 125A3).

The working space in Fig. 125-1 must be guarded to avoid encroachment by others into the equipment or into the service person when the working space is in an open area or passageway. There must be at least one entrance for access into the working space. See Rule 113 for exit requirements. The working space must not be used for storage. See Fig. 125-3.

125B. Working Space over 600 V.   For voltages above 600 V, the working space is provided in accordance with **NESC** Table 124-1. If the horizontal clearance of unguarded parts from **NESC** Table 124-1 is used as the working space around equipment, additional working room may be needed for tools such as hot sticks. A note to Rule 125B states that the minimum approach distances in Rule 441 should be considered. Examples of working space in areas with equipment over 600 V are shown in Figs. 125-4 and 125-5.

126. EQUIPMENT FOR WORK
ON ENERGIZED PARTS

This rule indirectly ties the work rules of Part 4 into the working space of Rule 125. If a worker is within the guard zone of **NESC** Table 124-1, Column 4, then

Switchgear

VIOLATION!
Working space must not
be used for storage.

STORAGE

STORAGE

STORAGE

STORAGE

See
Photo(s)

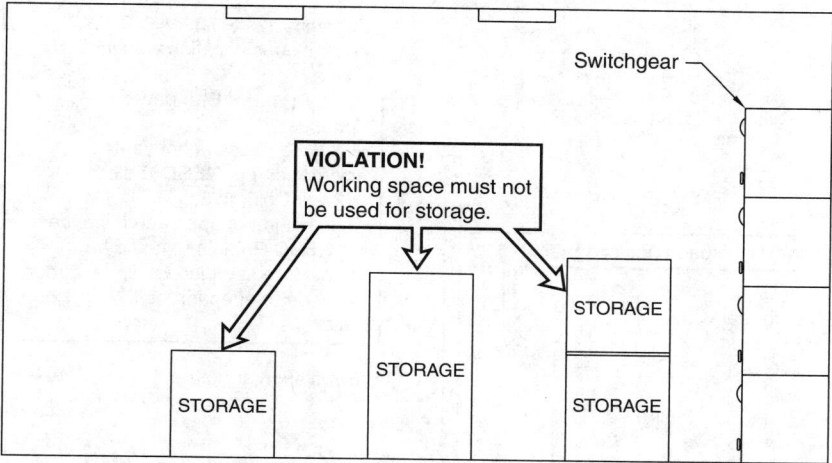

Fig. 125-3. Storage materials must not be in the working space (Rule 125A1).

Working space not less than 3'-6".
In this example, the safety clearance
zone to the fence or wall was calculated
to be less than the working space per
Rule 110A2, Exception 1.

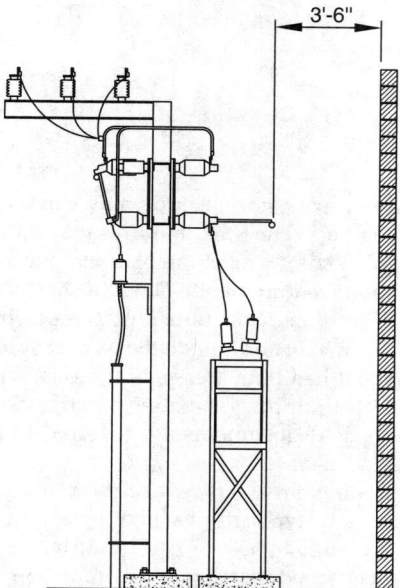

3'-6"

Example: 12.47/7.2 kV, 3Ø, 4W
electric supply station working
space.
• 15 kV (110 kV BIL) between
  phases.
• Fence or wall without holes or
  openings.
• Working space is horizontal
  clearance per **NESC** Table 124-1,
  Column 3.
• The minimum approach distances
  in **NESC** Rule 441 should be
  considered. Additional working room
  may be needed for tools (e.g., hot
  sticks).

Fig. 125-4. Example of working space over 600 V (Rule 125B).

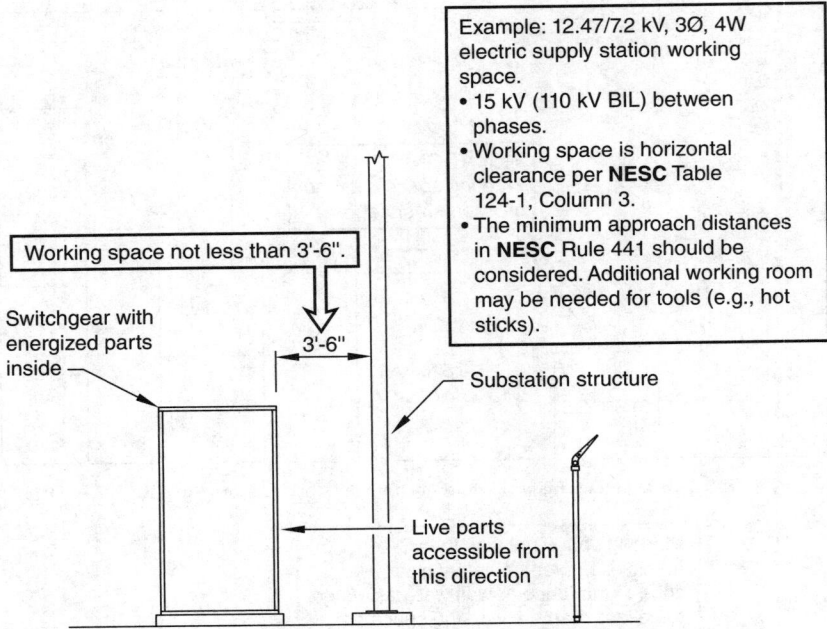

Example: 12.47/7.2 kV, 3Ø, 4W electric supply station working space.
• 15 kV (110 kV BIL) between phases.
• Working space is horizontal clearance per **NESC** Table 124-1, Column 3.
• The minimum approach distances in **NESC** Rule 441 should be considered. Additional working room may be needed for tools (e.g., hot sticks).

Working space not less than 3'-6".

Switchgear with energized parts inside

3'-6"

Substation structure

Live parts accessible from this direction

Fig. 125-5. Example of working space over 600 V (Rule 125B).

the worker must utilize protective equipment that is properly tested and rated for the voltage involved. The Work Rules in Part 4, specifically Rule 441, contain minimum approach distances to live parts.

127. CLASSIFIED LOCATIONS

Classified locations are locations where fire or explosion hazards may exist due to flammable gases, vapors, liquids, dust, or fibers. The **NESC** requires that classified locations in the vicinity of electric supply stations meet the National Electrical **Code** (NEC) hazardous (classified) locations requirements. The NEC has very detailed requirements related to classified locations, and rather than repeating them, the **NESC** requires that the NEC rules be met. In addition to the NEC requirements, the **NESC** provides requirements for coal-handling areas and various other hazardous locations primarily related to electric supply generating stations. Several National Fire Protection Association (NFPA) documents are required to be referenced for specific types of installations.

Separation from or ventilation of a hazardous area can reduce the classified area requirements. High-voltage facilities are typically located away from classified areas. Low-voltage equipment (i.e., under 600 V) can be installed in explosionproof enclosures and located in a classified location in accordance with the NEC rules.

128. IDENTIFICATION

This rule requires identification or labeling of equipment and devices in the electric supply station. This includes indoor switchboards and outdoor equipment. The identification must be uniform throughout any one station and not be placed on removable covers or doors that could be interchanged, therefore creating a labeling error. The **Code** is not specific as to the type of identification, but equipment must be sufficiently labeled for safe use and operation. Identification may include voltage levels, nameplate information, phase color coding, feeder numbering, etc. Rule 170 addresses identification of circuit breakers, reclosers, switches, and fuses. An example of identification in a substation is shown in Fig. 128-1.

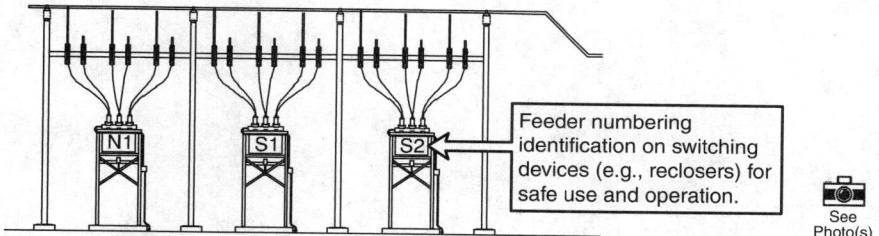

Feeder numbering identification on switching devices (e.g., reclosers) for safe use and operation.

See Photo(s)

Fig. 128-1. Example of identification in a substation (Rule 128).

129. MOBILE HYDROGEN EQUIPMENT

Hydrogen is primarily used in generating stations. Hydrogen gas is highly flammable. Bonding of mobile hydrogen equipment will reduce the chance for a difference in potential, which will reduce the chance of sparks that may cause an explosion.

Section 13

Rotating Equipment

130. SPEED CONTROL AND STOPPING DEVICES

Section 13 deals with generators, motors, motor generators, and rotary converters in electric supply stations. **NESC** Sec. 02, "Definitions of Special Terms," defines an electric supply station as a generating station or substation. Rotating equipment is more applicable to generating stations, as substations typically do not involve rotating equipment.

Section 13 only provides general rules related to basic safety features. Other IEEE standards are available, but not referenced in the **Code**, for generation station design. Using IEEE Standards and the National Electrical Code (NEC) for specific design features are good applications of Rule 012C, which requires using accepted good practice for particulars not specified in the **NESC**.

Rule 130 requires overspeed trip of prime movers (i.e., diesel engines, turbines, etc.) in addition to governors. Manual stopping devices are also required. Speed-limiting devices are required for separately excited AC motors and series motors, as they typically have problems with overspeed or runaway. Adjustable-speed motors must also have a speed-limiting device and must be equipped or connected to avoid weak fields that produce overspeed. Mechanical protection of the control circuits for stopping and speed limiting is also required.

131. MOTOR CONTROL

Rule 131 requires that motors do not restart automatically after a power outage if the unexpected starting could create injury to personnel. Two types of basic motor starting circuits are used in the motor control industry, low-voltage protection and low-voltage release.

Low-voltage protection (three-wire control) requires manual restarting by an equipment operator. This rule does provide an option to use low-voltage release (two-wire control with automatic restarting) if warning signals and time-delay features are designed into the control scheme. Examples of simple two-wire and three-wire motor control schemes are shown in Fig. 131-1.

132. NUMBER 132 NOT USED IN THIS EDITION

133. SHORT-CIRCUIT PROTECTION

This rule requires short-circuit protection for electric motors. Only a simple requirement is stated. Rule 012C, which requires accepted good practice, should be applied

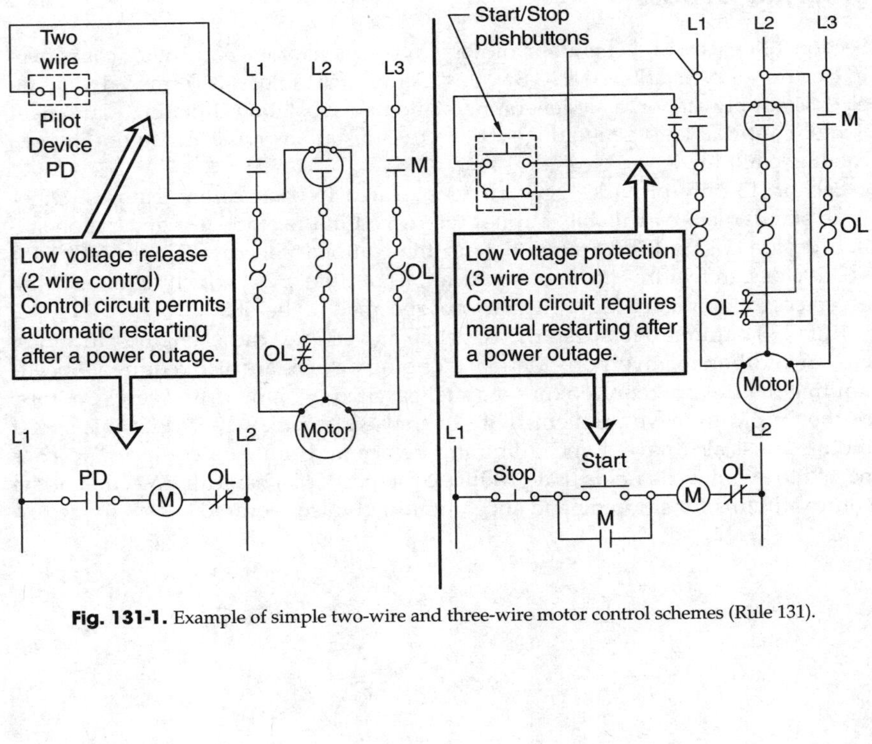

Fig. 131-1. Example of simple two-wire and three-wire motor control schemes (Rule 131).

for specific details. The National Electrical Code (NEC) specifies maximum fuse and circuit breaker sizes for various types and sizes of motors. The NEC is an excellent reference for accepted good practice in this case.

An example of motor protection components and terminology is shown in Fig. 133-1.

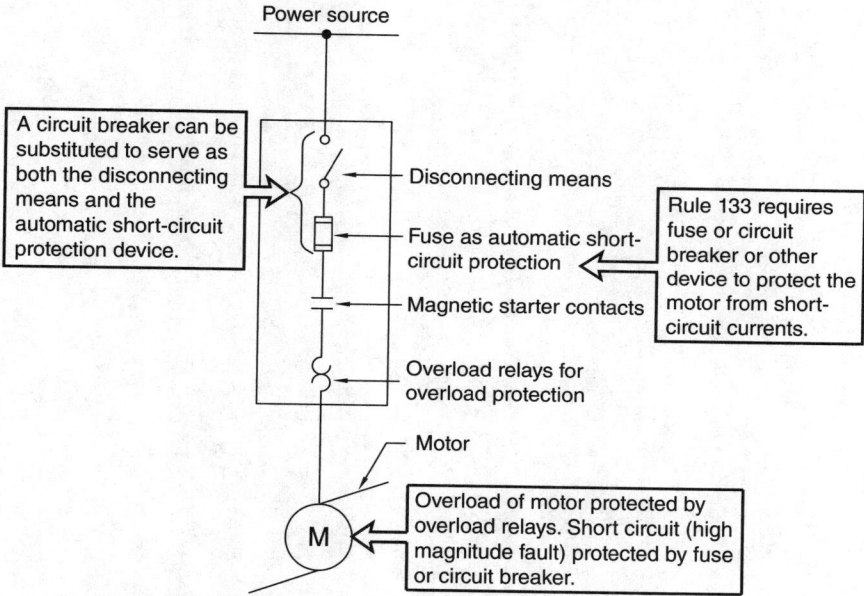

Fig. 133-1. Example of motor protection components and terminology (Rule 133).

Section 14

Storage Batteries

140. GENERAL

This section outlines the requirements for storage batteries used in electric supply stations. The battery system is typically used for DC control and DC operation of circuit breakers, circuit switchers, reclosers, etc. Batteries produce hydrogen gas, which is explosive, and electrolyte, which is acidic and corrosive. Sealed batteries can minimize these concerns. Typical battery systems used in substation applications include lead-acid and nickel-cadmium (NiCa). Batteries can be located with other equipment or in a dedicated battery room. The number of batteries and the amount of hydrogen produced by the batteries affect this decision. Rule 142 states that the natural or powered ventilation must limit the hydrogen accumulation to less than an explosive mixture. The **Code** does not provide a percent concentration that is considered safe. Battery manufacturers can provide guidelines for calculating hydrogen gas emissions and ventilation recommendations. The **NESC** outlines general, location, ventilation, rack, floor, illumination, and service facility requirements in Rules 140 through 146 (Rule 147 is not used). **Code** Rules 140 through 146 are outlined in Fig. 140-1.

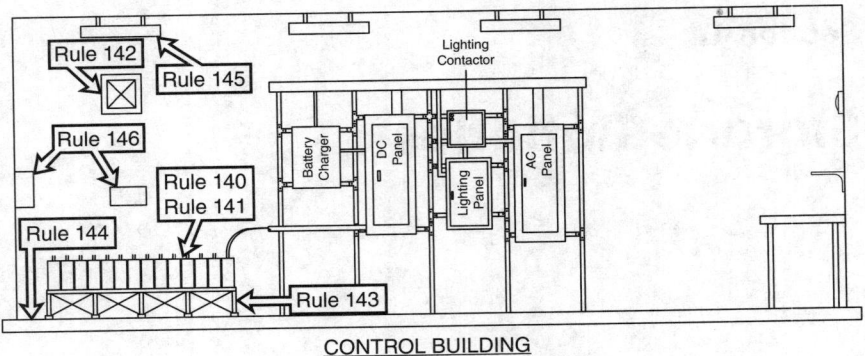

CONTROL BUILDING

Rule 140. General • Adequate space for operation and batteries.	**Rule 144. Floors in Battery Areas** • Acid-resistive floor. • Provisions to contain spilled electrolyte.
Rule 141. Location • Batteries located in a protective enclosure or area accessible only to qualified personnel.	**Rule 145. Illumination for Battery Areas** • Guard or isolate lighting fixtures (e.g., lenses or wire guards). • No receptacles or light switches in battery area.
Rule 142. Ventilation • Natural or powered ventilation. • Enunciate ventilation failure if required to keep hydrogen level below an explosive mixture.	**Rule 146. Service Facilities** • Goggles or face shield. • Acid-resistant gloves. • Protective aprons and overshoes. • Portable or stationary wash for eyes and skin. • Safety sign prohibiting smoking, sparks, or flames.
Rule 143. Racks • Racks firmly anchored, preferably to the floor. • Metal racks must be grounded.	

Fig. 140-1. Storage batteries (Rules 140 through 146).

See Photo(s)

141. LOCATION

See discussion in Rule 140 and Fig. 140-1.

142. VENTILATION

See discussion in Rule 140 and Fig. 140-1.

143. RACKS

See discussion in Rule 140 and Fig. 140-1.

144. FLOORS IN BATTERY AREAS

See discussion in Rule 140 and Fig. 140-1.

145. ILLUMINATION FOR BATTERY AREAS

See discussion in Rule 140 and Fig. 140-1.

146. SERVICE FACILITIES

See discussion in Rule 140 and Fig. 140-1.

Section 15

Transformers and Regulators

150. CURRENT-TRANSFORMER SECONDARY CIRCUITS PROTECTION WHEN EXCEEDING 600 V

This rule requires secondary circuits of current transformers (with primary circuits over 600 V) to be protected by conduit, covering, or some other protection. If the conduit or covering is metal, it must be effectively grounded and consideration must be given to circulating currents. Current transformers (CTs) must also have a provision for shorting the secondary wiring. These requirements are needed, as open or damaged current-transformer secondary circuits may cause a hazardous high voltage and arcing.

An example of protecting current-transformer secondary circuits is shown in Fig. 150-1.

151. GROUNDING SECONDARY CIRCUITS OF INSTRUMENT TRANSFORMERS

Instrument transformers consist of both voltage transformers (VTs) and current transformers (CTs). Voltage transformers are sometimes referred to as potential transformers (PTs). The secondaries of VTs and CTs must be effectively grounded. An example of basic VT and CT connections is shown in Fig. 151-1.

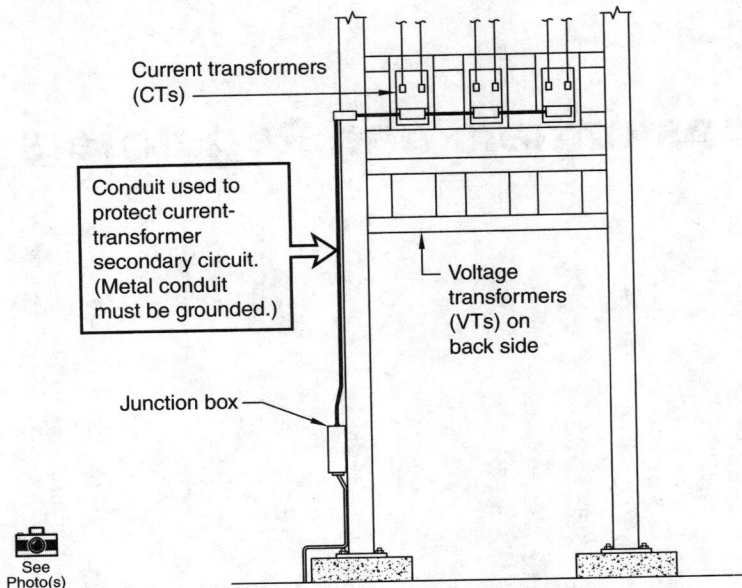

Current transformers (CTs)

Conduit used to protect current-transformer secondary circuit. (Metal conduit must be grounded.)

Voltage transformers (VTs) on back side

Junction box

See Photo(s)

Fig. 150-1. Example of protecting current-transformer secondary circuits (Rule 150).

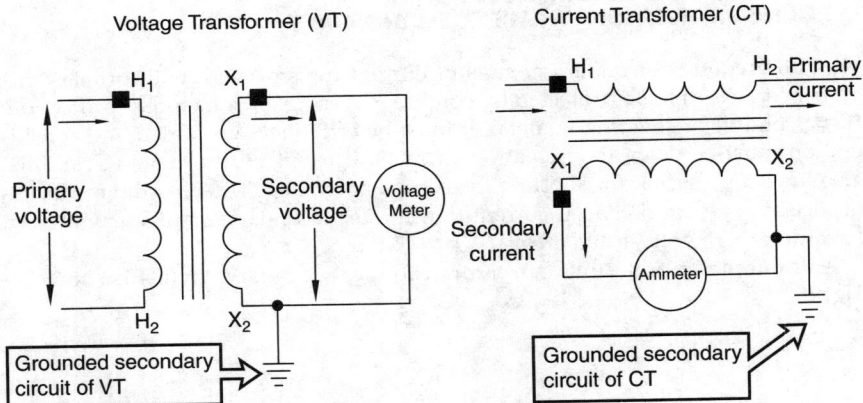

Voltage Transformer (VT)

H_1 X_1

Primary voltage

Secondary voltage

Voltage Meter

H_2 X_2

Grounded secondary circuit of VT

Current Transformer (CT)

H_1 H_2 Primary current

X_1 X_2

Secondary current

Ammeter

Grounded secondary circuit of CT

Fig. 151-1. Example of basic VT and CT connections (Rule 151).

152. LOCATION AND ARRANGEMENT OF POWER TRANSFORMERS AND REGULATORS

152A. Outdoor Installations. This rule requires that energized parts of power transformers be enclosed, guarded, or physically isolated as discussed in Rule 124. The case (enclosure) of a substation transformer and regulator must be effectively grounded or guarded.

The Code methods required for minimizing fire hazards related to liquid-filled power transformers installed in outdoor substations are outlined below (one or more methods must be used):

- Less flammable liquids
- Space separation
- Fire-resistant barriers
- Automatic extinguishing systems
- Absorption beds
- Enclosures

The Code requires that the degree of the fire hazard and the amount and characteristics of the liquid be considered when selecting a method, but no specifics are provided for the amount or type of oil, separation distance, etc. Rule 012C, which requires accepted good practice, must be applied. IEEE Standard 979, *IEEE Guide for Substation Fire Protection*, is an excellent reference for accepted good practice in this case.

The NESC does not address Spill Prevention Control & Countermeasure (SPCC) plans. SPCC plans are required by the Environmental Protection Agency (EPA). The amount of oil contained in the substation and the location of the substation relative to water are two of the many considerations that trigger the need for a substation SPCC plan.

There are times when an engineer or designer is trying to determine how close an oil-filled service transformer can be placed to a building. Rule 152 is in Part 1 and Part 1 only applies to electric supply stations, not to overhead or pad-mounted service transformers outside the substation fence. Part 2, Overhead Lines, and Part 3, Underground Lines, do not provide any additional rules related to locating oil-filled transformers near buildings. Part 2 does provide clearances to buildings for live parts and equipment cases, but these clearances are based on electrical safety, not fire safety. Rule 012C, which requires accepted good practice, must be applied in this instance. The National Electrical Code (NEC), Underwriters Laboratories (UL), and Factory Mutual (FM) are all good references for determining service transformer location standards.

152B. Indoor Installations. Transformers and regulators installed indoors require even greater consideration of fire hazards. Rule 152B is divided into three parts. The first paragraph covers traditional oil-filled transformers. The second paragraph covers dry-type transformers. The third paragraph covers oil-filled transformers with less flammable oil. Traditional oil-filled transformers use mineral oil, which is considered flammable. Less flammable oil is typically a fluid consisting of fire-resistant hydrocarbon. Dry-type transformers use no fluid oil for cooling. They are vented for air cooling. Dry-type transformers usually do not have the overload capability that oil-filled transformers have. The term "fire walls" is used but no specifics as to the fire wall ratings are provided (i.e., 1 hour rating, 2 hour rating, etc.). In addition to the accepted good practice documents discussed in Rule 152A, the Uniform Building Code (UBC) or other applicable building code document must be referenced for fire wall construction details. When an electric supply station is located indoors, for example, as part of a high-rise building in a large city, both the NESC and applicable building codes need to be addressed. NFPA 850, *Recommended Practice for Fire Protection for Electric Generating Plants and*

High Voltage Direct Current Converter Stations, can be referenced when designing generating stations.

The requirements for locating traditional oil-filled transformers indoors are outlined in Fig. 152-1.

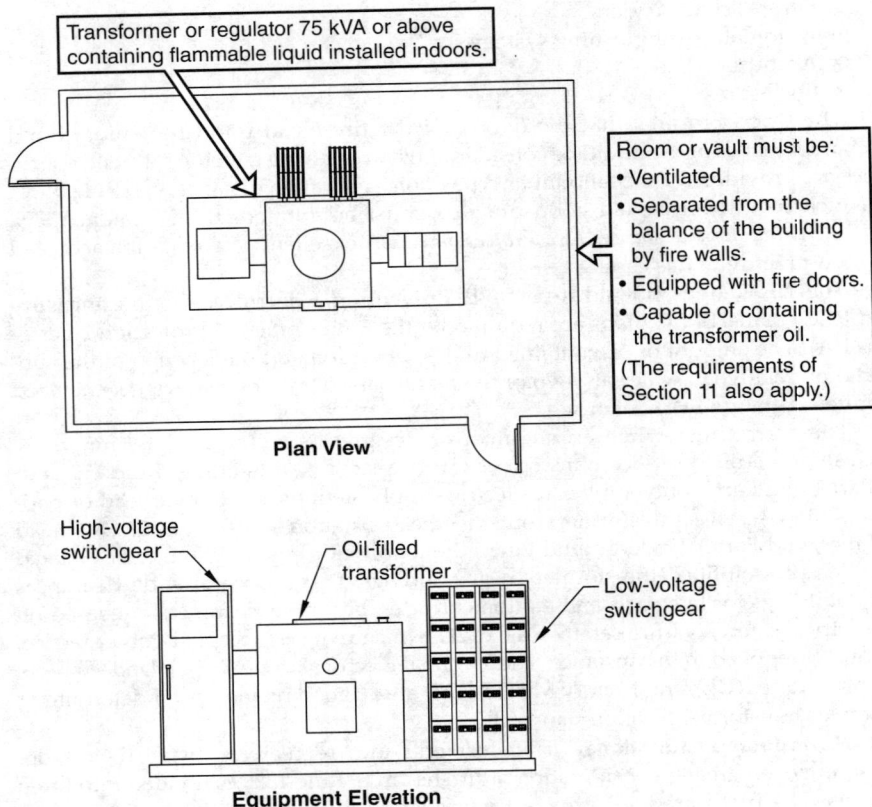

Fig. 152-1. Oil-filled transformer installed indoors (Rule 152B1).

The requirements for locating dry-type transformers indoors are outlined in Fig. 152-2.

The requirements for locating less flammable liquid transformers indoors are outlined in Fig. 152-3.

153. SHORT-CIRCUIT PROTECTION OF POWER TRANSFORMERS

Electric supply station power transformers are required to have short-circuit protection. A short circuit within the transformer provides a high-magnitude

Transformer or regulator of the dry type or containing
nonflammable liquid-or gas-installed indoors.

Plan View

Requirements:
- May be installed in a building (a separate room or vault may not be required depending on building codes).
- Energized parts must be enclosed and the case must be grounded.
- Pressure relief vents for nonbiodegradable liquids must have a method for absorbing toxic gases.

High-voltage switchgear — Dry-type transformer — Low-voltage switchgear

Equipment Elevation

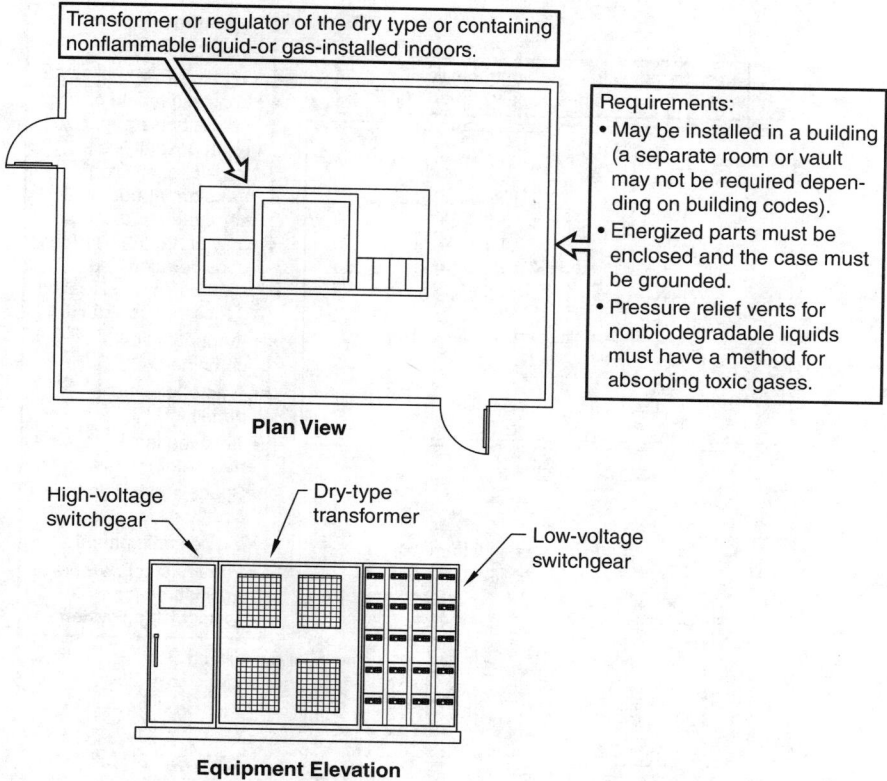

Fig. 152-2. Dry-type transformer installed indoors (Rule 152B2).

fault current. The short-circuit protection device typically will not be sized to protect transformer overload. Transformer overload is commonly monitored by metering the load and comparing the metered load to the transformer nameplate capacity.

This rule provides a list of automatically disconnecting short-circuit protection devices. Circuit breakers typically open all three phases of a three-phase line simultaneously. Fuses, on the other hand, open just the faulted phase. Removing just the faulted phase is acceptable. This rule does not apply to every transformer in the station, only to power transformers. For example, current transformers (CTs) are not typically fused. Voltage transformers (VTs) may be fused, but the Code does not require them to be fused. The choices for short-circuit protection of power transformers are outlined in Fig. 153-1.

Transformer with less flammable liquid installed indoors in a supply station building.

Requirements:
- Installation must be in a supply station building.
- Less flammable oil does not require a separate room or vault like a traditional oil-filled transformer but a separate room or vault may be the result of the consideration and selection of the location.

To locate, consider:
- Amount of liquid contained.
- Type of electrical protection.
- Tank venting.

Then select:
- Space separation from combustibles,
- Liquid confinement,
- Fire-resistant barriers or enclosures, or
- Extinguishing systems.

Plan View

High-voltage switchgear

Less flammable oil-filled transformer

Low-voltage switchgear

Equipment Elevation

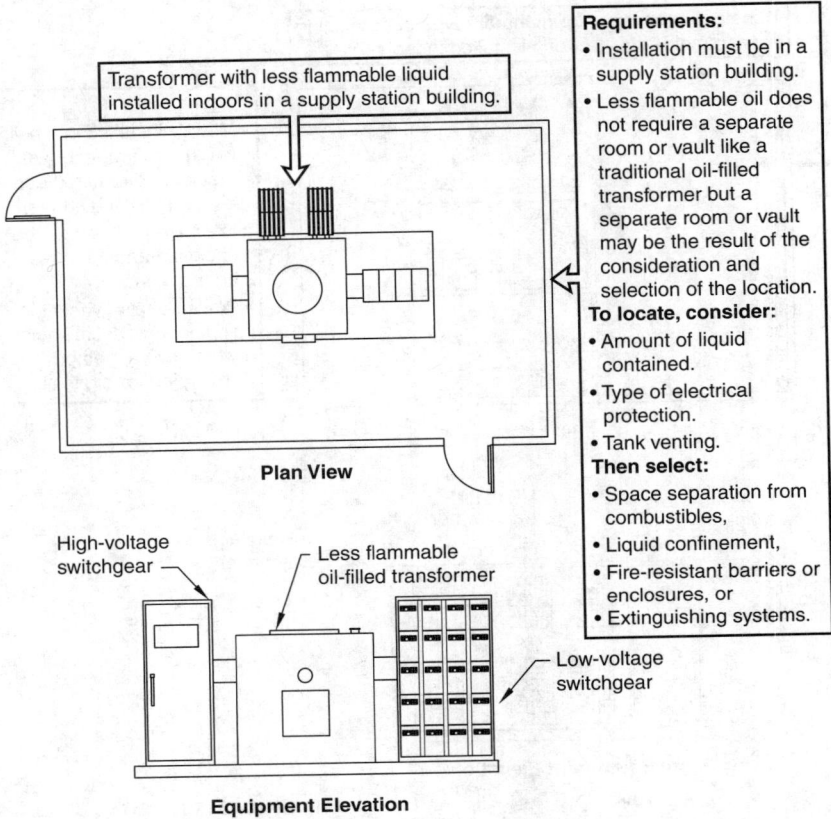

Fig. 152-3. Less flammable oil-filled transformer installed indoors (Rule 152B3).

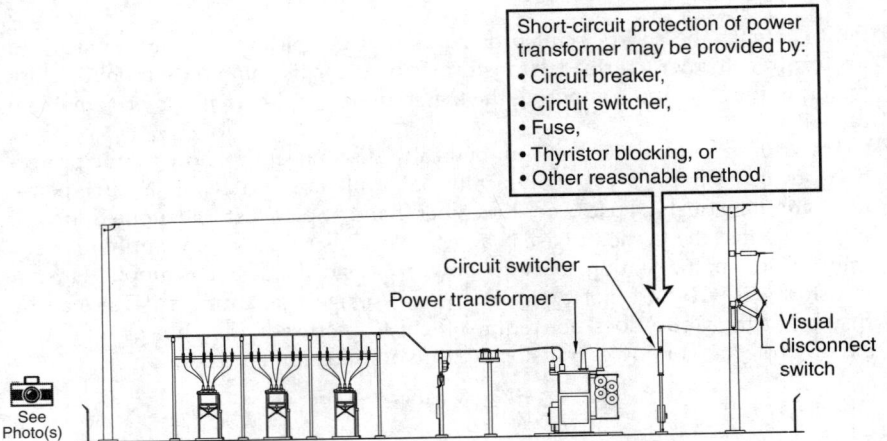

Short-circuit protection of power transformer may be provided by:
- Circuit breaker,
- Circuit switcher,
- Fuse,
- Thyristor blocking, or
- Other reasonable method.

Circuit switcher

Power transformer

Visual disconnect switch

See Photo(s)

Fig. 153-1. Choices for short-circuit protection of power transformers (Rule 153).

Section 16

Conductors

160. APPLICATION

Selecting conductors may appear to be a simple task but there are literally thousands to pick from. This rule requires that the following conductor features must be considered when choosing a conductor.

- Suitable for the location
- Suitable for the use
- Suitable for the voltage (insulation rating)
- Adequate ampacity (conductor size)

This section only applies to conductors in an electric supply station. Part 2 applies to overhead line conductors outside the substation and Part 3 applies to underground conductors outside the substation. Conductors in this rule, in Part 2, Overhead Lines, and in Part 3, Underground Lines, are discussed in general terms. The NESC does not specify conductor ampacity for various conductor sizes and types. To determine a conductor's ampacity, Rule 012C, which requires accepted good practice, must be applied. Ampacity tables can be found in the National Electrical Code (NEC) and in conductor manufacturers' literature. Both of these sources are excellent applications of Rule 012C. The NEC also specifies fuse and circuit breaker sizes for protecting 600 V conductors.

Conductors, by the definition in Sec. 02, include substation bus bars, as well as wires, which are typically bare, and cables, which are typically insulated.

Typical examples of conductors found in substations are shown in Fig. 160-1.

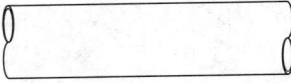

Bare 3" aluminum tube (bus bar)
- Voltage is dependent on the insulators that the bus bar is supported on.
- Ampacity (normal load current) is dependent on conductor material, size, and ambient temperature.
- Location: Above ground with proper clearance.

Insulated 1/0 aluminum URD cable (> 600 V)
- Voltage is dependent on insulation type and thickness.
- Ampacity (normal load current limit) is dependent on conductor material, size, insulation type, and conduit or burial method.
- Location: In conduit or direct buried at proper depths.

Insulated 1/0 copper (600 V or less)
- Voltage is dependent on insulation type and thickness.
- Ampacity (normal load current limit) is dependent on conductor material, size, insulation type, and conduit or burial method.
- Location: In conduit or direct buried at proper depths.

Fig. 160-1. Typical examples of conductors in substations (Rule 160).

161. ELECTRICAL PROTECTION

After the conductors are properly selected and applied, electrical protection of conductors is required. Overcurrent protection of conductors consists of two components, overload and short-circuit protection. Overload protection keeps the load current from exceeding the conductor's normal load current ampacity. Short-circuit protection keeps the fault current from exceeding the conductor's short-time fault current ampacity.

This rule requires overcurrent (both overload and short-circuit) protection by the following methods:
- Design of the system *and*
- Overcurrent devices *or*
- Alarm devices *or*
- Indication devices *or*
- Trip devices

A typical 120-V, 20-A, single-pole, thermal-magnetic circuit breaker protecting a No. 12 AWG copper, 600-V conductor provides both overload protection via the circuit breaker thermal element and short-circuit protection via the circuit breaker magnetic element. A typical 15-kV recloser protecting an overhead or underground distribution feeder may be set to only provide short-circuit protection. The overload protection of the feeder is monitored by checking percent conductor loading of the conductor during a system load study.

The electrical protection device may protect the conductor and a piece of equipment. For example, the high-side fuse of a substation transformer can provide short-circuit protection for the substation transformer and the bus bars. Overload protection of the transformer and bus bars can be provided by monitoring the load and comparing the load to the transformer and bus bar ratings.

Grounded conductors must be configured without overcurrent protection so there is not an interruption in the grounding protection. When a single-phase or three-phase circuit is fused or connected to a circuit breaker or recloser, the grounded conductor (neutral) must not be fused or connected to the circuit breaker or recloser contacts so that the continuity to ground is maintained.

Rule 161 applies to conductors and cables inside the substation. There is not a corresponding rule in Part 2 (Overhead Lines) or Part 3 (Underground Lines) requiring overcurrent protection for conductors outside the substation, therefore accepted good practice (Rule 012C) must be applied.

162. MECHANICAL PROTECTION AND SUPPORT

Conductors contained within the electric supply station must be adequately supported to withstand forces caused by the maximum short-circuit current that the conductor may experience. During short circuits, forces due to magnetic fields can occur on the substation conductors, including rigid bus bars. A rigid bus design must be adequately supported to withstand the short-circuit forces. The bus center-to-center spacing and the amount of short-circuit current both have an effect on the short-circuit forces. The strength of the substation structure and the strength of the insulators are typically based on the short-circuit forces and wind and ice loading, however for the conductors contained inside the substation fence, the Code does not specify wind and ice conditions. For wind and ice loads inside the substation, Rule 012C, which requires accepted good practice, must be applied. For overhead conductors that extend outside the electric supply station, the wind and ice loading and strength requirements in Part 2, Overhead Lines, Secs. 24, 25, 26, and 27 must be applied. For example, a transmission-deadend tower inside the substation that has overhead conductors that leave the substation must have the same loading and strength rules applied to it as the transmission pole outside the substation. Some utilities use a reduced tension span between the substation deadend tower and the first transmission pole outside the substation to reduce the loading and strength requirements on the substation deadend tower. This method typically requires guying on the first transmission pole outside the substation.

Commonly referenced design standards for substation conductor supports are IEEE Standard 605, *IEEE Guide for Design of Substation Rigid Bus Structures*, and ASCE 7, *Minimum Design Loads for Buildings* and *Other Structures*. The Uniform Building Code (UBC) is commonly referenced for seismic (earthquake) zones. The Rural Utilities Service (RUS) publishes Bulletin 1724E-300, *Design Guide for Rural Substations*, which contains detailed information on substation design calculations including calculations for short-circuit forces.

If a conductor, its insulation, or support is subject to mechanical damage, the following are required to limit the likelihood of damage or disturbance:

- Casing,
- Armor, or
- Other means

Typically in a substation, galvanized rigid steel conduit is used to protect and support insulated conductors subject to mechanical damage.

The mechanical protection and support requirements of this rule are outlined in Fig. 162-1.

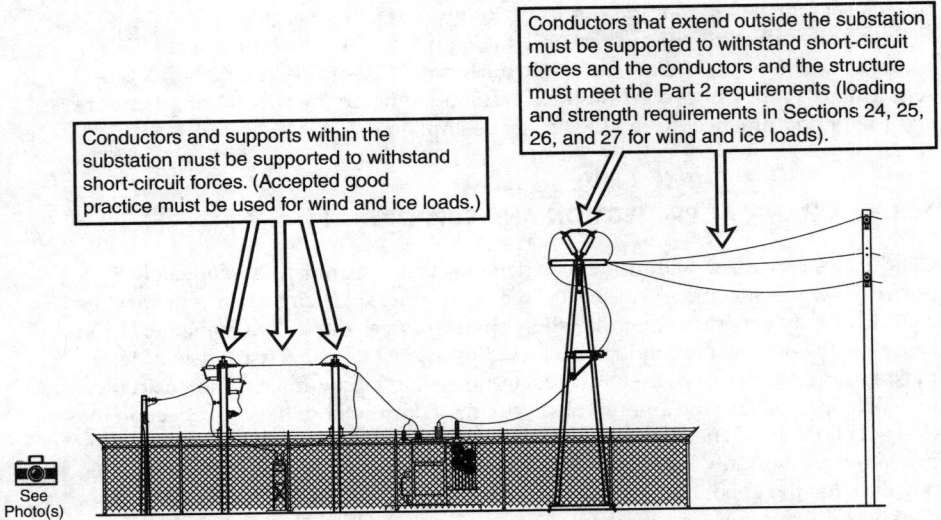

Conductors that extend outside the substation must be supported to withstand short-circuit forces and the conductors and the structure must meet the Part 2 requirements (loading and strength requirements in Sections 24, 25, 26, and 27 for wind and ice loads).

Conductors and supports within the substation must be supported to withstand short-circuit forces. (Accepted good practice must be used for wind and ice loads.)

See Photo(s)

Fig. 162-1. Substation conductor mechanical protection and support (Rule 162).

163. ISOLATION

Conductors inside the substation fence, when bare above 150 V to ground or insulated without a shield above 2500 V to ground, must be treated like the live parts in Rule 124. Nonshielded, insulated, and jacketed conductors may be installed in accordance with Rule 124C6, which discusses conductor guarding using insulation. The rules related to guarding a conductor with insulation are outlined in Fig. 163-1.

164. CONDUCTOR TERMINATIONS

The rules for conductor terminations in an electric supply substation are outlined in Fig. 164-1.

Bare conductors (bus bars) must meet
Rule 124 for clearance or guarding.

Per Rule 124C6a, an insulated
conductor above 2500 V to ground with a
grounded metallic shield (e.g., a
concentric neutral) is considered
guarded by insulation. The top of the
conduit does not have to meet any
vertical clearance requirements.

If the conductor is insulated but not
shielded, an appropriate barrier or
enclosure (e.g., a grounded rigid steel
conduit) is required per Rule 124C6b.
The top of the conduit must meet the
vertical clearance of **NESC** Table 124-1,
Column 3.

See
Photo(s)

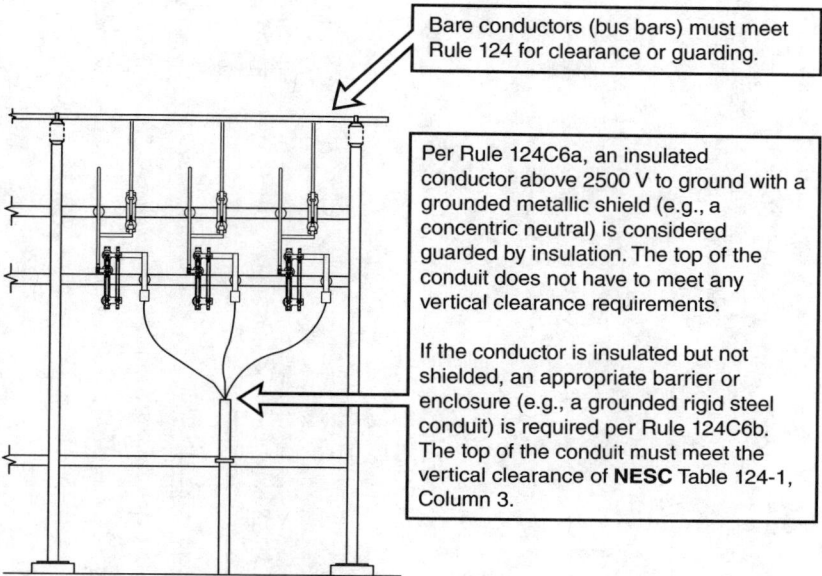

Fig. 163-1. Guarding a substation conductor with insulation (Rule 163).

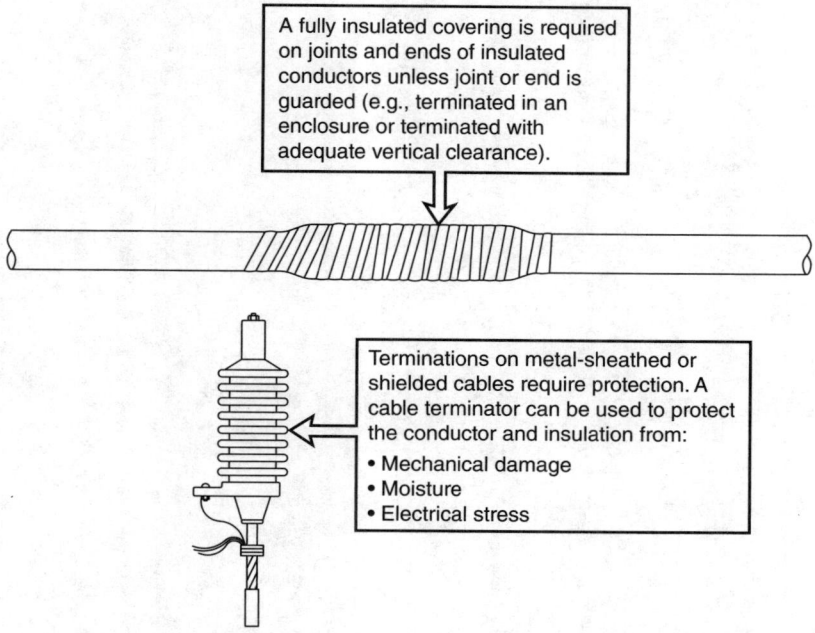

A fully insulated covering is required
on joints and ends of insulated
conductors unless joint or end is
guarded (e.g., terminated in an
enclosure or terminated with
adequate vertical clearance).

Terminations on metal-sheathed or
shielded cables require protection. A
cable terminator can be used to protect
the conductor and insulation from:

• Mechanical damage
• Moisture
• Electrical stress

Fig. 164-1. Conductor terminations (Rule 164).

Section 17

Circuit Breakers, Reclosers, Switches, and Fuses

170. ARRANGEMENT

This rule requires circuit breakers, reclosers, switches, and fuses be accessible only to qualified persons. Section 17 is part of Part 1, Electric Supply Stations; therefore, the rules of this section apply only to circuit breakers, reclosers, switches, and fuses located in electric supply stations. Circuit breakers, reclosers, switches, and fuses in electric supply stations are accessible only to qualified persons when Rule 110A is met.

To protect persons from energized parts or arcing, Rule 170 requires the following:
- Walls,
- Barriers,
- Latched doors,
- Location,
- Isolation, or
- Other means

The requirements of Rule 124, Guarding Live Parts, also apply.

Conspicuous and unique markings must be provided at the switching device or at any remote operating points to identify the equipment that is being controlled. The identification is for qualified personnel who are authorized to operate switching devices. See the discussion of identification and a related figure in Rule 128.

When the switch contacts are not normally visible (e.g., under oil, contained in a vacuum bottle, etc.), the switching device must be equipped with an operating position indicator. The work rules in Part 4 of the **NESC** provide switching control procedures and rules for de-energizing equipment or lines to protect employees. An example of a switch position indicator is shown in Fig. 170-1.

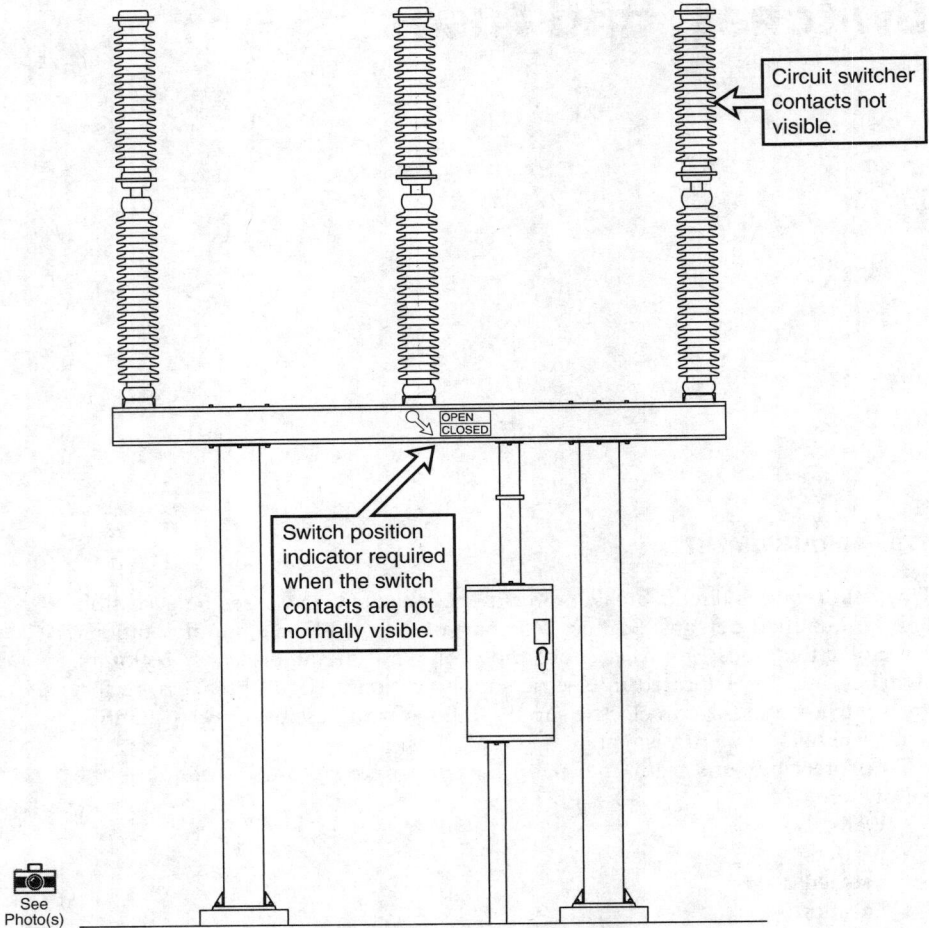

Circuit switcher contacts not visible.

OPEN
CLOSED

Switch position indicator required when the switch contacts are not normally visible.

See Photo(s)

Fig. 170-1. Example of a switch position indicator (Rule 170-1).

171. APPLICATION

The following ratings must be considered when applying a circuit breaker, circuit switcher, recloser, switch, or fuse:

- Voltage
- Continuous current

- Momentary current
- Short-circuit current interrupt rating

The maximum short-circuit current interrupt rating must be considered if the device is used to interrupt fault current.

When to apply ratings for momentary currents and interrupt currents is dependent on how the switch is used. If a switch is used as a disconnect only (i.e., it does not open under a fault condition), then it must be able to withstand a fault current flowing through it but it does not need to be rated to interrupt the fault current. The fault current it must withstand is termed the momentary fault current rating. If a switch is used to interrupt a fault, it must be rated to withstand the momentary fault current and interrupt the fault current without damage to the switch itself. A device that interrupts fault current must have a fault current interrupt rating.

It is important to note that the NESC does not address the selection of fuse, relay, or recloser settings such as time-current curves, delay intervals, number of operations to lockout, etc. These ratings are typically addressed in a system sectionalizing or coordination study. Careful selection of these and other settings is required for the safe and reliable operation of the electric system and accepted good practice must be applied in this case. The work rules in Part 4 of the NESC do address disabling a reclosing feature during work activities.

The interrupting capacity should be reviewed prior to each significant system change. For example, if a substation recloser is rated to interrupt 1250 A of short-circuit current and the substation transformer is replaced, the available fault current may increase to a value larger than 1250 A, which could cause damage to the recloser and personnel when the recloser operates to interrupt a fault. An example is given in Fig. 171-1.

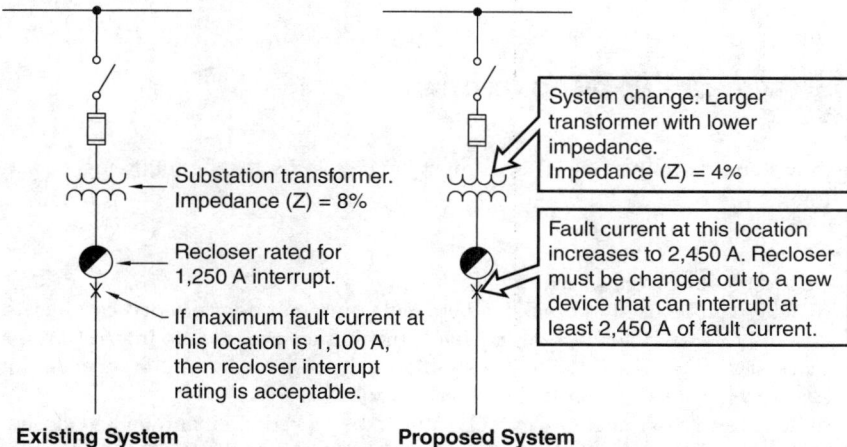

Existing System

Substation transformer. Impedance (Z) = 8%

Recloser rated for 1,250 A interrupt.

If maximum fault current at this location is 1,100 A, then recloser interrupt rating is acceptable.

Proposed System

System change: Larger transformer with lower impedance. Impedance (Z) = 4%

Fault current at this location increases to 2,450 A. Recloser must be changed out to a new device that can interrupt at least 2,450 A of fault current.

Fig. 171-1. Example of checking interrupting capacity prior to a significant system change (Rule 171).

172. CIRCUIT BREAKERS, RECLOSERS, AND SWITCHES CONTAINING OIL

Circuit breakers, reclosers, and switches containing oil receive special attention due to the flammability of the oil. Similar requirements are outlined in Rule 152 for oil-filled power transformers and regulators.

Segregation of oil-filled circuit-interrupting devices is required in the electric supply station. Segregation of oil-filled equipment minimizes fire damage to adjacent equipment or buildings. Segregation may be provided by the following methods:

- Spacing,
- Fire-resistant barrier walls, or
- Metal cubicles

Gas release vents require oil-separating devices or piping to a safe location. Means to control oil discharges from vents or tank rupture are required. Methods to control oil are:

- Absorption beds,
- Pits,
- Drains, or
- Any combination of the above

This list is slightly different from the requirements for oil-filled transformers and regulators in Rule 152 but the general intent is the same. IEEE Standard 979, *IEEE Guide for Substation Fire Protection,* is an excellent reference for fire protection requirements.

Buildings or rooms housing circuit breakers, reclosers, and switches containing flammable oil must be of fire-resistant construction (see Rule 152 for additional information). Not all circuit breakers, reclosers, and switches contain oil. Some are air break, some are vacuum break, and some are SF_6 gas insulated. This rule applies only to circuit breakers, reclosers, and switches that contain oil and are located in an electric supply station.

173. SWITCHES AND DISCONNECTING DEVICES

Switches and disconnecting devices must have capacity for the following system ratings:

- Voltage
- Current
- Load break current (if required)

Switches can be used to break load currents or open under no-load conditions. If required to break load current, the load current they are rated to interrupt must be marked on the switch. This value should not be confused with the short-circuit (fault) current-interrupt rating discussed in Rule 171.

Switches and disconnectors must be able to be locked open and locked closed, or plainly tagged where locks are not practical. Part 4 of the NESC specifies the work rules applicable to operating, locking, and tagging switches. Switches that are operated remotely and automatically must have a disconnecting means for the control circuit near the disconnecting apparatus to limit the likelihood of accidental operation of the switch.

174. DISCONNECTION OF FUSES

Disconnecting an energized fuse can be dangerous at any voltage. This rule requires fuses in circuits of more than 150 V to ground or more than 60 A (at any voltage) to be classified as disconnecting, or arranged to be disconnected from the source of power, or removed with insulating handles.

In most cases, fused cutout-type switches are configured such that the switchblade or fuse is "dead" when the switch is in the open position. Loop feeds can create energized blades when the switch is in the open position. The NESC does not have a requirement for how the switch blades or fuses are energized but does require the proper handling of the fuse. Proper consideration must be given to clearance requirements when fuse and switch blades are in the open position.

Section 18

Switchgear and Metal-Enclosed Bus

180. SWITCHGEAR ASSEMBLIES

180A. General Requirements for All Switchgear. This rule covers general requirements for all switchgear. Examples of switchgear found in electric supply stations are shown in Fig. 180-1.

The general requirements for all switchgear are outlined below:
- Secure to minimize movement.
- Support cables to minimize force on terminals.
- Locate away from liquid piping unless protected.
- Locate away from flammable gases or liquids.
- Install after general construction or provide temporary protection.
- Protect when doing maintenance in the area.
- Do not use as a physical support unless so designed.
- Interiors shall not be used for storage unless so designed.
- Metal instrument cases are to be grounded, enclosed in grounded metal covers, or made of insulating materials.

180B. Metal-Enclosed Power Switchgear. In addition to the general requirements for all switchgear, requirements exist for metal-enclosed power switchgear. The rules for metal-enclosed power switchgear are outlined in Fig. 180-2.

180C. Dead-Front Power Switchboards. Dead-front power switchboards with uninsulated rear connections must be installed in rooms capable of being locked with access limited to qualified personnel only. The uninsulated rear connections

125

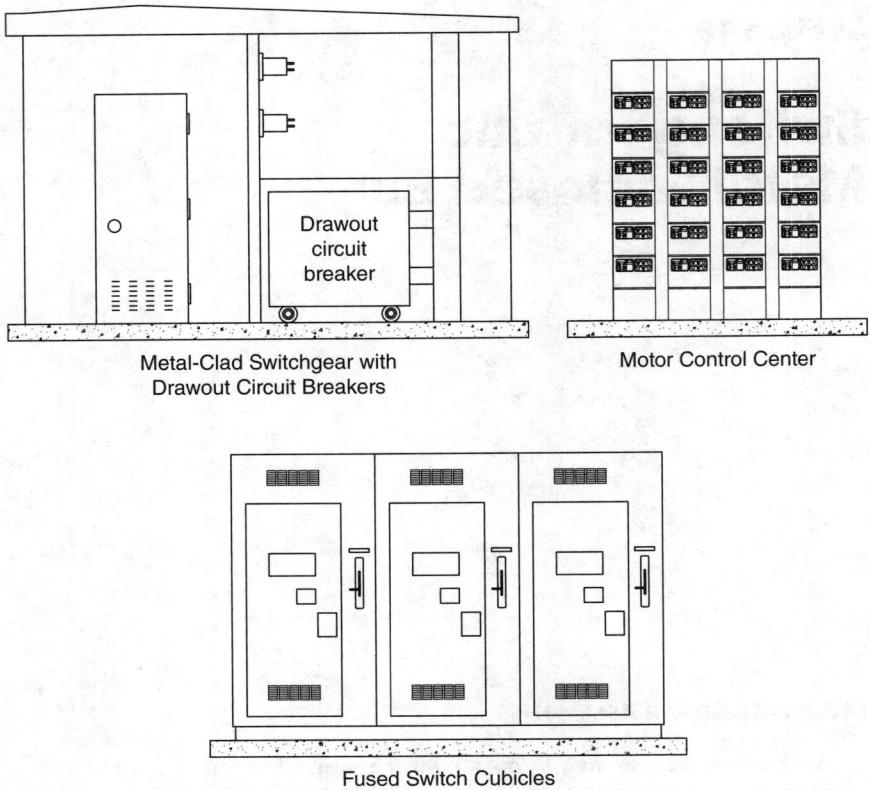

Metal-Clad Switchgear with
Drawout Circuit Breakers

Motor Control Center

Fused Switch Cubicles

Fig. 180-1. Examples of switchgear assemblies (Rule 180A).

typically will not meet the vertical and horizontal clearances of Rule 124; therefore, a separate locked room is required for these energized parts.

180D. Motor Control Centers. The fault current within the electric supply station can be very high due to low source impedance and little or no distribution-line impedance. The bus withstand ratings of motor control centers must therefore be given proper consideration. Bus bracing in low-voltage (e.g., 480-V) motor control centers commonly come in 50,000-, 100,000-, and 200,000-A values. This is a withstand rating, not an interrupt rating, as the bus must withstand the force of a momentary fault but not interrupt it. A fuse or circuit breaker in the motor control center will interrupt the fault. Current-limiting fuses can reduce the available fault current to which the motor control center bus and motor control center circuit breakers are exposed. Peak let-through currents for current-limiting fuses can be obtained from fuse manufacturers' literature. This rule also requires a safety sign for a motor control center cubicle having more than one voltage source. The rules regarding motor control center short-circuit ratings are outlined in Fig. 180-3.

Horizontal distance not less than 25' indoors or 10' outdoors. Exception: less distance with barrier. Rule does not apply to power transformer.

A safety sign must be placed on a cubicle containing more than one high-voltage source (e.g., transfer switch or bus tie breaker).

Container with flammable liquid or gas.

Low-voltage and high-voltage wiring in the switchgear cubicle must be separately routed and terminated.

Indoors: 25'

Outdoors: 10'

Cubicle #1   Cubicle #2   Cubicle #3

Control devices must be readily accessible.

Space in front of switchgear to remove and turn circuit breakers.

Switchgear enclosure must be designed for environment (e.g., indoor or outdoor).

Cubicle #4   Cubicle #5   Cubicle #6

Two means of exit, doors swing out with panic hardware. Exception: one door under certain conditions.

Nonhinged removable panel requires at least 3' of clear space. Working space requirements of Rule 125 also apply.

Panel boards, columns, etc., must not encroach the rear working space.

Hinge rear panel must swing open full 90°.

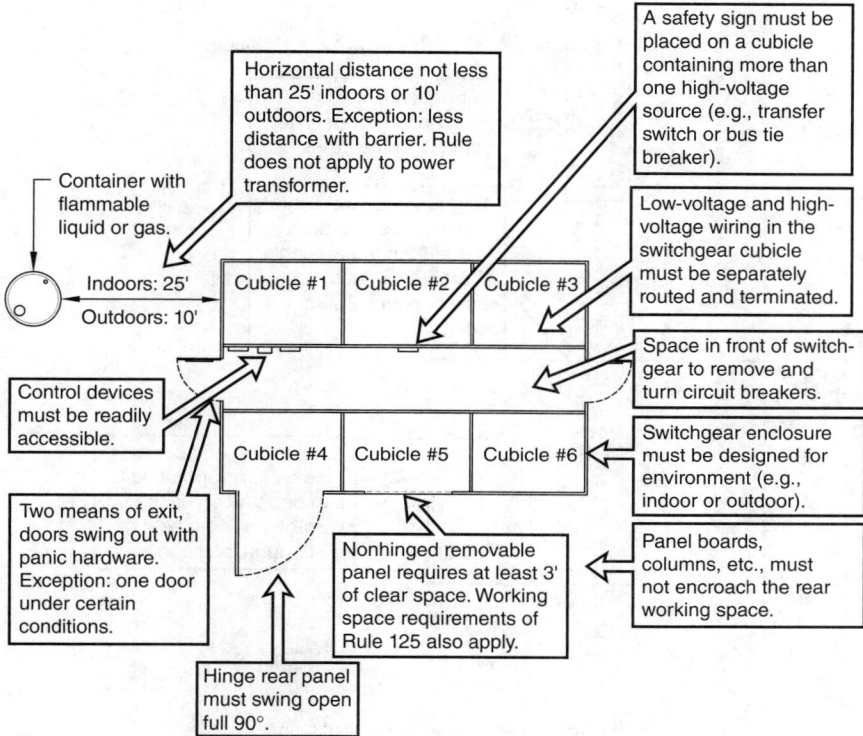

Fig. 180-2. Metal-enclosed power switchgear (Rule 180B).

180E. Control Switchboards. The requirements for control switchboards are outlined below:

- Control switchboards include cabinets with meters, relays, annunciators, computers, etc., for substation and generating station controls.
- Carpeting in control switchboard rooms must be antistatic and minimize toxic gas emissions under any conditions (e.g., water damage, fire damage, etc.).
- Adequate clearance in front and rear to read meters without stools.
- Personnel openings must be covered when not in use and must not be used for cable routing.

Reading control switchboard meters from stools produces a tripping or falling hazard. The rule related to adequate clearance for reading meters without stools is outlined in Fig. 180-4.

181. METAL-ENCLOSED BUS

181A. General Requirements for All Types of Bus.   The rules for metal-enclosed bus are outlined in Fig. 181-1.

Current-limiting fuse can be used to limit peak let-through fault current.

The motor control center bus must be braced higher than the available short-circuit (fault) current on the source side.

Bus bracing and circuit breaker interrupt rating can be reduced based on current-limiting fuse peak let-through fault current.

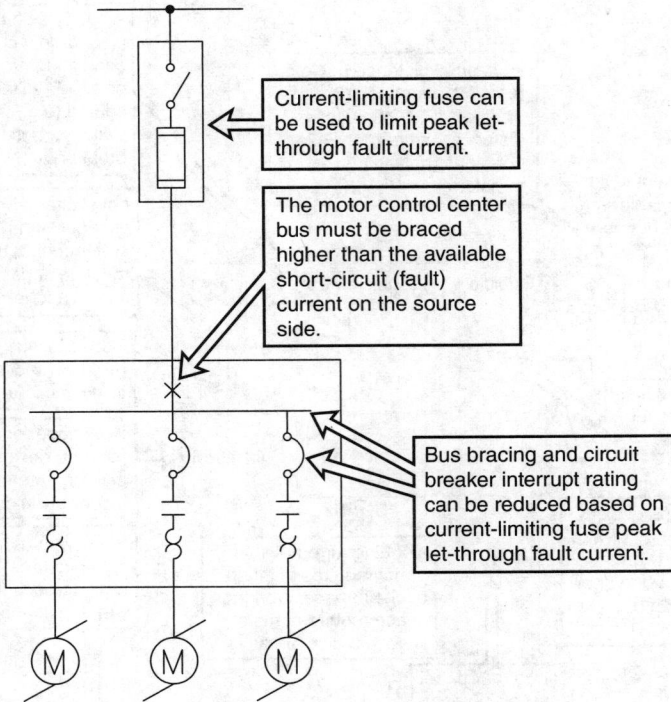

Fig. 180-3. Motor control center short-circuit ratings (Rule 180D).

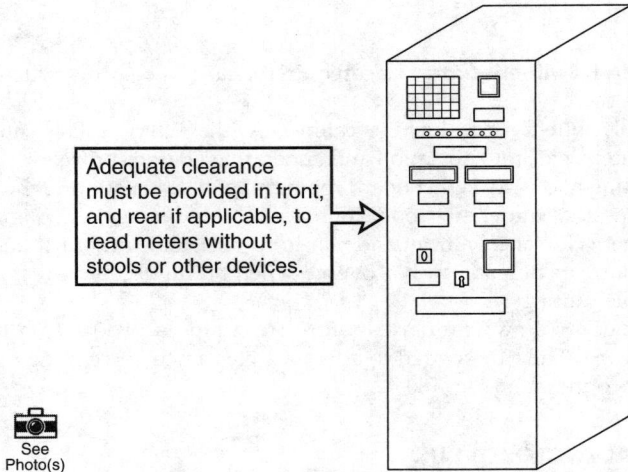

Adequate clearance must be provided in front, and rear if applicable, to read meters without stools or other devices.

See Photo(s)

Fig. 180-4. Adequate clearance for reading meters on control switchboards (Rule 180E3).

181B. Isolated-Phase Bus. Metal-enclosed bus is available in three common designs: segregated (isolated) phase bus bar, nonsegregated phase bus bar, and cable bus.

The use of isolated-phase bus requires the following special conditions:
- Clearance to magnetic material per manufacturer's recommendations
- Nonmagnetic conduit for alarm circuits
- Piping if enclosure drains are used
- Nonmagnetic wall plates
- Non-ferrous conduit for grounding conductors

Examples of metal-enclosed bus are shown in Fig. 181-2.

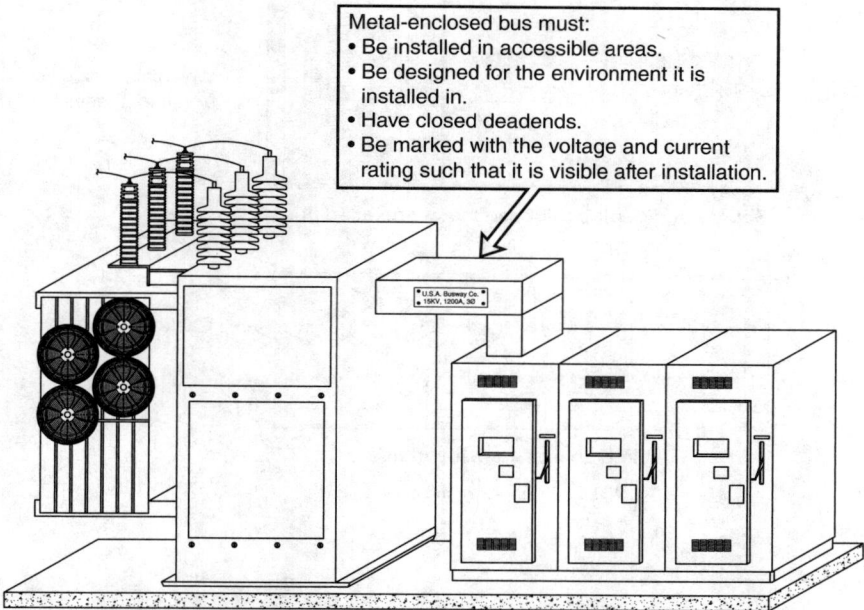

Metal-enclosed bus must:
- Be installed in accessible areas.
- Be designed for the environment it is installed in.
- Have closed deadends.
- Be marked with the voltage and current rating such that it is visible after installation.

Fig. 181-1. General requirements for metal-enclosed bus (Rule 181A).

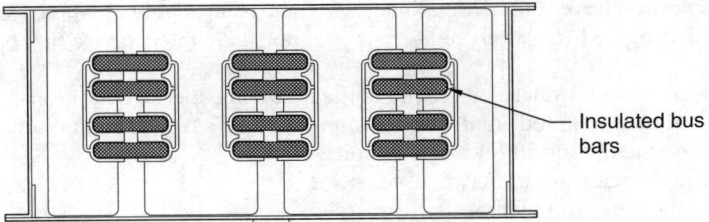

Nonsegregated phase metal-enclosed bus

Insulated bus bars

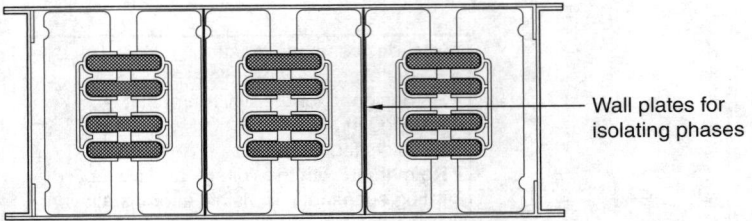

Segregated (isolated) phase metal-enclosed bus

Wall plates for isolating phases

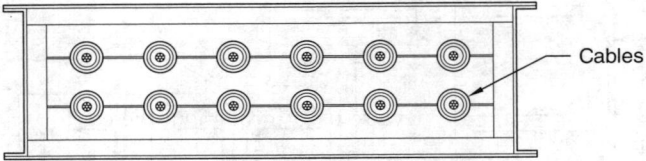

Cable type metal-enclosed bus

Cables

Fig. 181-2. Examples of metal-enclosed bus (Rule 181).

Section 19

Surge Arresters

190. GENERAL REQUIREMENTS

Surge arresters are used to protect equipment from overvoltage primarily due to switching surges and lighting. They are used throughout the electric utility system. The rules for surge arresters in Sec. 19 are for surge arresters installed in an electric supply station. Critical equipment is located in the electric supply station. The careful application of surge arresters is used to limit damage to expensive equipment such as the substation power transformer. Locating a surge arrester as close as practical to the equipment it protects limits voltage buildup across the equipment insulation. IEEE Standard C62.1 and IEEE Standard C62.11 are noted in the NESC. They are excellent references for surge arrester applications.

191. INDOOR LOCATIONS

Arresters can discharge hot gases and produce electric arcs. To avoid damage to its surroundings, if an arrester is located inside a building, it must be enclosed or located far away from passageways and combustible parts.

192. GROUNDING CONDUCTORS

Short conductive grounding leads are required to keep the voltage buildup across equipment insulation as low as possible. The grounding methods of Sec. 09 are referenced for additional information.

193. INSTALLATION

Surge arresters can be installed in various locations throughout the electric supply station. The rules regarding surge arrester installation in electric supply stations are outlined in Fig. 193-1.

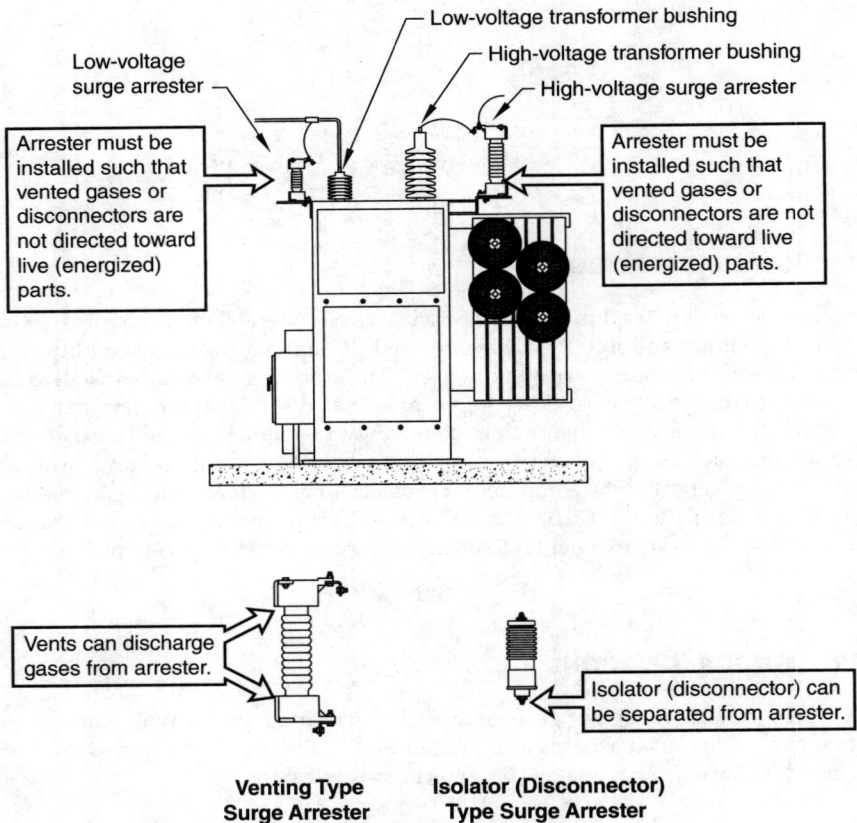

Low-voltage transformer bushing

High-voltage transformer bushing

High-voltage surge arrester

Low-voltage surge arrester

Arrester must be installed such that vented gases or disconnectors are not directed toward live (energized) parts.

Arrester must be installed such that vented gases or disconnectors are not directed toward live (energized) parts.

Vents can discharge gases from arrester.

Isolator (disconnector) can be separated from arrester.

See Photo(s)

Venting Type Surge Arrester

Isolator (Disconnector) Type Surge Arrester

Fig. 193-1. Arrester installation (Rule 193).

Part 2

Safety Rules for the Installation and Maintenance of Overhead Electric Supply and Communication Lines

GENERAL SECTIONS

01 INTRODUCTION   02 DEFINITIONS

03 REFERENCES   09 GROUNDING METHODS

GENERAL SECTIONS 01, 02, 03, 09

PART 1

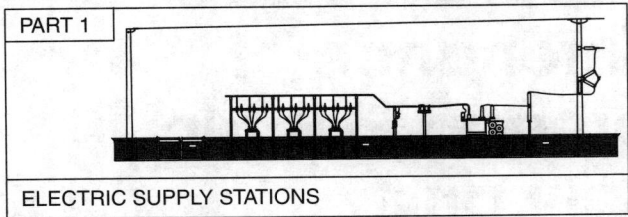

ELECTRIC SUPPLY STATIONS

PART 2

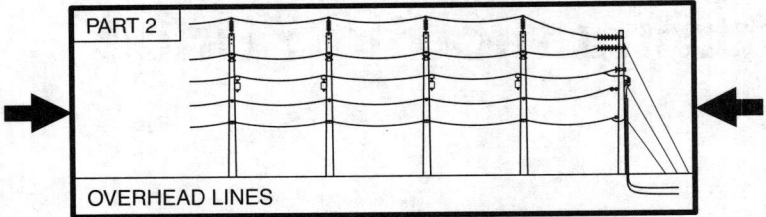

OVERHEAD LINES

PART 3

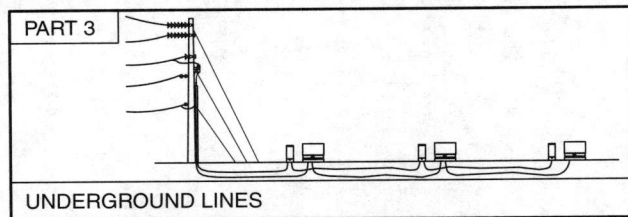

UNDERGROUND LINES

PART 4

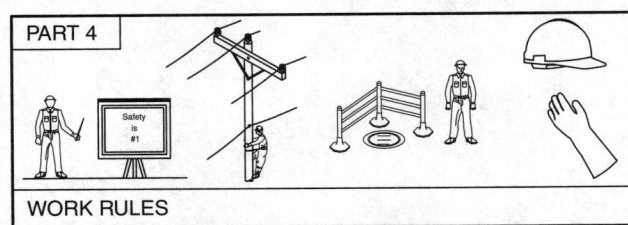

WORK RULES

Section 20

Purpose, Scope, and Application of Rules

200. PURPOSE

The purpose of Part 2, Overhead Lines, is similar to the purpose of the entire NESC outlined in Rule 010, except Rule 200 is specific to overhead supply and communication lines and equipment. Part 2 of the NESC focuses on the practical safeguarding of persons during the installation, operation, and maintenance of overhead supply (power) and communication lines and equipment.

201. SCOPE

The scope of Part 2, Overhead Lines, includes supply (power) and communication (phone, cable TV, fiber, etc.) conductors, cables, and equipment in overhead line applications. Separate rules apply to Electric Supply Stations (Part 1) and Underground Lines (Part 3). Part 2 covers overhead structural arrangements and extensions into buildings. Requirements are provided in Part 2 for:
- Spacing (center to center measurement)
- Clearances (surface to surface measurement)
- Strength of construction

There is some overlap between Overhead Lines (Part 2) and Underground Lines (Part 3). The overlap is at the underground riser on the overhead pole. Risers are covered in Sec. 23, Rule 239 of Part 2, and in Sec. 36 of Part 3. See Fig. 201-1.

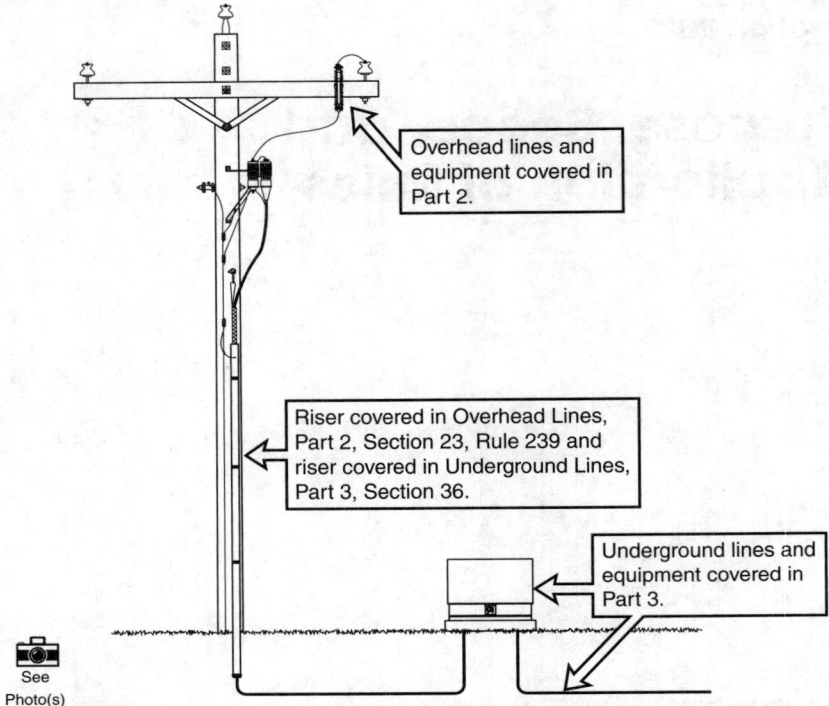

Fig. 201-1. Overlap between Overhead Lines (Part 2) and Underground Lines (Part 3) (Rule 201).

There is some overlap between Overhead Lines (Part 2) and Electric Supply Stations (Part 1). The overlap occurs when an overhead conductor extends outside the electric supply station fence. Conductors that extend outside the substation and their supports (e.g., poles, deadend towers, etc.) must meet the Part 2 requirements. Conductors and supports within the substation must meet the Part 1 requirements. See Fig. 201-2.

When overhead supply equipment is fenced, the rules of Part 1, Electric Supply Stations, may apply. See Rule 110A for a discussion.

The notes at the end of Rule 201 remind the reader that Part 4 of the NESC and the Occupational Health and Safety Administration (OSHA) regulations cover the work rules related to overhead lines. See Part 4, Work Rules, for a discussion.

See Rule 011, Scope, for additional information on the scope of the NESC.

202. APPLICATION OF RULES

Rule 202 references Rule 013 for the general application of Code rules. See Rule 013 for a discussion. Rule 202 has a "however" statement that is essentially an exception to Rule 013. If a supporting structure (e.g., pole) is being replaced,

Conductors that extend outside the substation must be supported to withstand short-circuit forces and the conductors and the structure must meet the Part 2 requirements (Loading and strength requirements in Secs. 24, 25, 26, and 27 for wind and ice loads) per Rule 162.

Conductors and supports within the substation must be supported to withstand short-circuit forces per Rule 162. (Accepted good practice must be used for wind and ice loads.)

See Photo(s)

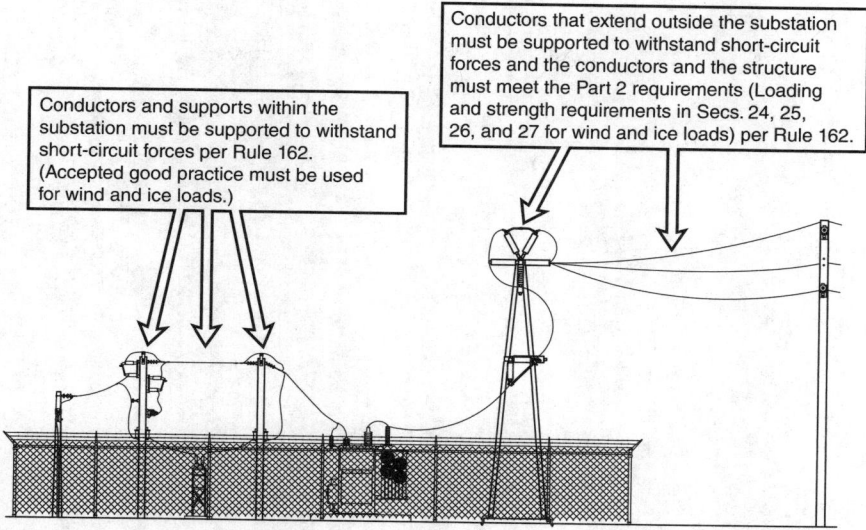

Fig. 201-2. Overlap between Overhead Lines (Part 2) and Electric Supply Stations (Part 1) (Rule 201).

the arrangement of equipment must conform to the current **NESC** requirements for vertical clearance between span wires or brackets carrying luminaires, traffic signals, or trolley conductors, and communications equipment covered in Rule 238C. No note or other explanation is provided in the **Code** as to why Rule 238C was singled out from all the overhead line rules.

Section 21

General Requirements

210. REFERENCED SECTIONS

This rule references four sections related to Part 2, Overhead Lines, so that rules do not have to be duplicated and the reader of the Code realizes that other sections are related to the information provided in Part 2. The related sections are:

- Introduction—Sec. 01
- Definitions—Sec. 02
- References—Sec. 03
- Grounding Methods—Sec. 09

The rules in Part 2, predominantly Rule 215, will provide the requirements for grounding overhead lines and equipment. The grounding methods are provided in Sec. 09.

211. NUMBER 211 NOT USED IN THIS EDITION

212. INDUCED VOLTAGES

Rule 212 recognizes that voltages induced by the supply line may affect the communication line in some manner. Noise on the communication line due to the supply line is one example of induced voltage problems. Specific requirements to deal with induced voltage are not specified other than cooperation and advance notice. The clearances provided in Part 2 between supply and communication lines and equipment are for safety purposes. They were not developed to mitigate induced voltage.

Induced voltage from a supply line, such as a high voltage transmission line, on to an adjacent fence line is not covered in the Code. Rule 012C, which requires accepted good practice, must be applied. Rule 232C1c does require limiting electrostatic effects between a supply line exceeding 98 kV to ground (exceeding 170 kV phase to phase) and a truck, vehicle, or equipment under the line. See Rule 232C for more information and a figure. Rules for induced voltages that occur when working on or near supply lines are covered in the Part 4 Work Rules.

213. ACCESSIBILITY

This rule generally states the requirement that overhead parts that need to be examined or adjusted during operation must be accessible to authorized supply and communication workers. This rule requires the following, all of which are discussed in more detail throughout Part 2:

- Climbing space
- Working space
- Working facilities
- Clearances between conductors

Climbing space is specifically discussed in Rule 236. Working space is covered in Rule 237. Clearances between conductors and between conductors and equipment are covered in various rules throughout Sec. 23.

214. INSPECTION AND TESTS OF LINES AND EQUIPMENT

The rules for inspection and testing on lines and equipment are broken down into two parts, when in-service (Rule 214A) and when out-of-service (Rule 214B). The open or closed status of a switch, fused cutout, recloser, etc., will determine if a line is in- or out-of-service. A line that is permanently abandoned must be removed or maintained in a safe condition. Opening the cutouts on a line and grounding the line may make it electrically safe but not structurally safe. If the permanently abandoned line does not meet the strength requirements in Sec. 26, it must be removed or maintained in a safe condition. The rules for inspecting and testing overhead lines and equipment are outlined in Fig. 214-1.

Rule 214 applies to overhead supply and communications lines. Similar inspection rules are outlined in Rule 121 for electric supply stations and Rule 313 for underground lines.

Rule 214 does not provide a specific requirement as to how often to inspect overhead lines. The inspection frequency must be determined by the utility "at such intervals as experience has shown to be necessary." This statement should consider what the local conditions are; for example, do poles rot faster in some areas due to acidic soil conditions, do insulators fail more in some areas due

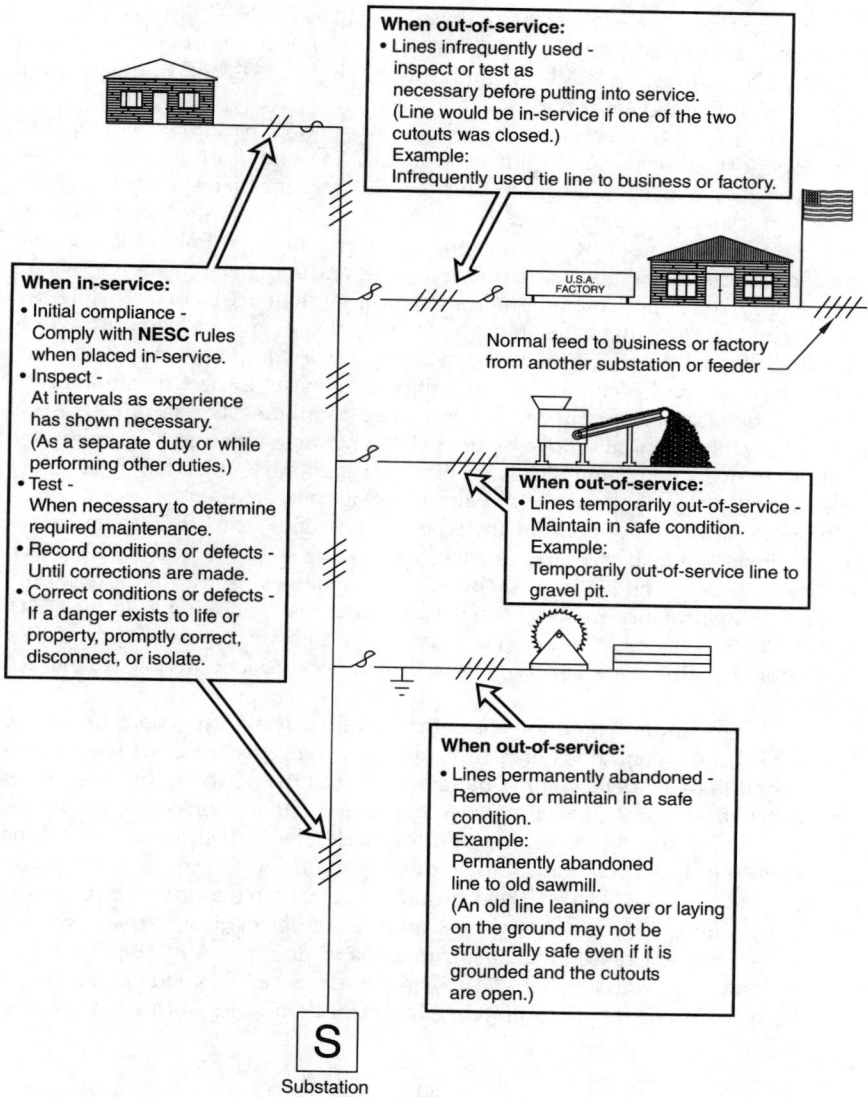

When out-of-service:
• Lines infrequently used -
  inspect or test as
  necessary before putting into service.
  (Line would be in-service if one of the two
  cutouts was closed.)
  Example:
  Infrequently used tie line to business or factory.

When in-service:
• Initial compliance -
  Comply with **NESC** rules
  when placed in-service.
• Inspect -
  At intervals as experience
  has shown necessary.
  (As a separate duty or while
  performing other duties.)
• Test -
  When necessary to determine
  required maintenance.
• Record conditions or defects -
  Until corrections are made.
• Correct conditions or defects -
  If a danger exists to life or
  property, promptly correct,
  disconnect, or isolate.

Normal feed to business or factory
from another substation or feeder

U.S.A.
FACTORY

When out-of-service:
• Lines temporarily out-of-service -
  Maintain in safe condition.
  Example:
  Temporarily out-of-service line to
  gravel pit.

When out-of-service:
• Lines permanently abandoned -
  Remove or maintain in a safe
  condition.
  Example:
  Permanently abandoned
  line to old sawmill.
  (An old line leaning over or laying
  on the ground may not be
  structurally safe even if it is
  grounded and the cutouts
  are open.)

S

Substation

See
Photo(s)

Fig. 214-1. Inspection and tests of lines and equipment when in- and out-of-service (Rule 214).

to salt fog, pollution from a nearby factory, lightning, etc. These considerations are part of applying accepted good practice for the given local conditions. See NESC Rule 012C. Most electric power utilities have an inspection program that has some type of record keeping and inspection frequency; however, the frequency of the inspection is not specified in the **NESC**. The authority having jurisdiction over utilities (i.e., a State Public Service Commission or State Public Utility Commission) sometimes sets a fixed inspection frequency for utilities in a particular state. It is important for utilities to check the authority having jurisdiction to determine if fixed inspection intervals apply.

A note in Rule 214A2 recognizes that inspections may be performed as a separate duty or performed while performing other duties. An example of an inspection performed as a separate operation is an electric utility sending an inspector to poles in a geographic area and testing them for rot. An example of an inspection performed while performing other duties is a communications utility technician looking for damaged hardware and low hanging wires while adding, changing, or removing a service drop. If a supply or communications employee notices a defect while performing another task, the defect must be addressed, not ignored. Typically supply (power) utility employees inspect power lines, power equipment, and associated power facilities and communications utility employees inspect communication lines, communication equipment, and associated communication facilities. Inspection of the pole is typically done by the pole owner. Inspection responsibilities on joint-use (power and communication) poles are commonly outlined in the joint-use agreement between the power and communication companies.

Rule 214 also does not provide any type of inspection checklist. The specifics as to what items are inspected and to what detail is not specified. Rule 012C, which requires accepted good practice, must be used. Sometimes electric utilities use both detailed and patrol inspections. These inspection terms are not found in the **NESC**. A State Public Service Commission or State Public Utility Commission may define detailed and patrol inspection tasks or accepted good practice (**NESC** Rule 012C) may be used.

Rule 214A4 requires conditions or defects found during inspections or tests to be recorded if the conditions or defects are not promptly corrected. The records of the conditions or defects must be maintained until the conditions or defects are corrected. The **NESC** does not provide a time frame for correcting conditions or defects found during inspections. For example, the correction could be done immediately or the correction could be done in the future during a pole replacement project. Per Rule 214A5, if the condition or defect can reasonably be expected to endanger human life or property, the condition or defect must be promptly corrected, disconnected, or isolated, but again the **NESC** does not provide a time frame for the prompt correction. Rule 012C, which requires accepted good practice, must be used to determine the correction time frame for both Rules 214A4 and 214A5.

215. GROUNDING OF CIRCUITS, SUPPORTING STRUCTURES, AND EQUIPMENT

Rule 215 is broken into three main paragraphs. Rule 215A reminds us that the methods of grounding are specified in Sec. 09 and the requirements of grounding are provided in this rule for circuits and non-current-carrying parts. Rule 215B focuses on the requirements for grounding circuits. See Sec. 02, Definitions, for a discussion of the term "effectively grounded." The circuit grounding requirements of Rule 215B are outlined in Fig. 215-1.

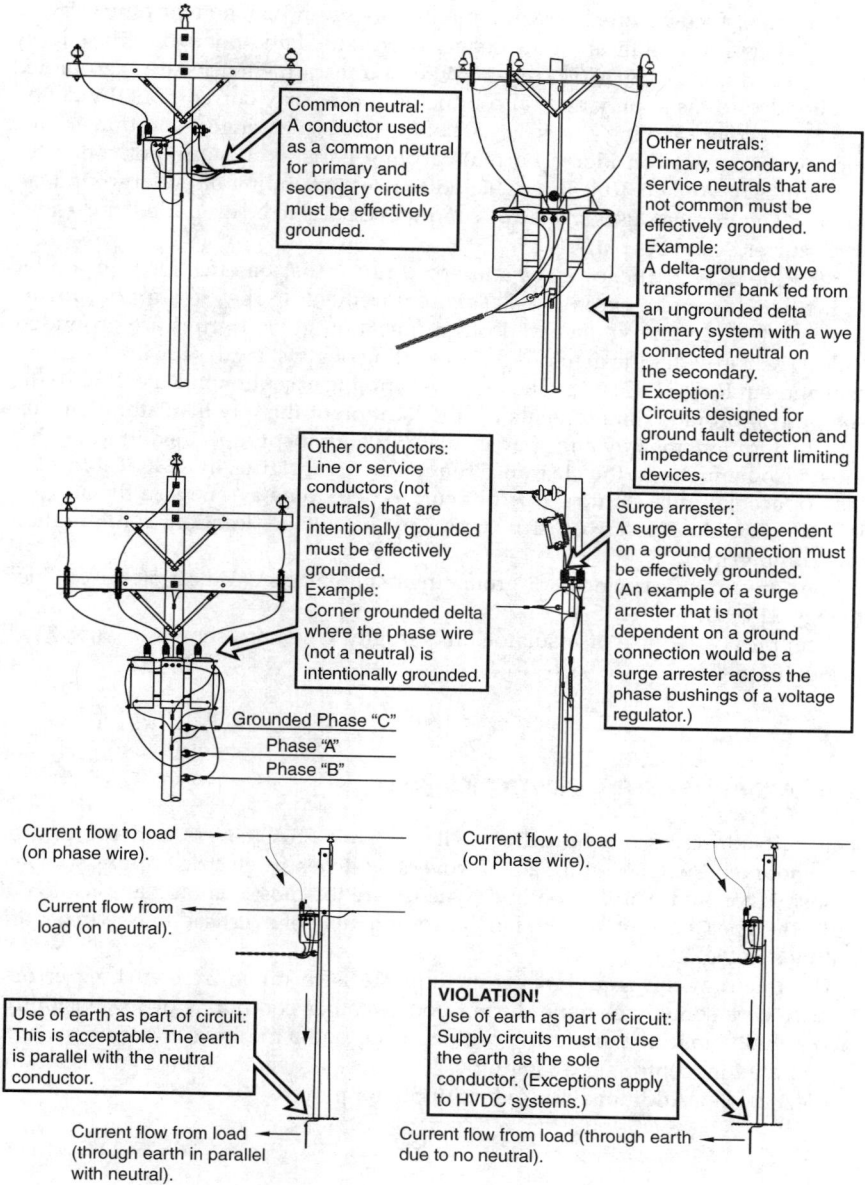

Common neutral:
A conductor used as a common neutral for primary and secondary circuits must be effectively grounded.

Other neutrals:
Primary, secondary, and service neutrals that are not common must be effectively grounded.
Example:
A delta-grounded wye transformer bank fed from an ungrounded delta primary system with a wye connected neutral on the secondary.
Exception:
Circuits designed for ground fault detection and impedance current limiting devices.

Other conductors:
Line or service conductors (not neutrals) that are intentionally grounded must be effectively grounded.
Example:
Corner grounded delta where the phase wire (not a neutral) is intentionally grounded.

Surge arrester:
A surge arrester dependent on a ground connection must be effectively grounded. (An example of a surge arrester that is not dependent on a ground connection would be a surge arrester across the phase bushings of a voltage regulator.)

Grounded Phase "C"
Phase "A"
Phase "B"

Current flow to load (on phase wire).

Current flow from load (on neutral).

Current flow to load (on phase wire).

Use of earth as part of circuit:
This is acceptable. The earth is parallel with the neutral conductor.

VIOLATION!
Use of earth as part of circuit: Supply circuits must not use the earth as the sole conductor. (Exceptions apply to HVDC systems.)

Current flow from load (through earth in parallel with neutral).

Current flow from load (through earth due to no neutral).

Fig. 215-1. Grounding of circuits (Rule 215B).

Rule 215C focuses on the requirements for grounding non-current-carrying parts. When a communication messenger is grounded and bonded to the supply neutral conductor, it can carry current. When a transformer enclosure is grounded and bonded to the supply neutral conductor, it can carry current. But, by definition of a current-carrying part in Sec. 02 of the **NESC**, messengers and transformer enclosures are not considered current-carrying parts as they are not connected to a source of voltage. Rule 314, which discusses grounding of underground circuits and equipment, uses the term "conductive parts to be grounded" instead of "non-current-carrying parts."

Per Rule 215C2, anchor guys and span guys must be effectively grounded or insulated (per the exceptions). This requirement applies to supply (power) guys supporting any voltage and communication guys. If guys are grounded, Rule 092C2 provides the methods for grounding of guys. If guys are insulated, the exception in Rule 215C2a (for anchor guys) and the exception in Rule 215C2b (for span guys) provide requirements for the location of the guy insulator or insulators. In addition, the guy and span insulator requirements specified in Rule 279A must also be met and the clearance requirements for guys in **NESC** Table 232-1 (as applicable) must be met. It is difficult to determine a "one size fits all" guy insulator standard as guyed structures vary in voltage, location, configuration, and complexity.

The non-current-carrying parts grounding requirements of Rule 215C are outlined in Figs. 215-2 through 215-4.

Examples of the use of insulator links in guy wires are shown in Figs. 215-5 through 215-8.

216. ARRANGEMENT OF SWITCHES

This rule addresses overhead line switches. Rule 381C covers underground or pad-mounted switches and Sec. 17 covers switches in electric supply stations. Transmission and distribution line switches are the most common application of Rule 216. The **Code** rules related to the arrangement of overhead line switches are outlined in Fig. 216-1.

The intent of the requirements in Rule 216 is to minimize switching errors. Switching is sometimes done during poor weather conditions or low visibility. Accessibility, marking, uniform positioning, etc., help minimize switching errors. The operation, control, and tagging of switches is covered in the Work Rules in Part 4, Sec. 44, "Additional Rules for Supply Employees."

217. GENERAL

Rule 217 provides general information regarding supporting structures (e.g., poles and towers) except for clearance to other objects. Clearance of a supporting structure to other objects is covered in the overhead line clearance rules in Sec. 23, Rule 231.

Per Rule 215C1, non-current-carrying metal parts that must be effectively grounded:
- Metal structures (e.g., poles)
- Metal-reinforced structures
- Lamp posts
- Conduits and raceways
- Cable sheaths
- Messengers
- Metal frames, cases, and hangers of equipment
- Metal switch handles and operating rods
Exceptions apply to:
- Certain parts 8' or higher and to parts that are otherwise isolated or guarded using a uniform practice over a well-defined area (e.g., a metal pin for a distribution insulator is typically not grounded).
- Equipment cases in certain specialized applications.
- Certain parts enclosing or supporting only communication conductors that are not exposed to open supply conductors.

- Per Rule 215C2, anchor guys and span guys must be effectively grounded per Rule 092C2.
- Exception: Anchor guys and span guys that are insulated per Rule 215C2 and Rule 279A.
- Additional requirements apply to span wires supporting luminaires or traffic signals and insulators in span wires supporting energized trolley or electric railroad contact conductors (Rule 215C3).
- Additional requirements apply to insulators used to limit galvanic corrosion (Rule 215C4).

Grounding of guy by using only the anchor does not meet grounding methods in Sec. 09. See Rule 092C.

Multiple communication messengers on the same structure exposed to power contacts, power induction, or lighting must be bonded together per Rule 215C5 and Rule 092C1. (Bonding each individual messenger to a pole ground is the common method of bonding multiple messengers together.)
Single messengers are required to be effectively grounded per Rule 215C1.

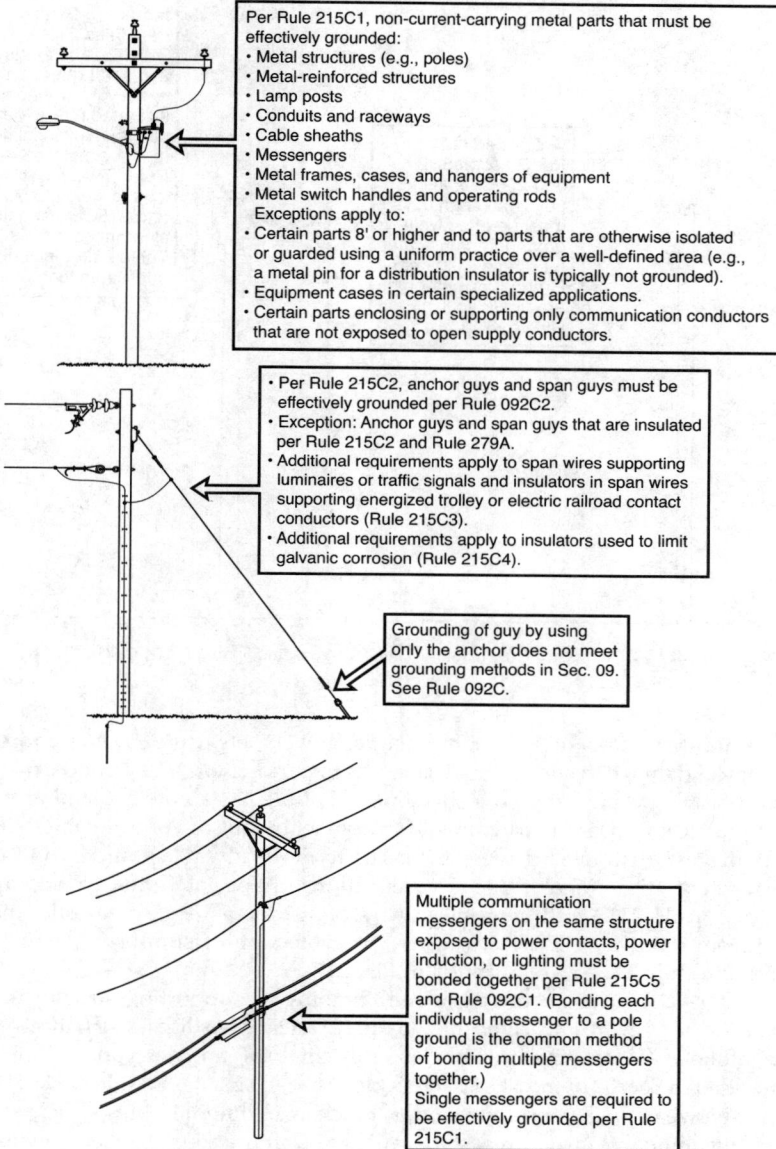

Fig. 215-2. Non-current-carrying parts to be grounded (Rule 215C).

Grounded Guy:
Per Rules 215C2a and 215C2b, anchor guys and span guys must be effectively grounded or insulated (per the exceptions). Grounding methods are provided in Rule 092C2.

Insulated Guy:
Per Rules 215C2a and 215C2b, anchor guys and span guys must be effectively grounded or insulated (per the exceptions). Guy insulators must be installed per Rule 279A, the exceptions to Rules 215C2a and 215C2b, and the clearances in **NESC** Table 232-1 including applicable footnotes.

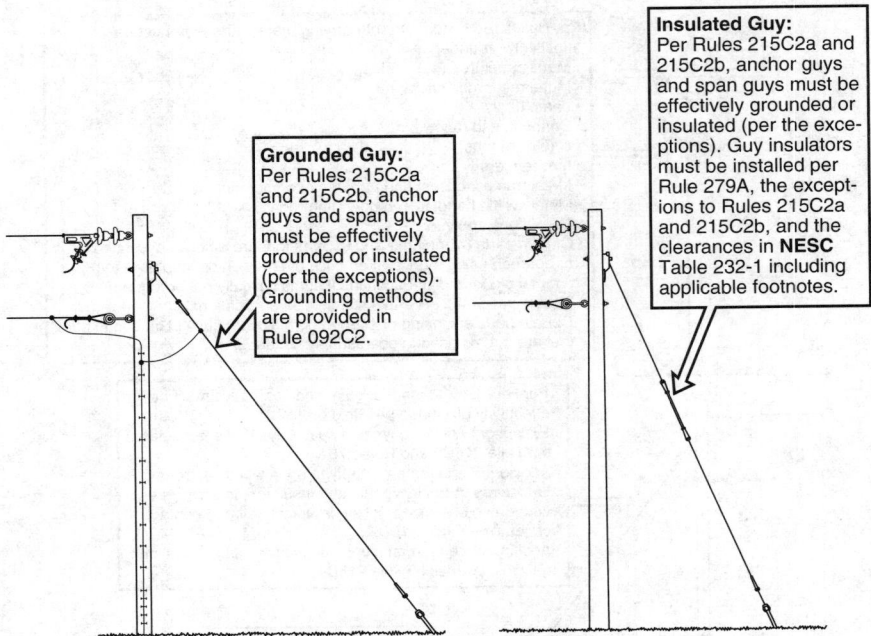

Fig. 215-3. Grounding (or insulating) anchor guys and span guys (Rule 215C2).

Supporting structures includes poles, lattice towers, etc. Rule 217A1a addresses mechanical damage to supporting structures in parking lots, alleys, and driveway areas subject to vehicular traffic abrasion. This Rule has a note explaining that it is not practical to protect structures from contact by out-of-control vehicles (e.g., drunk drivers, distracted drivers, etc.). The terms roadway, shoulder, and traveled way are all defined in Sec. 02, Definitions of Special Terms, to help apply this Rule. Rule 231B addresses clearances of structures (e.g., poles and anchor guys) from roadways. The rules related to protection of supporting structures covered in Rule 217A1 are outlined in Fig. 217-1.

Rule 217A2 focuses on unauthorized climbing of supporting structures. The Code uses the term "readily climbable" in Rule 217A2a when discussing lattice towers. A definition of "supporting structure" (readily climbable and not readily climbable) is provided in Sec. 02 of the NESC. The Code uses 8-ft spacing in Rules 217A2b and 217A2c between pole steps and standoff brackets to limit climbing. The concern of readily climbable structures, steps, and standoff brackets is to keep unauthorized people (e.g., the public) from climbing the pole or structure.

The rules related to climbing of supporting structures covered in Rule 217A2 are outlined in Figs. 217-2, 217-3, and 217-4.

Grounded Guy:
Per Rules 215C2a and 215C2b,
anchor guys and span guys must
be effectively grounded or insulated
(per the exceptions). Grounding
methods are provided in Rule 092C2.

If a span guy and its
associated anchor
guy are bonded
together, they may
be considered as
one guy.

Span guy

Anchor guy

Insulated Guy:
Per Rules 215C2a and 215C2b, anchor
guys and span guys must be effectively grounded
or insulated (per the exceptions). Guy insulators
must be installed per Rule 279A, the exceptions
to Rules 215C2a and 215C2b, and the clearances
in **NESC** Table 232-1 including applicable footnotes.

Span guy and
anchor guy are
not considered
as one guy.

Span guy

Anchor guy

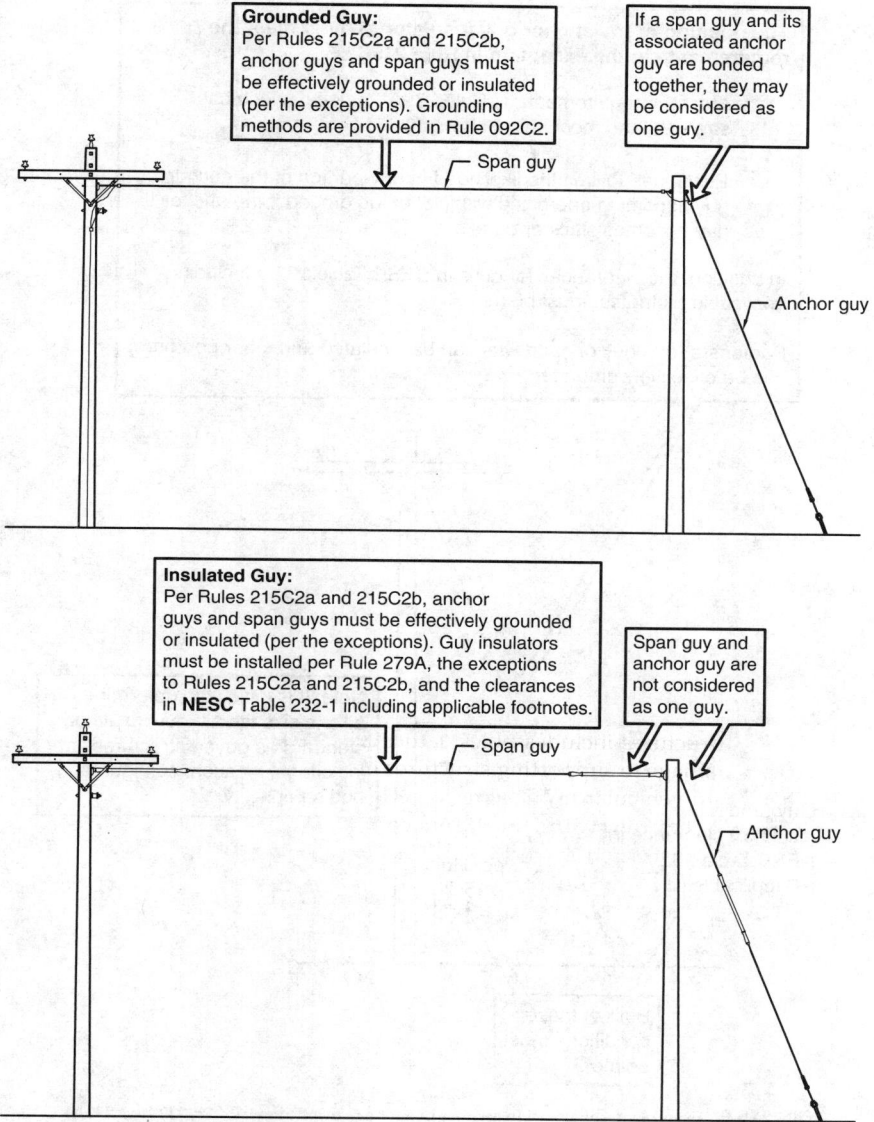

Fig. 215-4. Grounding (or insulating) anchor guys and span guys (Rule 215C2).

The position of the anchor guy insulator(s) must meet the requirements in the exception in Rule 215C2a:

- Meet the requirements of Rule 279A (material, electrical strength, and mechanical strength).

- Positioned to limit the likelihood of any portion of the anchor guy becoming energized within 8' of the ground if the anchor guy becomes slack or breaks.

In addition, the clearances for guys in **NESC** Table 232-1 including applicable footnotes, must be met.

Portions(s) of a guy or span wire can be insulated and other portion(s) can be effectively grounded.

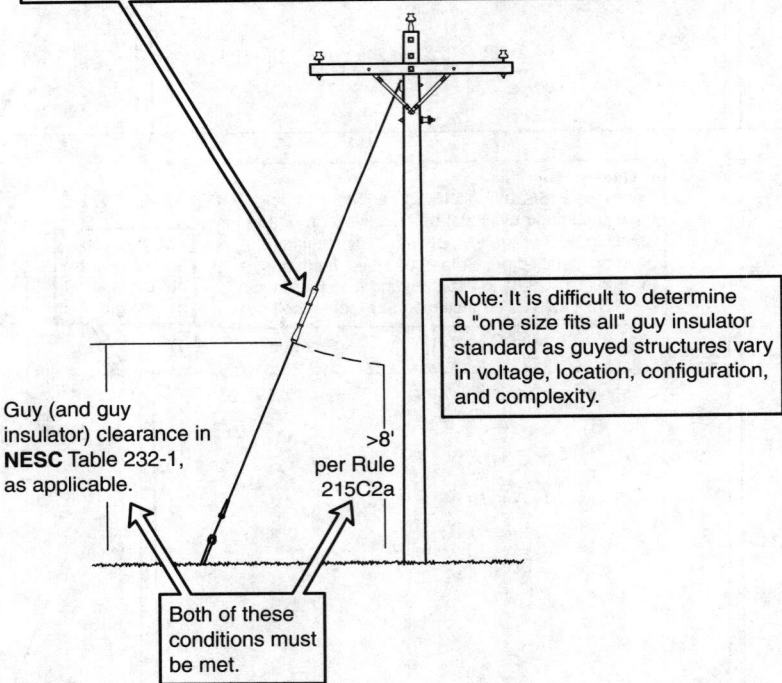

Note: It is difficult to determine a "one size fits all" guy insulator standard as guyed structures vary in voltage, location, configuration, and complexity.

Guy (and guy insulator) clearance in **NESC** Table 232-1, as applicable.

>8' per Rule 215C2a

Both of these conditions must be met.

Fig. 215-5. Example of the use of insulators in anchor guys (Rules 215C2a, 232, and 279A).

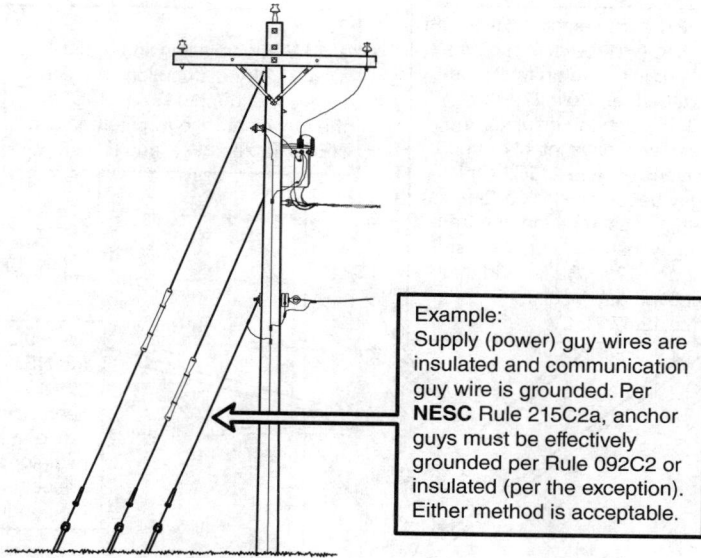

Example:
Supply (power) guy wires are insulated and communication guy wire is grounded. Per **NESC** Rule 215C2a, anchor guys must be effectively grounded per Rule 092C2 or insulated (per the exception). Either method is acceptable.

Fig. 215-6. Example of the use of insulators in anchor guys (Rules 215C2a, 232, and 279A).

Rule 217A3 provides requirements for identifying supporting structures (e.g., poles and towers) on overhead lines. Structure identification is required but not specifically limited to numbering. Location, construction, or marking can also be used to identify poles and towers for employees. Identification of switch poles (or the switches themselves) is especially critical so that switching operations discussed in Part 4, Work Rules, are done correctly. Some utilities number every pole and some only number corner, tap, or other selected poles. A small rural utility can use construction and location for identification (e.g., the transformer pole at the Smith Ranch). The method used to identify poles should consider the size and complexity of the utility's system. Rule 411A2 in Part 4, Work Rules, requires that diagrams (i.e., maps) be maintained on file for employees to use. Rule 220D applies to identification of the conductors on the supporting structures.

The rules related to identification of supporting structures covered in Rule 217A3 are outlined in Fig. 217-5.

Rule 217A4 covers attachments, decorations, and obstructions on supporting structures (e.g., poles). The supporting structure owner must agree (provide concurrence) to attachments made on structures of utility lines. This requirement is for any kind of attachment (utility or non-utility). The supporting structure owner is typically the power company or the traditional telephone company. The structure can also be owned by multiple parties. Structure ownership is typically defined in joint use agreements (contracts) between power and communication utilities. See Rule 222 for a discussion of joint use agreements. Typically, traditional Cable TV companies do not own poles; they only rent space on poles. The word

Per Rule 279A, the electrical strength design rating (voltage rating) of the guy insulator is based on multiples of the voltage (see Rule 279A for specifics) to which the insulator may be exposed with guys intact or under the conditions of Rule 215C2 (the anchor guy becomes slack or breaks). This example assumes that the transmission guy insulators are exposed to 115 kV and 12.47/7.2 kV and that the distribution guy insulator is exposed to 12.47/7.2 kV.

Example:
115 kV transmission line and a 12.47/7.2 kV distribution line. Guy links used to meet Rule 215C2a, the guy clearance requirements in **NESC** Table 232-1, and Rule 279A.

115 kV transmission line

See Rule 235E and **NESC** Table 235-6 for the clearance between the guy insulator and the adjacent power line conductor.

12.47/7.2 kV Distribution line

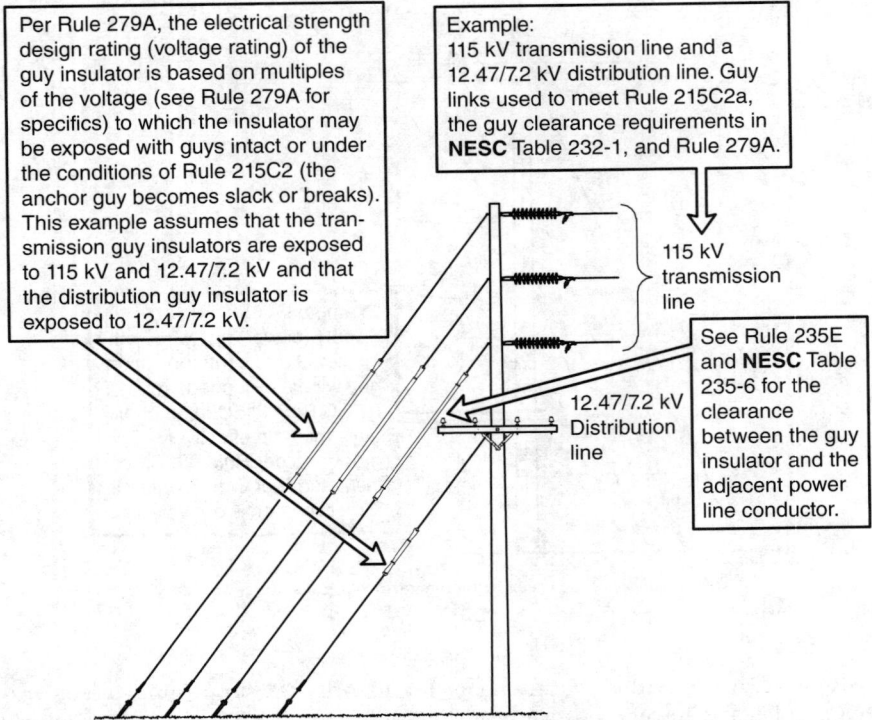

Fig. 215-7. Example of determining guy insulator link electrical strength (voltage) rating (Rules 215C2a, 232, and 279A).

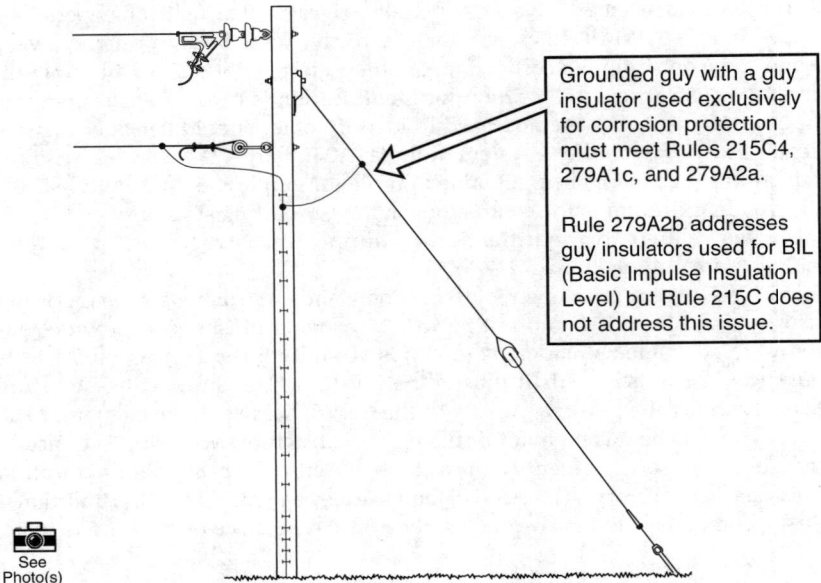

Grounded guy with a guy insulator used exclusively for corrosion protection must meet Rules 215C4, 279A1c, and 279A2a.

Rule 279A2b addresses guy insulators used for BIL (Basic Impulse Insulation Level) but Rule 215C does not address this issue.

See Photo(s)

Fig. 215-8. Example of a guy insulator used to limit galvanic corrosion (Rule 215C4).

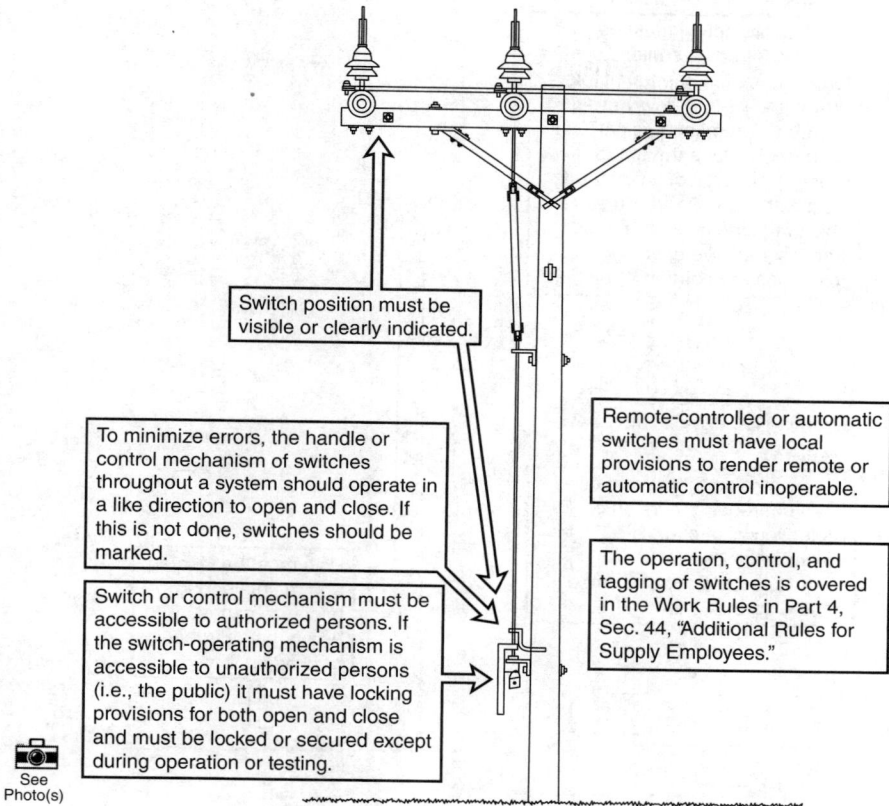

Switch position must be visible or clearly indicated.

To minimize errors, the handle or control mechanism of switches throughout a system should operate in a like direction to open and close. If this is not done, switches should be marked.

Switch or control mechanism must be accessible to authorized persons. If the switch-operating mechanism is accessible to unauthorized persons (i.e., the public) it must have locking provisions for both open and close and must be locked or secured except during operation or testing.

Remote-controlled or automatic switches must have local provisions to render remote or automatic control inoperable.

The operation, control, and tagging of switches is covered in the Work Rules in Part 4, Sec. 44, "Additional Rules for Supply Employees."

See Photo(s)

Fig. 216-1. Arrangement of overhead line switches (Rule 216).

typically is used in this paragraph as other ownership situations can exist. The word traditional is used in this paragraph to refer to telephone and cable utilities before the present day mixing of communication services. Nonutility attachments (e.g., a stop sign mounted on a wood utility pole) must have agreement (concurrence) of the occupant or occupants of the space in which the attachment is made (the supply or communications space) as well as the pole owner. For example, assume a wood pole is owned by the power company and has power, phone, and Cable TV lines attached to it. If a stop sign is mounted on the pole, the pole owner (the power company) and possibly the occupants of the communication space (the phone and Cable TV company) must both provide concurrence if the stop sign is in the communication space. In this example the stop sign would have to be above the lowest communication cable or conductor. See the definition of communication space in Sec. 02 of the **NESC**. Attachments must not result in violations (noncompliance) of the **NESC** rules. Rule 012C, which requires accepted good practice, may need to be applied as the **NESC** may not specifically address **Code** issues related to the attachment. For example, no clearance requirement exists between

Physical protection must be provided for poles subject to vehicle abrasion in parking lots, alleys, and next to driveways. The abrasion must be significant enough to reduce the pole's strength. No protection methods are listed in the **Code**. Rule 012C, which requires accepted good practice, must be used. Metal pole stubs are commonly applied.

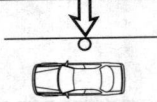

Parking lot, alley, or driveway

A pole outside the traveled way does not require marking or protection from contact by out-of-control vehicles.
(See Rule 231B for clearance of pole from a roadway.)

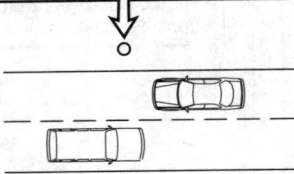

Pole must be located and maintained to minimize exposure to brush, grass, rubbish, or building fires.

Roadway

Waterway

Roadway   Bridge   Roadway

Overhead power line

Waterway

Supporting structure attached to bridge carrying open supply (power) conductors greater than 600 V must have a safety sign. (See Rule 110A for a discussion on safety signs.)

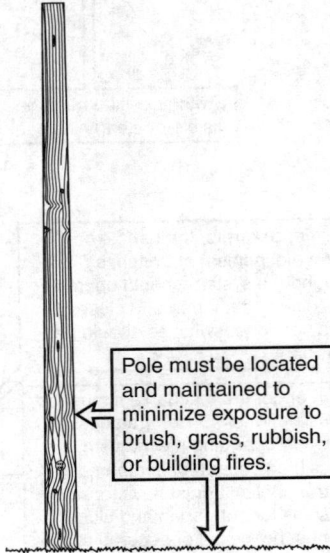

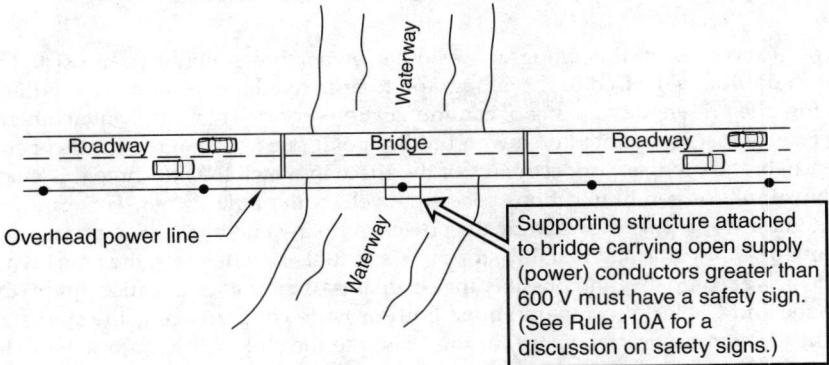

Fig. 217-1. Protection of supporting structures (Rule 217A1).

Condition:
Readily climbable closely latticed pole or tower (including bridge attachments) carrying open supply (power) conductors. Structure is adjacent to road, walkway, or gathering place such as a park, school, or playground. (See definition of "supporting structure, readily climbable and not readily climbable" in Sec. 02.)

Requirement:
Structure must be equipped with climbing barriers to inhibit climbing by the public or the structure must be posted with safety signs.
Exception:
This rule does not apply where access to the structure is limited by a fence meeting the height requirements of Rule 110A1.

See Photo(s)

Roadway   Sidewalk

Park/Playground/School

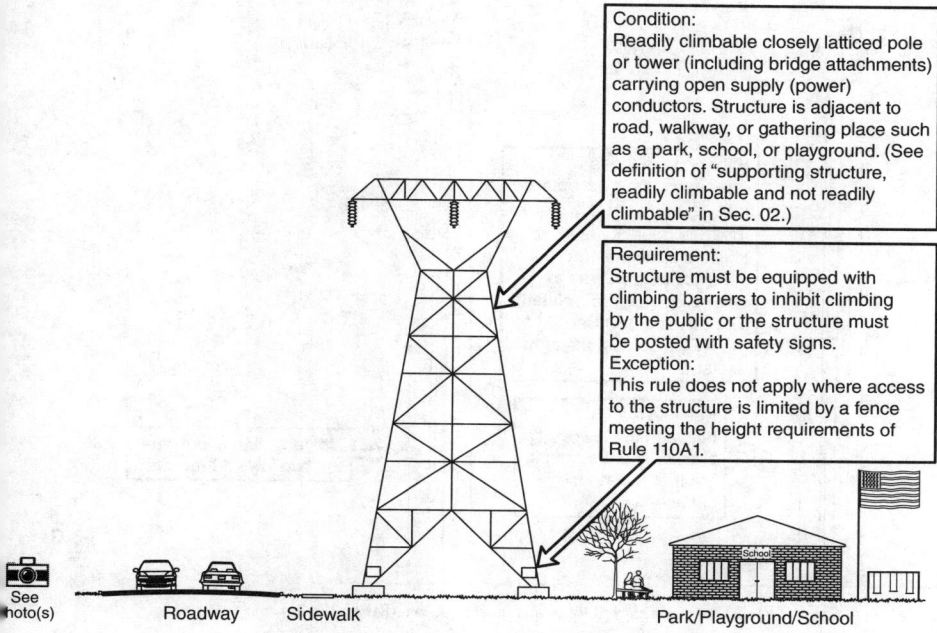

Fig. 217-2. Readily climbable supporting structures (Rule 217A2a).

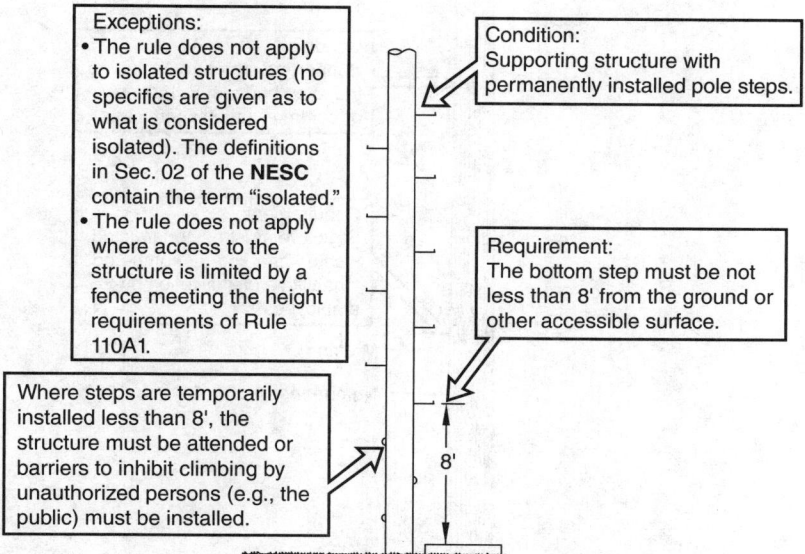

Exceptions:
• The rule does not apply to isolated structures (no specifics are given as to what is considered isolated). The definitions in Sec. 02 of the **NESC** contain the term "isolated."
• The rule does not apply where access to the structure is limited by a fence meeting the height requirements of Rule 110A1.

Condition:
Supporting structure with permanently installed pole steps.

Requirement:
The bottom step must be not less than 8' from the ground or other accessible surface.

Where steps are temporarily installed less than 8', the structure must be attended or barriers to inhibit climbing by unauthorized persons (e.g., the public) must be installed.

8'

See Photo(s)

Fig. 217-3. Permanently mounted and temporary steps on supporting structures (Rule 217A2b).

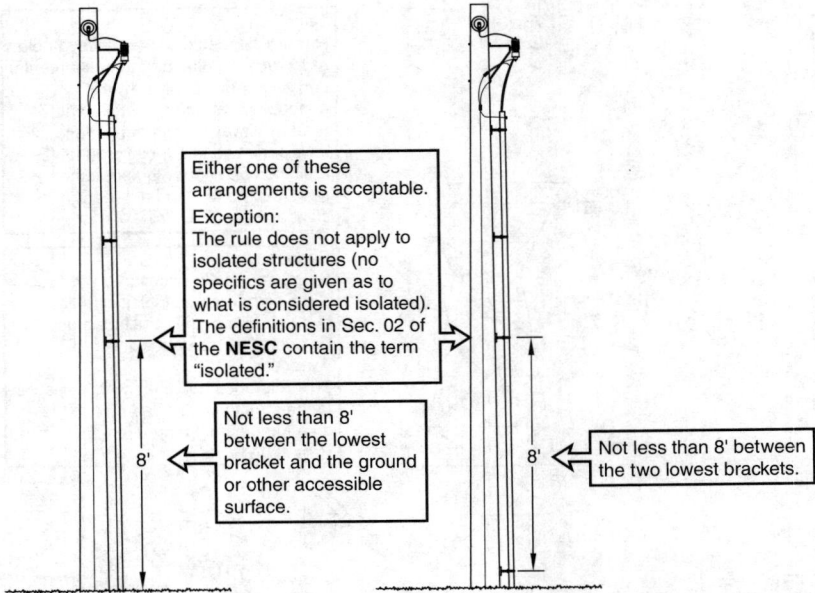

Either one of these arrangements is acceptable.

Exception:
The rule does not apply to isolated structures (no specifics are given as to what is considered isolated). The definitions in Sec. 02 of the **NESC** contain the term "isolated."

Not less than 8' between the lowest bracket and the ground or other accessible surface.

8'

8'

Not less than 8' between the two lowest brackets.

See Photo(s)

Fig. 217-4. Arrangement of standoff brackets (Rule 217A2c).

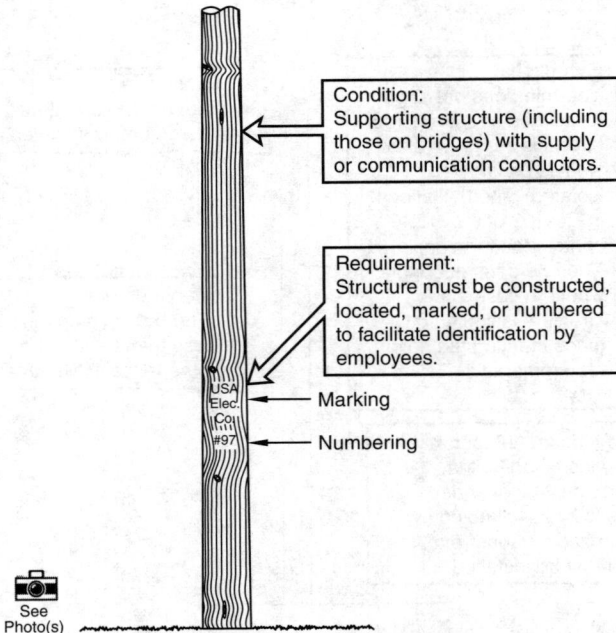

Condition:
Supporting structure (including those on bridges) with supply or communication conductors.

Requirement:
Structure must be constructed, located, marked, or numbered to facilitate identification by employees.

Marking

Numbering

See Photo(s)

Fig. 217-5. Identification of supporting structures (Rule 217A3).

a stop sign and a communications cable above it and no grounding requirements exist for stop signs. The wind load on the stop sign should be considered when the strength and loading rules are applied. Attachments must not obstruct the climbing space or present a climbing hazard to utility workers (linemen). See Rule 236 for a discussion of climbing space. Through-bolts must be properly trimmed but no dimensions are provided. Rule 012C, which requires accepted good practice, must be used. A through-bolt that is too long can be trimmed or replaced with a shorter bolt. Vines, nails, tacks, or other items that may interfere with climbing should be removed before climbing to avoid a slip or fall. The rules related to attachments, decorations, and obstructions on supporting structures are outlined in Fig. 217-6.

Rule 217B discusses unusual conductor supports. The **NESC** does not prohibit unusual conductor supports. It does require applying **Code** rules and additional precautions deemed necessary by the administrative authority to assure safety. For example, a power line supported on a bridge requires complying with the **NESC**

• Attachments must have concurrence of the pole owner (utility and nonutility attachments).
• Nonutility attachments must have concurrence of the occupants of the space (power or communications space).
• No attachment can create an **NESC** violation.
• Attachments must not obstruct the climbing space or present a climbing hazard.
• Through-bolts must be properly trimmed.
• Vines, nails, tacks, or other items which may interfere with climbing should be removed before climbing.

STOP

Garage
Sale →

See
ʰoto(s)

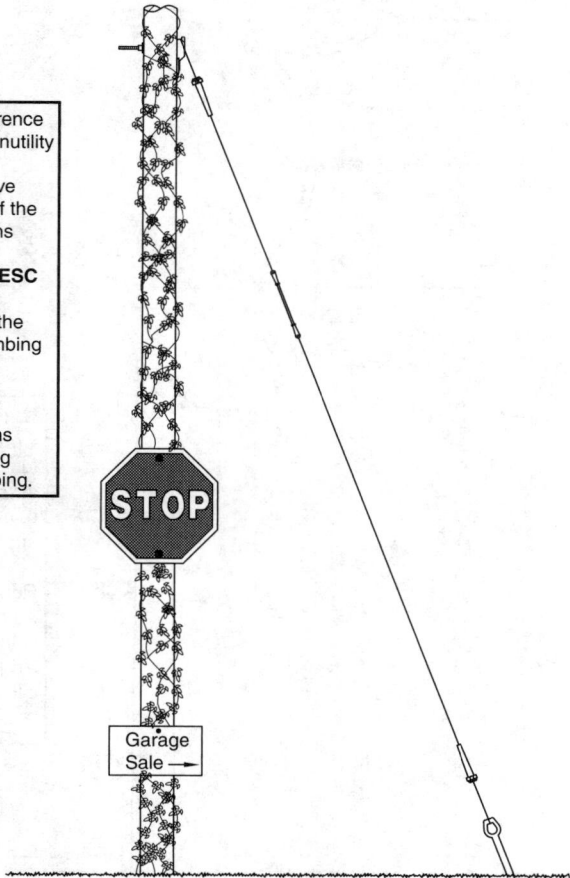

Fig. 217-6. Attachments, decorations, and obstructions on supporting structures (Rule 217A4).

rules for clearance to the bridge and the NESC rules for loading and strength of the power line structure and any additional rules required by the highway department or bridge authority. The rules required by the highway department or bridge authority may involve additional loading and strength requirements related to the power line being attached to the bridge and additional clearance requirements for persons using or maintaining the bridge. One of the most recognizable unusual conductor supports in the United Sates is the "Mickey Mouse" tower outside of Disney World in Orlando, FL. See Fig. 217-7.

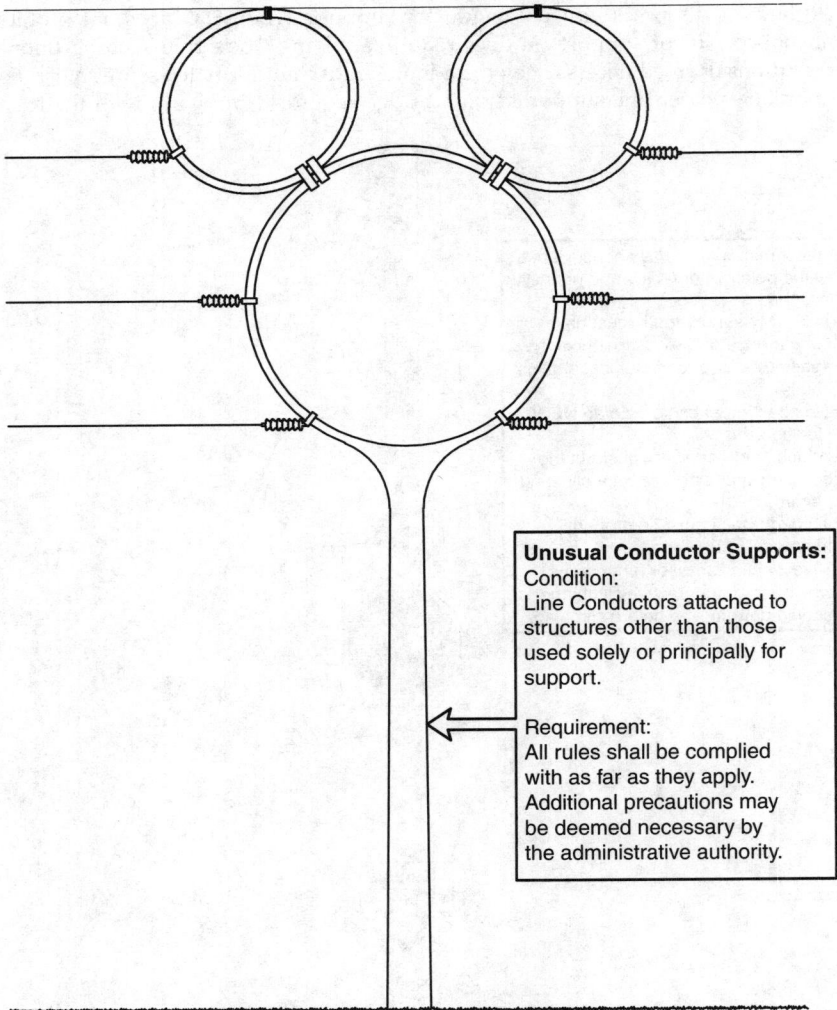

Unusual Conductor Supports:
Condition:
Line Conductors attached to structures other than those used solely or principally for support.

Requirement:
All rules shall be complied with as far as they apply. Additional precautions may be deemed necessary by the administrative authority.

Fig. 217-7. Unusual conductor supports (Rule 217B).

Rule 217B also addresses supporting conductors on trees and roofs. The **Code** states that supporting (attaching) overhead conductors to trees or roofs "should be avoided." The **Code** does not use stronger language like "shall not be attached." Even though "should" is used instead of "shall," everything practical must be done to avoid attaching conductors to trees or roofs. See Rule 015 for a discussion of the intent of the words "should" and "shall." The requirement to avoid supporting conductors on trees and roofs is shown in Fig. 217-8.

Rule 217C applies to the protection and marking of guys. The application of guy markers varies from utility to utility. Utilities are required to use guy markers on guys exposed to pedestrian traffic. A guy marker or protection must be used for guys in established parking areas. Some utilities use a guy marker on every guy but this is not a **Code** requirement. The note in Rule 217C1 clarifies that there is no intent to require guy markers on all anchor guys, only those required by the rule. Some utilities use bold fluorescent multicolored markers; others use conspicuous single color markers. Rule 217C2 clearly states that guys located outside the traveled way of a roadway or established parking areas do not require marking. Similar wording appears in Rule 217A1a for supporting structures. The note in Rule 217C2 explains that it is not practical to protect guys from contact by out-of-control vehicles (e.g., drunk drivers, distracted drivers, etc.). The terms roadway, shoulder, and traveled way are all defined in Sec. 02, Definitions of Special Terms, to help apply this Rule. Rule 231B addresses clearances of structures (e.g., poles and anchor guys) from roadways. Examples of using guy markers and protection are shown in Fig. 217-9.

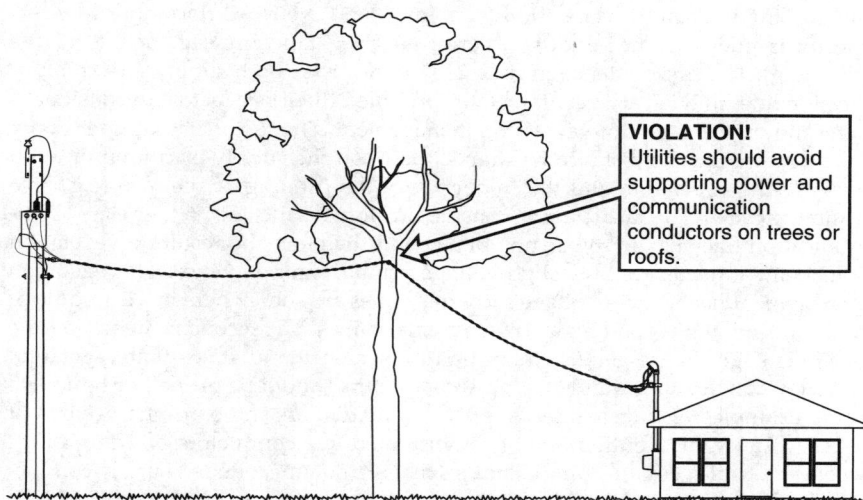

Fig. 217-8. Supporting conductors on trees and roofs (Rule 217B).

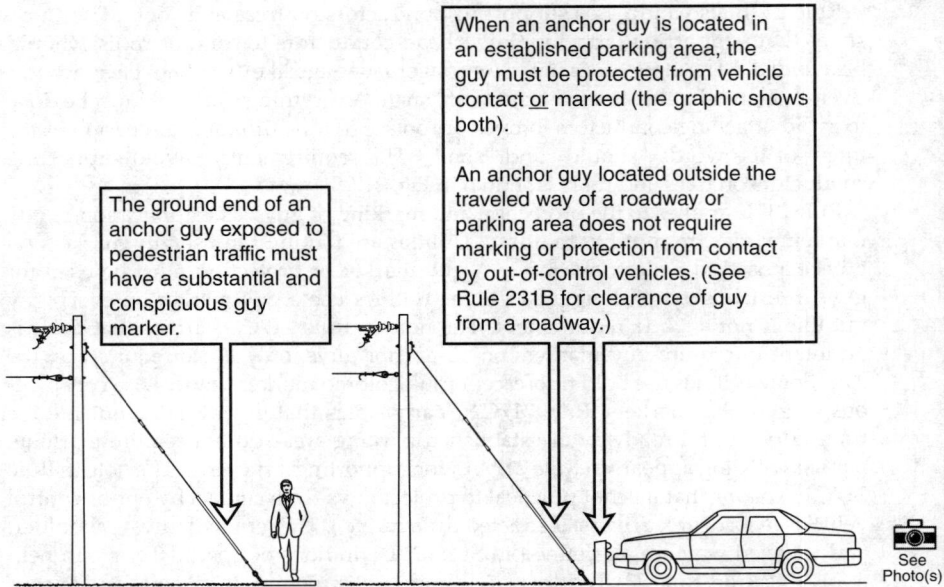

The ground end of an anchor guy exposed to pedestrian traffic must have a substantial and conspicuous guy marker.

When an anchor guy is located in an established parking area, the guy must be protected from vehicle contact or marked (the graphic shows both).

An anchor guy located outside the traveled way of a roadway or parking area does not require marking or protection from contact by out-of-control vehicles. (See Rule 231B for clearance of guy from a roadway.)

See Photo(s)

Fig. 217-9. Examples of protection and marking of guys (Rule 217C).

218. VEGETATION MANAGEMENT

The **NESC** does not provide a specific clearance from a supply or communication line (of any voltage) to vegetation (e.g., trees) and the **NESC** does not provide a specific frequency (time period) for performing vegetation management (e.g., tree trimming). The first sentence in Rule 218A1 addresses both supply and communication lines and is very general. Note 1 in Rule 218A1 lists factors to consider in determining the extent of vegetation management. The factors to consider apply to both selecting the distance (clearance) between the supply or communication line and the vegetation and the frequency of the pruning or removal (e.g., tree trimming cycle). The first factor to consider in Note 1 is the line voltage class. Communication lines have little or no voltage and therefore the resulting vegetation management needs are less. High voltage primary lines have greater vegetation management needs. Low voltage secondary lines are somewhere in the middle of communication lines and high voltage primary lines. The second sentence in Rule 218A1 is slightly less general than the first sentence and states that vegetation that may damage ungrounded supply conductors should be pruned or removed. Some examples of ungrounded supply conductors are the bare phase wires of a 12.47/7.2 kV distribution line or the insulated line conductors of a 120/240 V secondary triplex circuit. Some examples that are not ungrounded supply conductors are the effectively grounded neutral conductor associated with a 12.47/7.2 kV distribution line, the effectively grounded bare messenger of a 120/240 V secondary triplex circuit, and a communication cable. Since Rule 218 is not specific, Rule 012C, which requires accepted good practice, must be applied. It is common in the electric power utility industry to have some accepted good practice distance between a 12.47/7.2 kV power line and a tree and is common for there to be some tree trimming cycle for a 12.47/7.2 kV power line. It is common in the electric power utility industry to see a 120/240 V secondary triplex circuit going through a tree but to use some method (trimming or tree guard over the triplex cable) to prevent abrasion of the 120/240 V secondary triplex circuit. It is common in the communication utility industry to see a communications cable going through a tree. There is little or no electrical hazard in this case however, tree trimming for communication cables can be important to avoid damage to lashing wires which can cause the communication cable to separate from the messenger and to provide a clear working space for the communication lineman or technician during installation, operation, or maintenance of the communication line and equipment. Note 2 in Rule 218A1 provides a statement that it is not practical to prevent all tree-conductor contacts on overhead lines. It is possible for new vegetation growth to contact an ungrounded supply conductor but not damage the ungrounded supply conductor. To determine what amount of tree-conductor contact is practical to prevent and what amount of tree-conductor contact is not practical to prevent requires accepted good practice (Rule 012C). Note 2 should not be applied as an excuse for a poor vegetation management program, a utility needs to exercise due diligence when performing vegetation management and then if applicable, apply Note 2 on a case-by-case basis. Workers who trim trees in the vicinity of energized conductors must be qualified to do so.

The **Code** recognizes that at times tree trimming or removal may not be practical. If tree trimming or removal is not practical, the **Code** requires using methods to separate the conductor from the tree to avoid damage by abrasion and grounding. This requirement is very general. No specifics are provided as to how to accomplish the separation. One common example is using a factory made PVC tree guard over a 120/240-V, single-phase, three-wire triplex service drop in contact with a tree.

Although not stated in the **Code**, pruning or removal of vegetation near ungrounded supply conductors mitigates the chance of starting forest fires or wildfires and mitigates the chance of killing or injuring a person climbing a tree. The general **Code** rules for vegetation management are outlined in Fig. 218-1.

The North American Electric Reliability Corporation (NERC) standards apply to transmission line vegetation management. NERC Standard FAC-003-2/3 provides specific values for clearance between transmission line conductors and vegetation (trees). The NERC vegetation management standards were developed to address vegetation-related outages of transmission lines which can lead to cascading outages.

Line crossings, railroad crossings, limited-access highway crossings, and navigable waterways receive special considerations to minimize downed lines. Tree trimming for these areas should consider decaying trees or limbs and trees that are overhanging the line (sometimes referred to as danger trees). See Fig. 218-2.

• Vegetation management should be performed around supply and communication lines as experience has shown necessary.

• Vegetation that may damage ungrounded supply conductors should be pruned or removed.

• Where pruning or removal is not practical, the conductor should be separated from the tree with suitable materials or devices to avoid abrasion and grounding of the circuit through the tree.

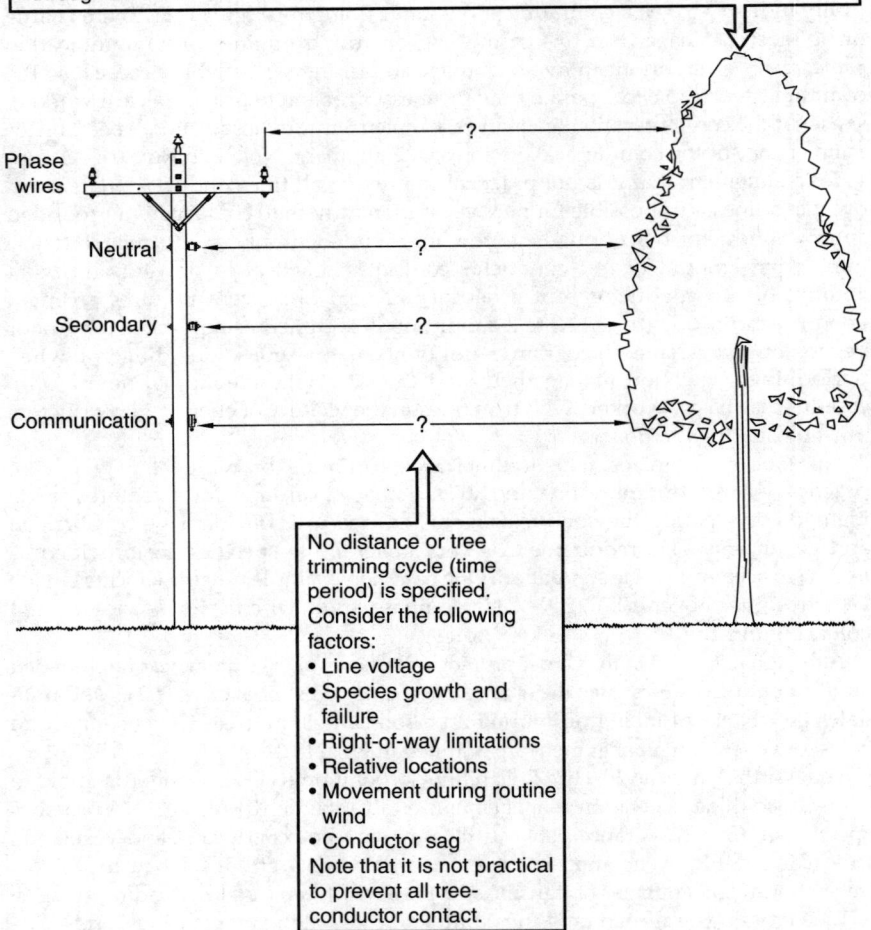

Phase wires

Neutral ?

Secondary ?

Communication ?

No distance or tree trimming cycle (time period) is specified. Consider the following factors:
• Line voltage
• Species growth and failure
• Right-of-way limitations
• Relative locations
• Movement during routine wind
• Conductor sag
Note that it is not practical to prevent all tree-conductor contact.

See Photo(s)

Fig. 218-1. General vegetation management requirements (Rule 218A).

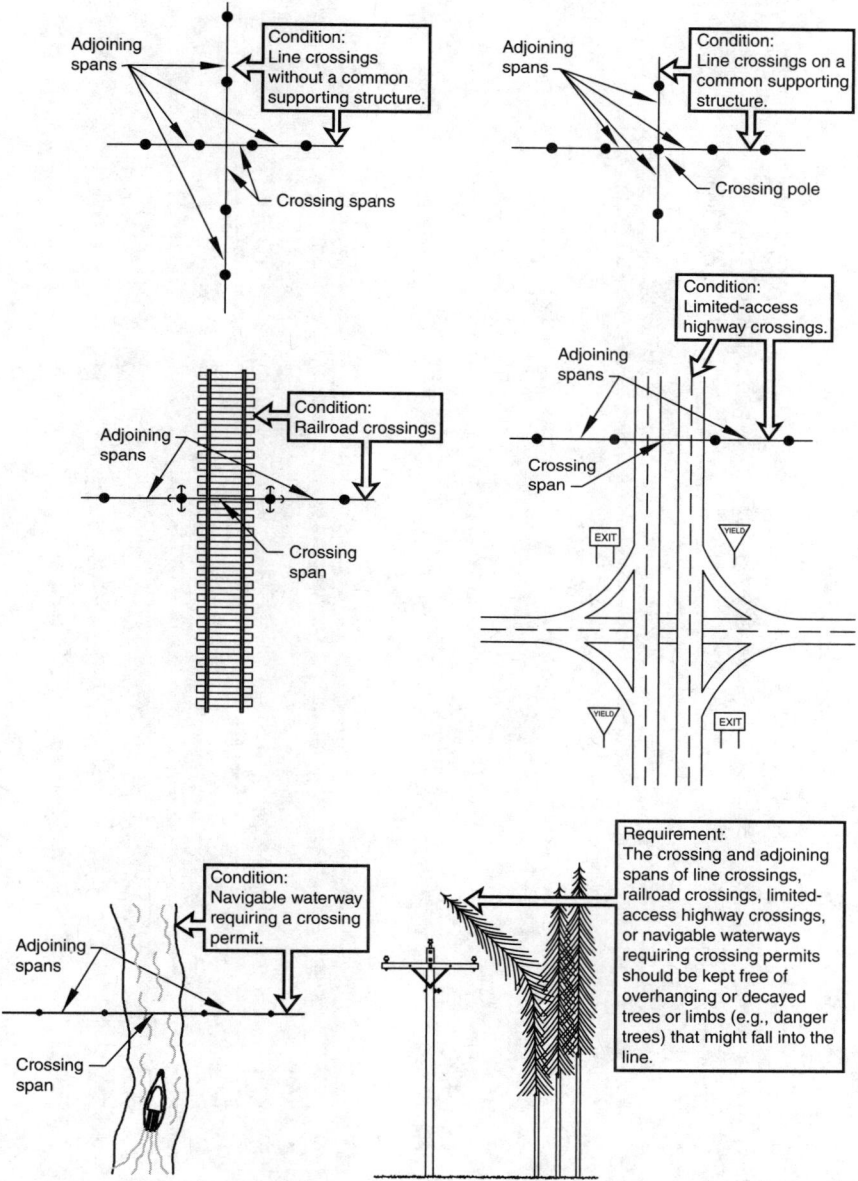

Fig. 218-2. Vegetation management at line, railroad, limited-access highway, and navigable waterway crossings (Rule 218B).

Section 22

Relations between Various Classes of Lines and Equipment

220. RELATIVE LEVELS

Rule 220A requires that the levels of different classes of conductors should be standardized by agreement of the utilities concerned. The **NESC** does not provide relative levels or specific locations for multiple communication circuits in the communications space. Years ago, telephone was the primary joint-use communications attachment on a power pole. Then came community antenna television systems (CATV), these systems were typically installed above telephone. Today, pole owners are receiving numerous requests for attachments by fiber utilities. **NESC** Rule 220A and Rule 235H (Clearance and Spacing between Communication Conductors, Cables, and Equipment) must both be referenced to determine the relative levels of communication cables in the communication space. Rules 220A and 235H require agreement between the utilities concerned or the parties involved. See Rule 235H for more information. Once standardization of levels have been agreed upon, there are several other **NESC** Rules that must be met including, but not limited to, clearance rules and strength and loading rules. A joint-use agreement between supply and communications utilities is one type of agreement that can be used to standardize utility locations on a pole. Rule 220B1 states that it is preferred that supply (power) conductors and equipment be located above communications conductors and equipment where practical. Four exceptions apply to Rule 220B1. See Fig. 220-1.

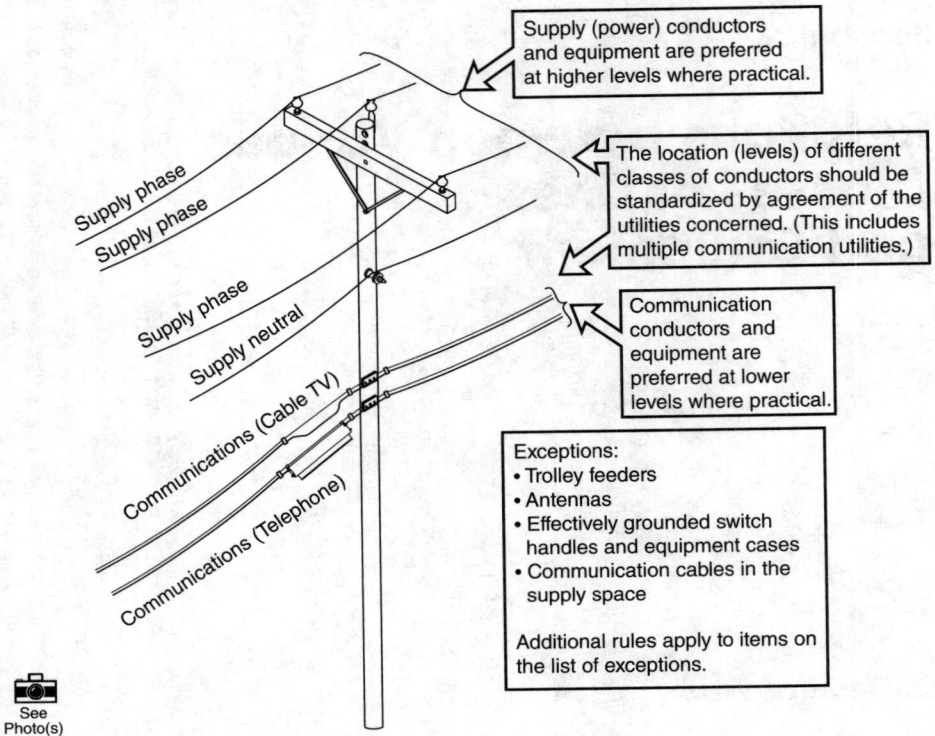

Supply (power) conductors and equipment are preferred at higher levels where practical.

The location (levels) of different classes of conductors should be standardized by agreement of the utilities concerned. (This includes multiple communication utilities.)

Communication conductors and equipment are preferred at lower levels where practical.

Exceptions:
• Trolley feeders
• Antennas
• Effectively grounded switch handles and equipment cases
• Communication cables in the supply space

Additional rules apply to items on the list of exceptions.

Supply phase
Supply phase
Supply phase
Supply neutral
Communications (Cable TV)
Communications (Telephone)

See Photo(s)

Fig. 220-1. Standardization of levels of supply and communication conductors (Rules 220A and 220B1).

Rule 220B2, Special Construction for Railroad Supply Circuits, of 600 Volts or Less and Carrying Power Not in Excess of 5 kW Associated with Railroad Communication Circuits, is a rule for special construction related to railroad signaling. There are seven conditions (paragraphs a through g) that must be met to apply this rule. Rule 220B2 does not apply to modern cable television and telephone circuits.

Rule 220C discusses where to position supply (power) lines of different voltages on overhead structures. The terms "crossings" and "conflicts" are used in Rule 220C1. Crossings are discussed in detail in Rule 241C. Conflicts are discussed in Rule 221 and a definition of "structure conflict" is provided in Sec. 02 of the NESC. Relative levels of supply lines at crossings and conflicts are shown in Figs. 220-2 and 220-3.

Rule 220C2 has requirements for structures used only for supply (power) conductors. Rule 220C2a covers structures with circuits owned by one utility and Rule 220C2b covers structures with circuits owned by separate supply utilities. See Figs. 220-4 and 220-5.

Transmission (higher
voltage) conductors above
distribution (lower voltage)
conductors at the crossing.

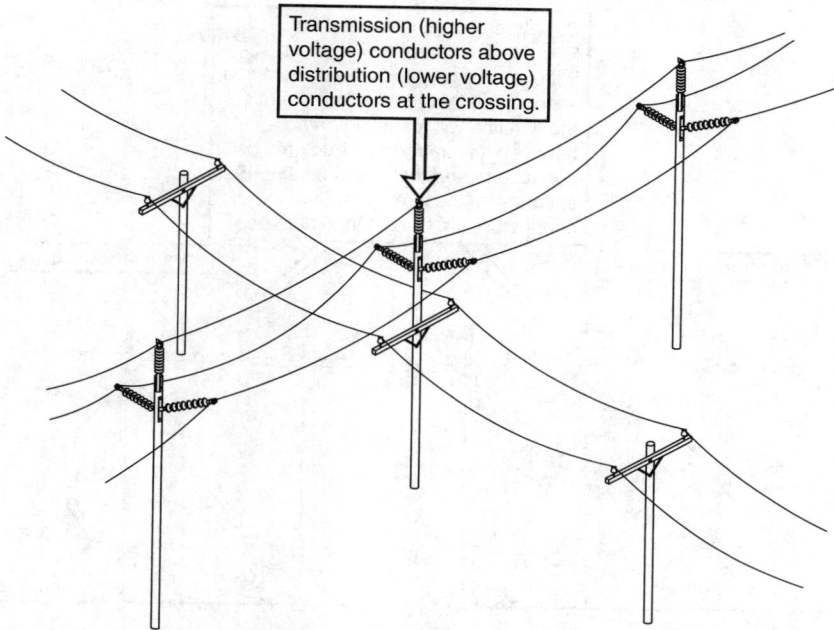

Fig. 220-2. Relative levels of supply lines of different voltages at crossings (Rule 220C1).

See
Photo(s)

Positioning a higher voltage line above a lower voltage line is a common construction practice. Clearance above ground is greater for higher voltage circuits, therefore having higher voltage circuits at higher positions permits greater clearance. Rule 220C states the relative levels of conductors, not the clearance between them. **NESC** Rule 235 is referenced in Rule 220C as this rule provides tables for horizontal and vertical clearances between conductors on a common supporting structure. Lower voltage circuits are typically worked on more than higher voltage circuits, so having lower voltage circuits at lower positions provides easier access.

Rules 220D and 220E require uniform positions of supply and communication conductors and cables (Rule 220D) and equipment (Rule 220E) or constructing, locating, marking, or numbering to facilitate identification by authorized employees who have to work on them. This rule is similar to Rule 217A3, Identification of Supporting Structures. Identifying overhead conductors by attachment to distinctive insulators or crossarms is an acceptable method. Using uniform positions of conductors does not prohibit changing locations systematically.

A neutral conductor, when on a crossarm with the phase conductors, is sometimes identified with a different color or style insulator or by labeling the crossarm with a letter "N" (Neutral) or letters "CN" (Common Neutral) below the neutral insulator. A pole with crossarms owned by multiple utilities can have the crossarms labeled (i.e., marked or numbered) with the utility name, or the crossarms

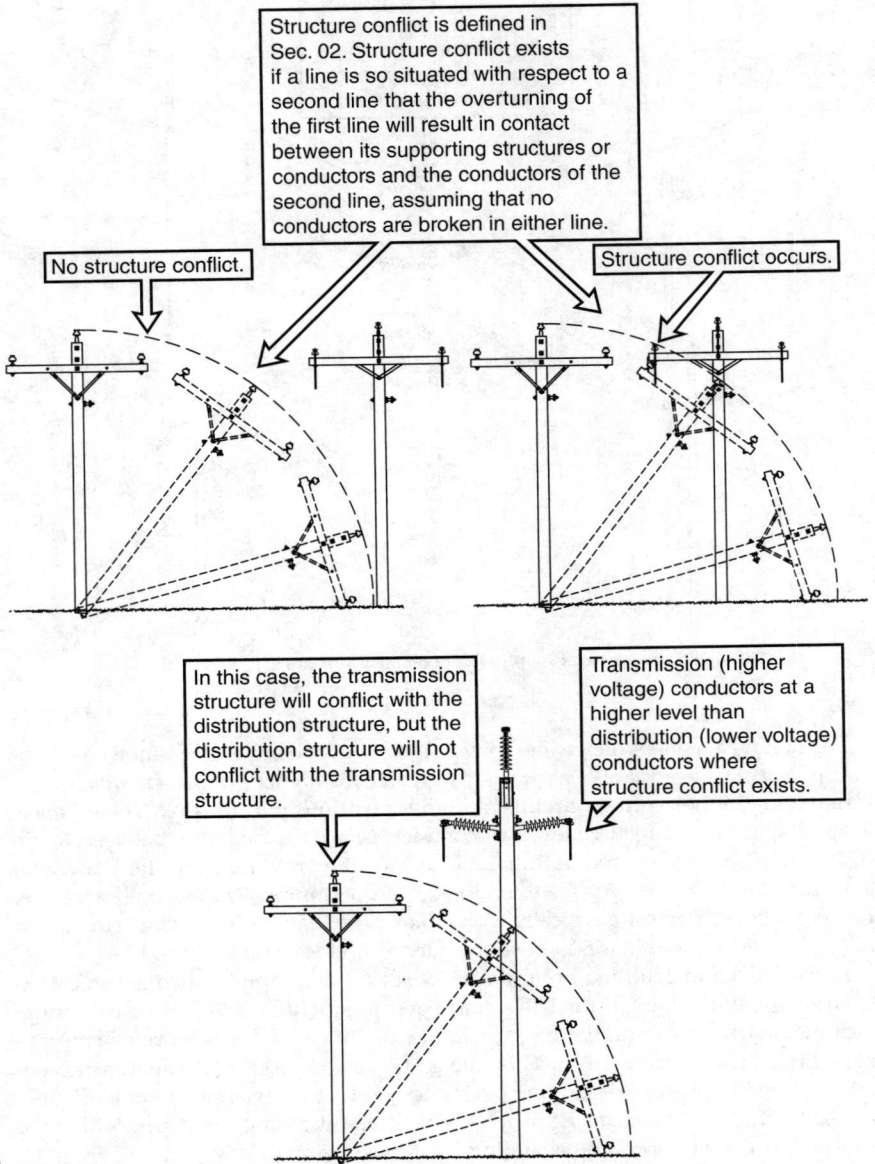

Structure conflict is defined in
Sec. 02. Structure conflict exists
if a line is so situated with respect to a
second line that the overturning of
the first line will result in contact
between its supporting structures or
conductors and the conductors of the
second line, assuming that no
conductors are broken in either line.

No structure conflict.

Structure conflict occurs.

In this case, the transmission
structure will conflict with the
distribution structure, but the
distribution structure will not
conflict with the transmission
structure.

Transmission (higher
voltage) conductors at a
higher level than
distribution (lower voltage)
conductors where
structure conflict exists.

See
Photo(s)

Fig. 220-3. Relative levels of supply lines of different voltages at structure conflict locations
(Rule 220C1).

Supply circuits of different
voltages owned by the same utility
should be arranged with the
higher voltage conductors above
the lower voltage conductors.

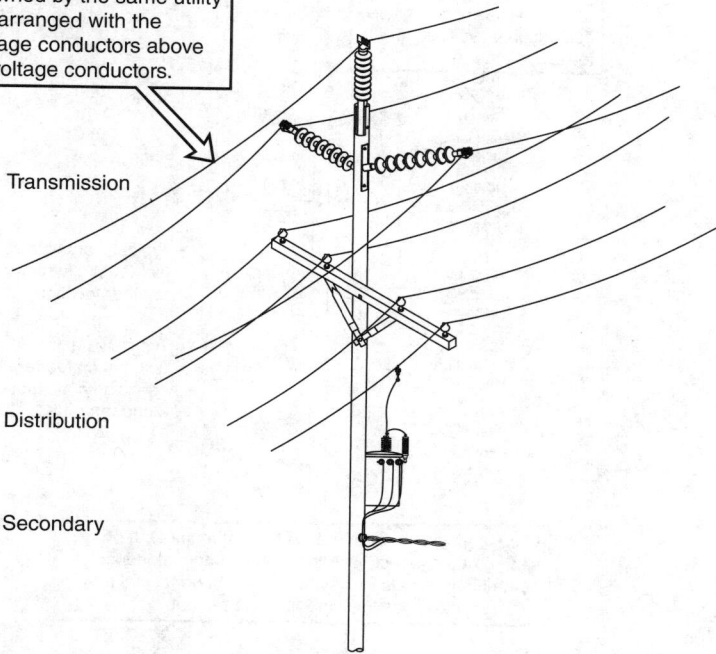

Transmission

Distribution

Secondary

See
Photo(s)

Fig. 220-4. Relative levels of supply circuits of different voltages owned by one utility (Rule 220C2a).

can be constructed or located such that the employees authorized to work on them
can recognize them as their own. Similarly, a pole with multiple communication
cable attachments can have the communication cables labeled (i.e., marked or
numbered) with color coding or a numbering system or the utility name, or the
cables can be constructed or located such that the employees authorized to work
on them can recognize them as their own. An example of identification of overhead conductors is shown in Fig. 220-6.

221. AVOIDANCE OF CONFLICT

The term "conflict" was introduced in Rule 220C. See Rule 220 for a discussion and
a figure related to structure conflict. Avoidance of conflict can be accomplished by
sufficient separation of lines, or by structure strength, or by combining the lines on
the same structure. See Fig. 221-1.

Many times, right-of-way constraints will prohibit two separate lines and collinear or joint-use construction will be required.

Supply circuits of different
voltages owned by separate
supply utilities can use the
basic highest voltage to
lowest voltage rule.

69 kV
Transmission
Utility #1

34.5 kV
Transmission
Utility #2

12.47 kV
Distribution
Utility #1

Typically an express
distribution feeder to
another location.

12.47 kV
Distribution
Utility #2

Typically the local
distribution feeder
servicing customers
along the route.

120/240 V
Secondary
Utility #2

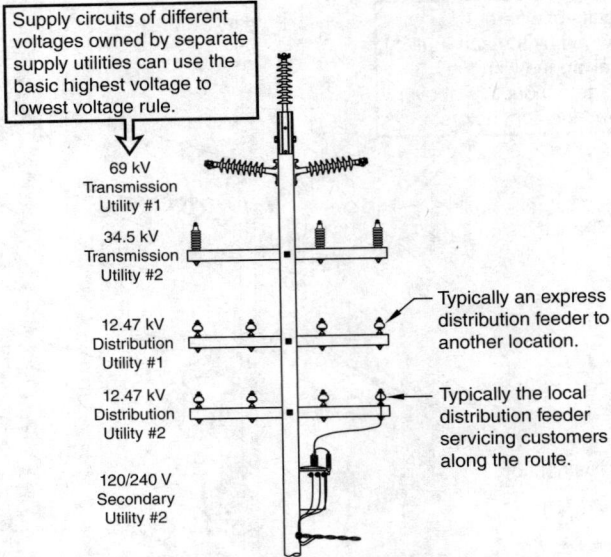

Grouping by utility is permitted but not required. If done,
each grouping must have the higher voltage conductors
at the higher levels in the group and the horizontal and
vertical clearances in Rule 235 must be met.

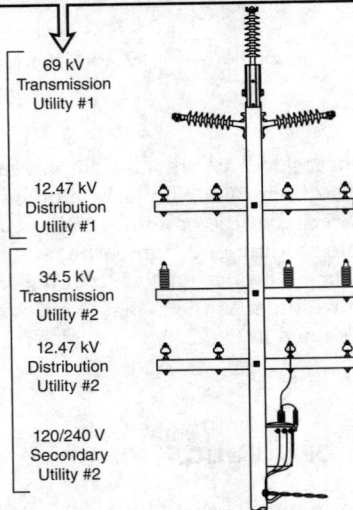

69 kV
Transmission
Utility #1

12.47 kV
Distribution
Utility #1

34.5 kV
Transmission
Utility #2

12.47 kV
Distribution
Utility #2

120/240 V
Secondary
Utility #2

Fig. 220-5. Relative levels of supply circuits of different voltages
owned by separate utilities (Rule 220C2b).

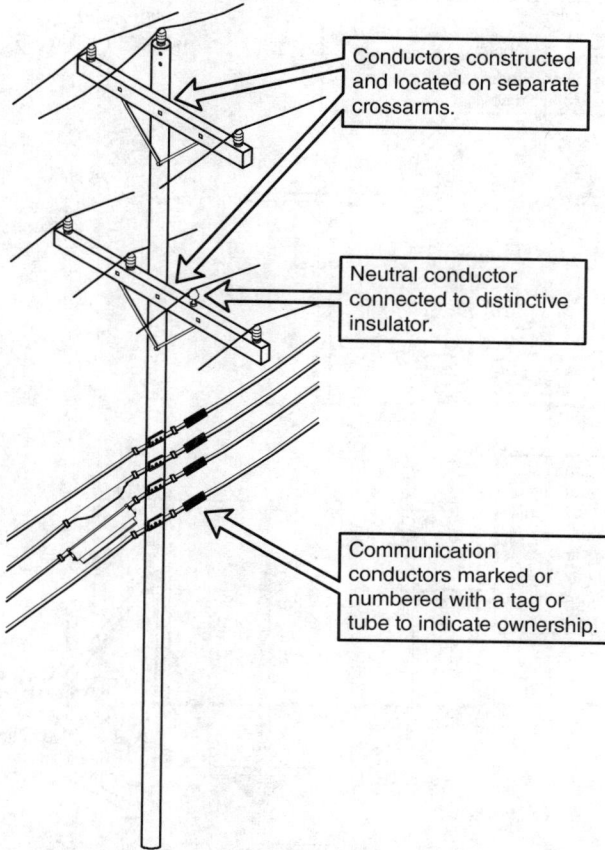

Conductors constructed
and located on separate
crossarms.

Neutral conductor
connected to distinctive
insulator.

Communication
conductors marked or
numbered with a tag or
tube to indicate ownership.

See
Photo(s)

Fig. 220-6. Example of identification of overhead conductors
(Rule 220D).

222. JOINT USE OF STRUCTURES

Per the definition of "joint use" in Sec. 02 of the **NESC**, joint use refers to two or
more utilities on the same structure which can be two or more of the same kind of
utility (e.g., two power utilities) or two or more of different kinds of utilities (e.g.,
power and communications). The **NESC** encourages considering the use of joint-
use construction along highways, roads, streets, and alleys. An example of a joint
use structure along a roadway is shown in Fig. 222-1.

To decide between joint use and separate lines along highways, roads, streets,
and alleys, the following must be considered by the utilities considering the joint
occupancy:

The basic rule...
Two separate lines either of which carries supply conductors (one could carry communication conductors) should be separated to avoid conflict.

If the basic rule is not practical...
The lines should be separated as far as practical and must be built to the grade of construction in Sec. 24 for a conflicting line. (Rule 233B applies to the minimum horizontal clearance required between adjacent lines.)

If the basic rule is not practical, an exception with two options is provided.

If the basic rule is not practical...
The lines must be combined on the same structure. (Rule 235 applies.)

See Photo(s)

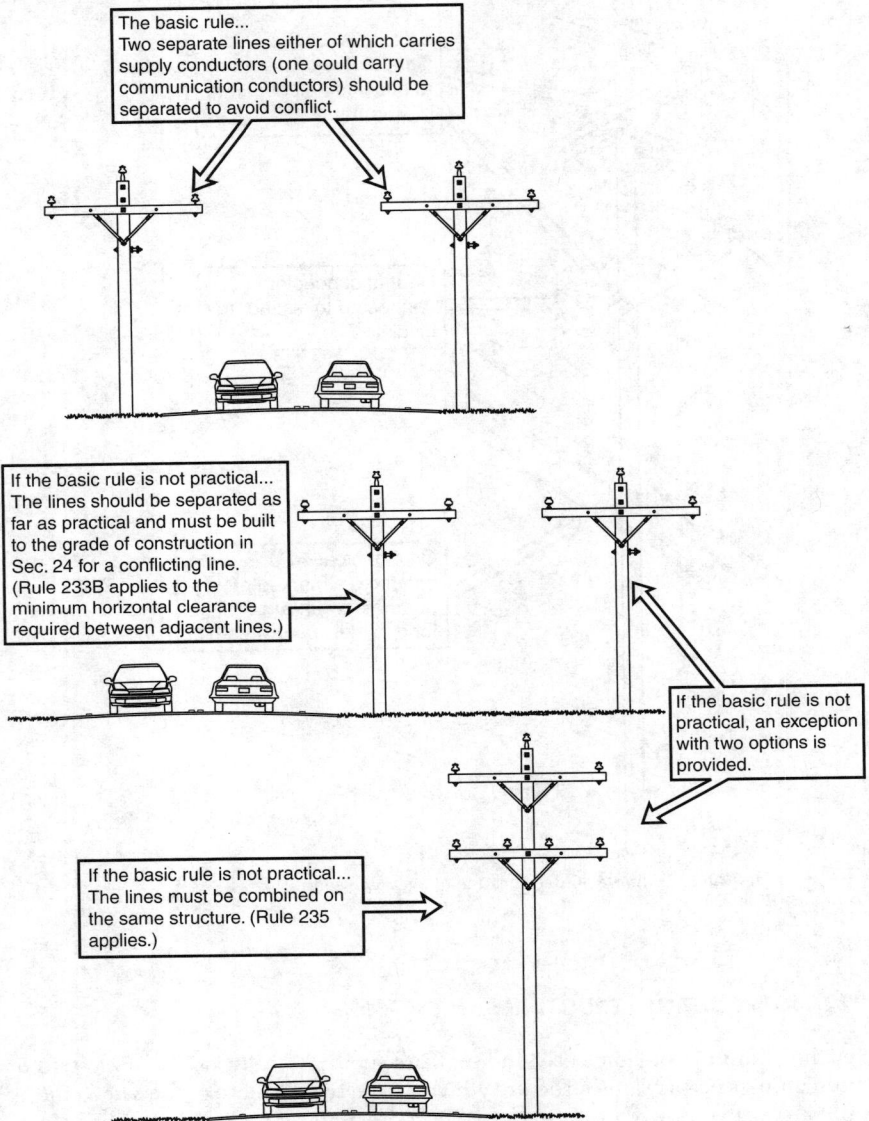

Fig. 221-1. Avoiding conflict between two separate lines (Rule 221).

- Character of circuits
- Worker safety
- Total number and weight of conductors
- Tree conditions
- Number of branches (taps)
- Number of service drops

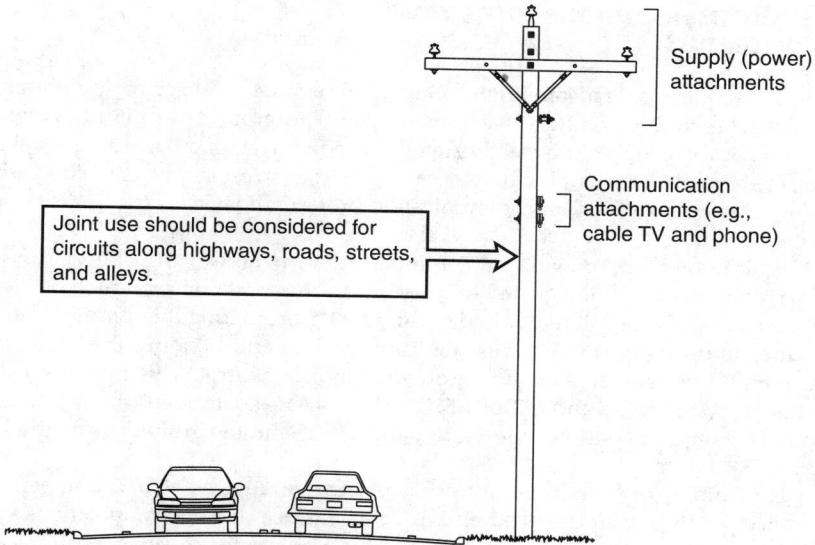

Fig. 222-1. Example of joint use of structures (Rule 222).

- Structure conflict
- Availability of right of way, etc.

When joint use is used, it must meet the grade of construction specified in Sec. 24. In addition to applying the appropriate grade of construction for loading and strength issues, numerous other rules apply to joint-use construction including the overhead line clearance rules in Part 2 and the underground separation rules in Part 3. Although not addressed in the **NESC**, a joint-use agreement is appropriate when joint-use construction is used. Joint-use agreements typically include wording regarding an attachment fee that is paid by the utility that is attaching to a pole owned by a different utility. Joint-use agreements also commonly include references to **NESC** requirements, methods to address payment for structures that must be replaced to meet joint-use requirements, and attachment locations. Some supply utilities reserve the top 10 or 12 ft of a distribution pole for supply attachments. This method assures supply space will be available for future distribution service transformers, primary taps, secondary services, etc. Other supply utilities permit communication attachments as long as space is available. This method may require that the communications utility vacate the pole or pay for a pole upgrade if distribution service transformers, primary taps, secondary services, etc. need to be added in the future and space is not available for these items. Similar approaches may be applied to underground ducts. Federal Communications Commission (FCC) or state regulations may specify pole attachment or underground duct procedures and rental fee methodologies. A joint-use agreement is appropriate for collecting pole rental fees, resolving make-ready issues, and assuring joint-use attachments are in accordance with the **NESC**.

223. COMMUNICATIONS PROTECTIVE REQUIREMENTS

Rule 223 requires that a communication apparatus that is subjected to lightning, contact with supply conductors exceeding 300 V to ground, a ground potential rise greater than 300 V, or a steady-state induced voltage of a hazardous level be protected by insulation and, where necessary, surge arresters in conjunction with fusible elements. Additional communication protective devices are also listed for severe conditions.

A typical joint-use (power and communication) overhead distribution line is subjected to the conditions listed in Rule 223A. The most common method used for the means of protection required in Rule 223B is insulating the communication conductors (in the form of a communication cable) and bonding the grounded communication messenger to the grounded supply neutral. This measure must be taken to satisfy Footnote 7 of NESC Table 242-1 and to meet the messenger grounding requirements in Rules 215C1 and 215C5 and the grounding method in Rule 097G.

It is common to find communication protective devices at building service entrance points and at the head end or central office of a communications system. The additional communication protective devices for severe conditions are typically applied to a communication line entering an electric supply station. The electric supply station typically has large fault current duties, which can severely damage a metallic communication cable. Commonly, isolation equipment or a fiber-optic communication cable is used to serve substations to mitigate this concern.

224. COMMUNICATION CIRCUITS LOCATED WITHIN THE SUPPLY SPACE AND SUPPLY CIRCUITS LOCATED WITHIN THE COMMUNICATION SPACE

Rule 224A applies to communication circuits located in the supply space. Examples of communication circuits located in the supply space are a communication circuit operated by a supply utility and used for communicating between supply stations or a communication circuit operated by a supply utility as a line of business to provide Internet, data, television, or telephone services. Another example is a communication utility locating a communications line in the supply space (as opposed to the communication space) using qualified supply line employees. The terms "supply space" and "communications space" are defined in Sec. 02 of the NESC. The "communication worker safety zone" is described in NESC Rules 235C4 and 238E.

If a communication circuit is located in the supply space (not the communication space), it must be installed and maintained by an employee qualified to work in the supply space per the work rules (Secs. 42 and 44) of the NESC. The employee in this case is typically a power lineman. If a communication circuit is located in the communication space (not in the supply space), it can be installed and maintained by an employee qualified to work in the communication space per the work rules (Secs. 42 and 43) of the **NESC**. The employee in this case is typically a communications lineman or communications technician.

Rule 224A2 states that an insulated communication cable supported by an effectively grounded messenger and located in the supply space must have the same clearance as neutrals meeting Rule 230E1 from communication circuits in the communication space and from supply conductors in the supply space. This requirement is also conveyed in **NESC** Table 235-5, Footnote 5 and Footnote 9. See Rules 235C and 238 for examples and additional information. Fiber-optic cables located in the supply space must meet Rule 230F. Rule 224 addresses communication circuits in the supply space. See Rule 235I for communication antennas in the supply space.

Examples of communication cables located in the supply space and the communication space are shown in Fig. 224-1.

Rule 224A3 provides conditions for locating a communication circuit in the supply space in one portion of the system and in the communication space in another portion of the system. The transition of the communication cable from the supply space to the communication space must occur on a single structure. See Fig. 224-2.

Rule 224B applies to special supply circuits used exclusively in the operation of communication circuits. Rule 224B1 applies to open wire (e.g., bare noninsulated) circuits. Rule 224B2 applies to a communication cable with a supply circuit embedded in it. Rule 224 addresses communication circuits in the supply space. See Rule 235I for communication antennas in the supply space.

225. ELECTRIC RAILWAY CONSTRUCTION

Electric railways can be in the form of electric locomotives on railroad tracks or electric trolleys on streetcar tracks. Rule 225 provides general information related to these types of installations. Rule 225 uses the term "electric railway," but there are times where trolley contact conductors are used for electric trolley buses that have tires on a roadway. The **NESC** addresses clearances of electric railway and trolley contact conductors in Rule 225 and throughout the clearance rules in Sec. 23. Size, strength, and loading issues related to electric railway and trolley contact conductors are addressed in Rule 261H3 and throughout Secs. 24, 25, 26, and 27.

Rule 225, as well as Rule 215C and other rules in Secs. 23, 24, 25, 26, and 27 address the span wires that support electric railway or trolley contact conductors. Span wires or brackets carrying trolley conductors may be located in the communication worker safety zone. See Rule 238 for more information.

Many times electric railway or electric trolley systems are operated at DC voltages. See Rule 230G for a discussion of DC vs. AC voltages and how to apply DC voltages to the clearance tables in the **NESC**. Part 4 of the **NESC** contains minimum approach distances for DC voltages in Sec. 44.

Communication cable located in the supply space (static wire with embedded fiber-optic communication).
This communication cable must be installed and maintained by an employee qualified to work in the supply space.

Transmission phase conductors in the supply space

Distribution phase and neutral conductor in the supply space

Supply space

See Rules 235C4 and 238E.

Communication worker safety zone

Communication space

Communication cable located in the supply space. A communication cable on an effectively grounded messenger per Rule 224A or a fiber-optic cable per Rule 230F. This communication cable must be installed and maintained by an employee qualified to work in the supply space.

Communication cable (cable TV) located in the communication space

Communication cable (telephone) located in the communication space

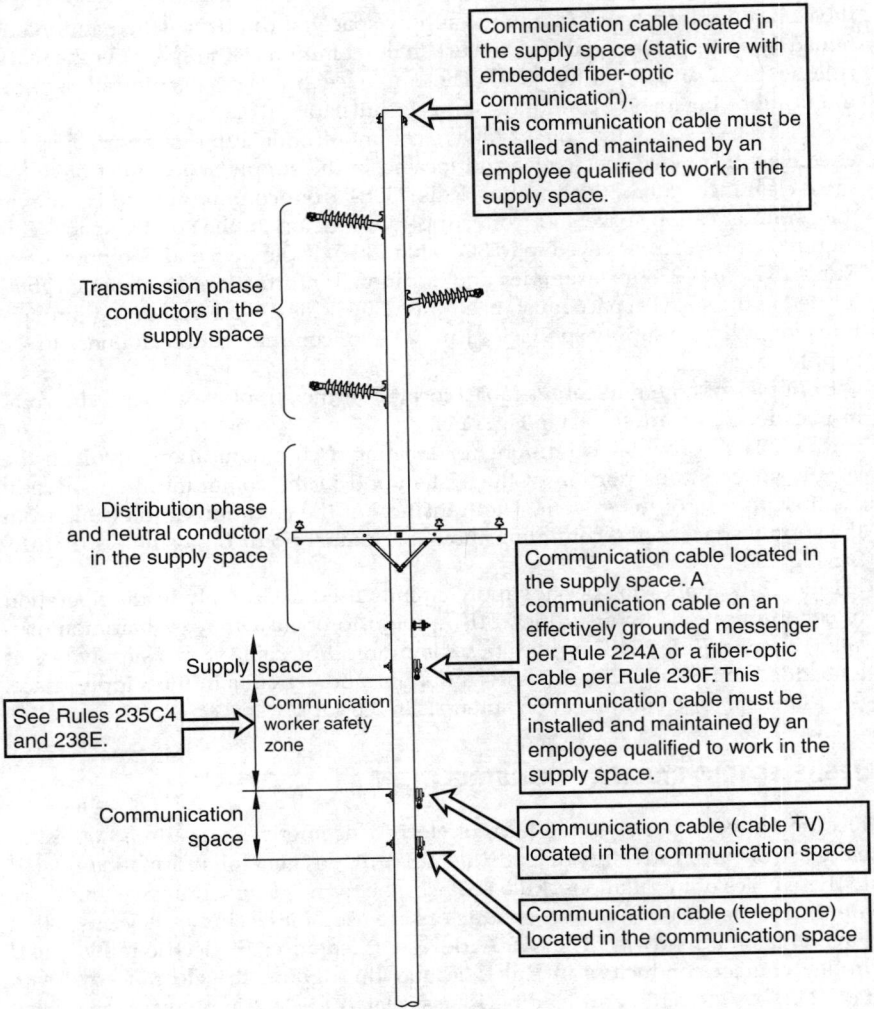

Fig. 224-1. Examples of communication cables located in the supply space and the communication space (Rule 224A).

Another reference for electric railway and electric trolley design is the American Railway Engineering and Maintenance-of-Way Association (AREMA). AREMA publishes a document titled *Manual for Railway Engineering.*

Additional terms specific to electric railway and trolley systems are pantograph, trolley pole, and third-rail. Examples of common terms used in electric railway and trolley systems are shown in Fig. 225-1.

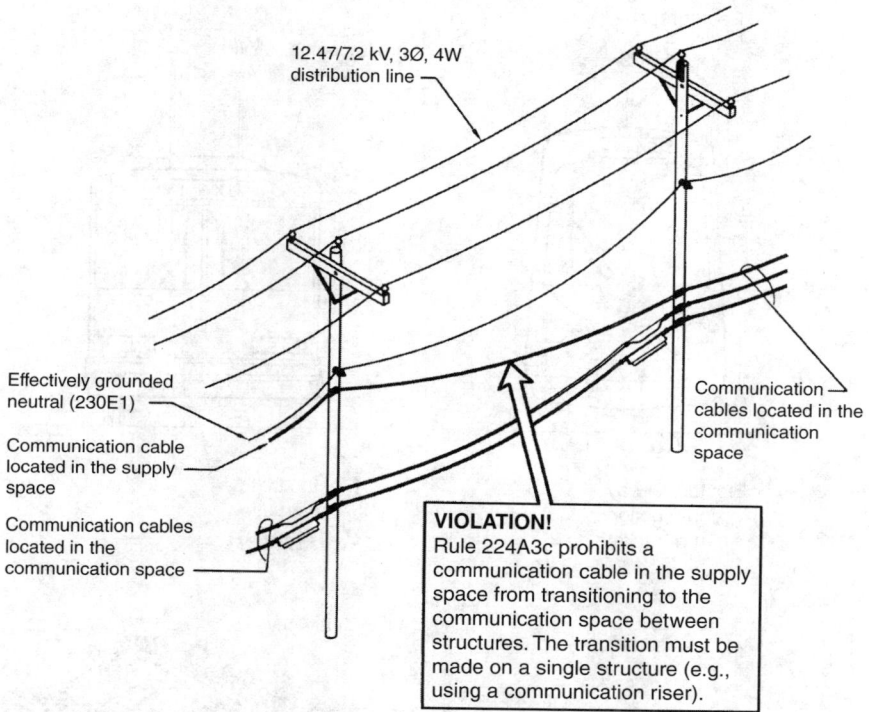

12.47/7.2 kV, 3Ø, 4W
distribution line

Effectively grounded
neutral (230E1)

Communication cable
located in the supply
space

Communication cables
located in the
communication space

Communication
cables located in the
communication
space

VIOLATION!
Rule 224A3c prohibits a
communication cable in the supply
space from transitioning to the
communication space between
structures. The transition must be
made on a single structure (e.g.,
using a communication riser).

Fig. 224-2. Example of a communication cable transition from the supply space to the communication space (Rule 224A3c).

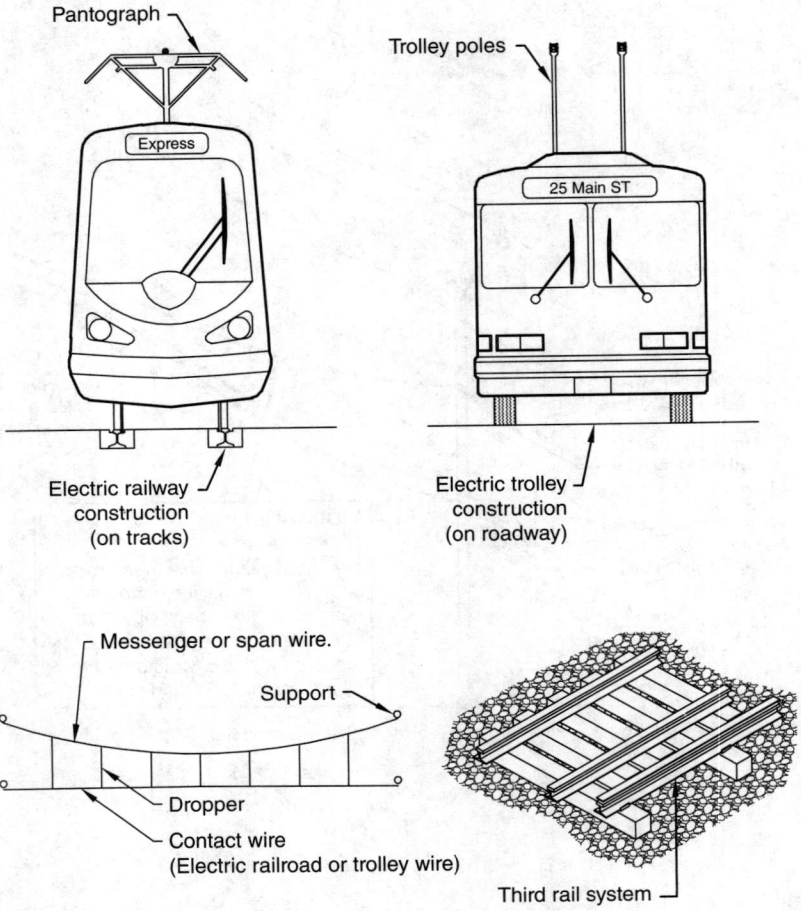

Pantograph

Express

Electric railway
construction
(on tracks)

Trolley poles

25 Main ST

Electric trolley
construction
(on roadway)

Messenger or span wire.

Support

Dropper

Contact wire
(Electric railroad or trolley wire)

Third rail system

See
Photo(s)

Fig. 225-1. Examples of common terms used in electric railway and trolley systems (Rule 225).

Section 23

Clearances

Section 23 is probably the most referenced section in the **NESC**. Rule 232, Vertical Clearance of Wires, Conductors, Cables, and Equipment above Ground, Roadway, Rail, or Water Surfaces, is probably the most referenced rule. Chances are that if a person only uses the **NESC** once a year, he or she will be using it to find a clearance value somewhere in this section. When an overhead line clearance question comes up, one of the first problems a person has is determining which rule applies to the **Code** clearance question at hand. An outline of each rule in Sec. 23 with an icon providing a graphical representation of what the rule covers and a column indicating if a sag chart is needed to determine clearance values is shown in Fig. 23-1.

Many of the clearances found in the **NESC** tables in this section are based on conductor sags. The **NESC** specifies maximum sags, minimum sags, sags with wind, sags with ice, sags at specified temperatures, initial sags, and final sags. It is important that the person using the **Code** to determine overhead line clearances understands how to apply a sag and tension chart. Important concepts that apply to sag and tension charts are outlined below:

- Sag and tension charts are created for a conductor type (e.g., 1/0 ACSR, 6/1 stranding, code word: Raven) at a ruling span (e.g., 300 ft) in a loading district (e.g., Medium Loading) and clearance zone (e.g., Zone 2). This information typically appears at the heading of the sag chart.
- A sag and tension chart describes what position a conductor will be in at various temperature and physical loading conditions. Examples of physical loading conditions are ice and wind.
- A ruling span is the span that "rules" or "governs" the behavior of all the spans between two conductor deadends.

INFO	RULE 230	General	Sag chart needed to check clearance? ☐ YES ☒ NO
	RULE 231	Clearances of supporting structures from other objects	☐ YES ☒ NO
	RULE 232	Vertical clearances of wires, conductors, cables, and equipment above ground, roadway, rail, or water surfaces	☒ YES ☐ NO
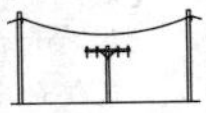	RULE 233	Clearances between wires, conductors, and cables carried on different supporting structures	☒ YES ☐ NO
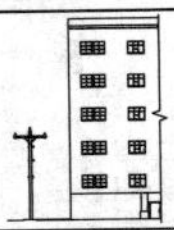	RULE 234	Clearances of wires, conductors, cables, and equipment from buildings, bridges, rail cars, swimming pools, and other installations	☒ YES ☐ NO
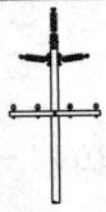	RULE 235	Clearance for wires, conductors, or cables carried on the same supporting structure	☒ YES ☐ NO
	RULE 236	Climbing space	☐ YES ☒ NO
	RULE 237	Working space	☐ YES ☒ NO
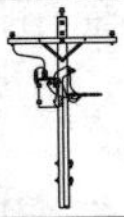	RULE 238	Vertical clearance between certain communications and supply facilities located on the same structure	☐ YES ☒ NO
	RULE 239	Clearance of vertical and lateral facilities from other facilities and surfaces on the same supporting structure	☐ YES ☒ NO

Fig. 23-1. Summary of overhead line clearance rules (Sec. 23).

- A ruling span is calculated between deadends by using the following formula:

$$RS = \sqrt{\frac{S_1^3 + S_2^3 + S_3^3 + \cdots + S_n^3}{S_1 + S_2 + S_3 + \cdots + S_n}}$$

Where RS = Ruling Span
$S_1, S_2, S_3 \cdots S_n$ = 1st, 2nd, 3rd $\cdots$ nth span lengths between the conductor deadends.

- The sags on a sag and tension chart are for a span length equal to the ruling span. To calculate sags for spans longer or shorter than the ruling span, the following formula can be used:

$$\text{Sag in feet} = \left(\frac{\text{span length}}{\text{ruling span length}}\right)^2 \times \text{ruling span sag in feet}$$

- The curved shape of a conductor suspended between two rigid supports is defined as a catenary curve. The catenary equation is a complex equation involving hyperbolic terms and is typically solved using computer programs. Manual sag and tension calculations are typically solved using a simple parabolic equation to approximate the more complex catenary equation.
- The temperature on a sag chart is the conductor temperature, not the ambient air temperature. The conductor temperature is based on the ambient air temperature, the cooling effect of the wind, the radiant heating effect of the sun, and the amount of electrical current flowing through the conductor. IEEE Standard 738 provides methods to calculate the current-temperature relationship of bare overhead conductors. Below are approximate examples for a 1/0 ACSR conductor:
  - ✓ 1/0 ACSR bare conductor has an approximate current-carrying capacity (ampacity) of 230 A. The 230 A is based on a conductor temperature of 167°F.
  - ✓ 1/0 ACSR bare conductor:
    Assume the conductor carries a small electrical load (e.g., 11.5 A, which is approximately 5 percent of the 230-A ampacity).
    During summer conditions: Assume a 104°F ambient temperature combined with the heating effect of the electrical current yields a 110°F conductor temperature.
    During winter conditions: Assume a −20°F ambient temperature combined with the heating effect of the electrical current yields a −15°F conductor temperature.
    The maximum sag conditions per Rule 232A would be:
    - 32°F with ice (e.g., 0.25 in of ice for Zone 2)
    - 120°F (since the maximum operating temperature from above is less than 120°F)
  - ✓ 1/0 ACSR bare conductor:
    Assume the conductor carries a large electrical load (e.g., 175 A, which is approximately 75 percent of the 230-A ampacity).

During summer conditions: Assume a 104°F ambient temperature combined with the heating effect of the electrical current yields a 167°F conductor temperature.

During winter conditions: Assume a −20°F ambient temperature combined with the heating effect of the electrical current yields a 32°F conductor temperature.

The maximum sag conditions per Rule 232A would be:
- 32°F with ice (e.g., 0.25 in of ice for Zone 2)
- 120°F
- 167°F (since the maximum operating temperature from above is greater than 120°F)

✓ 1/0 ACSR bare conductor:

Assume the conductor carries an emergency electrical load (e.g., 230 A, which is 100 percent of the 230-A ampacity).

During summer conditions: Assume a 104°F ambient temperature combined with the heating effect of the electrical current yields a 212°F conductor temperature.

During winter conditions: Assume a −20°F ambient temperature combined with the heating effect of the electrical current yields a 60°F conductor temperature.

The maximum sag conditions per Rule 232A would be:
- 32°F with ice (e.g., 0.25 in of ice for Zone 2)
- 120°F
- 212°F (since the maximum operating temperature from above is greater than 120°F)

- It is common to see conductor temperatures of 104°F, 167°F, and 212°F on a sag chart as they represent 40°C, 75°C, and 100°C, respectively.
- Electrical loads can peak in the summer or winter or both. Areas with a large amount of electric heating load tend to peak in the winter. Areas with a large amount of air conditioning load tend to peak in the summer.
- Sag and tension charts provide both initial and final sags and tensions. Initial values apply to the day a new conductor is strung up. Final values can occur anywhere in time after that point. For example, if the conductor is exposed to an ice storm the first week it is installed, the conductor could be at final sag. If the conductor was never exposed to ice and was exposed to very little wind, it would take many years of hanging under its own weight for the conductor to reach final sag.
- The additive constant for the heavy, medium, light, and warm islands loading districts is sometimes referred to as a K-factor. The additive constant is also used for clearance zones 1, 2, 3, and 4.
- The maximum sag on a sag chart typically occurs at one of the following conditions:
  ✓ 32°F with ice (e.g., 0.25 in of ice for Zone 2), final
  ✓ 120°F, final
  ✓ Greater than 120°F, final (e.g., 167°F, 212°F, etc.)

Larger sags may be observed at the following conditions, but these conditions are not required for checking vertical clearance:
  ✓ Heavy, medium, light, or warm islands loading (e.g., 0.25 in of ice, 4-lb/ft² wind, 0.20 lb/ft additive constant for medium loading)

- The heavy, medium, light, and warm island conditions (Rule 250B for loading) are also referred to as the Zone 1, 2, 3, and 4 conditions (Rule 230B for clearance).

- One of the Zone 1, 2, 3, and 4 conditions (e.g., Zone 2, 15°F, 0.25 in of ice, 4 lb/ft² wind, 0.20 lb/ft additive constant) must be applied to the conductor for physical loading purposes before the final sag is checked at 32°F with ice (e.g., 0.25 in of ice for Zone 2), 120°F, and greater than 120°F if so designed. Note that sag is not required to be checked at 15°F, 0.25 in of ice, 4 lb/ft² wind, 0.20 lb/ft additive constant. The heavy, medium, light, or warm islands (Zone 1, 2, 3, or 4) conditions must be applied to the conductor for physical loading purposes, but sag is checked at other conditions.

  ✓ 60°F with a 6-lb/ft² wind (this condition is required for conductor horizontal blowout, not vertical sag)

  ✓ 60°F with extreme wind (e.g., 17.76 lb/ft²) (this condition is required for conductor tension limits, not vertical sag)

  ✓ 15°F with extreme ice with concurrent wind (e.g., 0.25 in of ice with 2.30 lb/ft²) (this condition is required for conductor tension limits, not vertical sag)

- The minimum sag on a sag chart usually occurs at the initial sag of the coldest temperature without ice or wind loading (e.g., −20°F, initial).

- The tension must be checked at the heavy, medium, light, or warm islands loading condition (Rule 250B), the extreme wind condition (Rule 250C), if applicable, and the extreme ice with concurrent wind condition (Rule 250D), if applicable, to determine the maximum tension which is commonly referred to as the "design tension." It is possible for the maximum tension to occur at the minimum temperature condition.

- The minimum tension usually occurs at the maximum sag condition due to high temperatures at final tension.

- The sag and tensions on a sag chart can be varied by increasing tension which will decrease sag or by decreasing tension which will increase sag. Too much tension is not good for guying and structure strength and loading issues. Too much sag is not good for clearance issues. A balance must be reached between the two.

- The sag values on a sag and tension chart are used to calculate clearance. The tensions on a sag and tension chart are used to calculate strength and loading of structures and conductor supports.

- The following headings and abbreviations are used on a typical sag and tension chart:

  ✓ Conductor: Raven (this is the nickname for the conductor).

  ✓ 1/0 AWG, 6/1 stranding, ACSR: 1/0 AWG (American Wire Gauge) is the conductor size. ACSR stands for Aluminum Conductor, Steel Reinforced. The 6/1 represents six strands of aluminum twisted around one strand of steel.

  ✓ Area = cross-sectional area of the conductor

  ✓ Dia = diameter of the conductor

  ✓ Wt = weight of the conductor

  ✓ RTS = rated tensile strength of the conductor

✓ Span = ruling span
✓ The design condition is the limiting design condition entered into the program. It may be an **NESC** tension limit, a conductor manufacturer tension limit, or a user-defined sag or tension limit.

- Final sag values on the sag and tension chart include the effects of conductor inelastic deformation due to ice and wind loads and long-term material deformation (sometimes referred to as creep). **NESC** Appendix B provides additional information on conductor creep.
- **NESC**-defined tension limits (percentages) must be entered into the sag and tension program.
- User-defined sag and tension limits can be entered into a sag and tension program in addition to **NESC** design conditions. An example of a user-defined tension limit is 20 percent tension at the initial condition at the average annual minimum temperature (e.g., 20 percent tension at 0°F, initial). This limit is used to control aeolian vibration. Another example of a user-defined limit is specifying a conductor sag for a lower circuit to match the sag of a higher circuit. The sags of two different conductor sizes or types cannot perfectly match throughout the entire temperature range. Matching sags at 60°F at the final tension typically produces the closest matching sags throughout the entire temperature range. Other examples of user-defined sag and tension limits are 3 ft of sag at 60°F final, a sag of 1 percent of the ruling span at 60°F final, a 2000-lb maximum tension limit, etc. The conductor or cable manufacturer may also specify a sag or tension limit. When a user-defined tension is entered into a sag and tension program, the **NESC** design conditions must also be entered to verify that none of the **NESC** design conditions are exceeded.
- The phase and neutral conductor of the same circuit can carry different amounts of current and therefore operate at different temperatures. The neutral conductor may carry less current due to phase balance and neutral current cancellation or due to the fact that the earth and a communication messenger on a joint-use pole are in parallel with a multigrounded neutral. It is possible for the neutral and communication messenger on a joint-use (power and communications) line to operate at 120°F while the phase wires on the same line operate at temperatures above 120°F.
- If a line is existing, a sag chart can be created for the existing line by measuring the sag at a known conductor temperature and span length. This information is then entered into a program to create a sag chart. Another method is to find the third or fifth return wave for a known conductor temperature and span length. The return wave time can be converted to sag using the following formula:

$$\text{Sag in feet} = 4.025 \left(\frac{\text{time in seconds}}{2 \times \text{number of return waves}} \right)^2$$

In both methods described above, the line must be assumed to be in its initial or final sag condition. The return wave method is not accurate when used on deadend spans as the deadend insulator string absorbs the conductor wave and distorts the wave timing.

- Sag and tension charts must be run for secondary and communication conductors and cables to accurately check clearance and strength issues for these circuits.
- Sag and tension charts are used to create stringing charts for installing conductors in the field. Stringing charts typically provide sag in inches and third and fifth return wave time in seconds for various installation temperatures and span lengths contained by the ruling span.

Required sag and tension values in Rules 230, 232, 233, 234, 235, 250B, 250C, 250D, 251, and 261H are noted in the sample sag and tension chart provided in Fig. 23-2.

The sag and tension chart provides a sag value at the midspan or center of the span. There are times when sag needs to be checked somewhere else in the span, for example, if a rise in the ground line occurs around the quarter span instead of the midspan. Another example is if two lines cross in the span, one at 10% of the span distance, and one at 33% of the span distance. A graph can be used to estimate the percentage of midspan sag at various distances along the span length. See Fig. 23-3.

230. GENERAL

230A. Application. Rule 230A provides an introduction to Sec. 23. Section 23 covers clearances for overhead supply and communication lines. Burial depths for underground lines are covered in Part 3 of the **NESC** and clearances in electric supply stations are covered in Part 1 of the **NESC**.

Rule 230A provides a note regarding the development of clearance values and makes reference to Appendix A of the **NESC**, which outlines how **NESC** clearances are calculated starting with the 1990 edition of the **NESC**. Prior to 1990, clearance was calculated using a 60°F sag condition with a clearance adder for long spans instead of using a maximum sag condition. Therefore, clearance values prior to 1990 cannot be directly compared to today's clearances. The examples and discussion presented in this Handbook will all focus on using today's **Code** and today's clearance calculations. Reading and understanding prior methods may be needed if a person is trying to apply the "grandfather" requirements of Rule 013.

Temporary clearances are required to always be the same as permanent clearances. In other words, no clearance reductions exist for a temporary situation. This rule cannot be understated and is conveyed in Fig. 230-1.

Temporary installations are permitted to have lower grades of construction. See Rules 014 and 241.

Emergency installations do permit some slight decreases in clearance if certain conditions are met. Rule 014 is referenced as it provides general waiver information. Rule 230A2 provides the specifics, the most common of which are outlined in Fig. 230-2.

Emergency installations permit laying certain supply (power) and communication cables directly on the ground. This same permission can be found in Rule 311C. An example of supply and communication cables laid on the ground is shown in Fig. 230-3.

Where access is limited to qualified personnel only, as inside a properly fenced electric substation, no emergency clearance is specified.

Sample Sag and Tension Chart

Conductor RAVEN   #1/0 AWG 6/ 1 Stranding ACSR

Area= .0968 Sq. In Dia= .398 In Wt= .145 Lb/F RTS= 4380 Lb
Span= 300.0 Feet (NESC Medium Load Zone /NESC Clearance Zone 2)

	Design Points					Final			Initial		
	Temp	Ice	Wind	K	Weight	Sag	Tension	RTS	Sag	Tension	RTS
	F	In	Psf	Lb/F	Lb/F	Ft	Lb	%	Ft	Lb	%
Note 1	15.	.25	4.00	.20	.658	5.43	1366.	31.2	5.43	1366.	31.2
Note 2	15.	.25	2.30	.00	.387	4.31	1012.	23.1	4.07	1070.	24.4
Note 3	32.	.25	.00	.00	.346	4.49	868.	19.8	4.13	945.	21.6
Note 4	60.	.00	17.76	.00	.607	6.14	1114.	25.4	5.92	1154.	26.4
Note 5	60.	.00	6.00	.00	.246	4.66	595.	13.6	4.03	688.	15.7
Note 6	-20.	.00	.00	.00	.145	1.78	916.	20.9	1.61	1015.	23.2
	-10.	.00	.00	.00	.145	1.98	825.	18.8	1.73	945.	21.6
Note 12	0.	.00	.00	.00	.145	2.20	740.	16.9	1.86	876.	20.0*
	10.	.00	.00	.00	.145	2.46	663.	15.1	2.02	808.	18.5
Note 7	15.	.00	.00	.00	.145	2.60	628.	14.3	2.10	775	17.7
	20.	.00	.00	.00	.145	2.74	595.	13.6	2.20	743.	17.0
	30.	.00	.00	.00	.145	3.04	536.	12.2	2.40	680.	15.5
	32.	.00	.00	.00	.145	3.11	526.	12.0	2.44	668.	15.3
	40.	.00	.00	.00	.145	3.36	486.	11.1	2.62	622.	14.2
	50.	.00	.00	.00	.145	3.68	444.	10.1	2.87	569.	13.0
Note 8	60.	.00	.00	.00	.145	4.00	408.	9.3	3.14	520.	11.9
	70.	.00	.00	.00	.145	4.32	378.	8.6	3.42	478.	10.9
	80.	.00	.00	.00	.145	4.63	353.	8.1	3.70	441.	10.1
	90.	.00	.00	.00	.145	4.93	331.	7.6	4.00	408.	9.3
	100.	.00	.00	.00	.145	5.22	313.	7.1	4.29	380.	8.7
	104.	.00	.00	.00	.145	5.34	306.	7.0	4.41	370.	8.5
	110.	.00	.00	.00	.145	5.47	299.	6.8	4.58	356.	8.1
Note 9	120.	.00	.00	.00	.145	5.61	291.	6.7	4.84	335.	7.7
Note 10	167.	.00	.00	.00	.145	6.25	262.	6.0	6.14	266.	6.1
Note 11	212.	.00	.00	.00	.145	6.84	239.	5.5	6.76	242.	5.5

*Design Condition

Note 1:
15°F, 0.25" ice, 4 lb/ft^2 wind, 0.20 lb/ft K-factor (additive constant).
This condition (Medium Loading) is specified in Rules 250B and 251.
This condition (Zone 2) is specified in Rule 230B.
The tension from this line,1366 lb initial, is used for
guying and other structure strength and loading calculations.
This tension must be compared to the tension for extreme wind loading (Rule 250C), if
applicable, and the tension for extreme ice with concurrent wind loading (Rule 250D), if
applicable, to determine the maximum design tension. The maximum tension on the chart
is commonly referred to as the "design tension."
The percent rated tensile strength from this line, 31.2%, is used in Rule 261H1a.
Rule 261H1a specifies a 60% limit.

Note 2:
15°F, 0.25" ice, 2.30 lb/ft^2 wind.
This condition is specified in Rule 250D.
A 2.30 lb/ft^2 wind correspondents to a 30 mph wind in west-central Washington
State. 0.25"of ice is also required at this location.
The tension from this line, 1070 lb initial, is used for guying and
other structure strength and loading calculations when Rule 250D applies
(for structures and supported facilities over 60' above ground).
The percent rated tensile strength from this line, 24.4%, is used in Rule 261H1a.
Rule 261H1a specifies an 80% limit. This tension must be compared
to the tension for heavy, medium, light, and warm islands loading (Rule 250B) and the tension
for extreme wind loading (Rule 250C), if applicable, to determine the maximum design tension.

Fig. 23-2. Sample sag and tension chart (Sec. 23). (*Continued*)

Note 3: (Zone 2)
32°F, 0.25" ice.
This condition is specified in Rules 232A, 233A, 234A, and 235C2b(1)(c)(ii).
The sag from this line, 4.49' final, is used for clearance calculations.
This sag must be compared to the sag at 120°F, final, and greater than 120°F, final, if so designed, to determine the largest final sag.

Note 4:
60°F, 17.76 lb/ft^2 wind.
This condition is specified in Rule 250C. A 17.76 lb/ft^2 wind corresponds to an 85 mph wind in central Washington State. See Fig. 250-3 for the 17.76 lb/ft^2 calculation.
The tension from this line, 1154 lb initial, is used for guying calculations and other structure strength and loading calculations when Rule 250C applies (for structures and supported facilities over 60' above ground). The percent rated tensile strength from this line, 26.4%, is used in Rule 261H1a. Rule 261H1a specifies an 80% limit. This tension must be compared to the tension for heavy, medium, light, and warm islands loading (Rule 250B) and the tension for extreme ice with concurrent wind loading (Rule 250D), if applicable, to determine the maximum design tension.

Note 5:
60°F, 6 lb/ft^2 wind.
This condition is specified in Rules 233A and 234A.
The sags from this line, 4.66' final and 4.03' initial, are used for horizontal clearance calculations. The sags from this line are "resultant" sags in a diagonal position. Options typically exist in sag and tension programs to show both the horizontal and vertical components of the resultant sag value.

Note 6:
−20°F.
This condition is required in Rule 234A1d (A minimum conductor temperature for a specific area.)
The sag from this line, 1.61' initial, is used for vertical clearance under signs or buildings.

Note 7:
15°F (Medium Loading).
This condition is used in Rule 261H1c.
The percent rated tensile strengths from this line, 14.3% final, and 17.7% initial, are used in Rule 261H1c. Rule 261H1c specifies a 35% initial tension limit and a 25% final tension limit.

Note 8:
60°F.
This condition is used in Rules 233A, 235B, and 235C2b(3).
The sags from this line, 4.00' final and 3.14' initial, are used for vertical and horizontal clearance calculations.

Note 9:
120°F.
This condition is specified in Rules 232A, 233A, 234A, and 235C2b(1)(c)(i).
The sag from this line, 5.61' final, is used for clearance calculations.
The sag must be compared to the sag at 32°F, 0.25" ice, final, and greater than 120°F, final, if so designed, to determine the largest final sag.

Note 10:
167°F.
This condition is specified in Rules 232A, 233A, 234A, and 235C2b(1)(c)(i).
The sag from this line, 6.25' final, is used for clearance calculations.
This condition is only required if the line is designed to operate at a temperature greater than 120°F.

Fig. 23-2. Sample sag and tension chart (Sec. 23). (*Continued*)

Note 11:
212°F.
This condition is specified in Rules 232A, 233A, 234A, and 235C2b(1)(c)(i).
The sag from this line, 6.82' final, is used for clearance calculations.
This condition is only required if the line is designed to operate at a temperature greater than 120°F.

Note 12:
0°F initial, 20% tension limit.
This condition is not specified in the **Code**. It is a user-defined condition. For example, a 20% tension limit at the average minimum cold temperature may help prevent aeolian vibration. (See Note 3 in Rule 261H1c.) Other user-defined conditions can also be entered into a sag and tension program. The program will determine the most limiting condition of all the conditions entered. In this case the 20% tension limit at 0°F initial was the most limiting condition and the program flagged this value as the "*Design Condition."

Fig. 23-2. Sample sag and tension chart (Sec. 23). (*Continued*)

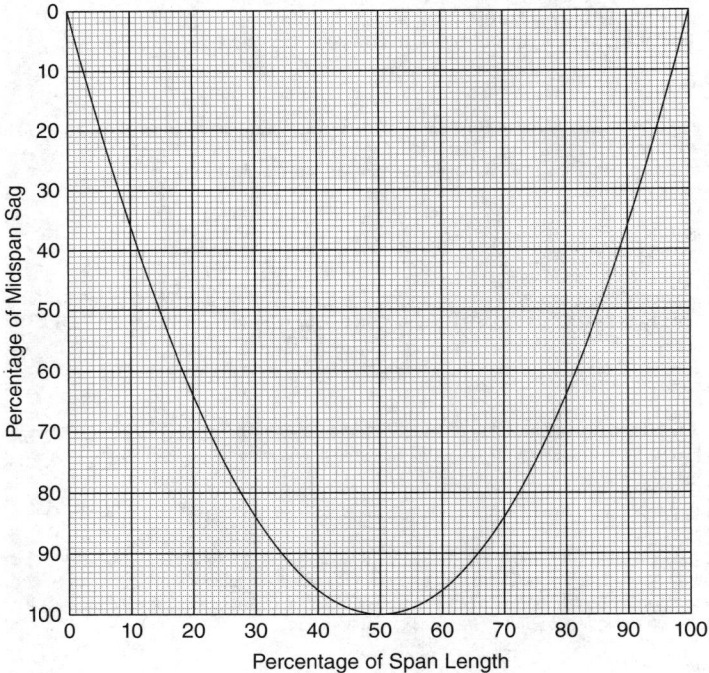

Fig. 23-3. Estimate of the percentage of midspan sag at various distances along the span length (Sec. 23).

TEMPORARY CLEARANCE ~~ALWAYS~~ PERMANENT CLEARANCE

Fig. 230-1. Relationship between temporary and permanent clearance (Rule 230A1).

The general requirements in Rule 014 require emergency installations to be removed, replaced, or relocated as soon as practical.

Rule 230A3 describes how clearance and spacing are measured. See Fig. 230-4.

Throughout Sec. 23 the term clearance is used more often than spacing. Part 3, Underground Lines, uses the term separation which is measured similar to clearance.

No exact values are specified. Horizontal clearances may also be reduced using accepted good practice.

CONDITION	TYPE OF CONDUCTOR		
	NEUTRAL (230E1) OR COMMUNICATIONS	SECONDARY DUPLEX, TRIPLEX, OR QUADRUPLEX (230C3)	PRIMARY VOLTAGE OPEN (NONINSULATED) SUPPLY CONDUCTORS 750 V TO 22 KV
Roads, streets, and other areas subject to truck traffic	Normal: 15.5' Emergency: 15.5'	Normal: 16.0' Emergency: 15.5'	Normal: 18.5' Emergency: No exact value is specified. Adjust for voltage and local conditions. Value must be greater than 15.5'.
Spaces and ways subject to pedestrians or restricted traffic only.	Normal: 9.5' Emergency: 9.0'	Normal: 12.0' Emergency: 9.0'	Normal: 14.5' Emergency: No exact value is specified. Adjust for voltage and local conditions. Value must be greater than 9.0'.

Caution: Use is very limited.

For the purposes of this rule a truck is defined as a vehicle exceeding 8' in height. An ambulance or fire rescue unit responding to an emergency could be 8' high or higher.

Fig. 230-2. Reductions of overhead clearances for emergency conditions (Rule 230A2).

EMERGENCY ONLY
Supply and communication cables may be laid directly on the grade if they are guarded or located so they do not obstruct pedestrians or vehicle traffic and if they are marked.

No specifics are provided on how to guard, locate, and mark the cables. Accepted good practice must be used.

Underground Residential Distribution (URD) cable (greater than 600 V) meeting Rule 230C or 350B laid on grade.

Secondary triplex 230C3 cable laid on grade.

Communications cable (Phone or Cable TV) laid on grade.

See Photo(s)

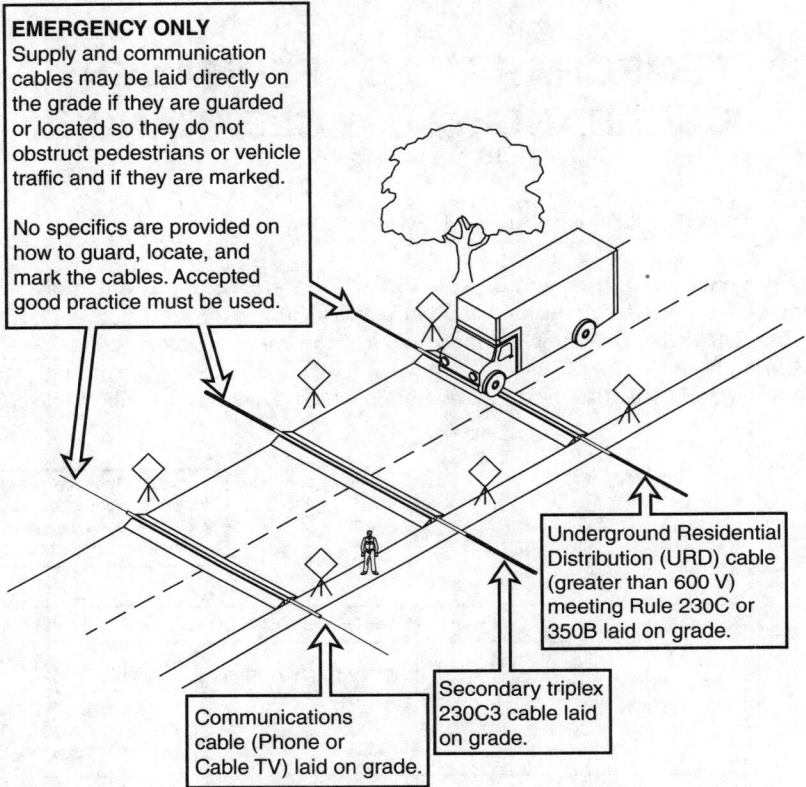

Fig. 230-3. Example of cables laid on grade for emergency conditions (Rules 230A2d and 311C).

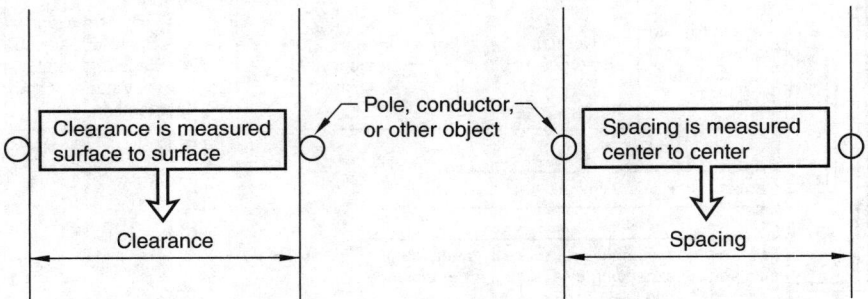

Clearance is measured surface to surface

Pole, conductor, or other object

Spacing is measured center to center

Clearance

Spacing

Fig. 230-4. Measurement of clearance and spacing (Rule 230A3).

Rule 230A4 provides rounding requirements specific to the clearance calculations in Sec. 23. In general, Rule 018 permits rounding "off" to the nearest significant
digit unless otherwise specified in applicable rules. One rule where other rounding is specified is Rule 230A4 which requires rounding "up" as Sec. 23 deals with
overhead line clearances which are typically specified as "not less than" clearances.
Rounding "off" follows the rules of traditional rounding learned in math class. An
example of rounding "off" is rounding 20.02 down to 20.0 or rounding 20.66 up
to 20.7. An example of rounding "up" for "not less than" clearance is rounding
20.02 to 20.1 because rounding "off" to 20.0 would not meet a clearance required
to be "not less than" 20.02. The number of decimal places for the final rounded
value depends on the number of decimal places for the original starting value from
the **Code** table or rule. An exception is provided for millimeters which must be
rounded up to the next multiple of 25 mm. For example, the clearance values in
NESC Table 232-1 have one decimal place (e.g., 18.5 ft). Therefore, a calculated
clearance for a voltage greater than 22 kV per Rule 232C would be rounded "up"
to the tenths place (e.g., 20.02 would be rounded up to 20.1). An exception and
an example are provided in Rule 230A4 requiring rounding down when determining clearance at specified conditions based on field measurements. The example
pertains to vertical clearance required in Rule 232 and **NESC** Table 232-1. The
example not only explains when rounding down is appropriate but also explains
the difference between field-measured sag and the maximum sag conditions that
are required per Rule 230A.

230B. Ice and Wind Loading for Clearances. Rule 230B specifies ice and wind
loading for clearance purposes. **NESC** Fig. 230-1 and **NESC** Tables 230-1 and 230-2
are the three basic references to determine ice and wind loading for Zone 1, Zone 2,
Zone 3, and Zone 4. Zone 4 has values for above and below 9,000 ft of elevation
as some warm islands can have high mountains with ice storms. **NESC** Fig. 230-1
is nearly identical to **NESC** Fig. 250-1. The only difference is Zone 1 in **NESC**
Fig. 230-1 is labeled "Heavy" in **NESC** Fig. 250-1, Zone 2 is labeled "Medium," Zone
3 is labeled "Light," and Zone 4 is labeled "Warm Islands." The values specified in
NESC Tables 230-1 and 230-2 are comparable to the values specified in **NESC** Tables
250-1 and 251-1. The values are identical; however, the presentation format varies
slightly. The requirements in Sec. 25 (**NESC** Fig. 250-1, and **NESC** Tables 250-1 and 251-1)
are for determining conductor tension and calculating strength and loading issues.
The requirements in Sec. 23 (**NESC** Fig. 230-1, and **NESC** Tables 230-1 and 230-2) are for
determining conductor sag and therefore calculating clearance. **NESC** Appendix B
provides additional information on applying ice and wind loads for clearance and
strength. The requirements in **NESC** Table 230-2 (e.g., Zone 2, 15°F, 0.25 in of ice,
4 lb/ft² wind, 0.20 lb/ft additive constant) must be applied to the conductor for physical loading purposes before the final sag is checked at 32°F with ice (e.g., 0.25 in of
ice for Zone 2), 120°F, and greater than 120°F if so designed (see Rule 232A). Note
that sag is not required to be checked at 15°F, 0.25 in of ice, 4 lb/ft² wind, 0.20 lb/ft
additive constant. This condition must be on the sag and tension chart (for clearance Zone 2 and Medium Loading) but clearance is checked at the temperature and
loading conditions in Rule 232A (or other applicable rule). The Zone 1, 2, 3, or 4
conditions (whichever is applicable) from Table 230-2 must be applied to the conductor for physical loading purposes, but sag is checked at the ice condition

(if applicable) in **NESC** Table 230-1 and high temperature conditions defined in the individual clearance rules (e.g., Rule 232A). More specifically, the values specified in **NESC** Table 230-2 for Zones 1, 2, 3, and 4 are for "clearance loading" in Sec. 23. Clearance measurement conditions in Section 23 are covered in Rules 232A, 233A, 234A, 235B, and 235C2b(1)(c). Rule 250C (extreme wind) and 250D (extreme ice with concurrent wind) can impose greater loads (and therefore more sag) than the loads in Zones 1, 2, 3, and 4. However, the loads specified in Rules 250C and 250D are for strength calculations, only the loads of Zones 1, 2, 3, and 4 are required for clearance purposes (see **NESC** Appendix B). Rule 230I, Maintenance of Clearances and Spacings, requires that conductors be resagged if an excessive ice or wind storm stretched conductors to the point of a clearance violation. Rules 230B3 and 230B4 are similar to Rules 251A and 251B. See Rules 251A and 251B for a discussion and related figures. The requirements for Zones 1, 2, 3, and 4 are outlined in Fig. 230-5.

230C. Supply Cables.   The terms 230C1, 230C2, and 230C3 cables will be used over and over throughout the rules of Sec. 23 and throughout the clearance tables in Sec. 23. The most common of these three cables used in construction today is the 230C3 cable, which is an overhead secondary duplex, triplex, or quadruplex cable. The rules defining the construction of 230C1, 230C2, and 230C3 supply cables are outlined in Fig. 230-6.

A bare messenger or neutral is required for 230C1, 230C2, and 230C3 cables. An insulated neutral is more common for underground secondary duplex, triplex, and quadruplex construction. The 230C1, 230C2, and 230C3 cables are relying on the effectively grounded bare messenger or neutral to carry fault current if an insulation failure occurs.

230D. Covered Conductors.   The covered conductors Rule 230D refers to are commonly called tree wire. Tree wire cables are not fully insulated like underground residential distribution (URD) cables or 230C cables. They are covered to limit the likelihood of a short circuit in case of momentary contact with a tree limb. Covered conductors can be attached to insulators on crossarms or used as part of a spacer cable system. Since 230D cables are not fully insulated, they must be considered bare conductors for clearance purposes except that the clearance between the conductors may be reduced when the conductors are owned, operated, or maintained by the same utility. See Fig. 230-7.

230E. Neutral Conductors.   The phrase "neutral conductors meeting 230E1" will be used over and over throughout the rules of Sec. 23 and throughout the clearance tables in Sec. 23. See Sec. 02, Definitions, for a discussion of the term "effectively grounded." The rules for a 230E1 neutral are outlined in Fig. 230-8.

230F. Fiber-Optic Cable.   Fiber-optic supply cable and fiber-optic communication cable are **NESC** terms to categorize where a fiber-optic cable is located. Examples of communication cables located in the supply space and in the communication space are discussed in Rule 224. Rule 224A2a applies to insulated communication circuits supported on an effectively grounded messenger. Rule 224A2b references Rule 230F for fiber-optic cable requirements.

Fiber-optic cables that are strung on overhead pole lines are supported on a messenger or they are contained in an All-Dielectric Self-Supporting (ADSS) cable. Fiber-optic cables can also be embedded in a metallic messenger or static wire. An example of an All-Dielectric Self-Supporting (ADSS) cable is shown in Fig. 230-9.

Clearance Zone	Ice and Wind Condition at a Specified Temperature (Clearance Loading)	Location
Zone 1	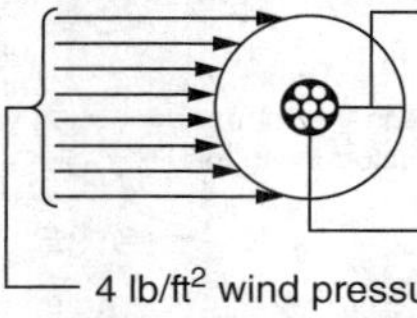 1/2" of radial ice Conductor at 0°F 4 lb/ft^2 wind pressure on conductor with ice Ice weight: 57 lb/ft^3 Additive constant: 0.30 lb/ft	See a map of the USA in **NESC** Fig. 230-1.
Zone 2	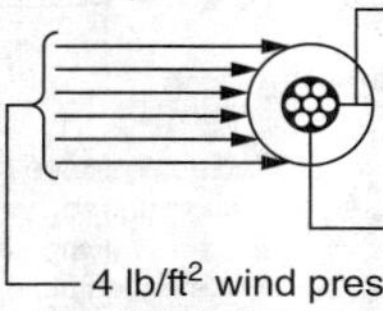 1/4" of radial ice Conductor at 15°F 4 lb/ft^2 wind pressure on conductor with ice Ice weight: 57 lb/ft^3 Additive constant: 0.20 lb/ft	See a map of the USA in **NESC** Fig. 230-1.
Zone 3	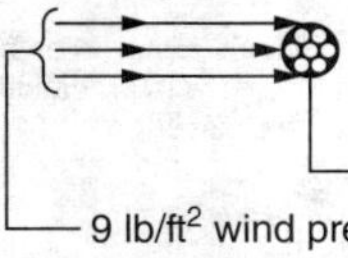 Conductor at 30°F 9 lb/ft^2 wind pressure on conductor without ice no ice Additive constant: 0.05 lb/ft	See a map of the USA in **NESC** Fig. 230-1.
Zone 4 (for altitudes of sea level to 9000 ft)	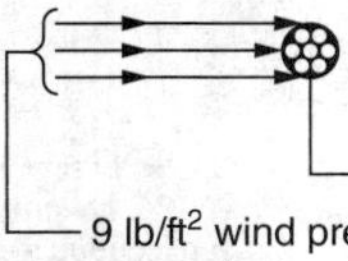 Conductor at 50°F 9 lb/ft^2 wind pressure on conductor without ice no ice Additive constant: 0.05 lb/ft	See a map of the USA in **NESC** Fig. 230-1 (American Samoa, Guam, Hawaii, Puerto Rico, Virgin Islands, and other islands located from latitude 25 degrees south through 25 degrees north).
Zone 4 (for altitudes above 9000 ft)	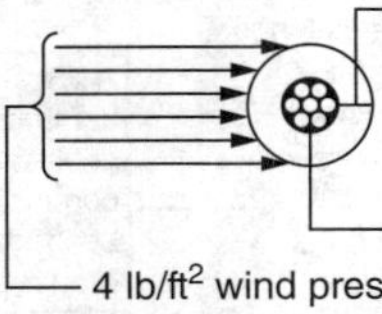 1/4" of radial ice Conductor at 15°F 4 lb/ft^2 wind pressure on conductor with ice Ice weight: 57 lb/ft^3 Additive constant: 0.20 lb/ft	See a map of the USA in **NESC** Fig. 230-1 (American Samoa, Guam, Hawaii, Puerto Rico, Virgin Islands, and other islands located from latitude 25 degrees south through 25 degrees north).

Note: Values shown above are for "clearance loading" in Sec. 23. Clearance measurement conditions in Sec. 23 are covered in Rules 232A, 233A, 234A, 235B, and 235C2b(1)(c).

Fig. 230-5. Zone 1, 2, 3, and 4 ice and wind loading for clearances (Rule 230B).

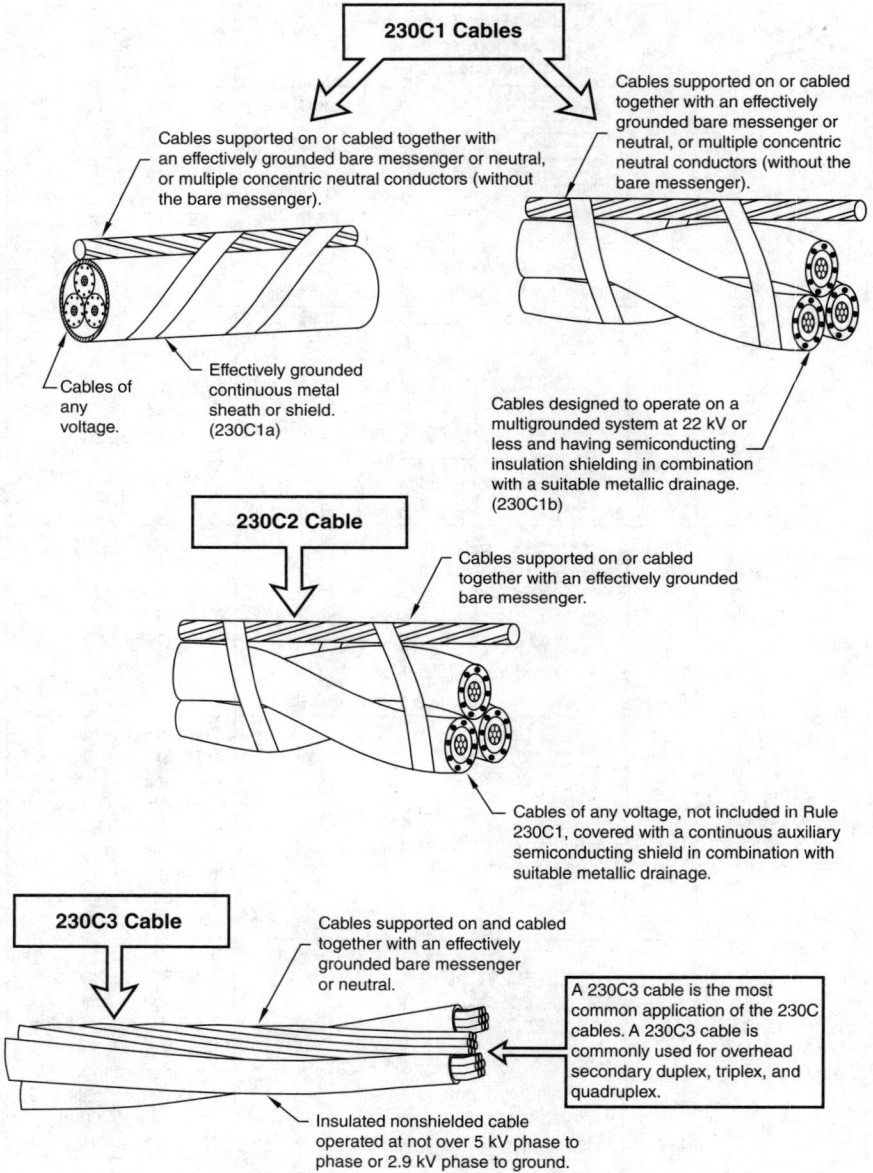

230C1 Cables

Cables supported on or cabled together with an effectively grounded bare messenger or neutral, or multiple concentric neutral conductors (without the bare messenger).

Cables supported on or cabled together with an effectively grounded bare messenger or neutral, or multiple concentric neutral conductors (without the bare messenger).

Cables of any voltage.

Effectively grounded continuous metal sheath or shield. (230C1a)

Cables designed to operate on a multigrounded system at 22 kV or less and having semiconducting insulation shielding in combination with a suitable metallic drainage. (230C1b)

230C2 Cable

Cables supported on or cabled together with an effectively grounded bare messenger.

Cables of any voltage, not included in Rule 230C1, covered with a continuous auxiliary semiconducting shield in combination with suitable metallic drainage.

230C3 Cable

Cables supported on and cabled together with an effectively grounded bare messenger or neutral.

A 230C3 cable is the most common application of the 230C cables. A 230C3 cable is commonly used for overhead secondary duplex, triplex, and quadruplex.

See Photo(s)

Insulated nonshielded cable operated at not over 5 kV phase to phase or 2.9 kV phase to ground.

Fig. 230-6. Construction of 230C1, 230C2, and 230C3 supply cables (Rule 230C).

Bare messenger

Covered "tree wire" conductor

Covered conductors must be considered bare conductors for all clearance requirements except clearance between conductors. (e.g., Clearance of the conductor to the ground or a roadway must be based on a bare open supply conductor.)

Clearance between covered conductors may be reduced below the requirements for open (bare) conductors when the conductors are owned, operated, or maintained by the same utility. Intermediate spacers out in the span may be used to maintain conductor clearance and provide support.

See Photo(s)

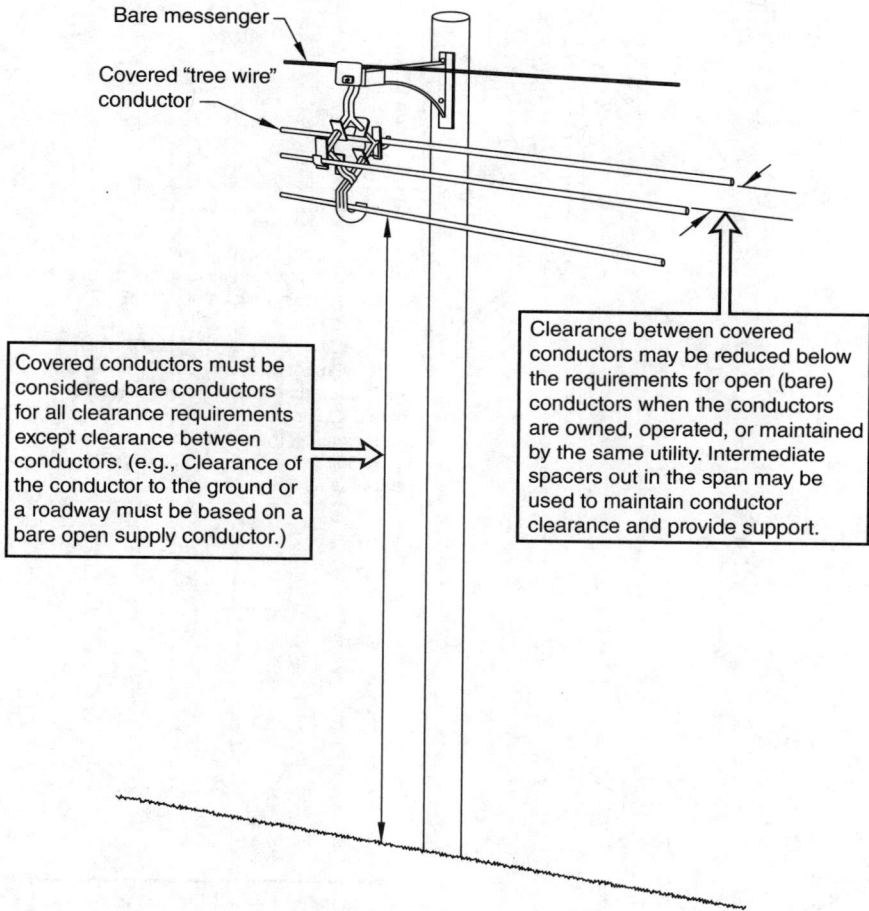

Fig. 230-7. Covered conductors (Rule 230D).

Fiber-optic cables that are located in the supply space and embedded in a messenger or conductor must have the same clearance from communication facilities as the messenger or conductor the fiber-optic cables are embedded in. Fiber-optic cables that are supported on an effectively grounded messenger or contained in an all-dielectric self-supporting cable are required to have the same clearance as neutrals meeting Rule 230E1 from communication facilities. This requirement can be applied to **NESC** Table 235-5, Footnotes 5 and 9. See Rules 235C and 238 for examples, figures, and additional information.

Fiber-optic cables located in the communication space, whether or not supported by a messenger, must have the clearance from supply facilities as required for a communication messenger.

230G. Alternating- and Direct-Current Circuits. The clearances in Sec. 23 apply to AC and DC circuits. The voltage used in a typical American home is 120-V alternating

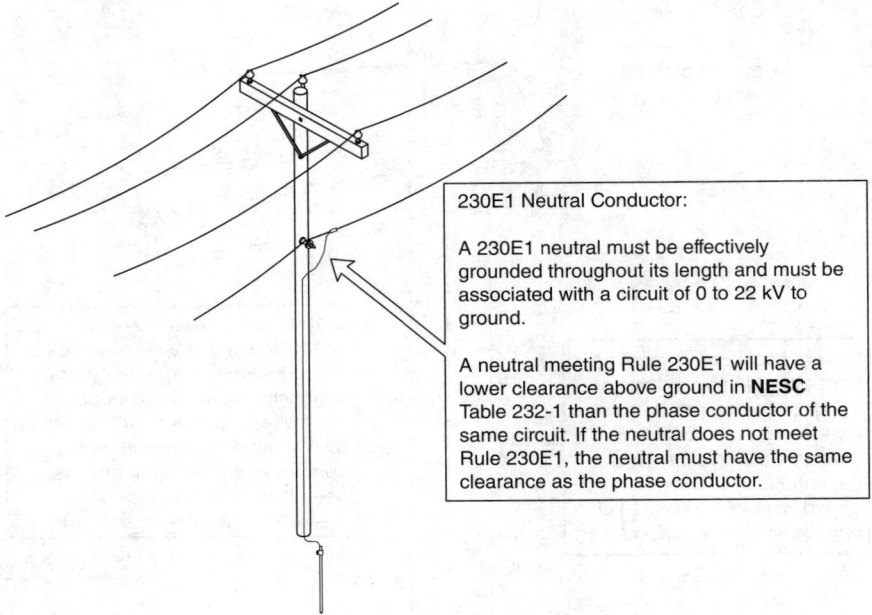

230E1 Neutral Conductor:

A 230E1 neutral must be effectively grounded throughout its length and must be associated with a circuit of 0 to 22 kV to ground.

A neutral meeting Rule 230E1 will have a lower clearance above ground in **NESC** Table 232-1 than the phase conductor of the same circuit. If the neutral does not meet Rule 230E1, the neutral must have the same clearance as the phase conductor.

See
Photo(s)

Fig. 230-8. Neutral conductors (Rule 230E).

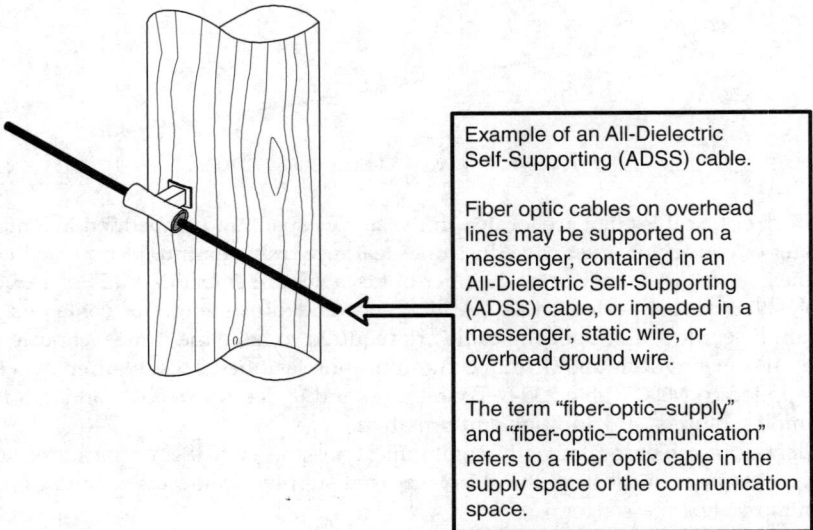

Example of an All-Dielectric Self-Supporting (ADSS) cable.

Fiber optic cables on overhead lines may be supported on a messenger, contained in an All-Dielectric Self-Supporting (ADSS) cable, or impeded in a messenger, static wire, or overhead ground wire.

The term "fiber-optic–supply" and "fiber-optic–communication" refers to a fiber optic cable in the supply space or the communication space.

See
Photo(s)

Fig. 230-9. Example of an All-Dielectric Self-Supporting (ADSS) cable (Rule 230F).

current at 60 Hz (60 cycles per second). The 120-V value is a root mean square (rms) measurement. Another term used for the rms voltage is the effective voltage. If an oscilloscope were plugged into a home receptacle, the scope would show that the top of the sine wave for the 120-V circuit actually crests at $120 \times \sqrt{2}$ or 169.7 V. This same discussion is true for 7.2 kV, 115 kV, 500 kV, etc. Per the definition of voltage in Sec. 02, the clearance tables in Sec. 23 apply to the rms voltage (e.g., 120 V) not the crest voltage (e.g., 169.7 V). The **NESC** does use the crest value when applying the formula for alternate clearances in Rule 232D and in other alternate clearance calculations in Sec. 23.

Values for DC, AC crest, and AC rms are outlined in Fig. 230-10.

For three-phase circuits, it is important to note the difference between phase-to-phase and phase-to-ground voltage. The phase-to-ground voltage is equal to the phase-to-phase voltage divided by the square root of 3. An example of phase-to-phase and phase-to-ground voltages is shown in Fig. 230-11.

Many of the clearance tables in Sec. 23 use phase-to-ground values; however, some tables that involve voltage between conductors will use phase-to-phase values.

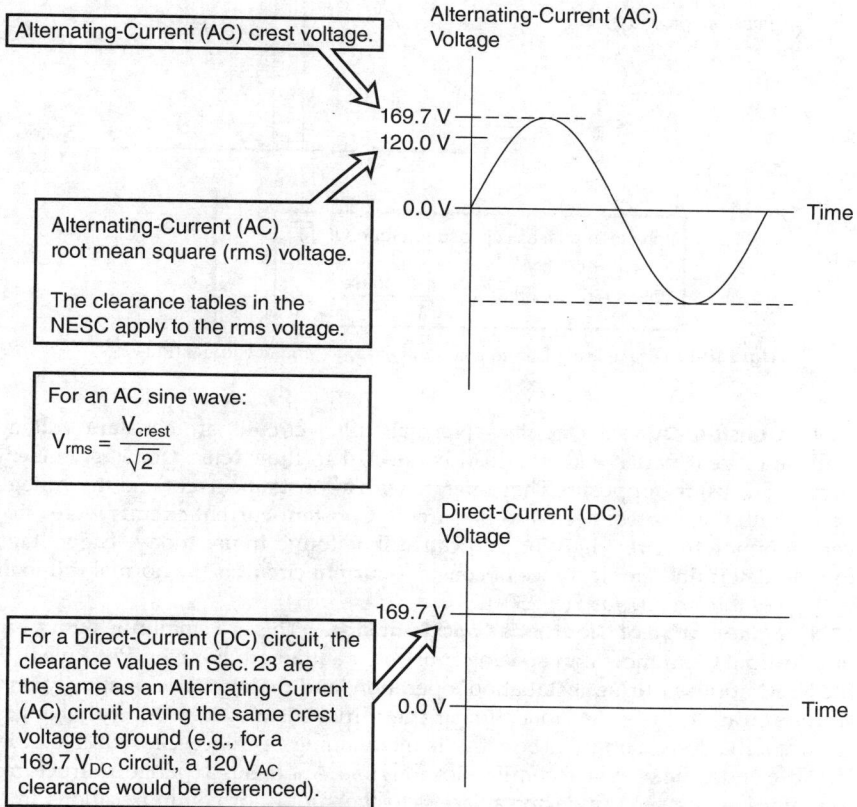

Alternating-Current (AC) crest voltage.

Alternating-Current (AC) Voltage

169.7 V

120.0 V

Alternating-Current (AC) root mean square (rms) voltage.

The clearance tables in the NESC apply to the rms voltage.

0.0 V — Time

For an AC sine wave:

$$V_{rms} = \frac{V_{crest}}{\sqrt{2}}$$

Direct-Current (DC) Voltage

169.7 V

For a Direct-Current (DC) circuit, the clearance values in Sec. 23 are the same as an Alternating-Current (AC) circuit having the same crest voltage to ground (e.g., for a 169.7 V_{DC} circuit, a 120 V_{AC} clearance would be referenced).

0.0 V — Time

Fig. 230-10. Alternating-Current (AC) and Direct-Current (DC) circuits (Rule 230G).

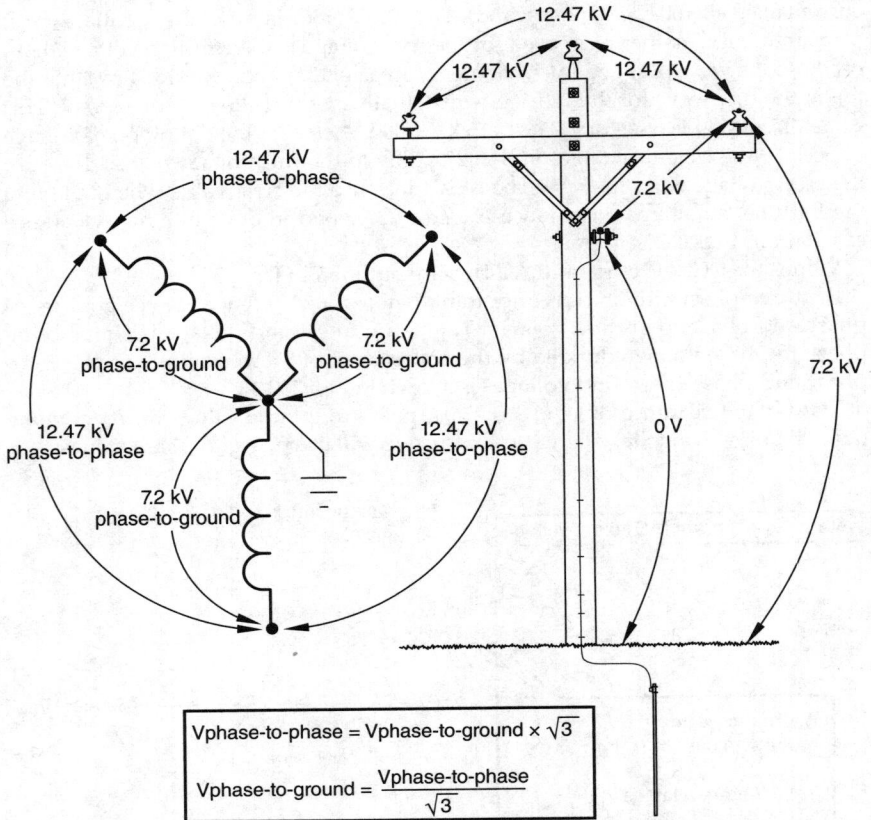

$$V\text{phase-to-phase} = V\text{phase-to-ground} \times \sqrt{3}$$

$$V\text{phase-to-ground} = \frac{V\text{phase-to-phase}}{\sqrt{3}}$$

Fig. 230-11. Example of phase-to-phase and phase-to-ground voltages (Rule 230G).

230H. Constant-Current Circuits. Normal utility circuits are constant voltage and the current varies with the load connected to the circuit. Constant-current circuits are just the opposite. They operate with a constant current and the voltage varies with the load connected to the circuit. Constant-current circuits were once very common for street lighting and can still be found in use today. The voltage used to determine the clearance of constant-current circuit is the normal full-load voltage of the circuit. See Fig. 230-12.

230I. Maintenance of Clearances and Spacings. This is a small rule with a big implication. Clearances and spacing must be maintained. Forever. Per Rule 010, the **NESC** applies during installation, operation, and maintenance of supply and communication lines, not just during the initial installation. Several changes can occur that force a utility to be active in maintaining clearance. One is a change in land use under the supply or communications line. Another is a change in structures (buildings, signs, etc.) under or adjacent to the supply or communications line. Another is excessive sag due to a major ice- or wind-storm. In each of these cases it

Current varies with the load.

$I = I_A + I_B + I_C + I_D$

Constant-voltage source (e.g., 120 V).

Voltage is held constant. Current varies as more load is added in parallel. Each load is rated for a constant voltage.

I_A I_B I_C I_D

Normal electric supply circuits are constant voltage. Constant-current circuits (e.g., series street lighting) must have clearance based on the normal full-load voltage which is the maximum voltage that the constant-current source is rated for.

Voltage varies with the load.

Constant-current source (e.g., 10 A).

Current is held constant. Voltage varies as more load is added in series. Each load is rated for a constant current.

V_A V_B

$V = V_A + V_B + V_C + V_D$

V_C V_D

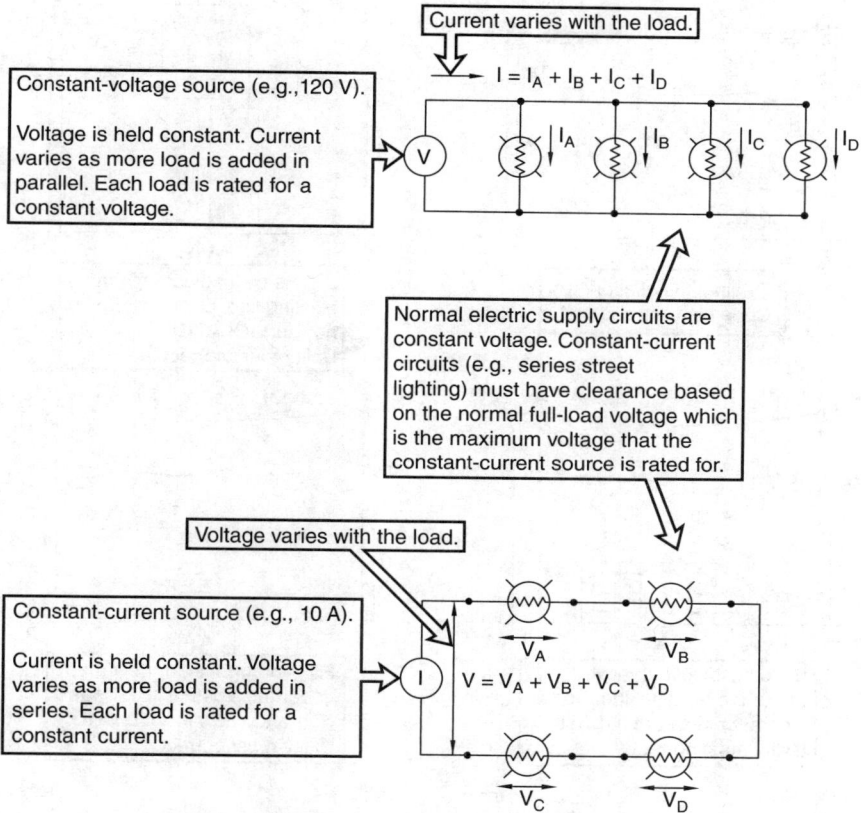

Fig. 230-12. Constant-current circuits (Rule 230H).

is the responsibility of the utility to maintain clearance. The note in Rule 230I that references Rule 013 is a reminder that when maintaining a line, the "grandfather clause" of Rule 013B may be applied and the inspection rules and work rules in the current **Code** edition must be applied per Rule 013C. Applying Rule 013B can help determine which edition of the **Code** is applicable. Applying Rule 013B does not excuse a utility from correcting a violation once the applicable edition is determined. See Rule 013 for additional information. Examples of maintenance of clearances and spacings are shown in Fig. 230-13.

In the case of the excessive sag due to a major ice or wind storm, Rule 230I recognizes that clearance cannot be maintained during or after an abnormal event of this nature. There is some reasonable time period for storm damage work to be repaired; however, the **NESC** does not address the time period. This same discussion applies to an abnormal event such as a car hitting a power pole and knocking down the wires or reducing the clearance of the wires. In this instance, clearance cannot be maintained during or sometime after the accident.

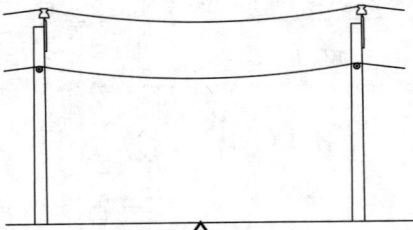

Original Installation.
Code clearance is met.

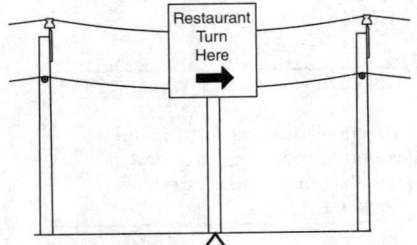

Restaurant
Turn
Here
➡

Sign is built adjacent to the
existing line. Code clearance
now in **VIOLATION.**
Utility must correct.

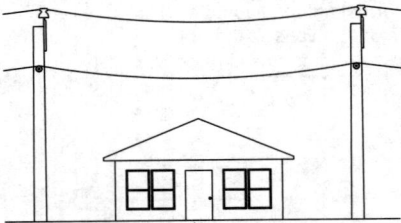

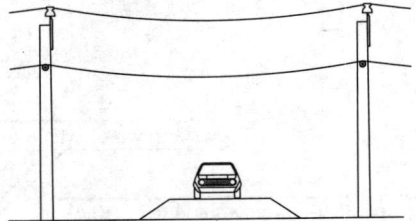

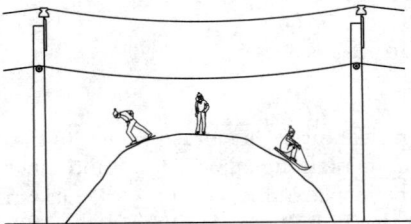

Building is built or moved
under the existing line. Code
clearance now in **VIOLATION.**
Utility must correct.

Fill is brought in for new road
under the existing line. Code
clearance now in **VIOLATION.**
Utility must correct.

Snow is plowed under the
existing line. No reductions
exist for temporary clearance.
Code clearance now in
VIOLATION. Utility must
correct.

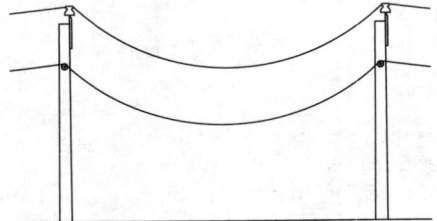

Existing conductor was
excessively stretched due to
major ice or wind storm. Code
clearance now in **VIOLATION.**
Utility must correct.

Fig. 230-13. Examples of maintenance of clearances and spacings (Rule 230I).

There is some reasonable time period for repairing damage to the line caused by the car hitting the pole; however, the **NESC** does not address the time period.

A utility must correct a **Code** clearance violation, even when the line was built with the proper **Code** clearance. The **NESC** does not address who pays for the correction. The utility may consider billing a third party for the correction, but the utility is responsible for maintaining clearance or, in other words, fixing the problem. Some utilities establish clearance values by using the **Code** clearance plus an adder. See Rule 010 for a discussion of clearance adders. The line inspection requirements of Rule 214 provide a means to bring clearance issues to the utility's attention and noncompliant conditions (violations) must be addressed per Rules 214A4 and 214A5, whichever is applicable.

231. CLEARANCES OF SUPPORTING STRUCTURES FROM OTHER OBJECTS

Rule 231 applies to supporting structures. The most common supporting structure is a pole but since other structures exist (e.g., lattice towers) the **NESC** uses the term supporting structure, not pole. Rule 231 does not apply to conductors, only to the supporting structure. A simple title for this section could be "Where can I set my pole?"

Included with the supporting structure are the support arms, anchor guys, equipment, and braces. The clearance requirements of this section are between the nearest parts of the objects concerned.

231A. From Fire Hydrants. Rule 231A is needed to provide the fire department crews adequate space to connect wrenches, hoses, and equipment to the fire hydrant. The rule for clearance of a supporting structure from a fire hydrant is outlined in Fig. 231-1.

231B. From Streets, Roads, and Highways. Rule 231B provides clearances between supporting structures (poles) and roads. Avoiding vehicle contact with poles can be just as important as maintaining clearance to energized conductors. Although not referenced, Rules 217A and 217C relate to Rule 231B. The note in Rules 217A1a and 217C2 regarding out-of-control vehicles helps distinguish between an ordinary vehicle using the road and an out-of-control vehicle that cannot be planned for.

Rule 231B1 applies to roads with curbs. No specific dimensions are specified for the distance that a supporting structure must be behind the curb. The rules for clearance of supporting structures from roads that have curbs are outlined in Fig. 231-2.

Rule 231B2 applies to roads without curbs. No specific dimensions are specified, just the requirement to locate the pole "a sufficient distance from the roadway to avoid contact by ordinary vehicles using and located on the traveled way." The **NESC** has definitions in Sec. 02 for the terms shoulder, roadway, and traveled way to clarify the requirements in this rule. The rules for clearance of supporting structures from roads that do not have a curb are outlined in Fig. 231-3.

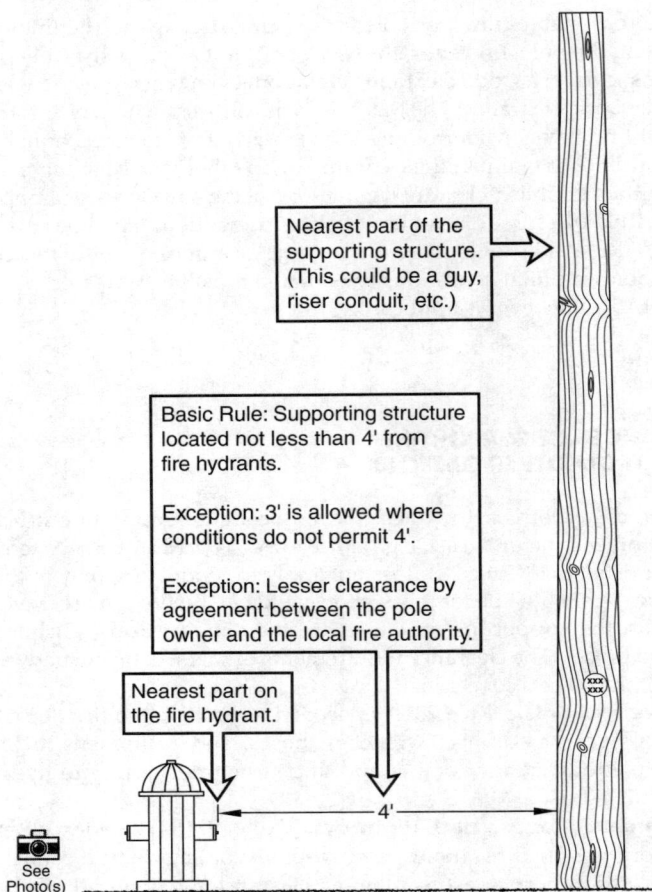

Fig. 231-1. Clearance of a supporting structure from a fire hydrant (Rule 231A).

Rule 231B3 recognizes various utilities compete for narrow rights-of-way along roads, streets, and highways. Special cases may exist and must be resolved using accepted good practice for the conditions at hand. Protection of supporting structures in parking lots, alleys, and next to driveways is addressed in Rule 217A. Protection and marking of guys is addressed in Rule 217C.

Rule 231B4 recognizes that in many cases a state highway department, county road department, city, or other governmental authority may require a permit to place poles in a right of way. The permit may have special requirements for pole placement.

231C. From Railroad Tracks. Rule 231C applies to supporting structures (poles) on lines paralleling or crossing railroad tracks. Four exceptions to Rules 231C1 and 231C2 permit reductions to the basic clearance provided in Rule 231C1.

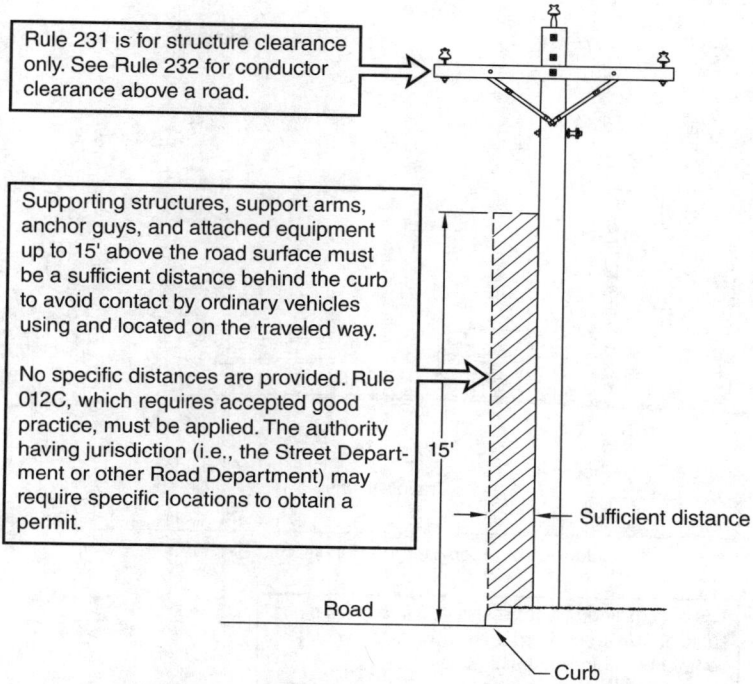

Rule 231 is for structure clearance only. See Rule 232 for conductor clearance above a road.

Supporting structures, support arms, anchor guys, and attached equipment up to 15' above the road surface must be a sufficient distance behind the curb to avoid contact by ordinary vehicles using and located on the traveled way.

No specific distances are provided. Rule 012C, which requires accepted good practice, must be applied. The authority having jurisdiction (i.e., the Street Department or other Road Department) may require specific locations to obtain a permit.

15'

Sufficient distance

Road

Curb

See Photo(s)

Fig. 231-2. Clearance of a supporting structure from a curb (Rule 231B1).

Many times when applying for a permit to parallel or cross a railroad right-of-way, the railroad company will require greater clearances than listed in this rule. The Code makes a reference to Rule 234I, which covers clearance of wires, conductors, and cables to rail cars. Rule 232, which applies to conductor clearance over railroad tracks, should also be checked. Rule 231C only applies to structures, not conductors. The rules for clearance of supporting structures from railroad tracks are outlined in Fig. 231-4.

232. VERTICAL CLEARANCES OF WIRES, CONDUCTORS, CABLES, AND EQUIPMENT ABOVE GROUND, ROADWAY, RAIL, OR WATER SURFACES

Rule 232 is probably referenced more than any other rule in the NESC. NESC Table 232-1 is probably referenced more than any other table. One problem that can occur is referencing NESC Table 232-1 without reading and understanding all the related rules that apply to the table. For example, Rules 230A through 230I define general clearance requirements and types of conductors used in NESC Table 232-1. Rule 232A describes the conductor temperature and loading (sag) conditions that

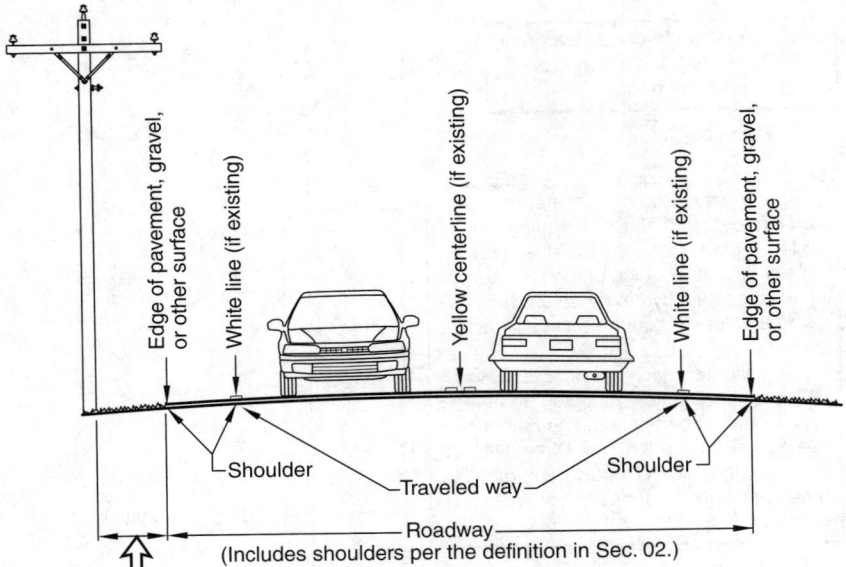

The supporting structure should be located a
sufficient distance from the roadway to avoid
contact by ordinary vehicles using and
located on the traveled way.

No specific distances are provided. Rule 012C,
which requires accepted good practice, must be
applied. The authority having jurisdiction (i.e., the
Highway Department or other Road Department)
may require specific locations to obtain a permit.

See
Photo(s)

Fig. 231-3. Clearance of a supporting structure from a roadway without a curb (Rule 231B2).

apply to **NESC** Table 232-1. Just as important as the related rules is the use and understanding of a sag and tension chart. The basics of how to use a sag and tension chart and a sample sag and tension chart are provided at the beginning of Sec. 23 of this Handbook.

232A. Application.   Rule 232A states the temperature and loading conditions that apply to the clearances in **NESC** Table 232-1. Three conditions are specified, and the condition that produces the largest final sag must be used. The conditions are:

- 120°F, no wind displacement, final sag.
- Greater than 120°F, if so designed, no wind displacement, final sag.
- 32°F, no wind displacement, final sag, with ice from the Zone specified in Rule 230B (e.g., Zone 2 requires 0.25 in of ice).

The conditions above are outlined on the sample sag and tension chart at the beginning of Sec. 23.

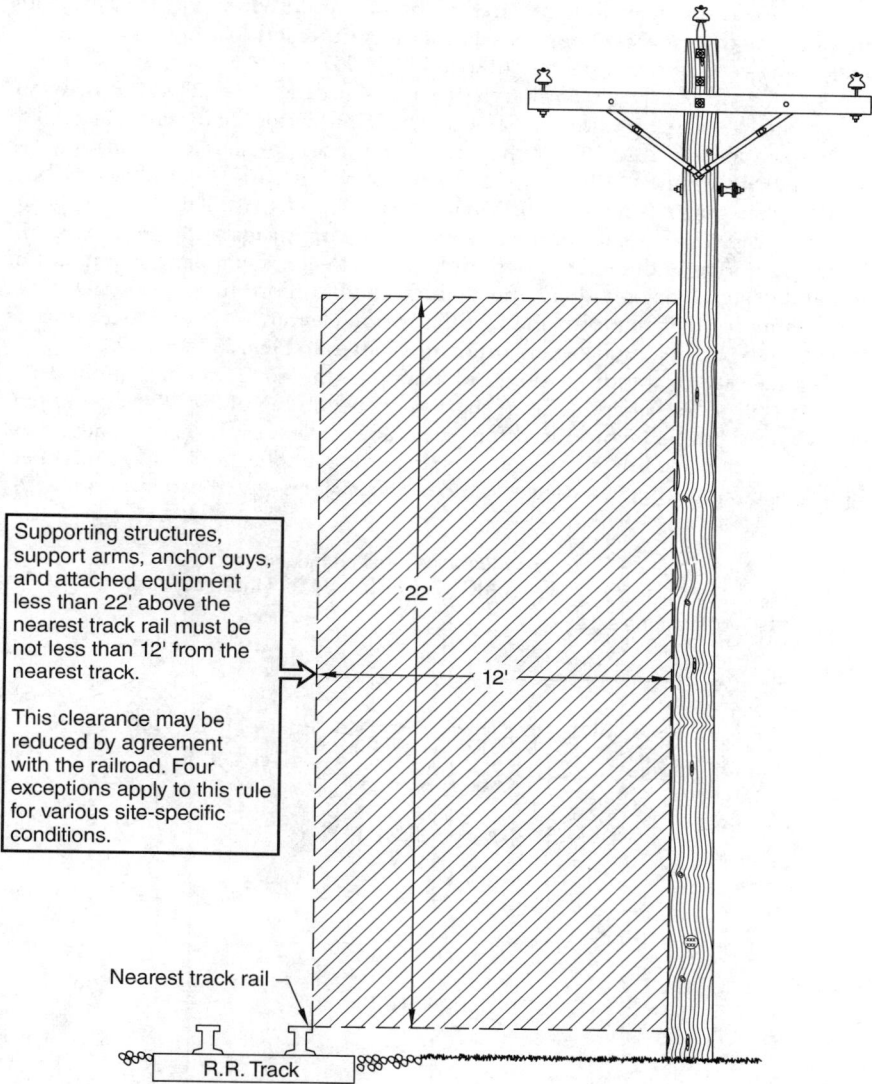

Supporting structures, support arms, anchor guys, and attached equipment less than 22' above the nearest track rail must be not less than 12' from the nearest track.

This clearance may be reduced by agreement with the railroad. Four exceptions apply to this rule for various site-specific conditions.

22'

12'

Nearest track rail

R.R. Track

See Photo(s)

Fig. 231-4. Clearance of a supporting structure from railroad tracks (Rule 231C).

The conductor conditions specified in Rule 232A are used to check clearance, not conductor tension. Conductor tension limits for open supply conductors are provided in Rule 261H.

The temperatures listed in Rule 232A are the conductor temperatures, not the ambient air temperature. See the beginning of Sec. 23 for additional information on conductor versus air temperature. The term loading in Rule 232A refers to the physical loads on the conductors in the form of ice and wind, not electrical loads in kilowatts or amperes. The conductor temperature and loading conditions for measuring vertical clearance are outlined in Fig. 232-1.

Using the sample sag and tension chart at the beginning of Sec. 23, 212°F, no wind, no ice, final tension, produces the largest sag (6.84 ft) of the conditions that need to be checked in Rule 232A. If the line was not designed to operate at temperatures above 120°F, then the 120°F, no wind, no ice, final value (5.61 ft) would be used as it is larger than the 32°F, no wind, 0.25 in ice, final value (4.49 ft). This example is not true for every case. Some conductors will have larger sags with ice, others will have larger sags with high temperatures. The wire size, span length, amount of ice, and design tension will all affect which condition produces the largest final sag. Examples that reinforce the fact that **Code** clearance for overhead lines is based on the largest final sag condition are shown in Figs. 232-2 and 232-3.

The majority of the discussion and examples in Sec. 23 revolve around the sample 1/0 ACSR sag and tension chart at the beginning of Sec. 23. Sag and tension charts can also be produced for secondary duplex, triplex, and quadruplex, communication cables on messengers (including phone, cable TV, and fiber-optic), self-supporting all-dielectric fiber-optic cables, overhead ground wires

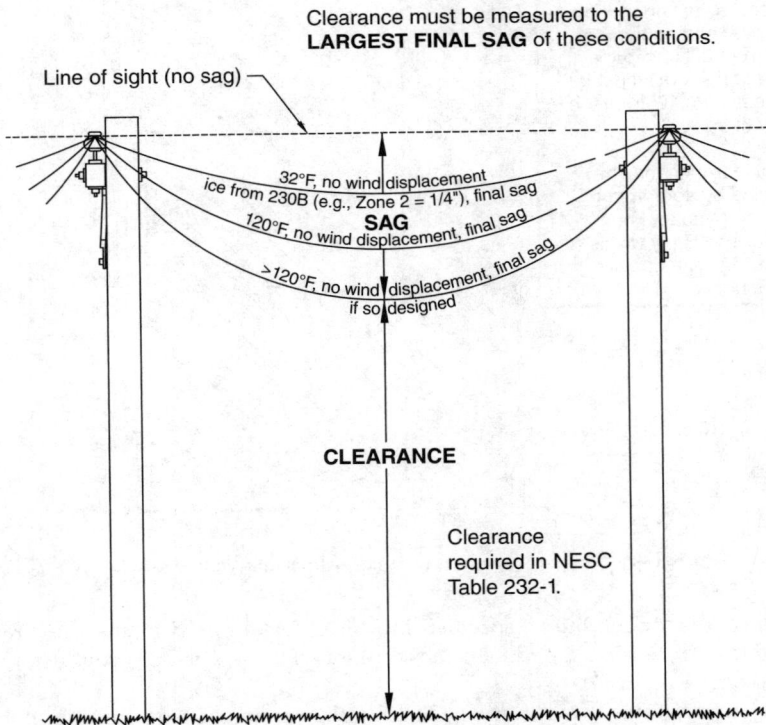

Fig. 232-1. Conductor temperature and loading conditions for measuring vertical clearance (Rule 232A).

Example:
- 300' span of 1/0 ACSR conductor.
- Assume the calculated ruling span is 300'.
- The clearance measurement is taken on a 50°F day; the conductor has a light electrical load all year long.
- Assume the conductor temperature is 60°F.
- Assume the maximum temperature the line is designed to operate at is 120°F.
- The midspan clearance measurement to a phase wire of 12.47/7.2 kV, 3Ø 4W effectively grounded circuit is 20'.
- Assume the line was strung using the sag and tension chart at the beginning of Sec. 23.
- Assume the line is at the final sag condition.

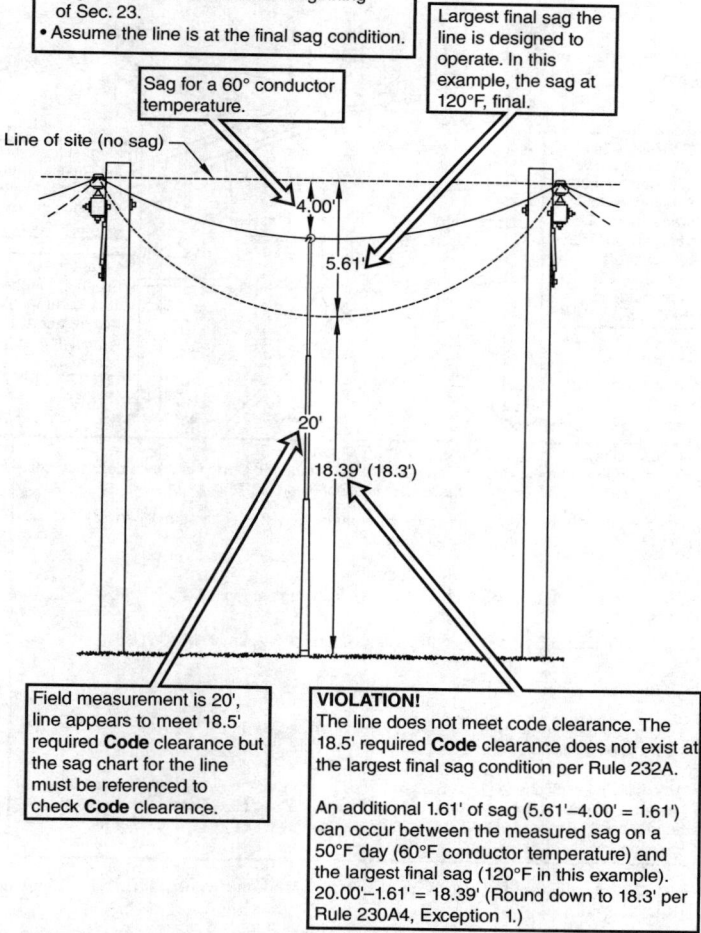

Largest final sag the line is designed to operate. In this example, the sag at 120°F, final.

Sag for a 60° conductor temperature.

Line of site (no sag)

4.00'

5.61'

20'

18.39' (18.3')

Field measurement is 20', line appears to meet 18.5' required **Code** clearance but the sag chart for the line must be referenced to check **Code** clearance.

VIOLATION!
The line does not meet code clearance. The 18.5' required **Code** clearance does not exist at the largest final sag condition per Rule 232A.

An additional 1.61' of sag (5.61'–4.00' = 1.61') can occur between the measured sag on a 50°F day (60°F conductor temperature) and the largest final sag (120°F in this example). 20.00'–1.61' = 18.39' (Round down to 18.3' per Rule 230A4, Exception 1.)

Fig. 232-2. Example of overhead clearance based on the largest final sag condition (Rule 232A).

Example:
• 400' span of 1/0 ACSR conductor.
• The 400' span is one span in a series of five spans from deadend to deadend.
• Assume the calculated ruling span is 300'.
• The clearance measurement is taken on a 50°F day; the conductor has a light electrical load all year long.
• Assume conductor temperature is 60°F.
• Assume the maximum temperature the line is designed to operate at is 120°F.
• The quarter span clearance measurement to phase wire of a 12.47/7.2 kV, 3Ø 4W effectively grounded circuit is 20'.
• The quarter span rise is 5'.
• Assume the line was strung using the sag and tension chart at the beginning of Sec. 23.
• Assume the line is at the final sag conditions.

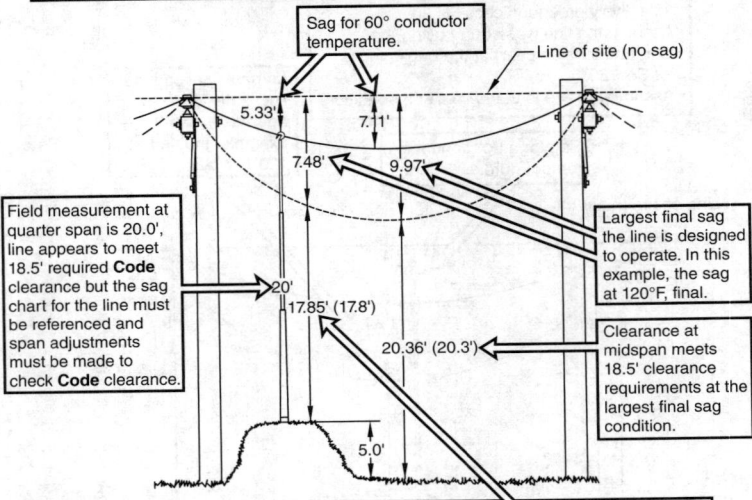

Sag for 60° conductor temperature.

Line of site (no sag)

5.33'   7.11'

7.48'   9.97'

Field measurement at quarter span is 20.0', line appears to meet 18.5' required **Code** clearance but the sag chart for the line must be referenced and span adjustments must be made to check **Code** clearance.

20'

17.85' (17.8')

20.36' (20.3')

5.0'

Largest final sag the line is designed to operate. In this example, the sag at 120°F, final.

Clearance at midspan meets 18.5' clearance requirements at the largest final sag condition.

VIOLATION!
The line does not meet **Code** clearance. The 18.5' required **Code** clearance does not exist at the quarter span location at the largest final condition per Rule 232A.

The sag on a 50° day (60° conductor temperature) for a 400' span at midspan is:

$$\left(\frac{400}{300}\right)^2 \times 4' = 7.11'$$

The sag at the quarter span is 75% of the midspan sag per Fig. 23-3.
7.11' × 0.75 = 5.33'

The largest final sag (at 120°F) in this example for a 400' span at midspan is:

$$\left(\frac{400}{300}\right)^2 \times 5.61' = 9.97'$$

The sag at the quarter span is 75% of the midspan sag per Fig. 23-3.
9.97 × 0.75 = 7.48'

An additional 2.15' of sag (7.48'–5.33') can occur between the measured sag on a 50°F day (60°F conductor temperature) and the largest final sag (120°F in this example) at the quarter span location. 20.00'–2.15' = 17.85' (Round down to 17.8' per Rule 230A4, Exception 1.)

Fig. 232-3. Example of overhead clearance based on the largest final sag condition (Rule 232A).

(static wires), and virtually any other type of overhead power and communication conductor or cable.

The exception to Rule 232A recognizes that electric railroad and trolley car conductors experience different conditions than typical power and communication conductors. Finally, a note to Rule 232A reminds us that the phase and neutral conductor may not operate at the same temperatures. The neutral conductor may carry less current due to phase balance and neutral current cancellation or due to the fact that the earth and a communication messenger on a joint use pole are in parallel with a multigrounded neutral.

Prior to 1990, a 60°F sag was used to determine clearance plus additional clearance was required for long spans. In 1990, the **Code** clearance rules experienced major revisions and the method to determine **Code** clearance was changed from using the 60°F sag condition to using the maximum sag conditions in Rule 232A. This change partly came about due to the fact that computer programs became available to generate detailed sag and tension charts like the sample at the beginning of Sec. 23 of this Handbook.

232B. Clearance of Wires, Conductors, Cables, Equipment, and Support Arms Mounted on Supporting Structures.   Rule 232B references the most commonly used table in the **NESC**, Table 232-1. Table 232-1 cannot be used without first reading and understanding the temperature and loading conditions in Rule 232A and without understanding the application of a sag and tension chart.

NESC Table 232-1 covers vertical clearance of wires, conductors, and cables above ground, roadway, rail, and water surfaces. **NESC** Table 232-2 covers unguarded rigid live parts of equipment. The basic difference between these two tables is that Table 232-1 addresses items that vary due to sag and Table 232-2 addresses items that are fixed or rigid and do not vary with sag. Equipment cases fall under Table 232-2. Secondary drip loops are included in Table 232-1 even though they do not vary much with sag.

An example of how clearance values are determined in **NESC** Tables 232-1 and 232-2 is shown in Fig. 232-4.

One of the main differences between **NESC** Table 232-1 (conductors) and **NESC** Table 232-2 (rigid parts) is that the mechanical and electrical component of clearance typically is reduced by 0.5 ft for rigid parts as they are not subject to sag variations. This is evident when comparing similar clearance conditions in **NESC** Tables 232-1 and 232-2.

NESC Tables 232-1 and 232-2 cover multiple conditions under the line, but the tables cannot cover every specific condition. For conditions not covered, Rule 012C, which requires accepted good practice, must be applied. The NOTE (not footnote) at the bottom of **NESC** Table 232-1 and **NESC** Table 232-2 explains how the clearances on the table were determined. Footnote 26 of **NESC** Table 232-1 and Footnote 3 of **NESC** Table 232-2 explains how to increase **Code** clearance for oversized vehicles. For example, if the truck in Fig. 232-4 was a 22-ft-high mining truck on an industrial site, the engineer or designer of the line is required to use the 4.5-ft (conductor) and 4.0-ft (rigid part) mechanical and electrical component of clearance for 12.47/7.2 kV (Rules 232C and 232D discuss when additional clearance is required for higher voltages). A 22-ft height must be substituted for the 14-ft reference component. See Fig. 232-5.

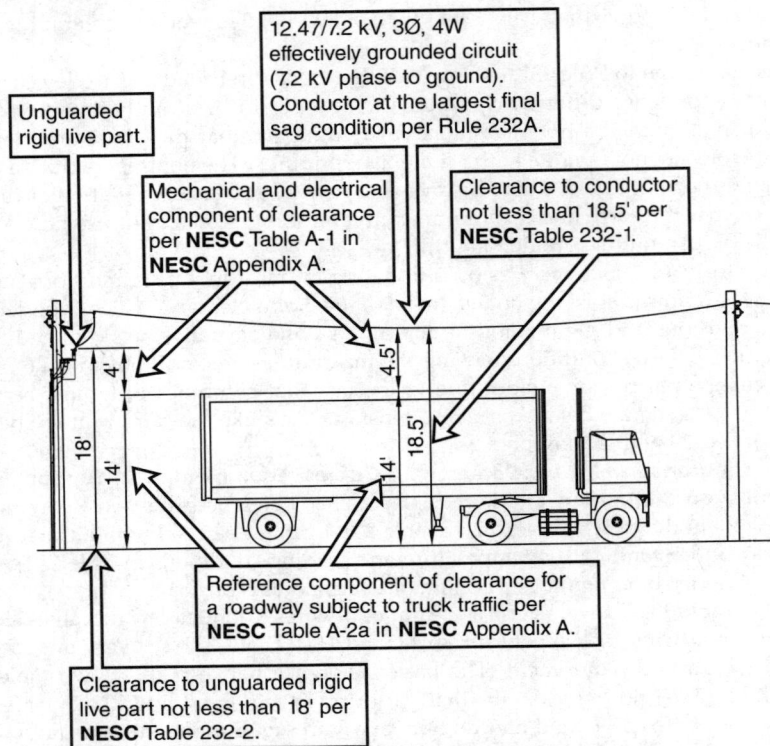

Fig. 232-4. Example of how clearance values are determined (Rule 232B).

Footnote 26 of **NESC** Table 232-1 and Footnote 3 of **NESC** Table 232-2 are also applicable to oversized farming vehicles. Footnote 26 of **NESC** Table 232-1 and Footnote 3 of **NESC** Table 232-2 can also be used to calculate the clearance for house moves or other oversized vehicles that exceed 14 ft in height but are traveling on roadways or highways that normally have trucks not exceeding 14 ft. The **NESC** does not address who pays the construction costs for increasing power line clearance due to oversized vehicles. OSHA Standard 1910.333 has a 4 ft clearance requirement for vehicular and mechanical equipment in transit with its structure lowered. This OSHA requirement is in addition to the **NESC** requirements. Mobile cranes used in the vicinity of power lines are addressed in the OSHA regulations in OSHA 1926.1408 which is part of OSHA 1926, Subpart CC.

The **NESC** clearance tables provide a wealth of information. They must be carefully reviewed before selecting the proper value from the table. Many of the clearances in **NESC** Table 232-1 are the same except that different footnotes apply to each. An explanation of how **NESC** Table 232-1 is formatted is shown in Fig. 232-6.

All of the clearances in **NESC** Table 232-1 are "not less than" values. See Rule 010 for a discussion of using **Code** clearance values plus an adder. Using the **Code** clearance plus an adder can help meet and maintain clearance over the life of an

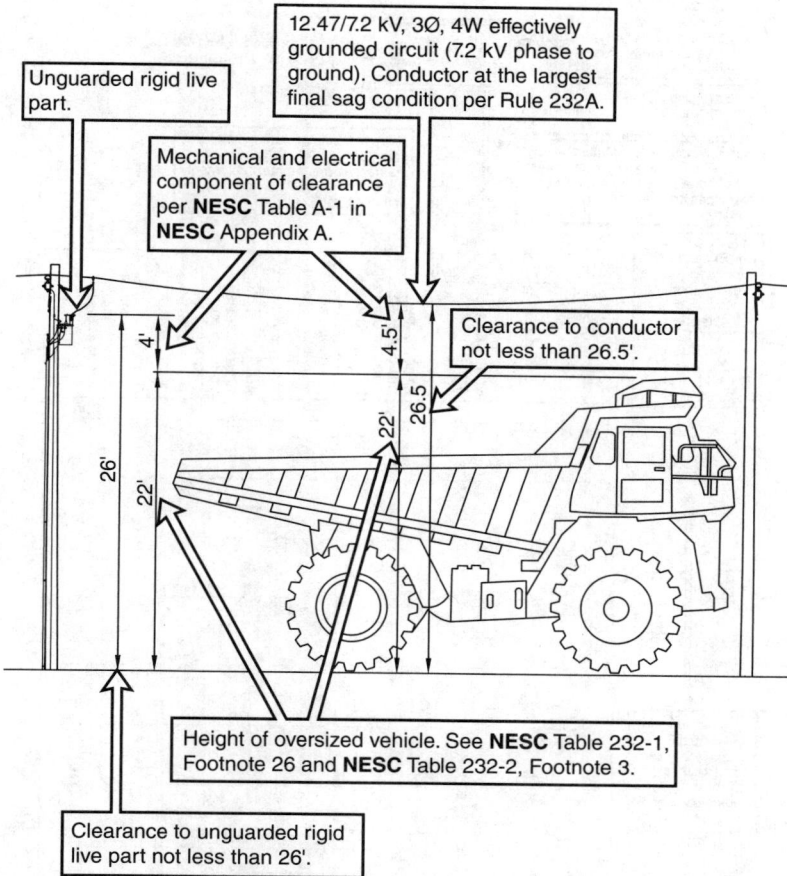

Fig. 232-5. Example of how to calculate clearance for an oversized vehicle (Rule 232).

installation as required in Rule 230I. If the authority issuing an occupancy permit (e.g., a railroad company, highway department, etc.) requires greater clearance than the **NESC**, the greater clearance required in the occupancy permit will take precedence over the **NESC** clearance values. The phase conductor, neutral conductor, secondary cable (0 to 750 V), and insulated communication cable clearances for the 10 land use categories in **NESC** Table 232-1 are shown in Figs. 232-7 through 232-16.

NESC Table 232-1 also addresses trolley and electrified railroad contact conductors and associated span or messenger wires. See Rule 225 for a discussion and a figure related to electric railway construction and trolley lines.

NESC Table 232-2 covers vertical clearance to equipment cases and unguarded rigid live parts above ground, roadway, and water surfaces. The land use categories are similar to **NESC** Table 232-1. The clearances in **NESC** Table 232-2 are

Table number corresponds with **NESC** rule number. The table cannot be used without reading the associated rules.

Table title gives a basic indication of what clearances are covered in the table.

Information in the parentheses discusses if table voltages are phase to ground or phase to phase and references the rule applicable to the table.

Specific **Code** wording and **Code** references are used for each conductor category.

Specific **Code** wording is used for the surface under the line.

These four rows have the same basic clearance values except different footnotes apply to each.

TABLE 232-1

Vertical Clearances of Wires, Conductors, and Cables Above Ground, Roadway, Rail, or Water Surfaces

(Voltages are ... See Rule...)

Neutral, insulated communications, secondary triplex, and phase conductors are the most commonly referenced items. They are located in these three columns.

	NEUTRAL (230E1) AND INSULATED COMMUNICATIONS	SECONDARY DUPLEX, TRIPLEX, AND QUADRUPLEX 0–750 V (230C3)	This column has miscellaneous items including open wire secondary.	PHASE CONDUCTORS UP TO 22 KV TO GROUND (e.g., 34.5/19.9 kV).	These columns have trolley and electric railroad values.
1. Track rails...					
2. Roads, streets,...	15.5'	16.0'		18.5'	
3. Driveways, parking...	15.5'	16.0'		18.5'	
4. Other areas...	15.5'	16.0'		18.5'	
5. Spaces and...					
6. Water areas...					
7. Water areas...					
a. <20 acres					
b. >20 to 200					
c. >200 to 2000					
d. >2000 acres					
8. Established boat ramps...					
Where wires,...					
9. Roads, streets,...	15.5'	16.0'		18.5'	
10. Roads where it is unlikely...					

1. Footnote ◄ Table footnotes to provide additional information (they carry the full force of the **Code**). See Rule 015.

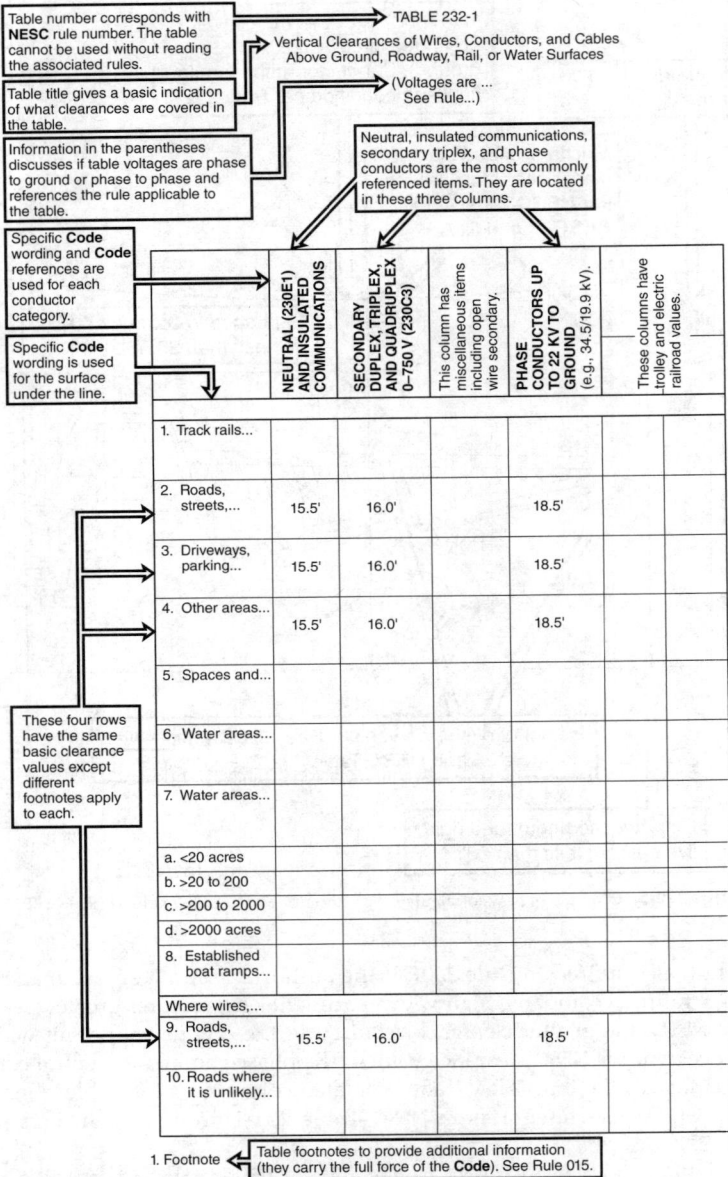

Fig. 232-6. Explanation of how **NESC** Table 231-1 is formatted (Rule 232).

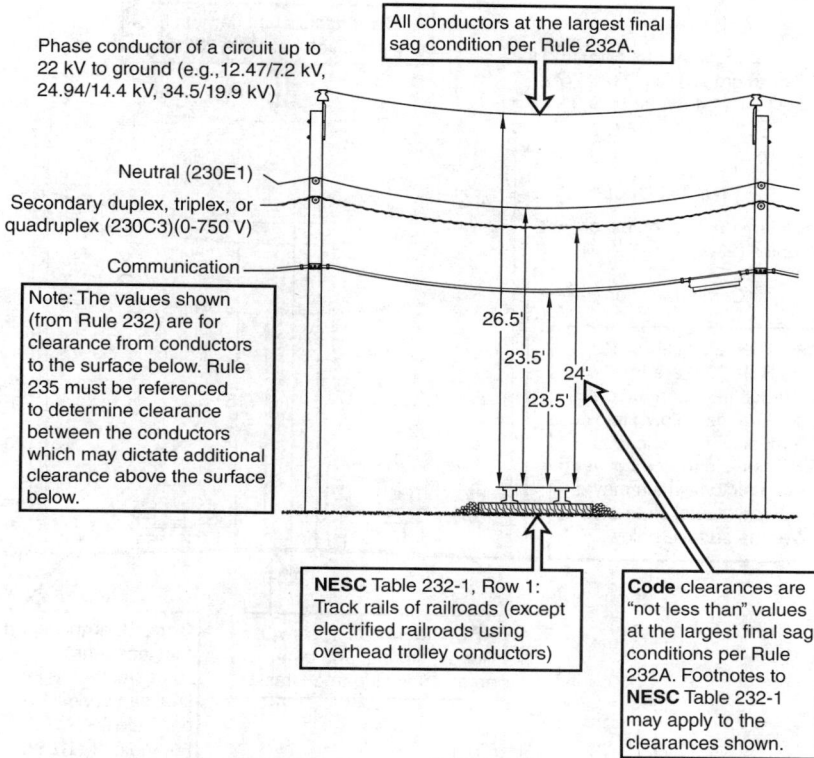

Phase conductor of a circuit up to 22 kV to ground (e.g.,12.47/7.2 kV, 24.94/14.4 kV, 34.5/19.9 kV)

All conductors at the largest final sag condition per Rule 232A.

Neutral (230E1)

Secondary duplex, triplex, or quadruplex (230C3)(0-750 V)

Communication

Note: The values shown (from Rule 232) are for clearance from conductors to the surface below. Rule 235 must be referenced to determine clearance between the conductors which may dictate additional clearance above the surface below.

26.5'

23.5'

24'

23.5'

NESC Table 232-1, Row 1: Track rails of railroads (except electrified railroads using overhead trolley conductors)

Code clearances are "not less than" values at the largest final sag conditions per Rule 232A. Footnotes to NESC Table 232-1 may apply to the clearances shown.

See Photo(s)

Fig. 232-7. Common clearance values from NESC Table 232-1 (Rule 232B1).

commonly $1/2$ ft less than the corresponding clearance in NESC Table 232-1 as the rigid parts are not subject to variations in sag. Rules 232B2 and 232B3 both reference NESC Table 232-2. An example of clearance to equipment cases and ungrounded rigid live parts is shown in Fig. 232-17.

Rule 232B4 includes vertical clearance requirements for effectively grounded and ungrounded conductive parts of luminaires (light fixtures). Modern luminaires are effectively grounded and operate on a constant voltage system. An example of clearance to an effectively grounded street light is shown in Fig. 232-18.

Per the title of Rule 232 and the titles of NESC Tables 232-1 and 232-2, the clearances specified in Rule 232 are vertical clearances. Per the note at the beginning of Rule 232B, Rule 232 does not provide horizontal or diagonal clearances to land surfaces. Rule 234 provides horizontal clearances to buildings and other structures but not to a land surface. To determine the horizontal or diagonal clearance to a land surface, Rule 012C, which requires accepted good practice, must be applied. See Fig. 232-19.

The Code clearances in Rule 232 and clearance values elsewhere in the Code are "not less than" values. Clearance greater than the Code requires is certainly

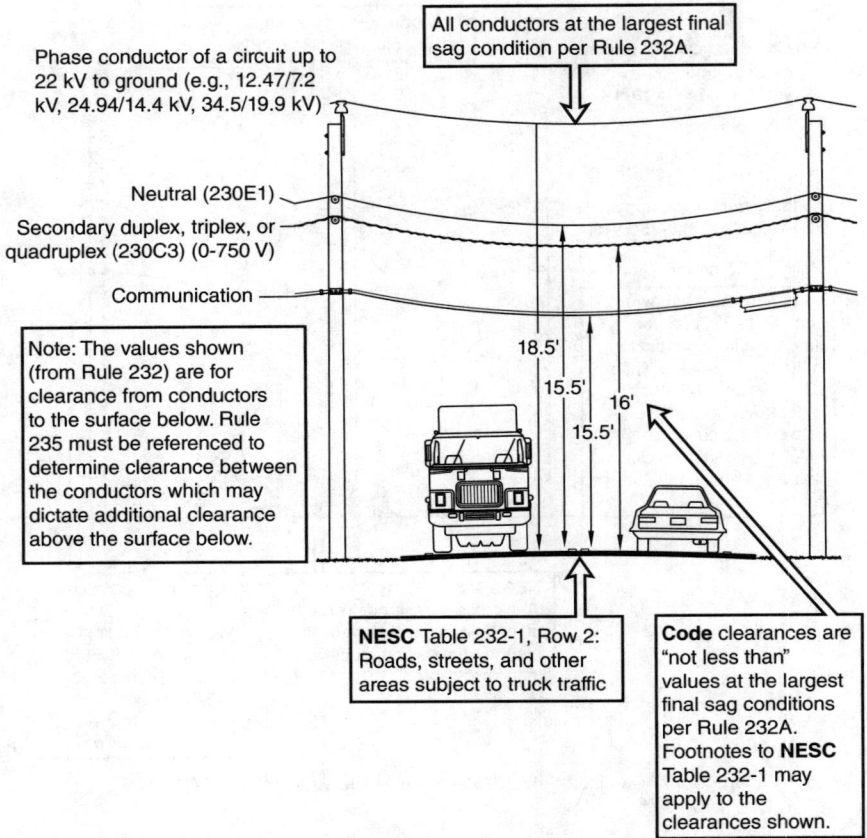

Fig. 232-8. Common clearance values from NESC Table 232-1 (Rule 232B1).

acceptable and may be needed to maintain the "not less than" clearance over the life of the line, but greater clearance is not required by the **NESC**. Certain agencies like the State Highway Department, railroad company, U.S. Army Corps of Engineers (water areas), or other agency issuing a crossing or occupancy permit may require greater clearance than the **NESC**. These agencies are considered the authority having jurisdiction. See Fig. 232-20.

The figures in this Handbook typically show the phase wire at the pole top position and the neutral wire at a lower position on the pole or on the same crossarm as the phase wire. It is not a **Code** violation to position the neutral wire at the top of the pole and the phase wire below the neutral on a crossarm or on the pole as long as the appropriate "not less than" clearances for each conductor in Rule 232 is met. **NESC** Table 235-5, Footnote 12, also applies to neutrals at the top position on a pole. This construction method is sometimes done in high

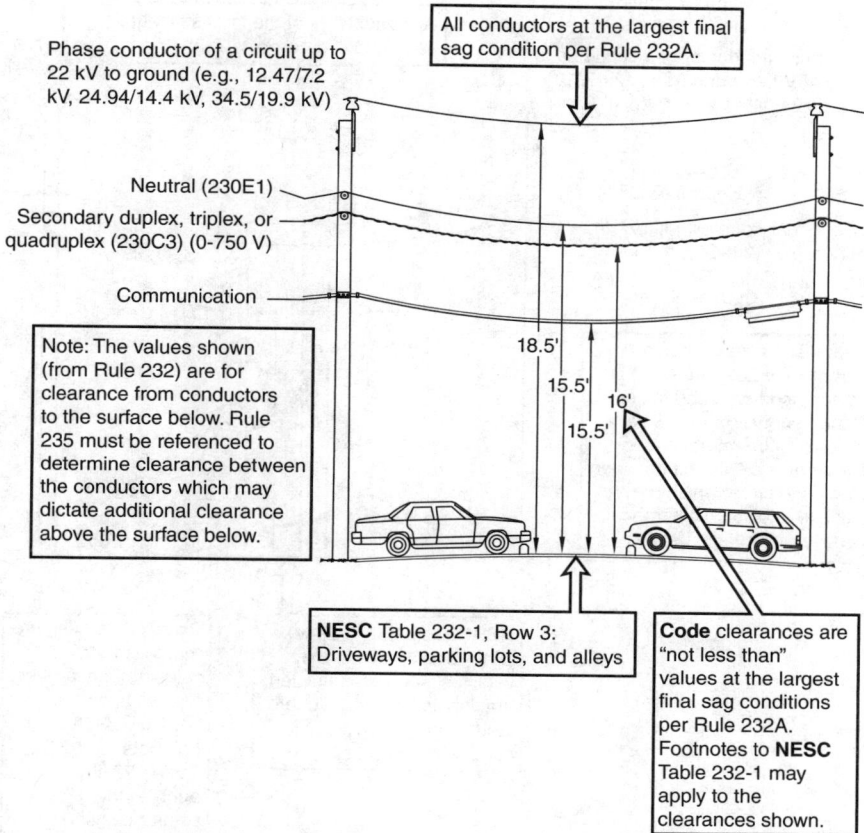

Phase conductor of a circuit up to 22 kV to ground (e.g., 12.47/7.2 kV, 24.94/14.4 kV, 34.5/19.9 kV)

All conductors at the largest final sag condition per Rule 232A.

Neutral (230E1)

Secondary duplex, triplex, or quadruplex (230C3) (0-750 V)

Communication

Note: The values shown (from Rule 232) are for clearance from conductors to the surface below. Rule 235 must be referenced to determine clearance between the conductors which may dictate additional clearance above the surface below.

18.5'
15.5'
16'
15.5'

NESC Table 232-1, Row 3: Driveways, parking lots, and alleys

Code clearances are "not less than" values at the largest final sag conditions per Rule 232A. Footnotes to NESC Table 232-1 may apply to the clearances shown.

Fig. 232-9. Common clearance values from NESC Table 232-1 (Rule 232B1).

lightning areas and is sometimes referred to as "neutral high" construction. Communication workers installing communication cables under the bottom phase wire need to be able to recognize the differences between a neutral wire and a phase wire. See Fig. 232-21.

There is one agency that gets involved if too much clearance exists above ground or near an airport. That agency is the Federal Aviation Administration (FAA). The NESC does not address placing marker balls on conductors or marking of tall structures. FAA Advisory Circular AC 70/7460-1K and AC 70/7460-2K address obstruction marking and lighting for structures including overhead power lines. See Fig. 232-22.

232C. Additional Clearances for Wires, Conductors, Cables, and Unguarded Rigid Live Parts of Equipment. The clearance values in Tables 232-1 and 232-2 must be increased for the following reasons:

- The voltage of the line exceeds 22 kV.

Phase conductor of a circuit up to
22 kV to ground (e.g., 12.47/7.2
kV, 24.94/14.4 kV, 34.5/19.9 kV)

All conductors at the largest final
sag condition per Rule 232A.

Neutral (230E1)

Secondary duplex, triplex, or
quadruplex (230C3) (0-750 V)

Communication

Note: The values shown
(from Rule 232) are for
clearance from conductors
to the surface below. Rule
235 must be referenced to
determine clearance between
the conductors which may
dictate additional clearance
above the surface below.

18.5'

15.5'

16'

15.5'

NESC Table 232-1, Row 4:
Other areas traversed by
vehicles, such as cultivated,
grazing, forest, and orchard
lands, industrial sites,
commercial sites, etc.

Code clearances are
"not less than"
values at the largest
final sag conditions
per Rule 232A.
Footnotes to **NESC**
Table 232-1 may
apply to the
clearances shown.

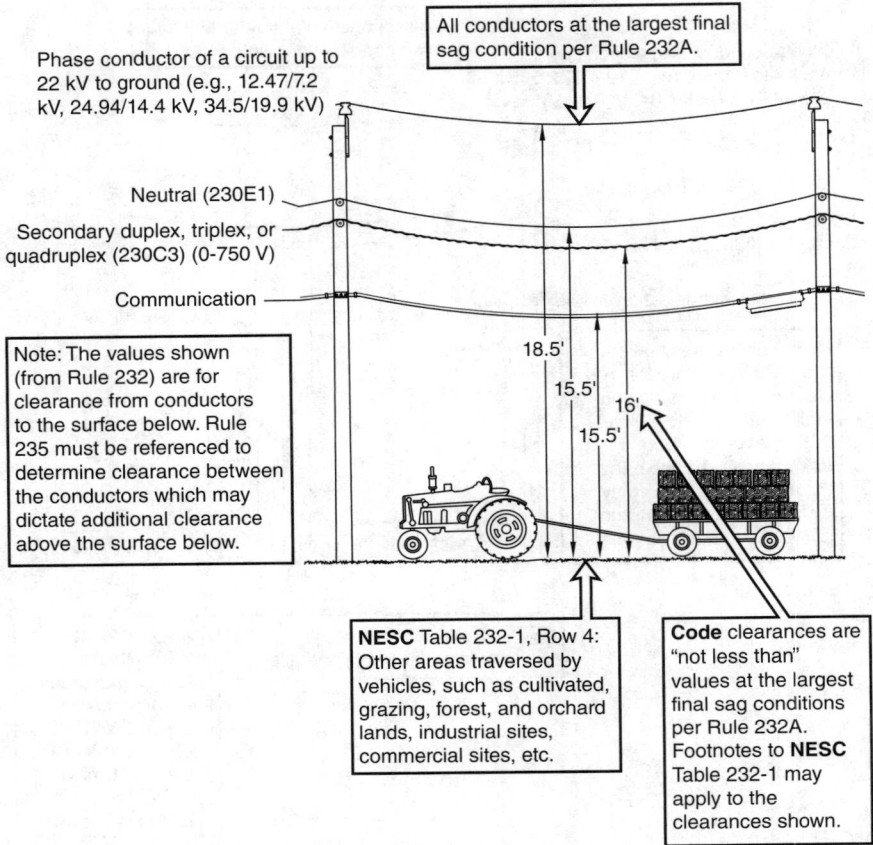

Fig. 232-10. Common clearance values from NESC Table 232-1 (Rule 232B1).

✓ 22 kV to ground for an effectively grounded circuit. The wording in parentheses under the titles of **NESC** Tables 232-1 and 232-2 permits effectively grounded circuits and those other circuits where all ground faults are cleared by promptly de-energizing the faulted section, both initially and following subsequent breaker operations.

✓ 22 kV phase-to-phase for an ungrounded (e.g., delta) circuit.

✓ The maximum operating voltage, typically 1.05 times the nominal operating voltage, must be used for lines over 50 kV. ANSI C84.1, American National Standard for Electric Power Systems and Equipment—Voltage Ratings (60 Hz), provides a table listing nominal voltage levels and the corresponding maximum voltage level.

✓ The clearance adder is 0.4 in per kilovolt in excess of 22 kV.

• The voltage of the line exceeds 50 kV and the line is above an elevation of 3300 ft.

Phase conductor of a circuit up to 22 kV to ground (e.g., 12.47/7.2 kV, 24.94/14.4 kV, 34.5/19.9 kV)

All conductors at the largest final sag condition per Rule 232A.

Neutral (230E1)

Secondary duplex, triplex, or quadruplex (230C3) (0-750 V)

Communication

Use extreme caution before applying pedestrian-only clearances. Areas accessible to horseback riders do not meet the pedestrian-only criteria.

14.5'

9.5'

12'

9.5'

Note: The values shown (from Rule 232) are for clearance from conductors to the surface below. Rule 235 must be referenced to determine clearance between the conductors which may dictate additional clearance above the surface below.

NESC Table 232-1, Row 5: Spaces and ways subject to pedestrians or restricted traffic only

Code clearances are "not less than" values at the largest final sag conditions per Rule 232A. Footnotes to NESC Table 232-1 may apply to the clearance shown.

Rules of the NESC and the National Electric Code (NEC) overlap at the service point. The authority having jurisdiction (e.g., a state or local electrical inspector) may require use of the NEC clearances at the service point.

12'

One of the most common pedestrian-only applications is a clearance of not less than 12' from the bottom of a secondary service (230C3 cable) drip loop. The drip loop fittings must be insulated per Rule 234C3a(1)

See Photo(s)

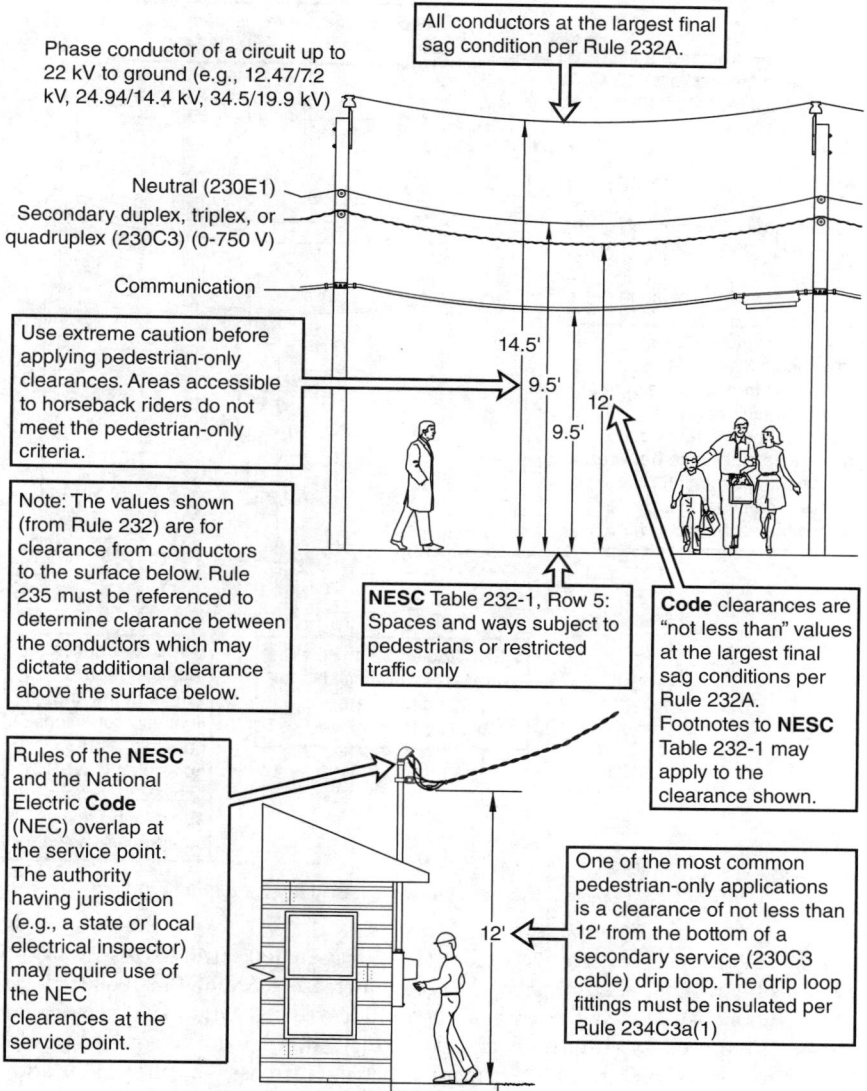

Fig. 232-11. Common clearance values from NESC Table 232-1 (Rule 232B1).

✓ 50 kV to ground for an effectively grounded circuit. The wording in parentheses under the titles of NESC Tables 232-1 and 232-2 permits effectively grounded circuits and those other circuits where all ground faults are cleared by promptly de-energizing the faulted section, both initially and following subsequent breaker operations.

✓ 50 kV phase-to-phase for an ungrounded (e.g., delta) circuit.

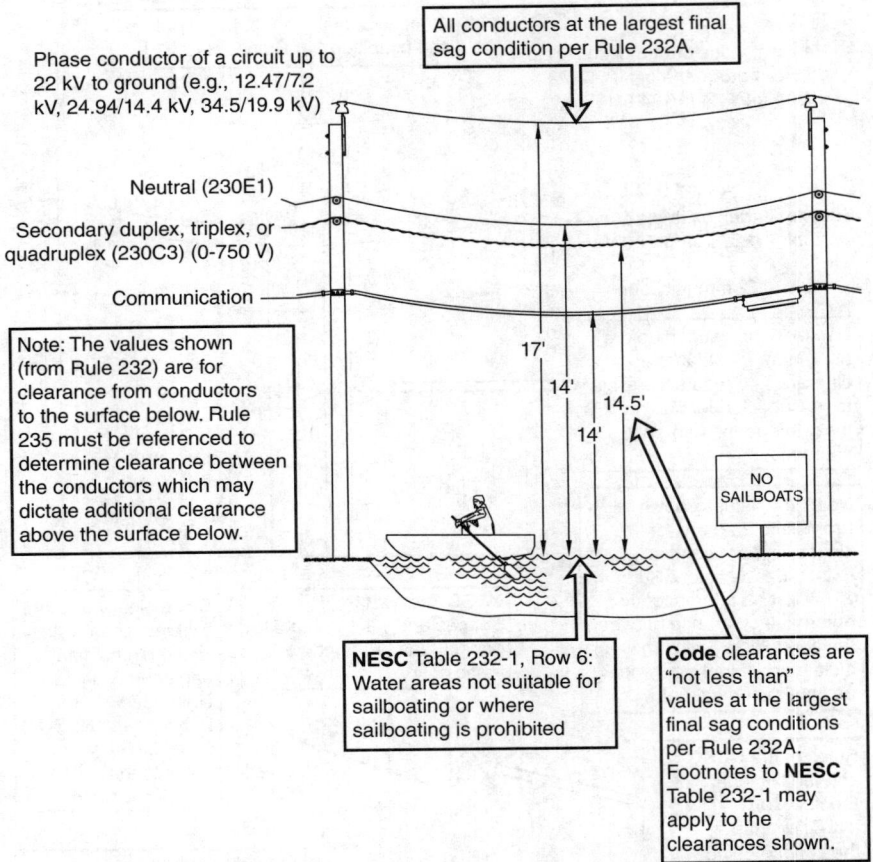

Fig. 232-12. Common clearance values from NESC Table 232-1 (Rule 232B1).

✓ A clearance adder of 3 percent (.03) for each 1000 ft in excess of 3300 ft of elevation. Assuming the 3 percent increase is applied in 1000 ft blocks (or steps), 3 percent is required for altitudes over 3300 ft to 4300 ft, 6 percent is required for altitudes over 4300 ft to 5300 ft, 9 percent is required for altitudes over 5300 ft to 6300 ft, etc. The percent increase is applied to the additional clearance required for lines above 22 kV, not to the total clearance.

✓ No increase in clearance is required for elevations above 3300 ft if the line is rated 50 kV or less.

• For voltages exceeding 98 kV AC to ground, additional clearance or other means are required to limit electrostatic field effects to 5 mA or less.

• For voltages over 470 kV, the clearance must be determined using the formulas in Rule 232D. Rule 232D can also be used as an alternate clearance method for lines over 98 kV AC to ground or 189 kV DC to ground.

• The "not less than" rounding requirements of Rule 230A4 apply.

Phase conductor of a circuit up to 22 kV to ground (e.g., 12.47/7.2 kV, 24.94/14.4 kV, 34.5/19.9 kV)

All conductors at the largest final sag condition per Rule 232A.

Neutral (230E1)

Secondary duplex, triplex, or quadruplex (230C3) (0-750 V)

Communication

Note: The values shown (from Rule 232) are for clearance from conductors to the surface below. Rule 235 must be referenced to determine clearance between the conductors which may dictate additional clearance above the surface below.

A B C D

Code clearances are "not less than" values at the largest final sag conditions per Rule 232A. Footnotes to NESC Table 232-1 may apply to the clearances shown.

NESC Table 232-1, Row 7:
Water areas suitable for sailboating including lakes, ponds, reservoirs, tidal waters, rivers, streams, and canals with an unobstructed surface area of:

	A	B	C	D
Less than 20 acres	20.5'	17.5'	17.5'	18.0'
Over 20 to 200 acres	28.5'	25.5'	25.5'	26.0'
Over 200 to 2000 acres	34.5'	31.5'	31.5'	32.0'
Over 2000 acres	40.5'	37.5'	37.5'	38.0'

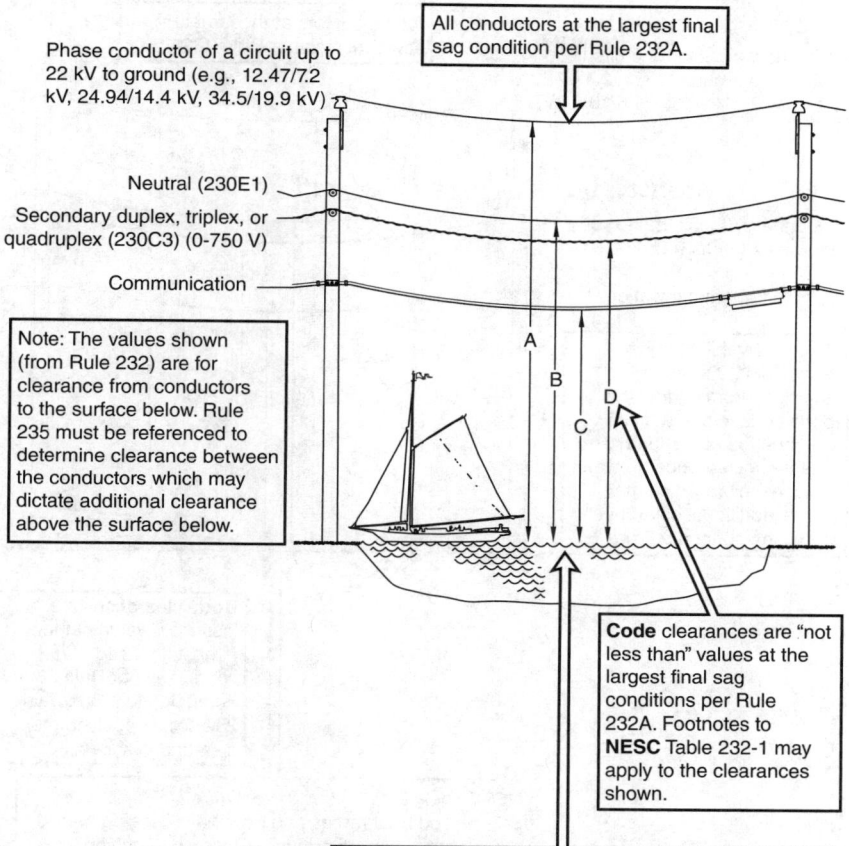

Fig. 232-13. Common clearance values from NESC Table 232-1 (Rule 232B1).

Phase conductor of a circuit up to
22 kV to ground (e.g., 12.47/7.2
kV, 24.94/14.4 kV, 34.5/19.9 kV)

All conductors at the largest final
sag condition per Rule 232A.

Neutral (230E1)

Secondary duplex, triplex, or
quadruplex (230C3) (0-750 V)

Communication

Note: The values shown
(from Rule 232) are for
clearance from conductors
to the surface below. Rule
235 must be referenced to
determine clearance between
the conductors which may
dictate additional clearance
above the surface below.

Code clearances are "not
less than" values at the
largest final sag
conditions per Rule 232A.
Footnotes to **NESC** Table
232-1 may apply to the
clearances shown.

NESC Table 232-1, Row 8:
Established boat ramps and associated rigging areas;
areas posted with sign(s) for rigging or launching
sailboats

	A	B	C	D
Less than 20 acres	25.5'	22.5'	22.5'	23.0'
Over 20 to 200 acres	33.5'	30.5'	30.5'	31.0'
Over 200 to 2000 acres	39.5'	36.5'	36.5'	37.0'
Over 2000 acres	45.5'	42.5'	42.5'	43.0'

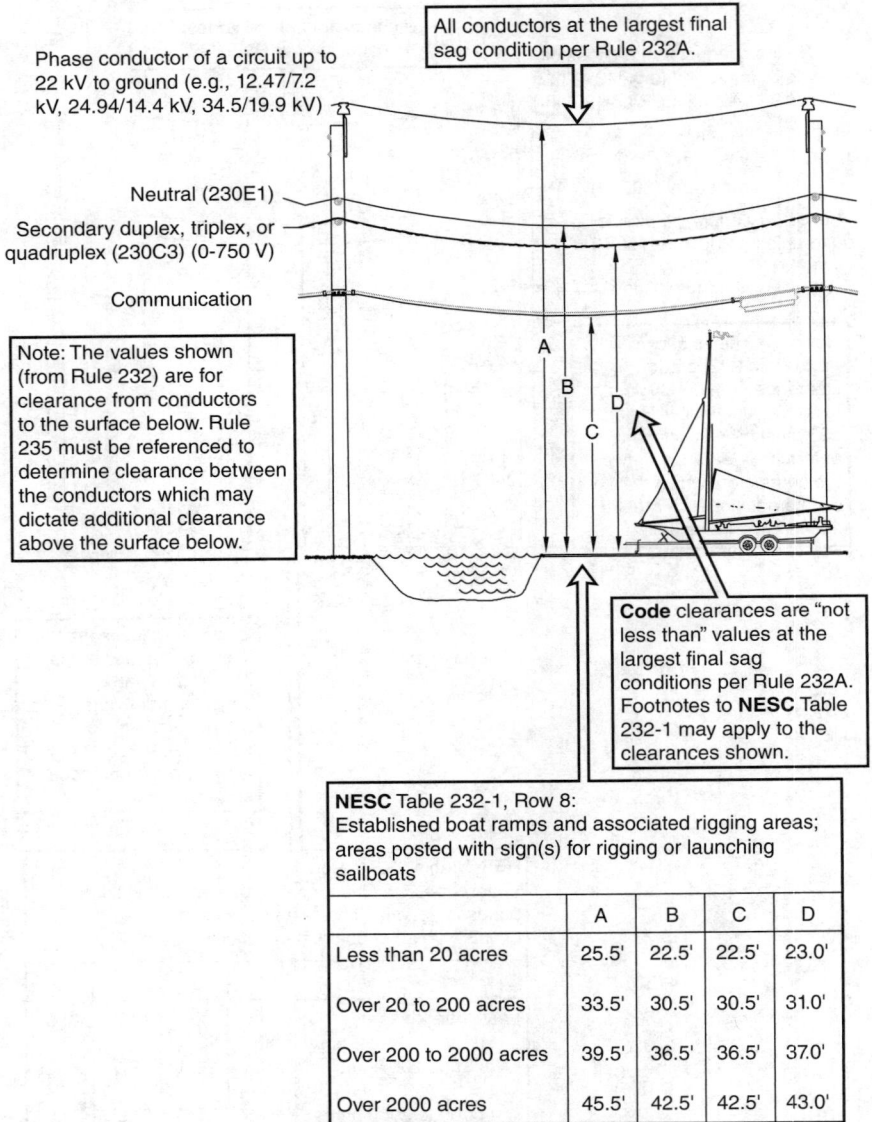

Fig. 232-14. Common clearance values from NESC Table 232-1 (Rule 232B1).

An example of an additional clearance calculation is shown in Fig. 232-23.

Rule 232C1c requires that lines exceeding 98 kV AC to ground have additional clearance or other means to reduce electrostatic field effects to 5 mA or less. High-voltage transmission lines can induce a voltage and therefore induce currents into metal objects like a truck or railroad car under the line. The average adult

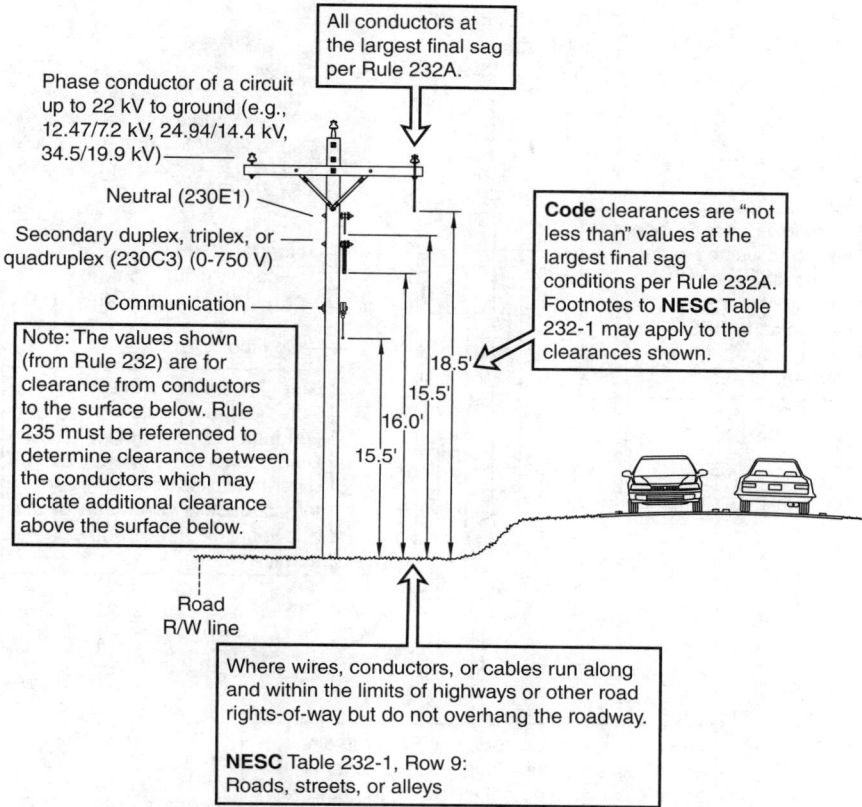

Fig. 232-15. Common clearance values from NESC Table 232-1 (Rule 232B1).

human body can detect an electrical shock at currents around 2 mA. Currents above 5 mA can cause pain and can be harmful to the body. Since no specific clearance adders or other methods are provided, Rule 012C, which requires accepted good practice, must be applied. Accepted good practice in this case may be an engineering analysis or field testing. The rules related to limiting electrostatic field effects to 5 mA are outlined in Fig. 232-24.

232D. Alternate Clearances for Voltages Exceeding 98 kV AC to Ground or 139 kV DC to Ground. The use of Rule 232D will permit reduced clearances for some voltages exceeding 98 kV to ground alternating current (AC) and 138 kV to ground direct current (DC). See Rule 230G for a discussion of AC and DC circuits. The best method for checking the application of Rule 232D is to use the formulas in Rule 232D and check the calculated results to NESC Table 232-4. If the calculated result does not match the examples provided in Table 232-4 for the same input conditions, a value was incorrectly entered or the formula was misapplied. Rule 232D also requires application of NESC Table 232-3, which lists reference heights similar to NESC Table A-2a in Appendix A of the NESC.

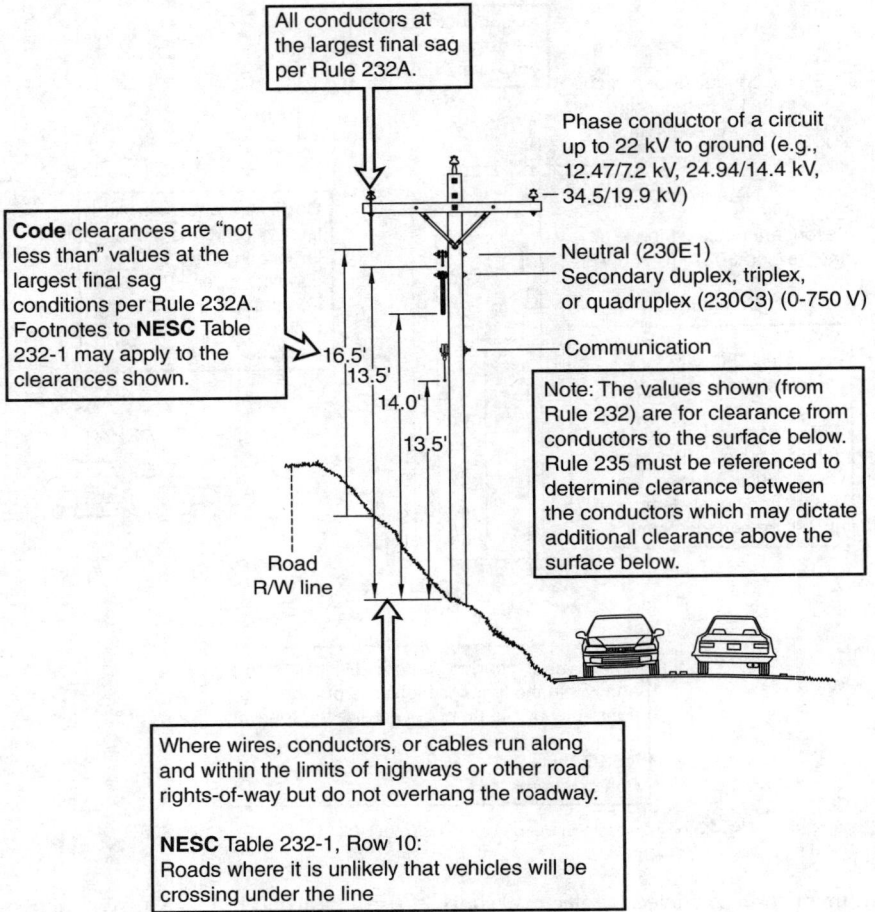

All conductors at the largest final sag per Rule 232A.

Phase conductor of a circuit up to 22 kV to ground (e.g., 12.47/7.2 kV, 24.94/14.4 kV, 34.5/19.9 kV)

Code clearances are "not less than" values at the largest final sag conditions per Rule 232A. Footnotes to **NESC** Table 232-1 may apply to the clearances shown.

Neutral (230E1)
Secondary duplex, triplex, or quadruplex (230C3) (0-750 V)

Communication

Note: The values shown (from Rule 232) are for clearance from conductors to the surface below. Rule 235 must be referenced to determine clearance between the conductors which may dictate additional clearance above the surface below.

16.5'
13.5'
14.0'
13.5'

Road R/W line

Where wires, conductors, or cables run along and within the limits of highways or other road rights-of-way but do not overhang the roadway.

NESC Table 232-1, Row 10:
Roads where it is unlikely that vehicles will be crossing under the line

Fig. 232-16. Common clearance values from **NESC** Table 232-1 (Rule 232B1).

A list of important factors to consider when using Rule 232D is outlined below:
- The alternate clearances are for lines exceeding 98 kV (AC) to ground and 139 kV (DC) to ground only.
- The alternate clearance method must be used for voltages over 470 kV.
- **NESC** Table 232-3 reference heights are to be used.
- Crest voltage (not rms voltage) must be used. See Rule 230G for a discussion.
- The maximum operating voltage, typically 1.05 times the nominal voltage, must be used. ANSI C84.1, American National Standard for Electric Power Systems and Equipment—Voltage Ratings (60 Hertz), provides a table listing nominal voltage levels and the corresponding maximum voltage level.
- Maximum switching surge factors must be known.

Clearance to a 7.2 kV phase to ground rigid live part is not less than 18'.

18'

15'

18'

15'

Clearance to an effectively grounded equipment case must be not less than 15'. This same clearance applies to platforms.

NESC Table 232-2, Row 1c: Other areas traversed by vehicles such as cultivated, grazing, forest, and orchard lands, industrial areas, commercial areas, etc.

See Photo(s)

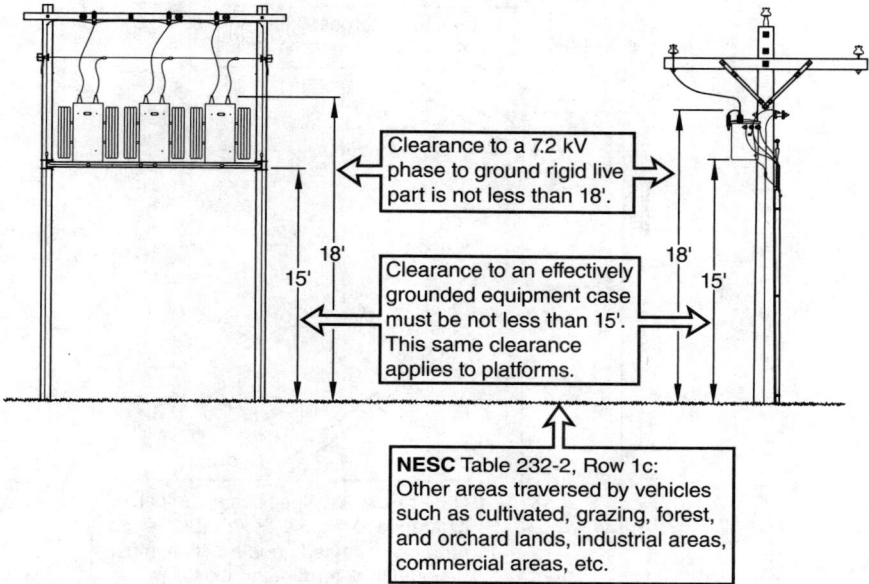

Fig. 232-17. Example of clearance to equipment cases and unguarded rigid live parts (Rules 232B2 and 232B3).

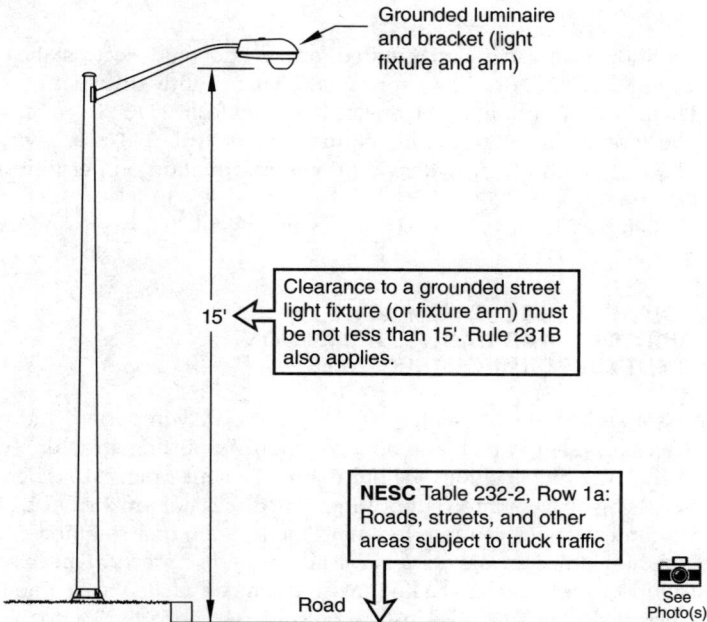

Grounded luminaire and bracket (light fixture and arm)

15'

Clearance to a grounded street light fixture (or fixture arm) must be not less than 15'. Rule 231B also applies.

NESC Table 232-2, Row 1a: Roads, streets, and other areas subject to truck traffic

See Photo(s)

Road

Fig. 232-18. Example of clearance to street lighting (Rule 232B4).

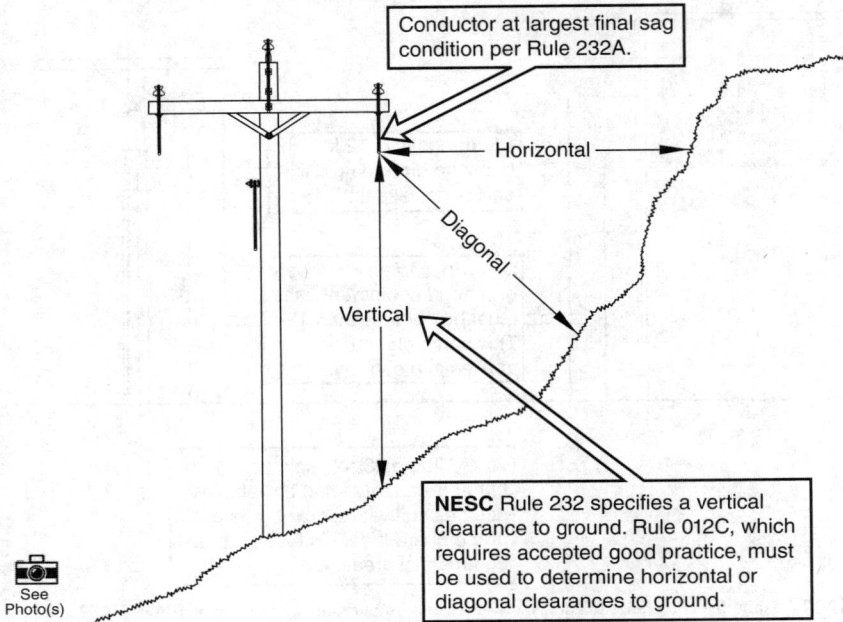

Conductor at largest final sag condition per Rule 232A.

Horizontal

Diagonal

Vertical

NESC Rule 232 specifies a vertical clearance to ground. Rule 012C, which requires accepted good practice, must be used to determine horizontal or diagonal clearances to ground.

See Photo(s)

Fig. 232-19. Vertical clearance to land surface (Rule 232).

- Additional clearances are required for elevation and electrostatic effects similar to Rule 232C. (Increases for elevation are slightly different.)
- There is a lower limit for clearance based on Rule 232C.
- The "not less than" rounding requirements of Rule 230A4 apply.
- **NESC** Appendix D provides additional information on per-unit overvoltage factors.
- **NESC** Table 232-4 can be used to verify that calculations were done correctly.

233. CLEARANCES BETWEEN WIRES, CONDUCTORS, AND CABLES CARRIED ON DIFFERENT SUPPORTING STRUCTURES

233A. General.   The first sentence of Rule 233 is very important and very powerful. "Crossings should be made on a common supporting structure, where practical." The wording includes "should" and "where practical." There are times when crossing on a common supporting structure is not practical. Turning lanes at roadway intersections can interfere with the location of a common crossing pole. Transmission lattice towers do not typically provide a practical means of attaching a distribution line crossing. If a line crossing can be attached to a common supporting structure, Rule 235 applies instead of Rule 233. See Rule 233B for a comparison of Rules 233B, 233C, 235B, and 235C. Rule 235, Clearance for Wires, Conductors, or

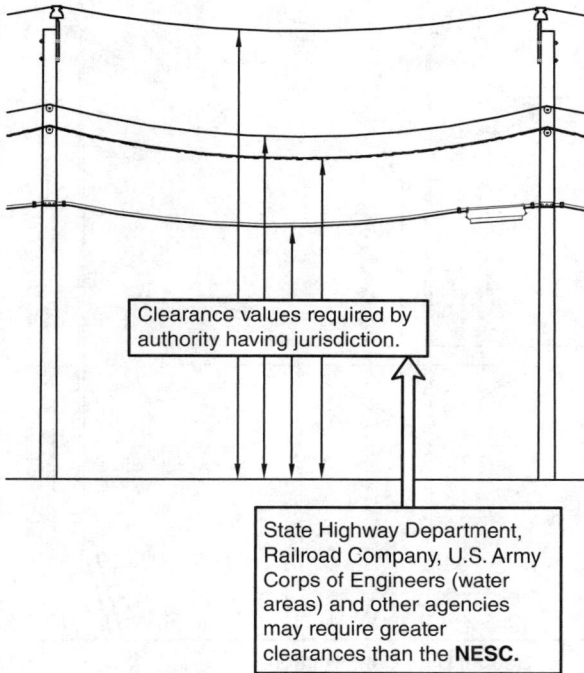

Clearance values required by
authority having jurisdiction.

State Highway Department,
Railroad Company, U.S. Army
Corps of Engineers (water
areas) and other agencies
may require greater
clearances than the **NESC.**

Fig. 232-20. Authority having jurisdiction (Rule 232).

Cables Carried on the Same Supporting Structure, is simpler to apply at a crossing because the conductors are tied onto the structure, not moving around in a span as they are in Rule 233. Examples of line crossings with and without a common supporting structure are shown in Fig. 233-1.

Two "envelopes" are addressed in this rule, the conductor movement envelope in Rule 233A1 that outlines an area of where the conductor can be positioned at various temperature and loading conditions; and the clearance envelope in Rule 233A2 that outlines the horizontal and vertical clearance area that must be maintained between the two conductors.

The conductor movement envelope in Rule 233A1a(3) has the same largest final sag conditions described in Rule 232A. See Rule 232A for a discussion. In addition to the largest final sag condition, Rule 233A1a(1) requires a 60°F, no wind displacement, initial and final sag condition and Rule 233A1a(2) requires a 60°F, 6-lb/ft² wind, initial and final sag condition. These values are outlined on the sample sag and tension chart at the beginning of Sec. 23.

The temperatures listed in Rule 233A are the conductor temperatures, not the ambient air temperature. See the beginning of Sec. 23 for additional information on conductor versus air temperature. The term loading in Rule 233A refers to the physical loads on the conductors in the form of ice and wind, not electrical loads in kilowatts or amperes. The conductor temperature and

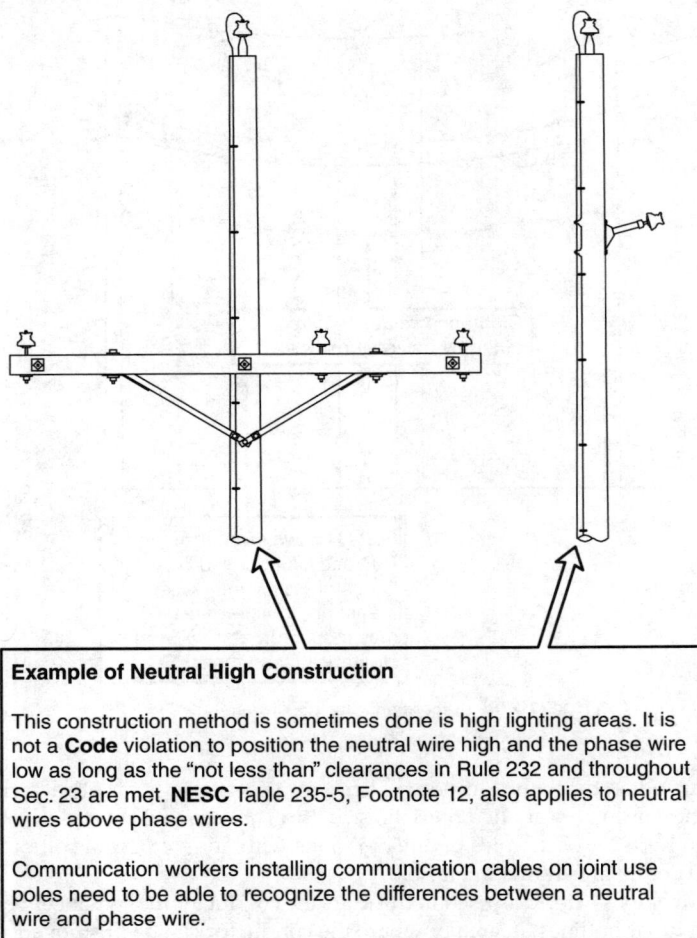

Example of Neutral High Construction

This construction method is sometimes done is high lighting areas. It is not a **Code** violation to position the neutral wire high and the phase wire low as long as the "not less than" clearances in Rule 232 and throughout Sec. 23 are met. **NESC** Table 235-5, Footnote 12, also applies to neutral wires above phase wires.

Communication workers installing communication cables on joint use poles need to be able to recognize the differences between a neutral wire and phase wire.

Fig. 232-21. Examples of neutral high construction (Rule 232).

loading conditions for measuring the conductor movement envelope are outlined in Fig. 233-2.

Using the sample sag and tension chart at the beginning of Sec. 23, 212°F, no wind, no ice, final produces the largest sag (6.84 ft) of the conditions that need to be checked in this rule. If the line were not designed to operate at high temperatures, then the 120°F, no wind, no ice, final value (5.61 ft) would be used as it is larger than the 32°F, no wind, 0.25 in ice, final value (4.49 ft). The 60°F, no wind, no ice, initial condition produces 3.14 ft of sag. The 60°F, no wind, no ice, final condition produces 4.00 ft of sag. The 60°F, 6 lb/ft^2 wind, no ice, initial condition produces 4.03 ft of sag. The 60°F, 6 lb/ft^2 wind, no ice, final condition produces 4.66 ft of sag. The 60°F conditions with wind have more sag than without wind but

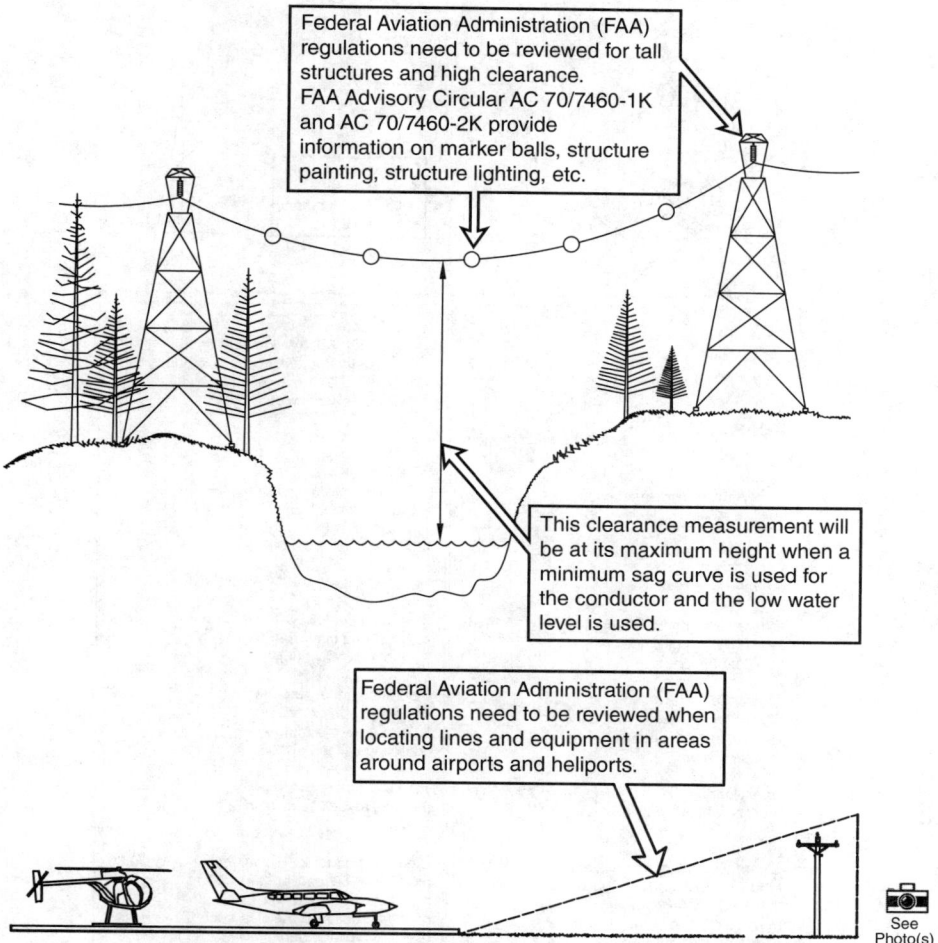

Federal Aviation Administration (FAA) regulations need to be reviewed for tall structures and high clearance. FAA Advisory Circular AC 70/7460-1K and AC 70/7460-2K provide information on marker balls, structure painting, structure lighting, etc.

This clearance measurement will be at its maximum height when a minimum sag curve is used for the conductor and the low water level is used.

Federal Aviation Administration (FAA) regulations need to be reviewed when locating lines and equipment in areas around airports and heliports.

See Photo(s)

Fig. 232-22. Federal Aviation Administration (FAA) requirements (Rule N/A).

the wind-blown conductor position is to the side of the span. The **NESC** does not provide the details of how to calculate the position of the blownout conductor at 60°F with a 6-lb/ft² wind. A transmission or distribution line design manual will typically include formulas to perform this calculation. See Rule 240 for a discussion of line design manuals. A 6-lb/ft² wind pressure on a cylindrical surface like a conductor, cable, or pole is equal to approximately a 50 mile per hour wind. The deflection of suspension insulators and flexible structures must be considered in the wind blowout calculation. See Fig. 233-3.

The footnotes to **NESC** Fig. 233-2 provide additional information on how to apply the conductor movement envelope. Footnote 1 requires that different wind directions be considered. The direction that produces the minimum

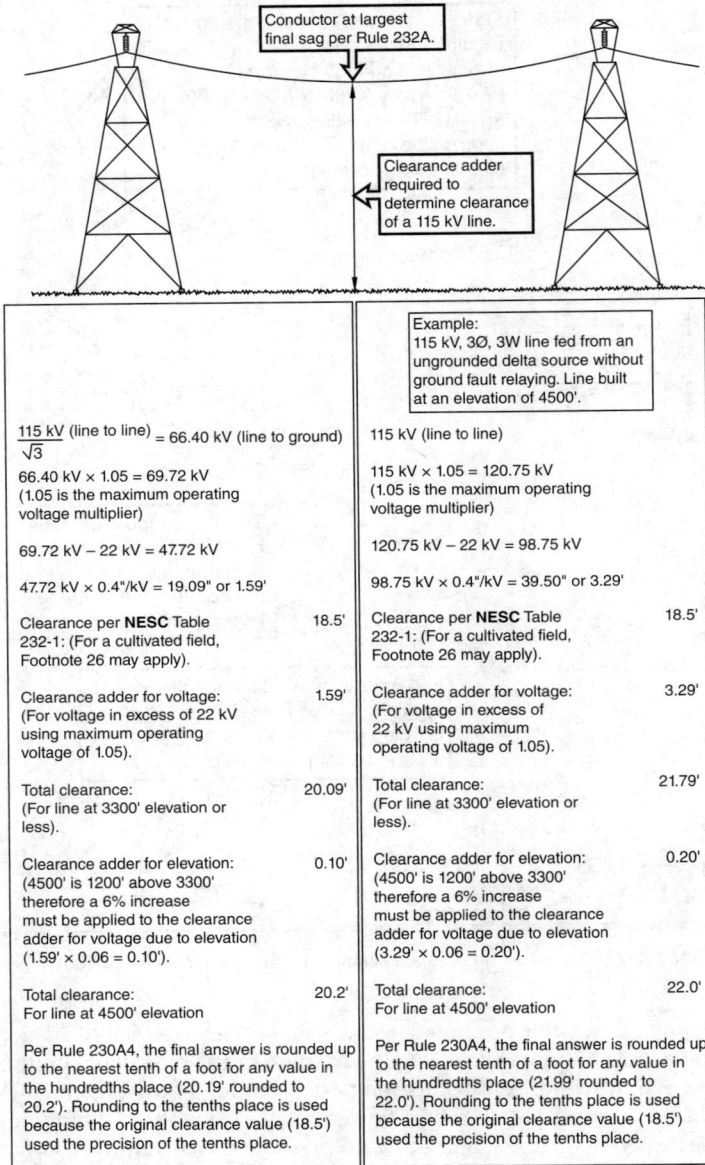

Conductor at largest final sag per Rule 232A.

Clearance adder required to determine clearance of a 115 kV line.

Example:
115 kV, 3Ø, 3W line fed from an ungrounded delta source without ground fault relaying. Line built at an elevation of 4500'.

$\dfrac{115 \text{ kV (line to line)}}{\sqrt{3}} = 66.40$ kV (line to ground)

66.40 kV × 1.05 = 69.72 kV
(1.05 is the maximum operating voltage multiplier)

69.72 kV − 22 kV = 47.72 kV

47.72 kV × 0.4"/kV = 19.09" or 1.59'

Clearance per **NESC** Table 232-1: (For a cultivated field, Footnote 26 may apply).	18.5'
Clearance adder for voltage: (For voltage in excess of 22 kV using maximum operating voltage of 1.05).	1.59'
Total clearance: (For line at 3300' elevation or less).	20.09'
Clearance adder for elevation: (4500' is 1200' above 3300' therefore a 6% increase must be applied to the clearance adder for voltage due to elevation (1.59' × 0.06 = 0.10').	0.10'
Total clearance: For line at 4500' elevation	20.2'

Per Rule 230A4, the final answer is rounded up to the nearest tenth of a foot for any value in the hundredths place (20.19' rounded to 20.2'). Rounding to the tenths place is used because the original clearance value (18.5') used the precision of the tenths place.

115 kV (line to line)

115 kV × 1.05 = 120.75 kV
(1.05 is the maximum operating voltage multiplier)

120.75 kV − 22 kV = 98.75 kV

98.75 kV × 0.4"/kV = 39.50" or 3.29'

Clearance per **NESC** Table 232-1: (For a cultivated field, Footnote 26 may apply).	18.5'
Clearance adder for voltage: (For voltage in excess of 22 kV using maximum operating voltage of 1.05).	3.29'
Total clearance: (For line at 3300' elevation or less).	21.79'
Clearance adder for elevation: (4500' is 1200' above 3300' therefore a 6% increase must be applied to the clearance adder for voltage due to elevation (3.29' × 0.06 = 0.20').	0.20'
Total clearance: For line at 4500' elevation	22.0'

Per Rule 230A4, the final answer is rounded up to the nearest tenth of a foot for any value in the hundredths place (21.99' rounded to 22.0'). Rounding to the tenths place is used because the original clearance value (18.5') used the precision of the tenths place.

Fig. 232-23. Example of an additional clearance calculation (Rules 232C1a and 232C1b).

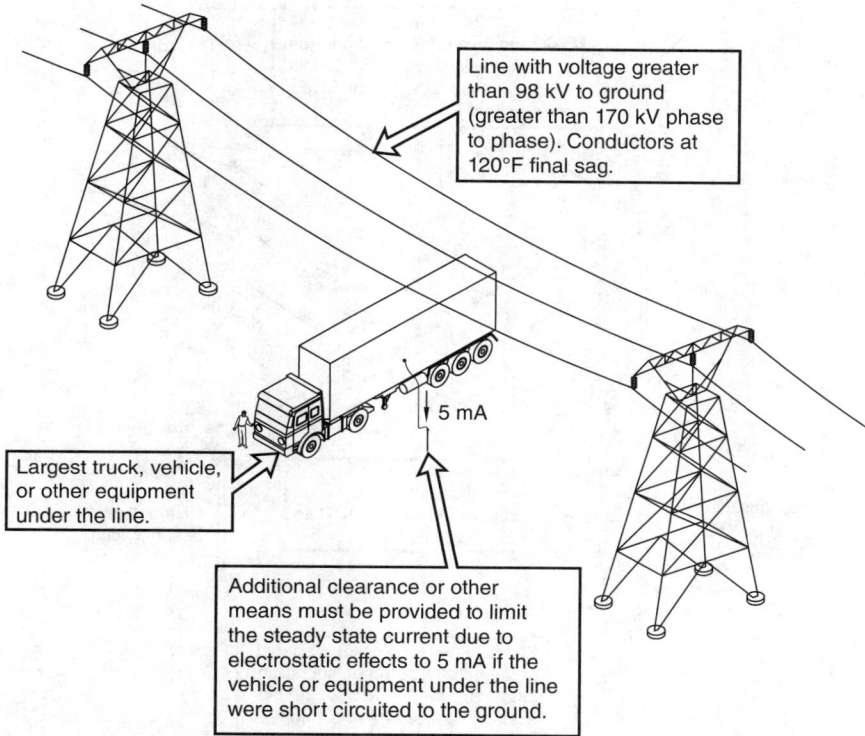

Fig. 232-24. Electrostatic field requirements (Rule 232C1c).

distance between conductors must be used. Footnote 5 applies when one line is above the other. When the largest final sag of the upper conductor is determined, the lower conductor temperature must be equal to the ambient air temperature used to determine the largest final sag of the upper conductor. This requirement is similar to, but worded slightly different than, Rule 235C2b(1)(c). Rule 235C2b(1)(c) requires checking vertical clearances during two conditions, when the top conductor is at a maximum operating temperature (120°F, or greater if so designed) and when the top conductor is at a cold temperature with ice (32°F, with ice from Rule 230B). Footnote 3 of **NESC** Fig. 233-2 only requires checking vertical clearance at one of these two conditions. Both conditions are checked to determine the greatest sag of the top conductor, but then vertical clearance is only required to be checked at the condition that produces the largest sag of the top conductor. The condition that produces the largest sag of the top conductor may not be the condition that produces the greater vertical clearance at the structure. For a thorough review, both conditions can be checked as required in Rule 235C2b(1)(c). See Rule 235C and the beginning of Sec. 23 for discussions related to conductor temperature and ambient temperature.

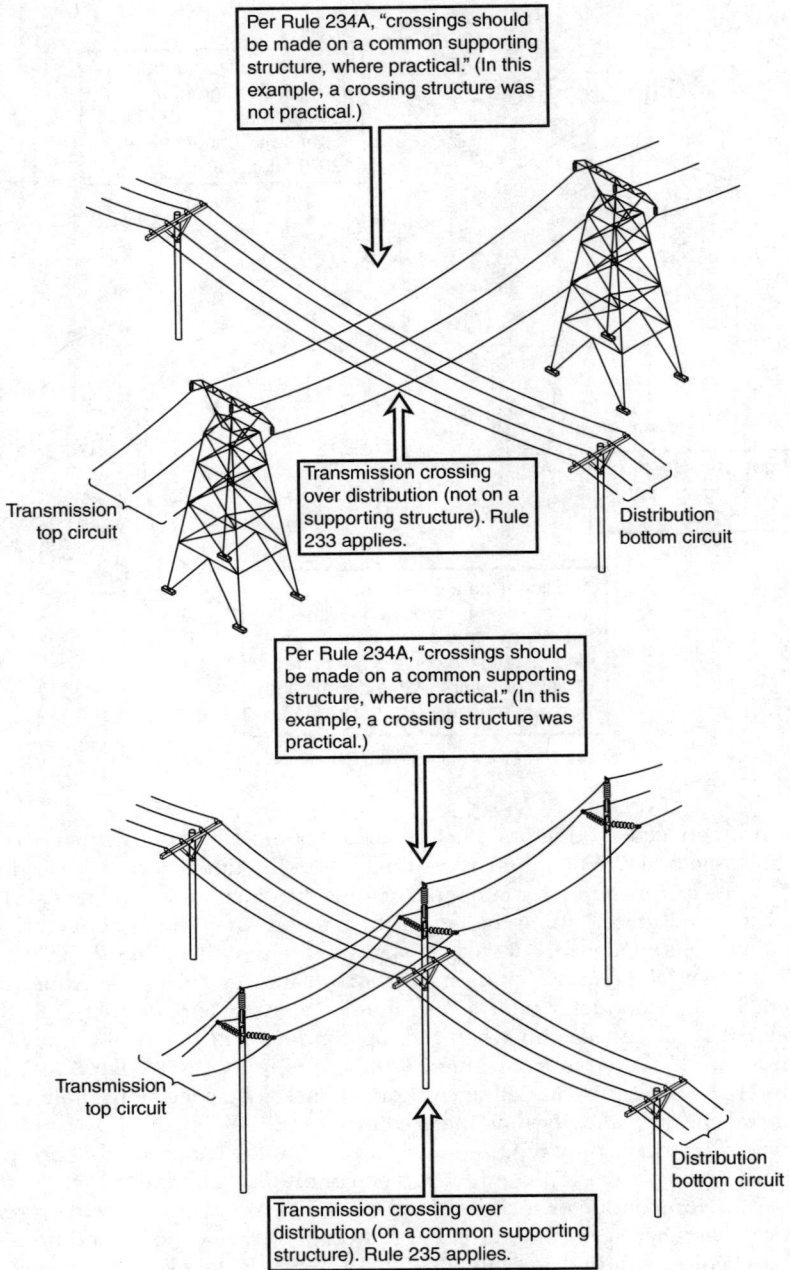

Per Rule 234A, "crossings should be made on a common supporting structure, where practical." (In this example, a crossing structure was not practical.)

Transmission top circuit

Transmission crossing over distribution (not on a supporting structure). Rule 233 applies.

Distribution bottom circuit

Per Rule 234A, "crossings should be made on a common supporting structure, where practical." (In this example, a crossing structure was practical.)

Transmission top circuit

Transmission crossing over distribution (on a common supporting structure). Rule 235 applies.

Distribution bottom circuit

Fig. 233-1. Examples of line crossings with and without a common supporting structure (Rule 233A).

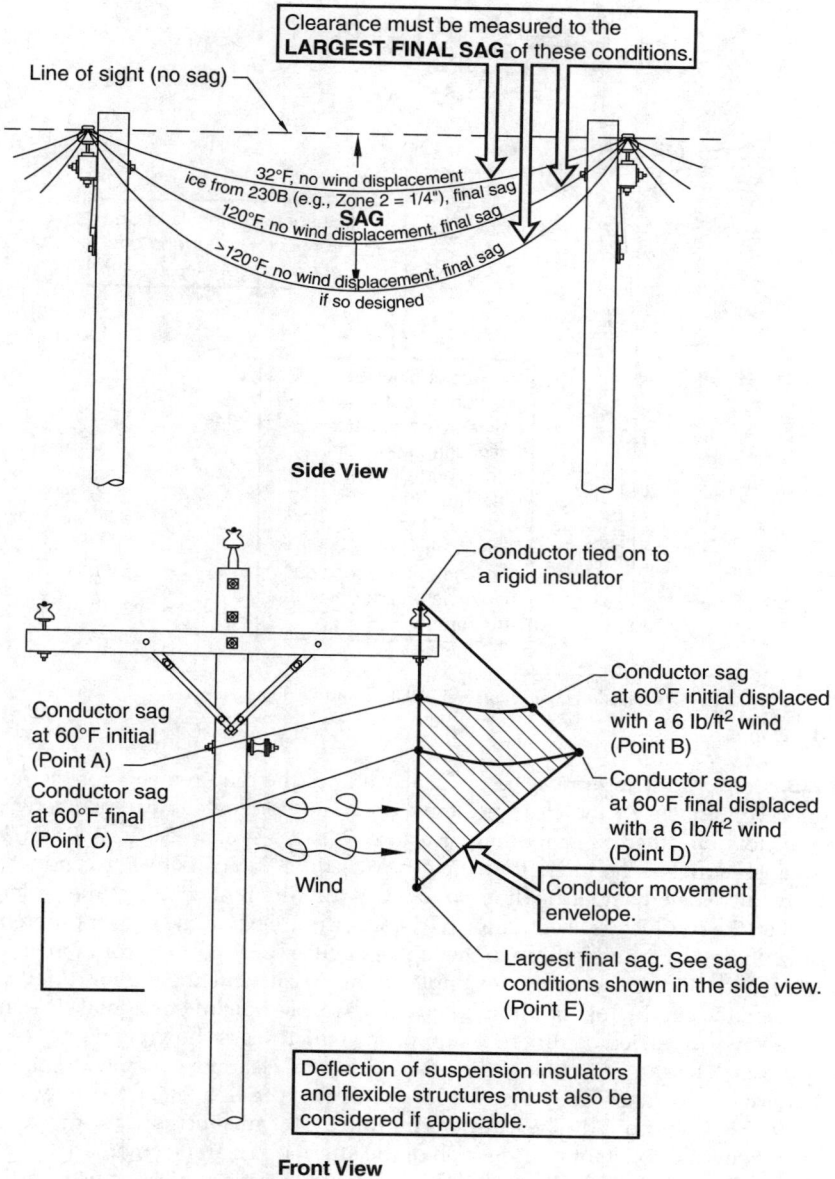

Clearance must be measured to the **LARGEST FINAL SAG** of these conditions.

Line of sight (no sag)

32°F, no wind displacement
ice from 230B (e.g., Zone 2 = 1/4"), final sag
SAG
120°F, no wind displacement, final sag
>120°F, no wind displacement, final sag
if so designed

Side View

Conductor tied on to a rigid insulator

Conductor sag at 60°F initial displaced with a 6 lb/ft² wind (Point B)

Conductor sag at 60°F final displaced with a 6 lb/ft² wind (Point D)

Conductor sag at 60°F initial (Point A)

Conductor sag at 60°F final (Point C)

Wind

Conductor movement envelope.

Largest final sag. See sag conditions shown in the side view. (Point E)

Deflection of suspension insulators and flexible structures must also be considered if applicable.

Front View

Fig. 233-2. Conductor movement envelope (Rule 233A).

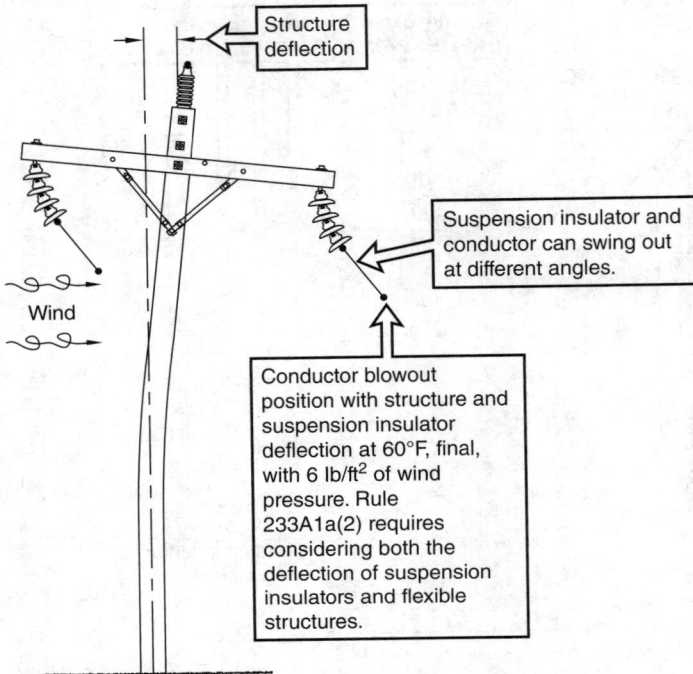

Fig. 233-3. Deflection of suspension insulators and flexible structures [Rule 233A1a(2)].

233B. Horizontal Clearance. Rule 233B outlines the requirements for the horizontal component of the clearance envelope. The horizontal clearance between conductors on different supporting structures is not less than 5.0 ft. A 0.4-in/kV clearance adder is required for voltages exceeding 22 kV between conductors involved. When comparing line conductors of different circuits, the voltage between the conductors is determined by using the greater of the phasor difference or the phase-to-ground voltage of the higher circuit. See Rule 235A for more information and a figure showing an example of how to calculate the phasor difference between line conductors of different circuits. An example of horizontal clearance between wires carried on different supporting structures is shown in Fig. 233-4.

Rule 233B has an exception to the 5.0-ft horizontal clearance for anchor guys.

Alternate clearances may be used for voltages exceeding 98 kV AC to ground or 139 kV DC to ground. The formulas in Rule 235B3 must be used for this calculation. See Rule 232D for a discussion of the alternate clearance method.

Rules 233B, 233C, 235B, and 235C are related to each other but use slightly different methods for determining clearance values. A comparison of the methods used to determine clearance in Rules 233B, 233C, 235B, and 235C is shown below:

- **Rule 233B.** Horizontal Clearance Between Wires, Conductors, and Cables Carried on Different Supporting Structures.

Example: Two parallel 12.47/7.2 kV effectively grounded distribution lines of the same height.

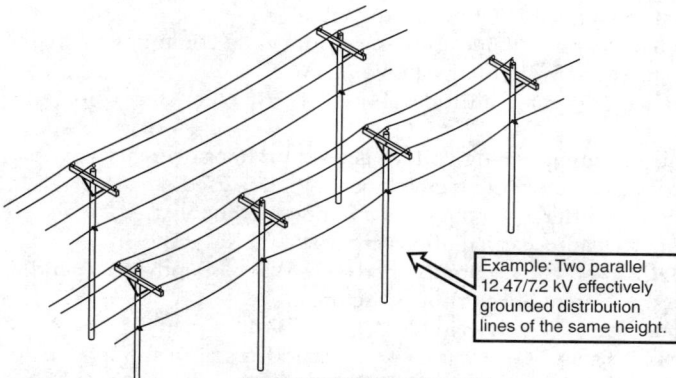

Wind must be considered in both directions. The lines will be closer with wind in this direction as opposed to wind in the opposite direction.

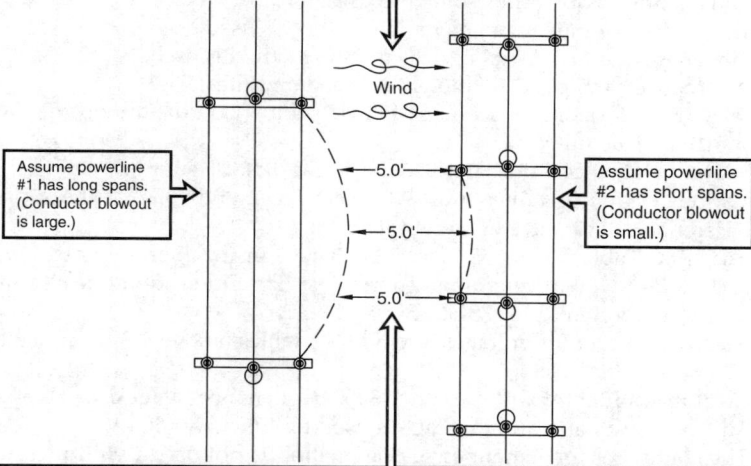

Wind

Assume powerline #1 has long spans. (Conductor blowout is large.)

—5.0'—

—5.0'—

—5.0'—

Assume powerline #2 has short spans. (Conductor blowout is small.)

The horizontal clearance between the lines must not be not less than 5.0' for voltages up to 22 kV between conductors per Rule 233B. A clearance adder of 0.4" per kV must be applied over 22 kV. In this example, the phasor difference between the line conductors is 7.2 kV + 7.2 kV = 14.4 kV (assuming a 180° phasor relationship). Therefore, no additional clearance for voltage applies and the maximum operating voltage and additional clearance for elevation requirements do not apply.

The clearance must be measured after considering the conductor movement envelopes of each conductor.
The horizontal clearance between the conductor of one line and the supporting structure of the other line must also be checked (Rule 234B).

See Photo(s)

Fig. 233-4. Example of horizontal clearance between wires carried on different supporting structures (Rule 233B).

✓ Clearance value of 5.0 ft in rule (no table).
✓ The 5.0 ft clearance value applies to supply and communication lines.
✓ Clearance adder for voltage above 22 kV.
✓ Clearance adder for altitude above 3300 ft when the voltage exceeds 50 kV.
✓ The maximum operating voltage is used for lines over 50 kV.
✓ The conductor movement envelope method is used.
✓ The phasor difference voltage calculation method is used.
✓ Alternate clearance calculations are based on Rule 235B3.

- **Rule 233C.** Vertical Clearance Between Wires, Conductors, and Cables Carried on Different Supporting Structures.
  ✓ Various clearance values in **NESC** Table 233-1.
  ✓ The **NESC** Table 233-1 clearance values address both supply and communication lines.
  ✓ Clearance adder for voltage above 22 kV.
  ✓ Clearance adder for altitude above 3300 ft when the voltage exceeds 50 kV.
  ✓ The maximum operating voltage is used for lines over 50 kV.
  ✓ The conductor movement envelope method is used.
  ✓ The phasor difference voltage calculation method is used.
  ✓ Alternate clearance calculations are based on Rule 233C3.

- **Rule 235B.** Horizontal Clearance Between Line Conductors on the Same Supporting Structure.
  ✓ Various clearance values in **NESC** Table 235-1 at the support.
  ✓ Various clearance values in **NESC** Tables 235-2 and 235-3 at the support based on voltage and sag out in the span.
  ✓ The **NESC** Table 235-1, 235-2, and 235-3 clearance values apply to supply lines and open wire communication lines. Communication line clearance is also covered in Rule 235H.
  ✓ Clearance adder for voltage shown in the tables (above 8.7 kV and above 50 kV).
  ✓ Clearance adder for altitude above 3300 ft when the voltage exceeds 50 kV.
  ✓ The maximum operating voltage is used for lines over 50 kV.
  ✓ The conductor movement envelope method is not used (wind pressures and sag conditions are specified in individual rules).
  ✓ The phasor difference voltage calculation method is used.
  ✓ Alternate clearance calculations are based on Rule 235B3.

- **Rule 235C.** Vertical Clearance Between Conductors at the Support on the Same Supporting Structure.
  ✓ Various clearance values in **NESC** Table 235-5 at the support.
  ✓ Various clearance values in Rule 235C2b at the support based on voltage and sag out in the span.
  ✓ The **NESC** Table 235-5 clearance values address both supply and communication lines. Clearance between supply and communication lines is also covered in Rule 235C4. Communication line clearance is also covered in Rule 235H. Communication antenna clearances are also covered in Rule 235I.

✓ Clearance adder for voltage shown in the table and in the rule (above 8.7 kV in the table and above 50 kV in the rule).

✓ Clearance adder for altitude above 3300 ft when the voltage exceeds 50 kV.

✓ The maximum operating voltage is used for lines over 50 kV.

✓ The conductor movement envelope method is not used (wind pressures and sag conditions are specified in individual rules).

✓ The phasor difference voltage calculation method is used.

✓ Alternate clearance calculations are based on Rule 233C3.

233C. Vertical Clearance.   Rule 233C outlines the requirements for the vertical component of the clearance envelope. **NESC** Table 233-1 is used to find the vertical clearance between conductors on different supporting structures. Rule 233C applies to crossing and adjacent wires, conductors, and cables carried on different support-ing structures. See Rule 233B for a comparison of Rules 233B, 233C, 235B, and 235C. Examples of vertical clearance between wires carried on different supporting struc-tures are shown in Figs. 233-5, 233-6, and 233-7.

Rule 233C has an exception that states that no vertical clearance is required between wires, conductors, or cables that are electrically interconnected at the crossing. An example of the exception to vertical clearance between wires that are electrically connected at a crossing is shown in Fig. 233-8.

See Rule 235A for more information and a figure showing an example of how to calculate the phasor difference between line conductors of different circuits.

Alternate clearances may be used for voltages exceeding 98 kV AC to ground or 139 kV DC to ground per Rule 233C3. The formulas in Rule 233C3 must be used for this calculation. The values in **NESC** Table 233-2 can be used to check the proper application of the formulas. See Rule 232D for a discussion of the alternate clearance method.

234. CLEARANCE OF WIRES, CONDUCTORS, CABLES, AND EQUIPMENT FROM BUILDINGS, BRIDGES, RAIL CARS, SWIMMING POOLS, AND OTHER INSTALLATIONS

234A. Application.   Rule 234 contains both vertical and horizontal clearance requirements. When horizontal clearance is specified in Rules 234B, 234C, and 234D, the clearance must be checked when the conductor is at rest and when the conductor is displaced by wind. Both conditions must be met, not one or the other. Rule 234A defines the conductor temperature and loading conditions that must be checked before applying the rules and tables in Sec. 234. The largest final sag con-ditions in Rule 234A are the same conditions provided in Rule 232A. See Rule 232A for a discussion.

Rule 234A2 provides a wind condition for checking horizontal clearance with wind. The wind condition in Rule 234A2 is the same as in Rule 233A except Rule 234A2 only requires checking the 60°F final sag condition. Rule 233A requires

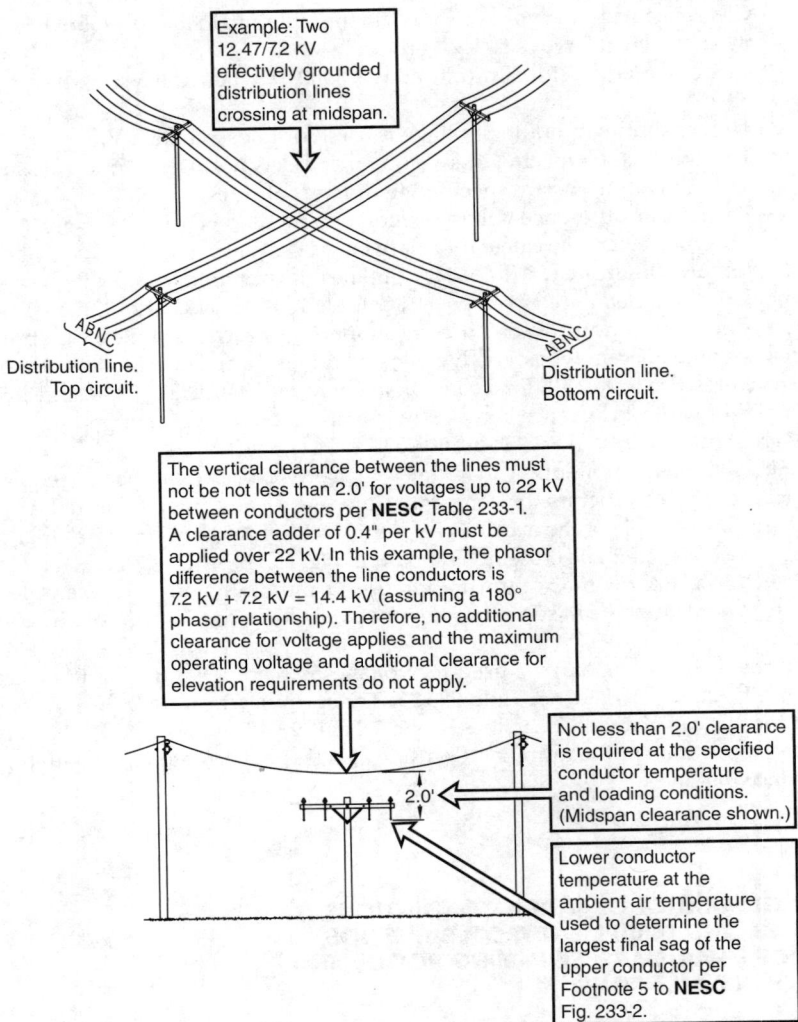

Example: Two 12.47/7.2 kV effectively grounded distribution lines crossing at midspan.

Distribution line.
Top circuit.

Distribution line.
Bottom circuit.

The vertical clearance between the lines must not be not less than 2.0' for voltages up to 22 kV between conductors per **NESC** Table 233-1. A clearance adder of 0.4" per kV must be applied over 22 kV. In this example, the phasor difference between the line conductors is 7.2 kV + 7.2 kV = 14.4 kV (assuming a 180° phasor relationship). Therefore, no additional clearance for voltage applies and the maximum operating voltage and additional clearance for elevation requirements do not apply.

Not less than 2.0' clearance is required at the specified conductor temperature and loading conditions. (Midspan clearance shown.)

2.0'

Lower conductor temperature at the ambient air temperature used to determine the largest final sag of the upper conductor per Footnote 5 to **NESC** Fig. 233-2.

Fig. 233-5. Example of vertical clearance between wires carried on different supporting structures (Rule 233C).

checking both the 60°F final and initial conditions. See Rule 233A for a discussion. The conductor temperature and loading conditions for measuring vertical and horizontal clearance of conductors to buildings and other installations are shown in Fig. 234-1.

The horizontal clearances specified in Rule 234 must be checked at rest and after applying a 6-lb/ft² wind at 60°F, final sag. The wind pressure may be reduced to 4 lb/ft² in sheltered areas but trees are not considered a shelter to a line. The deflection of suspension insulators must be considered. The deflection of flexible structures

Vertical clearance from **NESC** Table 233-1 and Rule 233C2:

Upper conductor:

$\frac{115}{\sqrt{3}}$ kV (line to line) = 66.40 kV (line to ground)

64.40 kV × 1.05 = 69.72 kV
(1.05 is the maximum operating voltage multiplier)

Lower conductor:

7.2 kV phase wire

Total clearance:

2.0' (from **NESC** Table 233-1)

76.92 kV (from phasor difference calculation in Fig. 235-2)

2.0' + 0.4"/kV over 22 kV (from Rule 233C2)

2.0' + 0.4"(76.92 - 22)

2.0' + 21.968"

2.0' + $\frac{21.968"}{12"}$

2.0' + 1.83'

3.83'

Round to 3.9'

Example: 12.47/7.2 kV effectively grounded distribution line crossing midspan under a 115 kV, 3Ø, 3W transmission line fed from a grounded wye source with ground fault relaying at an elevation of 3000'.

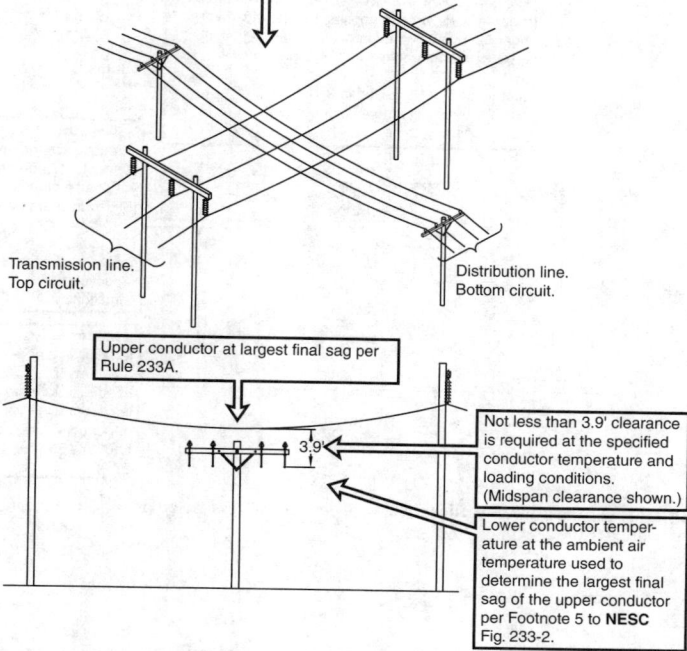

Transmission line. Top circuit.

Distribution line. Bottom circuit.

Upper conductor at largest final sag per Rule 233A.

3.9'

Not less than 3.9' clearance is required at the specified conductor temperature and loading conditions. (Midspan clearance shown.)

Lower conductor temperature at the ambient air temperature used to determine the largest final sag of the upper conductor per Footnote 5 to **NESC** Fig. 233-2.

See Photo(s)

Fig. 233-6. Example of vertical clearance between wires carried on different supporting structures (Rule 233C).

Example:
A communication cable on an effectively
grounded messenger crossing midspan under a
12.47/7.2 kV effectively grounded distribution line.

Distribution line.
Top circuit.

Communication
cable on a
messenger.
Bottom circuit.

The vertical clearance between the lines must not
be not less than 5.0' for voltages up to 22 kV
between conductors per Rule 233B. A clearance
adder of 0.4" per kV must be applied over 22 kV.
In this example, the phasor difference between the
line conductors is 7.2 kV. Therefore, no additional
clearance for voltage applies and the maximum
operating voltage and additional clearance for
elevation requirements do not apply.

(Footnote 5 of **NESC** Table 233-1
permits a reduction to 4.0' for certain
conditions.)

Not less than 5.0' clearance
is required at the specified
conductor temperature
and loading conditions.
(Midspan clearance shown.)

(Footnote 5 of **NESC**
Table 233-1 permits a
reduction to 4.0' for certain
conditions.)

5.0'

Lower conductor
temperature at the
ambient air temperature
used to determine the
largest final sag of the
upper conductor per
Footnote 5 of **NESC** Fig.
233-2.

See
Photo(s)

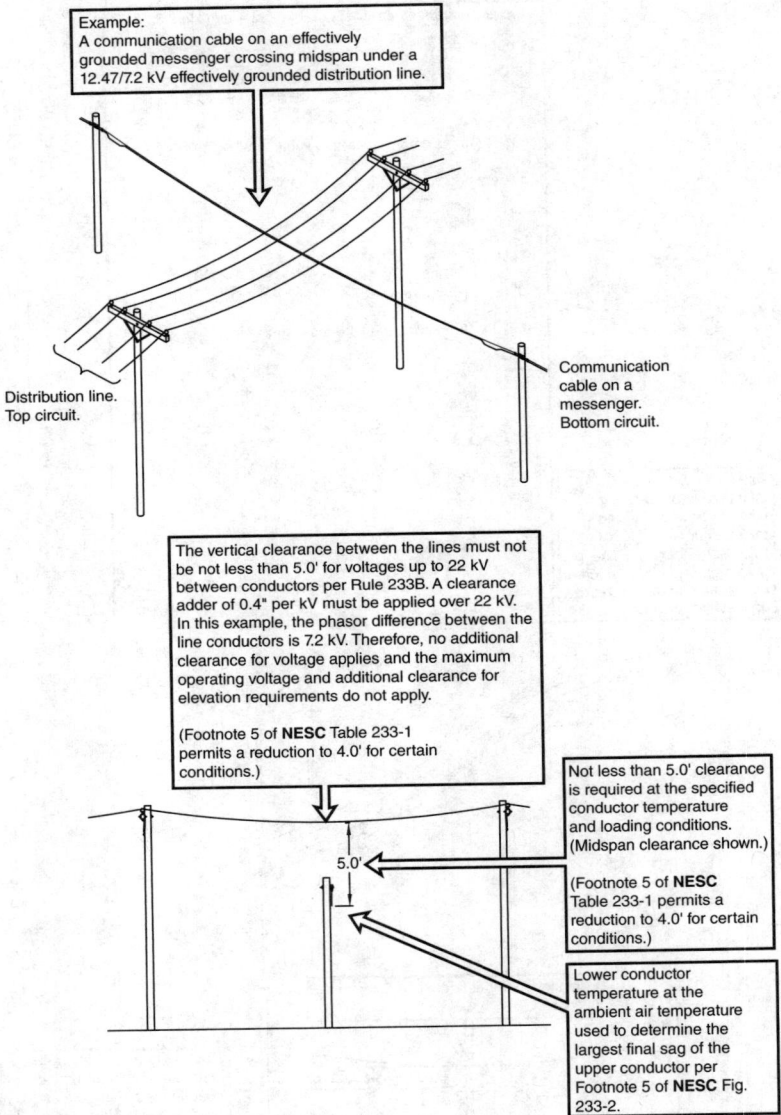

Fig. 233-7. Example of vertical clearance between wires carried on different supporting
structures (Rule 233C).

No vertical clearance is required between wires that are electrically connected at a crossing (e.g., a flying tap). But, in this example, vertical clearance would still be required as Phase A of the top circuit does not connect to Phase B or C or the neutral of the bottom circuit.

Distribution line.
Top circuit.

Distribution line.
Bottom circuit.

See
Photo(s)

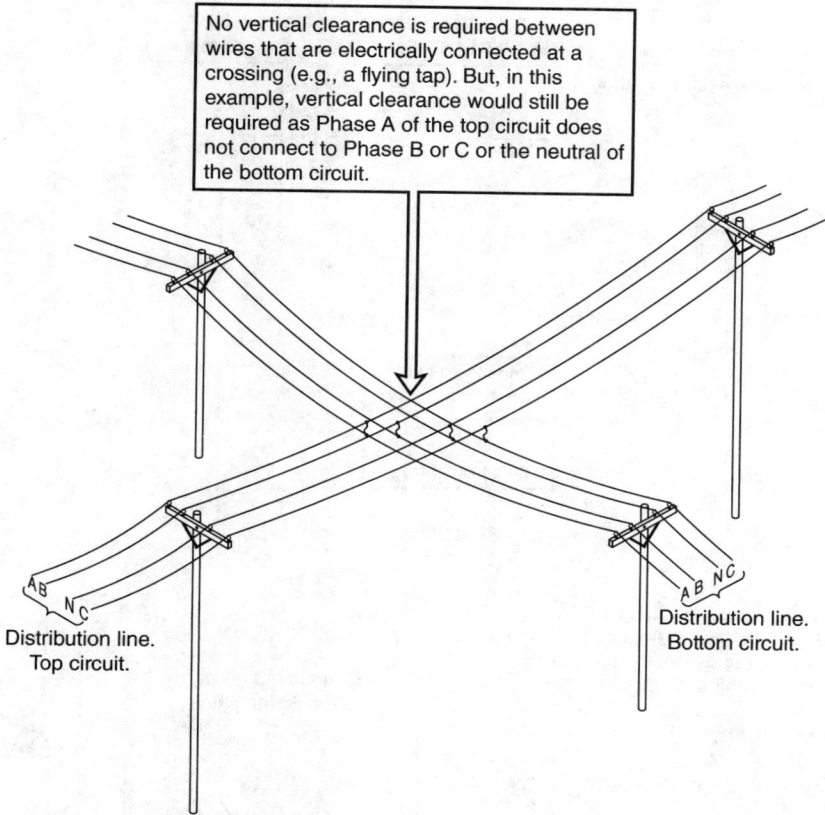

Fig. 233-8. Example of the exception to vertical clearance between wires that are electrically interconnected at a crossing (Rule 233C).

must be considered if the highest wire, conductor, or cable attachment is 60 ft or more above grade. Rule 233A1a(2) contains similar requirements for suspension insulators and flexible structures, but unlike Rule 234A2, Rule 233A1a(2) does not specify the 60-ft height requirement. See Fig. 234-2.

In addition to the largest final sag conditions, Rule 234A1d requires a minimum conductor sag be considered at the minimum conductor temperature with no wind at initial sag. The minimum conductor temperature will be based on the expected minimum ambient temperature. Using the sample sag and tension chart at the beginning of Sec. 23, −20°F, no wind, no ice, initial tension produces a minimum sag of 1.61 ft. This is the smallest sag on the sample sag and tension chart. The clearance under and alongside a building or sign must be checked when the conductor is in this minimum sag position. An example of checking clearance using a minimum sag condition is shown in Fig. 234-3.

Rule 234A3 describes the method that is used to Transition (T) between Horizontal (H) and Vertical (V) clearances. The T, H, and V values are shown in **NESC**

Clearance must be measured to the
LARGEST FINAL SAG of these conditions.

Line of sight (no sag)

32°F, no wind displacement
ice from 230B (e.g., Zone 2 = 1/4"), final sag
SAG
120°F, no wind displacement, final sag
≥120°F, no wind displacement, final sag
if so designed

A minimum sag
condition must be
used to check
clearance under a
sign or building.

Side View

Deflection of suspension
insulators and flexible
structures must also be
considered if applicable.

Conductor tied on
to a rigid insulator.

Conductor at 60°F,
final sag, displaced
by a 6 lb/ft^2 wind.

Wind

Largest final sag. See
sag conditions shown
in the side view.

Top View

Front View

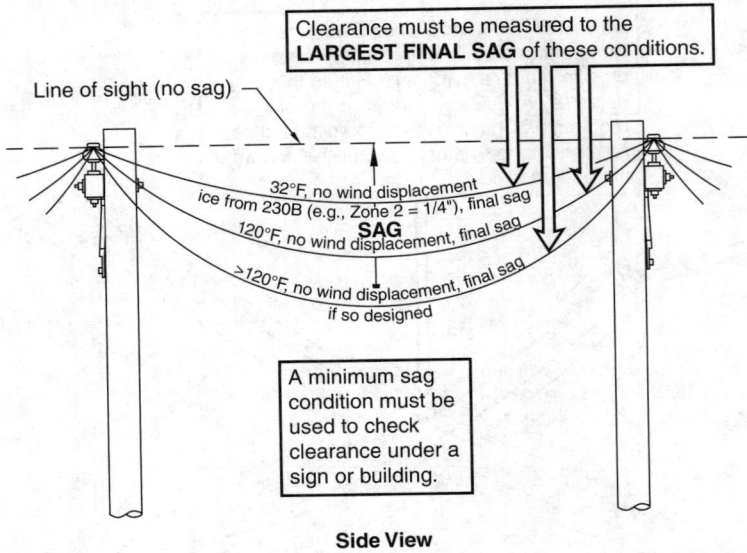

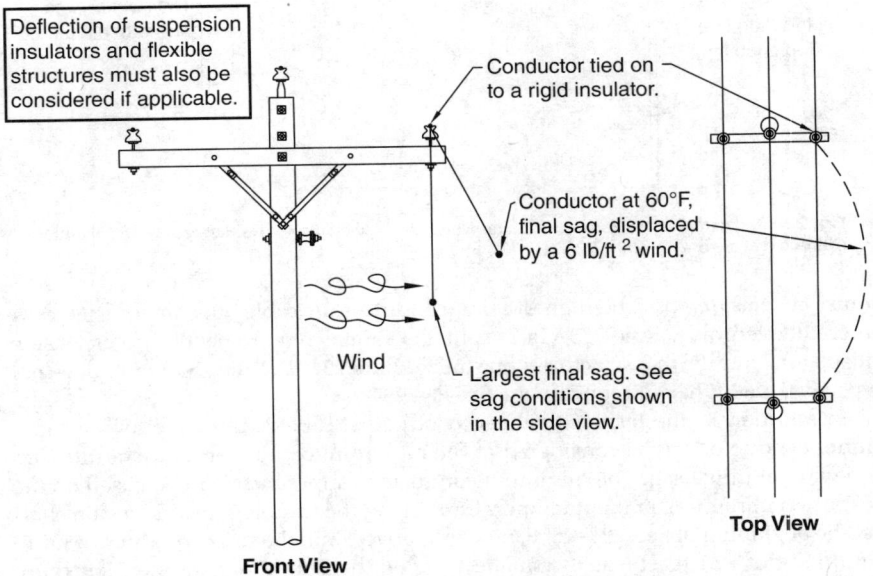

Fig. 234-1. Conductor temperature and loading conditions for measuring vertical and horizontal clearance to buildings and other installations (Rule 234A).

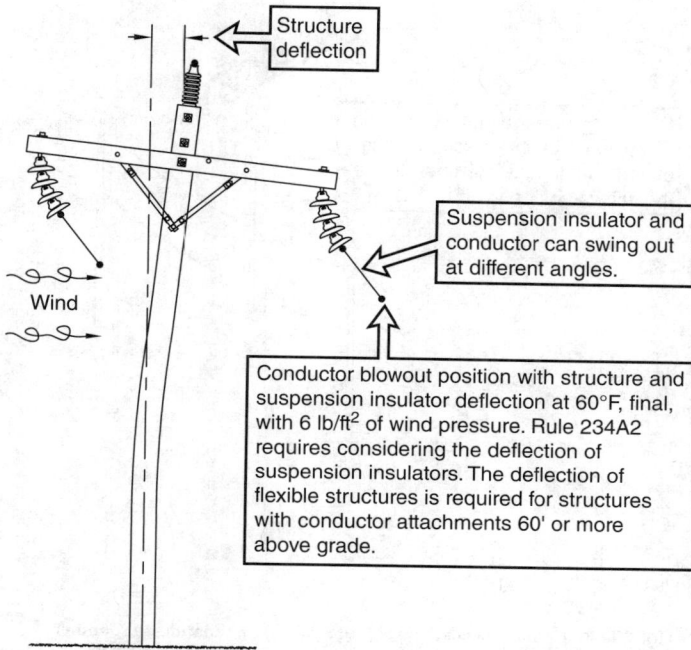

Fig. 234-2. Deflection of suspension insulators and flexible structures (Rule 234A2).

Figs. 234-1(a), (b), and (c). This is one of the few rules in the NESC that has graphics to help convey the Code requirements. NESC Figs. 234-1(a), (b), and (c) are used in conjunction with NESC Table 234-1.

234B. Clearances of Wires, Conductors, and Cables from Other Supporting Structures.   Rule 234B is used to check clearance of wires, conductors, and cables to street lighting poles, traffic signal poles, or a pole of another power line or communication line. In these cases, the line in question is not attached to the pole in question. It is important to note that the clearance is from "any part of such structure." Examples of parts that extend from street light and traffic signal pole structures are a photocell mounted on a street light pole and a video camera detector mounted on a traffic signal pole. OSHA requires non-qualified electrical workers to stay 10 ft or more away from power lines, this is commonly referred to as the "OSHA 10 ft Rule" (OSHA 1910.333). There are instances where the NESC clearance requirements of Rule 234B permit less than 10 ft of clearance between a street light and a power line. Where this occurs, the worker changing the light bulb on the street light needs to be a qualified electrical worker trained in the minimum approach distances in NESC Sec. 44 or OSHA 1910.269 or some other applicable standard. Examples of a 12.47/7.2-kV, three-phase, four-wire distribution line adjacent to a street light pole and a traffic signal pole are shown in Figs. 234-4 and 234-5.

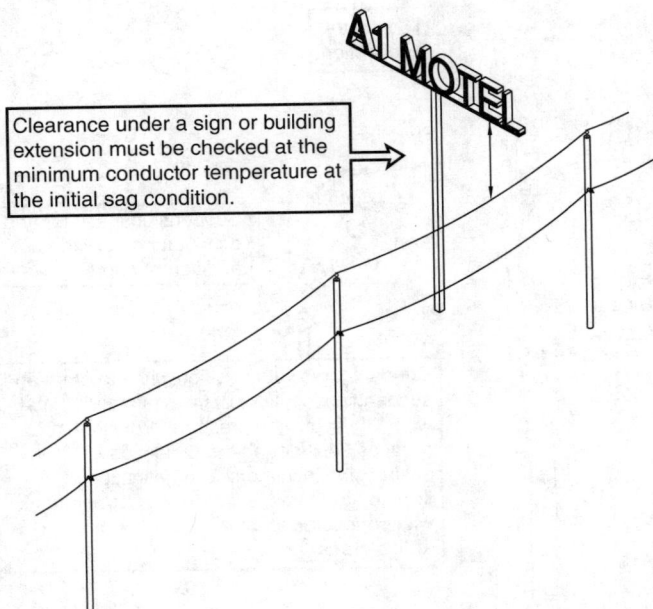

Fig. 234-3. Example of checking clearance using a minimum sag condition (Rule 234A).

There are times when a power line and street lighting compete for the same right of way and the only solution to meeting Rule 234B is bending the street light poles. See Fig. 234-6.

Another application of Rule 234B is a transmission line with a distribution underbuild that has "skip-span" construction. Skip-span construction is also sometimes used on a distribution line with a joint-use communication line attached to it. An example of a transmission line with a distribution underbuild that has skip-span construction is shown in Fig. 234-7.

The clearance adders for voltages above 22 kV are provided in Rule 234G. Alternate clearances for voltages exceeding 98 kV AC to ground or 139 kV DC to ground are provided in Rule 234H. The horizontal clearance in Rule 234B1 requires clearance adders for voltages greater than 22 kV for conductors at rest and greater than 22 kV for conductors displaced by wind. The horizontal wind table in Rule 234B1 does not have a table number. The table does not address 230E1 (neutral conductors), 230C3 cables below 750 V (e.g., 120/240 V secondary triplex conductors), or communication cables. The vertical clearance in Rule 234B2 also requires clearance adders for voltages greater than 22 kV.

234C. Clearances of Wires, Conductors, Cables, and Rigid Live Parts from Buildings, Signs, Billboards, Chimneys, Radio and Television Antennas, Tanks, Flagpoles and Flags, Banners, and Other Installations Except Bridges. Clearance from an energized conductor to a building or other installation is just as important as clearance to the surface below the line. As the title to Rule 234C states, this rule

Per Rule 234B1b, the horizontal clearance between the phase conductor of a 12.47/7.2 kV, 3∅, 4W distribution line and a street light pole must be not less than 4.5' with a 6 lb/ft² wind applied to the conductor at 60°F final sag. No wind clearance is specified for an effectively grounded neutral (230E1), a secondary (230C3) cable below 750 V, or a communication cable. The clearance is measured after the line is placed in the blowout position.

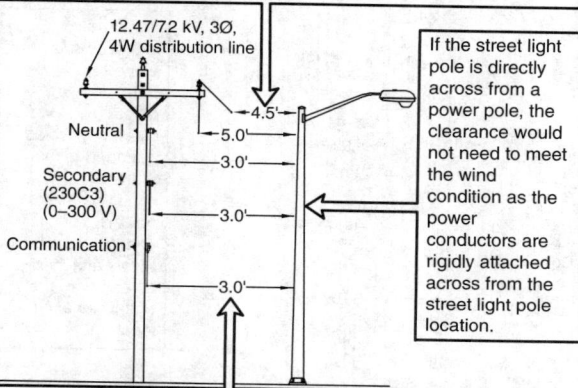

12.47/7.2 kV, 3∅, 4W distribution line

4.5'

Neutral — 5.0'

3.0'

Secondary (230C3) (0–300 V)

3.0'

Communication

3.0'

If the street light pole is directly across from a power pole, the clearance would not need to meet the wind condition as the power conductors are rigidly attached across from the street light pole location.

Per Rule 234B1a, the horizontal clearance between the phase conductor of a 12.47/7.2 kV, 3∅, 4W distribution line and a street pole (without wind) must be not less than 5.0'.

An exception permits the neutral (230E1), secondary (230C3) up to 300 V, and communication cable to have a horizontal clearance (without wind) to be not less than 3.0'.

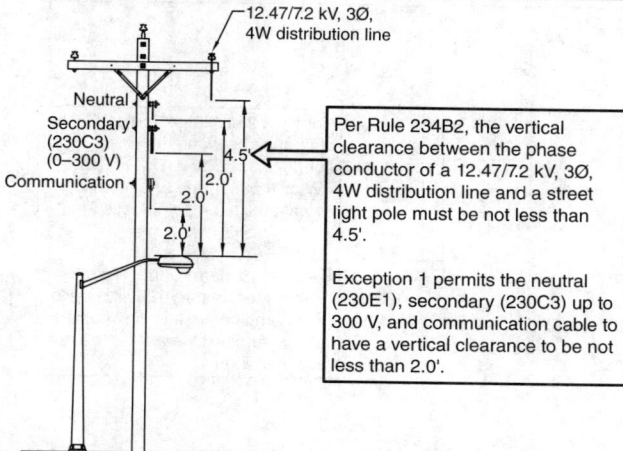

12.47/7.2 kV, 3∅, 4W distribution line

Neutral

Secondary (230C3) (0–300 V)

Communication

4.5'

2.0'

2.0'

2.0'

Per Rule 234B2, the vertical clearance between the phase conductor of a 12.47/7.2 kV, 3∅, 4W distribution line and a street light pole must be not less than 4.5'.

Exception 1 permits the neutral (230E1), secondary (230C3) up to 300 V, and communication cable to have a vertical clearance to be not less than 2.0'.

See Photo(s)

Fig. 234-4. Example of clearance of a conductor to a street lighting pole (Rule 234B).

Per Rule 234B1b, the horizontal clearance between the phase conductor of a 12.47/7.2 kV, 3Ø, 4W distribution line and a traffic signal pole must be not less than 4.5' with a 6 lb/ft^2 wind applied to the conductor at 60°F final sag. No wind clearance is specified for an effectively grounded neutral (230E1), a secondary (230C3) cable below 750 V, or a communication cable. The clearance is measured after the line is placed in the blowout position.

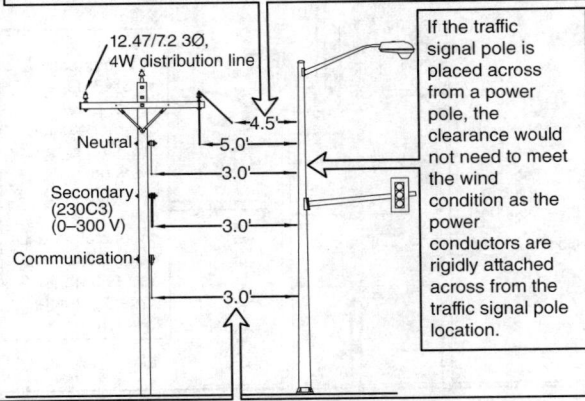

12.47/7.2 3Ø,
4W distribution line

←—4.5'—→

Neutral — 5.0'

3.0'

If the traffic signal pole is placed across from a power pole, the clearance would not need to meet the wind condition as the power conductors are rigidly attached across from the traffic signal pole location.

Secondary (230C3) (0–300 V) — 3.0'

Communication — 3.0'

Per Rule 234B1a, the horizontal clearance between the phase conductor of a 12.47/7.2 kV, 3Ø, 4W distribution line and a traffic signal pole (without wind) must be not less than 5.0'.

An exception permits the neutral (230E1), secondary (230C3) up to 300 V, and communication cable to have a horizontal clearance (without wind) to be not less than 3.0'.

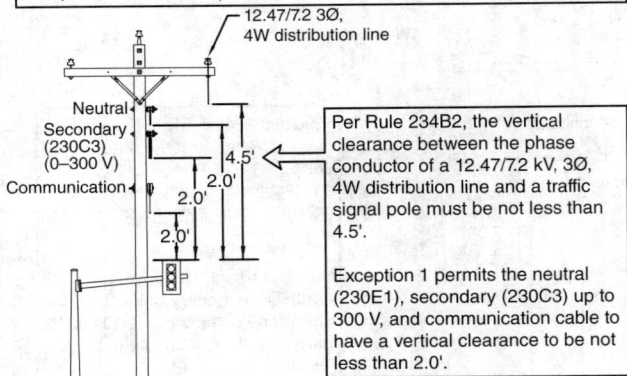

12.47/7.2 3Ø,
4W distribution line

Neutral
Secondary (230C3) (0–300 V)
Communication

4.5'
2.0'
2.0'
2.0'

Per Rule 234B2, the vertical clearance between the phase conductor of a 12.47/7.2 kV, 3Ø, 4W distribution line and a traffic signal pole must be not less than 4.5'.

Exception 1 permits the neutral (230E1), secondary (230C3) up to 300 V, and communication cable to have a vertical clearance to be not less than 2.0'.

See Photo(s)

Fig. 234-5. Example of clearance of a conductor to a traffic signal pole (Rule 234B).

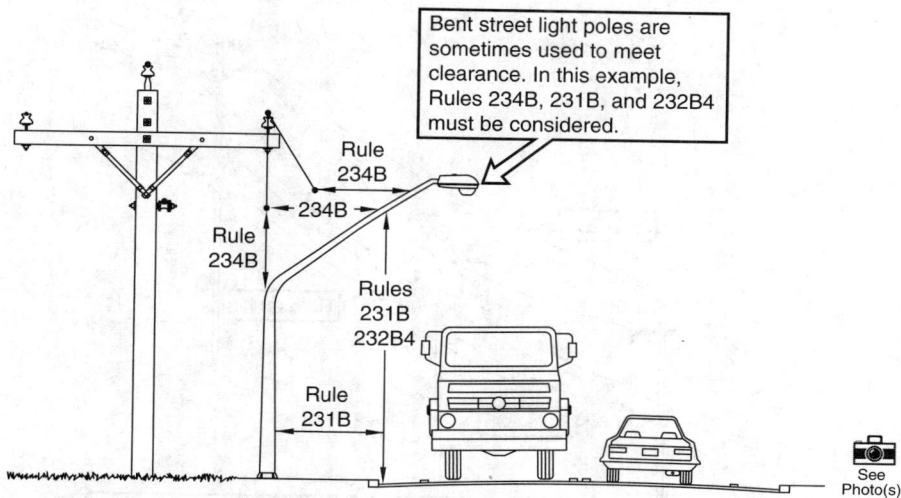

Fig. 234-6. Application of Rules 234B, 231B, and 232B4 to bent street lighting poles (Rule 234B).

includes not only clearance to buildings, but also signs, billboards, chimneys, radio and television antennas, tanks, flagpoles and flags, banners, and other installations except bridges. This is one of the few rules in the **NESC** that has graphics to help convey the **Code** requirements.

The first complete sentence of Rule 234C has four very important words that are not in the title of Rule 234C. The words are "...and any projections therefrom." This statement means that clearance must be maintained not just to the building structure, but to a gutter, awning, or any other projection from a building, sign, billboard, chimney, radio or television antenna, tank, flagpole and flag, banner, or other installation except bridges. The "except bridges" statement is due to the fact that bridges have their own rule, Rule 234D.

The vertical and horizontal clearances for Rule 234C are determined by using **NESC** Table 234-1 with the aid of **NESC** Figs. 234-1(a), 234-1(b), and 234-1(c). Normally, the **Code** requires you to meet all applicable rules, not one rule or another. However, in the case of building clearances, it is evident from the H, V, and T clearance envelope around the building in **NESC** Fig 234-1(a) that a power or communication line that meets the required vertical clearance above a building does not need to meet any required horizontal clearance to the side of a building and a power or communication line that meets the required horizontal clearance to the side of a building does not need to meet any required vertical clearance above a building. Other issues such as aerial trespass above a building or property line setback from the side of a building may arise when positioning power and communication lines adjacent to buildings but these issues are not **NESC** issues. Practical methods for meeting **NESC** clearance requirements to buildings include, but are not limited to, taller poles, using alley arm (side arm) framing, using bridge arm construction, using covered conductors, using covered bundled conductors,

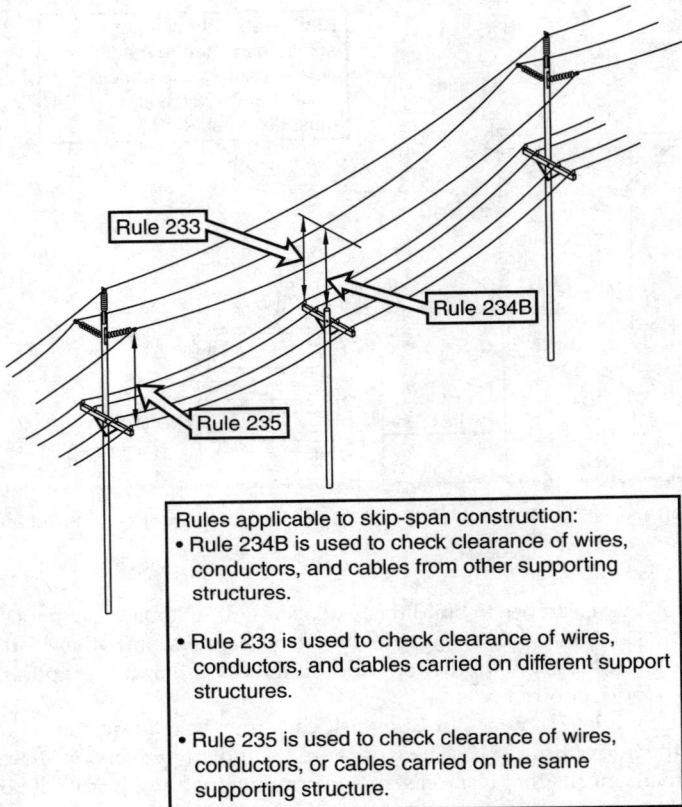

Rule 233

Rule 234B

Rule 235

Rules applicable to skip-span construction:
• Rule 234B is used to check clearance of wires, conductors, and cables from other supporting structures.

• Rule 233 is used to check clearance of wires, conductors, and cables carried on different support structures.

• Rule 235 is used to check clearance of wires, conductors, or cables carried on the same supporting structure.

See Photo(s)

Fig. 234-7. Example of skip-span construction (Rule 234B).

using aerial insulated conductors, increasing conductor tension for wind related clearances, rerouting the line, and burying the line. **NESC** Table 234-1 that is used in conjunction with **NESC** Fig. 234-1(a) address both wires (e.g., a 12.47 kV phase wire) and unguarded rigid live parts (e.g., an energized transformer bushing) in the same table.

It is important to carefully read the title of **NESC** Table 234-1, "Clearance of Wires, Conductors, Cables, and Unguarded Rigid Live Parts Adjacent but Not Attached to Buildings and Other Installations Except Bridges." The important wording is "...adjacent but not attached to..." This table is for conductors passing by a building, not a service that is attached to a building. Building services are covered in Rule 234C3.

The clearances in **NESC** Table 234-1 for conductors vertically over a building or other installation must be measured with the conductors at the largest final sag condition per Rule 234A. The vertical clearance for conductors under a sign or building projection must be measured with the conductors at the minimum sag condition as defined by Rule 234A. The clearances in **NESC** Table 234-1 for conductors to the side of (horizontal to) a building or other installation are at rest (no wind displacement) values. The horizontal clearance must also be checked using the wind condition in Rule 234A2. The horizontal clearances to conductors displaced by wind are provided in a small table in Rule 234C1. The table does not have an **NESC** table number. The table does not address 230E1 (neutral conductors), 230C3 cables below 750 V (e.g., 120/240 V secondary triplex conductors), or communication cables.

Examples of building clearance, billboard clearance, and flagpole clearance are shown in Figs. 234-8 through 234-12.

Rule 234C2 allows guarding as an option where **NESC** Table 234-1 clearances cannot be obtained. The note to this rule states that 230C1a cables are considered guarded. See the figure in Rule 230C for the details of a 230C1a cable.

Footnote 2 to **NESC** Table 234-1 permits reduced clearance to buildings. Footnote 2 permits clearance reductions for covered conductors. Covered conductor or "tree wire" is discussed in Rule 230D. Covered conductor applications along streets and in alleyways have become popular due to cramped space.

The **NESC** does not address clearance to buildings under construction. The Occupational Safety and Health Administration (OSHA) regulations apply to building construction. See the discussion in Rule 232B for buildings being transported on a roadway (e.g., house moves). Mobile cranes used in the vicinity of power lines are addressed in the OSHA regulations in OSHA 1926.1408 which is part of OSHA 1926 Subpart CC. OSHA requires nonqualified electrical workers to stay 10 ft or more away from power lines, this is commonly referred to as the "OSHA 10 ft Rule" (OSHA Standard 1910.333). The **NESC** permits a 12.47/7.2 kV distribution line to be not less than 7.5 ft from a building. The 7.5 ft clearance is the "at rest" value from **NESC** Table 234-1, meeting the "with wind" clearance from Rule 234C1b may increase the "at rest" requirement to greater than 7.5 ft. There can be instances where the **NESC** permits less than 10 ft of clearance to a building or sign. If a nonqualified electrical worker, for example a painter, is painting a building and is working more than 10 ft away from the power line, both the **NESC** and OSHA requirements are met. If the painter needs to work closer than 10 ft to the power line, the painter must contact the electric utility to work out a plan. The electric utility may decide to de-energize the power line, reposition the power line on temporary arms, or select some other method for the painter to work in accordance with the OSHA 10 ft requirement. The OSHA 10 ft Rule is a work rule that applies to the painter and the **NESC** 7.5 ft clearance (or greater with wind) is a clearance rule that applies to the electric utility. The OSHA 10 ft Rule does not require an electric utility to increase the **NESC** clearance and it does not excuse an electric utility from meeting the **NESC** clearance. Some States have high voltage safety acts with wording similar to the OSHA 10 ft requirement.

In addition to the **NESC** clearances for "tanks" in **NESC** Table 234-1, the Liquefied Petroleum Gas Code (NFPA 58) specifies clearances to LP-Gas (propane) tanks that are located under power lines. See Fig. 234-13.

The **NESC** does not specifically address clearance between a power line and irrigation equipment. **NESC** Table 234-1 can be used if the irrigation equipment is considered an "other installation not classified as buildings or bridges." In addition to the clearance between a power line and the irrigation piping, questions commonly arise as to the required clearance between a power line and the irrigation

Per Rule 234C1b, the horizontal clearance between the phase conductor of a 12.47/7.2 kV, 3Ø, 4W distribution line and a building must be not less than 4.5' with a 6 lb/ft^2 wind applied to the conductor at 60°F final sag. No wind clearance is specified for an effectively grounded neutral (230E1), a secondary (230C3) cable below 750 V, or a communication cable. The wind clearance is measured after the line is placed in the blowout position.

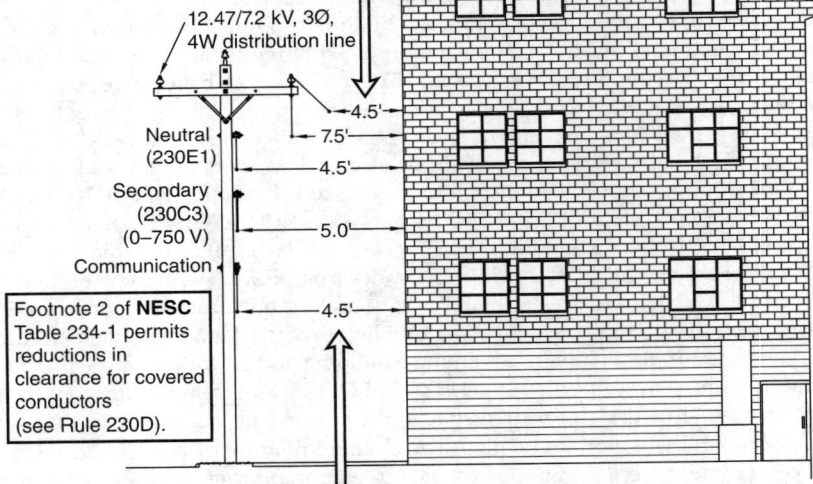

12.47/7.2 kV, 3Ø, 4W distribution line

4.5'

Neutral (230E1)   7.5'

4.5'

Secondary (230C3) (0–750 V)   5.0'

Communication

4.5'

Footnote 2 of **NESC** Table 234-1 permits reductions in clearance for covered conductors (see Rule 230D).

Per **NESC** Table 234-1, the horizontal clearance between the phase conductor of a 12.47/7.2 kV, 3Ø, 4W distribution line and a building (without wind) must be not less than 7.5'.

The neutral (230E1) and communication cable horizontal clearance (without wind) must be not less than 4.5'. The secondary (230C3) horizontal clearance (without wind) must be not less than 5.0'.

See Photo(s)

Fig. 234-8. Example of horizontal clearance of a conductor to a building (Rule 234C1).

water stream. Two documents are available for use as accepted good practice for determining the clearance between a power line and the irrigation water stream. They are "Guidelines for the Installation and Operation of Irrigation Systems Near High Voltage Transmission Lines" published by the Bonneville Power Administration (BPA) and "IEEE Guide for Installation, Maintenance and Operation of

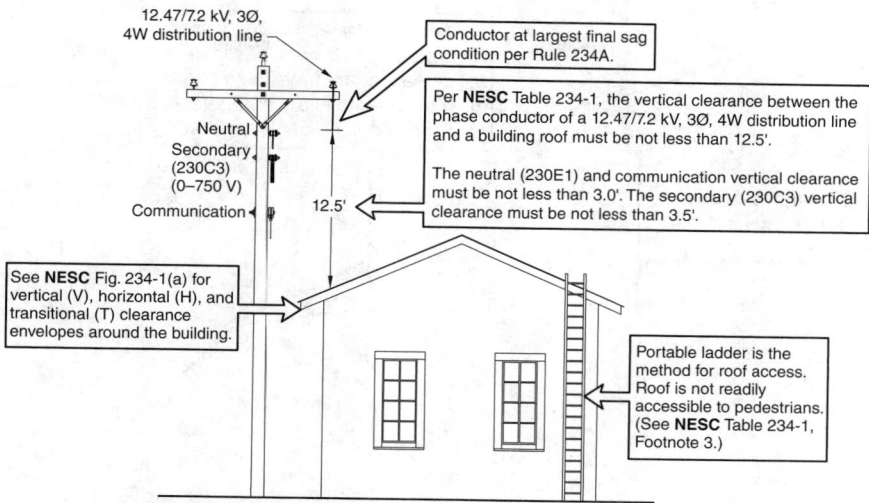

12.47/7.2 kV, 3Ø,
4W distribution line

Conductor at largest final sag
condition per Rule 234A.

Per **NESC** Table 234-1, the vertical clearance between the
phase conductor of a 12.47/7.2 kV, 3Ø, 4W distribution line
and a building roof must be not less than 12.5'.

Neutral
Secondary
(230C3)
(0–750 V)
Communication

The neutral (230E1) and communication vertical clearance
must be not less than 3.0'. The secondary (230C3) vertical
clearance must be not less than 3.5'.

12.5'

See **NESC** Fig. 234-1(a) for
vertical (V), horizontal (H), and
transitional (T) clearance
envelopes around the building.

Portable ladder is the
method for roof access.
Roof is not readily
accessible to pedestrians.
(See **NESC** Table 234-1,
Footnote 3.)

Fig. 234-9. Example of vertical clearance of a conductor to a building (Rule 234C1).

Irrigation Equipment Located Near or Under Power Lines" published by the Institute of Electrical and Electronics Engineers (IEEE Std. 1542). See Fig. 234-14.

The **NESC** does not specifically address clearance over fences or walls that are not part of a building structure. **NESC** Table 234-1 can be used if the fence or wall is considered an "other installation not classified as buildings or bridges." For clearance over a wall that is wide enough to stand on, the vertical clearance over "surfaces upon which personnel walk" is appropriate. For clearance over a fence that is not wide enough to stand on, the vertical clearance over "other portions of such installations" is appropriate. The clearances for wires, conductors, and cables above ground, roadway, rail, or water surfaces adjacent to the fence must also be met per **NESC** Table 232-1. **NESC** Table 234-1, Footnote 14 (which applies to conductors adjacent but not attached to buildings) and **NESC** Rule 234C3d(1), Exception 1 (which applies to supply conductors attached to buildings) address clearance above railings, walls, or parapets around balconies, decks, or roofs. See Fig. 234-15.

Rule 234C3 covers supply conductors attached to buildings. The supply conductors attached to the building must connect to a service entrance into the building. The **NESC** and the National Electrical Code (NEC) overlap at the service entrance point. The NEC has similar rules for service entrance conductors and depending on the authority having jurisdiction (e.g., a city or state electrical inspector), the NEC rules may govern. Examples of supply conductors attached and connecting to a service entrance into a building are shown in Figs. 234-16, 234-17, and 234-18.

Rule 234C4 permits communication conductors to be attached directly to buildings without any specific clearance requirements. Rule 234C4 does not state that the rule applies only to an attachment necessary for an entrance into the building as Rule 234C3 does for supply conductors. **NESC** Table 234-1 does require clearances from insulated communication cables to buildings for communication cables that are not attached to the building.

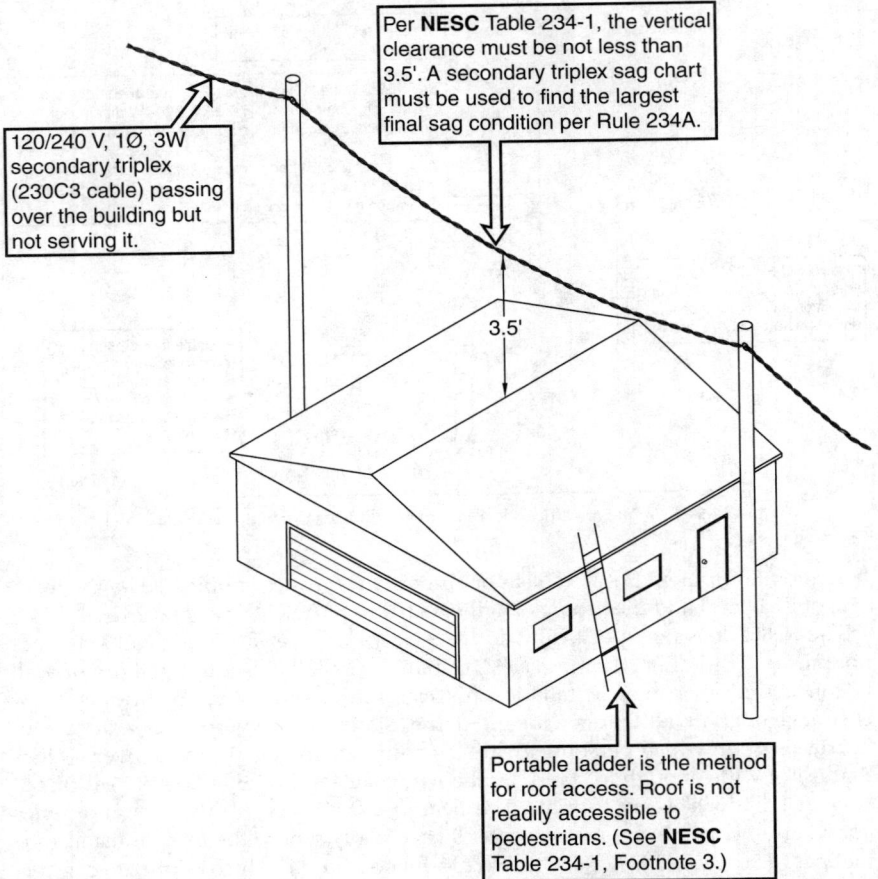

Per **NESC** Table 234-1, the vertical clearance must be not less than 3.5'. A secondary triplex sag chart must be used to find the largest final sag condition per Rule 234A.

120/240 V, 1Ø, 3W secondary triplex (230C3 cable) passing over the building but not serving it.

3.5'

Portable ladder is the method for roof access. Roof is not readily accessible to pedestrians. (See **NESC** Table 234-1, Footnote 3.)

See Photo(s)

Fig. 234-10. Example of vertical clearance of a supply service drop conductor over a building but not serving it (Rule 234C1).

Rule 234C5 requires a clear space for fire-fighting ladders for buildings exceeding three stories or 50 ft in height. A 6-ft (minimum) wide zone should exist adjacent to the building or beginning not over 8 ft from the building. Traditionally, the measurement for the clear space or zone that is at least 6 ft adjacent to the building is from the end of the crossarm to the building and the measurement for the 8 ft distance is from the opposite end of the crossarm to the building. The 6 ft zone beginning not over 8 ft from the building is not practical if 8 ft crossarms are used for a three-phase distribution line and a 7.5 ft clearance is required between the conductor near the edge of the crossarm and the building. Traditionally, this rule assumes the line is paralleling the building. An exception applies if the fire department does not use ladders in alleys near supply conductors. It is a good practice for supply utilities and the local fire department to communicate with each other to discuss safety issues related to overhead lines.

Per Rule 234C1b, the horizontal clearance between the phase conductor of a 12.47/7.2 kV, 3Ø, 4W distribution line and a billboard must be not less than 4.5' with a 6 lb/ft^2 wind applied to the conductor at 60°F final sag. No wind clearance is specified for an effectively grounded neutral (230E1), a secondary (230C3) cable below 750 V, or a communication cable. The clearance is measured after the line is placed in the blowout position.

If the billboard is directly across from a power pole, the clearance would not need to meet the wind condition as the power conductors are rigidly attached across from the billboard location.

Per **NESC** Table 234-1, the horizontal clearance between phase conductor of a 12.47/7.2 kV, 3Ø,4W distribution line and a billboard (without wind) must be not less than 7.5'.

The neutral (230E1) and communication horizontal clearance (without wind) must be not less than 3.0'. The secondary (230C3) horizontal clearance (without wind) must be not less than 3.5'.

Billboard not readily accessible to pedestrians.

Access is by a portable ladder (See **NESC** Table 234-1, Footnote 3)

Per **NESC** Table 234-1, the vertical clearance between the phase conductor of a 12.47/7.2 kV, 3Ø, 4W distribution line and a billboard must be not less than 8.0'.

The neutral (230E1) and communication vertical clearance must be not less than 3.0'. The secondary (230C3) vertical clearance must be not less than 3.5'.

The vertical clearance over the catwalk to the phase conductor must be not less than 13.5'.

The neutral (230E1) and communication cable vertical clearance must be not less than 10.5'. The secondary (230C3) vertical clearance must be not less than 11.0'.

Billboard not readily accessible to pedestrians.

Access is by a portable ladder (See **NESC** Table 234-1, Footnote 3)

See Photo(s)

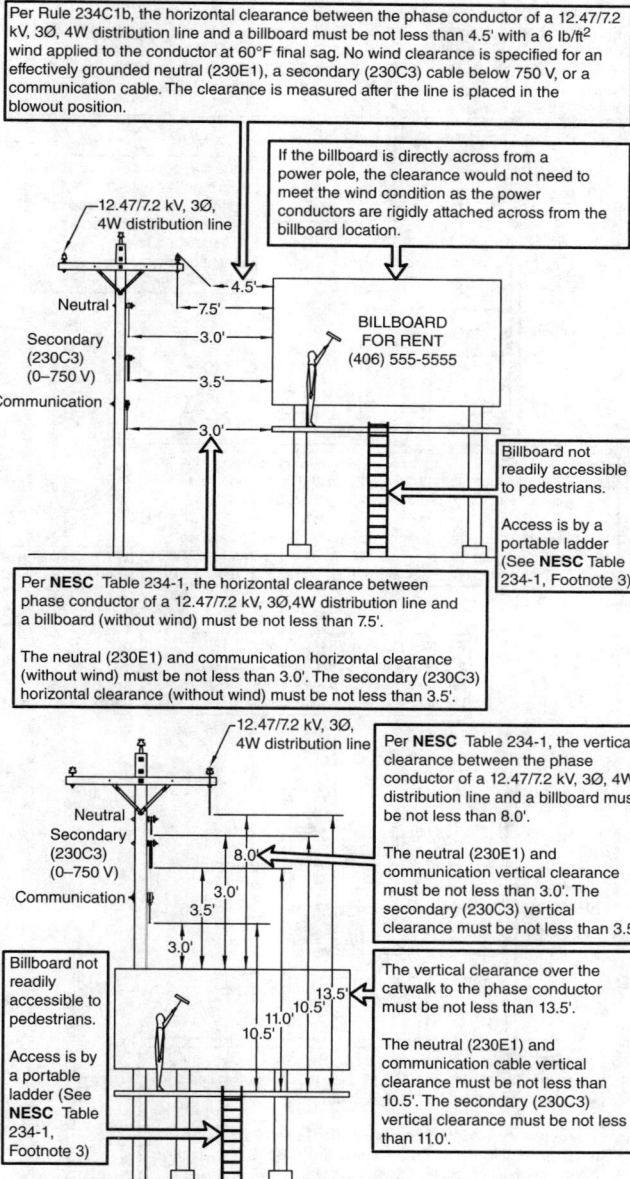

Fig. 234-11. Example of clearance of a conductor to a billboard (Rule 234C1).

Per Rule 234C1b, the horizontal clearance between the phase conductor of a
12.47/7.2 kV, 3Ø, 4W distribution line and a flag pole must be not less than 4.5' with
a 6 lb/ft^2 wind applied to the conductor at 60°F final sag. No wind clearance is
specified for an effectively grounded neutral (230E1), a secondary (230C3) cable
below 750 V, or a communication cable. The clearance is measured after the line is
placed in the blowout position.

12.47/7.2 kV, 3Ø,
4W distribution line

If the flag pole is directly across
from a power pole, the
clearance would not need to
meet the wind condition as the
power conductors are rigidly
attached across from the flag
pole location.

Wind

Neutral — 4.5'
— 7.5'

Secondary — 3.0'
(230C3)
(0–750 V) — 3.5'

Communication — 3.0'

Per **NESC** Table 234-1, the horizontal clearance between the
phase conductor of a 12.47/7.2 kV, 3Ø, 4W distribution line and a
flag pole (without wind) must be not less than 7.5'.

The neutral (230E1) and communication horizontal clearance
(without wind) must be not less than 3.0'. The secondary (230C3)
horizontal clearance (without wind) must be not less than 3.5'.

Both conditions must be met
(not one or the other). The
size of the flag will determine
which condition is greater.

Per **NESC** Table 234-1, Footnote 17,
the flag or banner is presumed to be
fully extended (by a light wind) but the
conductors are not displaced.

12.47/7.2 kV, 3Ø
4W distribution line

Neutral — 7.5'

Secondary — 3.0'
(230C3)
(0–750 V) — 3.5'
Wind

Communication — 3.0'

MONTANA

Per **NESC** Table 234-1, the horizontal clearance between the
phase conductor of a 12.47/7.2 kV, 3Ø, 4W distribution line and
a flag or banner must be not less than 7.5'

The neutral (230E1) and communication horizontal clearance
must be not less than 3.0'. The secondary (230C3) horizontal
clearance must be not less than 3.5'.

See
Photo(s)

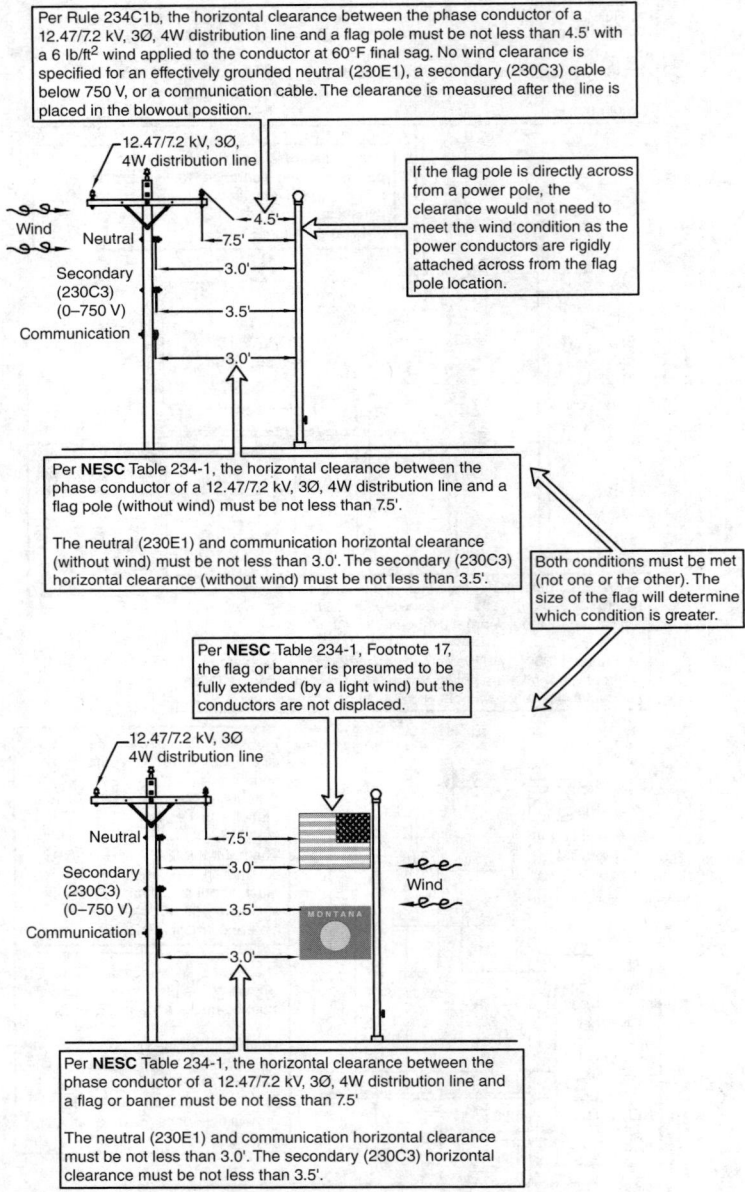

Fig. 234-12. Example of horizontal clearance of a conductor to a flagpole and flag
(Rule 234C1).

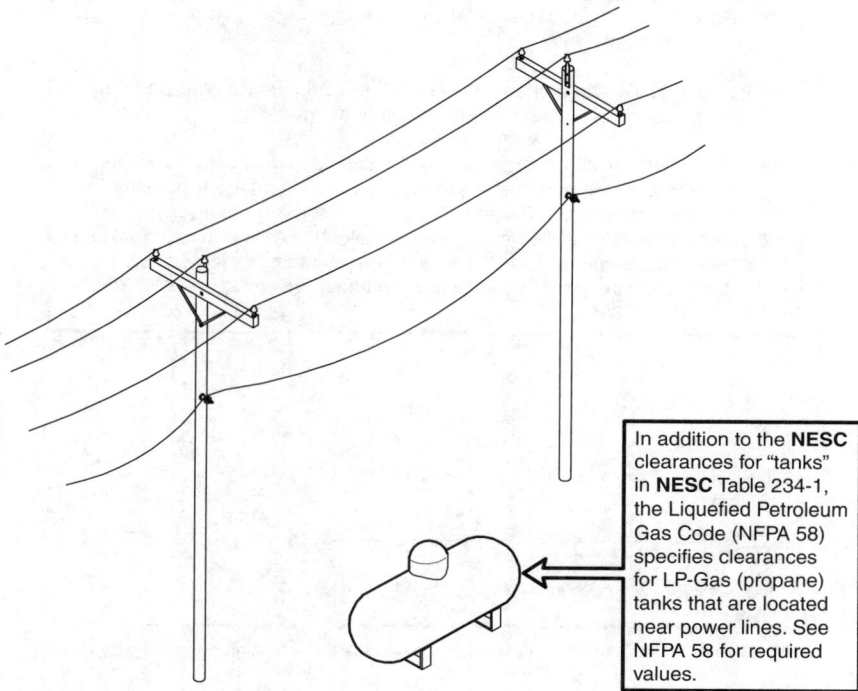

In addition to the **NESC** clearances for "tanks" in **NESC** Table 234-1, the Liquefied Petroleum Gas Code (NFPA 58) specifies clearances for LP-Gas (propane) tanks that are located near power lines. See NFPA 58 for required values.

Fig. 234-13. Liquefied Petroleum Gas Code (NFPA 58) requirements (Rule N/A).

The clearance adders for voltages above 22 kV are provided in Rule 234G. Alternate clearances for voltages exceeding 98 kV AC to ground or 139 kV DC to ground are provided in Rule 234H.

234D. Clearance of Wires, Conductors, Cables, and Unguarded Rigid Live Parts from Bridges.   Depending on the size and design of a bridge structure, power and communication lines and equipment may be located adjacent to or within a bridge structure. NESC Table 234-2 applies to bridge structures. NESC Table 234-2 addresses clearances over bridges when conductors are attached and when not attached. The table also addresses clearance beside, under, and within the bridge structure. The clearances in Table 234-2 are for the at-rest (no wind) conditions outlined in Rule 234A. The main concerns addressed in NESC Table 234-2 are outlined in Fig. 234-19.

NESC Figs. 234-1(a), (b), and (c), which focus on horizontal, vertical, and transitional clearances for signs and buildings, are also applicable to bridges. Bridges, unlike buildings and signs, do not have any required clearance to insulated communication cables and neutrals meeting Rule 230E1.

The horizontal clearance to bridges with wind displacement is outlined in a small table in Rule 234D. The table does not have an NESC table number. The table does not address 230E1 (neutral conductors), 230C3 cables below 750 V (e.g., 120/240-V secondary triplex conductors), or communication cables.

The **NESC** does not specifically address clearance between a power line and irrigation equipment.

NESC Table 234-1 can be used if the irrigation equipment is considered an "other installation not classified as buildings or bridges".

NESC Rule 012, which requires accepted good practice, must be used for the water stream. Common references are "Guidelines for the Installation and Operation of Irrigation Systems Near High Voltage Transmission Lines" published by the Bonneville Power Administration (BPA) and "IEEE Guide for Installation, Maintenance and Operation of Irrigation Equipment Located Near or Under Power Lines" published by the Institute of Electrical and Electronics Engineers (IEEE Std. 1542).

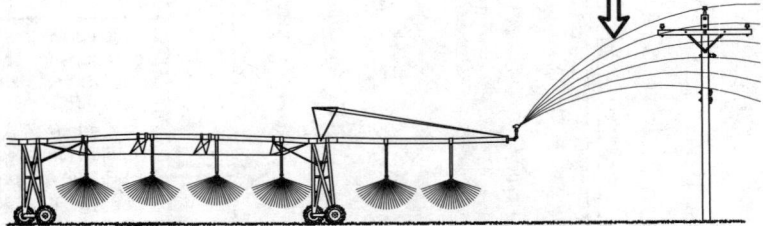

See Photo(s)

Fig. 234-14. Example of clearance between a power line and irrigation equipment (Rule 234C).

The clearance adders for voltages above 22 kV are provided in Rule 234G. Alternate clearances for voltages exceeding 98 kV AC to ground or 139 kV DC to ground are provided in Rule 234H.

Underground lines that have conduits attached to bridges are covered in Rules 320A, 322B, and 351C.

234E. Clearance of Wires, Conductors, Cables, or Unguarded Rigid Live Parts Installed over or near Swimming Areas with No Wind Displacement. Rule 234E applies not only to swimming pools but also to swimming areas and beaches. This rule applies to both in-ground swimming pools and permanently installed aboveground swimming pools. **NESC** Table 234-3 and **NESC** Figs. 234-3(a), 234-3(b), and 234-3(c) are provided in the **Code** to check clearance to both in-ground and aboveground swimming pools. Note 1 in Rule 234E clarifies the application of permanently installed aboveground swimming pools. The main concerns addressed in **NESC** Table 234-3 and **NESC** Figs. 234-3(a), 234-3(b), and 234-3(c) are outlined in Fig. 234-20.

A fully enclosed indoor pool does not apply to Rule 234E1 per Exception 1. A fully enclosed indoor pool is considered to be in a building therefore Rule 234C is applicable.

Exception 2 to Rule 234E1 exempts communication conductors and cables, effectively grounded surge-protection wires, neutral conductors meeting Rule 230E1, guys and messengers, supply cables meeting Rule 230C1, and supply cables of 0 to 750 V meeting Rule 230C2 or 230C3 when these facilities are 10 ft or more horizontally

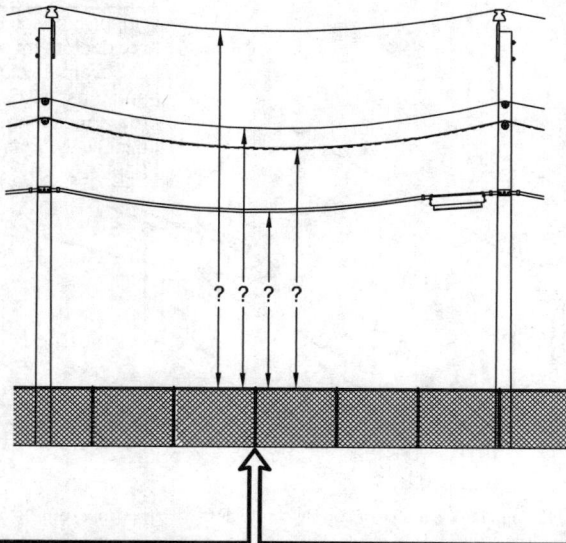

Fig. 234-15. Example of clearance of power and communication lines above a fence (Rule 234C).

> The **NESC** does not specifically address clearance over fences or walls that are not part of a building structure.
>
> **NESC** Table 234-1 can be used if the fence or wall is considered "other installations not classified as buildings or bridges." For clearances over a wall that is wide enough to walk on, the vertical clearance over "surfaces upon which personnel walk" is appropriate. For clearances over a fence that is not wide enough to walk on, the vertical clearance over "other portions of such installations" is appropriate.
>
> The clearances for wires, conductors, and cables above ground, roadway, rail, or water surfaces adjacent to the fence must also be met per **NESC** Table 232-1.
>
> **NESC** Table 234-1, Footnote 14 (which applies to conductors adjacent but not attached to buildings) and **NESC** Rule 234C3d(1), Exception 1 (which applies to supply conductors attached to buildings) address clearances above railings, walls, or parapets around balconies, decks, or roofs.

from the edge of the pool, diving platform, diving tower, water slide, or other fixed pool-related structures. At that location, Rule 232 clearances apply.

The use of underground power is not required but it is certainly an acceptable design practice for routing lines near swimming pools. If underground lines are used, Sec. 35, "Direct-buried cable and cable in duct not part of a conduit system," has rules for underground lines near swimming pools.

Note 2 in Rule 234E1 clarifies the clearance requirements for items that are similar to swimming pools but are not suitable for swimming and therefore do not

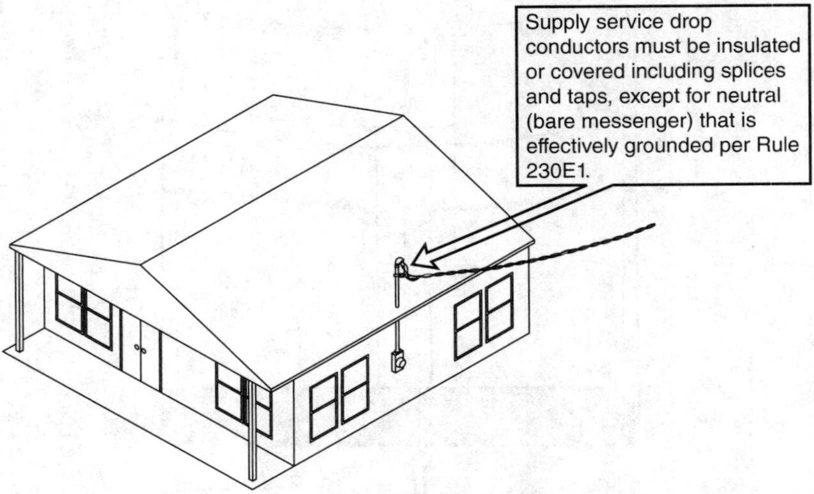

Supply service drop conductors must be insulated or covered including splices and taps, except for neutral (bare messenger) that is effectively grounded per Rule 230E1.

Fig. 234-16. Example of clearance of supply conductors (service drops) attached to buildings (Rule 234C3).

have skimmer and rescue poles associated with them. These items include whirlpools, hot-tubs, jacuzzis, spas, and wading pools. The note provides guidance as to how to treat each item for clearance purposes.

Swimming areas that are not pools (e.g., beaches, etc.) and have lifeguards using rescue poles must also comply with Table 234-3. The clearances developed for swimming pools were based on the use of the long aluminum rescue and skimmer poles that are used near poolsides. If these poles are also used at a beach area, Table 234-3 applies. If these poles are not used at beach areas, the clearances in Rule 232 (clearance above ground or water) apply. This is also true for water skiing areas. Rule 232 focuses on clearances above water as well as land and considers boating.

The clearance adders for voltages above 22 kV are provided in Rule 234G. Alternate clearances for voltages exceeding 98 kV AC to ground or 139 kV DC to ground are provided in Rule 234H.

234F. Clearance of Wires, Conductors, Cables, and Rigid Live Parts from Grain Bins.   Grain bins have unique features that segregate them from Rule 234C (buildings and other installations) and place them in their own dedicated rule. The building clearance rules do apply to grain bins with some very specific modifications. Grain bins are separated into two types, grain bins loaded by permanently installed augers, conveyers, or elevators (Rule 234F1), and grain bins loaded by portable augeurs, conveyers, or elevators (Rule 234F2). **NESC** Figs. 234-4(a) and 234-4(b) are provided in the **Code** to check clearance to grain bins. The main concerns that are addressed in **NESC** Figs. 234-4(a) and 234-4(b) are outlined in Fig. 234-21.

The horizontal clearance to a grain bin loaded by a portable auger can become quite large for tall grain bins as the horizontal clearance is based on the height of the grain bin. See Fig. 234-22.

Exception:
For a duplex, triplex, or quadruplex (230C3 cable) not exceeding
750 V, the clearance may be not less than 3' if the roof is not
readily accessible. The basic clearance is 10' for readily accessible
roofs.

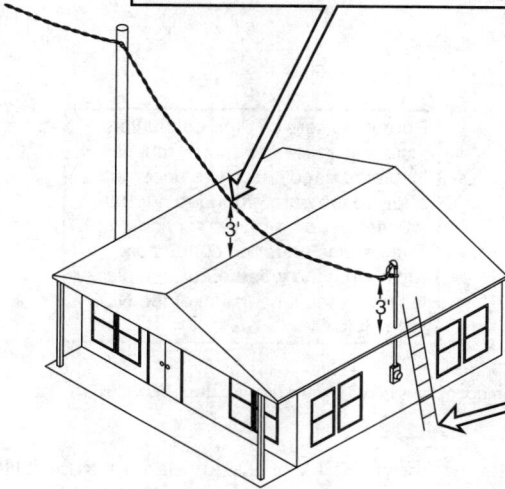

3'

3'

Portable ladder is the method
for roof access. Roof is not
readily accessible to
pedestrians. (See **NESC**
Table 234-1, Footnote 3.)

Exception:
For a duplex, triplex, or quadruplex
(230C3 cable) not exceeding 750 V, the
clearance may be not less than 18" for a
specific zone around the conduit mast,
which is shown in **NESC** Fig. 234-2 and
NESC Table 234-6.

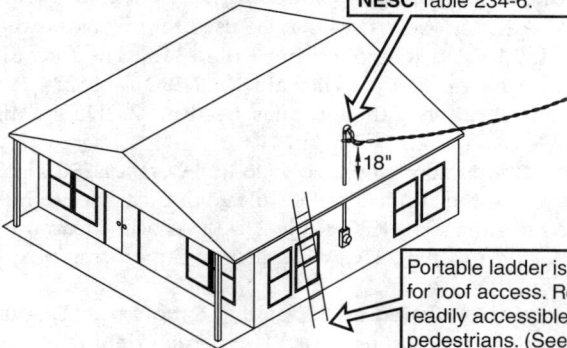

18"

Portable ladder is the method
for roof access. Roof is not
readily accessible to
pedestrians. (See **NESC**
Table 234-1, Footnote 3.)

See
Photo(s)

Fig. 234-17. Example of clearance of supply conductors (service drops) attached to buildings
(Rule 234C3).

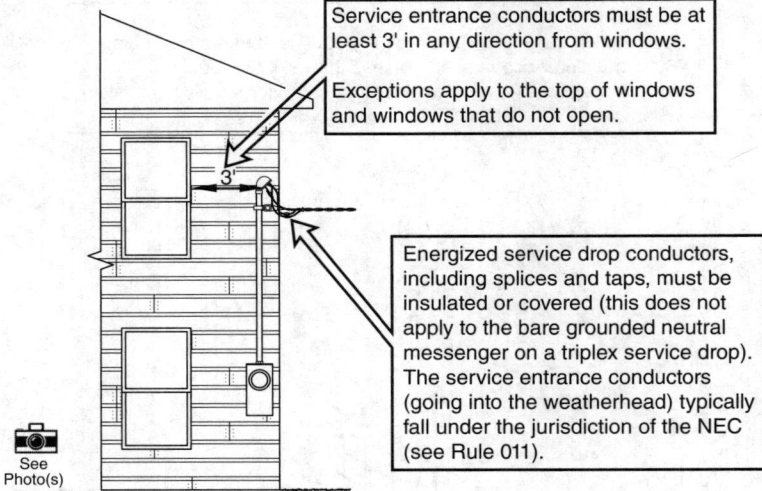

Service entrance conductors must be at least 3' in any direction from windows.

Exceptions apply to the top of windows and windows that do not open.

Energized service drop conductors, including splices and taps, must be insulated or covered (this does not apply to the bare grounded neutral messenger on a triplex service drop). The service entrance conductors (going into the weatherhead) typically fall under the jurisdiction of the NEC (see Rule 011).

See Photo(s)

Fig. 234-18. Example of clearance of supply conductors (service drops) attached to buildings (Rule 234C3).

The clearance adders for voltages above 22 kV are provided in Rule 234G. Alternate clearances for voltages exceeding 98 kV AC to ground or 139 kV DC to ground are provided in Rule 234H.

234G. Additional Clearances for Voltages Exceeding 22 kV for Wires, Conductors, Cables, and Unguarded Rigid Live Parts of Equipment. The clearance adders in Rule 234G which apply to Rules 234B, 234C, 234D, 234E, 234F, and 234J are similar to the clearance adders used in Rule 232C. See the discussion in Rule 232C.

234H. Alternate Clearances for Voltages Exceeding 98 kV AC to Ground or 139 kV DC to Ground. Alternate clearances may be used for voltages exceeding 98 kV AC to ground or 139 kV DC to ground per Rule 234H. The formulas in Rule 234H must be used for this calculation. The values in **NESC** Table 234-4 can be used to check the proper application of the formulas. See Rule 232D for a discussion of the alternate clearance method.

234I. Clearance of Wires, Conductors, and Cables to Rail Cars. NESC Fig. 234-5 is provided in the Code to aid the application of Rule 234I. Clearance to rail cars is also indirectly covered in Rule 232, **NESC** Table 232-1, Vertical Clearance over Track Rails of Railroads, and in Rule 231C, Clearance of Supporting Structures from Railroad Tracks.

234J. Clearance of Equipment Mounted on Supporting Structures. Unguarded rigid live parts are not subject to variations in sag. Unguarded rigid live parts are required to meet the clearances provided in Rule 234C or 234D, as applicable. Equipment cases, if effectively grounded, may be located on or adjacent to buildings, bridges, or other structures. A grounded meter enclosure is a common example of a grounded equipment case mounted on or adjacent to a building. If an equipment case is not effectively grounded, the clearances in Rules 234C or 234D, as applicable, must be applied. See Rule 215 for overhead equipment grounding requirements and see Sec. 02 for a

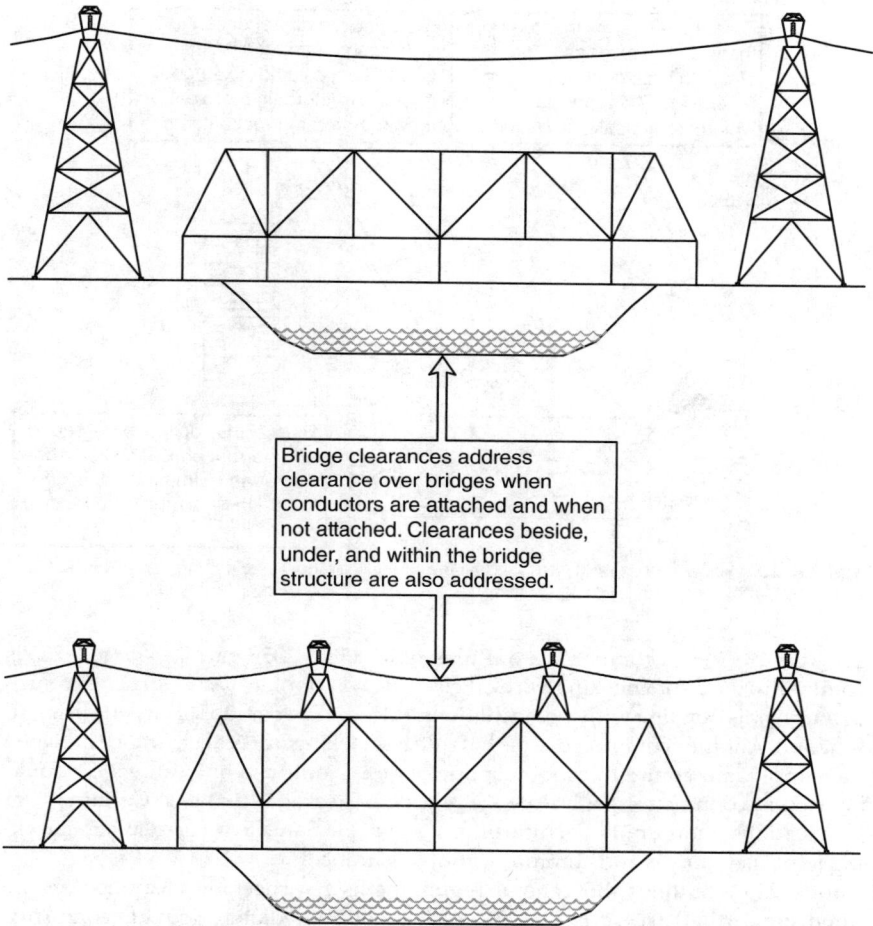

Bridge clearances address clearance over bridges when conductors are attached and when not attached. Clearances beside, under, and within the bridge structure are also addressed.

See
Photo(s)

Fig. 234-19. Concerns that are addressed when providing clearance to bridges (Rule 234D).

discussion of the term "effectively grounded." The **NESC** does not address the clearance between a pole and a building. An example of a pole, unguarded rigid live part, and grounded equipment case adjacent to a building is shown in Fig. 234-23.

235. CLEARANCE FOR WIRES, CONDUCTORS, OR CABLES CARRIED ON THE SAME SUPPORTING STRUCTURE

235A. Application of Rule.   Rule 235 is the opposite of Rule 233, Clearances between Wires, Conductors, and Cables Carried on Different Supporting Structures. Rule 235 addresses conductors on the same supporting structure (pole).

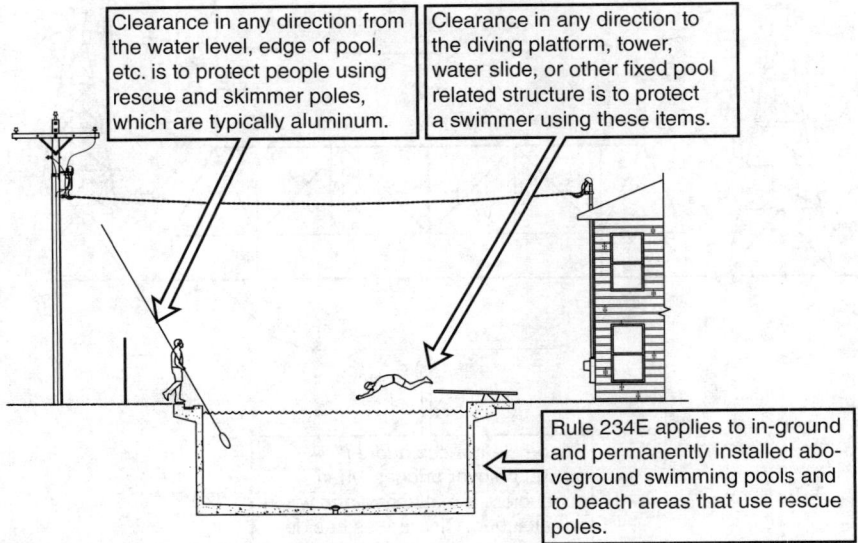

Fig. 234-20. Concerns that are addressed when providing clearance to swimming pools (Rule 234E).

See Rule 233B for a comparison of Rules 233B, 233C, 235B, and 235C. Rule 235 is commonly used for checking vertical clearance on double circuit structures such as a transmission line with a distribution underbuild or a double circuit distribution line. Another common use of this rule is checking vertical clearance between the supply and communication conductors on a joint-use line. Rule 235 should be used in conjunction with Rule 238, Vertical Clearance between Certain Communications and Supply Facilities Located on the Same Structure, when checking joint-use (power and communication) clearances.

Rule 235A outlines the general requirements for checking clearance on the same supporting structure. Rule 235A does not immediately provide conductor temperature and loading conditions as Rules 232A, 233A, and 234A do. The conductors in Rule 235 are tied onto a structure. Conductor temperature and loading conditions will be needed to check conditions out in the span. Conductor temperature and loading conditions are provided in Rule 235B1b for horizontal clearance and Rule 235C2b(1)(c) for vertical clearance.

A multiconductor cable meeting Rule 230C or 230D is considered a single conductor for the purposes of this rule. See Fig. 235-1.

When comparing line conductors of different circuits, the voltage between the conductors is determined using the greater of the phasor difference or the phase-to-ground voltage of the higher circuit. See Fig. 235-2.

235B. Horizontal Clearance between Line Conductors. Rule 235B applies to the horizontal clearance between line conductors attached to fixed supports (rigid insulators) on the same supporting structure. The horizontal clearance requirements apply between conductors of the same circuit or different circuits

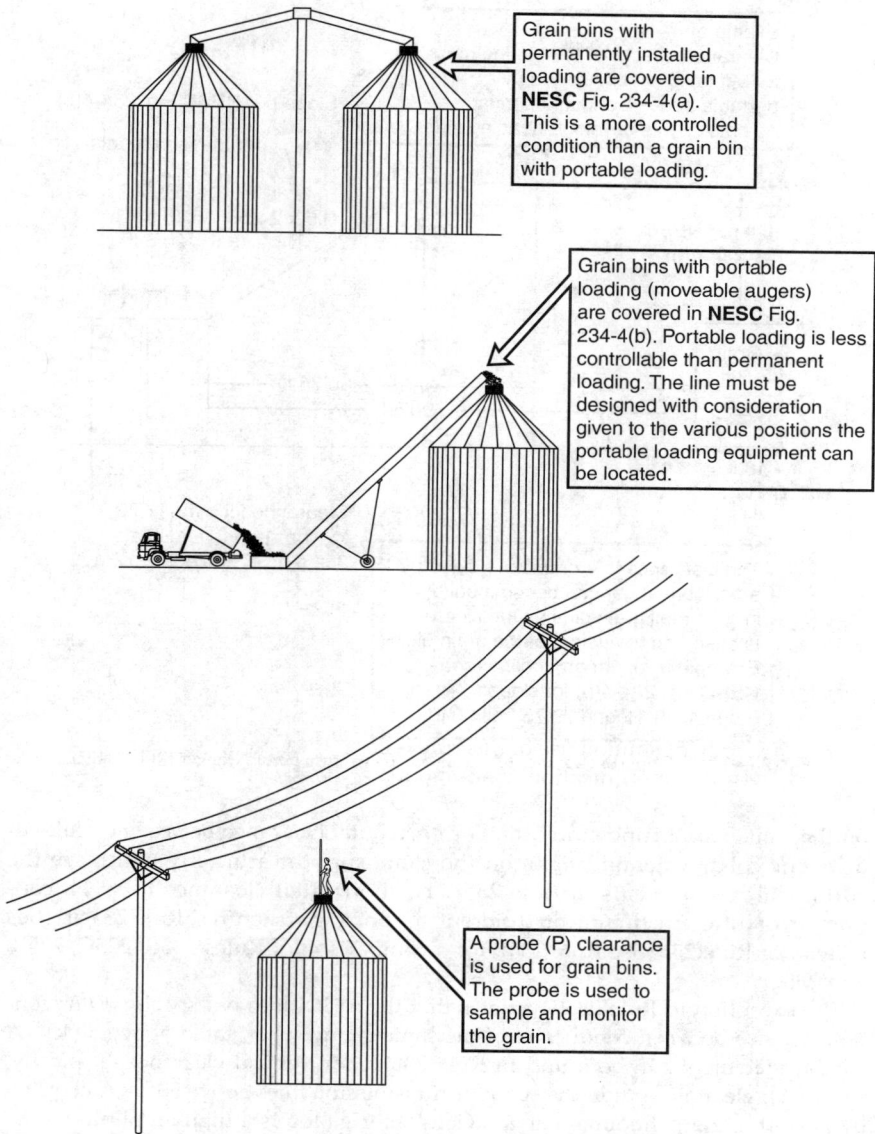

Grain bins with permanently installed loading are covered in **NESC** Fig. 234-4(a). This is a more controlled condition than a grain bin with portable loading.

Grain bins with portable loading (moveable augers) are covered in **NESC** Fig. 234-4(b). Portable loading is less controllable than permanent loading. The line must be designed with consideration given to the various positions the portable loading equipment can be located.

A probe (P) clearance is used for grain bins. The probe is used to sample and monitor the grain.

See Photo(s)

Fig. 234-21. Concerns that are addressed when providing clearance to grain bins (Rule 234F).

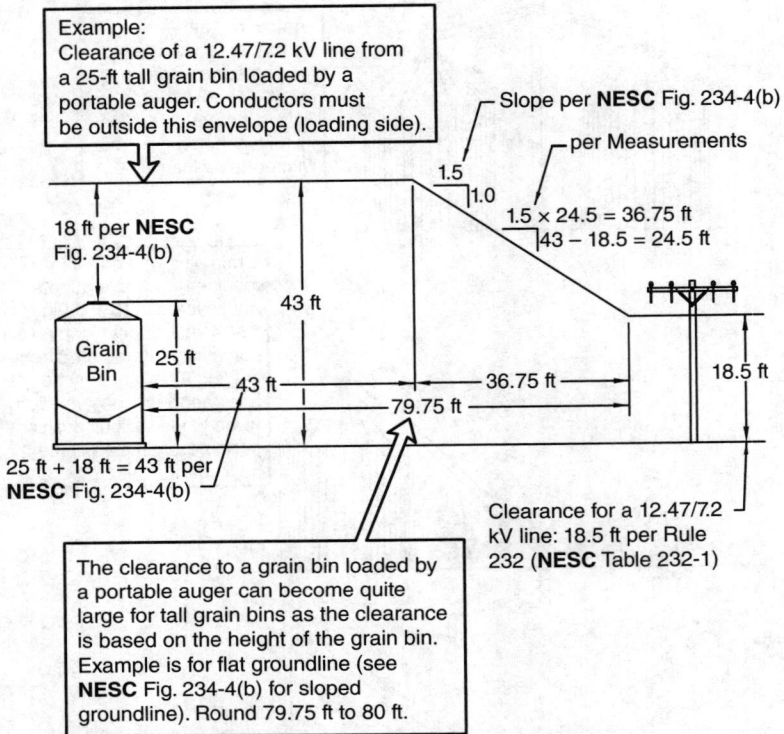

Example:
Clearance of a 12.47/7.2 kV line from a 25-ft tall grain bin loaded by a portable auger. Conductors must be outside this envelope (loading side).

Slope per **NESC** Fig. 234-4(b)

per Measurements

1.5

1.0

$1.5 \times 24.5 = 36.75$ ft
$43 - 18.5 = 24.5$ ft

18 ft per **NESC** Fig. 234-4(b)

43 ft

Grain Bin   25 ft

43 ft

36.75 ft

18.5 ft

79.75 ft

25 ft + 18 ft = 43 ft per **NESC** Fig. 234-4(b)

Clearance for a 12.47/7.2 kV line: 18.5 ft per Rule 232 (**NESC** Table 232-1)

The clearance to a grain bin loaded by a portable auger can become quite large for tall grain bins as the clearance is based on the height of the grain bin. Example is for flat groundline (see **NESC** Fig. 234-4(b) for sloped groundline). Round 79.75 ft to 80 ft.

Fig. 234-22. Example of clearance to a grain bin loaded by a portable auger (Rule 234F2).

on the same supporting structure. For horizontal clearance between circuits of different voltage classifications on the same support arm (crossarm), see the additional requirements in Rule 235F. For horizontal clearance between conductors of different circuits on different supporting structures, Rule 233 applies instead of Rule 235. See Rule 233B for a comparison of Rules 233B, 233C, 235B, and 235C.

The exception to Rule 235B1b states that the **NESC** does not specify a horizontal clearance between conductors of the same circuit when rated above 50 kV. A similar statement can be found in Rule 235C1 for vertical clearance. Typically, conductor clearance out in the span for transmission lines above 50 kV is checked by performing a galloping analysis. Conductor gallop is a high-amplitude, low-frequency oscillation. See Fig. 235-3.

The horizontal clearance must be checked at the support (crossarm) using **NESC** Table 235-1 and out in the span based on sag using either **NESC** Table 235-2 or **NESC** Table 235-3 depending on the conductor size. **NESC** Tables 235-2 and 235-3 have a formula at the bottom of each table. The table or the formula can be used. The formulas are also presented in Rule 235B1b. The horizontal clearance at the structure will typically be controlled by the clearance required out in the span based on sag. See Fig. 235-4.

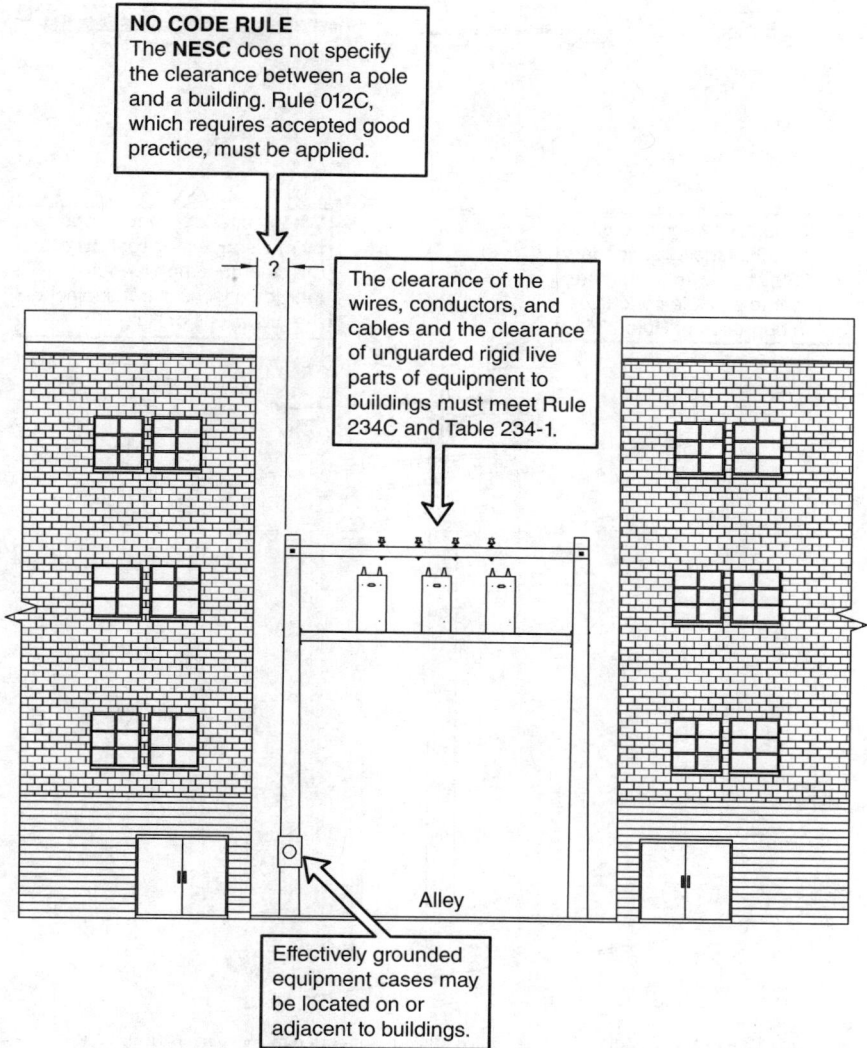

NO CODE RULE
The **NESC** does not specify the clearance between a pole and a building. Rule 012C, which requires accepted good practice, must be applied.

The clearance of the wires, conductors, and cables and the clearance of unguarded rigid live parts of equipment to buildings must meet Rule 234C and Table 234-1.

Alley

Effectively grounded equipment cases may be located on or adjacent to buildings.

Fig. 234-23. Example of a pole, conductors, unguarded rigid live parts, and grounded equipment case adjacent to a building (Rule 234J).

The values found in **NESC** Tables 235-1, 235-2, and 235-3 are clearances between conductors at supports, not spacing between supports. See Rule 230A3 for a discussion of spacing and clearance. Live metallic hardware electrically connected to the conductor, like a conductor tie wire on an insulator, must be addressed for the clearance at the structure but not for the clearance at the structure based on the sag out in the span. See Fig. 235-5.

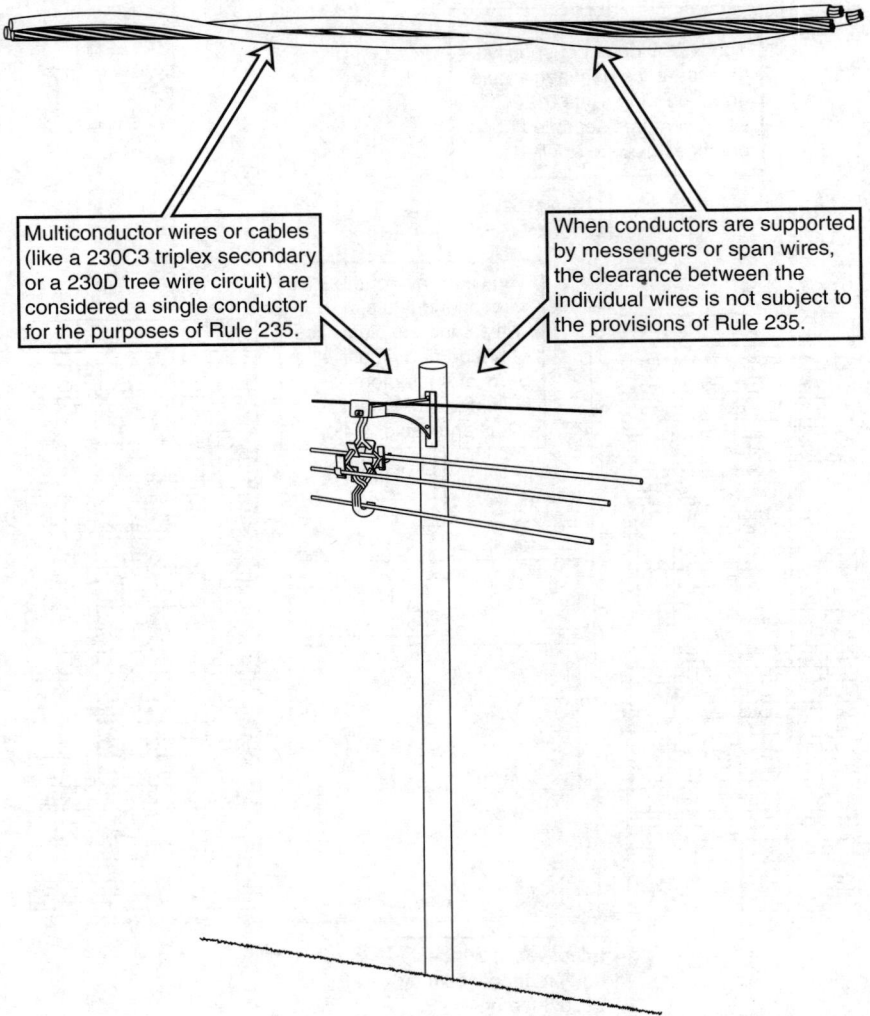

Multiconductor wires or cables (like a 230C3 triplex secondary or a 230D tree wire circuit) are considered a single conductor for the purposes of Rule 235.

When conductors are supported by messengers or span wires, the clearance between the individual wires is not subject to the provisions of Rule 235.

Fig. 235-1. Multiconductor cables and conductors supported by messengers (Rule 235A).

Rule 235B2 has requirements for suspension insulators because **NESC** Table 235-1 is for fixed supports. The rules for suspension insulators require applying a 6-lb/ft^2 wind to the conductor that the suspension insulator is holding. The wind pressure may be reduced to 4 lb/ft^2 in sheltered areas but trees are not considered a shelter to a line. The deflection of flexible structures and fittings must also be considered if the deflection reduces the clearance in question. See Rules 233A1a(2) and 234A2 for similar requirements.

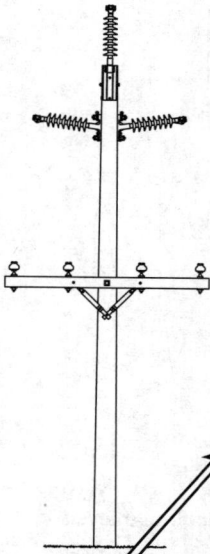

Top circuit:
115 kV transmission line fed from a
3Ø, 4W-wire grounded Y source
with ground fault relaying at an
elevation of 3000'.

$\dfrac{115 \text{ kV (line to line)}}{\sqrt{3}}$ = 66.40 kV (line to ground)

66.40 kV × 1.05 $\left(\begin{smallmatrix}\text{maximum operating}\\\text{voltage multiplier}\end{smallmatrix}\right)$ = 69.72 kV
Per Rule 235C2a(3).

(If 115 kV were fed from a 3Ø, 3 wire
delta source, the full 115 kV voltage
must be used.)

Bottom circuit:
12.47/7.2 kV 3Ø, 4W
effectively grounded
distribution line.

(1) **The phasor difference:**

Graphically:

Use 180° phasor relationship where
actual phasor relationship is not known.

Mathematically:

7.2 ∠ 180°   69.72 ∠ 0°

69.72 ∠ 0°–7.2 ∠ 180°

(69.72 + j0) – (–7.2 + j0)

Voltage between
circuits must be the
greater of (1) or (2). In
this case the value
76.92 is greater.

76.92 + j0

76.92 ∠ 0° kV

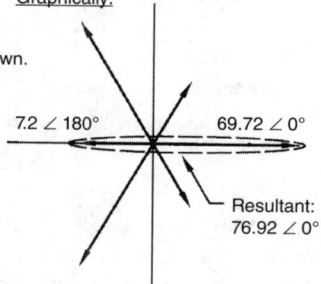

Resultant:
76.92 ∠ 0°

(2) **The phase-to-ground voltage of the higher circuit:**

$\dfrac{115}{\sqrt{3}}$ × 1.05 $\left(\begin{smallmatrix}\text{maximum operating}\\\text{voltage multiplier}\end{smallmatrix}\right)$ = 69.72 kV

If both circuits were 115 kV then either
circuit could be considered the higher
voltage circuit.

Fig. 235-2. Example of the voltage between line conductors of different circuits (Rule 235A).

Rule 235B3 provides alternate clearances for voltages exceeding 98 kV AC to
ground or 139 kV DC to ground. The formulas in Rule 235B3 must be used for
this calculation. The values in **NESC** Table 235-4 can be used to check the proper
application of the formulas. See Rule 232D for a discussion of alternate clearances.

Rule 235B can be applied using two different approaches. One approach is to
find the required horizontal clearance of a particular span. Another approach is
to determine a maximum span based on the horizontal clearance for a particular
structure type, conductor size, conductor sag, and circuit voltage.

> No horizontal clearance is specified between conductors of the same circuit out in the span when the voltage is above 50 kV. Rule 012C, which requires accepted good practice, must be applied. Typically, a galloping analysis is done to check horizontal and vertical clearances.

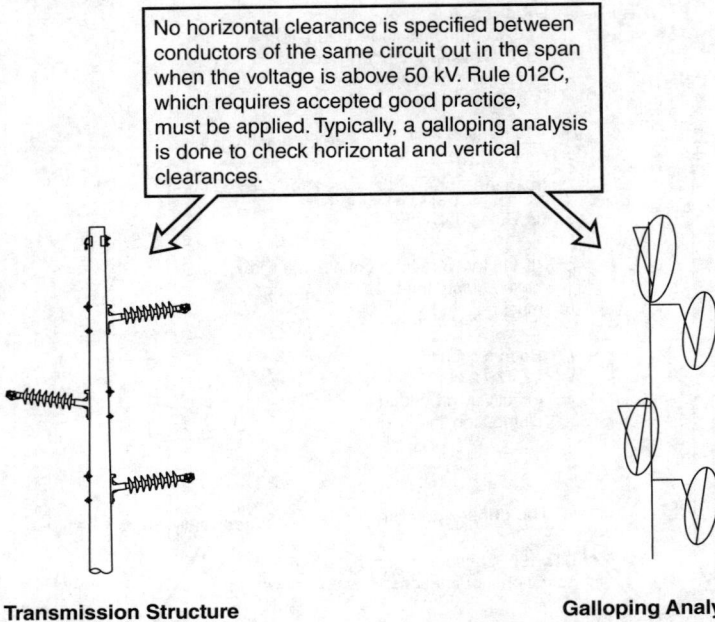

Transmission Structure **Galloping Analysis**

Fig. 235-3. Horizontal clearance between conductors of the same circuit rated above 50 kV (Rule 235B1b).

235C. Vertical Clearance at the Support for Line Conductors and Service Drops. Rule 235C applies to the vertical clearance between conductors at the support on the same supporting structure (e.g., pole). Rule 235C applies to line conductors and service drops. The terms line conductors (conductors, line) and service drop are both defined in **NESC** Sec. 02, Definitions of Special Terms. The vertical clearance requirements apply to conductors of the same and different circuits on the same supporting structure. For vertical clearance between conductors of different circuits on different supporting structures, Rule 233 applies instead of Rule 235. See Rule 233B for a comparison of Rules 233B, 233C, 235B, and 235C. Rule 235C1a describes two items that the rule does not cover and therefore **NESC** Table 235-5 does not apply to. First, the **NESC** does not specify a vertical clearance between conductors of the same circuit when rated above 50 kV. A similar statement can be found in an exception to Rule 235B1b for horizontal clearance out in the span. Second, Rule 235C1a does not specify a vertical clearance between ungrounded open supply conductors (e.g., bare noninsulated phase wires) of the same phase and circuit of the same utility (e.g., a tap off the main line of 0-50 kV). See Fig. 235-6.

The vertical clearance must be checked at the support (e.g., pole) using **NESC** Table 235-5 and out in the span based on sag using Rule 235C2b. Typically, the vertical clearance at the structure will be controlled by the clearance required out in the span based on the upper and lower conductor sags.

NESC Table 235-1 is used to determine the required clearance at the supports, but the clearance at the supports may need to be greater than the value in NESC Table 235-1 to meet NESC Table 235-2 or 235-3.

Example: Horizontal clearance of 1/0 ACSR conductors on a 12.47/7.2 kV 3Ø, 4W circuit.

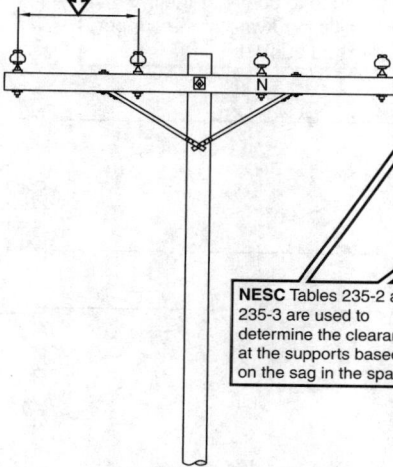

NESC Tables 235-2 and 235-3 are used to determine the clearance at the supports based on the sag in the span.

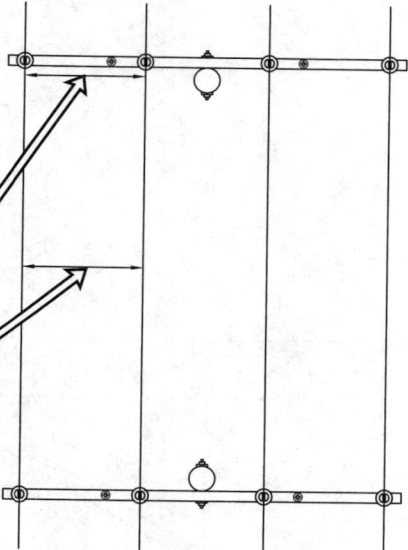

Using NESC Table 235-1

- Voltage between conductors involved is 12.47 kV for the two insulator pins on the left side of the crossarm.

- Per **NESC** Table 235-1 the required horizontal clearance is:

  12" + 0.4" per kV over 8.7 kV
  12" + 0.4" (12.47 − 8.7)
  12" + 0.4" (3.77)
  12" + 1.508"
  13.5"
  Round to 14"

Using NESC Table 235-3
(AWG #2 or larger conductor)

- Using the 1/0 ACSR sag chart at the beginning of Sec. 23, the sag at 60°F final with no wind is 4.00'.

- Per **NESC** Table 235-3, clearance required at the support for 12.47 kV between conductors, with 48" of sag, is 20".

The larger clearance (20") is required at the supports. Using this value limits the structure to a 300' maximum span as the sag value used in the calculation was from a sag chart with a 300' ruling span. A larger clearance is typically used between 12.47 kV insulators to permit longer span lengths.

Fig. 235-4. Example of horizontal clearance between line conductors (Rule 235B).

Per Rule 235C2b(1)(c), the upper conductor must be checked at 120°F or the maximum operating temperature at final sag while the lower conductor is at the same ambient temperature as the upper conductor without electrical loading at final sag. The upper conductor must also be checked at 32°F with ice from the Zone in Rule 230B at final sag, while the lower conductor is at the same ambient temperature as the upper conductor without electrical loading or ice at final sag.

Clearance between conductors at supports (Rule 235B1a) must consider live metallic hardware (e.g., conductor ties) electrically connected to the conductor per Rule 230A3. The clearance between conductors according to sags (out in the span) does not need to consider the live metallic hardware at the support per Rule 235B1b.

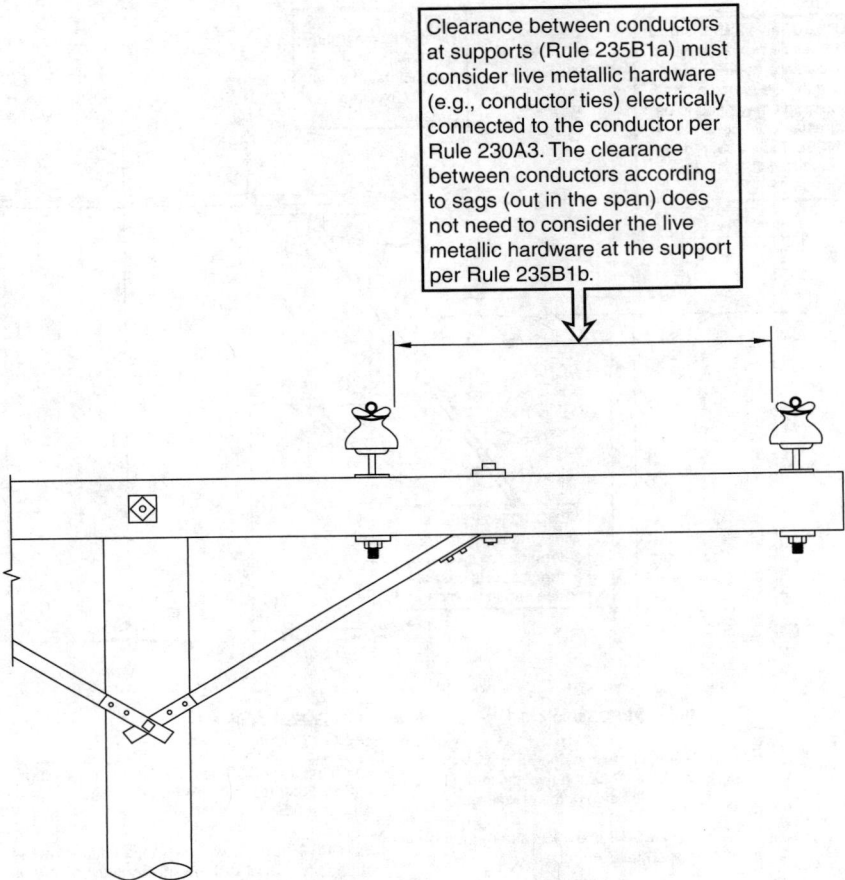

Fig. 235-5. Horizontal clearance measurements between conductors at supports (Rule 235B).

Using the 1/0 ACSR sample sag and tension chart at the beginning of Sec. 23, assume the following for an example of applying Rule 235C2b(1)(c)i:

- 1/0 ACSR phase conductor of a 12.47/7.2 kV, 3Ø, 4W, effectively grounded line is loaded to approximately 75% capacity (175 A) at both winter and summer peak periods, medium loading, clearance Zone 2, 300-ft ruling span. See the discussion of this example at the beginning of Sec. 23.
- Upper phase conductor at 167°F, final. Sag is 6.25 ft.
- Lower neutral conductor at 104°F, final. Sag is 5.34 ft.

If the top and bottom conductors were attached at the same height, the top conductor would be sagging 0.91 ft (6.25 ft – 5.34 ft = 0.91 ft) below the bottom conductor. The top conductor attachment must be raised 0.91 ft (11 in) to make the sags level plus 12 in (16 in × 0.75 = 12 in per **NESC** Table 235-5 and Rule 235C2b(1)(a)) to meet the required clearance in the span for a 7.2-kV phase conductor over a

No vertical clearance is specified between conductors of the same circuit when the voltage is above 50 kV. Rule 012C, which requires accepted good practice, must be applied. Typically, a galloping analysis is done to check horizontal and vertical clearances.

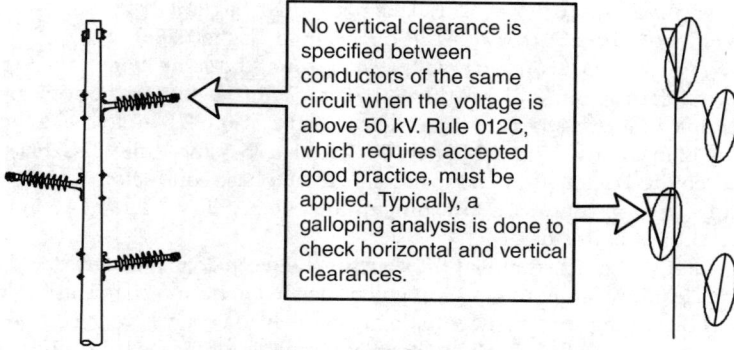

Conductors of same circuit above 50 kV

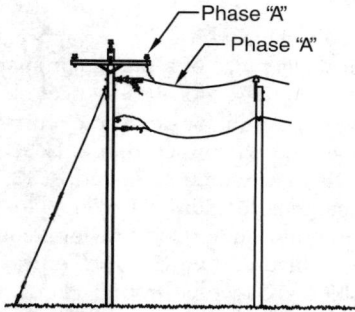

Phase "A"

Phase "A"

Conductors of same phase and circuit, 0–50 kV

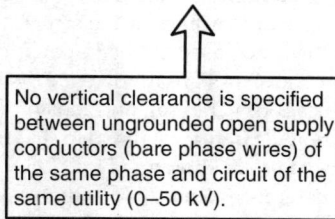

No vertical clearance is specified between ungrounded open supply conductors (bare phase wires) of the same phase and circuit of the same utility (0–50 kV).

Fig. 235-6. Items not covered under vertical clearance between conductors (Rule 235C1a).

neutral conductor. The top conductor attachment must be 23 in (11 in + 12 in = 23 in) higher than the bottom conductor.

Using the 1/0 ACSR sample sag and tension chart at the beginning of Sec. 23, assume the following for an example of applying Rule 235C2b(1)(c)ii:

• 1/0 ACSR phase conductor of a 12.47/7.2 kV, 3Ø, 4W, effectively grounded line is loaded to approximately 75% capacity (175 A) at both winter and summer peak periods, medium loading, clearance Zone 2, 300-ft ruling span. See the discussion of this example at the beginning of Sec. 23.

- Upper phase conductor at 32°F, 0.25 in ice, final. Sag is 4.49 ft.
- Lower neutral conductor at −20°F, no ice, final. Sag is 1.78 ft.

If the top and bottom conductors were attached at the same height, the top conductor would be sagging 2.71 ft (4.49 ft − 1.78 ft = 2.71 ft) below the bottom conductor. The top conductor attachment must be raised 2.71 ft (32.5 in) to make the sags level plus 12 in (16 in × 0.75 = 12 in per **NESC** Table 235-5 and Rule 235C2b(1)(a)) to meet the required clearance in the span for a 7.2-kV phase conductor over a neutral conductor. The top conductor attachment must be 32.5 in + 12 in = 44.5 in higher than the bottom conductor.

The greater vertical clearance at the structure between 23 in and 44.5 in is 44.5 in. A vertical clearance larger than 44.5 in will be needed to permit span lengths longer than the 300-ft ruling span in the sample sag and tension chart.

Exceptions apply to Rule 235C2b(1)(c) that permit ignoring the rule under certain conditions.

Examples of vertical clearance between line conductors are shown in Figs. 235-7 and 235-8.

Rule 235C can be applied using two different approaches. One approach is to find the required vertical clearance of a particular span. Another approach is to determine a maximum span based on the vertical clearance for a particular structure type, conductor size, conductor sag, and circuit voltage.

In addition to powerline-to-powerline clearance, **NESC** Table 235-5 applies to power-to-communication clearance. Rule 235C and **NESC** Table 235-5 specify the conductor-to-conductor clearance. In addition to conductor-to-conductor clearance, a joint-use (power and communication) **Code** review requires hardware-to-hardware clearance, which is specified in Rule 238. The vertical clearance requirements in Rule 238B and **NESC** Table 235-5 revolve around 40 in. Less than 40 in (e.g., 30 in) is acceptable for certain grounded supply facilities. Greater than 40 in is required for supply voltages above 8.7 kV to ground. The vertical clearance must be checked at the support (e.g., pole) using **NESC** Table 235-5 and out in the span based on sag using Rule 235C2b. Typically, the vertical clearance at the structure will be controlled by the clearance required out in the span based on the upper and lower conductor sags. Depending on what type of conductors are involved, a power conductor sag and tension chart and a communications cable sag and tension chart will be needed to check supply-to-communication clearance at midspan. Examples of joint-use (supply and communication) structures are shown in Figs. 235-9 through 235-14.

Exception 1 of Rule 235C1 references Rule 235G which covers conductor spacing on vertical racks or separate brackets. Exceptions 2 and 3 of Rule 235C1 apply to joint-use (supply and communication) construction. See Fig. 235-15.

In addition to the sag-related clearance checks in Rule 235C2b(1), Rule 235C2b(2) requires making sag adjustments when necessary to maintain clearance. When sag is reduced, tension is increased. If sag reductions are made to maintain clearance, the tension limits in Rule 261H1 must not be exceeded. Rule 235C2b(3) requires an additional special clearance check for joint-use (supply and communication) structures. See Fig. 235-16.

Alternate clearances may be used for voltages exceeding 98 kV AC to ground or 139 kV DC to ground. The formulas in Rule 233C3 must be used for this calculation. See Rule 232D for a discussion of the alternate clearance method.

Example:
Vertical clearance between 1/0 ACSR phase and
neutral conductors on a 12.47/7.2 kV, 3Ø, 4W circuit.

NESC Table 235-5 is used to determine the
required clearance at the supports, but the
clearance at the supports may need to be
greater than the value in **NESC** Table 235-5
to meet Rule 235C2b(1)(a).

The vertical clearance between
the outer phase conductor and
the neutral does not need to be
checked if adequate horizontal
clearance exists per Rule 235B.

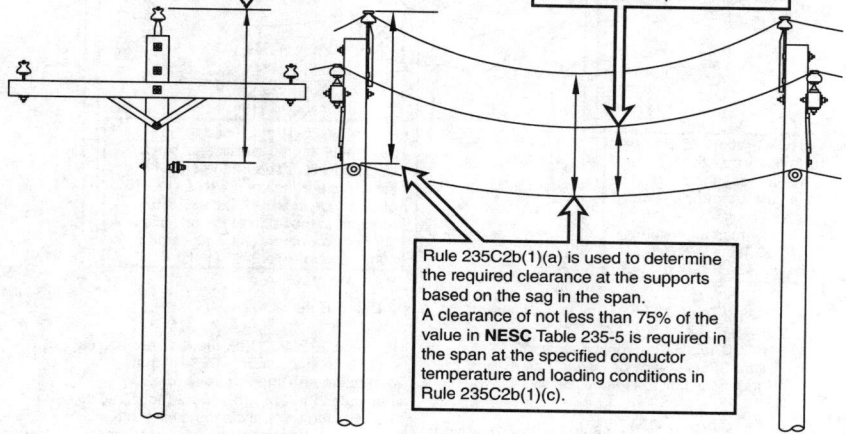

Rule 235C2b(1)(a) is used to determine
the required clearance at the supports
based on the sag in the span.
A clearance of not less than 75% of the
value in **NESC** Table 235-5 is required in
the span at the specified conductor
temperature and loading conditions in
Rule 235C2b(1)(c).

Using NESC Table 235-5

• Conductor at upper level is 7.2 kV phase
  to ground.

• Conductor at lower level is a 230E1 neutral.

• Per **NESC** Table 235-5, the required vertical
  clearance at the support is 16".

Using Rule 235C2b (1)(a)

• Per Rule 235C2b(1)(a), the required vertical clearance
  in the span is 75% of that required at the support.
  .75 × 16" = 12"

The 12" clearance applies at the specified
conductor temperature and loading
conditions per Rule 235C2b(1)(c).

The 16" vertical clearance is at the support and the 12" vertical
clearance is out in the span. The conductor attachment locations
at the support must be determined to maintain 12" of vertical
clearance out in the span. The greater of the two values (16" of
vertical clearance at the support or the conductor attachment
locations to maintain 12" of vertical clearance out in the span)
must be used. Typically, more than 16" of clearance will be
needed at the structure to maintain 12" of clearance out in the
span.

Fig. 235-7. Example of vertical clearance between line conductors (Rule 235C).

Example:
Vertical clearance between a 115 kV transmission line with ground fault relaying and a 12.47/7.2 kV effectively grounded distribution line. The lines are owned by the same utility and the line is built at an elevation of 3000'. Assume the circuits are 180° out of phase.

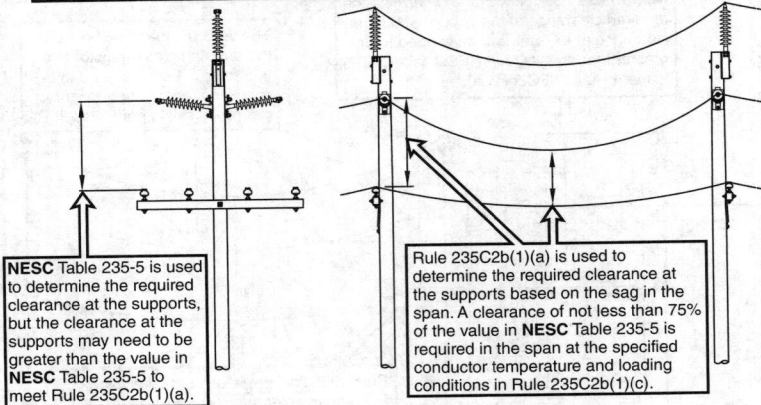

NESC Table 235-5 is used to determine the required clearance at the supports, but the clearance at the supports may need to be greater than the value in **NESC** Table 235-5 to meet Rule 235C2b(1)(a).

Rule 235C2b(1)(a) is used to determine the required clearance at the supports based on the sag in the span. A clearance of not less than 75% of the value in **NESC** Table 235-5 is required in the span at the specified conductor temperature and loading conditions in Rule 235C2b(1)(c).

Using Table 235-5

- **NESC** Table 235-5 only goes to 50 kV. Rule 235C2a provides clearance adders for voltages above 50 kV.

- $\frac{115 \text{ kV (line to line)}}{\sqrt{3}} = 66.40$ kV (line to ground)

  66.40 kV × 1.05 = 69.72 kV
  ($\begin{smallmatrix}1.05 \text{ is the maximum operating}\\ \text{voltage multiplier}\end{smallmatrix}$)

- Conductor at lower level is 7.2 kV phase to ground.

- The greater of the phasor difference or the phase to ground voltage of the higher circuit is 76.92 kV. (See Fig. 235-2)

- Per **NESC** Table 235-5, vertical clearance at the support is:

  16" + 0.4" per kV over 8.7 kV
  16" + 0.4" (50 − 8.7)
  16" + 16.52"
  32.52"a

Using Rule 235C2a

32.52" + 0.4" (76.92 − 50)
32.52" + 10.77"
43.29"
Round to 44"

Using Rule 235C2b

- Per Rule 235C2b(1)(a), the required vertical clearance in the span is 75% of that required at the support up to 50 kV. Per Rule 235C2b(1)(b), 100% (not 75%) of the required vertical clearance is required in the span for greater than 50 kV (exception apply). (0.75 × 32.52) + 10.77" = 35.16" Round to 36"

The 36" clearance applies at the specified conductor temperature and loading conditions per Rule 235C2b(1)(c).

The 44" vertical clearance is at the support and the 36" vertical clearance is out in the span. The conductor attachment locations at the support must be determined to maintain 36" of vertical clearance out in the span. The greater of the two values (44" of vertical clearance at the support or the conductor attachment locations to maintain 36" of vertical clearance out in the span) must be used. Typically, more than 44" of clearance will be needed at the structure to maintain 36" of clearance out in the span.

See Photo(s)

Fig. 235-8. Example of vertical clearance between line conductors (Rule 235C).

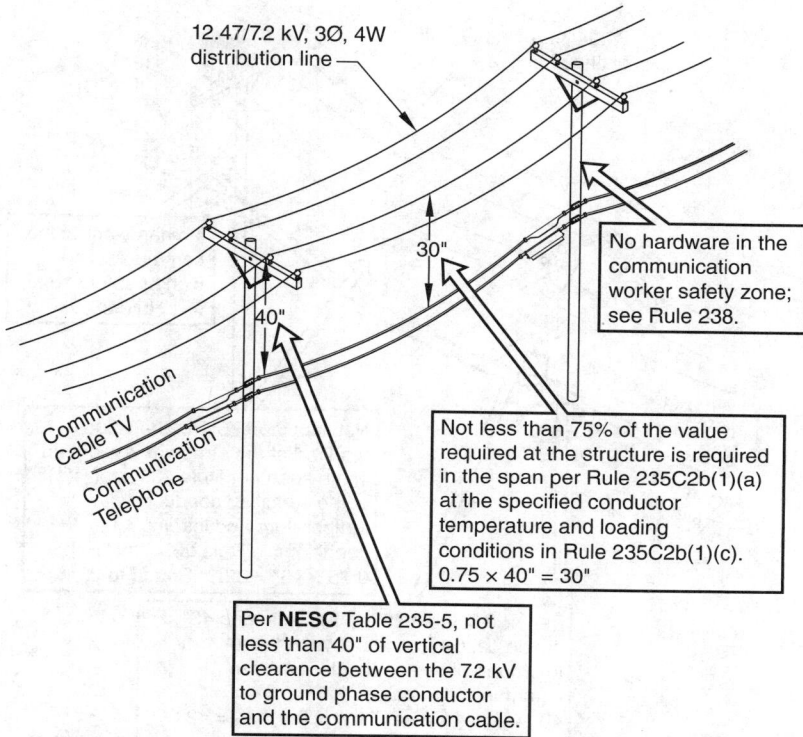

12.47/7.2 kV, 3Ø, 4W
distribution line

30"

No hardware in the
communication
worker safety zone;
see Rule 238.

40"

Communication
Cable TV

Communication
Telephone

Not less than 75% of the value
required at the structure is required
in the span per Rule 235C2b(1)(a)
at the specified conductor
temperature and loading
conditions in Rule 235C2b(1)(c).
0.75 × 40" = 30"

Per **NESC** Table 235-5, not
less than 40" of vertical
clearance between the 7.2 kV
to ground phase conductor
and the communication cable.

Fig. 235-9. Example of vertical clearance between joint-use (supply and communication) conductors (Rule 235C).

Rule 235C4 labels the area between the supply space and the communication space both at the structure and out in the span. The name for this area is the "Communication Worker Safety Zone." The work rules in Part 4 of the **NESC** provide the qualifications of a supply employee and a communication employee. Since the communication employee is not trained to work on supply lines, a safety zone exists for the communication employee's protection. If a communication line is positioned below a supply line on an overhead structure, but the required communication worker safety zone clearance does not exist, the communication employee can correct the violation if the communication employee does not violate the minimum approach distances and other requirements in Sec. 43. If the communication employee cannot maintain minimum the approach distances in Sec. 43, the communication employee must contact a supply employee to correct the violation. See Sec. 43 for additional information. The communication worker safety zone is defined by both Rule 238 and Rule 235C, not one rule or the other but the combined effect of both rules. Only a few select items are permitted in the communications worker safety zone. The exceptions are the items specified in Rules 238C, 238D, and 239. See the communication worker safety zone discussion and figure in Rule 238E.

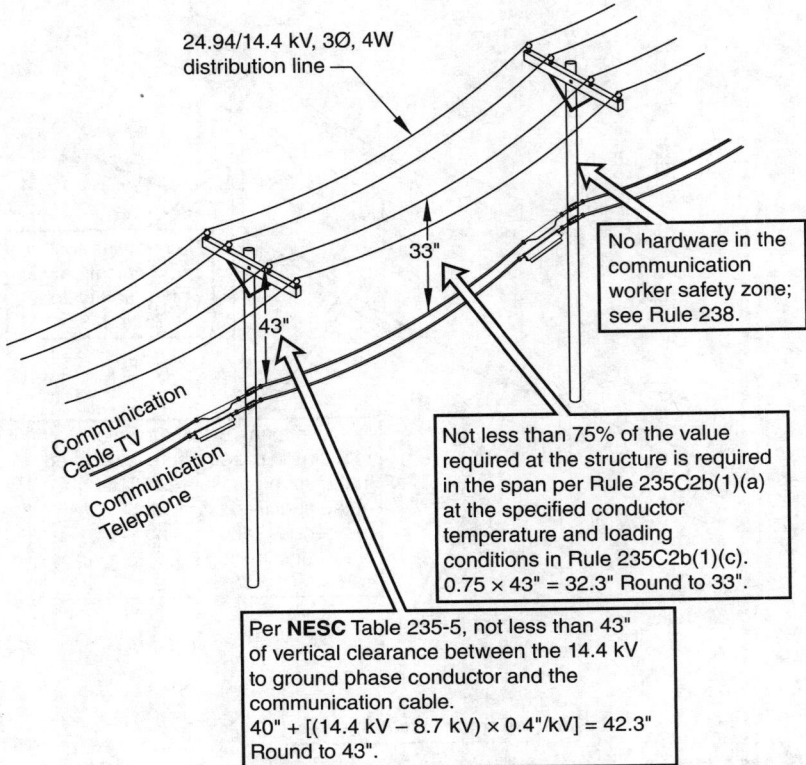

24.94/14.4 kV, 3Ø, 4W
distribution line —

33"

43"

Communication
Cable TV

Communication
Telephone

No hardware in the
communication
worker safety zone;
see Rule 238.

Not less than 75% of the value
required at the structure is required
in the span per Rule 235C2b(1)(a)
at the specified conductor
temperature and loading
conditions in Rule 235C2b(1)(c).
0.75 × 43" = 32.3" Round to 33".

Per **NESC** Table 235-5, not less than 43"
of vertical clearance between the 14.4 kV
to ground phase conductor and the
communication cable.
40" + [(14.4 kV − 8.7 kV) × 0.4"/kV] = 42.3"
Round to 43".

Fig. 235-10. Example of vertical clearance between joint-use (supply and communication) conductors (Rule 235C).

235D. Diagonal Clearance between Line Wires, Conductors, and Cables Located at Different Levels in the Same Supporting Structure. Diagonal clearances are not specified in Rule 235, only horizontal and vertical clearances. NESC Fig. 235-1 is provided in the Code for determining diagonal clearances. If both a horizontal and vertical clearance apply to a conductor, a diagonal clearance can be determined by drawing a box with the required horizontal clearance from Rule 235B and vertical clearance from Rule 235C.

235E. Clearances in Any Direction at or Near a Support from Line Conductors to Supports, and to Vertical or Lateral Conductors, Service Drops, and Span or Guy Wires Attached to the Same Support. NESC Table 235-6 is used to find clearance in any direction (not just horizontal or vertical) from line conductors (that are at or near a support) to supports and to vertical or lateral conductors, service drops, span or guy wires, and to communication antennas attached to the same support. The term line conductors refers to the conductors spanning from pole to pole. Vertical and lateral conductors are discussed in Rule 239. See Rule 239A for a figure showing vertical and lateral conductors. **NESC** Table 235-6 is commonly used for

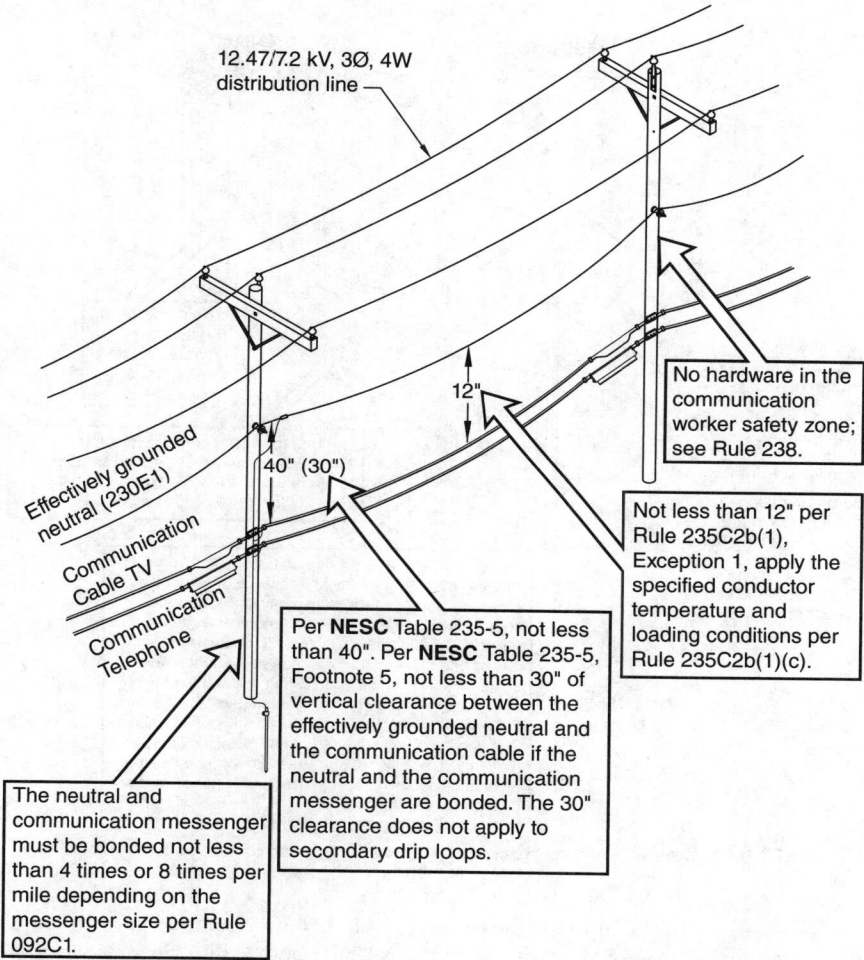

12.47/7.2 kV, 3Ø, 4W
distribution line

Effectively grounded
neutral (230E1)

Communication
Cable TV

Communication
Telephone

12"

40" (30")

No hardware in the
communication
worker safety zone;
see Rule 238.

Not less than 12" per
Rule 235C2b(1),
Exception 1, apply the
specified conductor
temperature and
loading conditions per
Rule 235C2b(1)(c).

Per **NESC** Table 235-5, not less
than 40". Per **NESC** Table 235-5,
Footnote 5, not less than 30" of
vertical clearance between the
effectively grounded neutral and
the communication cable if the
neutral and the communication
messenger are bonded. The 30"
clearance does not apply to
secondary drip loops.

The neutral and
communication messenger
must be bonded not less
than 4 times or 8 times per
mile depending on the
messenger size per Rule
092C1.

See
Photo(s)

Fig. 235-11. Example of vertical clearance between joint-use (supply and communication) conductors
(Rule 235C).

checking clearance to guys on complicated guying structures. Another common
application is clearance of a conductor to the supporting structure (e.g., pole). See
Fig. 235-17.

NESC Table 235-6 addresses clearance of supply and communication service
drops to line conductors. Service drop clearances are also addressed in Rules 232B
and 235C. Occasionally, a power service drop angles down past a communications
line conductor. An example of a power service drop angling down past a communications line conductor is shown in Fig. 235-18.

12.47/7.2 kV, 3Ø, 4W
distribution line

Effectively grounded
neutral (230E1)

Communication cable
located in the supply
space per Rule 224A2a or
Rule 230F1a or 230F1b.

Communications (Cable
TV) located in the
communication space

Communications
(Telephone) located in
the communication space

40" (30")

12"

No hardware
in the comm-
unication worker
safety zone; see
Rule 238.

Not less than 12" per Rule 235C2b(1),
Exception 1, apply the specified conductor
temperature and loading conditions per
Rule 235C2b(1)(c).

Per **NESC** Table 235-5, not less than 40".
Not less than 30" of vertical clearance between the
communication cable in the supply space and the
communication cable in the communication space
per **NESC** Table 235-5, Footnote 5. Footnotes 9 and
10 apply to the vertical clearance between the neutral
conductor and the communication cable in the supply
space. Bonding per Rule 097G is required to use the
30" and 12" clearance on 224A2a and 230F1a cables.
Bonding is not required for entirely dielectric cables
meeting Rule 230F1b. See Rule 224 for transitioning
a communication cable from the supply space to the
communication space.

See
Photo(s)

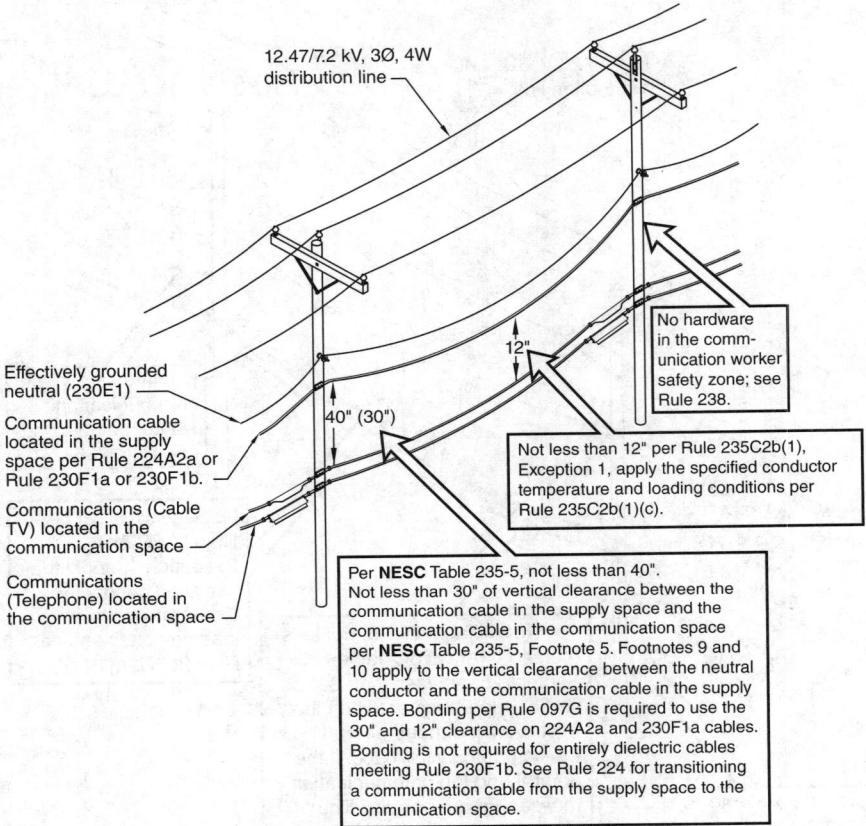

Fig. 235-12. Example of vertical clearance between joint-use (supply and communication) conductors
(Rule 235C).

The notes in Rule 235E1 direct the reader to Rule 235I for clearance to communication antennas in the supply space and to Rules 236D1 and 238 for clearance to communication antennas in the communication space. See Rules 235I, 236, and 238B for additional information.

Rule 235E2 has requirements for suspension insulators because **NESC** Table 235-6 is for fixed supports. The rules for suspension insulators require applying a 6-lb/ft² wind to the conductor that the suspension insulator is holding. The wind pressure may be reduced to 4 lb/ft² in sheltered areas, but trees are not considered a shelter to a line. The deflection of flexible structures and fittings must also be considered if the deflection reduces the clearance in question. See Rules 233A1(a)(2) and 234A2 for similar requirements.

Rule 235E3 provides alternate clearances for voltages exceeding 98 kV AC to ground or 139 kV DC to ground. The values in **NESC** Table 235-7 can be used to check the proper application of the formulas. See Rule 232D for a discussion of alternate clearances.

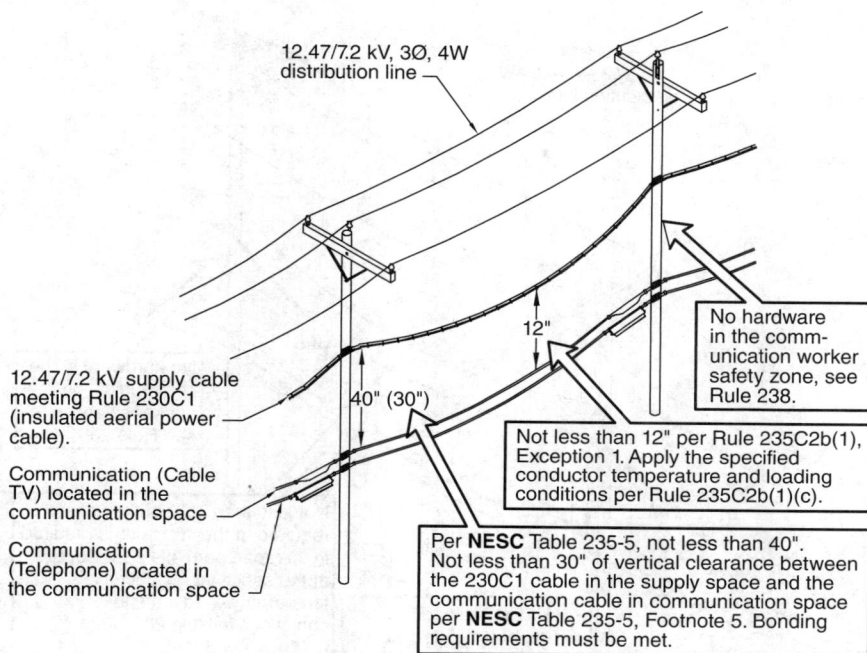

12.47/7.2 kV, 3Ø, 4W
distribution line

12.47/7.2 kV supply cable
meeting Rule 230C1
(insulated aerial power
cable).

Communication (Cable
TV) located in the
communication space

Communication
(Telephone) located in
the communication space

12"

No hardware
in the comm-
unication worker
safety zone, see
Rule 238.

40" (30")

Not less than 12" per Rule 235C2b(1),
Exception 1. Apply the specified
conductor temperature and loading
conditions per Rule 235C2b(1)(c).

Per **NESC** Table 235-5, not less than 40".
Not less than 30" of vertical clearance between
the 230C1 cable in the supply space and the
communication cable in communication space
per **NESC** Table 235-5, Footnote 5. Bonding
requirements must be met.

See
Photo(s)

Fig. 235-13. Example of vertical clearance between joint-use (supply and communication) conductors (Rule 235C).

235F. Clearances between Circuits of Different Voltage Classifications Located in the Supply Space on the Same Support Arm. Rule 235F recognizes that circuits of different voltage levels are not always separated vertically on a structure. Rule 220C assumes circuits of different voltage levels will be separated vertically at different heights and does not comment on horizontal separation. Rule 235F must be used in conjunction with Rule 235B, Horizontal Clearance between Line Conductors, and in some cases Rule 236, Climbing Space.

The NESC uses the term support arm for a crossarm. The terms bridge arm and sidearm are also used in this rule. A bridge arm is an arm that spans between two poles. A sidearm is an arm to one side of the pole. A sidearm is commonly called an alley arm. Examples of clearance between supply circuits of different voltages on the same support arm are outlined in Fig. 235-19.

235G. Conductor Spacing: Vertical Racks or Separate Brackets. Reduced vertical clearance between conductors is permitted when vertical racks are used. The rules for conductor spacing on vertical racks are outlined in Figs. 235-20 and 235-21.

235H. Clearance and Spacing between Communication Conductors, Cables, and Equipment. Rule 235H1 addresses the clearance and spacing between communication cables. The rule does not state if the clearance and spacing values are

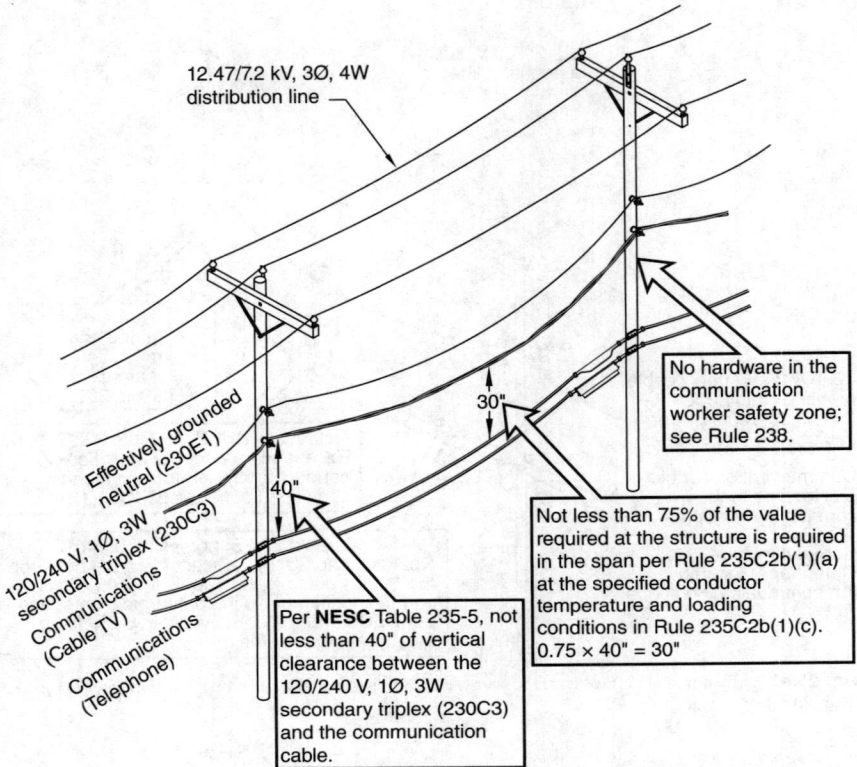

12.47/7.2 kV, 3Ø, 4W
distribution line

No hardware in the communication worker safety zone; see Rule 238.

30"

Effectively grounded neutral (230E1)

40"

120/240 V, 1Ø, 3W secondary triplex (230C3)

Communications (Cable TV)

Communications (Telephone)

Not less than 75% of the value required at the structure is required in the span per Rule 235C2b(1)(a) at the specified conductor temperature and loading conditions in Rule 235C2b(1)(c). 0.75 × 40" = 30"

Per **NESC** Table 235-5, not less than 40" of vertical clearance between the 120/240 V, 1Ø, 3W secondary triplex (230C3) and the communication cable.

See Photo(s)

Fig. 235-14. Example of vertical clearance between joint use (supply and communication) conductors (Rule 235C).

vertical, horizontal, or radial. Commonly, communication cables are attached directly to a pole and separated vertically on the pole, but they can also be separated horizontally by placing the cables on opposite sides of the pole or on a communications crossarm. A not less than 12-in spacing is required between messengers supporting communication cables. The 12-in value is a spacing dimension, not a clearance dimension. See Rule 230A3 for a discussion of clearance versus spacing. The 12-in spacing dimension allows through-bolts for supporting communication messengers to be drilled 12 in apart, center-to-center.

Communication cables supported on messengers are installed by first attaching the messenger to the pole and then lashing the communication cable to the messenger. The lashing machine travels down the messenger and needs the 12-in spacing for proper operating room. The 12-in spacing between messengers may be reduced by agreement between the parties involved. The parties involved include the communication cable attachees and the pole owner or owners.

Rule 235H2 requires a clearance (not spacing) between cables and equipment of different communication utilities anywhere in the span to be not less than 4 in.

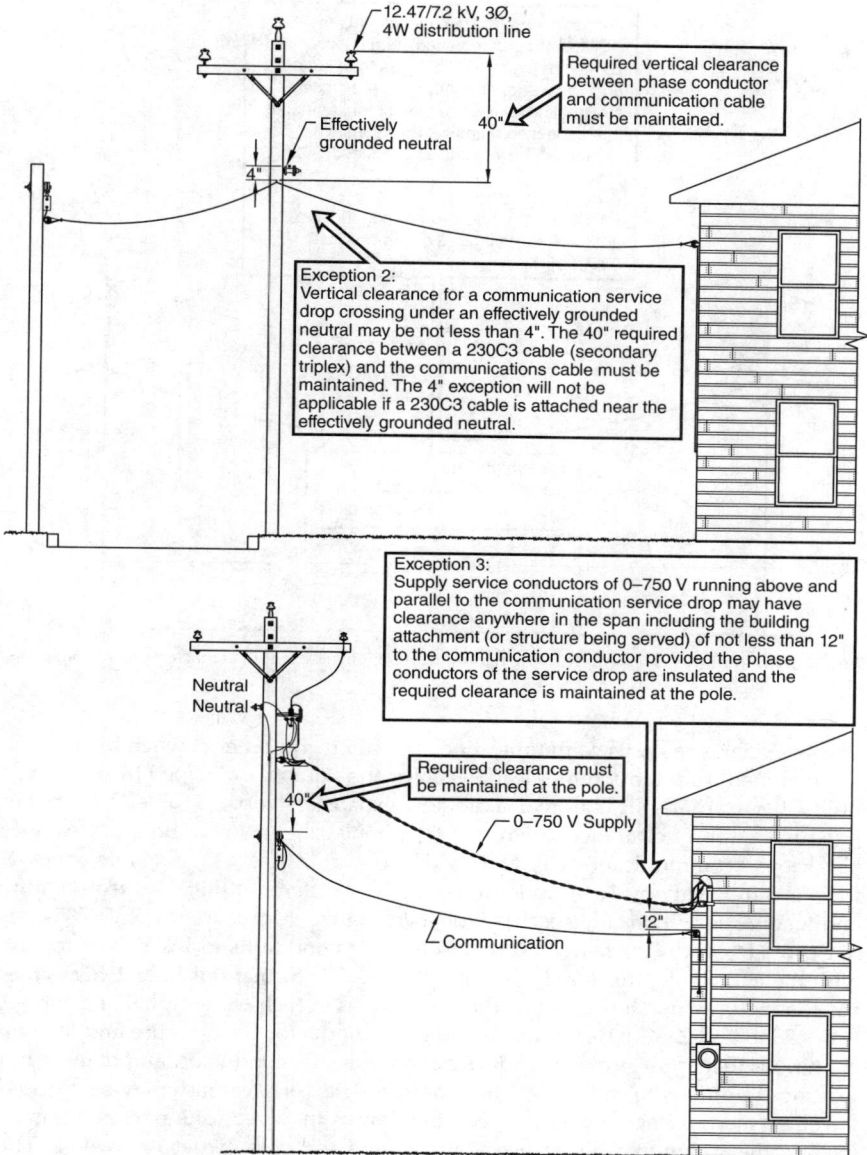

Fig. 235-15. Exceptions to vertical clearance requirements for joint-use (supply and communication) construction (Rule 235C1).

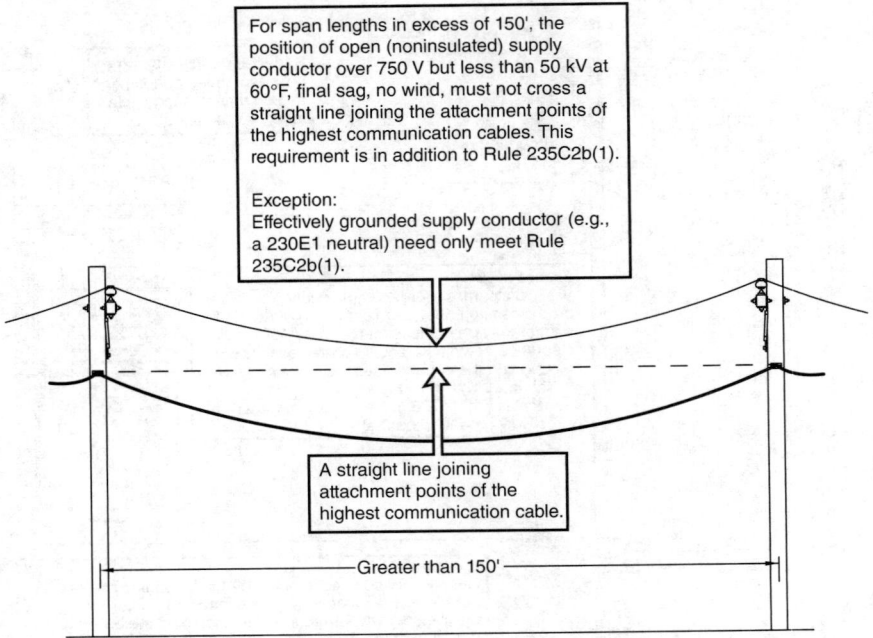

For span lengths in excess of 150', the position of open (noninsulated) supply conductor over 750 V but less than 50 kV at 60°F, final sag, no wind, must not cross a straight line joining the attachment points of the highest communication cables. This requirement is in addition to Rule 235C2b(1).

Exception:
Effectively grounded supply conductor (e.g., a 230E1 neutral) need only meet Rule 235C2b(1).

A straight line joining attachment points of the highest communication cable.

Greater than 150'

Fig. 235-16. Additional vertical clearance requirement for joint-use (supply and communication) lines [Rule 235C2b(3)].

The 4-in clearance keeps communication cables and equipment owned by one utility separated from the communication cables and equipment owned by a different utility. For example, the bottom of a cable TV expansion loop or a cable TV amplifier must have 4-in of clearance down to a telephone cable below. The 4-in clearance also keeps a communication circuit owned by one utility from sagging into or below the a communication circuit under it owned by a different utility. No temperature and loading conditions are specified for checking the 4-in clearance. Rule 235 does not provide conductor temperature and loading conditions in Rule 235A for use with the entire rule. Rules 232, 233, and 234 use this format but Rule 235 specifies the conductor temperature and loading conditions in each paragraph. For example, Rules 235B and 235C have within the rule the conductor temperature and loading conditions that apply. Rule 235H does not provide any conductor temperature and loading conditions to check the 4-in clearance. The 4-in clearance may be reduced (or eliminated) by agreement between the parties involved. The parties involved include the communication cable attachees and the pole owner or owners. The **Code** rules related to clearance and spacing between communication lines are outlined in Fig. 235-22.

If the parties agree to less than a 12 in spacing between communication messengers, the through bolt spacing on a pole can be less than 12 in. The **NESC** does not specify a minimum distance between through bolts on a wood pole. Rule 012C, which requires accepted good practice, must be used. See Fig. 235-23.

Example:
Double circuit
12.47/7.2 kV, 3Ø,
4W distribution
line (not joint use)

Rules 235B, 235C,
and 236 also apply.

235C

236

235B

Clearance to
anchor guy with
a guy strain
insulator must
be not less than
6".

Clearance to the surface
of the structure (pole)
must be not less than 4".
(Greater than 4" is
required for joint use
poles and less than 4"
is permitted when
Footnote 10 applies).
However, this value may
not provide adequate
climbing space per
Rule 236.

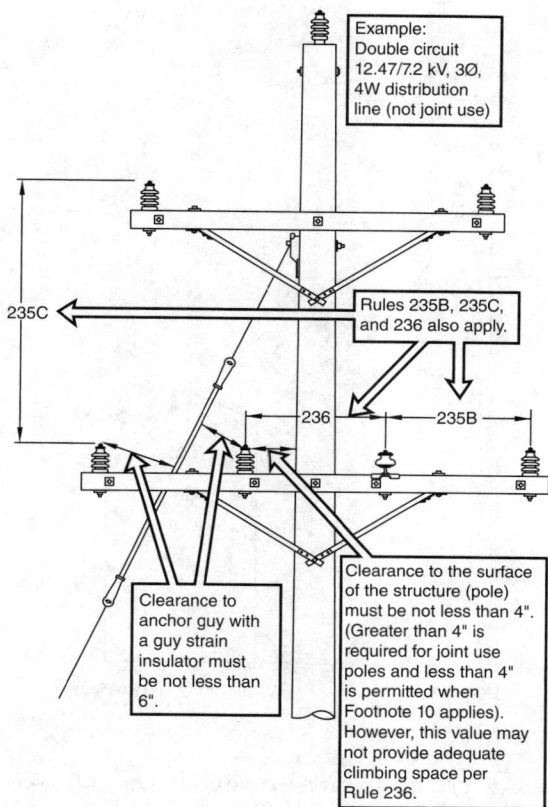

Using Table 235-6

- Table voltages are phase to phase.

- Using over 8.7 to 50 kV column and
  row 2b (anchor guys), the clearance
  required to the anchor guy is
  6" + 0.25" per kV over 8.7 kV.
  6" + 0.25" (12.47 − 8.7) = 6.94"
  Round to 7"

- If a guy strain insulator is used,
  Footnote 11 applies and a 25%
  reduction to 6.94" is applicable.
  7"− [7" × 0.25] = 5.25"
  Round to 6"

- Using over 8.7 to 50 kV column and
  row 4b (all other), the clearance
  required to the surface of the
  structure (pole) is 3" + 0.2" per kV
  over 8.7 kV.
  3" + 0.2" (12.47 − 8.7) = 3.75"
  Round to 4"

- Greater than 4" is required to the
  surface of the structure for joint use
  poles. Less than 4" is permitted when
  Footnote 10 is applicable.

- Rule 235B must be used to check
  the horizontal clearance between
  conductors. Rule 235C must be
  used to check the vertical clearance
  between conductors. Rule 236 must
  be checked for climbing space.

See
Photo(s)

Fig. 235-17. Example of clearance of a conductor to an anchor guy and a conductor to a support
(Rule 235E).

**235I. Communication Antenna Clearances in Any Direction from Supply and
Communication Lines Antennas in the Supply Space Attached to the Same
Supporting Structure.** Rule 235I provides clearance between supply conductors
and communication antennas (and associated conductive mounting hardware)
and communication antenna equipment cases (that support or are adjacent to
the antenna) mounted in the supply space in Rules 235I2a and 235I3. Rule 235I2b
addresses clearances of communication antennas located in the communication
space. Rules 235H, 238A, and 238B address clearance for communication antennas mounted in the communication space. Rule 235H addresses equipment, but
does not specifically use the term antenna. Rules 235A and 235B specifically
address antennas as equipment for clearance purposes. Rule 235I does not treat
the antenna as equipment. Rule 235I1 states that the antenna functions as a
rigid (vertical or lateral) open wire communication conductor for determining
clearances. Rules 235H and 235I2b do not specifically address a communication

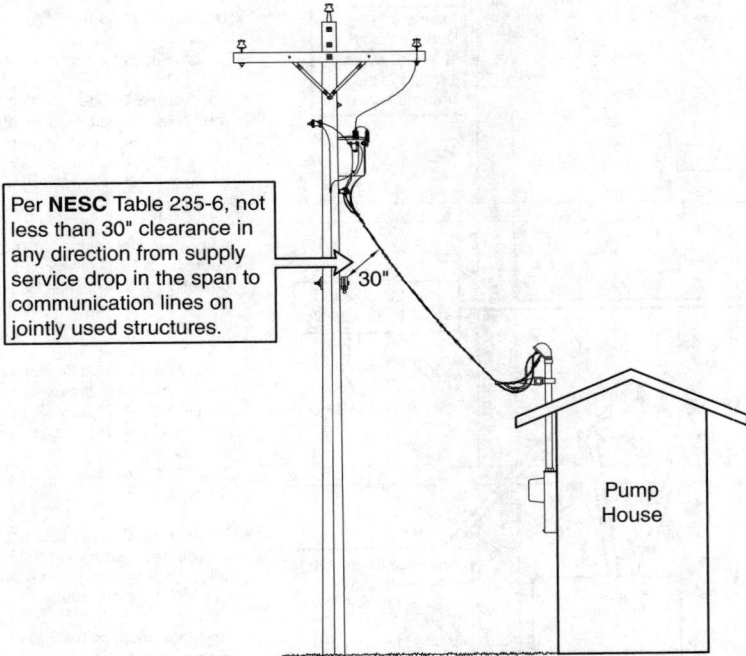

Per **NESC** Table 235-6, not less than 30" clearance in any direction from supply service drop in the span to communication lines on jointly used structures.

30"

Pump House

Fig. 235-18. Example of clearance of a supply service drop to a communication line conductor (Rule 235E).

antenna mounted to the strand of a communication conductor which is commonly referred to as a "strand mount antenna." In these applications the antenna owner or operator and the communications cable strand owner or operator are typically the same utility company. See Rules 224A, 235H, 238A, and 238B for additional discussion and figures. A communications antenna mounted in the supply space is typically mounted at the top of the pole above the supply conductors. Rule 235I references Rule 224A and specific rows in **NESC** Table 235-5 and **NESC** Table 235-6 to obtain the required clearance values. Normally, a communication vertical riser would be located on the pole that has a communication antenna. Requirements for vertical communication conductors passing through the supply space on jointly used structures (poles) are found in Rule 239H. When a communication antenna is located in the supply space, the worker installing or maintaining the communication antenna must be qualified to work in the supply space per the work rules (Secs. 42 and 44) of the **NESC**. In other words, a communication antenna installed in the supply space (e.g., at the pole top position) must be installed by a power lineman, not a communications lineman. A note in Rule 235I2 provides information that the clearances from **NESC** Table 235-6 are for electrical clearance, not for radio frequency radiation. Additional clearance may be needed to maintain a safe working distance when radiation exposure levels are considered. See Rule 420Q. A common reference for determining clearance to

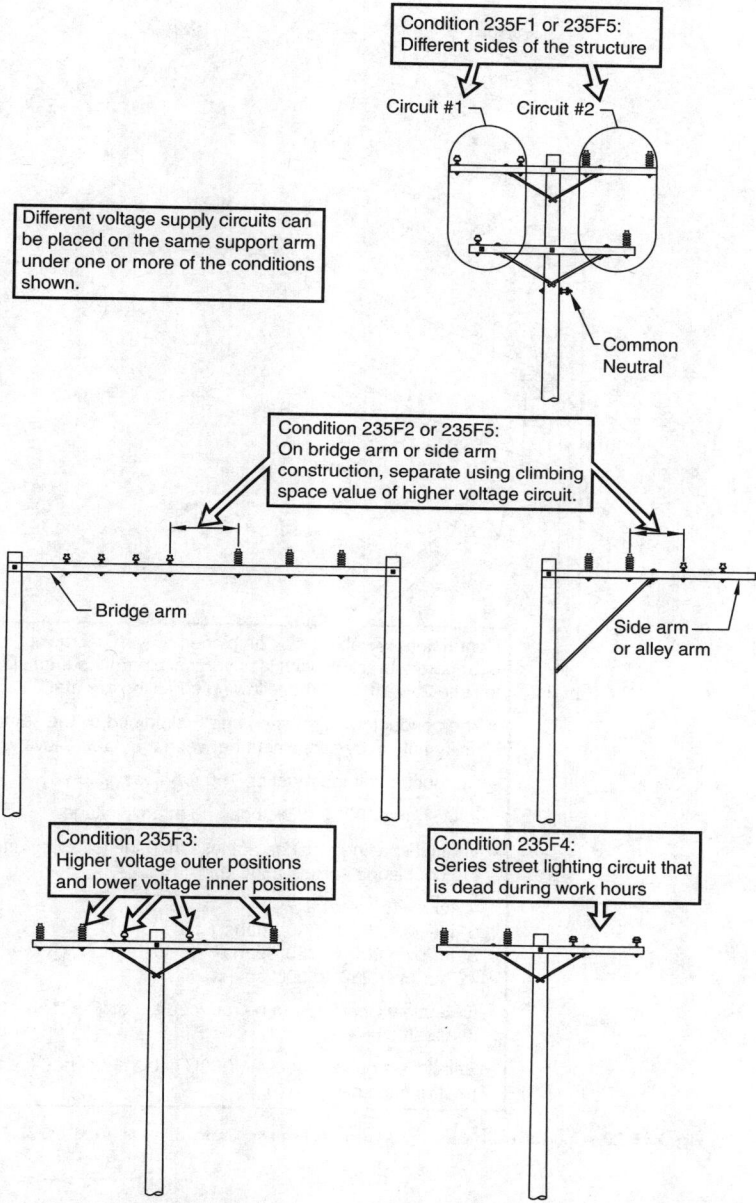

Condition 235F1 or 235F5:
Different sides of the structure

Circuit #1   Circuit #2

Common
Neutral

Different voltage supply circuits can be placed on the same support arm under one or more of the conditions shown.

Condition 235F2 or 235F5:
On bridge arm or side arm construction, separate using climbing space value of higher voltage circuit.

Bridge arm

Side arm or alley arm

Condition 235F3:
Higher voltage outer positions and lower voltage inner positions

Condition 235F4:
Series street lighting circuit that is dead during work hours

See
Photo(s)

Fig. 235-19. Examples of clearance between supply circuits of different voltages on the same support arm (Rule 235F).

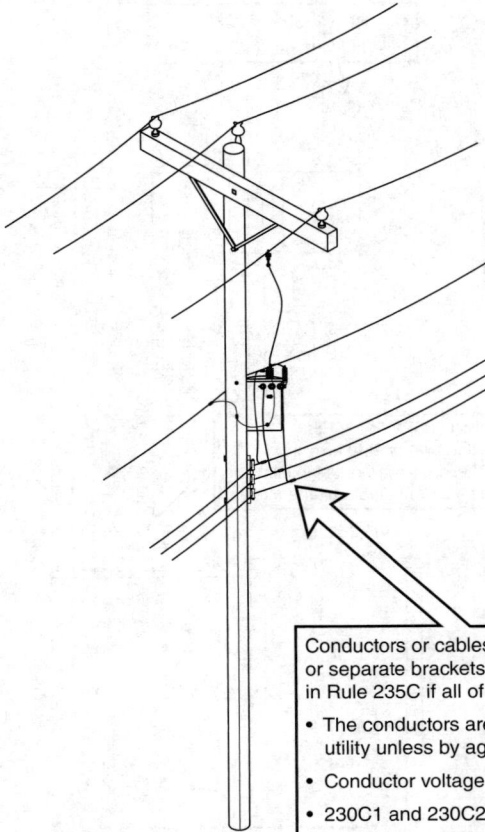

Conductors or cables may be placed on vertical racks
or separate brackets with less vertical clearance than specified
in Rule 235C if all of the following condition are met:

• The conductors are owned and maintained by the same
  utility unless by agreement between the parties involved.

• Conductor voltage must be 750 V or less.

• 230C1 and 230C2 cables can be at any voltage.

• Vertical spacing must be not less than the following under
  the conditions in Rule 235C2b(1)(c):

  4" for 0'–150' spans
  6" for over 150' to 200' spans
  8" for over 200' to 250' spans
  12" for over 250' to 300' spans

• Exception: Spacing can be reduced to not less than 4" with
  midspan spacers.

• Special exceptions apply to 230C cables attached to 230E1
  neutral brackets.

See
Photo(s)

Fig. 235-20. Conductor spacing on vertical racks or separate brackets (Rule 235G).

a communications antenna based on radio frequency radiation exposure limits is
the FCC (Federal Communications Commission) OET (Office of Engineering and
Technology) Bulletin 65. FCC OET Bulletin 65 is titled "Evaluating Compliance
with FCC Guidelines for Human Exposure to Radiofrequency Electromagnetic

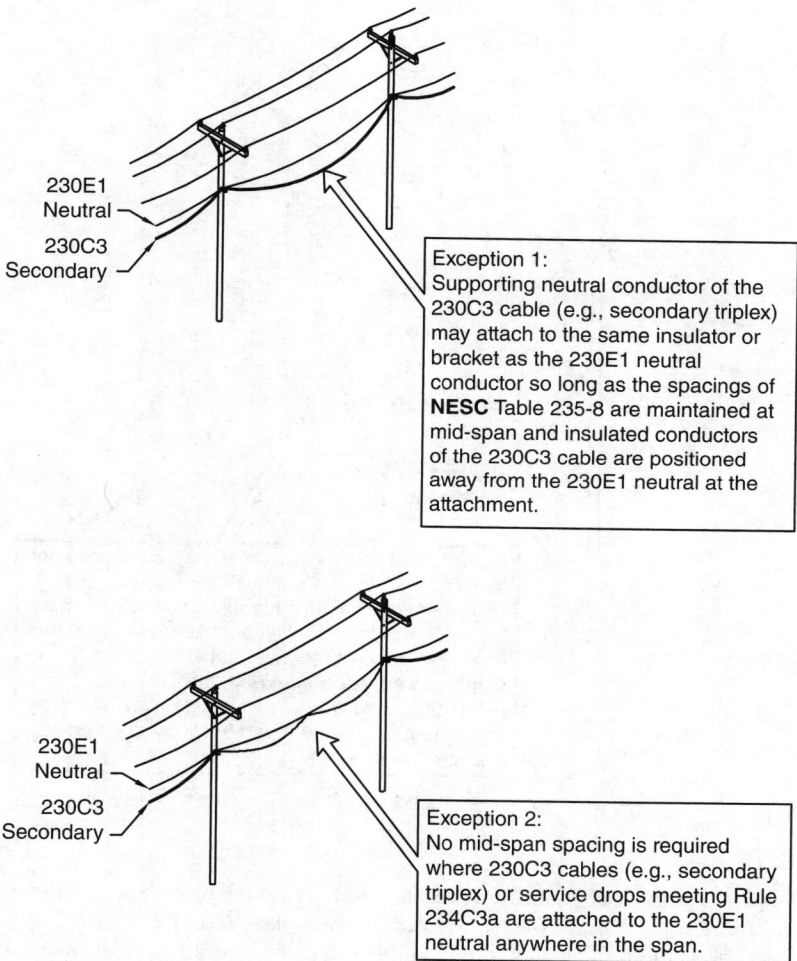

230E1
Neutral

230C3
Secondary

Exception 1:
Supporting neutral conductor of the
230C3 cable (e.g., secondary triplex)
may attach to the same insulator or
bracket as the 230E1 neutral
conductor so long as the spacings of
NESC Table 235-8 are maintained at
mid-span and insulated conductors
of the 230C3 cable are positioned
away from the 230E1 neutral at the
attachment.

230E1
Neutral

230C3
Secondary

Exception 2:
No mid-span spacing is required
where 230C3 cables (e.g., secondary
triplex) or service drops meeting Rule
234C3a are attached to the 230E1
neutral anywhere in the span.

Fig. 235-21. Exceptions to conductor spacing on vertical racks or separate brackets (Rule 235G).

Fields." Examples of clearance between supply lines and communication antennas
in the supply space are shown in Fig. 235-24.

236. CLIMBING SPACE

Rule 236, Climbing Space, and Rule 237, Working Space, may appear complicated
but they revolve around one simple idea. The supply and communication worker
must have adequate space to climb and work on a pole and its conductors, supports, and equipment.

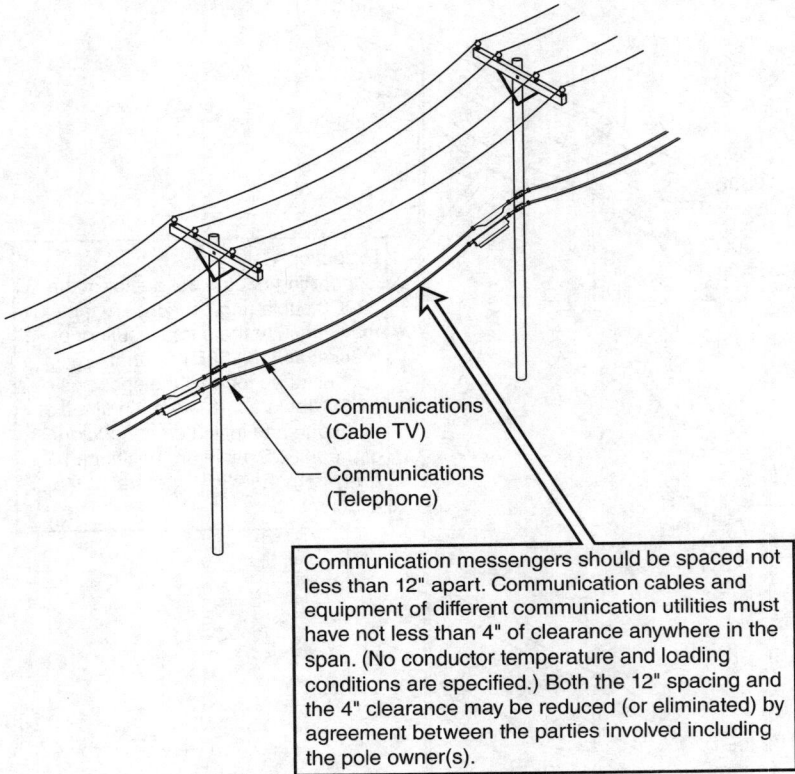

Communications
(Cable TV)

Communications
(Telephone)

Communication messengers should be spaced not
less than 12" apart. Communication cables and
equipment of different communication utilities must
have not less than 4" of clearance anywhere in the
span. (No conductor temperature and loading
conditions are specified.) Both the 12" spacing and
the 4" clearance may be reduced (or eliminated) by
agreement between the parties involved including
the pole owner(s).

See
Photo(s)

Fig. 235-22. Clearance and spacing between communication lines (Rule 235H).

The first sentence of Rule 236 is critical to applying (or not applying) the entire
rule. Rule 236 only applies if workers climb the pole. Wood poles are commonly
climbed by some utilities and rarely climbed by other utilities. Steel distribution
poles without steps are not intended to be climbed. If the pole is worked on from a
bucket truck, the climbing space rules do not apply. See Fig. 236-1.

Climbing space may be provided by temporarily moving the line conductors
using live line tools per Exception 3 of Rule 236E. The **NESC** work rules in Part 4
must be applied when climbing structures and moving live conductors.

Rule 236 provides requirements for measuring the climbing space in Rule 236B,
positioning arms in the climbing space in Rule 236C, and locating equipment
outside the climbing space in Rule 236D.

Rule 236E provides the basic horizontal clearances between conductors
bounding the climbing space. **NESC** Table 236-1 provides horizontal clearances
between conductors. The horizontal clearances in **NESC** Table 236-1 are intended
to provide 24 in of clear climbing space while the conductors bounding the
climbing space are covered with temporary insulation (e.g., insulating line hose,

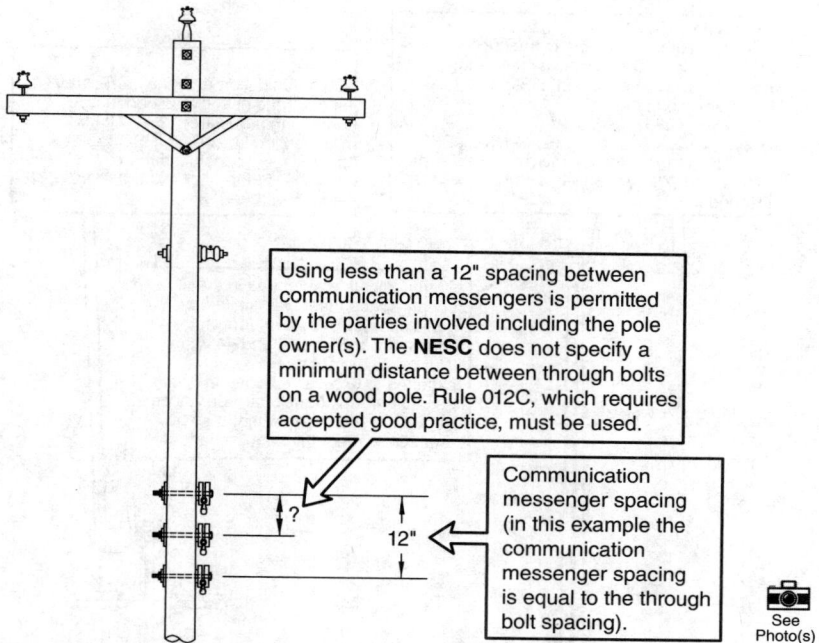

Fig. 235-23. Example of using less than a 12 inch spacing between communication messengers (Rule 235H).

insulating blankets, insulator hoods, etc.). For example, a 30-in horizontal clearance between conductors bounding the climbing space in NESC Table 236-1 (for 12.47/7.2 kV) assumes the temporary insulation on both sides of the climbing space is 6 in, leaving 24 in of clear climbing space. As the voltages on the table increase, the clearances increase as the temporary insulation is assumed to be thicker and bulkier. Examples of climbing space between conductors are shown in Figs. 236-2 and 236-3.

Methods to provide climbing space on buckarm construction are provided in Rule 236F. A buck arm is a crossarm used to change the direction of all or a part of the conductors on the line. A buck arm is generally placed at right angles to the line arm.

Rule 236G provides rules for climbing space past longitudinal runs not on support arms. The location, size, and quantity of the longitudinal runs not on support arms (e.g., communication cables attached to the pole) must be considered when determining if a qualified worker can climb past them. The NESC does not define a specific quantity or size of cables that a qualified climber can climb past. On a typical joint-use (power and communications) pole, it is the power lineman that has to climb past the communication cables (not the communication lineman), therefore, it is appropriate for the power lineman (or power company) to work with the communication lineman (or communications company) to determine if the location, size, and quantity of the cables permit

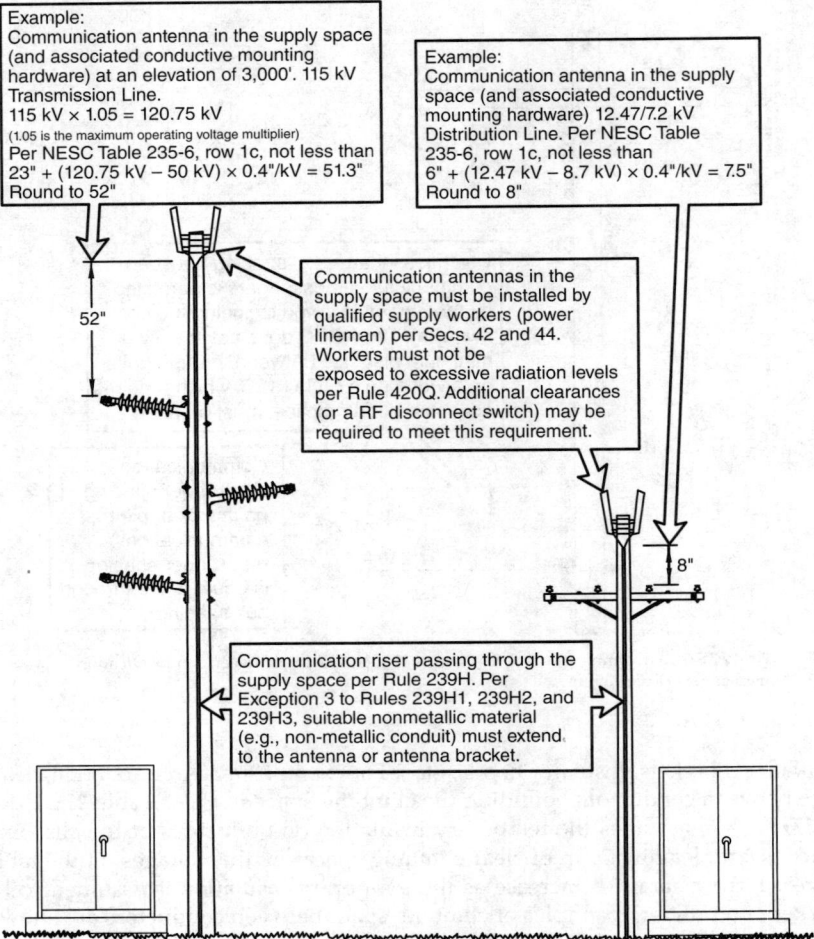

Fig. 235-24. Examples of clearance between supply lines and communication antennas in the supply space (Rule 235I).

qualified workers to climb past them. Keeping one side of the pole open and free from communication attachments is one way of assuring that adequate climbing space exists, however, there are times when communication attachments are made to both sides of the pole which is commonly referred to as pole boxing. Pole boxing is not specifically addressed in the **NESC**. The utility responsible for pole replacements typically does not prefer pole boxing (locating communication conductors on both sides of the pole) as it makes the pole replacement more time consuming. Commonly, communication companies want to use pole boxing to avoid a power-to-communications clearance violation or a

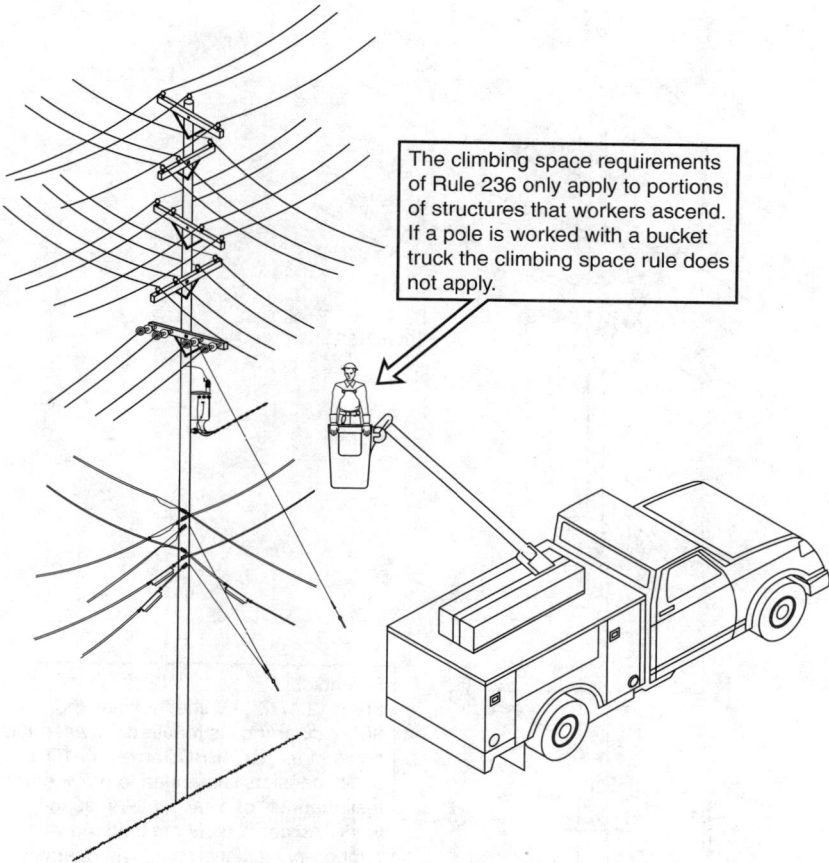

The climbing space requirements of Rule 236 only apply to portions of structures that workers ascend. If a pole is worked with a bucket truck the climbing space rule does not apply.

Fig. 236-1. Climbing space application (Rule 236).

communications-to-groundline clearance violation. If it is determined that the location, size, and quantity of the communication cables do not permit qualified workers to climb past them, a communications crossarm can be used to provide adequate climbing space between the communication conductors or adequate climbing space can be provided by having the communication cables attached directly to the pole on one side of the pole and placed on a standoff bracket on the other side of the pole. The crossarm method and standoff bracket method are not needed if it is determined that the location, size, and quantity of the cables (longitudinal runs not on support arms) permit qualified workers to climb past them. Secondary service drops are also addressed in Rule 236G, and **NESC** Fig. 236-1 is provided as a guide.

Per Rule 236H, vertical conductors in conduit (risers) that are securely attached to the pole without spacers (offset brackets) are not considered an obstruction to

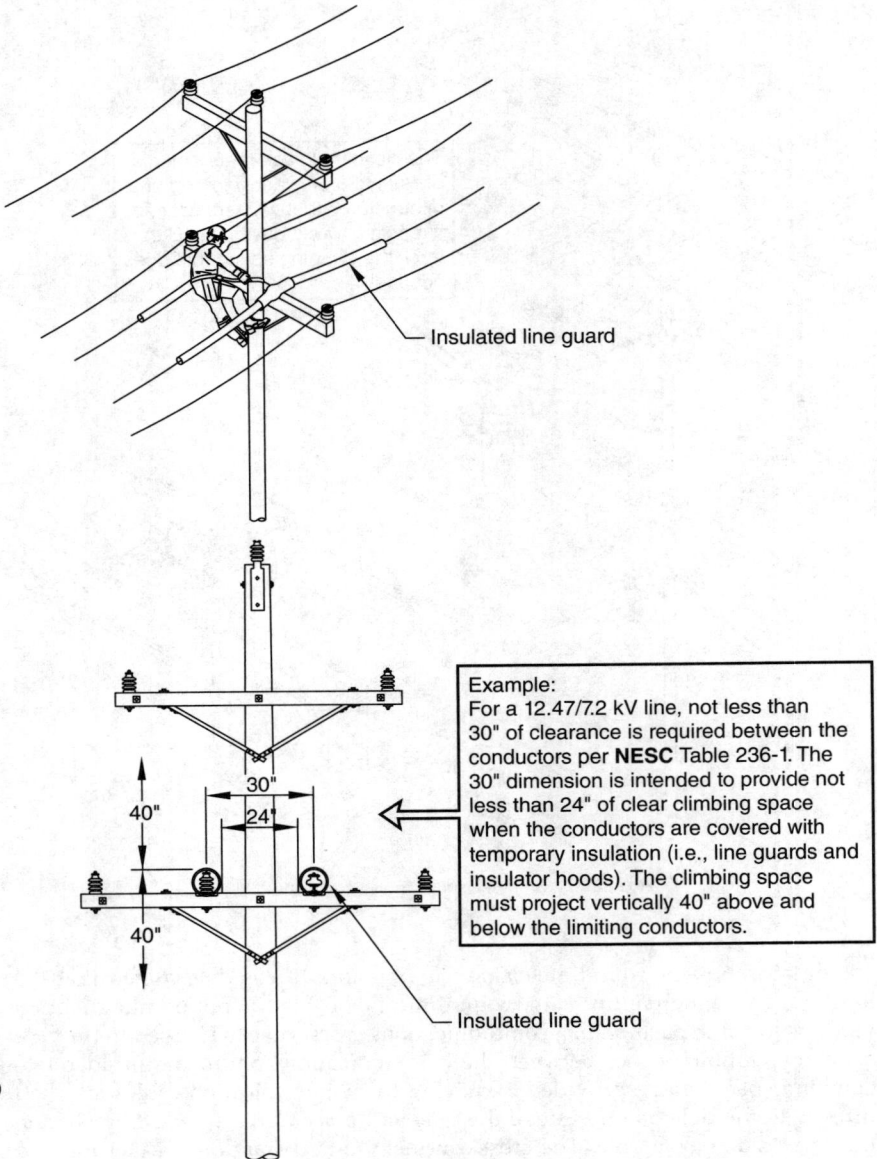

Insulated line guard

Example:
For a 12.47/7.2 kV line, not less than
30" of clearance is required between the
conductors per **NESC** Table 236-1. The
30" dimension is intended to provide not
less than 24" of clear climbing space
when the conductors are covered with
temporary insulation (i.e., line guards and
insulator hoods). The climbing space
must project vertically 40" above and
below the limiting conductors.

Insulated line guard

See
Photo(s)

Fig. 236-2. Examples of climbing space between conductors (Rule 236E).

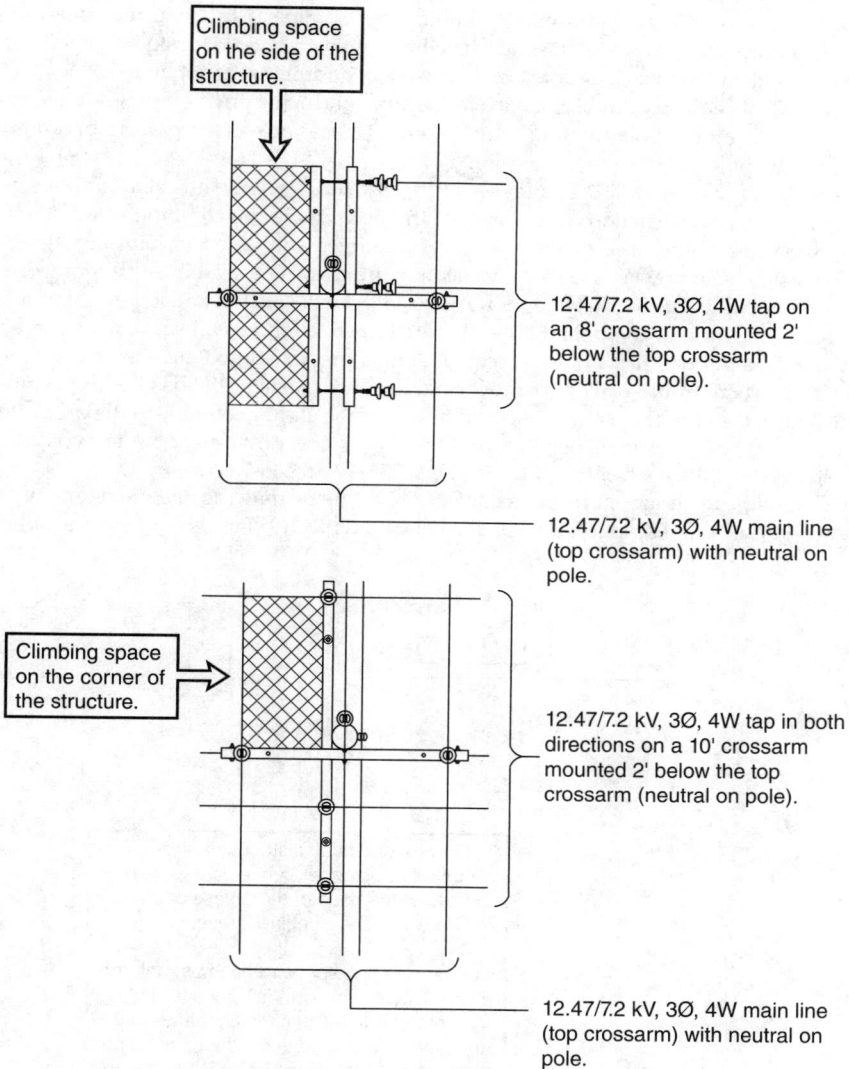

Climbing space on the side of the structure.

12.47/7.2 kV, 3Ø, 4W tap on an 8' crossarm mounted 2' below the top crossarm (neutral on pole).

12.47/7.2 kV, 3Ø, 4W main line (top crossarm) with neutral on pole.

Climbing space on the corner of the structure.

12.47/7.2 kV, 3Ø, 4W tap in both directions on a 10' crossarm mounted 2' below the top crossarm (neutral on pole).

12.47/7.2 kV, 3Ø, 4W main line (top crossarm) with neutral on pole.

Fig. 236-3. Examples of climbing space between conductors (Rule 236E).

the climbing space. When using the clearances in **NESC** Table 236-1, a conduit can be located within the clearances provided. Rule 362B, in Sec. 36, Risers, provides an additional requirement stating that the number, size, and location of the risers must be limited to allow adequate access for climbing. No specifics are provided as to the number or size of the conduits, Rule 012C, which requires accepted good practice, must be used. A conduit mounted on offset brackets is considered an obstruction to the climbing space. However, conduit offset brackets are used to organize the conduit risers on a pole and make climbing easier, especially when multiple conduits or large conduits are present. The climbing space is then located on a side or corner of the pole that does not contain the riser conduits on offset brackets.

Rule 236I requires climbing space to the top conductor position where the center phase is on the pole top and the two outer phases are on a crossarm mounted below the pole top.

One item that the NESC does not specifically address, but indirectly implies, is the location of obstructions at the base of the pole where the climbing space starts. Obstructions at the base of the pole are sometimes addressed in joint-use agreements between power and communication utilities. For example, the joint-use agreement may specify that telephone pedestals can be located outside the climbing space at the base of the pole. Other obstructions like fences can sometimes not be avoided, as many times a pole is located along a property line where a fence exists. If an obstruction like a telephone pedestal or fence exists at the base of the climbing space, the climbing space rules can be met if the climber climbs on the opposite side of the pole or rotates position on the way up and down the pole to avoid the obstruction at the base of the pole. See Figs. 236-4 and 236-5.

The fall protection requirements in Rule 420K necessitate the need for some type of wood pole fall restriction device. This device typically clamps or cinches around

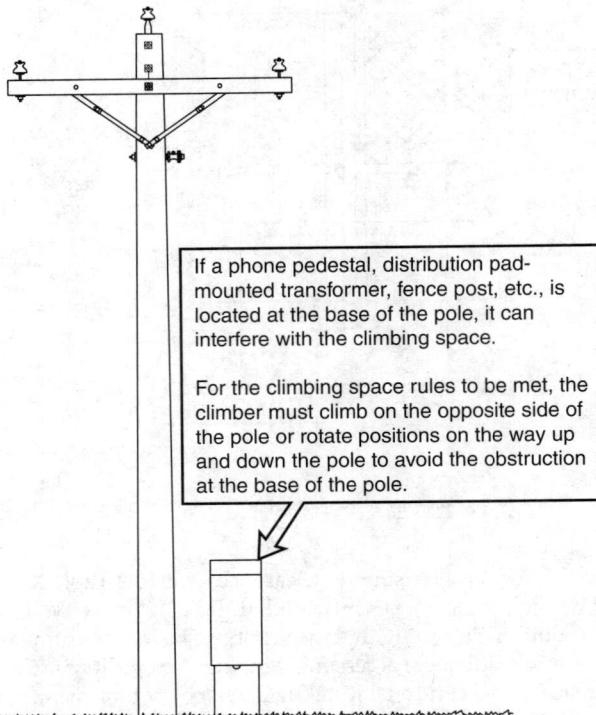

If a phone pedestal, distribution pad-mounted transformer, fence post, etc., is located at the base of the pole, it can interfere with the climbing space.

For the climbing space rules to be met, the climber must climb on the opposite side of the pole or rotate positions on the way up and down the pole to avoid the obstruction at the base of the pole.

See Photo(s)

Fig. 236-4. Example of obstructions at the base of the climbing space (Rule 236).

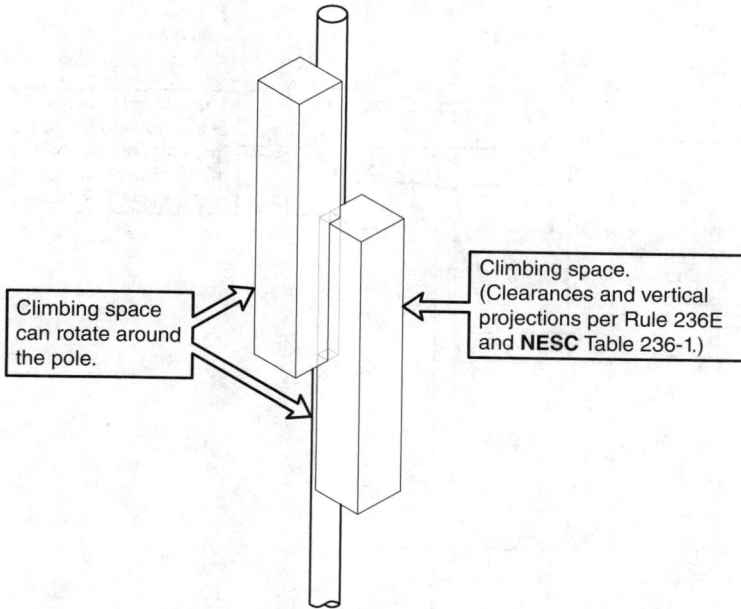

Fig. 236-5. Example of rotating the climbing space around a pole (Rule 236).

a wood pole when the pole is being climbed. Rule 420K also applies to transitioning across obstacles on the pole such as power and communication equipment. Power and communication equipment must be mounted outside of the climbing space. Although not required by Rule 236, stand-off brackets can be used to mount the power or communications equipment to provide a method for the pole climber to meet Rule 420K. In other words, the pole climber will only need to transition the wood pole fall restriction device across the stand-off bracket, not the full height of the power or communication equipment.

237. WORKING SPACE

Rule 237, Working Space, and Rule 236, Climbing Space, may appear complicated but they revolve around one simple idea. The supply and communication worker must have adequate space to climb and work on a pole and its conductors, supports, and equipment.

The working space rules are tied into the climbing space rules in Rule 236A. The working space dimensions required in Rule 237B are obtained from the climbing space requirements in Rule 236E and **NESC** Table 236-1. The working space dimension rules are outlined in Fig. 237-1.

Rule 237C provides requirements for working space relative to vertical and lateral conductors. Conductor jumpers and vertical risers that are not in conduit must be located outside the working space.

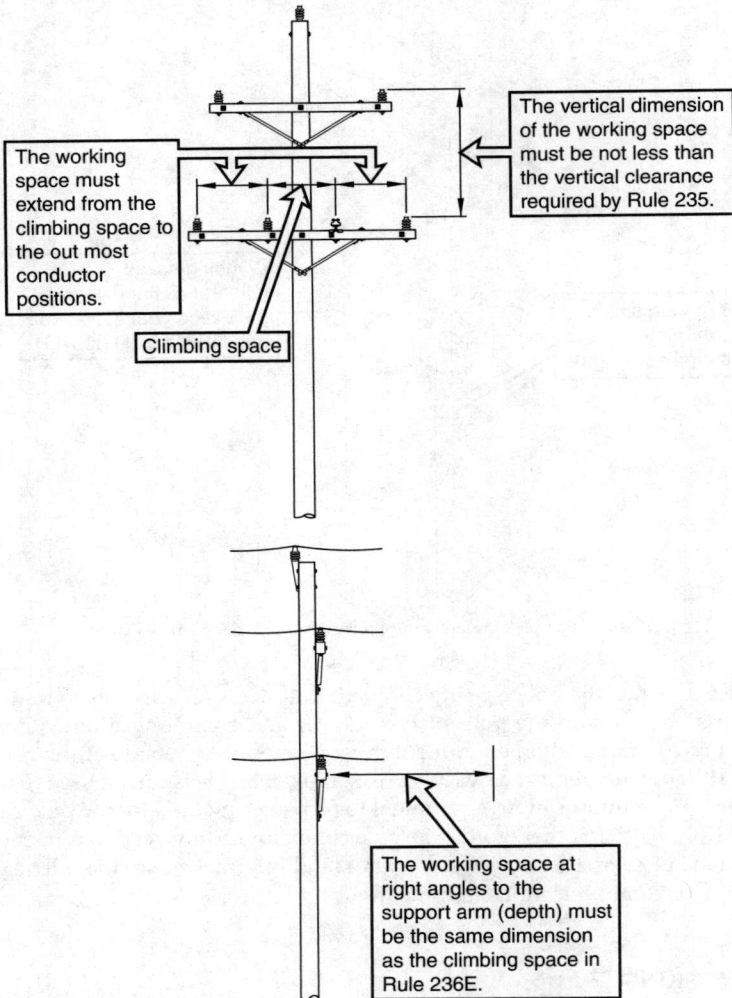

The working space must extend from the climbing space to the out most conductor positions.

The vertical dimension of the working space must be not less than the vertical clearance required by Rule 235.

Climbing space

The working space at right angles to the support arm (depth) must be the same dimension as the climbing space in Rule 236E.

Fig. 237-1. Working space dimensions (Rule 237B).

Rule 237D provides rules for working space relative to buckarm construction. This rule is to be used with the corresponding climbing space requirements in Rule 236F. **NESC** Fig. 237-1 is provided as a guide.

Energized equipment must be guarded to avoid contact if all the conditions listed in Rule 237E apply.

Rule 237F ties the working space requirements of Rule 237 to the work rules of Part 4. An example of working clearances from energized equipment is shown in Fig. 237-2.

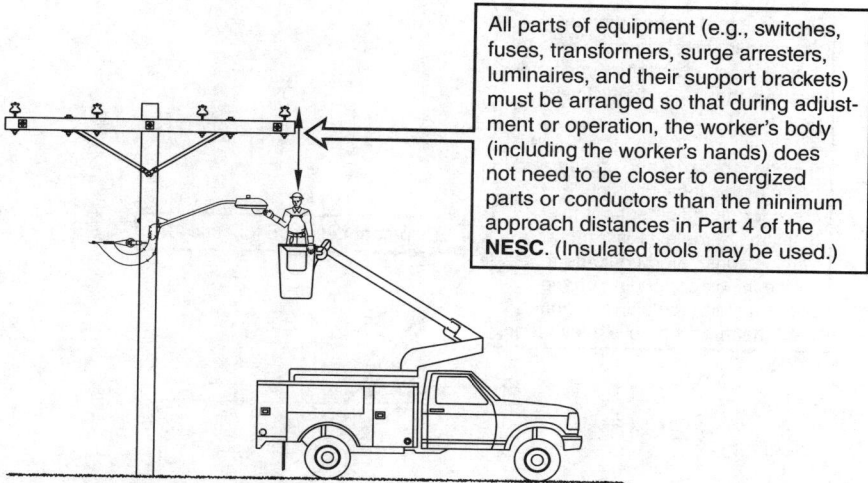

All parts of equipment (e.g., switches, fuses, transformers, surge arresters, luminaires, and their support brackets) must be arranged so that during adjustment or operation, the worker's body (including the worker's hands) does not need to be closer to energized parts or conductors than the minimum approach distances in Part 4 of the **NESC**. (Insulated tools may be used.)

See Photo(s)

Fig. 237-2. Example of working clearances from energized equipment (Rule 237F).

238. VERTICAL CLEARANCE BETWEEN SPECIFIED COMMUNICATION AND SUPPLY FACILITIES LOCATED ON THE SAME STRUCTURE

Rule 238 is a small rule in terms of the amount of text, but it has a big impact on clearance between supply (power) and communication facilities.

238A. Equipment. A definition is provided in Rule 238A for equipment as it applies in Rule 238. Since the definition is specific to this rule, it is provided in Rule 238 instead of Sec. 02, "Definitions of Special Terms." The definition of equipment on a typical joint-use (supply and communication) pole is outlined in Fig. 238-1.

A wood crossarm and crossarm brace (for supply conductors) are not considered equipment per the definition of equipment in this rule. A metal crossarm brace (for supply conductors) may be considered equipment per the definition depending on the position of the metal brace and what the metal brace is attached to. Wood and metal crossarm braces supporting communication cables are always considered equipment per the definition of equipment in this rule.

238B. Clearances in General. The vertical clearance between the following joint use facilities must be considered:

- Supply conductors and communication equipment
- Communication conductors and supply equipment
- Supply and communication equipment

NESC Table 238-1 is used to establish the vertical clearance requirements. NESC Table 238-1 provides vertical clearances only, not horizontal or diagonal clearances. The vertical clearance required in Rule 238 is in addition to the vertical clearance required in Rule 235C. The clearance requirements in both Rules 235C and 238 must be met. The clearances required in Rule 235 and 238 may be the

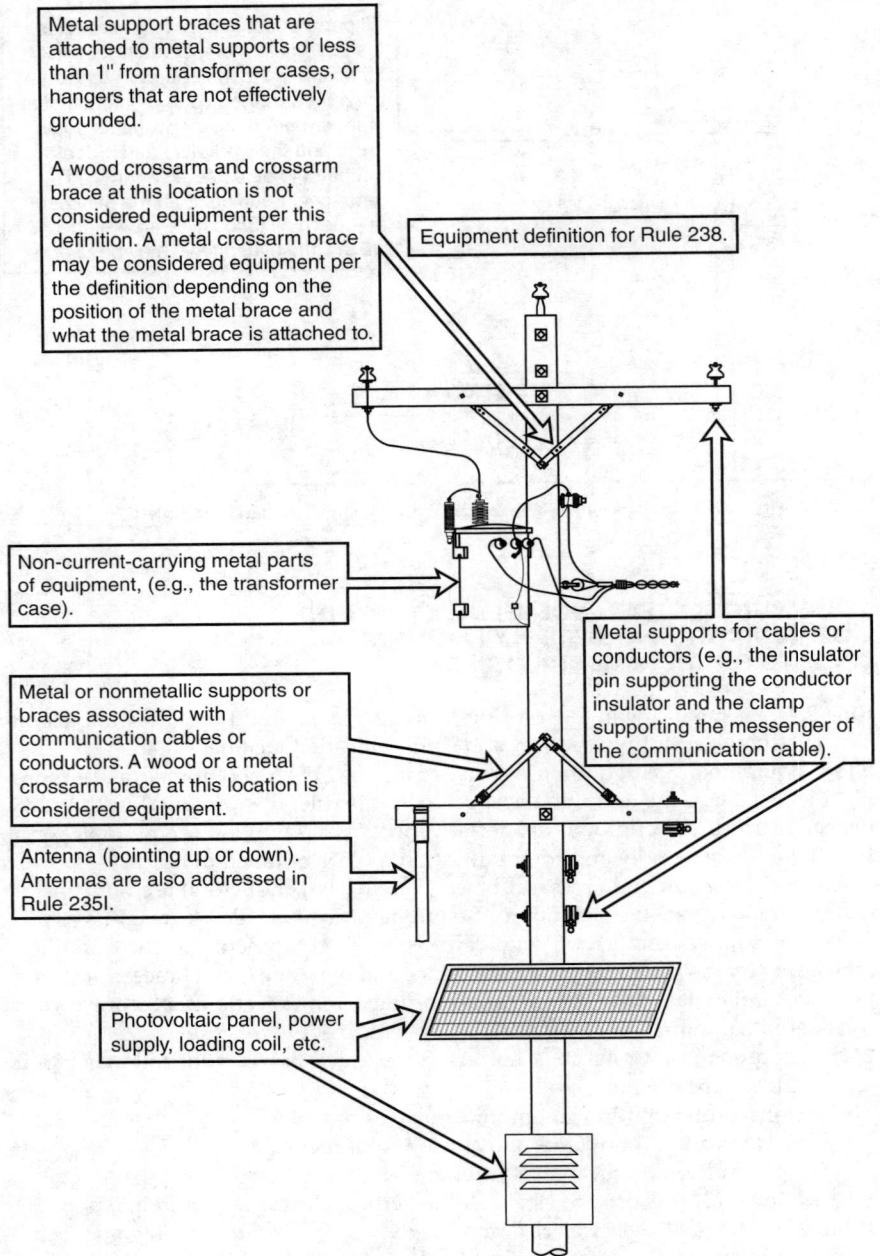

Metal support braces that are attached to metal supports or less than 1" from transformer cases, or hangers that are not effectively grounded.

A wood crossarm and crossarm brace at this location is not considered equipment per this definition. A metal crossarm brace may be considered equipment per the definition depending on the position of the metal brace and what the metal brace is attached to.

Equipment definition for Rule 238.

Non-current-carrying metal parts of equipment, (e.g., the transformer case).

Metal supports for cables or conductors (e.g., the insulator pin supporting the conductor insulator and the clamp supporting the messenger of the communication cable).

Metal or nonmetallic supports or braces associated with communication cables or conductors. A wood or a metal crossarm brace at this location is considered equipment.

Antenna (pointing up or down). Antennas are also addressed in Rule 235I.

Photovoltaic panel, power supply, loading coil, etc.

See Photo(s)

Fig. 238-1. Definition of equipment as it applies to vertical clearance between communication and supply facilities on the same structure (Rule 238A).

same (e.g., 40 in), but the clearance is measured at different locations. In Rule 238, a supply equipment to communication equipment measurement is required. In Rule 235C, a supply conductor to communication cable measurement is required, both at the structure and out in the span. The vertical clearance requirements in Rule 238B and **NESC** Table 238-1 revolve around 40 in. Less than 40 in (e.g., 30 in) is acceptable for certain grounded supply facilities. Greater than 40 in is required for supply voltages above 8.7 kV to ground. Exceptions to the values in **NESC** Table 238-1 are provided in Rules 238C and 238D. Examples of vertical clearance between supply and communication equipment on the same structure are shown in Figs. 238-2 through 238-11.

NESC Table 238-1 does not use an upper level/lower level format. **NESC** Table 235-5 does use an upper level/lower level format but Footnote 12 to **NESC** Table 235-5 addresses locating facilities in opposite positions.

The installation of a solar (photovoltaic) panel on a pole may involve both power and communication utilities depending on the location of the solar panel. If the solar panel is located in the supply space, the communication utilities on

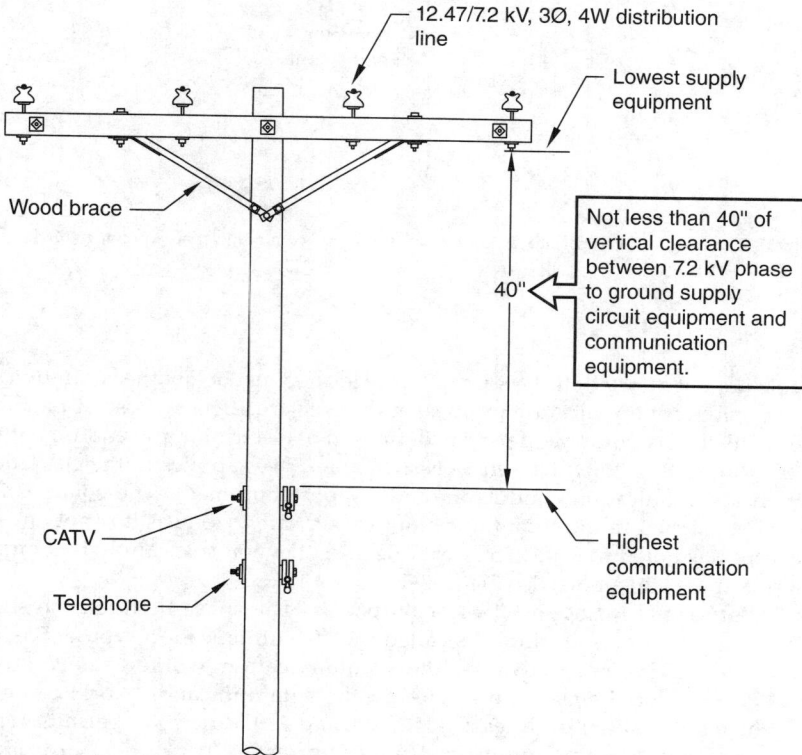

Fig. 238-2. Example of vertical clearance between supply and communication equipment on the same structure (Rule 238B).

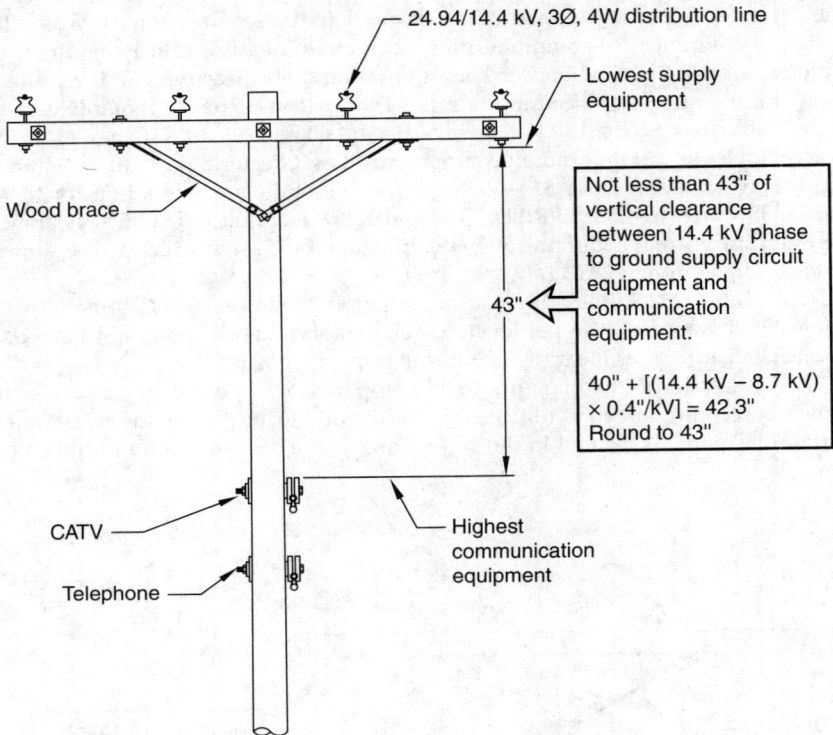

24.94/14.4 kV, 3Ø, 4W distribution line

Lowest supply equipment

Wood brace

Not less than 43" of vertical clearance between 14.4 kV phase to ground supply circuit equipment and communication equipment.

$$40" + [(14.4\ kV - 8.7\ kV) \times 0.4"/kV] = 42.3"$$
Round to 43"

43"

CATV

Telephone

Highest communication equipment

Fig. 238-3. Example of vertical clearance between supply and communications equipment on the same structure (Rule 238B).

the pole are not affected. If the solar panel is located just below the communication space, the communication utilities on the pole are affected. A solar panel is considered supply equipment per the definition of electric supply equipment in Sec. 02. Rule 238B, which references **NESC** Table 238-1, applies to the clearance between communication conductors and supply equipment. The solar panel cannot be installed in the communication worker safety zone as it is not one of the exceptions noted in Rule 235C4 or Rule 238E. Several rules apply to mounting a solar panel on a pole. See Fig. 238-12.

238C. Clearances for Span Wires or Brackets.   The space between the supply and communication facilities is called the "Communication Worker Safety Zone." Rule 238E which defines the communication worker safety zone allows a few select items to be located in the communication worker safety zone which are covered in Rules 238C, 238D, and 239. Rule 238C permits span wires or brackets carrying luminaires (light fixtures), traffic signals, or trolley conductors to be located in the communication worker safety zone. The **Code** recognizes that lighting fixtures and traffic signals provide their own safety

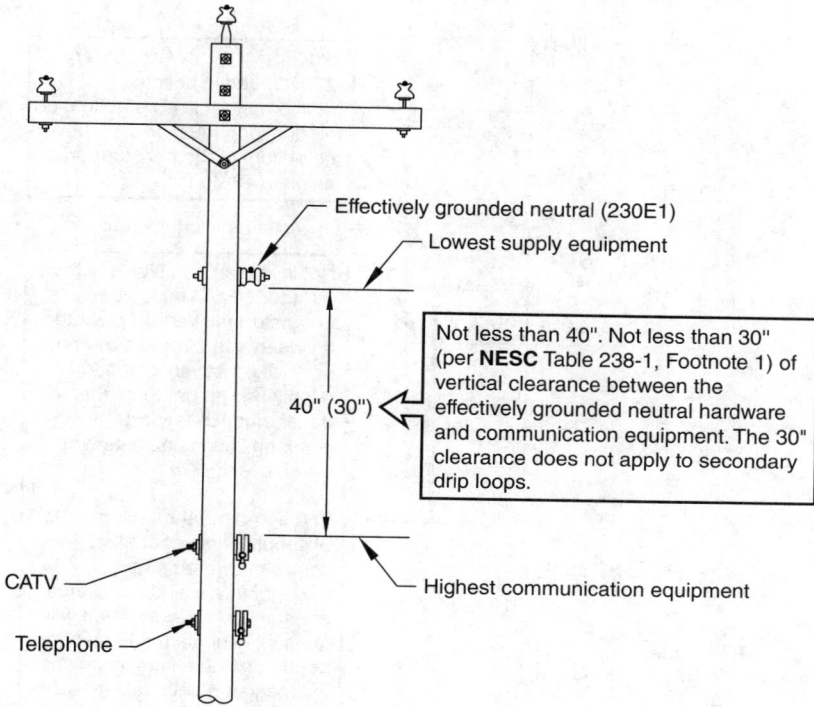

Effectively grounded neutral (230E1)

Lowest supply equipment

40" (30")

Not less than 40". Not less than 30"
(per **NESC** Table 238-1, Footnote 1) of
vertical clearance between the
effectively grounded neutral hardware
and communication equipment. The 30"
clearance does not apply to secondary
drip loops.

CATV

Highest communication equipment

Telephone

See
Photo(s)

Fig. 238-4. Example of vertical clearance between supply and communication equipment on the
same structure (Rule 238B).

function and commonly the appropriate mounting height for span wires or
brackets carrying light fixtures, traffic signals, or trolley conductors is in the
space between the supply and communication facilities. Rule 238C references
NESC Table 238-2, which provides the required clearances for these items. If a
luminaire or traffic signal is located in the communication worker safety zone,
a drip loop is needed for service to the luminaire or traffic signal. Rule 238D
applies to the drip loop.

238D. Clearance of Drip Loops Associated with Luminaires and Traffic Signals
The space between the supply and communication facilities is called the "Communication Worker Safety Zone." Rule 238E which defines the communication worker
safety zone allows a few select items to be located in the communication worker
safety zone which are covered in Rules 238C, 238D, and 239. Rule 238D recognizes
that lighting fixtures, lighting fixture brackets, and traffic signal brackets are commonly fed from a drip loop positioned below the lighting or traffic signal bracket.
Special clearances for these drip loops permit them to be located in the communication worker safety zone. An exception applies to Rule 238D that permits the
clearance of a drip loop for a luminaire or traffic signal to be not less than 3 in

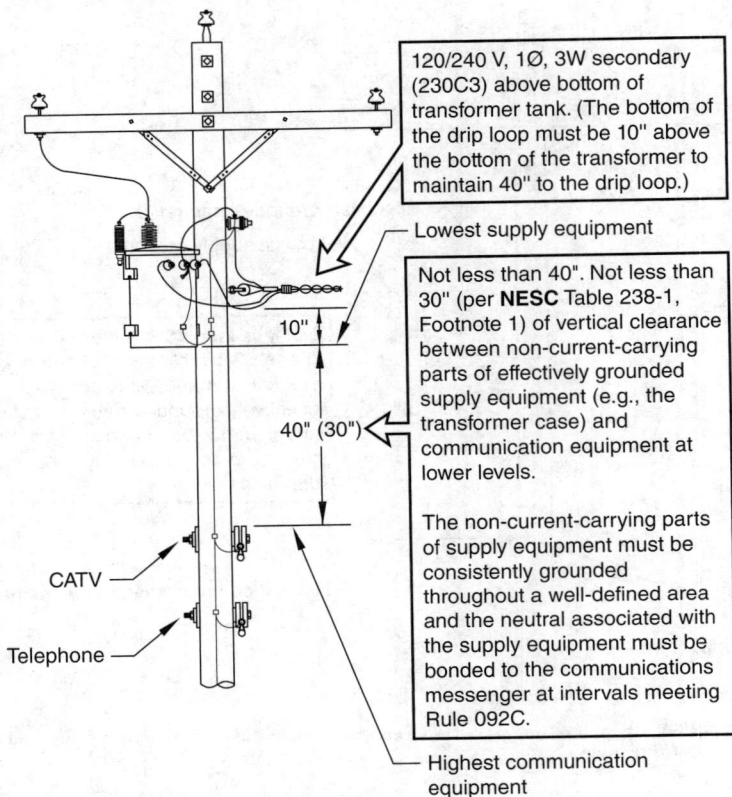

120/240 V, 1Ø, 3W secondary (230C3) above bottom of transformer tank. (The bottom of the drip loop must be 10" above the bottom of the transformer to maintain 40" to the drip loop.)

Lowest supply equipment

Not less than 40". Not less than 30" (per **NESC** Table 238-1, Footnote 1) of vertical clearance between non-current-carrying parts of effectively grounded supply equipment (e.g., the transformer case) and communication equipment at lower levels.

The non-current-carrying parts of supply equipment must be consistently grounded throughout a well-defined area and the neutral associated with the supply equipment must be bonded to the communications messenger at intervals meeting Rule 092C.

10"

40" (30")

CATV

Telephone

See Photo(s)

Highest communication equipment

Fig. 238-5. Example of vertical clearance between supply and communication equipment on the same structure (Rule 238B).

above the highest communication cable, through bolt, or other equipment if certain conditions apply. Examples of lighting fixtures mounted above, in, and below the communication worker safety zone, along with lighting fixture drip loops, are shown in Figs. 238-13 through 238-16.

It can be easy to lose sight of the fact that the reduced clearance to secondary drip loops only applies to drip loops feeding luminaires or traffic signals. For example, even though a 120/240 V circuit can be used to feed a house or a street lighting luminaire, the 12-in clearance only applies to the luminaire (street light) drip loop entering the luminaire or luminaire bracket. Luminaires and traffic signals serve their own safety functions so they merit special **Code** consideration. See Fig. 238-17.

238E. Communication Worker Safety Zone. Rule 238E labels the area between the supply space and the communication space both at the structure and out in the span. The name for this area is the "Communication Worker Safety Zone." The

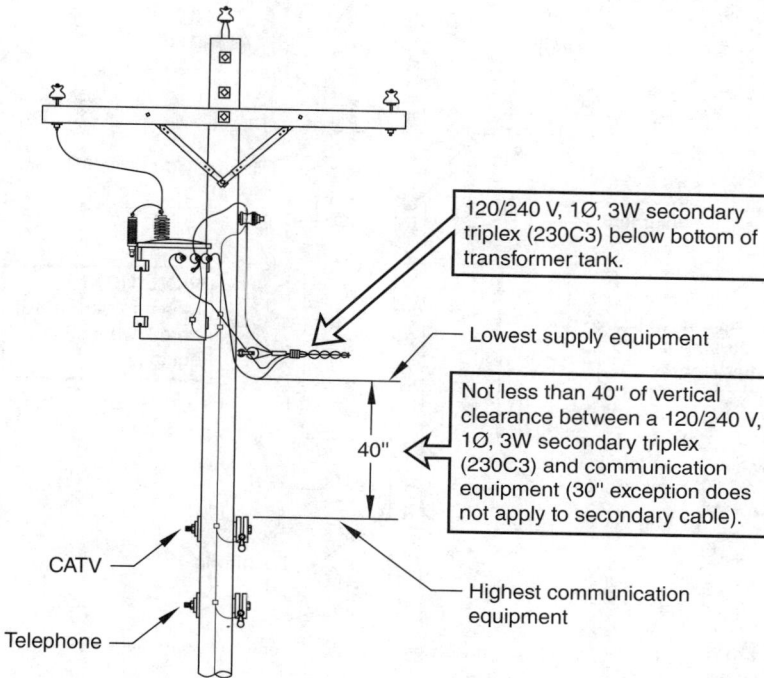

120/240 V, 1Ø, 3W secondary triplex (230C3) below bottom of transformer tank.

Lowest supply equipment

Not less than 40" of vertical clearance between a 120/240 V, 1Ø, 3W secondary triplex (230C3) and communication equipment (30" exception does not apply to secondary cable).

40"

CATV

Telephone

Highest communication equipment

See Photo(s)

Fig. 238-6. Example of vertical clearance between supply and communication equipment on the same structure (Rule 238B).

work rules in Part 4 of the NESC provide the qualifications of a supply employee and a communication employee. Since the communication employee is not trained to work on supply lines, a safety zone exists for the communication employee's protection. If a communication line is positioned below a supply line on an overhead structure, but the required communication worker safety zone clearance does not exist, the communication employee can correct the violation if the communication employee does not violate the minimum approach distances and other requirements in Sec. 43. If the communication employee cannot maintain the minimum approach distances and other requirements in Sec. 43, the communication employee must contact a supply employee to correct the violation. See Sec. 43 for additional information. The communication worker safety zone is defined by both Rule 238 and Rule 235C, not one rule or the other but the combined effect of both rules. The supply space above the Communication Worker Safety Zone and the communication space below the Communication Worker Safety Zone are defined in Sec. 02, Definitions of Special Terms. Sec. 02 also defines the terms electric supply equipment, communication equipment, conductor (multiple definitions) and lines (multiple definitions). Understanding these terms is helpful to applying the Communication Worker Safety Zone specified in both Rules 235C4 and 238E. The Communication Worker Safety Zone clearance

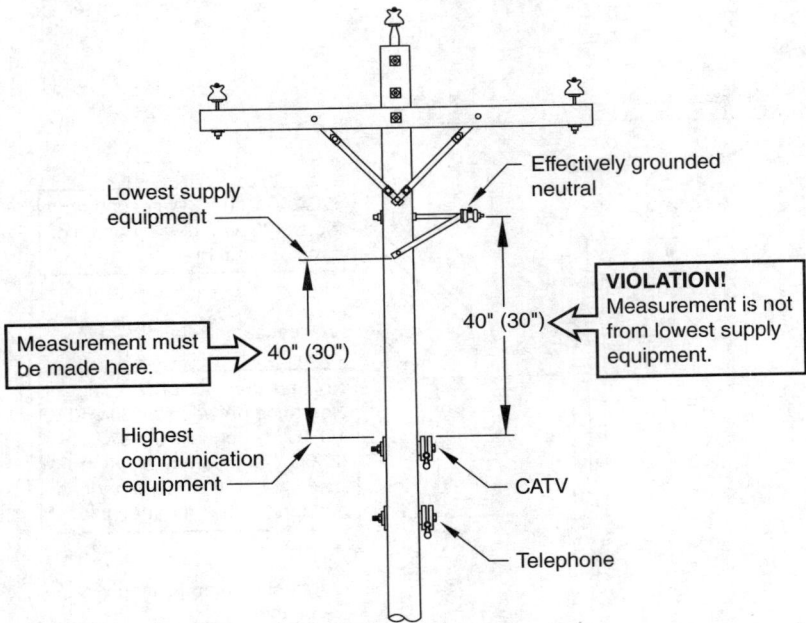

Fig. 238-7. Example of vertical clearance between supply and communication equipment on the same structure (Rule 238B).

revolves around 40 in, however, higher voltages (above 8.7 kV to ground) require greater than 40 in, exceptions may permit less than 40 in and knowing where to measure the upper and lower limits of the Communication Worker Safety Zone is critical to applying the rule. In addition to the items specified in Rules 238C and 238D, Rule 239 (Vertical Risers) are permitted to pass through the Communication Worker Safety Zone. The most common application in Rule 239 is Rule 239G, see Rule 239G for a discussion and a figure. It is possible to locate communication cables and communication antennas in the supply space. See Rules 224 and 235I. It is also possible to locate supply equipment below the communication space. See Rules 232 and 238. The appropriate paragraphs in Rule 239 need to be referenced for supply and communication risers going through the supply and communication spaces. See Fig. 238-18.

239. CLEARANCE OF VERTICAL AND LATERAL FACILITIES FROM OTHER FACILITIES AND SURFACES ON THE SAME SUPPORTING STRUCTURE

239A. General. Rule 239A1 provides a list of vertical and lateral conductors that may be placed directly on the supporting structure (pole). A conduit may also be placed directly on the pole. Rule 239D requires guarding of vertical conductors

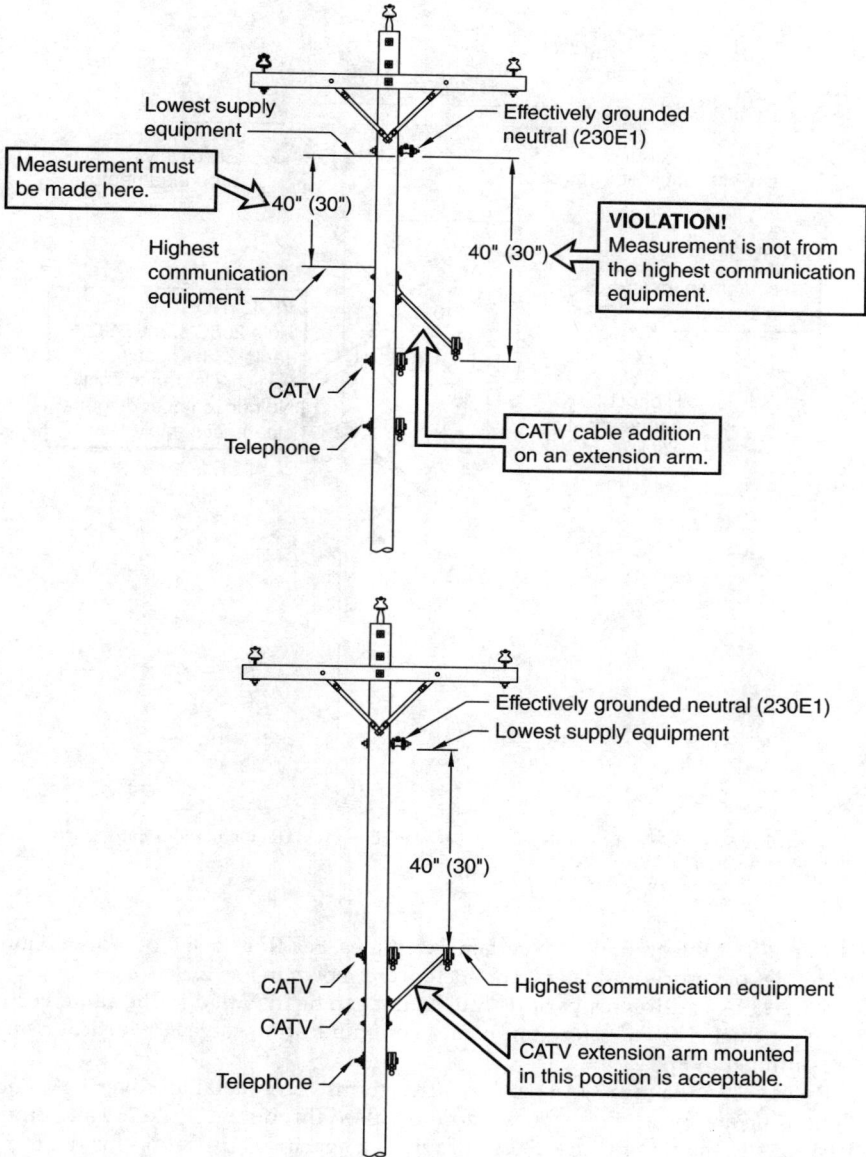

Fig. 238-8. Example of vertical clearance between supply and communication equipment on the same structure (Rule 238B).

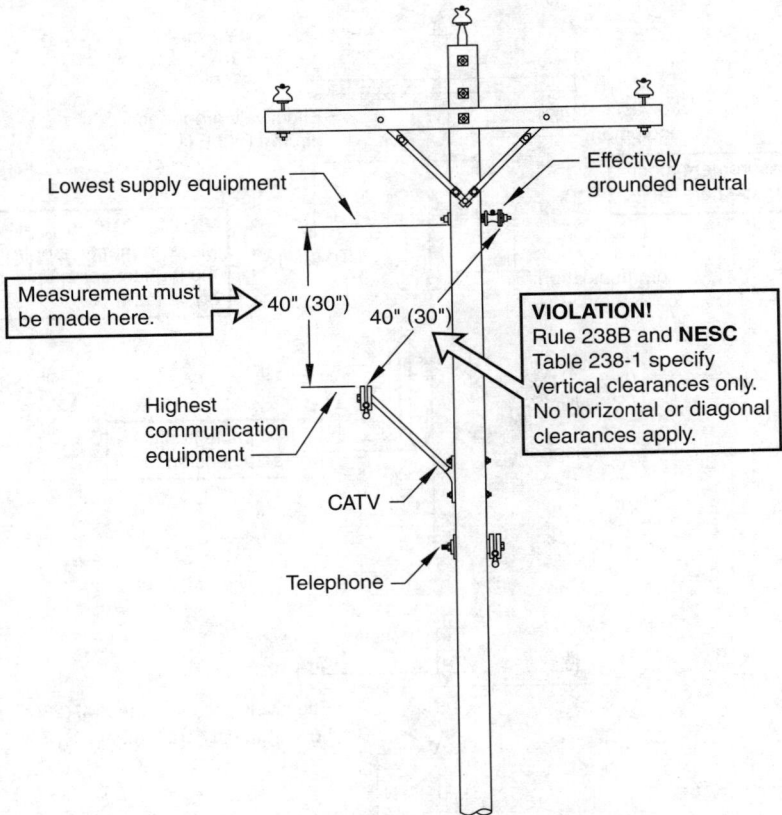

Fig. 238-9. Example of vertical clearance between supply and communication equipment on the same structure (Rule 238B).

within 8 ft of the ground with certain exceptions. See Rule 239D for a discussion. Examples of vertical and lateral conductors are shown in Fig. 239-1.

Rule 239A2 addresses various cables that can be installed in the same vertical riser duct. The rules for supply and communication cables in vertical risers are outlined in Fig. 239-2.

Rules 239A5 and 239A6 address the terms "nonmetallic covering" and "guarding and protection." These terms are used throughout Rule 239 and other locations in the Code. Rule 239A5 provides a specific clarification for the term "nonmetallic covering" when the term is used in Rule 239. The Code does not specify the type of conduit or U-guard that is required for risers. See Rule 352 for a discussion of various types of ducts (conduits).

239B. Location of Vertical or Lateral Conductors Relative to Climbing Spaces, Working Spaces, and Pole Steps. Climbing spaces, working spaces, and pole steps must not be obstructed with vertical and lateral conductors. Rule 239B

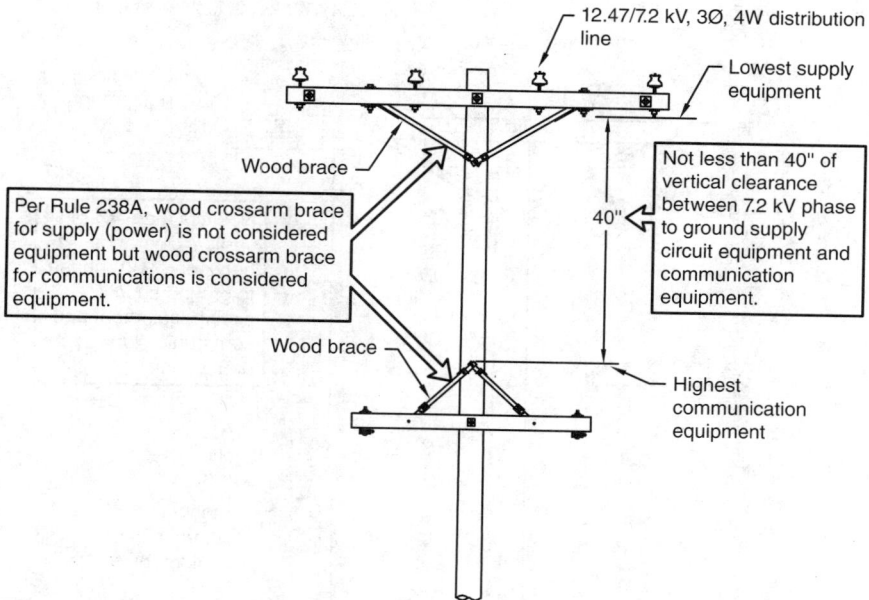

Per Rule 238A, wood crossarm brace for supply (power) is not considered equipment but wood crossarm brace for communications is considered equipment.

Not less than 40" of vertical clearance between 7.2 kV phase to ground supply circuit equipment and communication equipment.

12.47/7.2 kV, 3Ø, 4W distribution line

Lowest supply equipment

Wood brace

40"

Wood brace

Highest communication equipment

Fig. 238-10. Example of vertical clearance between supply and communication equipment on the same structure (Rule 238B).

applies to vertical conductors not in conduit. A note to this rule reminds the reader that vertical runs in conduit securely attached to the pole surface are not considered to obstruct the climbing space per Rule 236H.

239C. Conductors Not in Conduit. Vertical and lateral conductors not in conduit must have the same clearances from conduits (enclosing other conductors) as from other structure surfaces.

239D. Guarding and Protection Near Ground. Vertical conductors (risers) within 8 ft of the ground must be guarded. Conduit or U-guard are the most common types of guards used to protect the vertical conductors. An exception to Rule 239D1 provides a list of conductors and cables that do not require guarding. See Fig. 239-3.

Rule 239D is very similar to Rule 093D, Guarding and Protection, for grounding conductors. Rule 239D also overlaps with Rule 362, Pole Risers—Additional Requirements. See Rules 093D and 362 for additional information and related figures.

239E. Requirements for Vertical and Lateral Supply Conductors on Supply Line Structures or within Supply Space on Jointly Used Structures. The locations and conductors that apply to Rule 239E are graphically shown in Fig. 239-4.

The general clearances for open (e.g., bare noninsulated) vertical and lateral supply conductors from the surface of the support are provided in **NESC** Table239-1. NESC Table 239-1 also provides clearance to span, guy, and messenger wires. Clearances in Rule 235E (**NESC** Table 235-6) must also be met. **NESC** Table 235-6

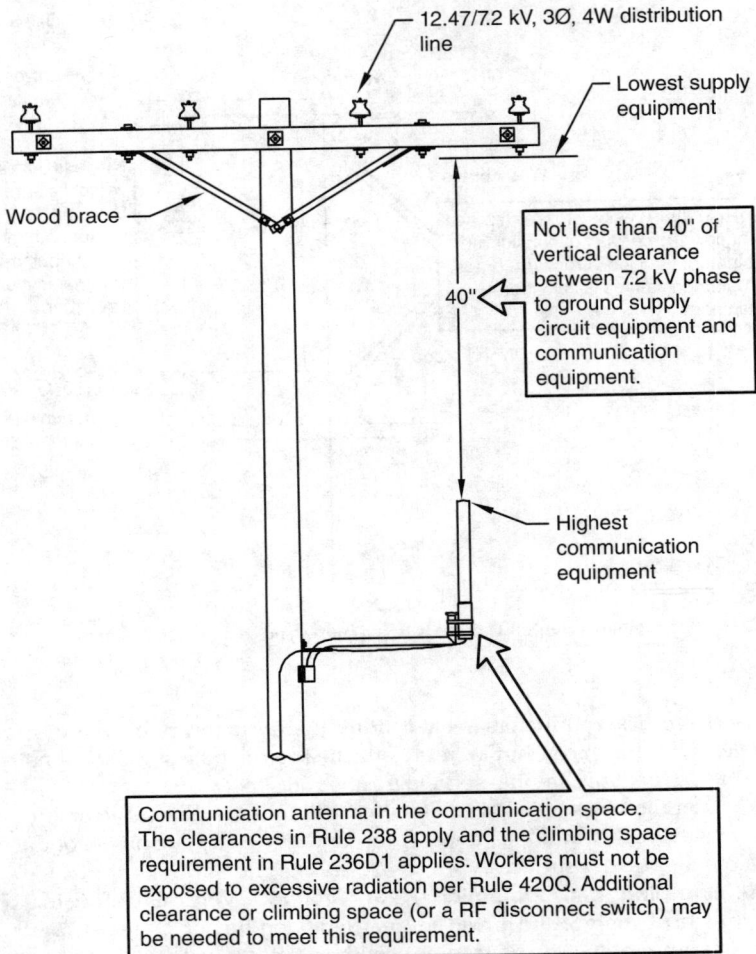

Fig. 238-11. Example of vertical clearance between supply and communication equipment on the same structure (Rule 238B).

provides clearances from line conductors (i.e., the conductors spanning from pole to pole) to vertical and lateral conductors attached to the same support (pole).

NESC Table 239-2 must be applied to special cases. If open line conductors (e.g., bare conductors spanning from pole to pole) are within 4 ft of the pole, vertical conductors must meet two special conditions. The special conditions in Rules 239E2a(1) and 239E2a(2) only apply if open line conductors are within 4 ft of the pole, which is very common, even on an 8-ft crossarm, and if workers climb to work on the line conductors while the vertical conductors in questions are energized. If the vertical conductors in question are de-energized while climbing, or

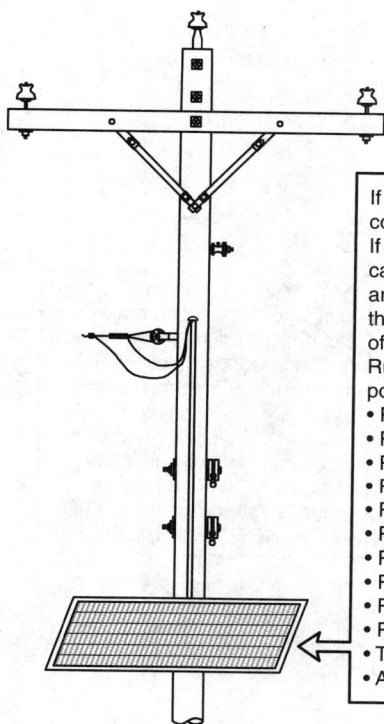

If the solar panel is located in the supply space, the communication utilities on the pole are not affected. If the solar panel is located just below the communication space, the communication utilities on the pole are affected. The solar panel can not be installed in the communication worker safety zone as it is not one of the exceptions noted in Rule 235C4 or Rule 238E. Rules that apply to a solar panel installation on a pole include:
- Rule 213 (Accessibility)
- Rule 215 and Sec. 09 (Grounding)
- Rule 217A4 (Attachments)
- Rule 220E (Identification)
- Rule 232 (Clearance)
- Rule 235 (Clearance)
- Rule 236 (Climbing Space)
- Rule 237 (Working Space)
- Rule 238 (Clearance)
- Rule 239 (Vertical Risers)
- The Work Rules of Part 4
- Any related rules

See Photo(s)

Fig. 238-12. Example of mounting a solar (photovoltaic) panel on a pole (Rule 238B).

if the pole is worked on from a bucket truck, then the special cases of Rule 239E2 do not apply. NESC Table 239-2 requires clearances larger than NESC Table 239-1 to provide room for workers to climb near energized vertical conductors. NESC Table 239-2 also provides a distance from above and below the line conductor where the clearances apply. If a conduit is used within the distance specified, it must be nonmetallic or protected by a nonmetallic covering to provide additional safety for a worker climbing the pole and working near an energized vertical conductor.

239F. Requirements for Vertical and Lateral Communication Conductors on Communication Line Structures or within the Communication Space on Jointly Used Structures. The locations and conductors that Rule 239F apply to are graphically shown in Fig. 239-5.

Uninsulated (open wire) vertical and lateral communication conductors are not as common as they once were. Open wire communication has been replaced with insulated communication cables. When an insulated communication cable riser exists on a joint-use (supply and communication) pole, the specified clearance to supply facilities in Rule 239F is essentially the same as required by Rules 235 and 238.

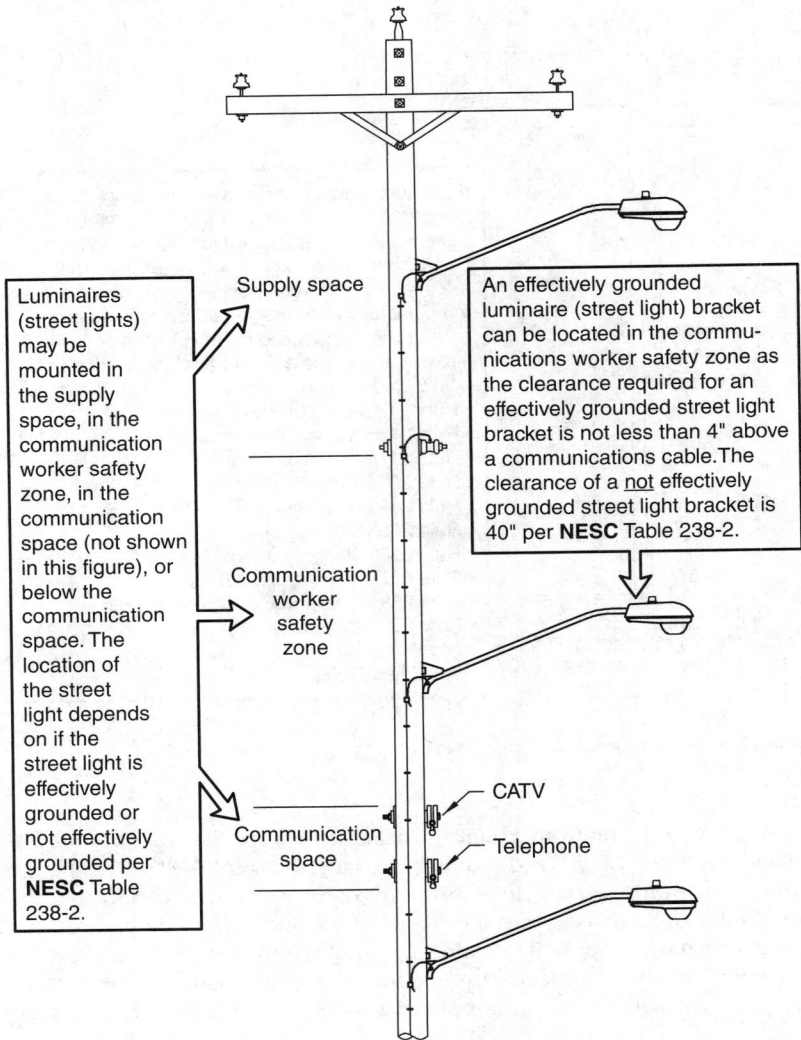

Luminaires (street lights) may be mounted in the supply space, in the communication worker safety zone, in the communication space (not shown in this figure), or below the communication space. The location of the street light depends on if the street light is effectively grounded or not effectively grounded per **NESC** Table 238-2.

Supply space

An effectively grounded luminaire (street light) bracket can be located in the communications worker safety zone as the clearance required for an effectively grounded street light bracket is not less than 4" above a communications cable. The clearance of a _not_ effectively grounded street light bracket is 40" per **NESC** Table 238-2.

Communication worker safety zone

Communication space

CATV

Telephone

Fig. 238-13. Example of locating a luminaire on a pole (Rule 238).

239G. Requirements for Vertical Supply Conductors and Cables Passing through Communication Space on Jointly Used Line Structures.   The locations and conductors to which Rule 239G applies are graphically shown in Fig. 239-6.

Rule 239G applies to a number of common joint-use (supply and communication) conditions. Providing a conduit over supply conductors 40 in above the highest communication attachment is a condition that relates to the communication worker safety zone discussed in Rule 238E. Examples of vertical supply conductors on joint-use (supply and communication) poles are shown in Figs. 239-7 and 239-8.

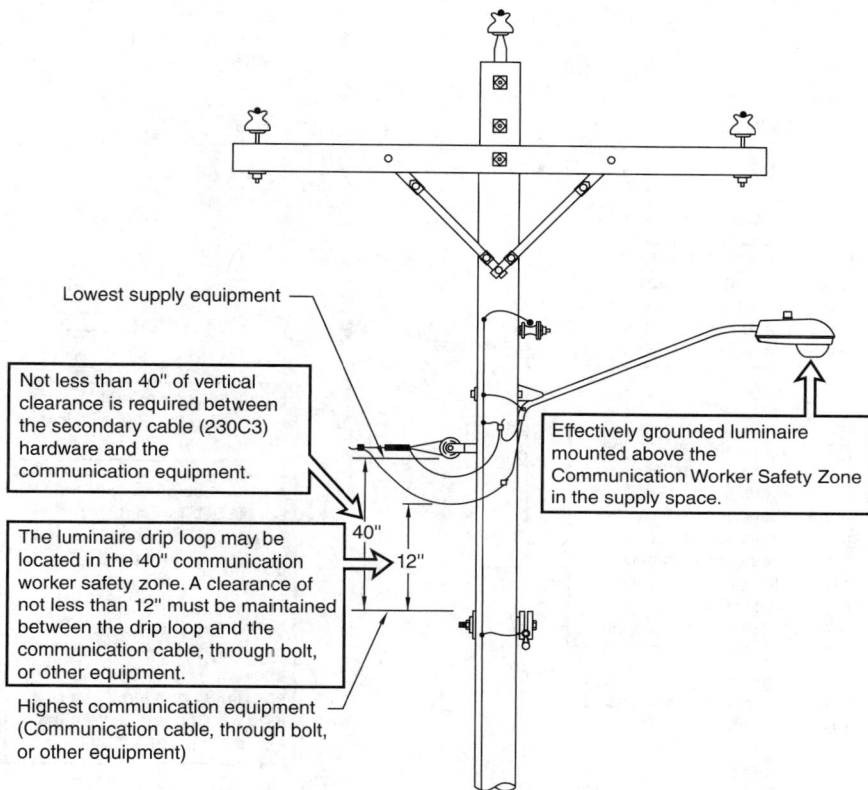

Lowest supply equipment ─

Not less than 40" of vertical clearance is required between the secondary cable (230C3) hardware and the communication equipment.

Effectively grounded luminaire mounted above the Communication Worker Safety Zone in the supply space.

The luminaire drip loop may be located in the 40" communication worker safety zone. A clearance of not less than 12" must be maintained between the drip loop and the communication cable, through bolt, or other equipment.

40"
12"

Highest communication equipment (Communication cable, through bolt, or other equipment)

See Photo(s)

Fig. 238-14. Example of vertical clearance between a drip loop feeding a luminaire and communication equipment (Rules 238C and 238D).

Exception 2 to Rule 239G1 permits omitting guarding from a supply grounding conductor (pole ground) on a joint-use (supply and communication) pole if certain conditions are met. See Fig. 239-9.

Rule 239G4 requires a supply cable service drop to be at least 40 in above the highest communications attachment. A similar requirement exists in Rules 235C and 238B. A common example is a 120/240 V, single-phase, three-wire service drop connection at the pole. Rule 239G4 also states that a supply cable service drop can be at least 40 in below the lowest communications attachment (in the communications space). This is a less common location for a supply cable service drop. An example of this installation is a tall pole for a railroad crossing with a railroad signal building located near the base of the tall pole. The service can be run from a transformer in the supply space, down the pole in conduit to at least 40 in below the lowest communications attachment, and then take off as a service drop to the railroad signal building. The supply service drop in the communications space must have insulated splices and connections for the energized phase conductors (not for the grounded neutral which typically serves as the messenger).

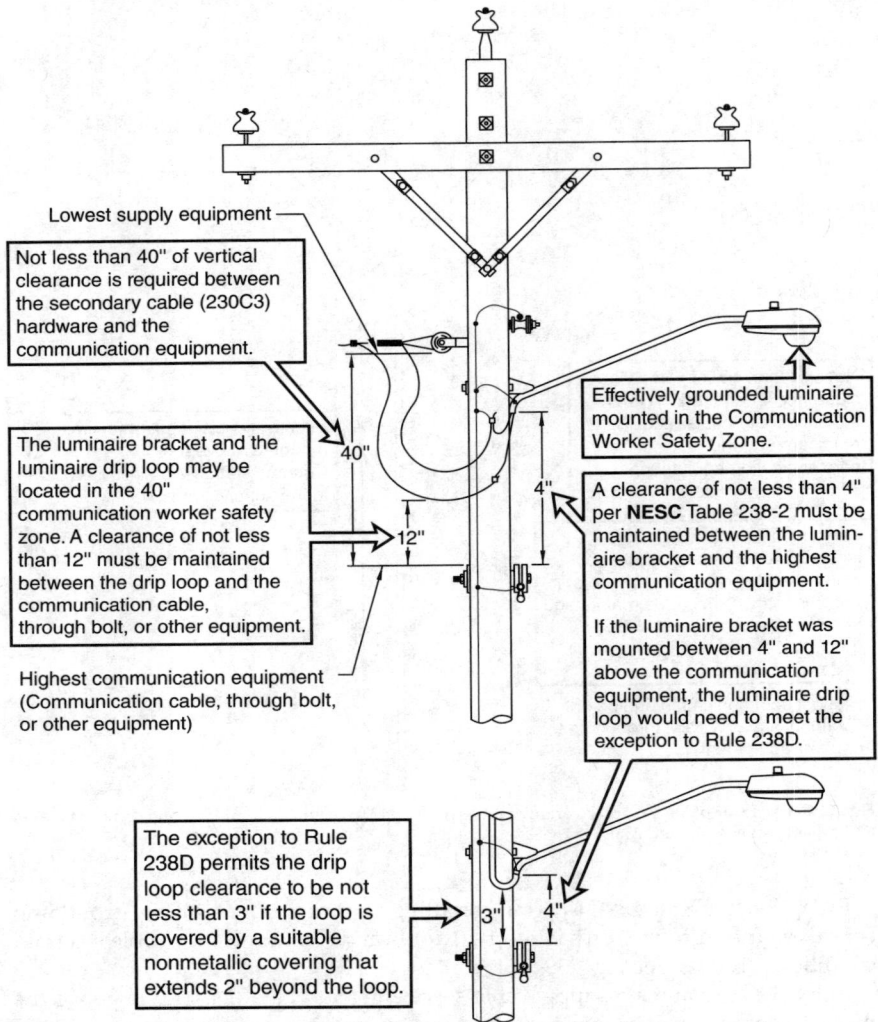

Lowest supply equipment ─

Not less than 40" of vertical clearance is required between the secondary cable (230C3) hardware and the communication equipment.

The luminaire bracket and the luminaire drip loop may be located in the 40" communication worker safety zone. A clearance of not less than 12" must be maintained between the drip loop and the communication cable, through bolt, or other equipment.

Highest communication equipment (Communication cable, through bolt, or other equipment)

Effectively grounded luminaire mounted in the Communication Worker Safety Zone.

A clearance of not less than 4" per **NESC** Table 238-2 must be maintained between the lumin-aire bracket and the highest communication equipment.

If the luminaire bracket was mounted between 4" and 12" above the communication equipment, the luminaire drip loop would need to meet the exception to Rule 238D.

The exception to Rule 238D permits the drip loop clearance to be not less than 3" if the loop is covered by a suitable nonmetallic covering that extends 2" beyond the loop.

See Photo(s)

Fig. 238-15. Example of vertical clearance between a drip loop feeding a luminaire and communication equipment (Rules 238C and 238D).

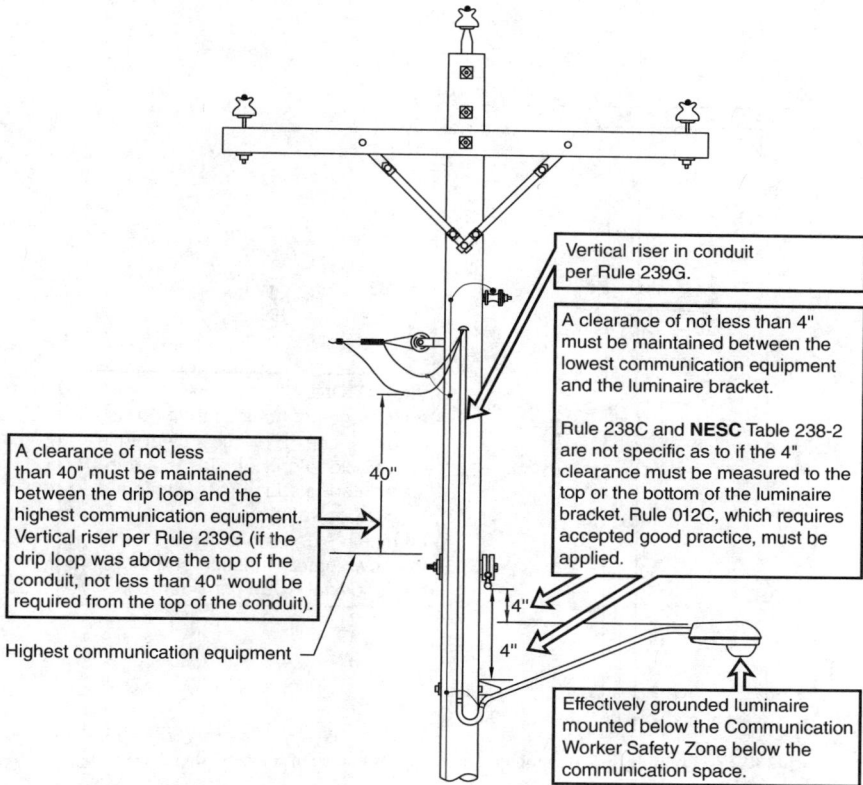

Vertical riser in conduit
per Rule 239G.

A clearance of not less than 4"
must be maintained between the
lowest communication equipment
and the luminaire bracket.

Rule 238C and **NESC** Table 238-2
are not specific as to if the 4"
clearance must be measured to the
top or the bottom of the luminaire
bracket. Rule 012C, which requires
accepted good practice, must be
applied.

A clearance of not less
than 40" must be maintained
between the drip loop and the
highest communication equipment.
Vertical riser per Rule 239G (if the
drip loop was above the top of the
conduit, not less than 40" would be
required from the top of the conduit).

Highest communication equipment

40"

4"

4"

Effectively grounded luminaire
mounted below the Communication
Worker Safety Zone below the
communication space.

Fig. 238-16. Example of vertical clearance between a drip loop feeding a luminaire and communication equipment (Rules 238C, 238D, and 239G).

239H. Requirements for Vertical Communication Conductors Passing through Supply Space on Jointly Used Structures. The locations and conductors to which Rule 239H applies are graphically shown in Fig. 239-10.

Communication conductors passing through the supply space can be seen on poles with an overhead ground wire (static) with an embedded fiber-optic communication cable and on supply poles with communication antennas on top. See Rule 235I for additional information on clearance to communication antennas in the supply space.

239I. Operating Rods. Operating rods for switches like a three-phase gang-operated distribution switch can pass through the communication space but they must be located outside the climbing space. See Rule 216 for a figure showing an overhead line switch operating rod. See Rule 236 for additional climbing space information.

239J. Additional Rules for Standoff Brackets. Standoff brackets may be used to support conduits. Both metallic and nonmetallic conduits may be supported.

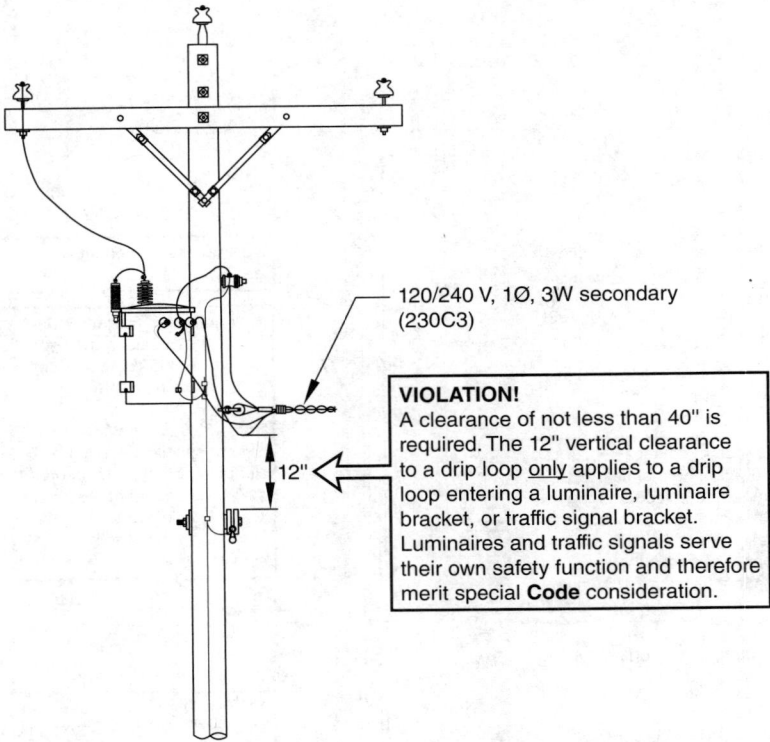

120/240 V, 1Ø, 3W secondary
(230C3)

VIOLATION!
A clearance of not less than 40" is required. The 12" vertical clearance to a drip loop <u>only</u> applies to a drip loop entering a luminaire, luminaire bracket, or traffic signal bracket. Luminaires and traffic signals serve their own safety function and therefore merit special **Code** consideration.

12"

Fig. 238-17. Example of a common joint-use (supply and communication) violation (Rule 238D).

The conductors inside the conduits must have sufficient insulation. Tree wire described in Rule 230D cannot be installed in a conduit riser as it is not fully insulated. Standoff brackets may be used to support cables not in conduit if the cable is a communication cable, 230C1a supply cable, or a supply cable less than 750 V with a single outer jacket or sheath. Traditional duplex, triplex, and quadruplex supply cables do not meet this criteria, nor do single insulated underground secondary conductors. The rules for positioning standoff brackets on a structure to inhibit climbing by unqualified persons are provided in Rule 217A2.

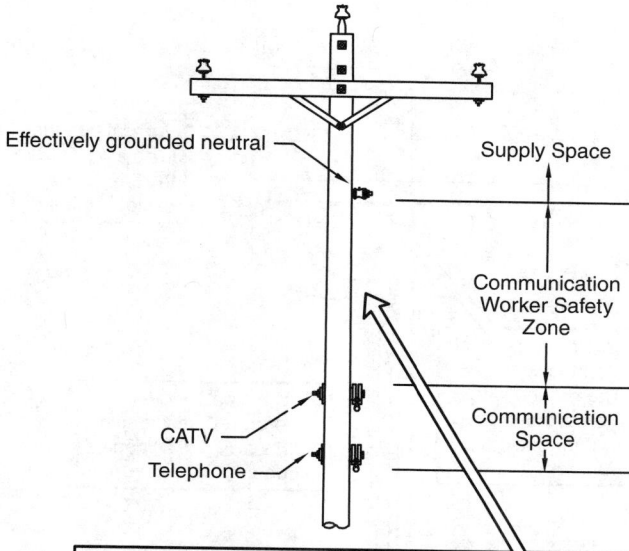

Effectively grounded neutral ─

Supply Space

Communication
Worker Safety
Zone

CATV ─

Communication
Space

Telephone ─

The "Communication Worker Safety Zone" is a space on the pole where nothing is located (with some exceptions). Both Rules 238 and 235C are used to define the space. Both rules must be met, not just one or the other.

Exceptions that are permitted in the communication worker safety zone are as follows:
• Span wires or brackets carrying luminaires (light fixtures), traffic signals, or trolley conductors (see Rule 238C).
• Drip loops associated with light fixtures or traffic signals (see Rule 238D).
• Vertical risers (see Rule 239).

See
Photo(s)

Fig. 238-18. Communication worker safety zone (Rule 238E).

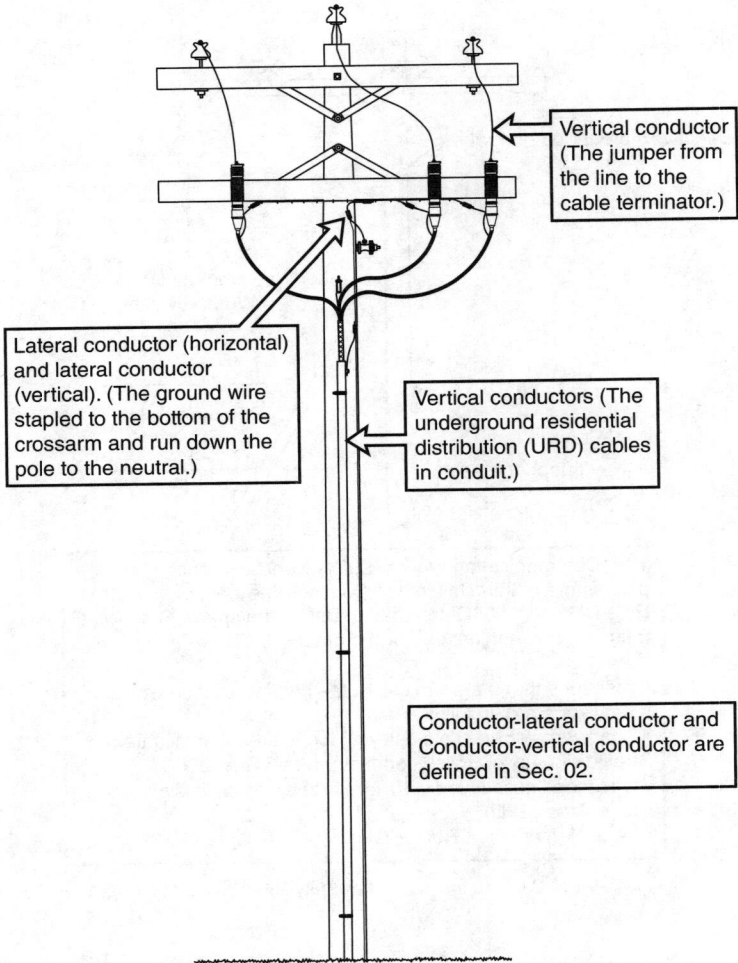

Vertical conductor (The jumper from the line to the cable terminator.)

Lateral conductor (horizontal) and lateral conductor (vertical). (The ground wire stapled to the bottom of the crossarm and run down the pole to the neutral.)

Vertical conductors (The underground residential distribution (URD) cables in conduit.)

Conductor-lateral conductor and Conductor-vertical conductor are defined in Sec. 02.

See Photo(s)

Fig. 239-1. Examples of vertical and lateral conductors (Rule 239A).

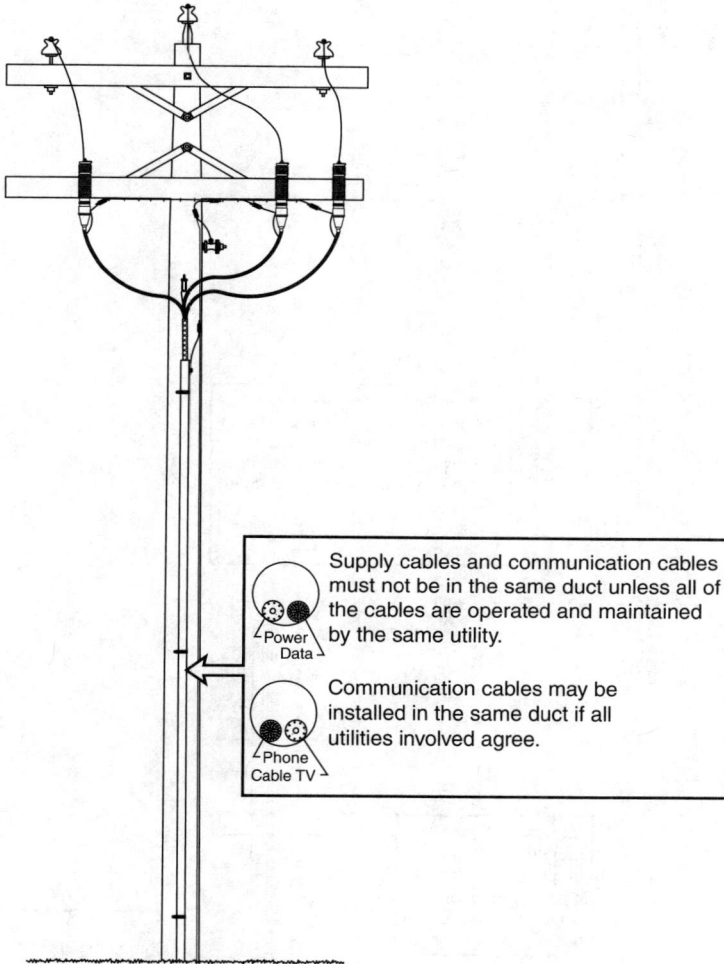

Fig. 239-2. Supply and communication cables in vertical risers (Rules 239A2d and 239A2e).

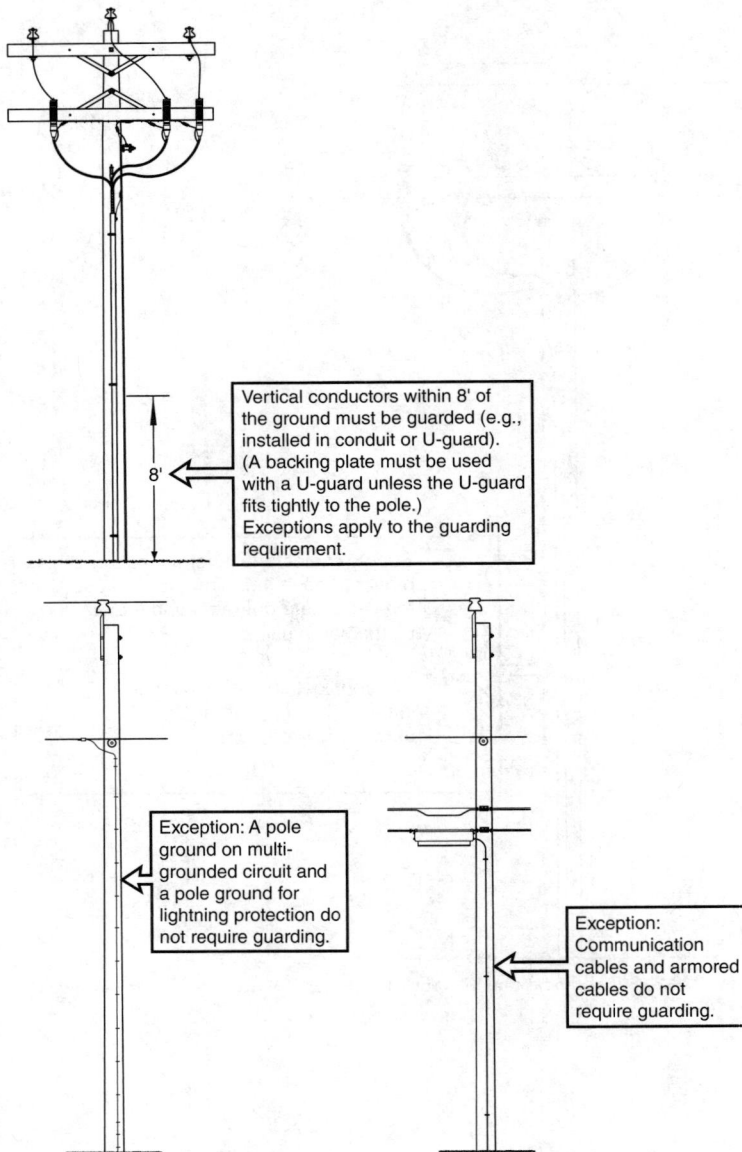

Vertical conductors within 8' of the ground must be guarded (e.g., installed in conduit or U-guard). (A backing plate must be used with a U-guard unless the U-guard fits tightly to the pole.) Exceptions apply to the guarding requirement.

8'

Exception: A pole ground on multi-grounded circuit and a pole ground for lightning protection do not require guarding.

Exception: Communication cables and armored cables do not require guarding.

See Photo(s)

Fig. 239-3. Guarding and protection near ground (Rule 239D).

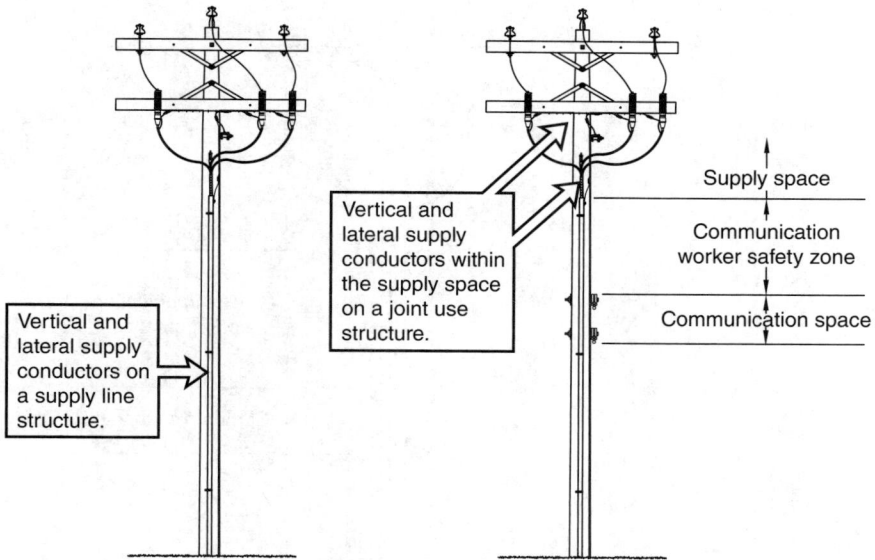

Fig. 239-4. Location of vertical and lateral supply conductors on supply line structures or within supply space on jointly used structures (Rule 239E).

Vertical and lateral supply conductors on a supply line structure.

Vertical and lateral supply conductors within the supply space on a joint use structure.

Supply space

Communication worker safety zone

Communication space

See Photo(s)

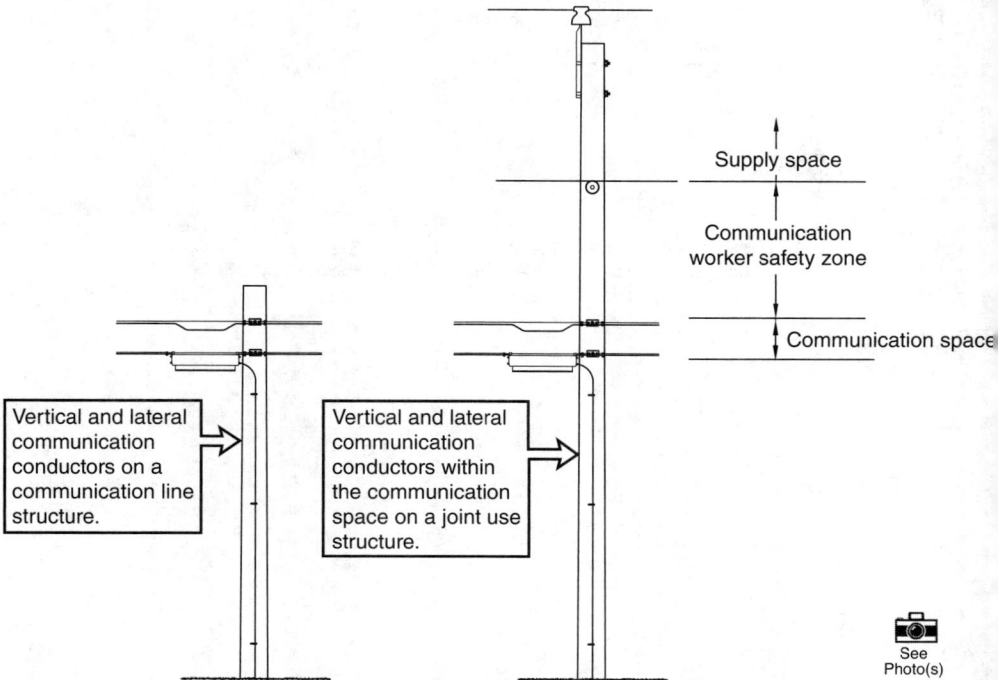

Fig. 239-5. Location of vertical and lateral communication conductors on communication line structures or within the communication space on jointly used structures (Rule 239F).

Supply space

Communication worker safety zone

Communication space

Vertical and lateral communication conductors on a communication line structure.

Vertical and lateral communication conductors within the communication space on a joint use structure.

See Photo(s)

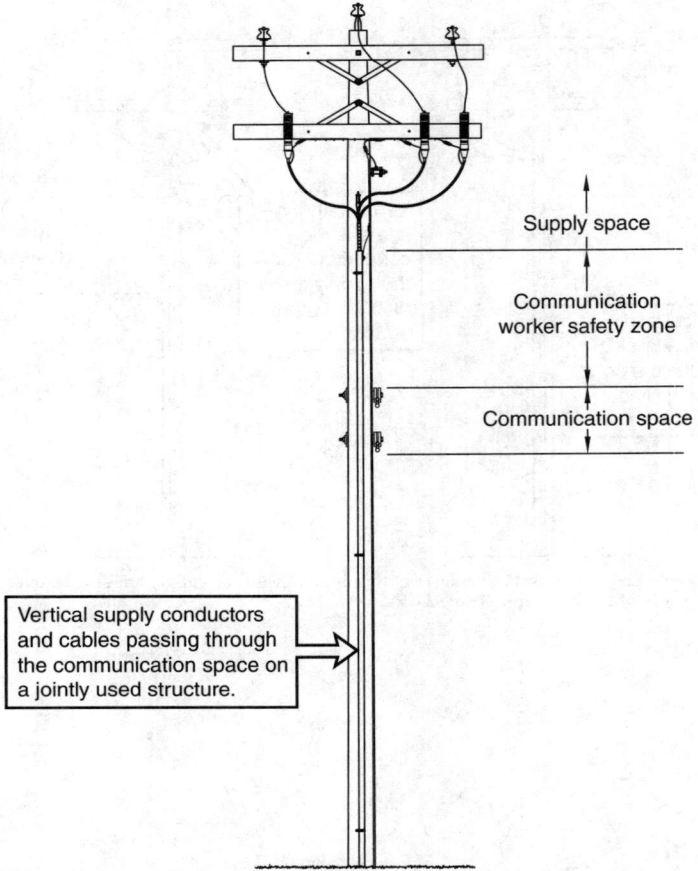

Fig. 239-6. Location of vertical supply conductors and cables passing through the communication space on jointly used line structures (Rule 239G).

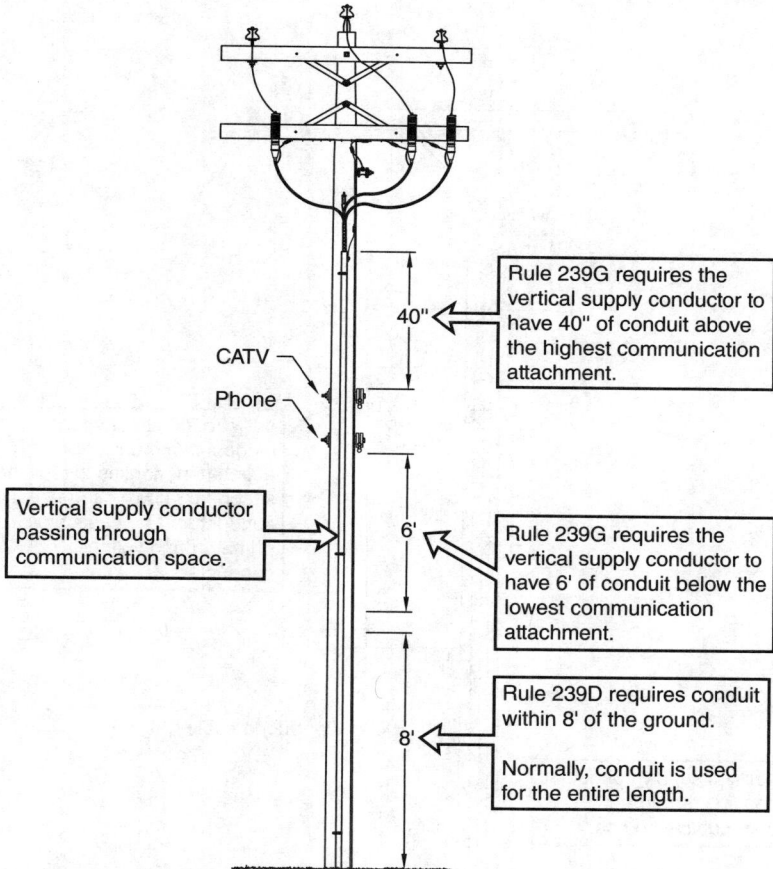

Rule 239G requires the vertical supply conductor to have 40" of conduit above the highest communication attachment.

CATV

Phone

40"

Vertical supply conductor passing through communication space.

6'

Rule 239G requires the vertical supply conductor to have 6' of conduit below the lowest communication attachment.

8'

Rule 239D requires conduit within 8' of the ground.

Normally, conduit is used for the entire length.

See Photo(s)

Fig. 239-7. Example of vertical supply conductors on a joint-use (supply and communication) pole (Rule 239G1).

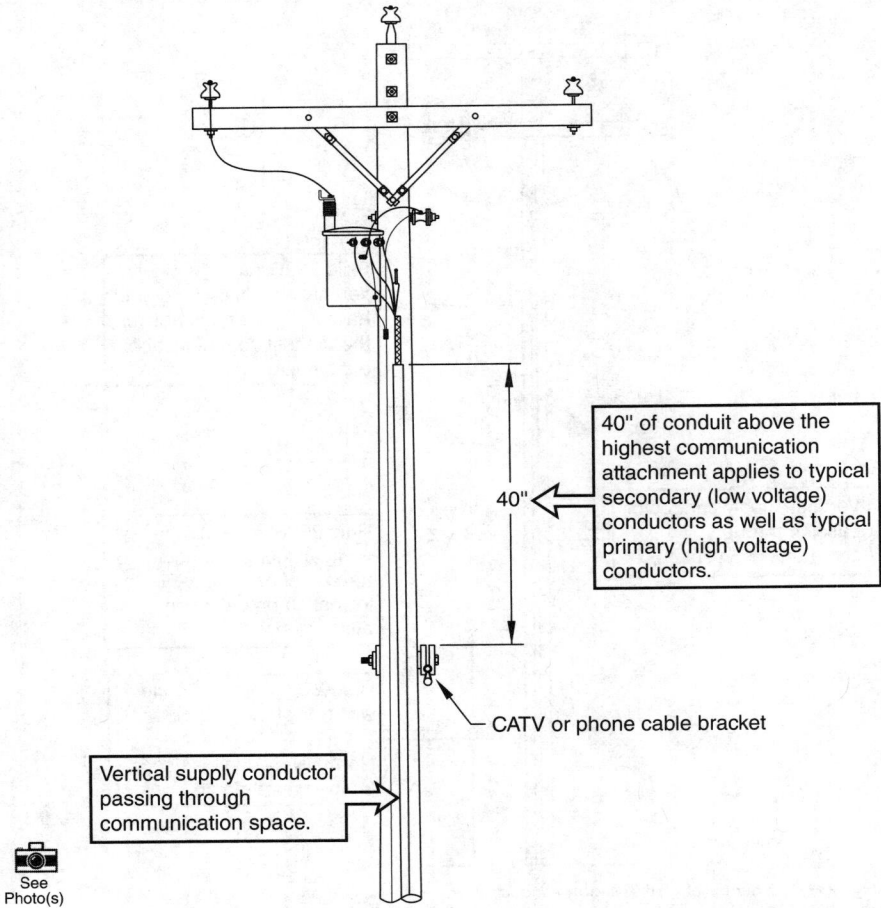

40" of conduit above the highest communication attachment applies to typical secondary (low voltage) conductors as well as typical primary (high voltage) conductors.

40"

CATV or phone cable bracket

Vertical supply conductor passing through communication space.

See Photo(s)

Fig. 239-8. Example of vertical supply conductors on a joint-use (supply and communication) pole (Rule 239G1).

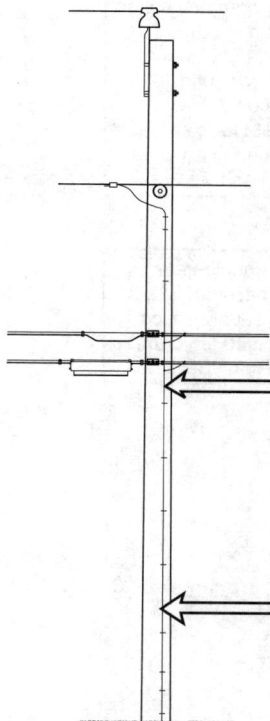

Exception 2 to Rule 239G1 permits omitting a
conduit or covering on the pole ground passing
through the communication space if:

• There are no trolley or ungrounded traffic signal
attachments, or ungrounded street light fixtures
below the communication attachment.

• The grounding conductor (pole ground) is
connected to a conductor that is part of an
effectively grounded system (e.g., an effectively
grounded neutral).

• The grounding conductor (pole ground) does not
have any connections to equipment between the
grounding electrode (ground rod) and the
effectively grounded conductor (neutral) unless
the equipment has additional connections to the
effectively grounded conductor (neutral).

• The grounding conductor (pole ground) is
bonded to the grounded communication
facilities (messenger) at that structure.
(See Rule 097G.)

Rules 093D and 239D allow guarding to be
omitted from the pole ground on multigrounded
circuits.

Fig. 239-9. Exception for a vertical supply grounding conductor on a joint-use (supply and communication) pole (Rule 239G1).

Vertical communication
conductors passing through
the supply space typically
connect to a communication
antenna or an overhead
ground wire (static) with an
embedded fiber-optic
communication cable.
See Rule 235I for clearance
to communication antennas
in the supply space.

Vertical communication
conductors passing
through the supply space
on a jointly used structure
must meet Rule 239H.

See
Photo(s)

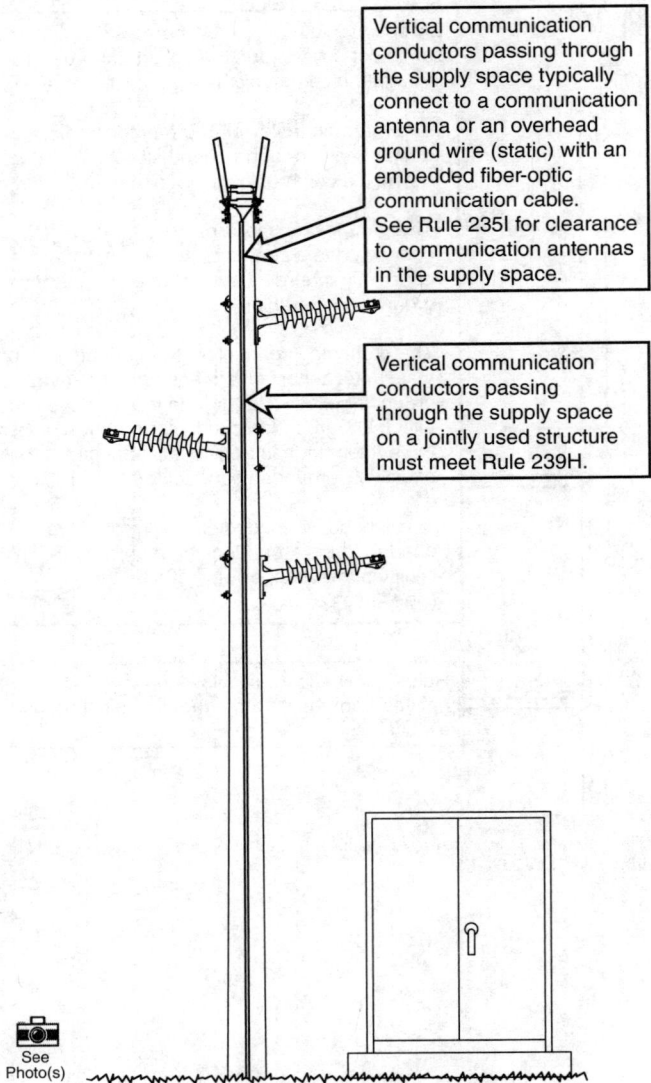

Fig. 239-10. Location of vertical communication conductors
passing through the supply space on jointly used structures
(Rule 239H).

Section 24

Grades of Construction

240. GENERAL

Sections 24, 25, and 26 are all interrelated. Section 24, "Grades of Construction," defines the required strength of overhead line construction for safety purposes. Section 25, "Loadings for Grades B and C," defines the physical loads (e.g., ice, wind, and temperature conditions) that overhead line construction must be able to withstand and the load factors that must be applied to the physical loads. Section 26, "Strength Requirements," defines the required strength of materials used in constructing overhead lines and the strength factors that must be applied to the materials. Line insulators are not addressed in Secs. 24, 25, or 26. Insulators are covered in their own section, Sec. 27.

The details of applying load factors and strength factors are covered in Secs. 25 and 26. The purpose of Sec. 24 is to define the grade of construction that is required for different situations.

This Handbook addresses the **Code** requirements in Secs. 24, 25, 26, and 27. Line design calculations are not presented in this Handbook. Line design calculations can be found in transmission and distribution line design manuals. Large utilities normally develop their own line design manuals. Small rural utilities typically use the transmission and distribution line design manuals published by the Rural Utilities Service (RUS), which in the past was referred to as the Rural Electrification Administration (REA). The RUS distribution line design manuals consist of RUS Bulletins 1724E-150 through 1724E-154. The RUS design manual for high voltage transmission lines is RUS Bulletin 1724E-200. The RUS also publishes design manuals for communication lines. This

Handbook will aid those individuals writing and using line design manuals and provide a deeper understanding of how the **Code** applies to line design calculations.

Typical strength and loading line design calculations include, but are not limited to, the following:

- Maximum wind span based on wind with ice on conductors
- Maximum weight span based on weight of conductors and ice
- Moment due to wind on pole
- Allowable resisting moment of pole
- Transverse, vertical, and total components of conductor loading
- Total ground line moment on pole
- Ruling span
- Diameter of pole at any point
- Dead-end guying strength
- Bisector guying strength
- Anchor strength
- Weak-link of guy attachment, guy wire, and anchor assembly
- Crossarm strength (vertical)
- Crossarm strength (longitudinal)
- Pole buckling
- Maximum line angle based on insulator strength
- Material deflection
- Equipment loading

Single pole structures are typically easier to analyze than multiple pole and lattice structures. Line design calculations also involve clearance calculations in addition to strength and loading calculations (e.g., clearance above ground, clearance to buildings and other structures, horizontal clearance between conductors, vertical clearance between conductors, etc.). The clearance calculations will affect the line design and therefore the selection of pole heights and pole classes. The sample sag and tension chart shown at the beginning of Sec. 23 provides sag for clearance calculations and conductor tension for strength and loading calculations. The "design tension" used for strength and loading calculations is the largest of the conductor tensions calculated after applying Rule 250B (heavy, medium, light, or warm islands loads), Rule 250C (extreme wind loads) if applicable, and Rule 250D (extreme ice with concurrent wind loads) if applicable. The 250B, 250C, and 250D loads are applied to conductors in accordance with Rule 251. The 250B, 250C, and 250D loads are applied to line supports (e.g., structures) in accordance with Rule 252. Rule 250A specifies that where all three rules apply (250B, 250C, and 250D), the required loading shall be the one that has the greatest effect. See Rules 250, 251, and 252 for additional information.

If more than one condition applies to the construction of an overhead line, the condition requiring the higher grade of construction is used. The order of grades of construction is discussed in Rule 241B. A note in Rule 240A references Rule 014 for grades of construction and strength for emergency and temporary installations. This issue is also addressed in **NESC** Table 261-1, Footnote 4.

Section 24 applies to both alternating current (AC) and direct current (DC) circuits. A DC circuit is considered equivalent to the root mean square (rms) values of an

AC circuit when applying Sec. 24. This is different from Sec. 23, which requires DC circuits to be equivalent to the AC crest value. See Rule 230G for a discussion of AC rms and crest values.

241. APPLICATION OF GRADES OF CONSTRUCTION TO DIFFERENT SITUATIONS

241A. Supply Cables. For the purposes of determining the grade of construction of a supply cable, supply (not communication) cables are broken into two types. Cable (Type 1) supply cables are cables meeting Rules 230C1, 230C2, and 230C3. See Rule 230C for a discussion of these types of supply cables. The 230C1, 230C2, and 230C3 cables are typically supported on messengers. The strength requirements for supply cable messengers are covered in Rule 261I.

Open (Type 2) supply cables are all other supply cables that do not meet the requirements for 230C1, 230C2, and 230C3 cables. Type 2 supply cables include open wire (e.g., bare) conductors and covered (e.g., tree wire) conductors.

241B. Order of Grades. Rule 241B outlines the order of grades. A general description of each grade of construction is outlined below:

- *Grade B.* The highest grade. The line will be "extra stout." Load and strength factors will be applied to make the strength of the line higher than any other grade. The required pole class will be larger than C and N grades (e.g., a Class 3 pole instead of a Class 5 pole) for the same span length or the required span length for Grade B will be shorter than grades C and N for the same pole class. Crossarms will be stronger than grades C and N, etc.
- *Grade C.* The middle grade. The line will be more "stout" than Grade N but not as "stout" as Grade B. Load and strength factors will be applied but they will not increase the strength of the line as much as Grade B. Grade C at crossings will be more "stout" than Grade C not at crossings.
- *Grade N.* The lowest grade. The line must be designed to support the anticipated loads but no load or strength factors are specified.

Figure 241-1 provides a graphical representation of the grades of construction.

241C. At Crossings. One overhead power line is considered to be at crossings when it crosses another line, a railroad track, a limited access highway, or a navigable waterway. An exception exists for joint use or collinear construction which in itself is not considered at crossings. See Figs. 241-2 through 241-4.

When two lines cross, the grade of construction for the conductors and supporting structures of the lower line can be determined without considering the upper line. In other words, the grade of construction of the lower line is based on the land use. This is because the lower line cannot fall into the upper line. The upper line, however, can fall into the lower line and cause problems. The conductors and supporting structures of the upper line at the crossing require careful application of Rule 241C3 (Multiple Crossings), Rule 242 (Grades of Construction for Conductors, **NESC** Table 242-1), and Rule 243 (Grades of Construction for Line Supports).

An example of supply lines at multiple crossings is shown in Fig. 241-5.

An example of communication lines at multiple crossings involving communications, supply, and railroad tracks is shown in Fig. 241-6.

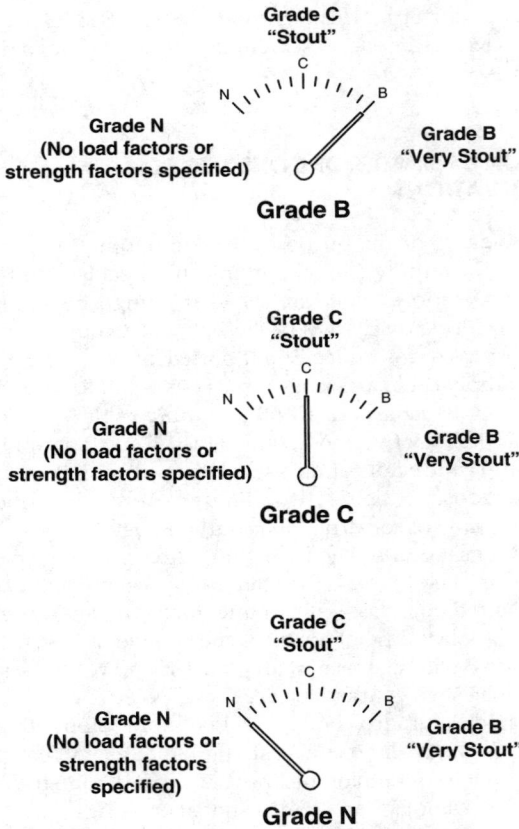

Grade C
"Stout"

Grade N
(No load factors or
strength factors specified)

Grade B
"Very Stout"

Grade B

Grade C
"Stout"

Grade N
(No load factors or
strength factors specified)

Grade B
"Very Stout"

Grade C

Grade C
"Stout"

Grade N
(No load factors or
strength factors
specified)

Grade B
"Very Stout"

Grade N

Fig. 241-1. Order of grades of construction (Rule 241B).

241D. Structure Conflicts.   Structure conflict exists when one line can fall down and conflict with another line. See Rules 220C and 221 for additional information and figures related to structure conflict. Rule 241D also references structure conflict in Rule 243A4 and in Sec. 02, Definitions. For the purposes of determining grades of construction, conflicting lines are treated the same as line crossings since one line can fall on the other.

242. GRADES OF CONSTRUCTION FOR CONDUCTORS

Rule 242 references **NESC** Table 242-1 for determining grades of construction for conductors (and cables). It is important to stress the term conductors. Rule 242

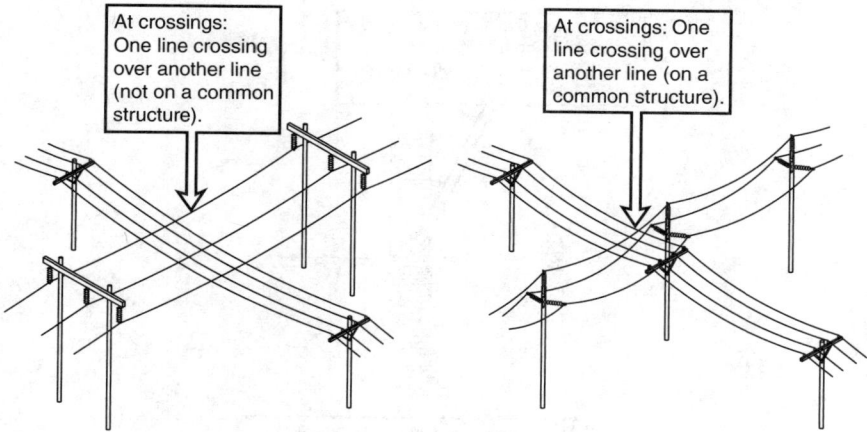

At crossings:
One line crossing
over another line
(not on a common
structure).

At crossings: One
line crossing over
another line (on a
common structure).

Fig. 241-2. Lines considered to be at crossings (Rule 241C1a).

and **NESC** Table 242-1 are used to determine grades of construction for conductors. Rule 243 is used to determine grades of construction for line supports. The grade of construction of a line support is based on the grade of construction of the conductors attached to the line support. Rules 242A through 242F provide specific information for special types of conductors.

There are several important items of discussion related to **NESC** Table 242-1.

- **NESC** Table 242-1 is for conductors and cables (supply and communication) alone, at crossing, or on the same structures with other conductors and cables.
- Supply or communication lines crossing over railroad tracks, limited access highways, or navigable waterways requiring waterway crossing permits must be built to Grade B construction.
- **NESC** Table 242-1 distinguishes between open and cable conductors. Cable conductors are 230C1, 230C2, and 230C3 cables. Open conductors are typically bare (noninsulated) conductors. See Rule 241A for a discussion.
- A 12.47/7.2-kV distribution line (open conductors) on an exclusive private right-of-way can be built to Grade N. This is not a common practice. Typically utilities build to Grade C construction for both common or public rights-of-way and exclusive private rights-of-way.
- An effectively grounded 115-kV transmission line (with ground fault relaying) on a public right-of-way may be built to Grade C. This is not a common practice. Typically utilities build transmission lines to Grade B construction.
- A 12.47/7.2-kV distribution line (open conductors) crossing another 12.47/7.2-kV distribution line (open conductors) must be built to Grade C construction. **NESC** Table 253-1 provides separate load factors for Grade C at crossings and elsewhere.

Per **NESC** Table 242-1, a 12.47/7.2-kV distribution line (open conductors) built joint use with a communication cable must be built to Grade B unless Footnote 6

At crossings: One line crossing over
or overhanging a railroad track.
Grade B construction required per
NESC Table 242-1.

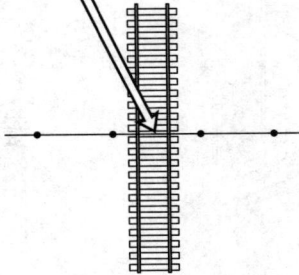

At crossings: One line crossing over or
overhanging the traveled way of a limited
access highway. Grade B construction
required per **NESC** Table 242-1.

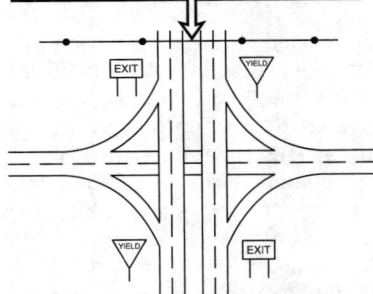

EXIT

YIELD

YIELD

EXIT

At crossings: One line crossing over or
overhanging a navigable waterway
requiring a waterway crossing permit.
Grade B construction required per **NESC**
Table 242-1.

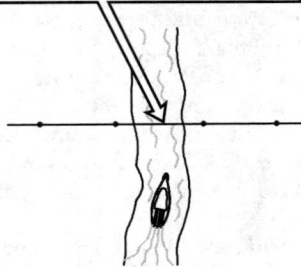

See
Photo(s)

Fig. 241-3. Lines considered to be at crossings
(Rule 241C1b).

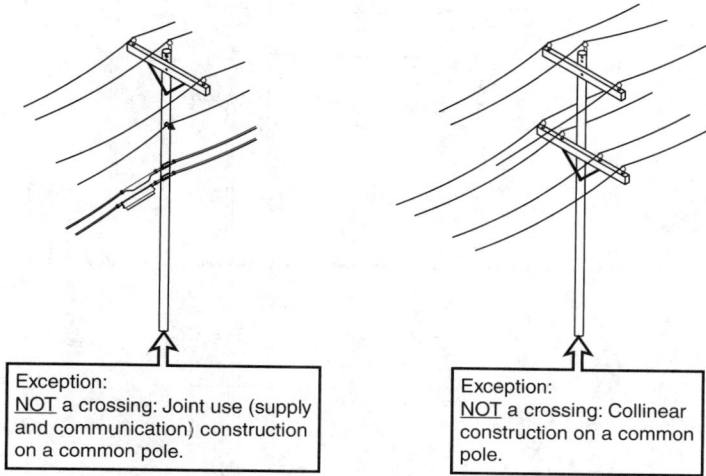

Exception:
NOT a crossing: Joint use (supply and communication) construction on a common pole.

Exception:
NOT a crossing: Collinear construction on a common pole.

Fig. 241-4. Exceptions to lines considered to be at crossings (Rule 241C1).

or 7 can be met. For a 12.47/7.2-kV line, Footnote 6 cannot be applied. Footnote 7 (Parts a and b) must then be met to construct a Grade C joint use (power and communication) line. Typically utilities build to Grade C construction for this application. The protection of communication equipment required in Footnote 7 is also required in Rule 223. The most common method used for the means of protection required in Footnote 7 is insulating the communication conductors (in the form of a communication cable) and bonding the grounded communication messenger to the grounded supply neutral. See Rule 223 for additional information.

A 120/240-V, single-phase, three-wire triplex cable meeting Rule 230C3 falls under "cable" in **NESC** Table 242-1 per Rule 241A. A 120/240-V, single-phase, three-wire triplex cable built joint use with a communication cable can be built to Grade N per **NESC** Table 242-1.

243. GRADES OF CONSTRUCTION FOR LINE SUPPORTS

Rule 242, Grades of Construction for Conductors, applies to grades of construction for conductors and this rule, Rule 243, applies to grades of construction for line supports. Line supports in this rule include structures, crossarms, insulators, pins, etc., basically the components that support the conductors.

243A. Structures. The structure (e.g., pole) grade of construction must be the same as the highest grade of conductor supported. In other words, if a pole has Grade B transmission conductors, Grade C distribution conductors, and Grade N

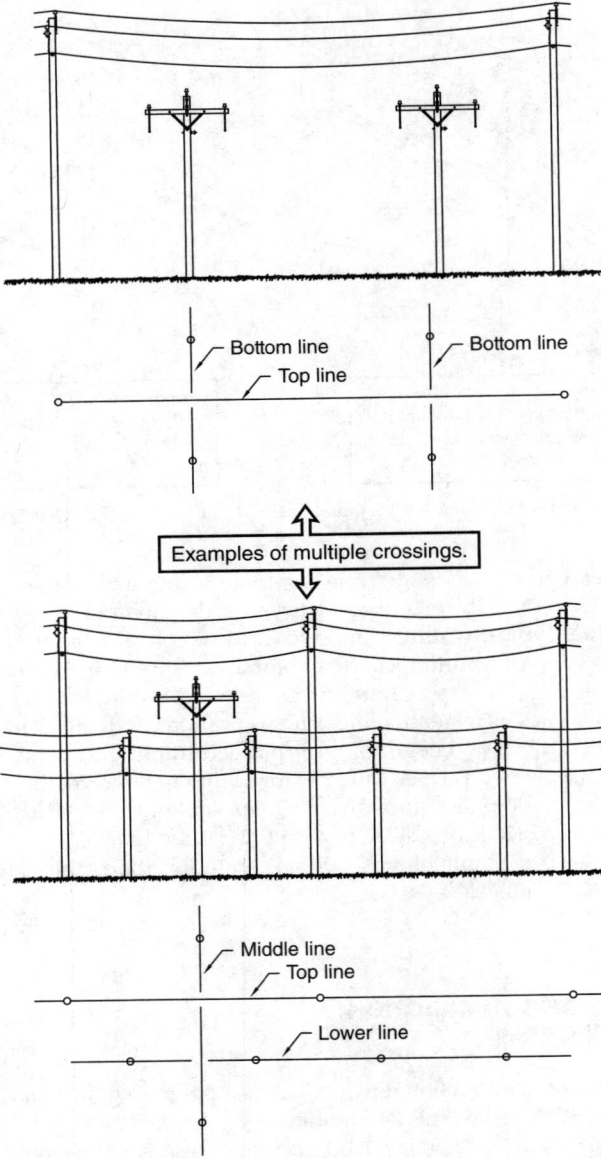

Fig. 241-5. Example of supply lines at multiple crossings (Rule 241C4a).

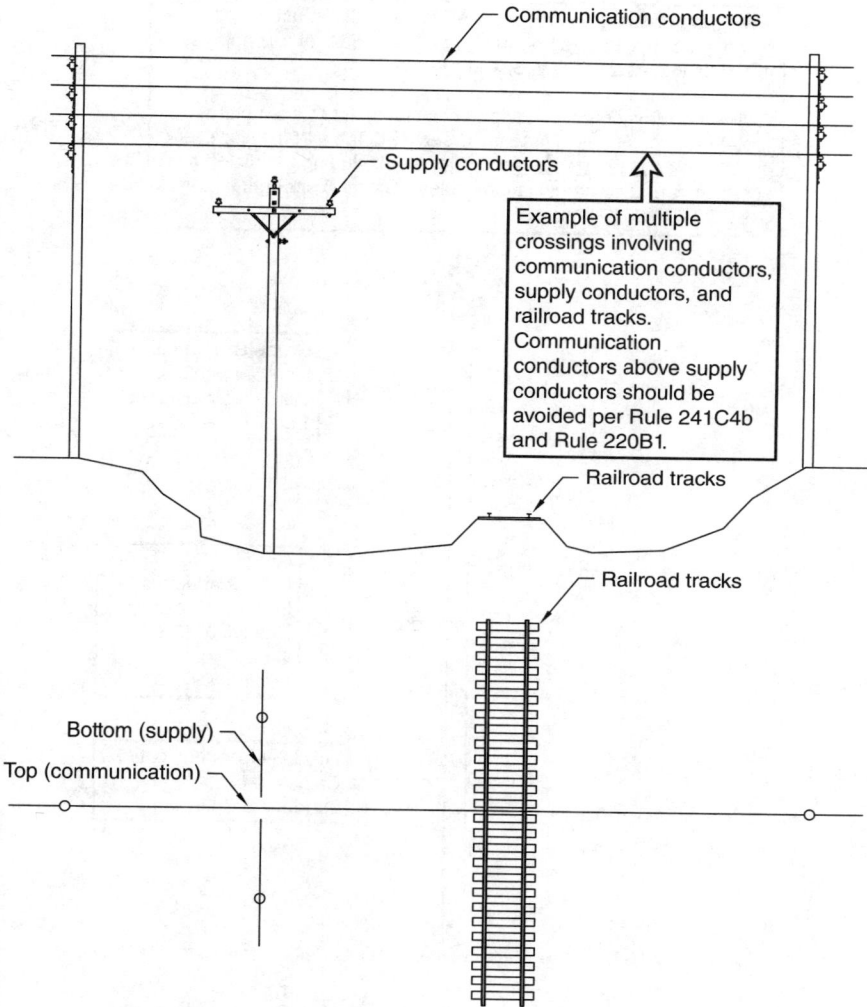

Fig. 241-6. Example of communication lines at multiple crossings (Rule 241C4b).

communication conductors, all attached to the same pole, then the pole itself must be designed to Grade B construction. There are four modifications to the general rule. See Fig. 243-1.

243B. Crossarms and Support Arms. The crossarm (or another type of support arm) grade of construction must be the same as the highest grade of conductor that the crossarm supports. If a crossarm carries Grade B conductors, the crossarm itself must be designed to Grade B. It is possible to have a Grade C crossarm on a Grade B structure (pole). For example, if a pole has Grade B transmission conductors

The grade of construction for the structure (pole) must be <u>the highest grade supported on the pole</u>. Four modifications apply to this Rule (Rules 243A1, 2, 3, and 4).

Example: A maximum wind span calculation for this structure (pole) requires Grade B load factors (**NESC** Table 253-1) <u>for every conductor and cable attached to the pole</u> (including the distribution conductors and communication cables). Grade B strength factors (**NESC** Table 261-1) must also be used.

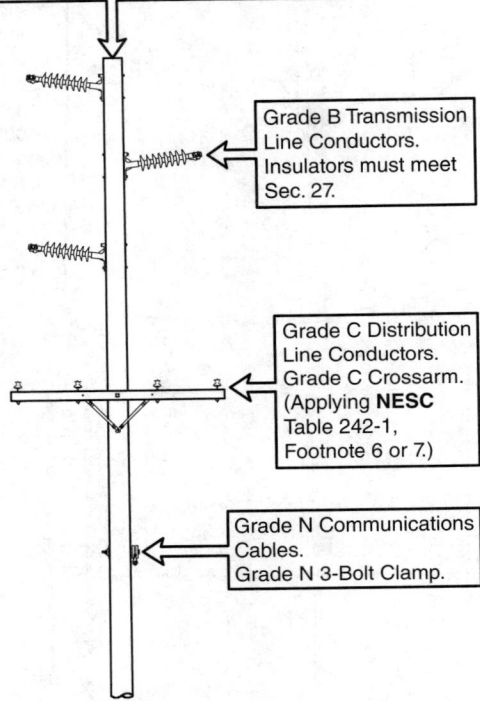

Grade B Transmission Line Conductors. Insulators must meet Sec. 27.

Grade C Distribution Line Conductors. Grade C Crossarm. (Applying **NESC** Table 242-1, Footnote 6 or 7.)

Grade N Communications Cables. Grade N 3-Bolt Clamp.

Fig. 243-1. Grades of construction for line supports (Rule 243A).

on a Grade B crossarm and Grade C distribution conductors on a Grade C crossarm, the structure (pole) must be Grade B per Rule 243A but the distribution crossarm can remain Grade C per Rule 243B. There are three modifications to the general rule.

243C. Pins, Armless Construction Brackets, Insulators, and Conductor Fastenings. The pins, armless construction brackets, insulators, and conductor fastenings grade of construction must be the same as the grade of their associated conductor. There are four modifications to the general rule. The most notable is Modification 4 (Rule 243C4). This modification states that insulators used on open conductor supply lines are covered in their own section, Sec. 27.

Section 25

Loadings for Grades B and C

250. GENERAL LOADING REQUIREMENTS AND MAPS

Sections 24, 25, and 26 are all interrelated. Section 24, "Grades of Construction," defines the required strength of overhead line construction for safety purposes. Section 25, "Loadings for Grades B and C," defines the physical loads (e.g., ice, wind, and temperature conditions) that overhead line construction must be able to withstand and the load factors that must be applied to the physical loads. Section 26, "Strength Requirements," defines the required strength of materials used in constructing overhead lines and the strength factors that must be applied to the materials. Line insulators are not addressed in Secs. 24, 25, or 26. Insulators are covered in their own section, Sec. 27.

The purpose of Sec. 24 is to define the grade of construction that is required for different situations. The details of applying load factors and strength factors are covered in Secs. 25 and 26.

This Handbook addresses the Code requirements in Secs. 24, 25, 26, and 27. Line design calculations are not presented in this Handbook. Line design calculations can be found in transmission and distribution line design manuals. Large utilities commonly develop their own line design manuals. Some utilities use the transmission and distribution line design manuals published by the Rural Utilities Service (RUS), which in the past was referred to as the Rural Electrification Administration (REA). The RUS distribution line design manuals consist of RUS Bulletins 1724E-150 through 1724E-154. The RUS design manual for high-voltage transmission lines is RUS Bulletin 1724E-200. The RUS also publishes design manuals for communication lines. This Handbook will aid those individuals writing and using line design manuals and provide a deeper understanding of how the **Code** applies to line design calculations.

Typical strength and loading line design calculations include, but are not limited to, the following:

- Maximum wind span based on wind with ice on conductors
- Maximum weight span based on weight of conductors and ice
- Moment due to wind on pole
- Allowable resisting moment of pole
- Transverse, vertical, and total components of conductor loading
- Total ground line moment on pole
- Ruling span
- Diameter of pole at any point
- Dead-end guying strength
- Bisector guying strength
- Anchor strength
- Weak-link of guy attachment, guy wire, and anchor assembly
- Crossarm strength (vertical)
- Crossarm strength (longitudinal)
- Pole buckling
- Maximum line angle based on insulator strength
- Material deflection
- Equipment loading

Single pole structures are typically easier to analyze than multiple pole and lattice structures. Line design calculations also involve clearance calculations in addition to strength and loading calculations (e.g., clearance above ground, clearance to buildings and other structures, horizontal clearance between conductors, vertical clearance between conductors, etc.). The clearance calculations will affect the line design and therefore the selection of pole heights and pole classes. The sample sag and tension chart shown at the beginning of Sec. 23 provides sag for clearance calculations and conductor tension for strength and loading calculations. The "design tension" used for strength and loading calculations is the largest of the conductor tensions calculated after applying Rule 250B (heavy, medium, light, or warm islands loads), Rule 250C (extreme wind loads) if applicable, and Rule 250D (extreme ice with concurrent wind loads) if applicable. The 250B, 250C, and 250D loads are applied to conductors in accordance with Rule 251. The 250B, 250C, and 250D loads are applied to line supports (e.g., structures) in accordance with Rule 252. Rule 250A specifies that where all three rules apply (250B, 250C, and 250D), the required loading shall be the one that has the greatest effect. See Rules 250, 251, and 252 for additional information.

250A. General.   If the grade of construction required in Sec. 24, "Grades of Construction," is B or C, Sec. 25 will define the physical loadings in the form of ice, wind, and load factors that increase the physical loads associated with overhead line construction. The terms "load specified in Rule 250," "Rule 250B loads," "Rule 250C loads," and "Rule 250D loads" will be used throughout Secs. 25 and 26.

When the loads in Rule 250B (heavy, medium, light, or warm islands loading), Rule 250C (extreme wind loading), and Rule 250D (extreme ice with concurrent wind) must all be considered, the required loading must be the one that has the greatest effect. Rule 250 also specifies that the intent is to apply wind in an essentially horizontal plane. In other words, the intent is not to apply upwinds or updrafts or downwinds or downdrafts. If some local terrain condition did warrant applying upwinds or downwinds, this practice is certainly not prohibited. A loading map for overhead lines in the United States is provided in **NESC** Fig. 250-1. The United States is divided into heavy, medium, light, and warm islands loading areas. Since only four loading areas are defined, the **NESC** recognizes that heavier loads than specified in the map may exist for a specific region. Using a higher loading than the map shows is certainly acceptable. Using a lower loading requires approval of an administrative authority (e.g., the State Public Service Committee).

The weather loadings specified in Rules 250B, 250C, and 250D may not be sufficient for the forces imposed during construction and maintenance. Additional loads may need to be considered to adjust for this condition.

Rule 250A4 states that if lines are designed to meet Sec. 25 (loadings for Grades B and C) and Sec. 26 (strength requirements), the lines will have sufficient capability to resist earthquake ground motions.

250B. Combined Ice and Wind District Loading. NESC Fig. 250-1 (map of the USA) and **NESC** Table 250-1 are the two basic references to determine ice and wind loading for heavy, medium, light, and warm islands loading. **NESC** Fig. 250-1 and **NESC** Table 250-1 provide a listing of the most predominate warm islands (Hawaii, Puerto Rico, Guam, Virgin Islands, and American Samoa). The statement "the loads of Rule 250B" is used throughout Secs. 25 and 26. When ice is specified, it is specified as a radial value. For example, a conductor with 1/4 in of radial ice (medium loading) adds 1/2 in to the diameter of the conductor (1/4 in of ice plus conductor diameter plus 1/4 in of ice).

The requirements for heavy, medium, light, and warm islands loading are outlined in Fig. 250-1.

250C. Extreme Wind Loading. NESC Fig. 250-2 (a through e) and **NESC** Tables 250-1, 250-2, and 250-3 are the basic references for determining extreme wind loading. **NESC** Fig. 250-2(b) provides a listing of the most predominate warm islands (Hawaii, Puerto Rico, Guam, Virgin Islands, and American Samoa) and their corresponding extreme wind loading values. The statement "the loads of Rule 250C" is used throughout Secs. 25 and 26.

Rule 250C only applies to structures (e.g., poles) and supported facilities (e.g., conductors, static wires, messengers, cables, etc.) more than 60 ft above ground or water level. See Fig. 250-2.

Rules 261A1c (metal, prestressed, and reinforced concrete poles), 261A2e (wood poles), and 261A3d (fiber-reinforced polymer poles) require the application of extreme wind to the pole, without conductors, for any height pole. Typically, applying the heavy, medium, light, or warm islands loading conditions to a pole and its conductors less than 60 ft above ground will result in a worse case design condition than a pole less than 60 ft above ground with extreme wind applied to the pole without conductors. The less than 60 ft requirement exists for special cases. If a pole is pre-engineered (e.g., a steel pole or a laminated wood pole), the design strength at the ground line could have a higher resisting moment in the transverse direction than in the longitudinal direction. For a line less than 60 ft above ground, the transverse loading requires application of Rule 250B (heavy, medium, light, or warm islands) loads. The transverse and longitudinal directions require application of Rule 261A1c, 261A2e, or 261A3d for extreme wind on the pole, without conductors.

Loading	Ice and Wind Condition at a Specified Temperature	Location
HEAVY	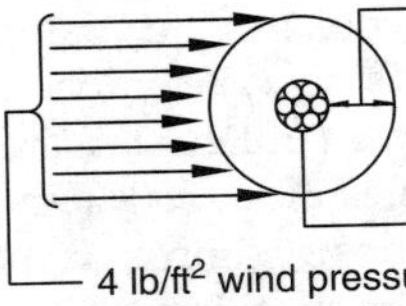 1/2" of radial ice Conductor at 0°F 4 lb/ft² wind pressure on conductor with ice Ice weight: 57 lb/ft³ Additive constant: 0.30 lb/ft	See a map of the USA in **NESC** Fig. 250-1.
MEDIUM	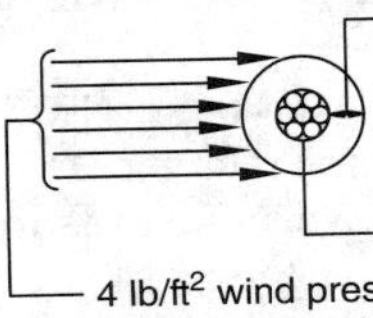 1/4" of radial ice Conductor at 15°F 4 lb/ft² wind pressure on conductor with ice Ice weight: 57 lb/ft³ Additive constant: 0.20 lb/ft	See a map of the USA in **NESC** Fig. 250-1.
LIGHT	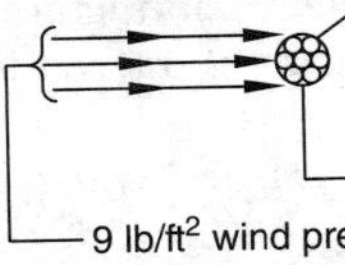 No ice Conductor at 30°F 9 lb/ft² wind pressure on conductor without ice No ice Additive constant: 0.05 lb/ft	See a map of the USA in **NESC** Fig. 250-1.
Warm Islands (for altitudes of sea level to 9000 ft)	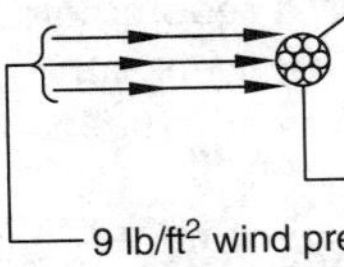 No ice Conductor at 50°F 9 lb/ft² wind pressure on conductor without ice No ice Additive constant: 0.05 lb/ft	See a map of the USA in **NESC** Fig. 250-1 (American Samoa, Guam, Hawaii, Puerto Rico, Virgin Islands, and other islands located from 0 to 25° latitude, north or south).
Warm Islands (for altitudes above 9000 ft)	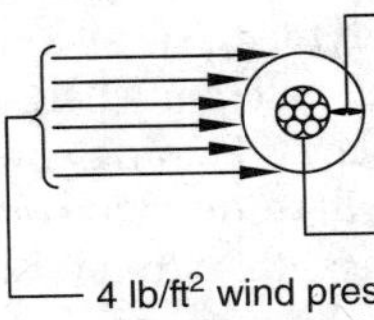 1/4" of radial ice Conductor at 15°F 4 lb/ft² wind pressure on conductor with ice Ice weight: 57 lb/ft³ Additive constant: 0.20 lb/ft	See a map of the USA in **NESC** Fig. 250-1 (American Samoa, Guam, Hawaii, Puerto Rico, Virgin Islands, and other islands located from 0 to 25° latitude, north or south).

Fig. 250-1. Heavy, medium, light, and warm islands ice and wind loading (Rule 250B).

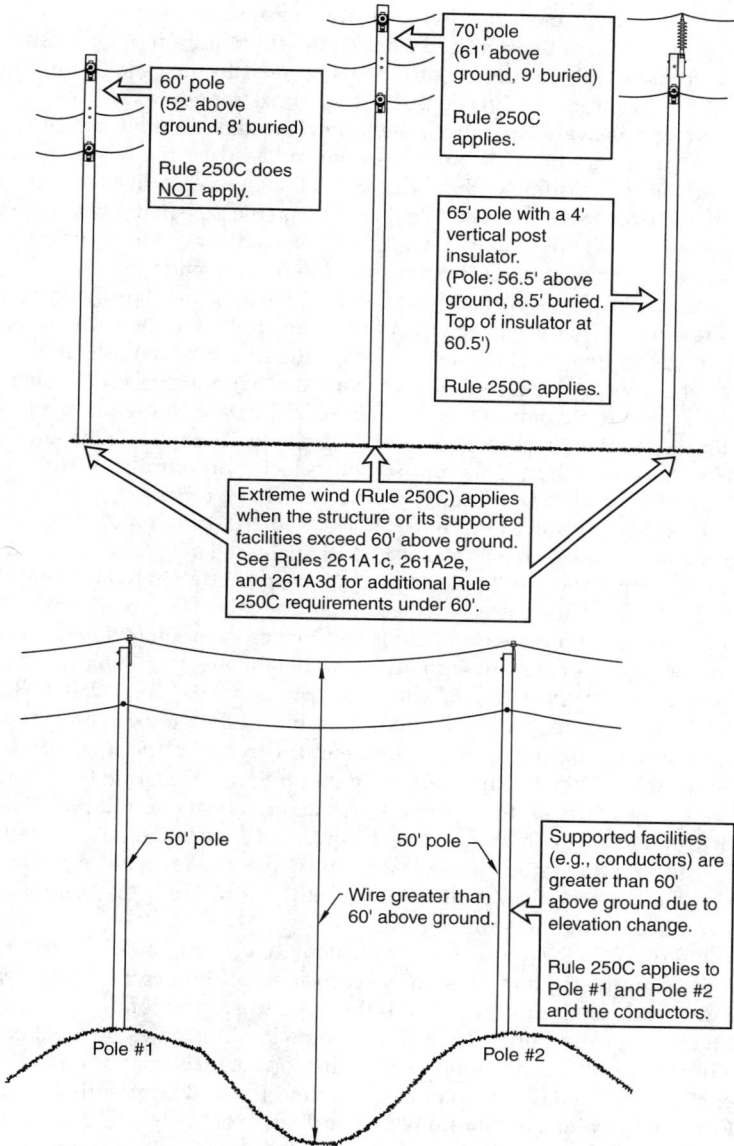

Fig. 250-2. Examples of facilities requiring extreme wind loading (Rule 250C).

Typically the heavy, medium, light, or warm islands loading requirements applied to conductors in the transverse direction will require more pole strength than the extreme wind loading requirements applied to the pole (without conductors) in the longitudinal direction. If a traditional round wood pole is used, the same strength is available in both the transverse and longitudinal directions. Assuming a traditional round wood pole is sized based on the heavy, medium, light, or warm islands loading with conductors in the transverse direction, adequate strength should exist for the extreme wind loading on the pole (without conductors) in any direction. Equal pole strength in all directions may not be accurate for a pre-engineered pole utilizing different strengths in different directions.

If it is determined that Rule 250C applies (i.e., the structures or the supported facilities are over 60 ft), the extreme wind must be applied to the entire structure and supported facilities, not just the portions of the structure above 60 ft. Extreme wind is applied without ice on the conductors and without ice on the structure at a 60°F temperature condition. **NESC** Table 250-1 specifies the temperature at which extreme wind is checked. A formula is provided to calculate the wind load on an object based on the basic wind speed, velocity pressure exposure coefficient, gust response factor, and force coefficient (shape factor).

The basic wind speed is found in **NESC** Fig. 250-2 (a through e). The velocity pressure exposure coefficient (k_z) increases with increased height due to wind friction near the earth's surface. Typical k_z values are provided in **NESC** Table 250-2. The formulas to calculate k_z are provided under the table. The gust response factor (G_{RF}) decreases with increased height and longer span lengths because wind velocity becomes more constant with increased height and gusts average out over longer span lengths. Typical G_{RF} values are provided in **NESC** Table 250-3. The formulas to calculate G_{RF} are provided under the table. Both k_z and G_{RF} have separate values for the structure height and wire height. Force coefficient (shape factor) is defined in Rule 252B; see Rule 252B for a discussion. A separate factor may be needed for the structure and the wire if the structure is not a round pole. The formulas for k_z and G_{RF} are complex and the heights and span lengths associated with k_z and G_{RF} can vary along the line. Rule 250C provides a note at the end of the rule to simplify the mathematical product of k_z and G_{RF} if desired. An example of an extreme wind calculation is shown in Fig. 250-3.

The value of 17.76 lb/ft^2 for the wire (calculated in Fig. 250-3) is entered at 60°F on the sample sag and tension chart located at the beginning of Sec. 23. This design condition is only required if the structure or the conductor is greater than 60 ft above ground or water level. A line with 45-ft poles on level ground would not require this condition to be on the conductor sag and tension chart. Additional extreme wind examples can be found in **NESC** Appendix C.

250D. Extreme Ice with Concurrent Wind Loading.   NESC Fig. 250-3 (a through f) and **NESC** Tables 250-1 and 250-4 are the basic references for determining extreme ice with concurrent wind loading. No extreme ice with concurrent wind loading is specified for warm islands. The statement "the loads of 250D" is used throughout Secs. 25 and 26. Rule 250D only applies to structures (e.g., poles) and supported facilities (e.g., conductors, static wires, messengers, cables, etc.) more than 60 ft above ground or water level. This same requirement can be found in Rule 250C. See the discussion and figure in Rule 250C for examples of facilities greater than 60' above ground.

Extreme wind

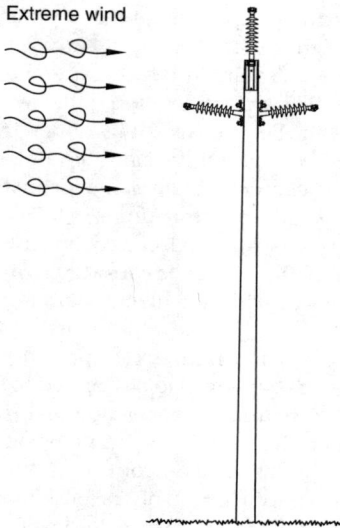

Example:
• Extreme wind in central Washington State.
  V = 85 miles/hour per **NESC** Figure 250-2(a).
• Pole height above ground line is 70'.
• Wire heights above ground line are 73' and 67'.
• Wind span length is 300'.
• The conductor and the pole have round shapes.

$$\text{Wind pressure} = 0.00256 \cdot (V)^2 \cdot k_z \cdot G_{RF} \cdot I \cdot C_f$$

Extreme wind pressure on the pole:	Extreme wind pressure on the wires:
• V = 85 miles/hour per **NESC** Figure 250-2a	• V = 85 miles/hour per **NESC** Figure 250-2a
• k_z structure = 1.10 per **NESC** Table 250-2	• k_z wire = 1.20 per **NESC** Table 250-2
• G_{RF} structure = 0.93 per **NESC** Table 250-3	• G_{RF} wire = 0.80 per **NESC** Table 250-3
• I = 1.0 for utility structures per **NESC** Rule 250C	• I = 1.0 for utility structures per **NESC** Rule 250C
• C_f = 1.0 for cylindrical structures per **NESC** Rule 252B2	• C_f = 1.0 for cylindrical shapes per **NESC** Rule 251A2
Wind pressure = 18.92 lb/ft²	Wind pressure = 17.76 lb/ft²

Fig. 250-3. Example of an extreme wind calculation (Rule 250C).

If it is determined that Rule 250D applies (i.e., the structure or the supported facilities are over 60 ft) the extreme ice with concurrent wind must be applied to the entire structure and supported facilities, not just the portions of the structure above 60 ft. Extreme ice with concurrent wind is applied to the conductor at a 15°F temperature condition. Only the wind portion of the extreme ice with concurrent wind needs to be applied to the structure (i.e., pole) as only the conductors are required to be covered with ice, not the pole.

   NESC Table 250-1 specifies the temperature at which extreme ice with concurrent wind is checked. **NESC** Table 250-4 provides horizontal wind pressures in lb/ft² based on wind speed in mph. Per the wording under the title of **NESC** Table 250-4, **NESC** Table 250-4 is only for use with Rule 250D, not Rule 250C. **NESC** Table 250-4 does not specify a force coefficient (shape factor). Although it is not stated in the **Code**, the horizontal wind pressure values in **NESC** Table 250-4 correspond to cylindrical surfaces. The appropriate force coefficients (shape factors) for flat or latticed structures in Rule 252B2 should be considered for non-cylindrical surfaces. The ice portion for extreme ice with concurrent wind is determined from **NESC** Fig. 250-3 (a through f) and applied to the conductors in accordance with Rule 251. See Rule 251 for a discussion and figures for applying ice and wind to conductors. Rule 250D2 allows the ice (not wind) to be derated to 0.80 for Grade C construction. An extreme ice with concurrent wind value of 15°F, 0.25 in of ice, and 2.30 lb/ft² of wind is entered on the sample sag and tension chart located at the beginning of Sec. 23. This value was chosen from west-central Washington State. This design condition on the conductor is only required if the structure or the conductor is greater than 60 ft above ground or water level. A line with 45-ft poles on level ground would not require this condition to be on the conductor sag and tension chart.

251. CONDUCTOR LOADING

251A. General.   The ice and wind loads in Rule 250, specifically Rules 250B, 250C, and 250D, must be applied to conductors. Rule 251 specifies how to apply the ice and wind loads to conductors.

   If the conductor consists of a cable on a messenger, the ice and wind loads are applied to both the cable and the messenger. Rule 251A2 specifies a method for determining wind loads on conductors and cables without an ice covering. Examples of wind loads on a stranded conductor and on communication cables supported on a messenger are shown in Fig. 251-1.

   Rule 251A3 specifies a method for determining ice loads on various conductor and cable configurations. Rule 251A4 requires that testing or a qualified engineering study be performed if the ice loading method in Rule 251A3 is reduced. Examples of ice and wind loads on a stranded conductor and on communication cables supported on a messenger are shown in Fig. 251-2.

251B. Load Components.   Rule 251B specifies the individual loads to consider when ice and wind are applied to a conductor. The load components are broken into three parts: vertical load, horizontal load, and total load.

   The vertical load must consider the weight of the conductor plus conductor spacers or any equipment that the conductor supports (e.g., Federal Aviation Administration marker balls). These items must be covered with ice to calculate weight for the medium, heavy, and high elevation warm islands loading districts per Rule 250B or the extreme ice loading per Rule 250D (if applicable).

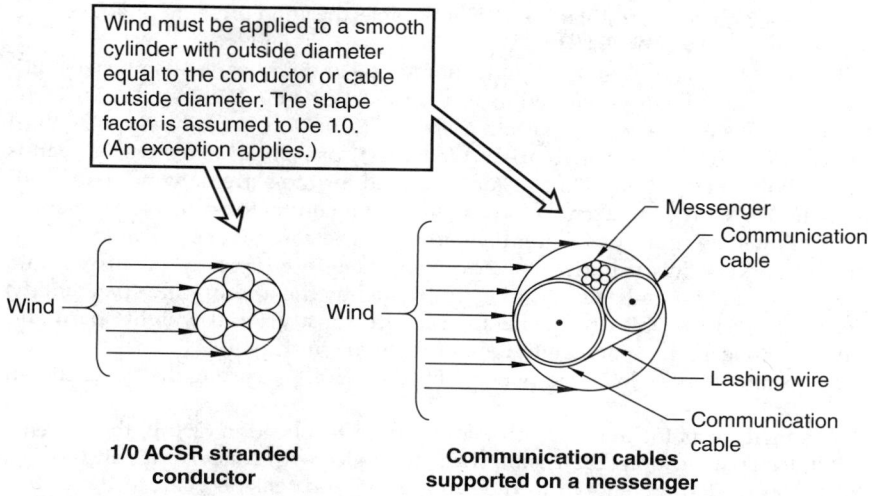

Fig. 251-1. Examples of wind loads on various conductor configurations (Rules 251A1 and 251A2).

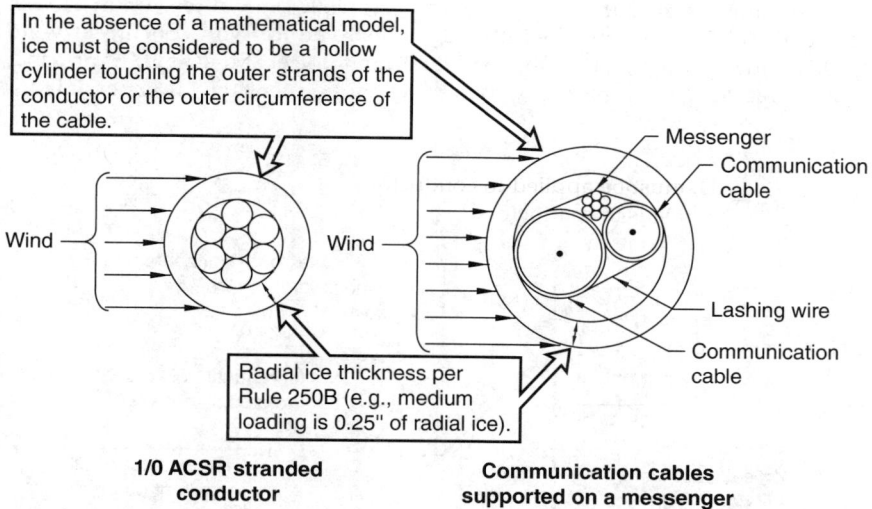

Fig. 251-2. Examples of ice and wind loads on various conductor configurations (Rule 251A3).

The horizontal load must consider wind on the conductors plus spacers and equipment that the conductor supports (e.g., Federal Aviation Administration marker balls). These items must be covered with ice to calculate the wind force for medium, heavy, and high elevation warm islands loading districts per Rule 250B or the extreme ice loading per Rule 250D (if applicable). The wind pressures in Rule 250B must be considered and, if applicable, the wind pressures in Rules 250C and 250D must be considered.

The total load is the resultant of the horizontal and vertical loads plus an additive constant (sometimes referred to as a K-factor) given in **NESC** Table 251-1. The additive constant varies by loading district (i.e., 0.30 lb/ft for heavy, 0.20 lb/ft for medium, 0.05 lb/ft for light, and 0.05 lb/ft or 0.20 lb/ft for warm islands depending on elevation). No constant is added for the extreme wind loading or extreme ice with concurrent wind loading. The conductor or cable messenger tension must be computed from the total load. The conductor vertical load component and conductor horizontal load component are typically calculated on a per foot of conductor basis and then multiplied by the appropriate span lengths described in Rule 252A (e.g., the span length for a vertical weight span) and Rule 252B4 (e.g., the span length for a horizontal wind span).

An example of the load components for the medium loading district is shown in Fig. 251-3.

The total loads for the medium loading design condition (250B), the extreme wind design condition (250C), and the extreme ice with concurrent wind design condition (250D) are shown in the sample sag and tension chart at the beginning of Sec. 23. The sample sag and tension chart indicates an initial tension of 1366 lb for 15°F, 0.25 in ice, 4 lb/ft^2 wind, and an additive constant of 0.20 lb/ft. This tension is higher than the extreme wind (250C) initial tension of 1154 lb for 60°F and 17.76 lb/ft^2 of wind and the extreme ice with concurrent wind (250D) initial tension of 1070 lb for 15°F, 0.25 in of ice, and 2.30 lb/ft^2 of wind. The 1366-lb tension is commonly called the "design tension." It is possible for

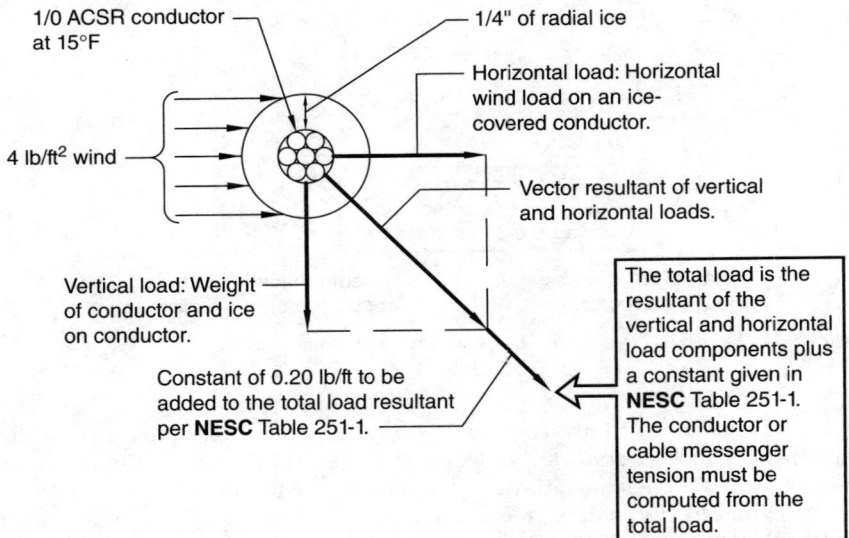

1/0 ACSR conductor at 15°F

1/4" of radial ice

Horizontal load: Horizontal wind load on an ice-covered conductor.

4 lb/ft^2 wind

Vector resultant of vertical and horizontal loads.

Vertical load: Weight of conductor and ice on conductor.

The total load is the resultant of the vertical and horizontal load components plus a constant given in **NESC** Table 251-1. The conductor or cable messenger tension must be computed from the total load.

Constant of 0.20 lb/ft to be added to the total load resultant per **NESC** Table 251-1.

Fig. 251-3. Example of load components for the medium loading district (Rule 251B).

the 250C (extreme wind) or the 250D (extreme ice with concurrent wind) loads to produce a higher conductor tension than the 250B (heavy, medium, light, or warm islands) loads. If the 250C or 250D loads are applicable (i.e., the structure or conductors are more than 60 ft above ground or water), then the "design tension" must be chosen from the largest conductor tension based on the 250B, 250C, and 250D loads. The 250B, 250C, and 250D loads are applied to conductors in accordance with Rule 251. The 250B, 250C, and 250D loads are applied to line supports (e.g., structures) in accordance with Rule 252. Rule 250A specifies that where all three rules apply (250B, 250C, and 250D), the required loading shall be the one that has the greatest effect.

Rule 261H1 requires the open (e.g., bare noninsulated) supply conductor tension at the Rule 251 condition (using the loads of Rule 250B) to be not more than 60 percent of the conductor rated breaking strength. Rule 261H also specifies tension limits if Rules 250C or 250D are applicable. The sample sag and tension chart at the beginning of Sec. 23 indicates a rated tensile strength of 31.2 percent for the 1366-lb design tension. This percentage meets the not more than 60 percent requirement based on Rule 250B loads. Rule 261H1 also specifies an initial and final tension limit at the applicable temperature in Table 251-1 without external load (i.e., without ice or wind). Rules 261I (supply cable messengers) and 261K2 (communication cable messengers) specify messenger tension limits for the 250B (heavy, medium, light, or warm islands loading), 250C (extreme wind loading), and 250D (extreme ice with concurrent wind loading) conditions applied in accordance with Rule 251.

252. LOADS ON LINE SUPPORTS

252A. Assumed Vertical Loads.   Ice and wind loads are first defined in Rule 250, specifically Rules 250B, 250C, and 250D. How to apply ice and wind loading to conductors is defined in Rule 251. How to apply ice and wind loading on line supports is specified in Rule 252. Rule 250A2 should be reviewed for proper consideration of construction and maintenance loads.

Rule 252A discusses how to apply vertical loads to line supports (i.e., poles, towers, foundations, crossarms, pins, insulators, and conductor fastenings). Rule 252A does not specify any load factors. The load factors are provided in Rule 253. An example of vertical loads on line supports is shown in Fig. 252-1.

A difference in elevation of supports must be considered as it will affect the length of the vertical (weight) span. An example of vertical loads on line supports with different elevations is shown in Fig. 252-2.

It is possible to have a negative vertical load on a line support if the conductor position is above the line support attachment position. This condition can occur on hilly terrain and is commonly referred to as uplift. Vertical loads on wood poles and structures due to guy tensions are discussed in Rule 261A2c.

252B. Assumed Transverse Loads.   Rule 252B discusses how to apply transverse (horizontal) loads to line supports (i.e., poles, towers, foundations, crossarms, pins, insulators, and conductor fastenings). Rule 252B does not specify any load factors. The load factors are provided in Rule 253. An example of transverse loads on line supports is shown in Fig. 252-3.

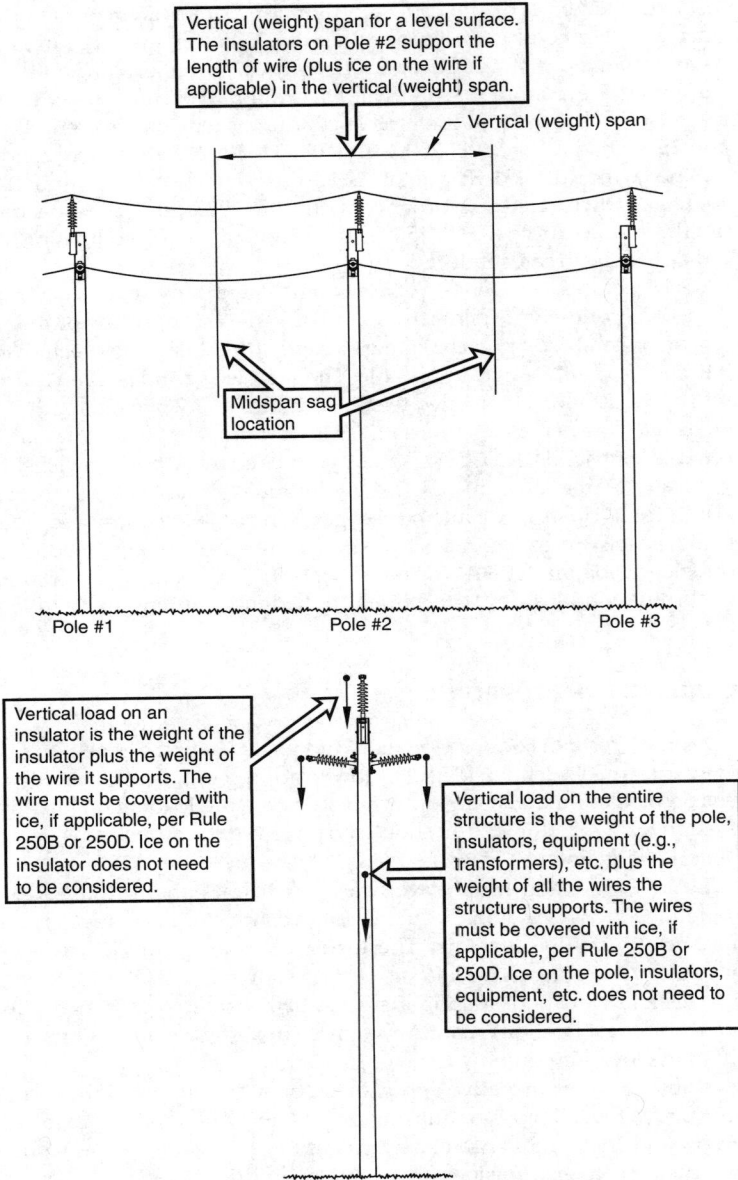

Vertical (weight) span for a level surface. The insulators on Pole #2 support the length of wire (plus ice on the wire if applicable) in the vertical (weight) span.

Vertical (weight) span

Midspan sag location

Pole #1   Pole #2   Pole #3

Vertical load on an insulator is the weight of the insulator plus the weight of the wire it supports. The wire must be covered with ice, if applicable, per Rule 250B or 250D. Ice on the insulator does not need to be considered.

Vertical load on the entire structure is the weight of the pole, insulators, equipment (e.g., transformers), etc. plus the weight of all the wires the structure supports. The wires must be covered with ice, if applicable, per Rule 250B or 250D. Ice on the pole, insulators, equipment, etc. does not need to be considered.

Fig. 252-1. Example of vertical loads on line supports (Rule 252A).

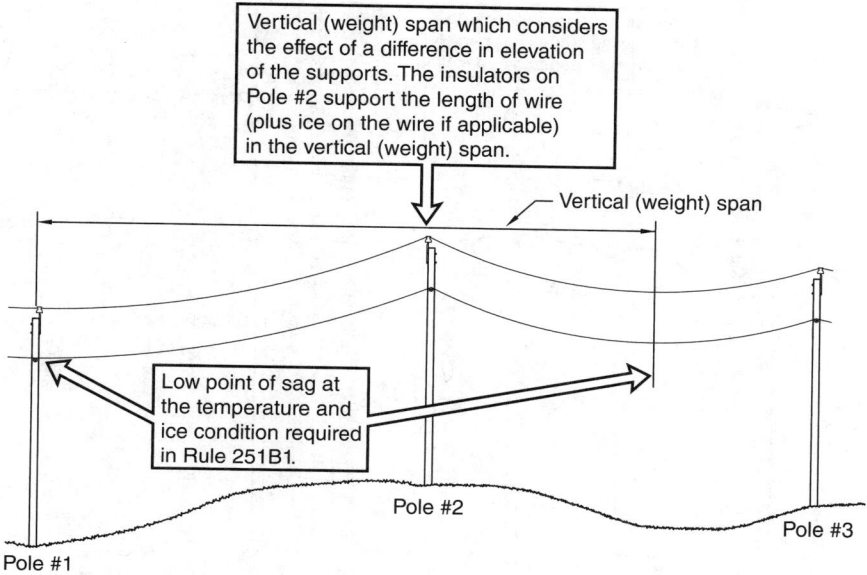

> Vertical (weight) span which considers the effect of a difference in elevation of the supports. The insulators on Pole #2 support the length of wire (plus ice on the wire if applicable) in the vertical (weight) span.

Vertical (weight) span

> Low point of sag at the temperature and ice condition required in Rule 251B1.

Pole #2

Pole #3

Pole #1

Fig. 252-2. Example of vertical loads on line supports with different elevations (Rule 252A).

Rule 252B2 specifies force coefficients (shape factors) to be used for various surface types. Transverse wind loads on structures are calculated without ice covering. Examples of wind load on various structure types are shown in Fig. 252-4.

Rule 252B3 requires angle structure to consider both the transverse wind load and the wire tension load. The wind direction must be applied to give the maximum resultant load. The angle of the line affects how to apply the wind. Normally, for small and medium angles, an oblique wind in the direction of the bisector of the angle will provide the maximum resultant load. Proper reductions are permissible due to the angularity of the wind on the wire. See Fig. 252-5.

Rule 252B4 states that the calculated transverse load must be based on the average of the two spans adjacent to the structure concerned. A note applies to using engineering judgment for special spans. Rule 252B4 does not require elevation to be considered as Rule 252A does for a vertical span.

252C. Assumed Longitudinal Loading.   Longitudinal loads are loads in line with the conductors. The most obvious longitudinal load is at a conductor dead-ended (single sided deadend) on a pole; however, longitudinal loading can occur on double deadends and on tangent structures under certain conditions. Rule 252C breaks longitudinal loading into six parts (252C1 through 252C6). Rule 252C does not specify any load factors. The load factors are provided in Rule 253.

Rule 252C1 specifies longitudinal tensions to be used when a Grade B line section is located within a line of a lower grade. A common example is a Grade C distribution line crossing a limited access highway, railroad tracks, or a navigable

Transverse (horizontal wind) span based on
the average of the two spans adjacent to
the structure concerned per Rule 252B4.

Transverse (horizontal wind) span

Pole #1   Pole #2   Pole #3

Transverse load on an insulator
is the horizontal wind on the
insulator plus the horizontal
wind on the wire it supports.
The wire must be covered with
ice, if applicable, per Rule 250B
or 250D. Ice on the insulator
does not need to be considered.

Wind

Transverse load on the entire
structure and equipment on
structure is the horizontal
wind on the pole, insulators,
equipment (e.g.,
transformers), etc., plus the
horizontal wind on all the
wires the structure supports.
The wires must be covered
with ice, if applicable, per Rule
250B or 250D. The wind
pressures in Rule 250B must be
considered and, if applicable,
the wind pressures in Rules
250C and 250D must be
considered. Ice on the pole,
insulators, equipment, etc. does
not need to be considered.

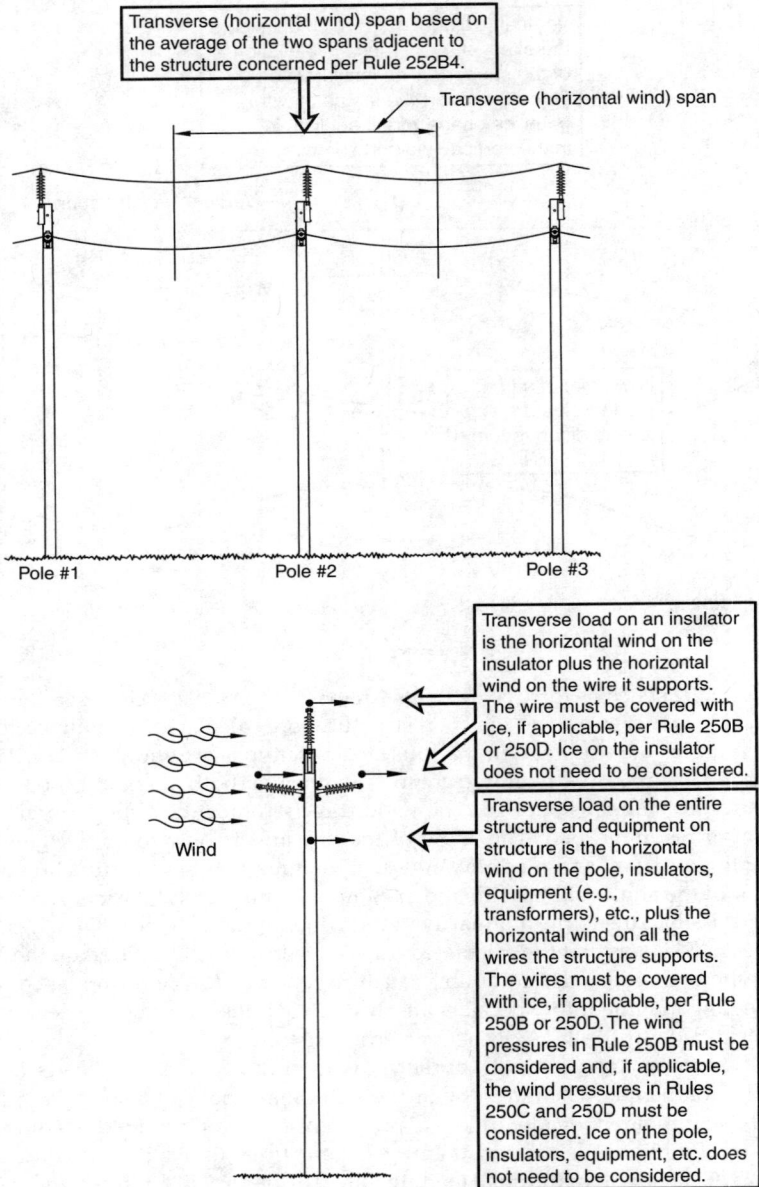

Fig. 252-3. Example of transverse loads on line supports (Rule 252B).

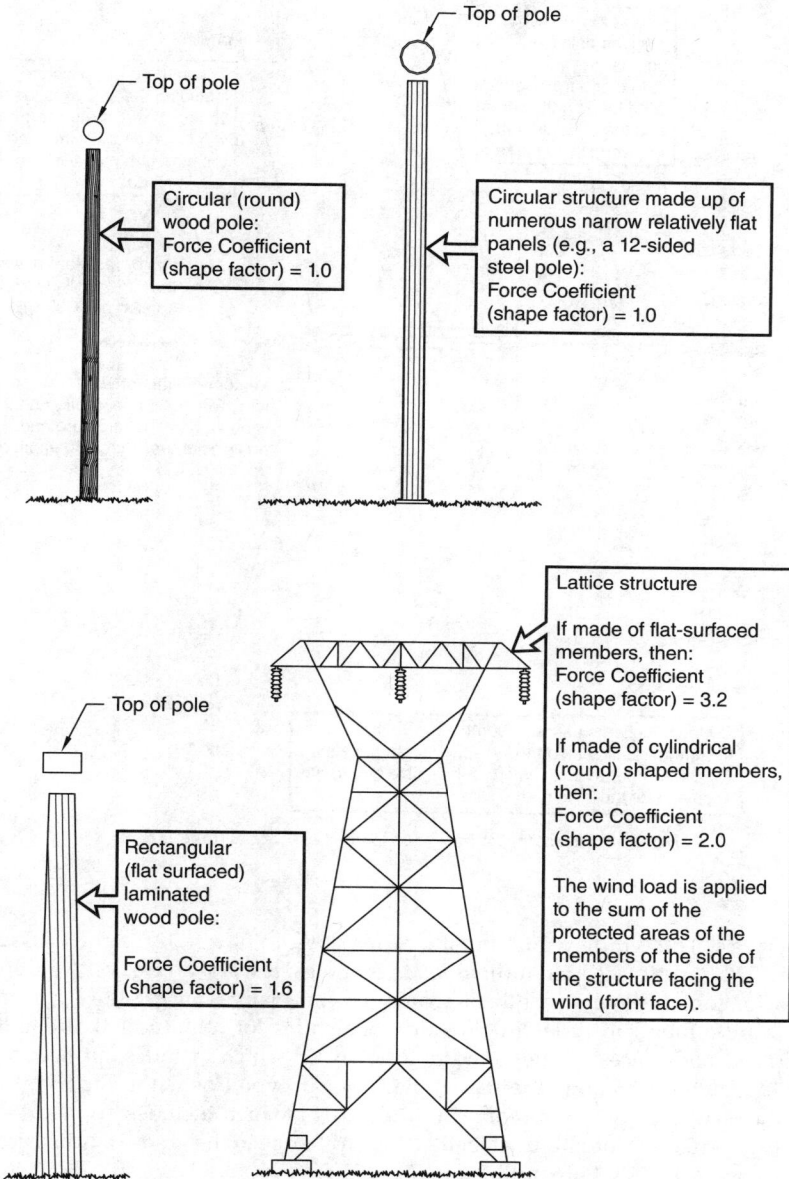

Fig. 252-4. Examples of wind load on various structure types (Rule 252B2).

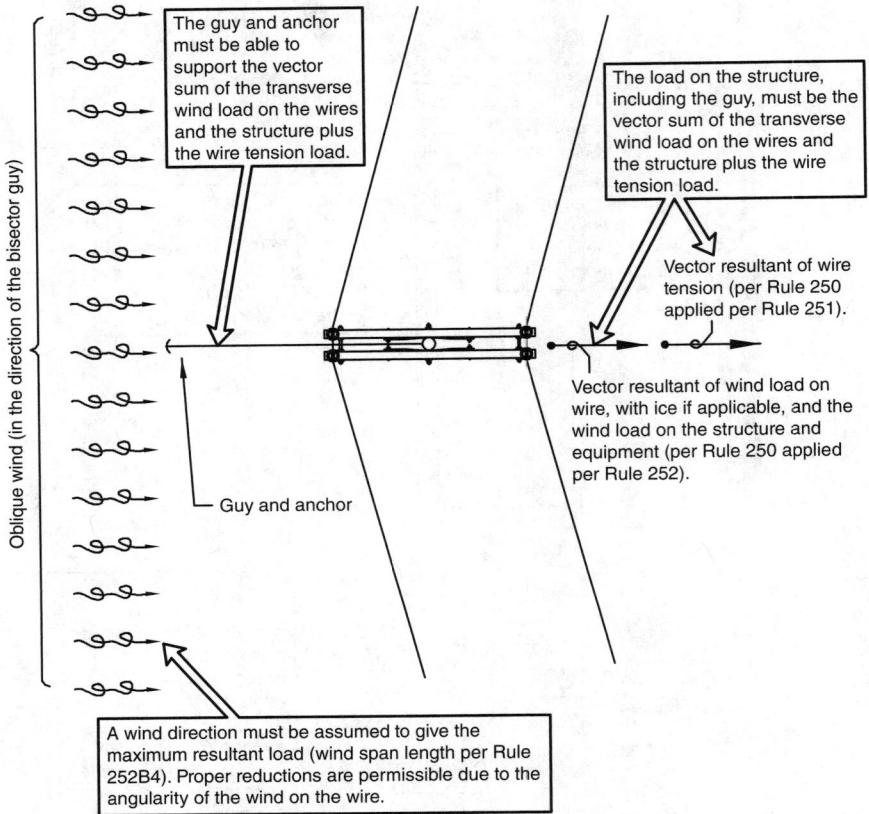

The guy and anchor must be able to support the vector sum of the transverse wind load on the wires and the structure plus the wire tension load.

The load on the structure, including the guy, must be the vector sum of the transverse wind load on the wires and the structure plus the wire tension load.

Vector resultant of wire tension (per Rule 250 applied per Rule 251).

Vector resultant of wind load on wire, with ice if applicable, and the wind load on the structure and equipment (per Rule 250 applied per Rule 252).

Guy and anchor

Oblique wind (in the direction of the bisector guy)

A wind direction must be assumed to give the maximum resultant load (wind span length per Rule 252B4). Proper reductions are permissible due to the angularity of the wind on the wire.

Fig. 252-5. Example of transverse loads at line angles (Rule 252B3).

waterway. The Grade C line must be increased to Grade B for the crossing. Rule 252C1 addresses longitudinal loads on poles, towers, and guys. Additional rules for longitudinal strength of structures for changes in grade of construction can be found in Rule 261A5. Additional rules for longitudinal strength of crossarms and braces can be found in Rule 261D5 which includes applied loads. Additional rules for longitudinal strength of pin-type (insulator supports) and conductor fastenings can be found in Rule 261F1 which includes applied loads. The Code rules for longitudinal loads where a change in the grade of construction occurs per Rule 252C1 are outlined in Fig. 252-6.

Rule 252C2 addresses the older open-wire communication circuits with multiple conductors on a crossarm. This construction has frequently been replaced with multiple pair insulated cables and fiber-optic cables.

Rule 252C3 provides rules for longitudinal loading on dead-end structures. See Fig. 252-7.

If conductors have a 3000 lb breaking strength or less (typically #4 ACSR and smaller) then the longitudinal load on the supporting structure is the tension of 2/3 but not fewer than two conductors in the direction of the higher grade section. Exception, the tension of one conductor can be used for circuits of one or two conductors.

If conductors have a breaking strength greater than 3000 lb (typically #2 ACSR and larger) then the longitudinal load on the support structure is the tension of one conductor when there are eight or fewer conductors (including an overhead ground wire). If there are more than eight conductors in the crossing, the tension of two conductors must be used.

Grade C distribution line

Grade B crossing

Grade C distribution line

Guy used to support longitudinal load.

Grade B section in a Grade C line.

Guy used to support longitudinal load.

The conductors selected must produce the maximum stress on the support.

Railroad tracks

See Photo(s)

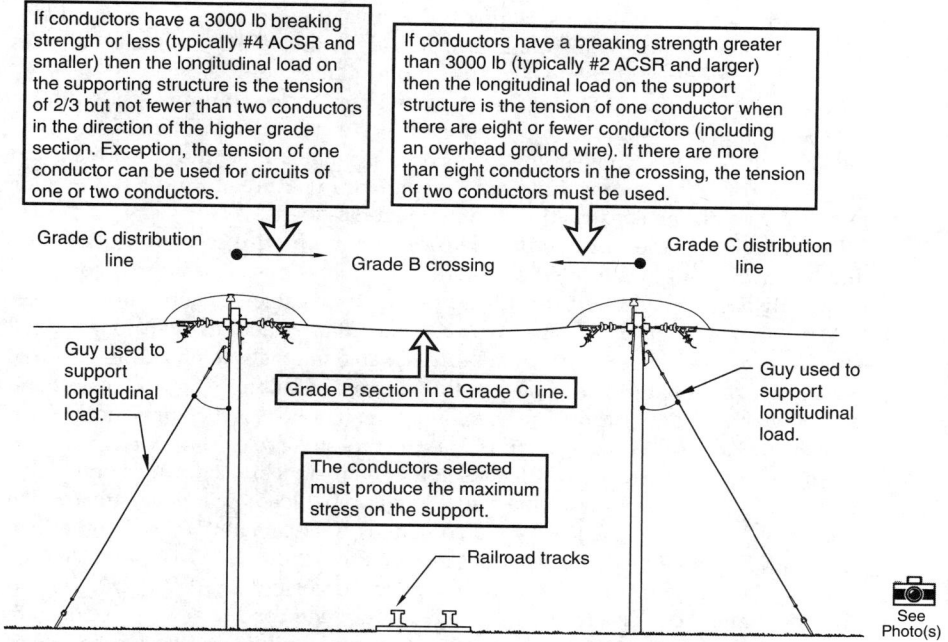

Fig. 252-6. Longitudinal loads at a change in grade of construction (Rule 252C1).

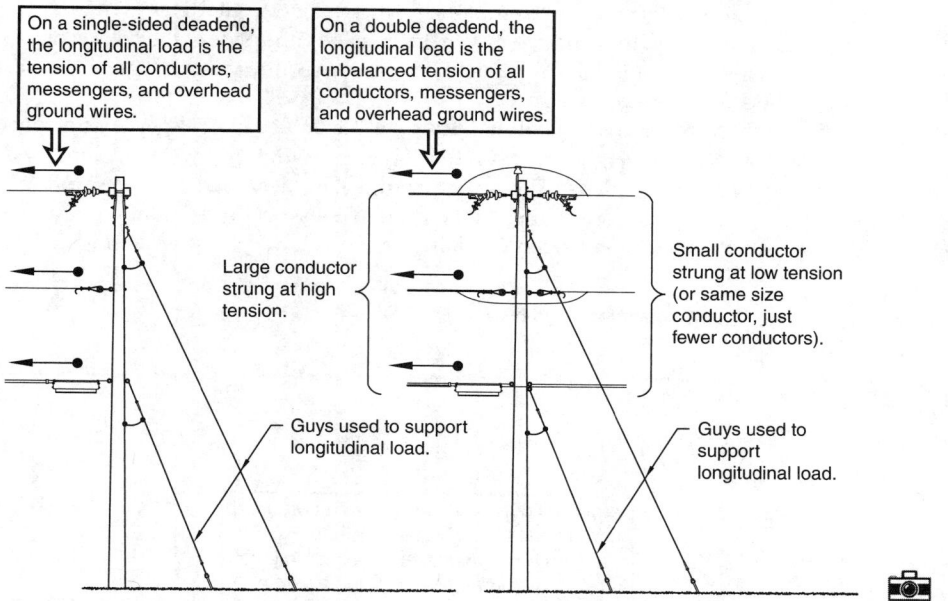

On a single-sided deadend, the longitudinal load is the tension of all conductors, messengers, and overhead ground wires.

On a double deadend, the longitudinal load is the unbalanced tension of all conductors, messengers, and overhead ground wires.

Large conductor strung at high tension.

Small conductor strung at low tension (or same size conductor, just fewer conductors).

Guys used to support longitudinal load.

Guys used to support longitudinal load.

Single-Sided Deadend

Double Deadend

See Photo(s)

Fig. 252-7. Longitudinal loads at deadends (Rule 252C3).

Rule 252C4 requires that a structure be capable of supporting unbalanced longitudinal loads created by unequal vertical loads or unequal spans. Extreme cases will require a double dead-end structure and guying.

Rule 252C5 requires that consideration be given to longitudinal loads during wire stringing operations. Temporary dead-end structures and temporary guying are typically used to meet this criterion if permanent facilities do not exist at the wire stringing locations. Rule 252C5 is similar in nature to Rule 250A2, which requires considering construction and maintenance loads.

Rule 252C6 specifies longitudinal tensions for open-wire communication conductors at railroad and limited access highway crossings. This construction has frequently been replaced with multiple pair insulated cables and fiber-optic cables.

A recommendation at the end of 252C states that structures with longitudinal strength capacity should be provided at reasonable intervals along a line. This rule is very general in nature. The intent of this recommendation is to avoid a "domino effect," which could occur in long stretches of straight lines if the wires in one span were to break. Some degree of longitudinal structure strength exists in a tangent wood pole designed for transverse wind loads. If the conductors break during a nonloaded condition (i.e., no ice or wind), the actual longitudinal loads will be less than if the conductors break during ice and wind conditions. A certain amount of longitudinal strength will also be provided by application of Rules 261A1c, 261A2e, and 261A3d. Typically the transverse loading and the application of Rules 261A1c, 261A2e, and 261A3d will not be as severe as a longitudinal dead-end conductor load. A line with angles, especially 90° turns, has longitudinal strength built in. A line that continues straight for a long distance (e.g., a 5 mile stretch) may require a double dead-end structure in the middle of the long straight stretch with enough longitudinal strength to act as a single dead-end if the conductors were to break on either side of the pole. When to insert a double dead-end pole in a straight line section requires accepted good practice as the Code does not specify a specific distance. See Fig. 252-8.

252D. Simultaneous Application of Loads. A structure may experience a combination of vertical (weight), transverse (horizontal wind and wire tension), and longitudinal (in-line) loads. The structure must be designed to withstand the simultaneous application of these loads when they occur simultaneously. An example of a structure experiencing simultaneous loads is shown in Fig. 252-9.

Double dead-end pole — Guy and anchor

Tangent pole

A double dead-end structure can be used to provide longitudinal strength to avoid a "domino effect" which could occur in long stretches of straight lines if the wires in one span were to break.

Fig. 252-8. Recommendation to provide longitudinal strength capability (Rule 252C6).

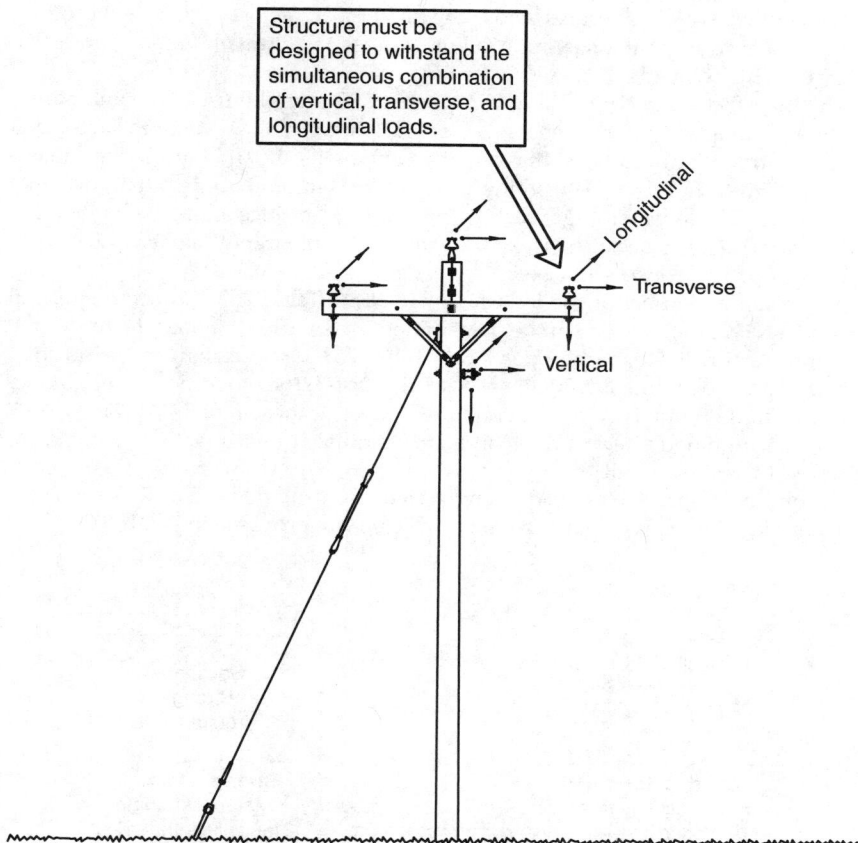

Fig. 252-9. Simultaneous application of loads (Rule 252D).

253. LOAD FACTORS FOR STRUCTURES, CROSSARMS, SUPPORT HARDWARE, GUYS, FOUNDATIONS, AND ANCHORS

Rule 253 requires load factors to be applied to the loads of Rules 250B, 250C, and 250D. This requirement is also stated in a similar fashion in Rules 261A1a and 261A1b for metal, prestressed-concrete, and reinforced-concrete structures, in Rules 262A2a and 262A2b for wood structures, and in Rules 261A3a and 261A3b for fiber-reinforced polymer structures. Load factors are provided in **NESC** Table 253-1. A load factor is a load multiplier that is greater than or equal to 1.0 with one exception, which is Grade C extreme wind (Rule 250C) loads. Grade C extreme ice with concurrent wind (Rule 250D) also has a load factor less than 1.0 for the ice portion only of extreme ice with concurrent wind. The Grade C load factor for Rule 250D is shown in Rule 250D2, not in Table 253-1. A load factor is applied to a base load to increase it. The load factors for Grade B are higher than Grade C with one exception, which is vertical loads. The higher load factors for Grade B produce a line that is stronger than a Grade C line because the Grade B line is required to withstand higher loads.

The load factors in **NESC** Table 253-1 in Sec. 25 must be used in combination with the strength factors in **NESC** Table 261-1 in Sec. 26. A strength factor is a material strength multiplier that is less than or equal to 1.0. It is applied to a material strength to reduce it. The strength factors for Grade B are less than or equal to Grade C. The lower strength factors for Grade B produce a line that is stronger than a Grade C line because less of the materials' full strength is permitted to be used for Grade B construction.

The combined effect of the load factor in **NESC** Table 253-1 and the strength factor in **NESC** Table 261-1 can be determined by dividing the load factor by the strength factor. For example, per **NESC** Table 253-1, the transverse wind load factor of a Grade B line is 2.50. Per **NESC** Table 261-1, the wood pole strength factor of a Grade B line is 0.65. The combined effect of these factors is 2.50 ÷ 0.65 = 3.85. The combined effect of the load factor and the strength factor is sometimes referred to as the overall "safety factor."

An outline of load factors and strength factors is provided in Fig. 253-1.

Examples of applying load and strength factors are provided in Fig. 253-2.

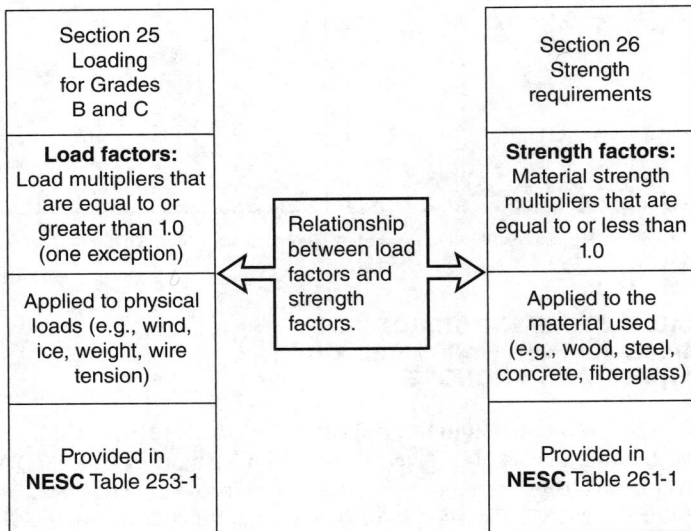

Fig. 253-1. Load and strength factors (Rules 253 and 261).

Example #1	Example #2	Example #3	Example #4
Conditions: - Transverse wind load on 1/0 ACSR wire - Medium loading district (Rule 250B) (Rules 250C and 250D are not applicable) - Wood poles - Grade B line	**Conditions:** - Transverse wind load on 1/0 ACSR wire - Medium loading district (Rule 250B) (Rules 250C and 250D are not applicable) - Wood poles - Grade C line	**Conditions:** - Transverse wind load on 1/0 ACSR wire - Medium loading district (Rule 250B) (Rules 250C and 250D are not applicable) - Steel poles - Grade B line	**Conditions:** - Transverse wind load on 1/0 ACSR wire - Medium loading district (Rule 250B) (Rules 250C and 250D are not applicable) - Steel poles - Grade C line
Load and Strength Factor: - Load factor from **NESC** Table 253-1: 2.50 - Strength factor from **NESC** Table 261-1: 0.65 (Footnote 2 applies to deterioration) (Footnote 4 applies to temporary service)	**Load and Strength Factor:** - Load factor from **NESC** Table 253-1: 2.20 (at crossings) 1.75 (elsewhere) - Strength factor from **NESC** Table 261-1: 0.85 (Footnote 2 applies to deterioration) (Footnote 4 applies to temporary service)	**Load and Strength Factor:** - Load factor from **NESC** Table 253-1: 2.50 - Strength factor from **NESC** Table 261-1: 1.0 (Footnote 6 applies to deterioration)	**Load and Strength Factor:** - Load factor from **NESC** Table 253-1: 2.20 (at crossings) 1.75 (elsewhere) - Strength factor from **NESC** Table 261-1: 1.0 (Footnote 6 applies to deterioration)

Example #5	Example #6	Example #7	Example #8
Conditions: - Longitudinal wire tension load at a 1/0 ACSR deadend - Medium loading district (Rule 250B) (Rules 250C and 250D are not applicable) - Wood poles - Grade B line	**Conditions:** - Longitudinal wire tension load at a 1/0 ACSR deadend - Medium loading district (Rule 250B) (Rules 250C and 250D are not applicable) - Wood poles - Grade C line	**Conditions:** - Longitudinal wire tension load at a 1/0 ACSR deadend - Medium loading district (Rule 250B) (Rules 250C and 250D are not applicable) - Steel poles - Grade B line	**Conditions:** - Longitudinal wire tension load at a 1/0 ACSR deadend - Medium loading district (Rule 250B) (Rules 250C and 250D are not applicable) - Steel poles - Grade C line
Load and Strength Factor: - Load factor from **NESC** Table 253-1: 1.65 (Footnote 2 applies to communications only) - Strength factor from **NESC** Table 261-1: 0.65 (Footnote 2 applies to deterioration) (Footnote 4 applies to temporary service)	**Load and Strength Factor:** - Load factor from **NESC** Table 253-1: 1.30 (Footnote 4 applies to metal guy wires) - Strength factor from **NESC** Table 261-1: 0.85 (Footnote 2 applies to deterioration) (Footnote 4 applies to temporary service)	**Load and Strength Factor:** - Load factor from **NESC** Table 253-1: 1.65 (Footnote 2 applies to communications only) - Strength factor from **NESC** Table 261-1: 1.0 (Footnote 6 applies to deterioration)	**Load and Strength Factor:** - Load factor from **NESC** Table 253-1: 1.30 (Footnote 4 applies to metal pole and metal guy wires) - Strength factor from **NESC** Table 261-1: 1.0 (Footnote 6 applies to deterioration)

Fig. 253-2. Examples of applying load and strength factors (Rules 253 and 261).

Section 26

Strength Requirements

260. GENERAL (SEE ALSO SECTION 20)

Sections 24, 25, and 26 are all interrelated. Section 24, "Grades of Construction," defines the required strength of overhead line construction for safety purposes. Section 25, "Loadings for Grades B and C," defines the physical loads (e.g., ice, wind, and temperature conditions) that overhead line construction must be able to withstand and the load factors that must be applied to the physical loads. Section 26, "Strength Requirements," defines the required strength of materials used in constructing overhead lines and the strength factors that must be applied to the materials. Line insulators are not addressed in Secs. 24, 25, or 26. Insulators are covered in their own section, Sec. 27.

The purpose of Sec. 24 is to define the grade of construction that is required for different situations. The details of applying load factors and strength factors are covered in Secs. 25 and 26.

This Handbook addresses the **Code** requirements in Secs. 24, 25, 26, and 27. Line design calculations are not presented in this Handbook. Line design calculations can be found in transmission and distribution line design manuals. Large Utilities commonly develop their own line design manuals. Some utilities use the transmission and distribution line design manuals published by the Rural Utilities Service (RUS), which in the past was referred to as the Rural Electrification Administration (REA). The RUS distribution line design manuals consist of RUS Bulletins 1724E-150 through 1724E-154. The RUS design manual for high voltage transmission lines is RUS Bulletin 1724E-200. The RUS also publishes design manuals for communication lines. This Handbook will aid those individuals writing and using line design manuals and provide a deeper understanding of how the **Code** applies to line design calculations.

Typical strength and loading line design calculations include, but are not limited to, the following:

- Maximum wind span based on wind with ice on conductors
- Maximum weight span based on weight of conductors and ice
- Moment due to wind on pole
- Allowable resisting moment of pole
- Transverse, vertical, and total components of conductor loading
- Total ground line moment on pole
- Ruling span
- Diameter of pole at any point
- Dead-end guying strength
- Bisector guying strength
- Anchor strength
- Weak-link of guy attachment, guy wire, and anchor assembly
- Crossarm strength (vertical)
- Crossarm strength (longitudinal)
- Pole buckling
- Maximum line angle based on insulator strength
- Material deflection
- Equipment loading

Single pole structures are typically easier to analyze than multiple pole and lattice structures. Line design calculations also involve clearance calculations in addition to strength and loading calculations (e.g., clearance above ground, clearance to buildings and other structures, horizontal clearance between conductors, vertical clearance between conductors, etc.). The clearance calculations will affect the line design and therefore the selection of pole heights and pole classes. The sample sag and tension chart shown at the beginning of Sec. 23 provides sag for clearance calculations and conductor tension for strength and loading calculations. The "design tension" used for strength and loading calculations is the largest of the conductor tensions calculated after applying Rule 250B (heavy, medium, light, or warm islands loads), Rule 250C (extreme wind loads) if applicable, and Rule 250D (extreme ice with concurrent wind loads) if applicable. The 250B, 250C, and 250D loads are applied to conductors in accordance with Rule 251. The 250B, 250C, and 250D loads are applied to line supports (e.g., structures) in accordance with Rule 252. Rule 250A specifies that where all three rules apply (250B, 250C, and 250D), the required loading shall be the one that has the greatest effect. See Rules 250, 251, and 252 for additional information.

260A. Preliminary Assumptions.   Rule 260A1 explains how to account for deformation, deflection, and displacement of parts of a structure when performing strength calculations. The **NESC** does not require every structure to have a deflection analysis. The **NESC** does say that when calculating stresses, allowance may be made for deformation, deflection, and displacement when the effects can be evaluated. The decision to consider deformation, deflection, and displacement when performing line design calculations may depend on the method used to perform the calculations. Typically, line design calculations performed by hand or using simple spreadsheets do not account for deformation, deflection, and displacement unless some percentage increase to the pole loading is applied to account for the deformation, deflection, and displacement. Line design calculations performed using sophisticated software programs make analyzing deformation, deflection, and displacement a simpler task. The term P-Delta is commonly used when referring to structure deformation, deflection, and displacement. Sophisticated software programs enable the user to perform a nonlinear P-Delta analysis for loading and strength from the ground line to the top of the pole (not just at the ground line). Sophisticated software programs can also apply fiber strength reductions due to the fiber strength height effect for wood poles (see Rule 261A2 for more information). The **NESC** provides a two-part approach to determine deformation, deflection, and displacement. First, the ice and wind loads of Rule 250 are applied to the structure without the load factors of Rule 253. These are the loads that are used to determine deformation, deflection, and displacement. Second, from the deformation, deflection, and displacement position, the structure strength is analyzed using the load factors in Rule 253. Examples of allowance for deformation, deflection, or displacement are shown in Fig. 260-1.

If allowance is made for deformation, deflection, or displacement at crossings or conflicts, the calculations used must be subject to mutual agreement of the parties involved.

The note in Rule 260A1 addresses the strength of a structure during the expected life of a structure. The note indicates that guying or bracing may be needed to meet the required strength of the structure over time due to everyday stresses on the structure produced by gravity or tension. For example, a 50-year-old wood dead-end pole that is not guyed may have sufficient strength for some portion of its service life but may not have sufficient strength toward the end of its service life due to the stress of everyday loads.

Per Rule 260A2, the **NESC** recognizes that new materials may become available that are not covered in the strength requirements section. Since the **NESC** is on a 5-year revision cycle, the **Code** permits trial installations of new materials. New materials must be tested and evaluated. See Rule 013 for additional requirements.

260B. Application of Strength Factors. Rule 260B1 explains how to apply **NESC** load and strength factors to line design calculations. An outline of load and strength factors is provided in Fig. 260-2.

NESC Table 261-1 is the main table in Sec. 26 for determining strength factors of various materials. It has strength factors for the loads of 250B (ice and wind in heavy, medium, light, and warm islands loading districts), for the loads of 250C (extreme wind), and for the loads of 250D (extreme ice with concurrent wind). **NESC** Table 261-1 must be used in combination with **NESC** Table 253-1. See Rule 253 for a discussion and examples of applying load and strength factors.

NESC Table 261-1 contains information relating to deterioration of structure strength under the title of the table and in the table footnotes. Per **NESC** Table 261-1, Footnote 2, for Rule 250B loads a wood pole must be replaced or rehabilitated when deterioration reduces the structure strength to 2/3 of that required when installed. The word "required" is used to distinguish between the required strength of a structure and the actual strength of a structure. If the wood pole class was required to be Class 5, but a Class 4 (larger circumference) pole was installed, the pole must be replaced or rehabilitated when deterioration reduces the structure strength to 2/3

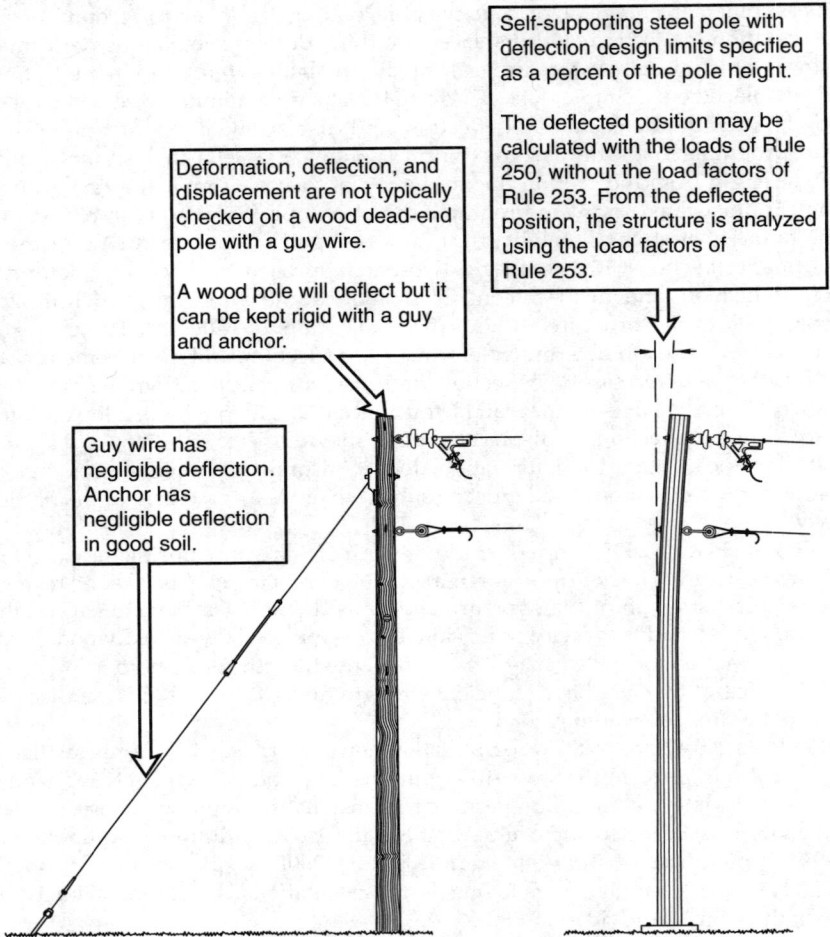

Fig. 260-1. Examples of allowance for deformation, deflection, or displacement (Rule 260A).

of the Class 5 rating (the required strength rating, not the actual strength rating). The most common wood pole deterioration is ground line rot. When wood distribution poles are inspected for ground line deterioration, it is common for the inspector to evaluate the actual strength rating of the pole unless extra steps are taken to determine the required strength rating of the pole. The most common rehabilitation method is pole stubbing. If a pole stub is added, the stubbed pole must have strength greater than 2/3 of that required when installed. If the pole is replaced instead of stubbed, the full strength requirements of **NESC** Table 261-1 apply. Footnote 3 of **NESC** Table 261-1 is similar to Footnote 2. Footnote 2 specifies a strength deterioration of 2/3 for Rule 250B (heavy, medium, light, and warm islands) loads. Footnote 3 specifies a deterioration of 3/4 for Rule 250C (extreme wind loads) and Rule 250D

Section 25 Loading for Grades B and C		Section 26 Strength Requirements
Load factors: Load multipliers that are equal to or greater than 1.0 (one exception)	**Relationship between load factors and strength factors.**	**Strength factors:** Material strength multipliers that are equal to or less than 1.0
Applied to physical loads (e.g., wind, ice, weight, wire tension)		Applied to the material used (e.g., wood, steel, concrete, fiberglass)
Provided in **NESC** Table 253-1		Provided in **NESC** Table 261-1

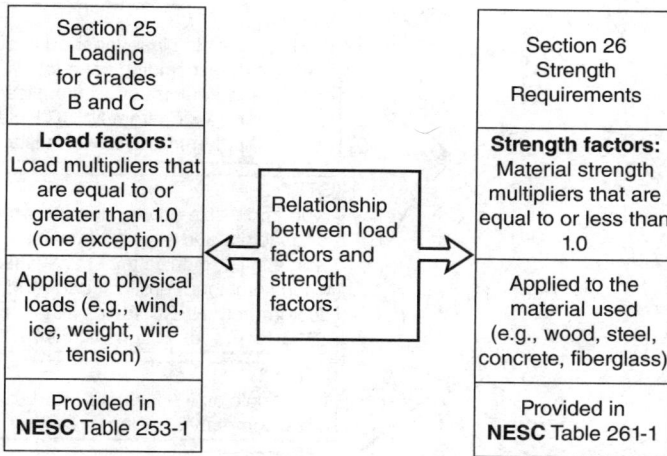

Fig. 260-2. Load and strength factors (Rule 260B).

(extreme ice with concurrent wind loads). Footnotes 2 and 3 state that the required strength of the structure (e.g., pole) must be based on revised loading when new or changed facilities modify the loads on an existing structure. The wording under the title of **NESC** Table 261-1 further defines how deterioration applies to existing poles that have new or changed lines or equipment added to them. Footnote 4 of **NESC** Table 261-1 describes the reduced strength of a structure (e.g., pole) built for temporary construction. This footnote does not use the same wording as Rule 014B which permits Grade N construction for temporary installations. Footnote 6 addresses deterioration for metal structures (e.g., steel poles). Per Footnote 6, no deterioration is permitted below the required strength of the structure. If deterioration is anticipated, a strength factor of less than 1.0 must be used. Pole owners typically test wood poles at or near the groundline for rot to determine if the pole strength is adequate or if the wood pole needs to be replaced or rehabilitated (e.g., using a steel truss or wood stub). Visual inspections can be performed for other structural members such as crossarms and pole hardware and the above ground portion of the pole is typically visually inspected. See Rule 214 for discussion of inspection and testing. Examples of how loads, materials, and load and strength factors affect the size of a wood pole are shown in Figs. 260-3 and 260-4.

Listings of common wood pole class sizes, examples of pole tags and pole branding information, and an example of a common wood pole burial depth formula are provided in Figs. 260-5 through 260-7.

Insulators are covered in their own section, Sec. 27, Line Insulation. **NESC** Table 261-1 does not apply to insulators. Rule 260B1 notes several standards for determining structure design capacity. See Rule 250 or 260 for a discussion of utility line design manuals.

Per Rule 260B2, if Rule 250C (extreme wind) or Rule 250D (extreme ice with concurrent wind) applies and the strength factor is not defined in Rule 261 or Table 261-1, a strength factor of 0.80 must be used for supported facilities.

STEP 1:
Wood pole size (class) needed to support loads on pole including conductors, cables, and equipment with applicable wind and ice loads from **NESC** Rule 250B (heavy, medium, light, or warm islands).

STEP 2:
Wood pole size (class) needed when load factors and strength factors (sometimes referred to as "safety factors" when combined) are included (during design) from **NESC** Tables 253-1 and 261-1 for grades B and C construction.

STEP 3:
Wood pole size (class) selected due to utility company standard inventory.

STEP 4:
Wood pole size (class) to be maintained. Pole must be replaced or rehabilitated (e.g., using a steel truss or wood stub) when deterioration reduces the structure strength to 2/3 of that required when installed (per **NESC** Table 261-1, Footnote 2). Deterioration can occur from the outside-in (external decay) or the inside-out (hollow heart). The wood pole size (class) selected in Step 3 can deteriorate more than **NESC** Table 261-1, Footnote 2 as the pole is sized above the **NESC** requirements.

See Photo(s)

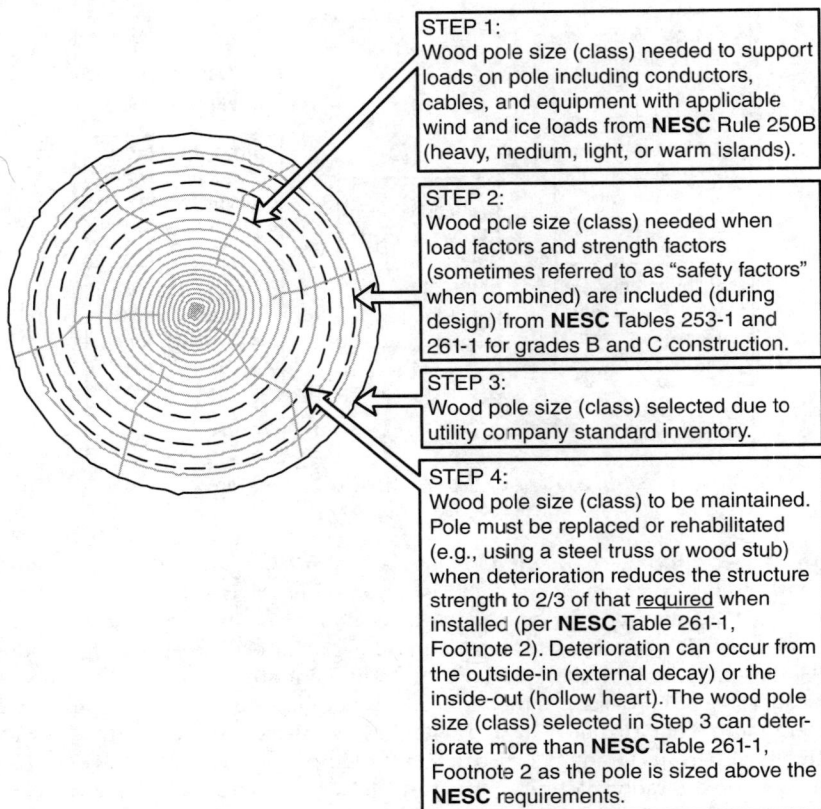

Fig. 260-3. Example of how loads, material, and load and strength factors affect the size of a wood pole (Rule 260B).

261. GRADES B AND C CONSTRUCTION

261A. Supporting Structures. The first sentence in Rule 261A states that the strength requirements of supporting structures (e.g., poles) may be provided by the structure alone or with the aid of guys or braces or both. See Fig. 261-1.

The strength requirements for guys and guyed structures are provided in Rules 261A2a (Exception 1), 261A2c, 261C, and 264. An example of structure strength using a structure alone or a structure with guys is shown in Fig. 261-2.

261A1. Metal, Prestressed-, and Reinforced-Concrete Structures. The Code rules related to strength requirements of metal, prestressed-, and reinforced-concrete structures are outlined in Fig. 261-3.

261A2. Wood Structures. The permitted stress level for natural wood poles can vary due to splices, wood grain, knots in the wood, moisture, etc. An ANSI standard, ANSI O5.1, must be used to determine the fiber stress to which strength factors in **NESC** Table 261-1 are applied. The Annex to the ANSI O5.1 Standard

STEP 3:
Wood pole size (class) selected due to utility company standard inventory.

STEP 2:
Wood pole size (class) needed when load factors and strength factors (sometimes referred to as "safety factors" when combined) are included (during design) from **NESC** Tables 253-1 and 261-1 for Grades B or C construction.

STEP 4:
Wood pole size (class) to be maintained. Pole must be replaced or rehabilitated (e.g., using a steel truss or wood stub) when deterioration red-uces the structure strength to 2/3 of that required when installed (per **NESC** Table 261-1, Footnote 2). Deterioration can occur from the outside-in (external decay) or the inside-out (hollow heart). The wood pole size (class) selected in Step 3 can deteriorate more than **NESC** Table 261-1, Footnote 2, as the pole is sized above **NESC** requirements.

STEP 1:
Wood pole size (class) needed to support loads on pole including conductors, cables, and equipment with applicable wind and ice loads from **NESC** Rule 250B (heavy, medium, light, or warm islands).

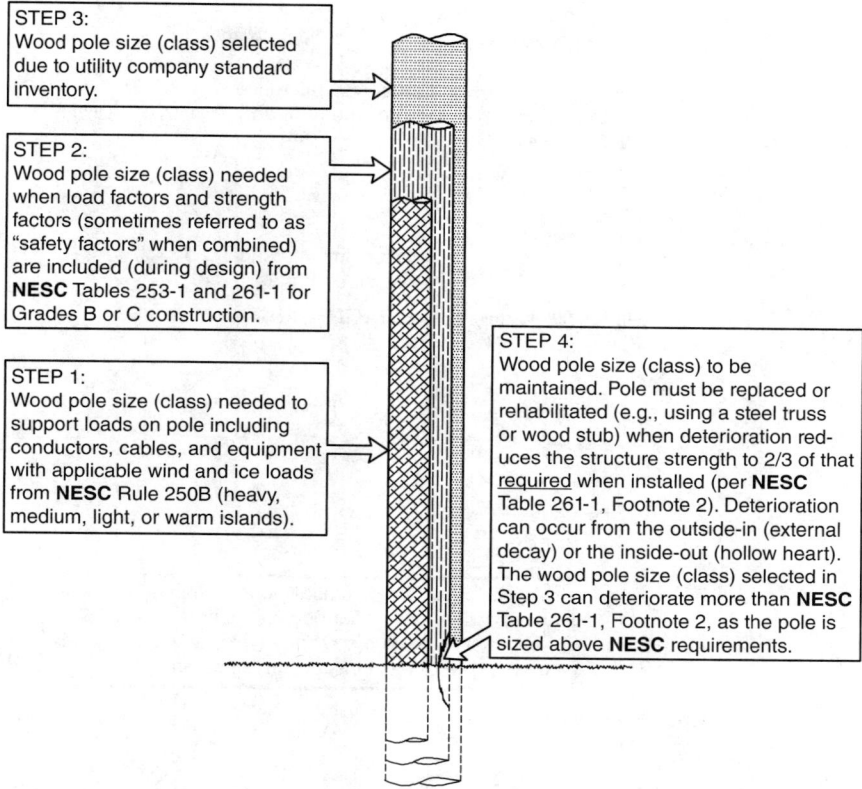

Fig. 260-4. Example of how loads, material, and load and strength factors affect the size of a wood pole (Rule 260B).

for wood poles contains fiber strength reduction factors due to the fiber strength height effect of natural wood poles. Two items can affect the pole-resisting moment along the pole: the circumference of the pole which decreases at points above the ground line due to the taper of the pole and the fiber strength of the pole which also decreases at points above the ground line. ANSI Standard O5.1 is referenced in Rule 261A2b(1) and therefore forms a part of the **NESC** (see the discussion in Sec. 03, References). However, the fiber strength reduction factors in ANSI O5.1 are listed in the Annex of the ANSI O5.1 document and the Annex is for information only and not part of the official Standard. In other words, the **NESC** requires the use of ANSI O5.1 but ANSI O5.1 does not require the use of the fiber strength reduction factors in the Annex. The Annex to the ANSI O5.1 Standard concluded that for poles 55 ft and shorter, the point of maximum bending stress is usually at or near the ground line. This conclusion corresponds to the 55 ft exception for Rule 261A2a. Using the ANSI O5.1 fiber strength reduction

COMMON POLE CLASS RATINGS
(using a 50' pole length)

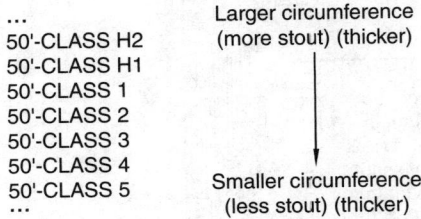

...
50'-CLASS H2   Larger circumference
50'-CLASS H1   (more stout) (thicker)
50'-CLASS 1
50'-CLASS 2
50'-CLASS 3
50'-CLASS 4
50'-CLASS 5   Smaller circumference
...   (less stout) (thicker)

Fig. 260-5. Listing of common wood pole classes (Rule 260).

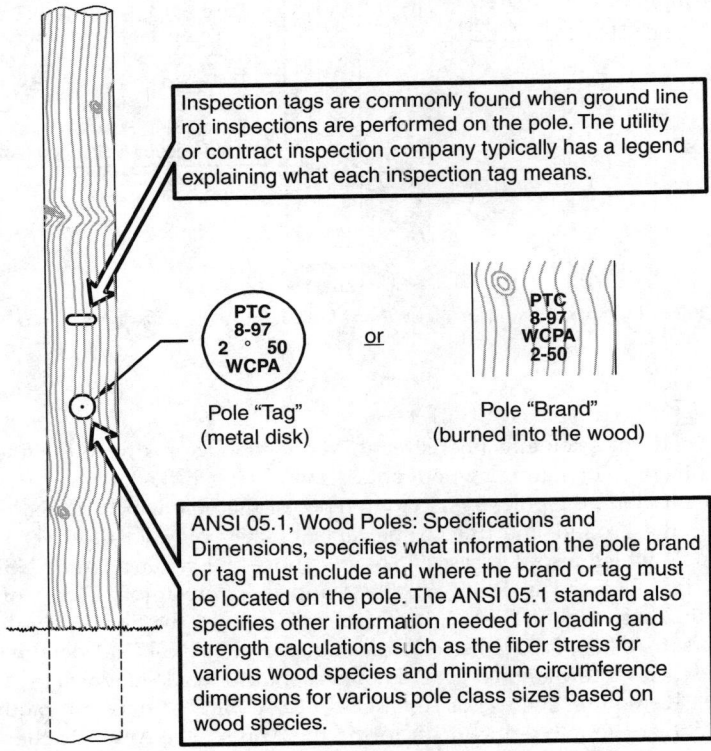

Inspection tags are commonly found when ground line rot inspections are performed on the pole. The utility or contract inspection company typically has a legend explaining what each inspection tag means.

PTC
8-97
2 ° 50
WCPA

or

PTC
8-97
WCPA
2-50

Pole "Tag"   Pole "Brand"
(metal disk)   (burned into the wood)

ANSI 05.1, Wood Poles: Specifications and Dimensions, specifies what information the pole brand or tag must include and where the brand or tag must be located on the pole. The ANSI 05.1 standard also specifies other information needed for loading and strength calculations such as the fiber stress for various wood species and minimum circumference dimensions for various pole class sizes based on wood species.

See
Photo(s)

Fig. 260-6. Examples of a pole tag and pole brand (Rule 260).

50' Pole

43'

The **NESC** does not specify a wood pole burial depth. Rule 012C, which requires accepted good practice, must be applied. A common wood pole burial depth formula is:

- 10% of the pole length plus 2' (10%+2')

- Example for a 50' pole: 0.10 (50') + 2' 5' + 2' = 7'

- Laminated wood pole manufacturers typically specify different formulas.

- Some utilities use lesser depths in rock or greater depths for soft soil.

The burial depth can affect clearances above ground and available strength at the ground line.

7'

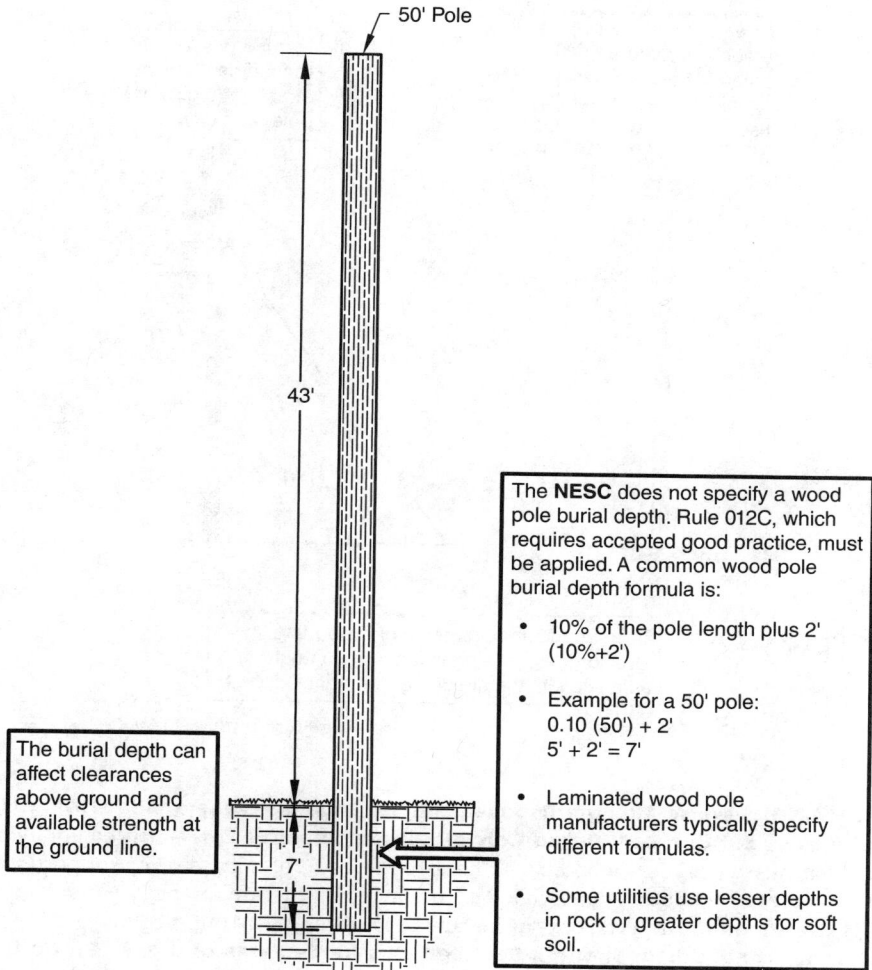

Fig. 260-7. Example of a common wood pole burial depth formula (Rule 260).

factors in line design calculations is not an **NESC** violation, they are not required but they are certainly not prohibited.

The permitted stress level of sawn or laminated wood structural members, cross-arms, and braces is determined by using the ultimate fiber stress of the material in ANSI O5.2 or ANSI O5.3 multiplied by the strength factors in **NESC** Table 261-1.

The **Code** rules related to strength requirements of wood structures are outlined in Figs. 261-4 through 261-7.

261A3. Fiber-Reinforced Polymer Structures.   The Code rules related to the strength requirements of fiber-reinforced polymer structures are outlined in Fig. 261-8.

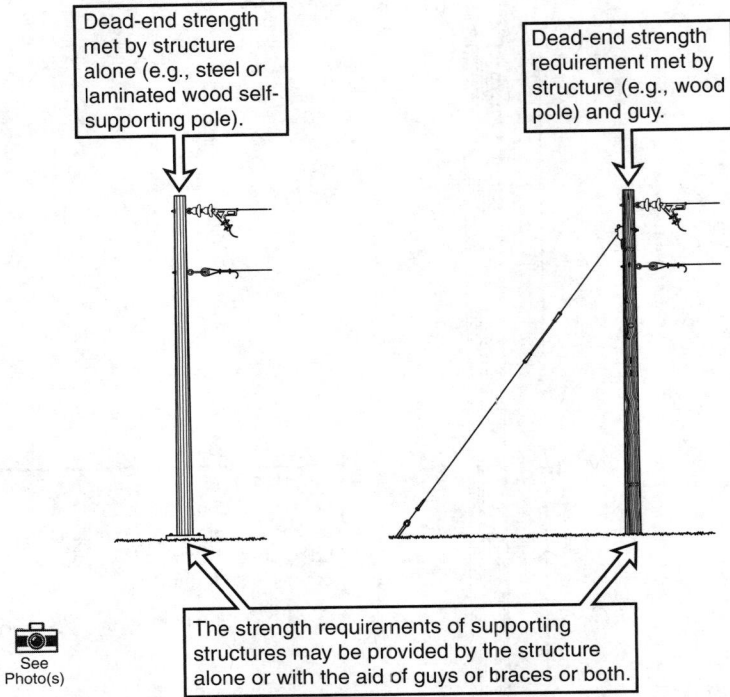

Fig. 261-1. Strength requirements of supporting structure (Rule 261A).

261A4. Transverse Strength Requirements for Structures Where Side Guying Is Required, But Can Be Installed Only at a Distance. Rule 261A4 can be applied to various types (e.g., wood, steel, fiber-reinforced polymer, etc.) of supporting structures. The Code rules related to transverse structure strength where side guys are required, but can only be installed at a distance, are outlined in Fig. 261-9.

261A5. Longitudinal Strength Requirements for Sections of Higher Grade in Lines of a Lower Grade Construction. Rule 261A5 can be applied to various types (e.g., wood, steel, fiber-reinforced polymer, etc.) of supporting structures. The Code rules related to longitudinal strength requirements for sections of higher grade in lines of a lower grade construction are outlined in Fig. 261-10.

Rule 252C1 provides longitudinal loading requirements for sections of Grade B construction when located in lines of lower than Grade B construction.

Rule 261A5b requires guying or increased clearance to overcome the increased sag that occurs when a structure is deflected.

261B. Strength of Foundations, Settings, and Guy Anchors. The Code rules for strength of foundations, settings, and guy anchors are outlined in Fig. 261-11.

Rule 261B states that design or experience may be used to determine the strength of foundations, settings, and guy anchors. The use of experience is commonly applied to local soil conditions within the service area of a utility. A foundation,

Grade B railroad crossing span using
strength of the structure alone (e.g., a heavier
pole class). The R.R. company issuing
a permit may require guys even if the
strength of the structure alone is sufficient.

Grade C
(Class 5 pole)

Grade B
(Class 3 pole)

Grade B
(Class 3 pole)

Grade C
(Class 5 pole)

Railroad tracks

Grade B railroad crossing
span using strength of the
structure with the aid of guys.

Grade C
(Class 5 pole)

Grade B
(Class 5 pole
with guys)

Grade B
(Class 5 pole
with guys)

Grade C
(Class 5 pole)

Railroad tracks

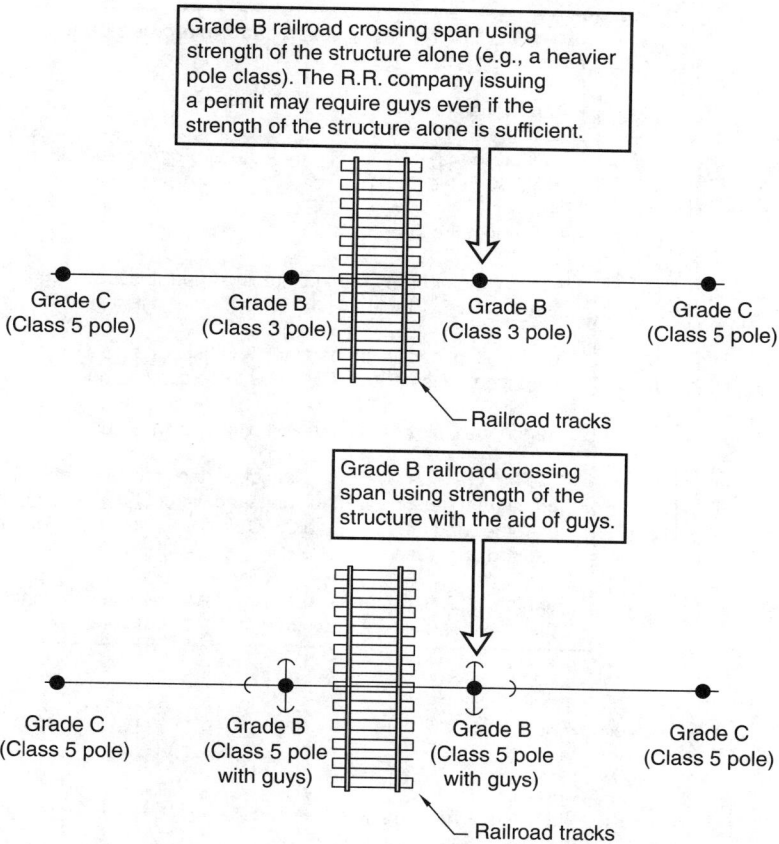

Fig. 261-2. Example of structure strength using a structure alone or a structure with guys
(Rule 261A).

setting, or guy anchor is only as strong as the soil it is set in. The **NESC** does not
specify wood pole burial depths, see Rule 260 for a discussion and a figure. The
strength of a guy anchor is determined by both the strength of the anchor rod
(see Rule 264F) and the holding capacity of the anchor based on a particular soil
classification. If soil types are unknown by experience, soil borings and a geotech-
nical evaluation can be done to determine soil conditions. Numerous guy anchor
methods exist including expanding, screw, plate, swamp, cone, rock, etc. Each
anchor type needs to be evaluated with the soil it is installed in. The **NESC** does not
specify a distance between guy anchors. Rule 012C, which requires accepted good
practice, must be used. Where two utilities or two separate circuits share a guy
anchor, the anchor in soil must be of sufficient strength for the combined loads of
the attached guy wires. Examples of multiple guy anchors and multiple guy wires
attached to a single guy anchor are shown in Fig. 261-12.

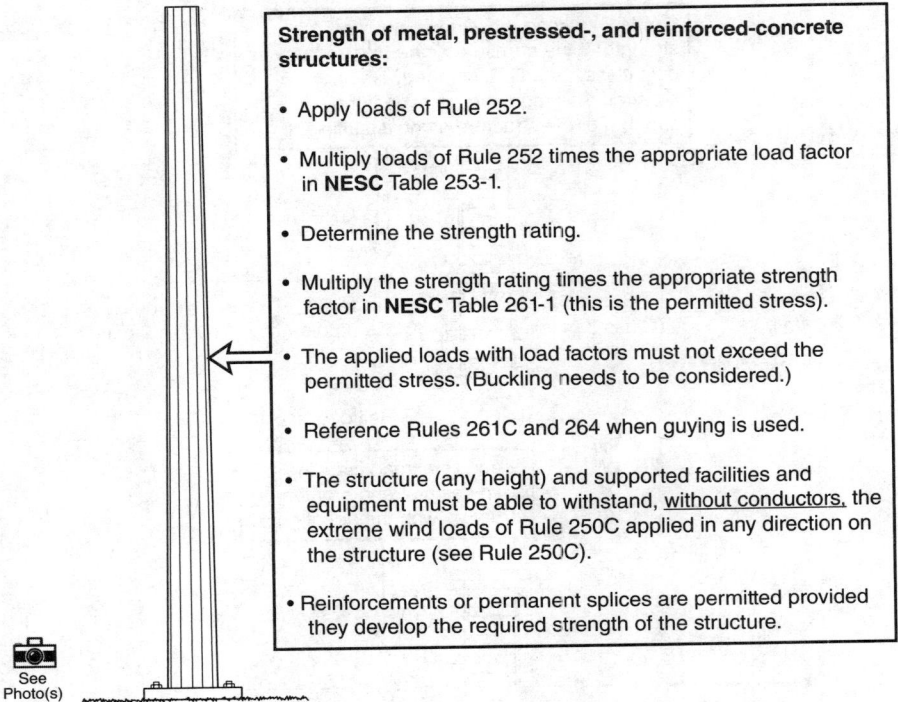

Strength of metal, prestressed-, and reinforced-concrete structures:

• Apply loads of Rule 252.

• Multiply loads of Rule 252 times the appropriate load factor in **NESC** Table 253-1.

• Determine the strength rating.

• Multiply the strength rating times the appropriate strength factor in **NESC** Table 261-1 (this is the permitted stress).

• The applied loads with load factors must not exceed the permitted stress. (Buckling needs to be considered.)

• Reference Rules 261C and 264 when guying is used.

• The structure (any height) and supported facilities and equipment must be able to withstand, <u>without conductors,</u> the extreme wind loads of Rule 250C applied in any direction on the structure (see Rule 250C).

• Reinforcements or permanent splices are permitted provided they develop the required strength of the structure.

See Photo(s)

Fig. 261-3. Strength requirements of metal, prestressed-, and reinforced-concrete structures (Rule 261A1).

The NOTES in Rule 261B provide a reminder that several factors may reduce clearance and structure strength. See Fig. 261-13.

261C. Strength of Guys and Guy Insulators. Guy strength is covered in Rule 264. Guy insulator strength is covered in Rule 279A1c. Rule 261C provides requirements relative to the integration of the guy with the structure to which the guy is attached. Rule 261A2a (Exception 1) and Rule 261A2c provide similar requirements for guys attached to wood poles only. See Fig. 261-14.

The NOTES in Rule 261C are similar to the NOTES in Rule 261B. Excessive movement of a guy (or movement of the anchor attached to the guy) can reduce clearance or structure capacity.

261D. Crossarms and Braces. Rule 261D covers crossarms and braces made of various materials.

261D1. Concrete and Metal Crossarms and Braces. The Code rules for concrete and metal crossarms and braces are outlined in Fig. 261-15.

261D2. Wood Crossarms and Braces. The loads on wood crossarms and braces and the permitted stress levels on wood crossarms and braces are covered in this rule. The permitted stress levels on wood crossarms and braces are also covered in Rule 261A2b(2). The Code rules for wood crossarms and braces in Rule 261D2 are outlined in Fig. 261-16.

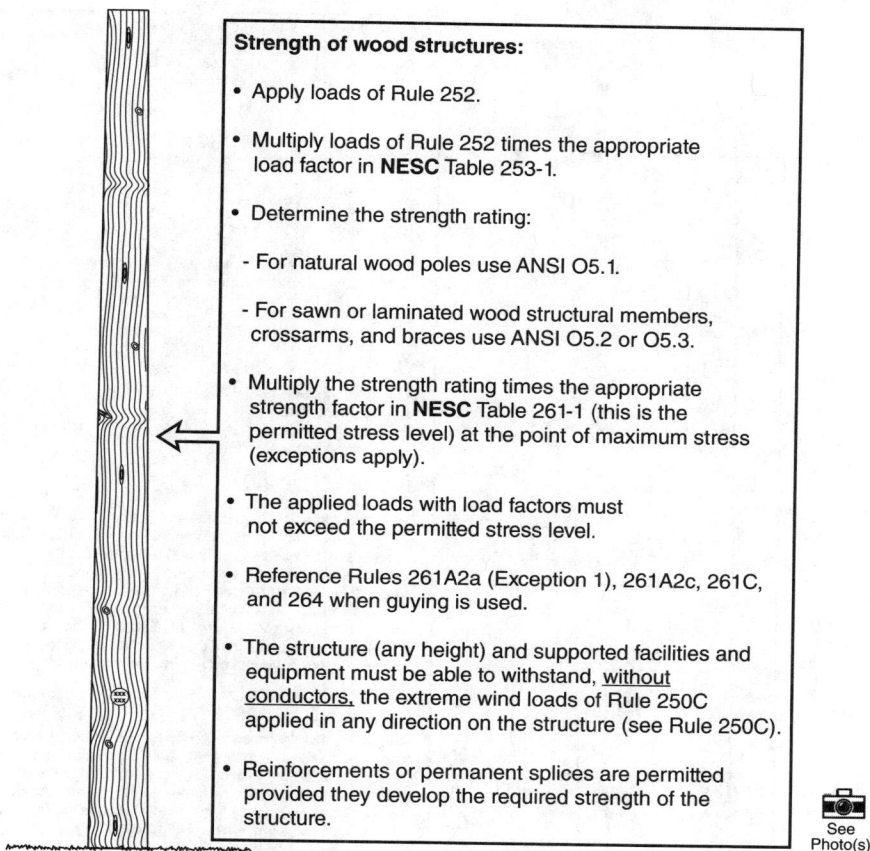

Strength of wood structures:

- Apply loads of Rule 252.

- Multiply loads of Rule 252 times the appropriate load factor in **NESC** Table 253-1.

- Determine the strength rating:

  - For natural wood poles use ANSI O5.1.

  - For sawn or laminated wood structural members, crossarms, and braces use ANSI O5.2 or O5.3.

- Multiply the strength rating times the appropriate strength factor in **NESC** Table 261-1 (this is the permitted stress level) at the point of maximum stress (exceptions apply).

- The applied loads with load factors must not exceed the permitted stress level.

- Reference Rules 261A2a (Exception 1), 261A2c, 261C, and 264 when guying is used.

- The structure (any height) and supported facilities and equipment must be able to withstand, <u>without conductors,</u> the extreme wind loads of Rule 250C applied in any direction on the structure (see Rule 250C).

- Reinforcements or permanent splices are permitted provided they develop the required strength of the structure.

See Photo(s)

Fig. 261-4. Strength requirements of wood structures (Rule 261A2).

In addition to the requirement to use **NESC** Table 253-1 with **NESC** Table 261-1, the Code also specifies minimum dimensions for select Southern Pine and Douglas Fir wood crossarms in **NESC** Table 261-2. Crossarms of other wood species may be used if they provide equal strength. See Fig. 261-17.

261D3. Fiber-Reinforced Polymer Crossarms and Braces. The Code rules for fiber-reinforced polymer crossarms and braces are outlined in Fig. 261-18.

261D4. Crossarms and Braces of Other Materials. Crossarms of other materials not specified in Rule 261D (i.e., other than concrete, metal, wood, and fiber-reinforced polymer) must meet the strength requirements of wood crossarms and braces.

261D5. Additional Requirements. Rule 261D5 provides additional rules for longitudinal strength of crossarms. Tension values and construction methods are specified. Using double wood crossarms is one method of meeting the longitudinal strength requirements of a Grade B line. A support assembly of equivalent strength (i.e., a pre-engineered crossarm assembly) is also acceptable.

- A moment is measure of rotation force about a point.

- The applied moment due to loads on the pole can be calculated by the following formula:

$$M = F \times d$$

  Where:

  M = Moment (ft-lb)
  F = Force (lb)
  d = Distance (ft)

- The resisting moment due to the strength of the pole can be calculated by the following formula:

$$M = K \times F \times C^3$$

  Where:

  M = Moment (ft-lb)
  K = Constant (ft/in)
  F = Fiber Stress (lb/in^2)
  C = Circumference (in)

- Basic Rule:

  Wood structures shall be designed to withstand the loads in Rule 252 multiplied by the appropriate load factor in **NESC** Table 253-1 (Applied Moment) without exceeding the fiber strength in ANSI O5.1 multiplied by the appropriate strength factor in **NESC** Table 261-1 (Resisting Moment). The Applied Moments from the groundline (Mgroundline) to the top of the pole (M_1, M_2, M_3, etc.) must be compared to the Resisting Moments from the groundline to the top of the pole to determine the point of maximum stress.

- Exception:

  Unguyed naturally grown wood poles (not laminated wood poles) 55 ft or less in total length (total pole length, not height above ground) acting as single-based structures or unbraced multiple-pole structures (single pole structures, not H-frame poles with X-braces) only require checking the stress point at the groundline (Mgroundline only).

  Guyed wood poles (any length) require checking the stress point at the points of attachment for guys and guy struts. See Rules 261A2c and 261C2 for strength requirements of guys and guyed wood poles.

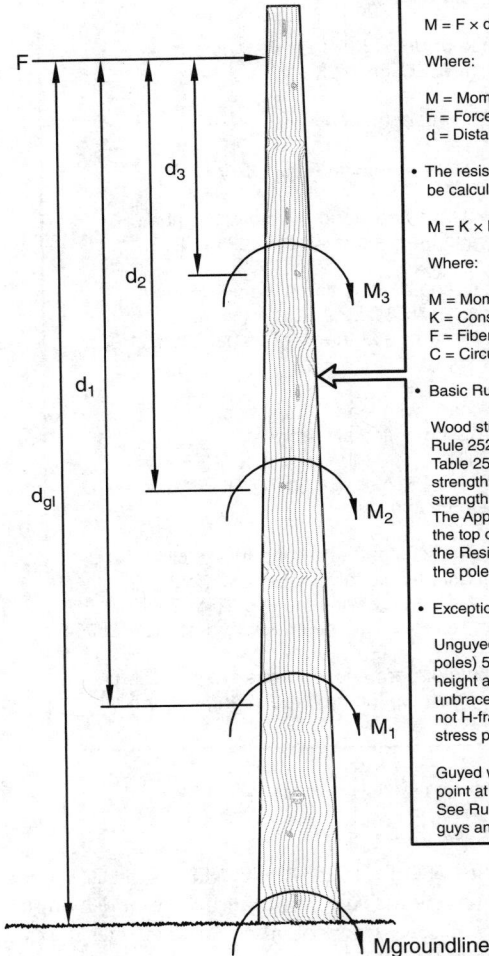

Fig. 261-5. Example of wood pole point of maximum stress (Rules 261A2a and 261A2b).

261E. Insulators.   Insulators are covered in Sec. 27. See Rule 277 for strengths of line insulators and Rule 279 for strengths of guy and span insulators.

261F. Strength of Pin-Type or Similar Construction and Conductor Fastenings.
Per Rule 261F1a, the longitudinal strength of insulator pins and conductor ties can be determined by applying the loads in Rule 252 multiplied by the load factors in Rule 253, or by applying 700 lb to the pin, whichever is greater. A tangent structure can have unbalanced longitudinal loading if the spans on each side of the structure

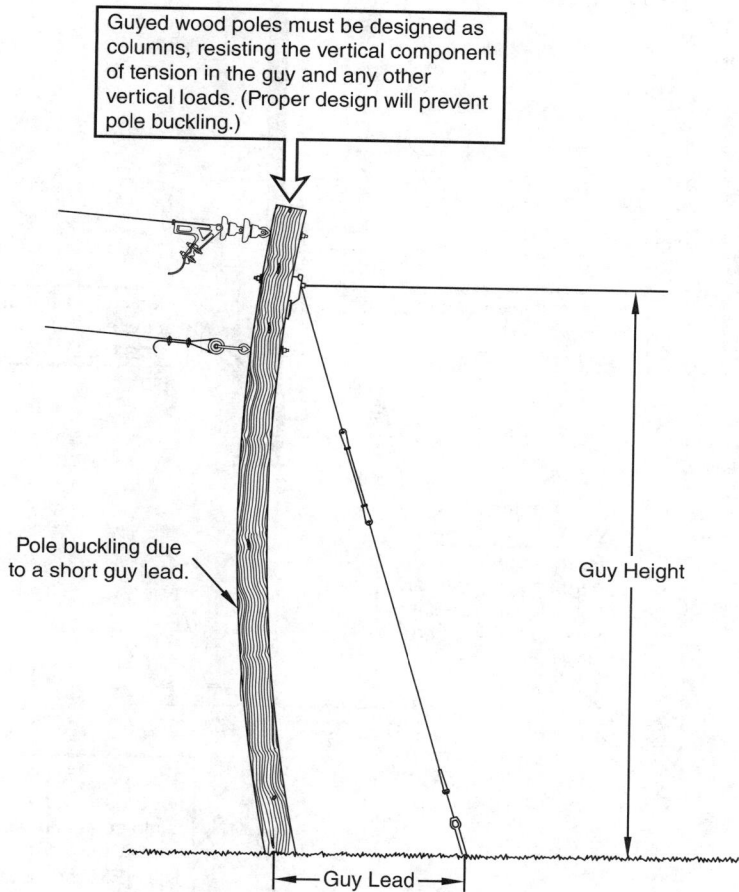

Guyed wood poles must be designed as columns, resisting the vertical component of tension in the guy and any other vertical loads. (Proper design will prevent pole buckling.)

Pole buckling due to a short guy lead.

Guy Height

Guy Lead

Fig. 261-6. Strength of guyed wood poles (Rule 261A2c).

are not equal or if the span on one side is loaded with ice and the span on the other side has dropped its ice. Rule 261F1b specifies a construction method for meeting Rule 261F1a. Using double wood pins and ties is a construction method that is acceptable for meeting the longitudinal strength requirements of Grade B construction. The application of double wood pins in Rule 261F1b corresponds with the application of double wood arms in Rule 261D5a(2). Both rules are for conductors with tensions limited to 2000 lb. Wood insulator pins were once commonly used and occasionally can still be found on older lines. Steel insulator pins are used in new construction. The **NESC** does not specify a construction application using steel pins. It is common to see steel pins used in place of wood pins on double crossarm construction. Rule 261F1c relates to Rule 261A5. Rule 261F1d relates to Rule 261A4. Rule 261F2 relates to Rule 261D5c. Rule 261F3 states that single conductor

Wood reinforcement
at the groundline
(wood stub)

Metal reinforcement
at the groundline
(steel truss)

Reinforcements or
permanent splices are
permitted provided they
develop the required
strength of the pole.

A pole top extension is
not considered a
reinforcement or splice.

The **NESC** does not
specifically address pole
top extensions. Rule
012C, which requires
accepted good practice,
must be applied. Use of
a pole top extension
requires careful review of
the strength and loading
line design calculations
used to size the original
pole.

Permanent splice
at any section
along the pole

See
Photo(s)

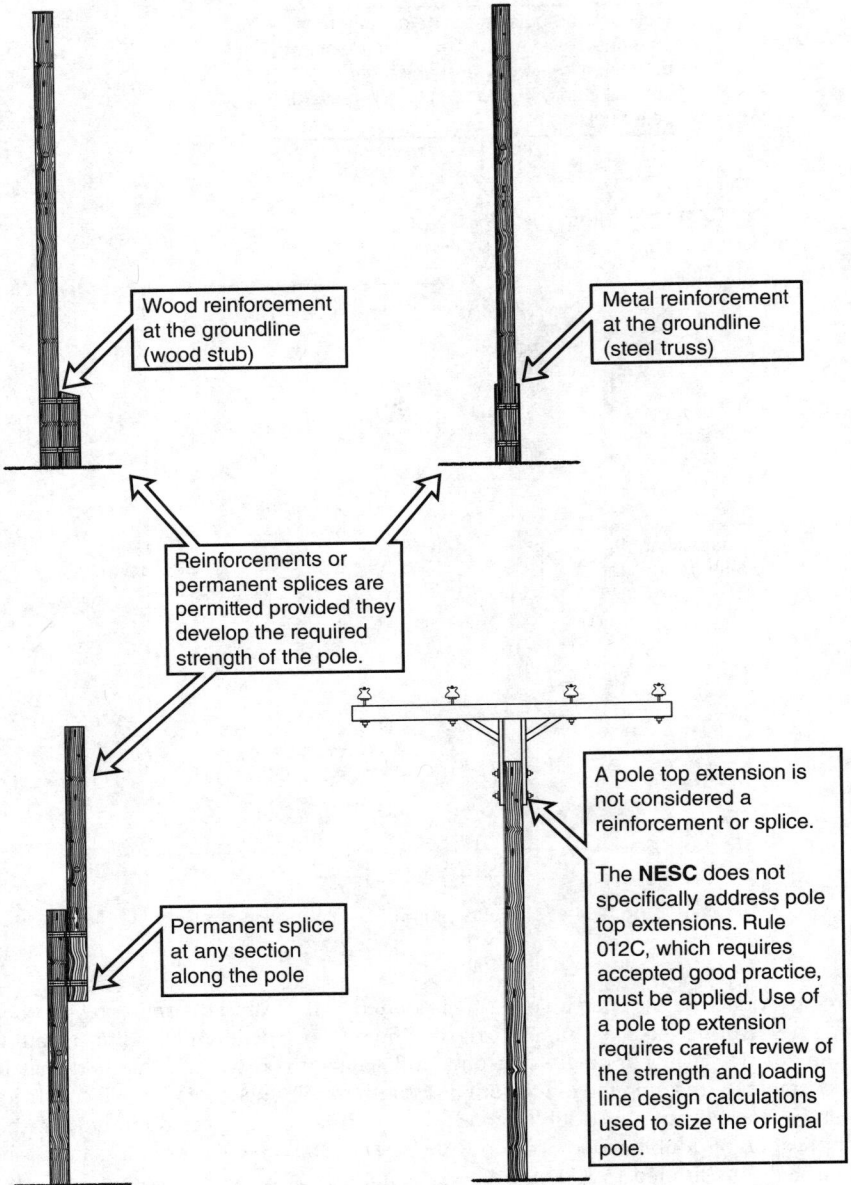

Fig. 261-7. Examples of spliced and reinforced wood poles (Rule 261A2d).

> **Strength of fiber-reinforced polymer structures:**
>
> • Apply loads of Rule 252.
>
> • Multiply loads of Rule 252 times the appropriate load factor in **NESC** Table 253-1.
>
> • Determine the strength rating (using 5th percentile strength or less).
>
> • Multiply the strength rating times the appropriate strength factor in **NESC** Table 261-1 (this is the permitted load).
>
> ← • The applied loads with load factors must not exceed the permitted load. (Buckling needs to be considered.)
>
> • Reference Rules 261C and 264 when guying is used.
>
> • The structure (any height) and supported facilities and equipment must be able to withstand, <u>without conductors</u>, the extreme wind loads of Rule 250C applied in any direction on the structure (see Rule 250C).
>
> • Reinforcements or permanent splices are permitted provided they develop the required strength of the structure.

See Photo(s)

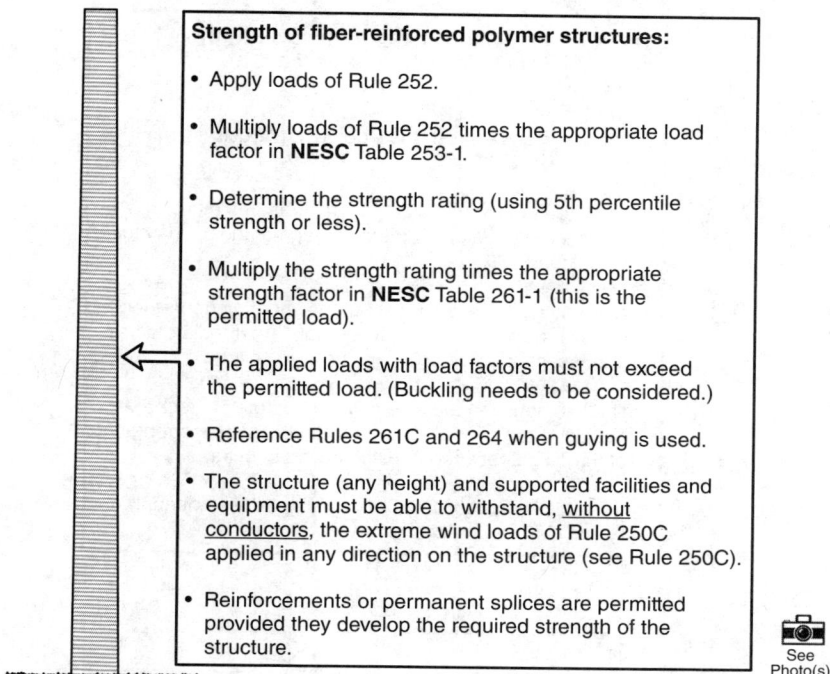

Fig. 261-8. Strength requirements of fiber-reinforced polymer structures (Rule 261A3).

supports used instead of double wood pins must have the equivalent strength of double wood pins. An example of wood pins is shown in Fig. 261-19.

261G. Armless Construction. The Code rules for armless construction are outlined in Fig. 261-20.

261H. Open Supply Conductors and Overhead Shield Wires. Rule 261H1 specifies tension limits for open (e.g., bare) supply conductors and overhead shield wires. The tension limits in Rule 261H1 should be entered into a conductor sag and tension program. Rule 261H1a(1) applies to the loaded condition of the conductor per Rule 250B which uses the heavy, medium, light, and warm island loading districts. Rule 261H1a(2) applies to the loaded condition of the conductor per Rules 250C (extreme wind) and 250D (extreme ice with concurrent wind). Rule 250B applies to all poles. The application of Rules 250C and 250D only applies to structures (e.g., poles) and supported facilities (e.g., conductors, static wires, messengers, cables, etc.) more than 60 ft above ground or water level. See Rules 250C and 250D for a discussion. Rule 261H1b addresses Aeolian vibration damage to conductors and related hardware. Aeolian vibration is a low-amplitude, high-frequency vibration that can cause failure of the conductor and related hardware. Rule 261H1b lists four Aeolian vibration mitigation methods. Three of the methods (vibration control devices, stress-reduction devices, and self damping conductors or vibration

**Transverse structure strength where side guys
are required but can only be installed at a
distance:**

- The distance between the side guyed structures must
  not be over 800'.

- The line between the side guyed structures must be
  substantially straight and the average of the spans must
  not exceed 150'.

- The side guyed poles must be designed to support the
  transverse load of the entire section. Intermediate poles
  are assumed not to carry any transverse load.

- The line between the side guyed structures must be
  designed to Grade B except for the transverse strength
  of the poles between the side guyed structures.

- Rule 261A4 applies to Grade B only; it does not apply to
  Grade C.

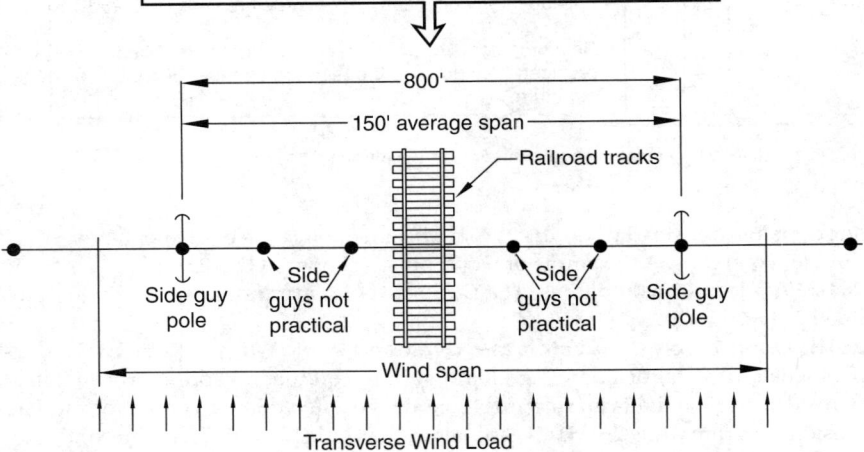

Fig. 261-9. Transverse structure strength where side guys are required but can only be installed at a distance (Rule 261A4).

resistant conductors) involve selecting special line hardware or special conductors. The fourth method in Rule 261H1b(4) involves reducing conductor tension and can be applied to various conductor types and does not involve special line hardware. If Rule 261H1b(4) is selected as the Aeolian vibration mitigation method, Rule 261H1c provides the tension limits to be used. The tension limits of Rules 261H1a and 261H1c are outlined in Fig. 261-21.

Longitudinal strength requirements for sections of higher grade in lines of a lower grade construction:

- The distance between structures with the required longitudinal strength must not be over 800'.

- The distance between the higher grade section and the structure with the required longitudinal strength must not exceed 500'.

- The line between the structures with the required longitudinal strength must be designed to Grade B transverse strength and stringing requirements.

- The line between the structures with the required longitudinal strength must be approximately straight or guyed.

- The structures with the required longitudinal strength may be guyed structures (or self-supporting structures).

- Rule 261A5 applies to Grade B only; it does not apply to Grade C.

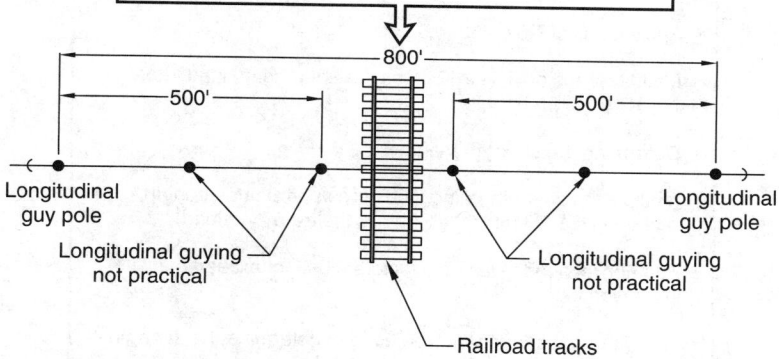

Fig. 261-10. Longitudinal strength requirements for sections of higher grade in lines of a lower grade construction (Rule 261A5).

The conductor manufacturer may also specify tension limits. See the sample sag and tension chart at the beginning of Sec. 23 for a discussion of **NESC** tension limits and user-defined design conditions.

Rule 261H2 specifies strength requirements for splices, taps, dead-end fittings, and associated hardware. Rules 261F and 261M provide similar requirements. Rule 261H2 provides special requirements for crossing spans and deadends. The requirements for location of splices and taps in crossing and adjacent spans are outlined in Fig. 261-22.

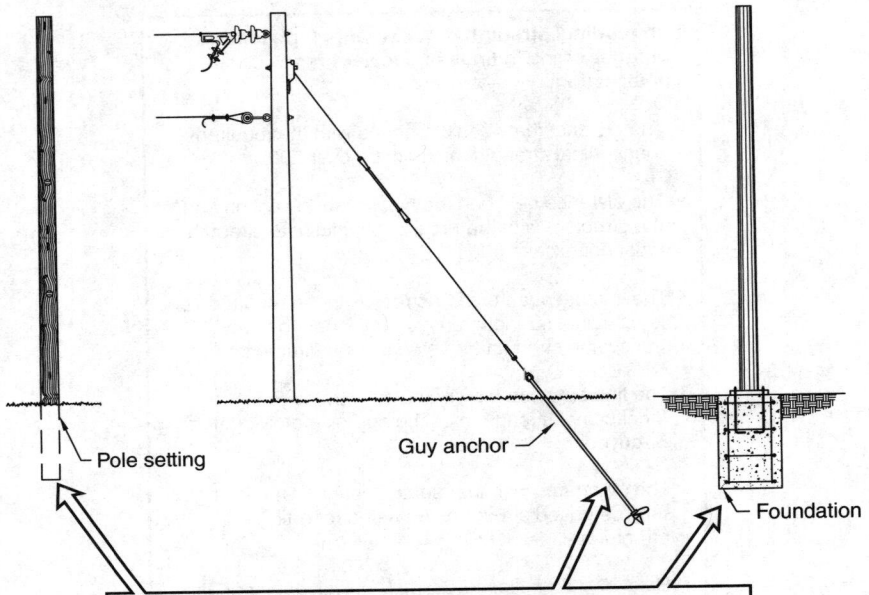

Strength of foundations, settings, and guy anchors:

- Apply loads of Rule 252.

- Multiply loads of Rule 252 times the appropriate load factor in **NESC** Table 253-1.

- Determine the strength rating.

- Multiply the strength rating times the appropriate strength factor in **NESC** Table 261-1 (this is the permitted load).

- The applied loads with load factors must not exceed the permitted load.

- Design or experience may be used to determine the strength of foundations, settings, and guy anchors.

Fig. 261-11. Strength of foundations, settings, and guy anchors (Rule 261B).

261I. Supply Cable Messengers.   The Code rules for strength of supply cable messengers are outlined in Fig. 261-23.

261J. Open-Wire Communication Conductors.   Open-wire communications circuits have typically been replaced with insulated communication cables or fiber-optic cables. The rules for open-wire communication circuits continue to remain in the Code. Wire tension limits for open-wire communications in Grade B and C construction are the same as the wire tension limits for supply conductors in Rule 261H1. An example of open-wire communication conductors is shown in Fig. 261-24.

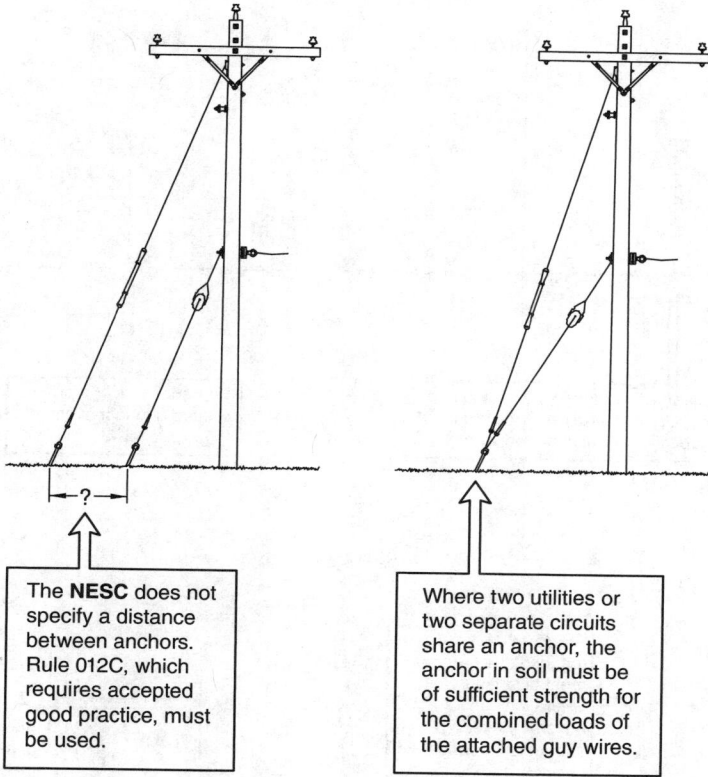

The **NESC** does not specify a distance between anchors. Rule 012C, which requires accepted good practice, must be used.

Where two utilities or two separate circuits share an anchor, the anchor in soil must be of sufficient strength for the combined loads of the attached guy wires.

Fig. 261-12. Examples of multiple guy anchors and multiple guy wires attached to a single guy anchor (Rule 261B).

261K. Communication Cables and Messengers. The Code rules for strength of communication cables and messengers are outlined in Fig. 261-25.

261L. Paired Metallic Communication Conductors. Rule 261L provides requirements for paired metallic communication conductors supported on messengers and paired metallic communication conductors not supported on messengers in addition to the requirements in Rule 261K. Paired metallic conductors are typically used by telephone utilities.

261M. Support and Attachment Hardware. Rule 261M applies to support and attachment hardware (e.g., nuts, bolts, etc.) not covered in Rule 261F (pin-type or similar construction and conductor fastenings) or Rule 261H2 (splices, taps, dead-end fittings, and associated attachment hardware). Appropriate loads, load factors, and strength factors apply to support hardware.

261N. Climbing and Working Steps and Their Attachments to the Structure. Rule 261N provides strength requirements for climbing devices (e.g., steps, ladders, platforms, and attachments). The load required is 300 lb for the weight of a

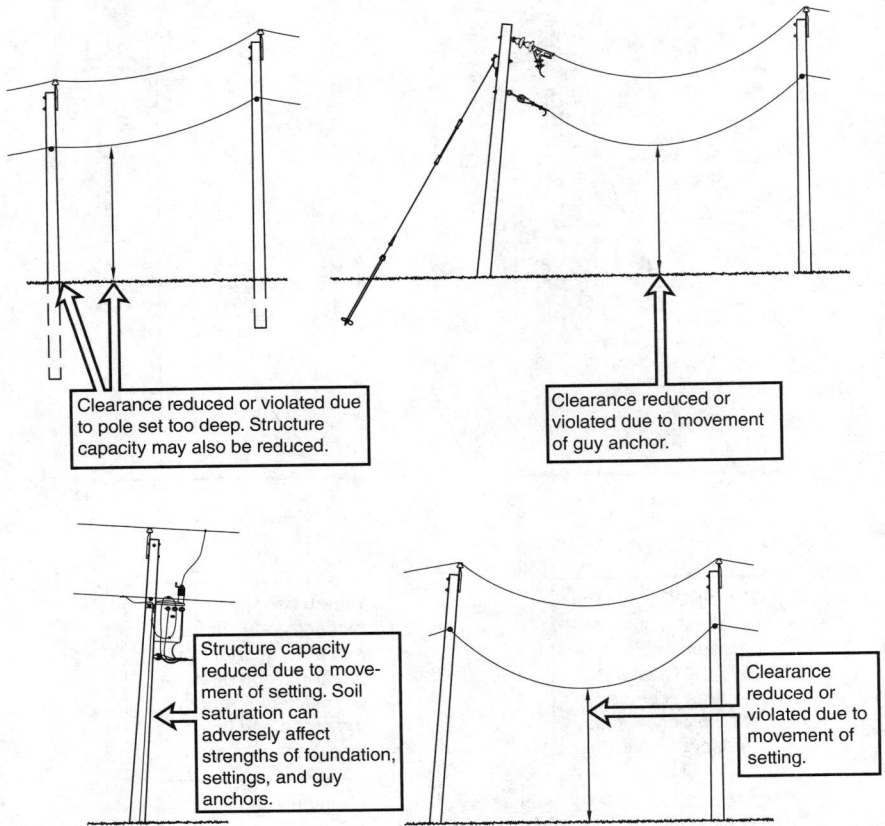

Fig. 261-13. Examples of excessive movement of foundations, settings, and guy anchors (Rule 261B).

lineworker including the weight of the items the lineworker carries on a structure unless the owner determines another value. Instead of derating the material strength of the steps, ladders, platforms, etc., a load factor of 2.0 is provided which increases the 300 lb load to 600 lb. The steps, ladders, platforms, etc. must be capable of supporting this load.

262. NUMBER 262 NOT USED IN THIS EDITION

263. GRADE N CONSTRUCTION

Grade N is the lowest grade of construction. Grade N supply and communication lines and structures have to withstand expected loads, including line personnel working on the structure. No load or strength factors are required for Grade N. Said another way, the line or structure is designed to withstand the expected loads and the load factors

Wood, Reinforced Concrete, or Fiber-Reinforced Polymer Structure:

Only the guy is assumed to restrain the dead-end wire tension as the wood, reinforced concrete, or fiber-reinforced polymer structure is assumed to be flexible and the guy is not. The pole acts as a strut (column) only unless the structure is sufficiently rigid (e.g., the pole has approximately the same deflection as the guy wire). Excessive movement of guys (or anchors) can reduce clearance or structure capacity.

Metal or Prestressed Concrete Structure:

The guy must be considered an integral part of the structure. Both pole and guy work together to restrain the dead-end wire tension.

Dead-end wire tension

Dead-end wire tension

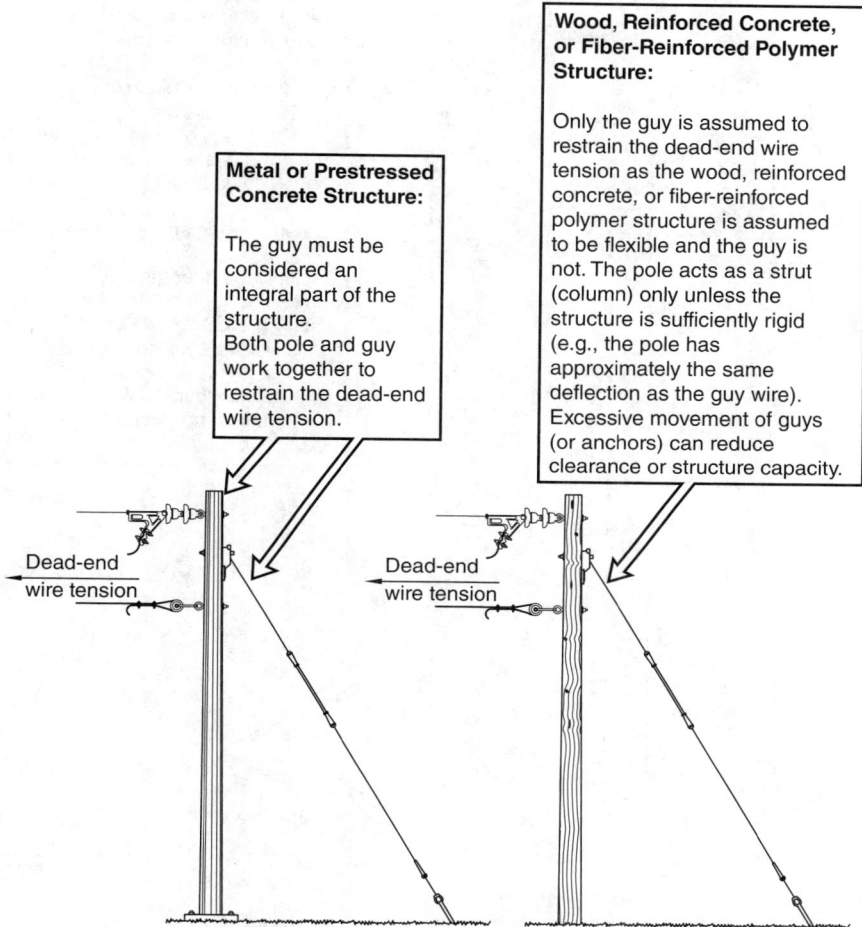

Fig. 261-14. Strength of guys and guy insulators (Rule 261C).

and strength factors are equal to 1.0. Rules 263A through 263I contain minimum conductor size requirements and requirements for typical construction practices.

264. GUYING AND BRACING

264A. Where Used. Guys and braces are used when a pole does not have sufficient strength alone. Guys and requirements for guyed poles are also covered in Rules 261A2a (Exception 1), 261A2c, and 261C. Guy insulators are covered in Rule 279. Conductors and power and communication cables on messengers cannot be used as guy wires. Conductors and power and communication cables supported

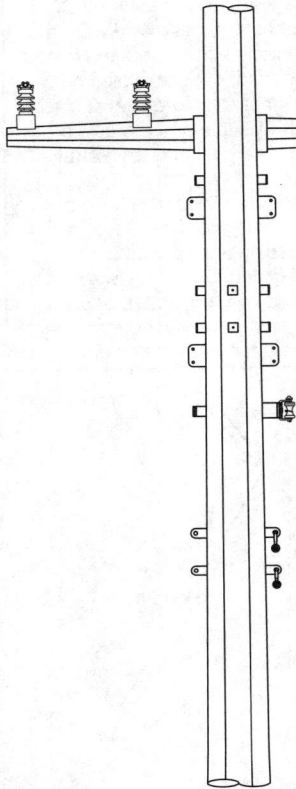

Strength of concrete and metal crossarms and braces:

- Apply loads of Rule 252.

- Multiply loads of Rule 252 times the appropriate load factor in **NESC** Table 253-1.

- Determine the strength rating.

- Multiply the strength rating times the appropriate strength factor in **NESC** Table 261-1 (this is the permitted load).

- The applied loads with load factors must not exceed the permitted load.

See Photo(s)

Fig. 261-15. Strength of concrete and metal crossarms and braces (Rule 261D1).

on messengers add vertical (weight) and horizontal (transverse wind) loads to structures and additional horizontal (transverse) loads at angle structures. It is possible for the loads due to line angles of multiple conductors on the same pole to cancel or offset each other but using conductors or power and communication cables on messengers as guy wires is not an acceptable practice. Guys are used to limit the increase in conductor sags (which can cause clearance to be decreased) and guys are used to provide support for unbalanced loads at the following locations:

- Corners
- Angles
- Deadends
- Large differences in span lengths
- Changes in grades of construction

Examples of guying and bracing are shown in Fig. 264-1.

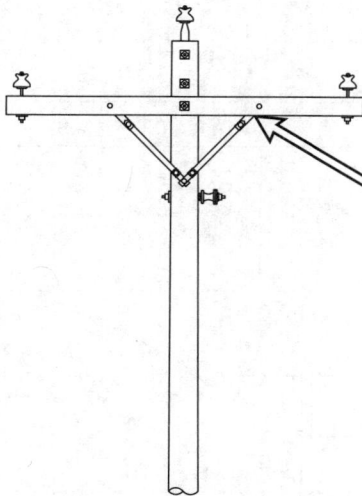

Strength of wood crossarms and braces:

- Apply loads of Rule 252.

- Multiply loads of Rule 252 times the appropriate load factor in **NESC** Table 253-1.

- Determine the strength rating:
  - For solid sawn or laminated wood, use the appropriate ultimate fiber strength of the material.

- Multiply the strength rating times the appropriate strength factor in **NESC** Table 261-1 (this is the permitted stress level).

- The applied loads with load factors must not exceed the permitted stress level.

- Crossarms of select Southern Pine or Douglas Fir must meet minimum **NESC** dimension values.

See Photo(s)

Fig. 261-16. Strength of wood crossarms and braces (Rule 261D2).

264B. Strength. Rule 264B applies to the strength of guys. Rule 217C addresses the protection and marking requirements of guys. The **Code** rules for guy strength are outlined in Fig. 264-2.

264C. Point of Attachment. The guy or brace should be attached to the structure as near as practical to the center of the conductor load. An individual guy for each conductor is typically not practical for small conductors or small angles but may be required for large conductors or large angles. Special consideration is given to insulation reduction for lines exceeding 8.7 kV. Guy insulators used exclusively for BIL insulation are discussed in Rule 279A2. The **Code** rules for the point of attachment of the guy or brace are outlined in Fig. 264-3.

264D. Guy Fastenings. The strength rating of a guy fastening is just as important as the strength rating of the guy wire itself. Numerous guy attachment methods exist including guy plates, pole bands, wrap guys, etc. The **Code** requires the use of guy thimbles and guy shims for certain guying conditions. The entire guy and anchor system, including guy fastenings, should be reviewed to determine the weak link in the guy and anchor system. A guy is only as strong as the weakest link connected to the guy wire. An example of the weakest link in a guy and anchor system is shown in Fig. 264-4.

264E. Electrolysis. Electrolysis (corrosion) of the anchor or anchor rod can jeopardize the strength of the structure. Corrosion normally occurs when two dissimilar metals exist in the soil (e.g., a copper ground rod and a steel anchor rod).

Crossarms of select Southern
Pine or Douglas Fir must meet
minimum **NESC** dimension values.
Crossarms of other species may
be used provided they have equal
strength.

Grade B	Grade C

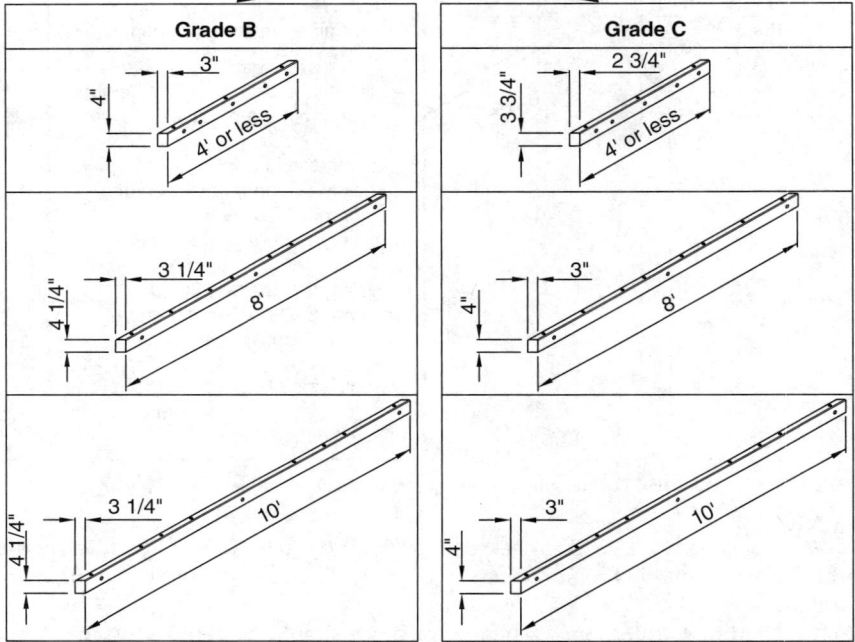

Fig. 261-17. Minimum wood crossarm dimensions for Southern Pine and Douglas Fir (Rule 261D2b).

The soil conditions can have an effect on the rate of corrosion. Typical solutions
to the problem are using guy insulators or using steel-coated copper ground rods.
Cathodic protection in the form of a sacrificial anode is another solution. Guy
insulators used exclusively for limiting corrosion are discussed in Rule 279A2.

264F. Anchor Rods. Rule 264F1 applies to the anchor rod only. Rule 264F2
applies to the anchor and rod assembly. Rules 261B and 264B also relate to the
strength of guy anchors. The **Code** rules for strength of anchor rods and anchors
provided in Rule 264F are outlined in Fig. 264-5.

Strength of fiber-reinforced polymer crossarms and braces:

- Apply loads of Rule 252.

- Multiply loads of Rule 252 times the appropriate load factor in **NESC** Table 253-1.

- Determine the strength rating (using 5th percentile or less).

- Multiply the strength rating times the appropriate strength factor in **NESC** Table 261-1 (this is the permitted load).

- The applied loads with load factors must not exceed the permitted load.

See Photo(s)

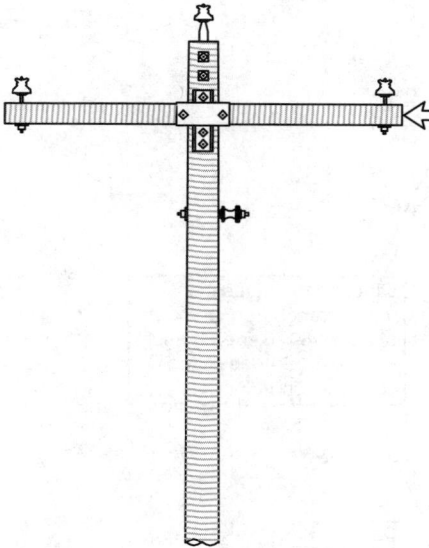

Fig. 261-18. Strength of fiber-reinforced polymer crossarms and braces (Rule 261D3).

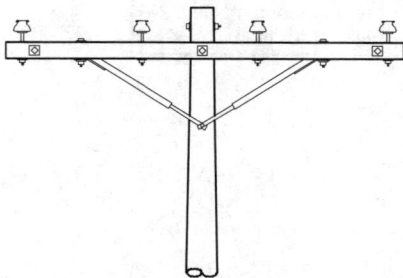

Example of steel insulator pin used in modern line construction.

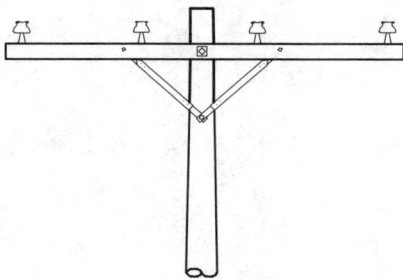

Example of wood insulator pin used in older line construction.

See Photo(s)

Fig. 261-19. Example of wood and steel pins (Rule 261F).

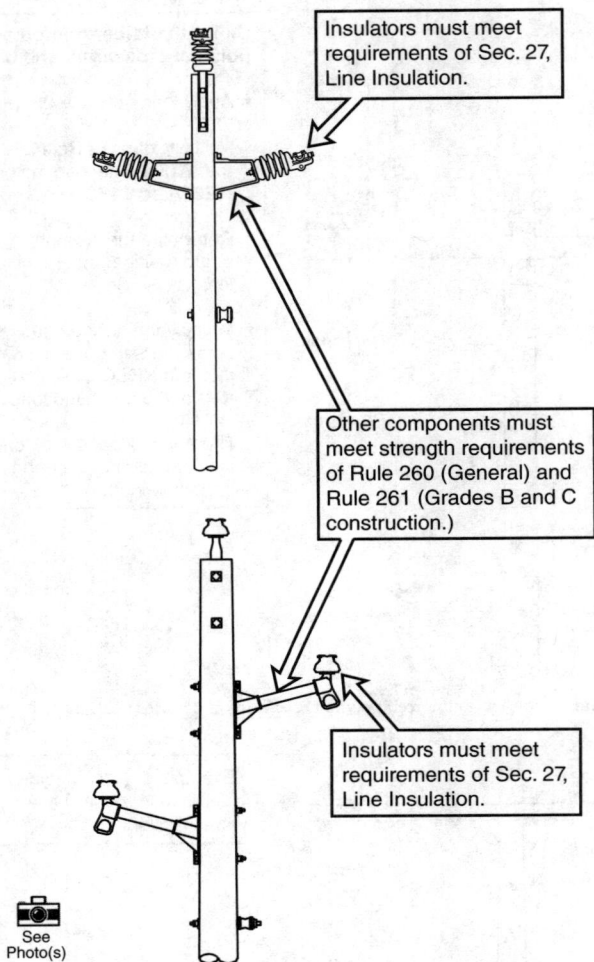

Insulators must meet requirements of Sec. 27, Line Insulation.

Other components must meet strength requirements of Rule 260 (General) and Rule 261 (Grades B and C construction.)

Insulators must meet requirements of Sec. 27, Line Insulation.

See Photo(s)

Fig. 261-20. Strength requirements for armless construction (Rule 261G).

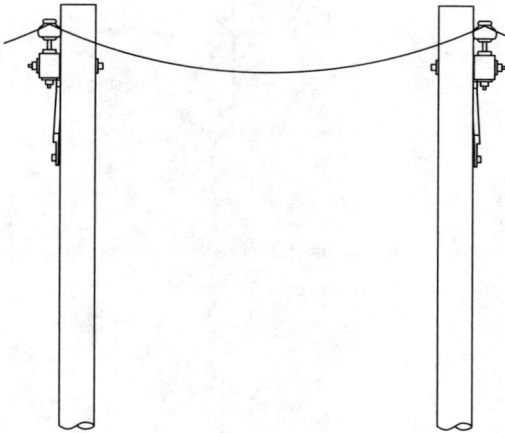

Tension limits for open supply conductors and overhead shield wires:

- The tension in the conductor must not be more than 60% of the rated breaking strength of the conductor when the loads of Rule 250B (heavy, medium, light, or warm islands) in Rule 251 are applied to the supply conductor using a load factor of 1.0.

- The tension in the conductor must not be more than 80% of the rated breaking strength of the conductor when the loads of Rules 250C (extreme wind) and 250D (extreme ice with concurrent wind) in Rule 251 are applied (if applicable) to the supply conductor using a load factor of 1.0.

- If reducing tension is used to mitigate Aeolian vibration, the tension in the conductor must not be more than 35% initial, at the applicable temperature in **NESC** Table 251-1 without external loading.

- If reducing tension is used to mitigate Aeolian vibration, tension in the conductor must not be more than 25% final, at the applicable temperature in **NESC** Table 251-1 without external loading.

- Notes are provided defining initial tension, final tension, and stating that the above limitations may not protect the conductor from damage due to Aeolian vibration.

Fig. 261-21. Tension limits for open supply conductors and overhead shield wires (Rules 261H1a and 261H1c).

Condition:
Lines crossing without a common supporting structure.

Requirement:
Splices should be avoided in crossing and adjacent spans. If impractical, specific strength rules must be met. Taps should be avoided in crossing spans. If required, specific strength rules must be met.

Condition:
Lines crossing on a common supporting structure.

Requirement:
Splices should be avoided in crossing and adjacent spans. If impractical, specific strength rules must be met. Taps should be avoided in crossing spans. If required, specific strength rules must be met.

Adjacent spans

Crossing spans

Adjacent spans

Crossing pole

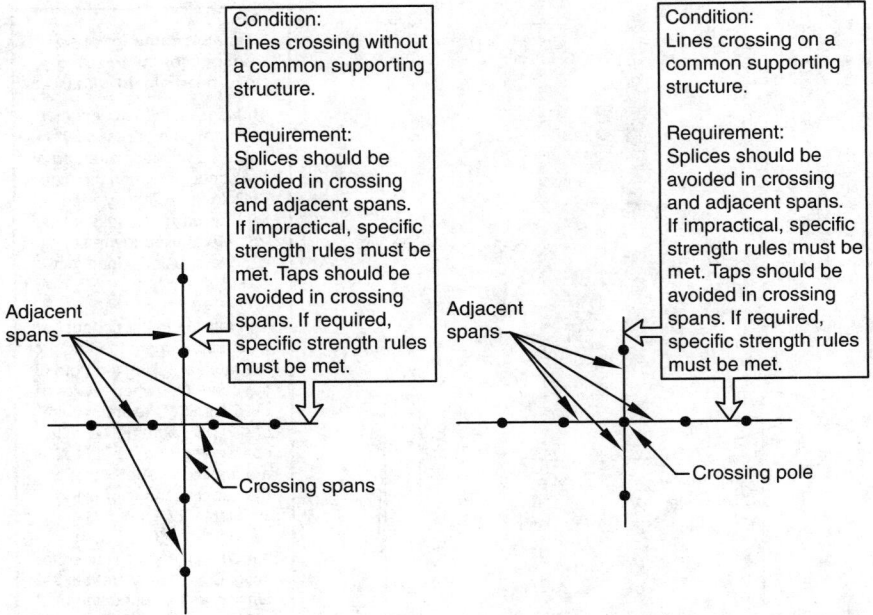

Fig. 261-22. Location of splices and taps in crossing and adjacent spans (Rule 261H2).

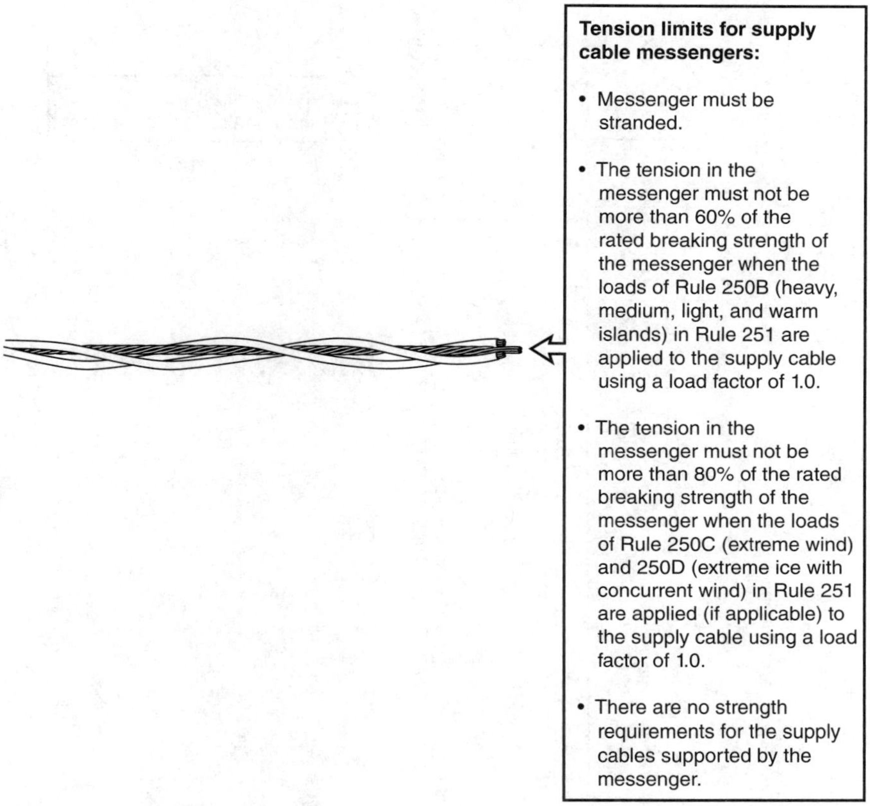

Tension limits for supply cable messengers:

- Messenger must be stranded.

- The tension in the messenger must not be more than 60% of the rated breaking strength of the messenger when the loads of Rule 250B (heavy, medium, light, and warm islands) in Rule 251 are applied to the supply cable using a load factor of 1.0.

- The tension in the messenger must not be more than 80% of the rated breaking strength of the messenger when the loads of Rule 250C (extreme wind) and 250D (extreme ice with concurrent wind) in Rule 251 are applied (if applicable) to the supply cable using a load factor of 1.0.

- There are no strength requirements for the supply cables supported by the messenger.

Fig. 261-23. Tension limits for supply cable messengers (Rule 261I).

Typical open-wire communication conductors.

See Photo(s)

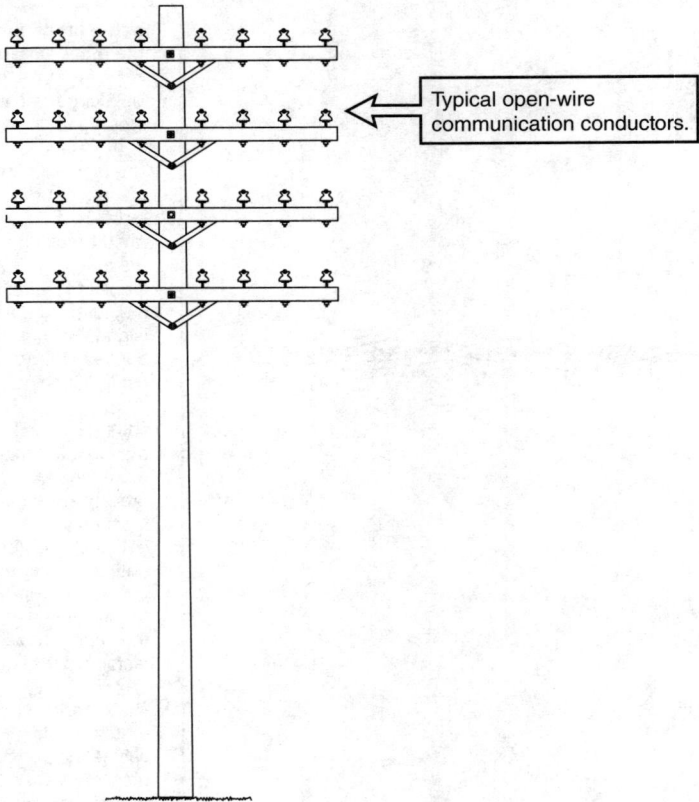

Fig. 261-24. Example of open-wire communication conductors (Rule 261J).

Tension limits for communication cables and messengers:

- There are no strength requirements for the communication cables supported by the messenger.

- The tension in the messenger must not be more than 60% of the rated breaking strength of the messenger when the loads of Rule 250B (heavy, medium, light, and warm islands) in Rule 251 are applied to the communications cable using a load factor of 1.0.

- The tension in the messenger must not be more than 80% of the rated breaking strength of the messenger when the loads of Rule 250C (extreme wind) and 250D (extreme ice with concurrent wind) in Rule 251 are applied (if applicable) to the communications cable using a load factor of 1.0.

- Self-supporting cables must meet the strength requirements for communication messengers. Note that lesser tensions may be needed to meet self-supporting fiber optic cable operational reliability.

- Metallic paired communication conductors have additional requirements in Rule 261L.

Fig. 261-25. Tension limits for communication cables and messengers (Rule 261K).

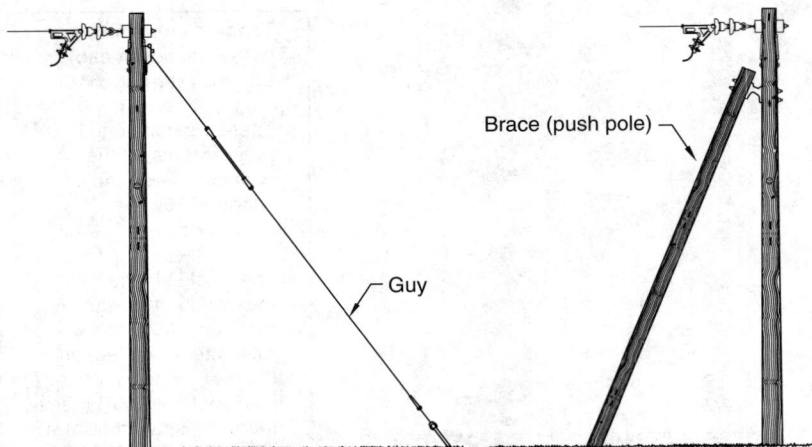

Fig. 264-1. Examples of guying and bracing (Rule 264A).

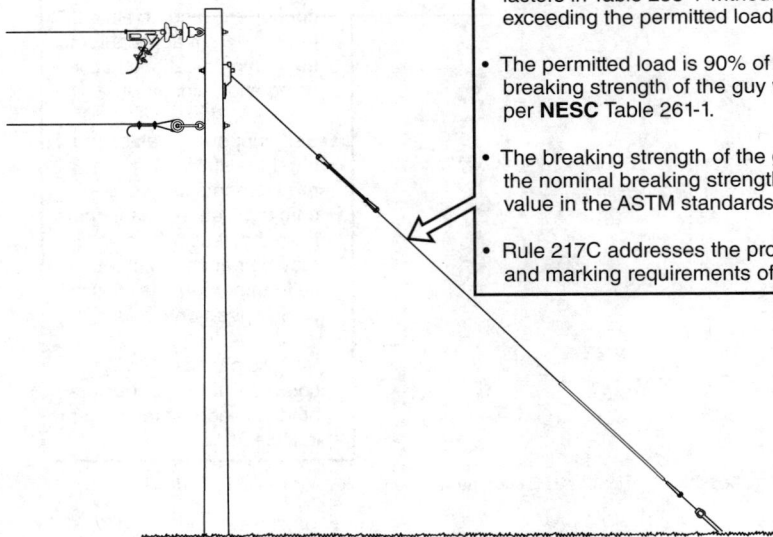

Strength of guys:

- Designed to withstand the loads in Rule 252 multiplied by the load factors in Table 253-1 without exceeding the permitted load.

- The permitted load is 90% of the breaking strength of the guy wire per **NESC** Table 261-1.

- The breaking strength of the guy is the nominal breaking strength value in the ASTM standards.

- Rule 217C addresses the protection and marking requirements of guys.

Fig. 264-2. Guy strength requirements (Rule 264B).

Point of attachment as near as practical to center of the conductor load. On lines exceeding 8.7 kV, locations may be adjusted to minimize the reduction of insulation values.

Point of attachment as near as practical to center of the conductor load. On lines exceeding 8.7 kV, locations may be adjusted to minimize the reduction of insulation valves.

See Photo(s)

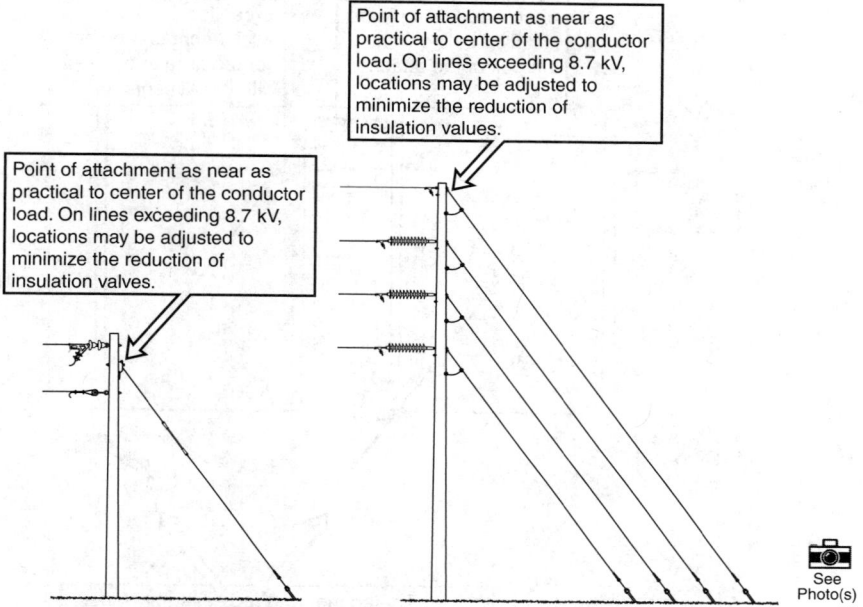

Fig. 264-3. Point of guy attachment (Rule 264C).

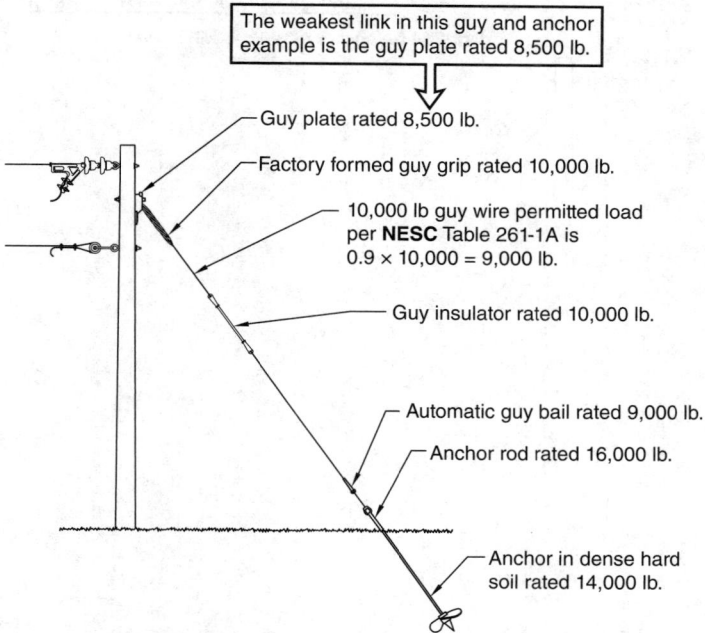

The weakest link in this guy and anchor example is the guy plate rated 8,500 lb.

Guy plate rated 8,500 lb.

Factory formed guy grip rated 10,000 lb.

10,000 lb guy wire permitted load per **NESC** Table 261-1A is $0.9 \times 10,000 = 9,000$ lb.

Guy insulator rated 10,000 lb.

Automatic guy bail rated 9,000 lb.

Anchor rod rated 16,000 lb.

Anchor in dense hard soil rated 14,000 lb.

Fig. 264-4. Example of the weakest link in a guy and anchor system (Rule 264D).

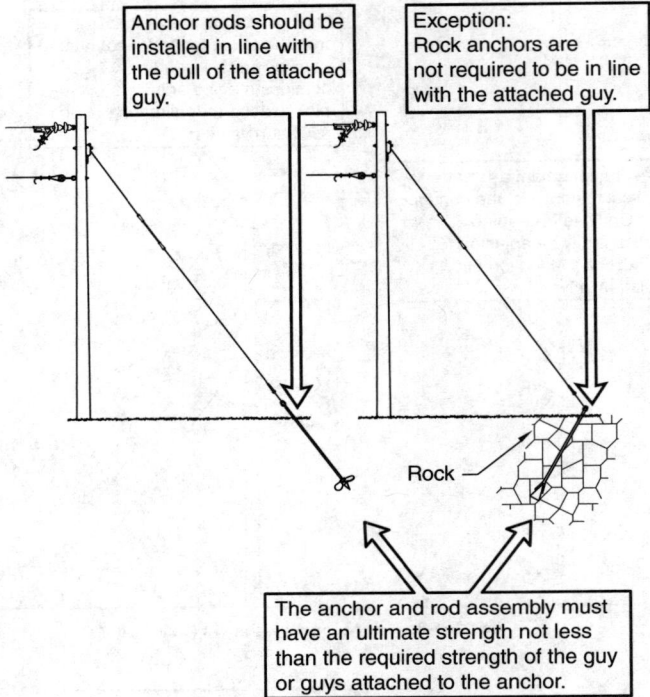

Anchor rods should be installed in line with the pull of the attached guy.

Exception:
Rock anchors are not required to be in line with the attached guy.

Rock

The anchor and rod assembly must have an ultimate strength not less than the required strength of the guy or guys attached to the anchor.

Fig. 264-5. Anchor rods (Rule 264F).

Section 27

Line Insulation

270. APPLICATION OF RULE

Line insulation in Sec. 27 applies to insulators for open conductor supply lines only, not for communication lines or secondary duplex, triplex, or quadruplex (230C3) cables. Section 27 does not apply to substation insulators as Sec. 27 is in Part 2, Overhead Lines. Note 1 references Rule 243C4. Rule 243C4 states that the strength requirements in Sec. 27 apply to all grades of construction. There are not separate Grade B and Grade C strength factors for insulators. Note 2 references Rule 242E for insulation requirements of neutral conductors. Rule 242E states that supply neutrals, which are effectively grounded and not located above supply conductors of more than 750 V to ground, do not need to meet any insulation requirements. Even though effectively grounded neutrals do not need to meet any insulation requirements, it is common to see neutral conductors attached to insulators for support and termination purposes. Also, even though the messengers of secondary duplex, triplex, and quadruplex (230C3) cables do not need to meet any insulation requirements, it is common to see the messengers of these cables attached to insulators for support and termination purposes. The Code rules related to the application of line insulation are outlined in Fig. 270-1.

271. MATERIAL AND MARKING

Supply circuit insulators must be manufactured and marked in accordance with the ANSI/NEMA C29 series of standards. An exception applies if performance and

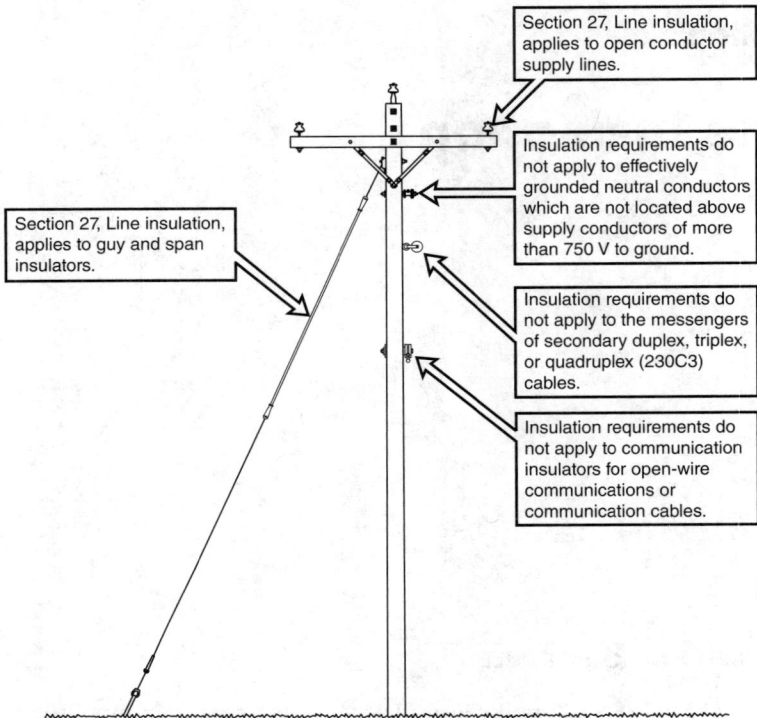

Fig. 270-1. Application of line insulation (Rule 270).

marking are appropriate for the application. Examples of supply circuit insulators are shown in Fig. 271-1.

Examples of identification marking on supply circuit insulators are shown in Fig. 271-2.

272. RATIO OF FLASHOVER TO PUNCTURE VOLTAGE

Rule 272 requires insulators to meet conditions and standards for flashover to puncture voltage ratios. A list of applicable standards is provided. When a standard does not exist for the flashover to puncture voltage ratio, Rule 272 specifies a not-to-exceed value of 75 percent with an exception that permits not more than 80 percent in areas of high atmospheric contamination. The flashover to puncture voltage ratio is used by insulator manufacturers but it is not always published in insulator catalog data. Insulator catalog data typically does include the low-frequency dry flashover value.

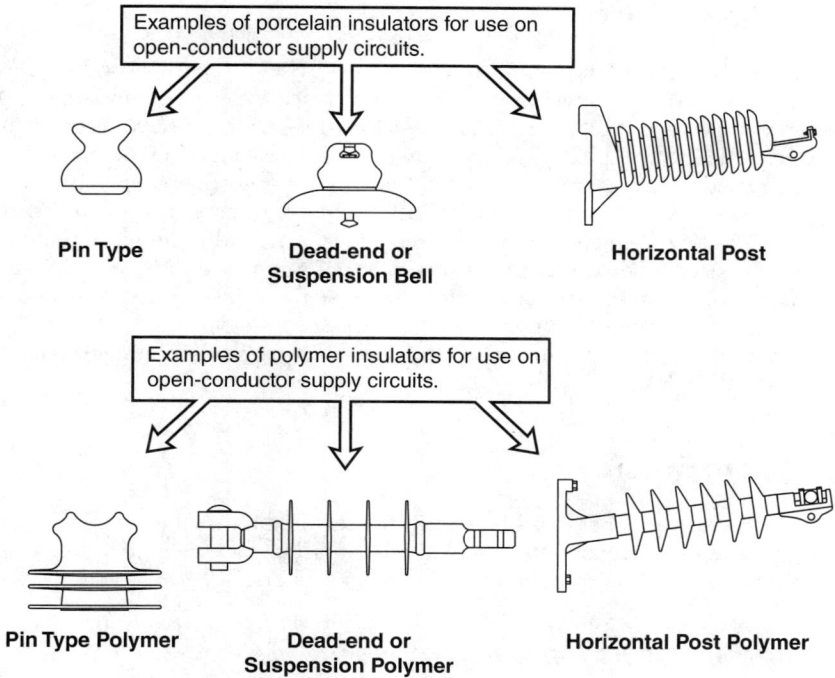

Examples of porcelain insulators for use on open-conductor supply circuits.

Pin Type

Dead-end or Suspension Bell

Horizontal Post

Examples of polymer insulators for use on open-conductor supply circuits.

Pin Type Polymer

Dead-end or Suspension Polymer

Horizontal Post Polymer

Fig. 271-1. Examples of supply circuit insulators (Rule 271).

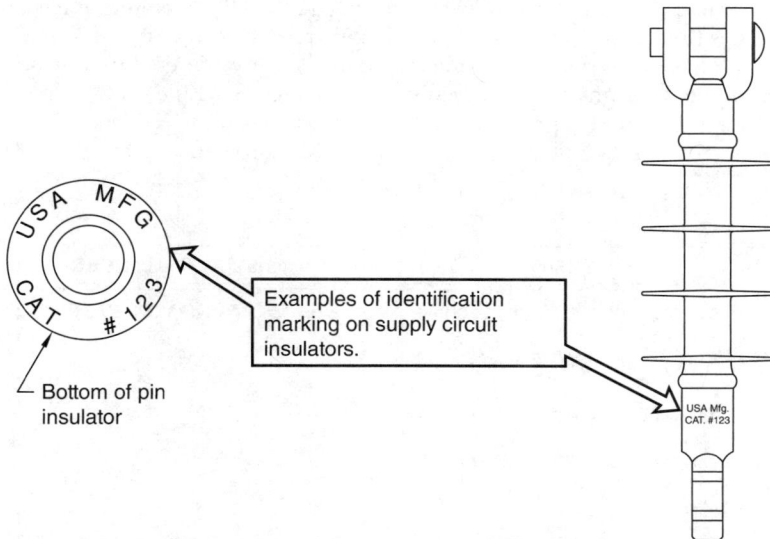

Examples of identification marking on supply circuit insulators.

Bottom of pin insulator

USA Mfg. CAT. #123

Fig. 271-2. Examples of identification marking on supply circuit insulators (Rule 271).

273. INSULATION LEVEL

Rule 273 specifies insulator dry flashover voltages in **NESC** Table 273-1 which must be compared to dry flashover tests done in accordance with ANSI Standard C29.1. Using dry flashover voltage ratings lower than the values in **NESC** Table 273-1 requires a qualified engineering study. Values higher than shown in **NESC** Table 273-1 must be used for areas with severe lightning, high atmospheric contamination (e.g., salt water fog, industrial plant pollution, etc.), or other unfavorable circumstances. The low-frequency dry flashover rating is typically the most common value referenced when comparing insulator specifications. Additional ratings are commonly published in insulator catalog sheets and in insulator specifications. Typical insulation level ratings are listed in Fig. 273-1.

In addition to electrical ratings, the **NESC** specifies mechanical strength ratings for insulators in Rule 277.

274. FACTORY TESTS

Insulator factory tests per ANSI Standards are required for each insulator or insulating part at or above 2.3 kV. Accepted good practice may be used where standards do not exist. An exception exists for guy insulators.

275. SPECIAL INSULATOR APPLICATIONS

Special consideration must be given to insulators for use on constant-current circuits (e.g., series street lighting). The voltage rating of the insulator must be based on the full load rated voltage of the supply transformer, not just the operating voltage of the circuit. See Rule 230H for a discussion and figure related to constant-current circuits.

Single-phase lines fed from three-phase lines are required to use insulators based on the three-phase line phase-to-phase voltage. See Rule 230G for an example and figure related to phase-to-phase and phase-to-ground voltages. Insulation requirements of single-phase circuits connected to three-phase circuits are shown in Fig. 275-1.

Rating	Values specified in **NESC**
• Low frequency 60 Hz dry flashover	Yes (see **NESC** Table 273-1)
• Low frequency 60 Hz wet flashover	No
• Critical impulse flashover - positive	No
• Critical impulse flashover - negative	No
• Total leakage distance	No

Fig. 273-1. Typical insulation level ratings (Rule 273).

The phase-to-phase voltage must be used to select the proper insulation level on both a three-phase circuit and a single-phase circuit fed from a three-phase line.

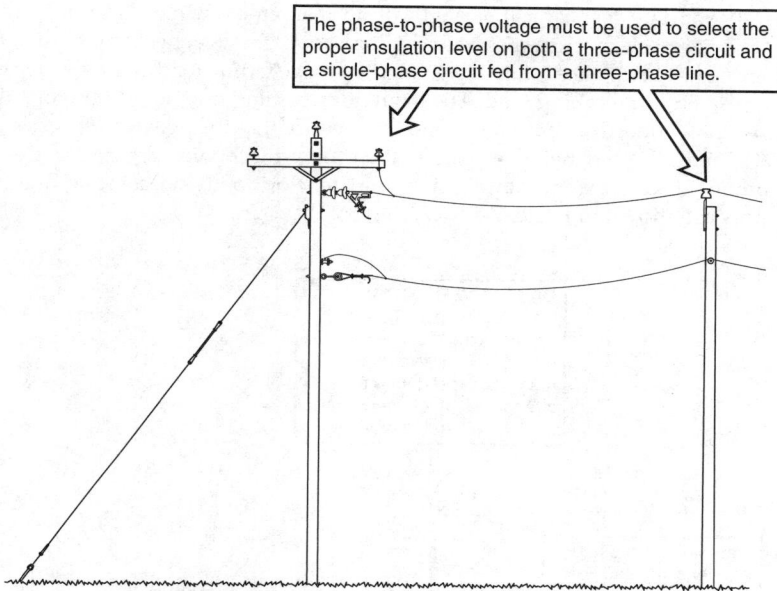

Fig. 275-1. Insulators of single-phase circuits connected to three-phase circuits (Rule 275).

276. NUMBER 276 NOT USED IN THIS EDITION

277. MECHANICAL STRENGTH OF INSULATORS

Rule 277 specifies the mechanical strength requirements for insulators. This is done in the form of strength factors that are applied to the insulators. Load factors for wire tension, wire vertical (weight) loads, and wire transverse (wind) loads are not applicable. Rule 277 requires insulators to withstand the loads in Rules 250, 251, and 252. Load factors are in Rule 253 and Rule 253 is not specified in Rule 277. Said another way, the load factor for the loads on insulators is 1.0. The strength factor percentages provided in **NESC** Table 277-1 do not distinguish between Grade B and Grade C construction. This is one reason why insulators have their own section (Sec. 27) rather than being included in Rule 261 and **NESC** Table 261-1. The hardware for insulators is covered in Rules 261M, 261F, 261H2, and **NESC** Table 261-1. The percent of strength ratings provided in **NESC** Table 277-1 are for use with Rule 250B (heavy, medium, light, and warm islands) loads and Rule 250C (extreme wind) and Rule 250D (extreme ice with concurrent wind) loads.

The strength factor percentages listed in **NESC** Table 277-1 are based on manufacturers ratings when determined in accordance with specific ANSI Standards for the type of insulator, the insulator material, and the type of load on the insulator. Per Footnote 4 to **NESC** Table 277-1, some insulators do not currently have a standard related to strength. In these cases, Rule 012C, which requires accepted good practice, must be applied.

The appropriate mechanical strength of an insulator that needs to be derated may not be very clear on an insulator catalog sheet. Catalog sheets may use terms like routine test load, maximum design load, average failing load, average breaking load, or maximum working load. The insulator manufacturer should be consulted or the ANSI Standards should be reviewed to verify that the percentages provided in **NESC** Table 277-1 are being applied to the proper mechanical strength rating.

Examples of cantilever, compression, and tension loads on ceramic line post insulators are shown in Fig. 277-1.

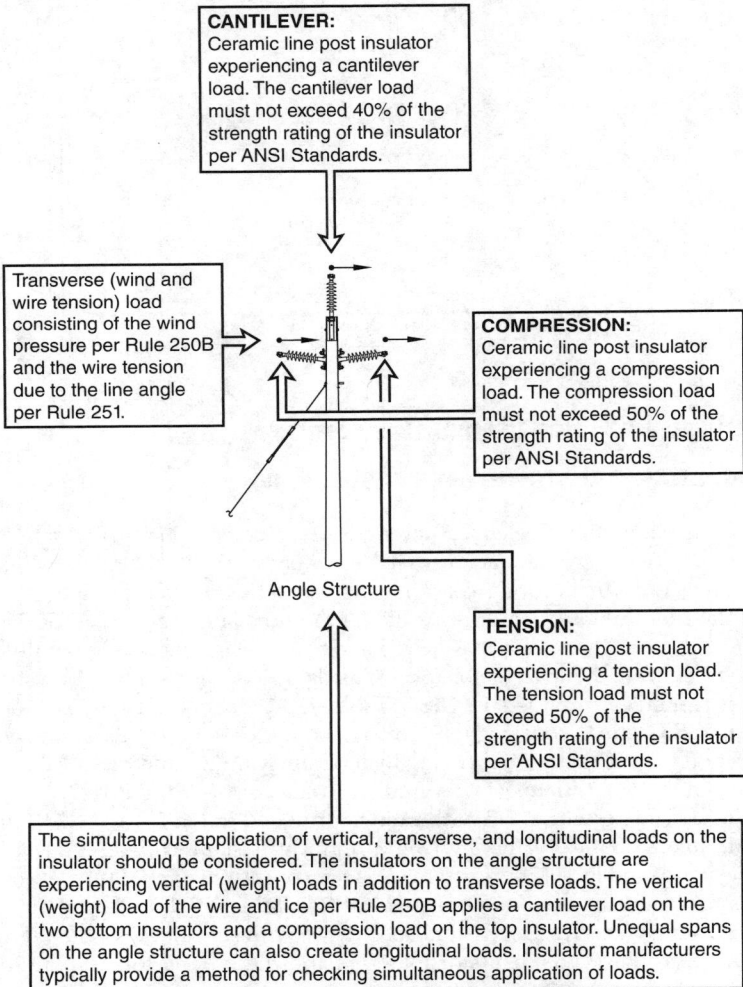

CANTILEVER:
Ceramic line post insulator experiencing a cantilever load. The cantilever load must not exceed 40% of the strength rating of the insulator per ANSI Standards.

Transverse (wind and wire tension) load consisting of the wind pressure per Rule 250B and the wire tension due to the line angle per Rule 251.

COMPRESSION:
Ceramic line post insulator experiencing a compression load. The compression load must not exceed 50% of the strength rating of the insulator per ANSI Standards.

Angle Structure

TENSION:
Ceramic line post insulator experiencing a tension load. The tension load must not exceed 50% of the strength rating of the insulator per ANSI Standards.

The simultaneous application of vertical, transverse, and longitudinal loads on the insulator should be considered. The insulators on the angle structure are experiencing vertical (weight) loads in addition to transverse loads. The vertical (weight) load of the wire and ice per Rule 250B applies a cantilever load on the two bottom insulators and a compression load on the top insulator. Unequal spans on the angle structure can also create longitudinal loads. Insulator manufacturers typically provide a method for checking simultaneous application of loads.

Fig. 277-1. Examples of cantilever, compression, and tension loads on ceramic line post insulators (Rule 277).

278. AERIAL CABLE SYSTEMS

Aerial cable systems primarily consist of covered conductors described in Rule 230D. The insulators supporting aerial cables must meet Rule 273. The covered conductors of aerial cable systems are considered bare conductors for electrical insulation requirements. The insulators and spacers of an aerial cable system must meet Rule 277. Rule 278B2 requires that the insulating spacers used in aerial cable systems withstand the loads specified in Sec. 25. Section 25 includes Rule 253, which contains load factors. The 50% of ultimate rated strength requirement in Rule 278B2 is for use with Rule 250B (heavy, medium, light, and warm islands) loads and the appropriate overload factors in Rule 253. The loads of 250C (extreme winds) and 250D (extreme ice with concurrent wind) are not addressed, therefore, Rule 012C, which requires accepted good practice, must be applied.

279. GUY AND SPAN INSULATORS

279A. Insulators.   Per Rules 215C2a and 215C2b, anchor and span guys must be effectively grounded or insulated (per the exceptions). If grounded, Rule 092C2 provides the methods for the grounding of guys. If insulated, the exceptions to Rules 215C2a and 215C2b provide requirements for the location of anchor and span guy insulators. See Fig. 279-1.

The Code rules for the material properties of guy insulators and the electrical and mechanical strength of guy insulators per Rule 279A1 are outlined in Fig. 279-2.

The requirements for mechanical strength of a guy insulator hinge on the word "required" strength of the guy. Guy insulators do not require the application of strength factors like line insulators in Rule 277. Guy insulators indirectly have load factors applied to them via the load factors that are applied to wire tension and wind loading. See Fig. 279-3.

Per Rule 279A2a, if a guy insulator is used exclusively for limiting corrosion of an anchor attached to a grounded guy on an effectively grounded system, then it is not classified as a guy insulator. Therefore, the electrical requirements in Rule 279A1b do not apply. The guy in this case is assumed to be grounded per Rule 215C2 using the grounding methods in Rule 092C2. The corrosion protection insulator must not reduce the mechanical strength of the guy. Additional requirements for guy insulators used to limit galvanic corrosion can be found in Rule 215C4. An example of a guy insulator used exclusively for limiting galvanic corrosion is shown in Fig. 279-4.

Per Rule 279A2b, if a guy insulator is used exclusively for meeting BIL requirements for a structure on an effectively grounded system, then it is not classified as a guy insulator. However, the strength requirement of the guy insulator must meet Rule 279A1c, and either a combination of Rules 215C2 and 279A1 or a combination of Rules 092C2 and 215C2 must be met. NESC Fig. 279-1 is provided in the Code as a reference for applying this rule.

The clearance requirement between a guy (with or without a guy insulator) and an energized conductor on the same structure (pole) is provided in NESC Table 235-6. See Rule 235E for a discussion.

Grounded Guy:
Per Rules 215C2a and 215C2b, anchor guys and span guys must be effectively grounded or insulated (per the exceptions). Grounding methods are provided in Rule 092C2.

Insulated Guy:
Per Rules 215C2a and 215C2b, anchor guys and span guys must be effectively grounded or insulated (per the exceptions). Guy insulators must be installed per Rule 279A, the exceptions to Rules 215C2a and 215C2b, and the clearances in **NESC** Table 232-1 including applicable footnotes.

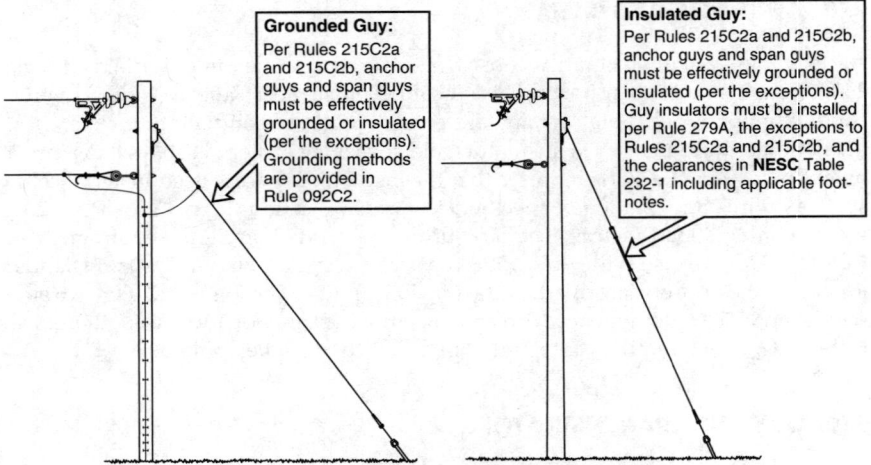

Fig. 279-1. Guy and span insulators (Rule 279).

279B. Span-Wire Insulators.   The requirements of span-wire insulators are similar to guy insulators. Rule 215C is referenced for application and Rule 274 is referenced for testing. The rated ultimate strength of a span-wire insulator is based on the required strength of the span wire in which it is located.

SECTION NUMBER 28 NOT USED IN THIS EDITION

SECTION NUMBER 29 NOT USED IN THIS EDITION

Material:

• Wet-process porcelain.

• Wood (not common in modern construction).

• Fiber reinforced polymer.

• Other material of suitable electrical and mechanical properties.

Electrical strength:

• The guy insulator may consist of one or more units.

• The design of the guy insulator dry flashover voltage must be at least double the voltage to which the insulator may be exposed with guys intact or under the conditions of Rule 215C2.

• The design of the guy insulator wet flashover voltage must be at least as high as the voltage to which the insulator may be exposed with guys intact or under the conditions of Rule 215C2.

• ANSI testing standards apply.

• UV protection is required.

Mechanical strength:

• The rated ultimate strength of the guy insulator must be at least equal to the required strength of the guy.

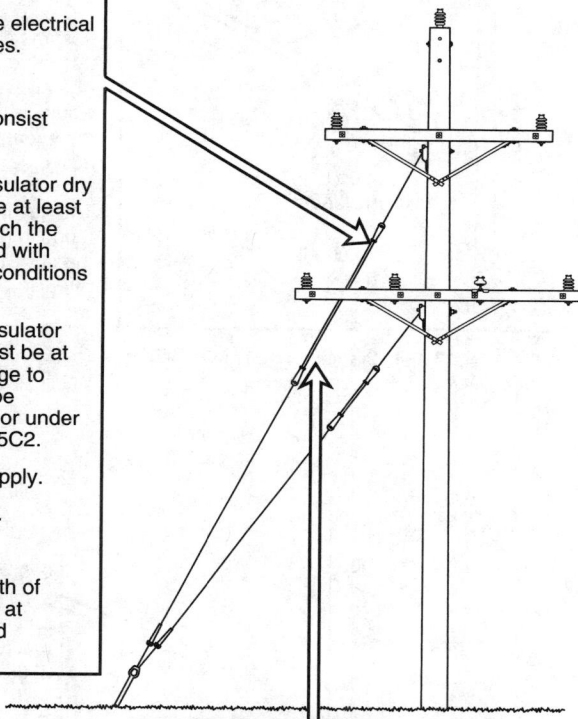

• See Rule 235E and **NESC** Table 235-6 for clearance between a guy (with or without a guy insulator) and an energized conductor.

• See Rules 215C2 and 092C2 for information on grounding vs. insulating guys and the location of the guy insulator

See Photo(s)

Fig. 279-2. Properties of guy insulators (Rule 279A1).

Example:

• Assume the structure has a small angle and the required guy tension is 2,000 lb (The 2,000 lb was calculated using the load factors in **NESC** Table 253-1.)

• Assume a 10,000 lb guy wire was used due to the fact it is a stock item (the allowable load on the guy wire would be 10,000 × .9 = 9,000 lb per the strength factors in **NESC** Table 261-1.)

• The guy insulator must be rated 2,000 lb (the required strength of the guy). It is not required to be rated 10,000 lb to match the guy wire.

10,000 lb guy insulator

2,000 lb guy load

10,000 lb guy wire

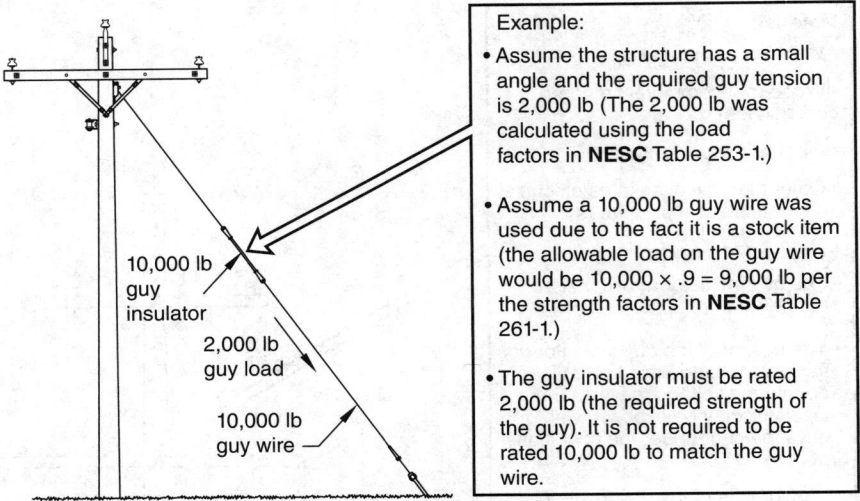

Fig. 279-3. Example of the required strength of a guy insulator (Rule 279A1c).

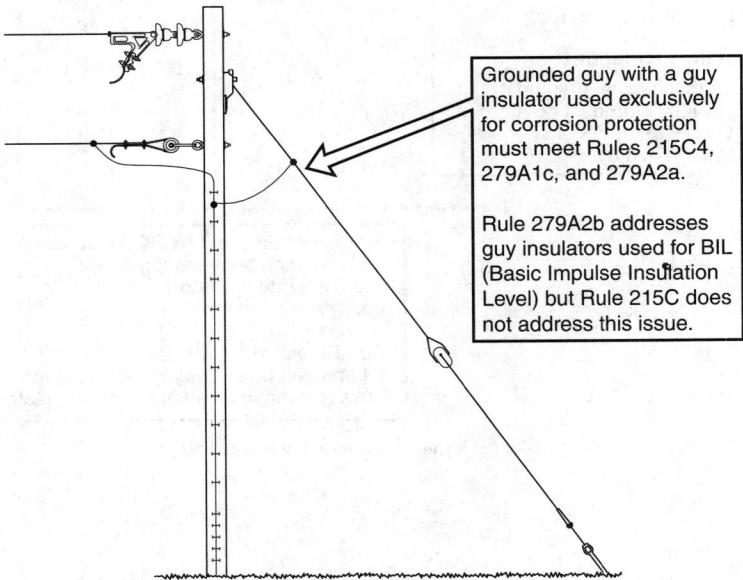

Grounded guy with a guy insulator used exclusively for corrosion protection must meet Rules 215C4, 279A1c, and 279A2a.

Rule 279A2b addresses guy insulators used for BIL (Basic Impulse Insulation Level) but Rule 215C does not address this issue.

See Photo(s)

Fig. 279-4. Example of a guy insulator used to limit galvanic corrosion (Rule 279A2a).

Part 3

Safety Rules for the Installation and Maintenance of Underground Electric Supply and Communication Lines

GENERAL SECTIONS

01 INTRODUCTION   02 DEFINITIONS

03 REFERENCES   09 GROUNDING METHODS

GENERAL SECTIONS 01, 02, 03, 09

PART 1

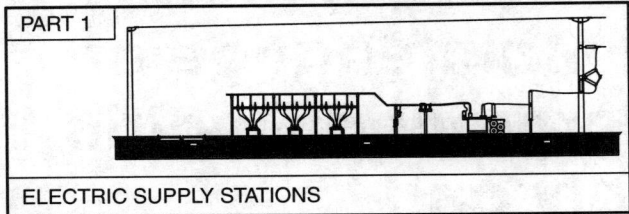

ELECTRIC SUPPLY STATIONS

PART 2

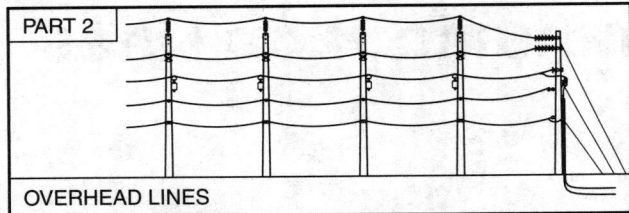

OVERHEAD LINES

PART 3

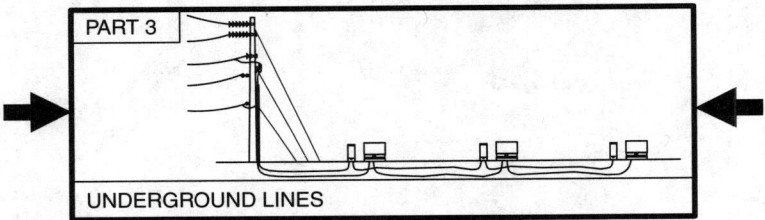

UNDERGROUND LINES

PART 4

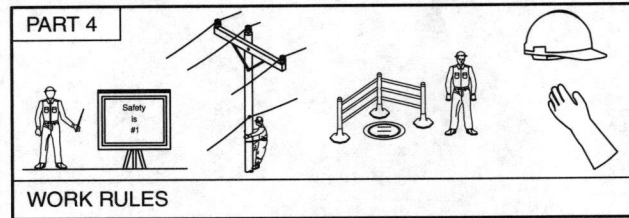

WORK RULES

Section 30

Purpose, Scope, and Application of Rules

300. PURPOSE

The purpose of Part 3, Underground Lines, is similar to the purpose of the entire NESC outlined in Rule 010, except Rule 300 which is specific to underground supply and communication lines and equipment. Part 3 of the NESC focuses on the practical safeguarding of persons during the installation, operation, and maintenance of underground supply (power) and communication lines and equipment.

301. SCOPE

The scope of Part 3, Underground Lines, includes supply (power) and communications (phone, cable TV, etc.) cables and equipment in underground or buried applications. The term underground equipment covers buried equipment and pad-mounted (aboveground) equipment used with underground conductors or cables. Separate rules apply to Electric Supply Stations (Part 1) and Overhead Lines (Part 2). Part 3 covers underground structural arrangements and extensions into buildings.

There is some overlap between Underground Lines (Part 3) and Overhead Lines (Part 2). The overlap is at the underground riser on the overhead pole. Risers are covered in Sec. 36 of Part 3 and in Sec. 23, Rule 239 of Part 2. See Fig. 301-1.

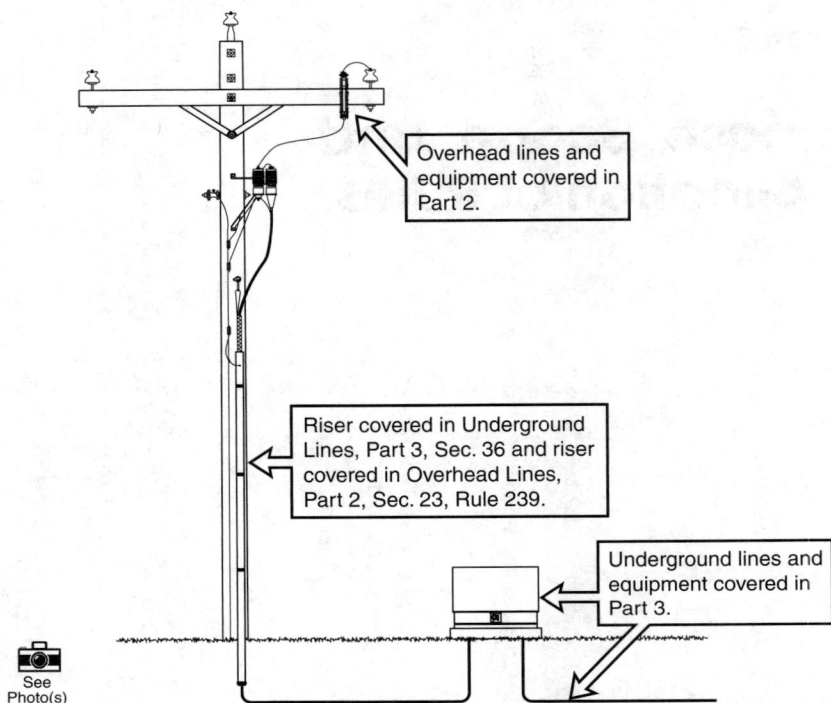

Fig. 301-1. Overlap between Underground Lines (Part 3) and Overhead Lines (Part 2) (Rule 301).

There are more distinct differences between Underground Lines (Part 3) and Electric Supply Stations (Part 1). Underground lines outside the electric supply station are covered in Part 3. Conductors and conduit inside the electric supply station are covered in Sec. 16 of Part 1. See Fig. 301-2.

The second to last sentence of Rule 301 clarifies the application of the **NESC** versus the National Electrical Code (NEC) to underground supply and communication lines. Rule 011 discusses the scope of the **NESC** and the NEC. The **NESC** covers conductors and equipment when they are serving a utility function (not an office building wiring function). The **NESC** underground lines, rules cover utility functions. There are instances when utilization wiring may be associated with the utility function of underground lines. For example, utilization wiring may be needed for lighting and ventilation in a vault. The **NESC** does not provide specific rules for utilization wiring. Rule 012C, which requires accepted good practice, must be applied when specific conditions are not covered. The NEC is an excellent reference for accepted good practice in this case.

When underground supply equipment is fenced, the rules of Part 1, Electric Supply Stations, may apply. See Rule 110A for a discussion.

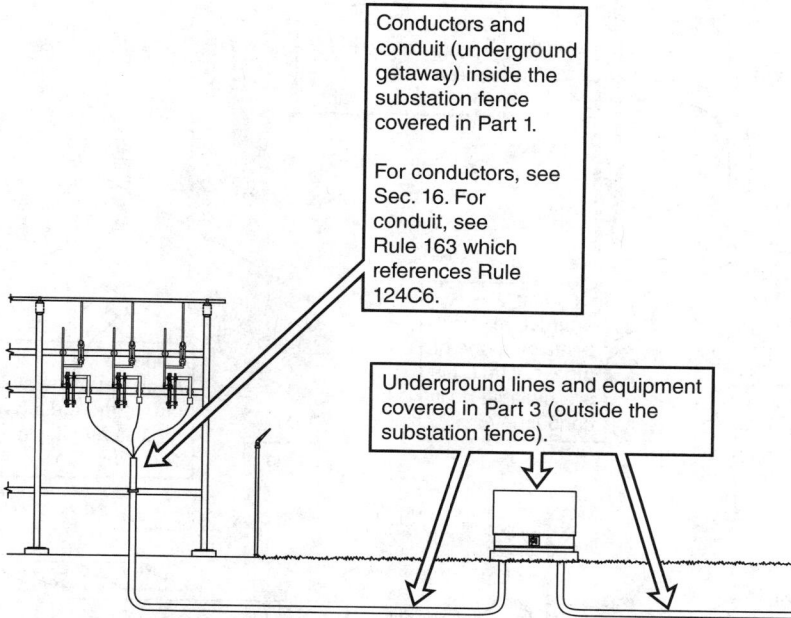

Fig. 301-2. Differences between Underground Lines (Part 3) and Electric Supply Stations (Part 1) (Rule 110).

302. APPLICATION OF RULE

Rule 302 references Rule 013 for the general application of **Code** rules. See Rule 013 for a discussion.

Rule 302 defines how the **NESC** uses the following terms in Part 3, Underground Lines:

- Duct
- Conduit
- Conduit system

The common trade use of these terms can be slightly different from the **NESC** definitions. The use of these terms as they apply to the **NESC** is shown in Fig. 302-1.

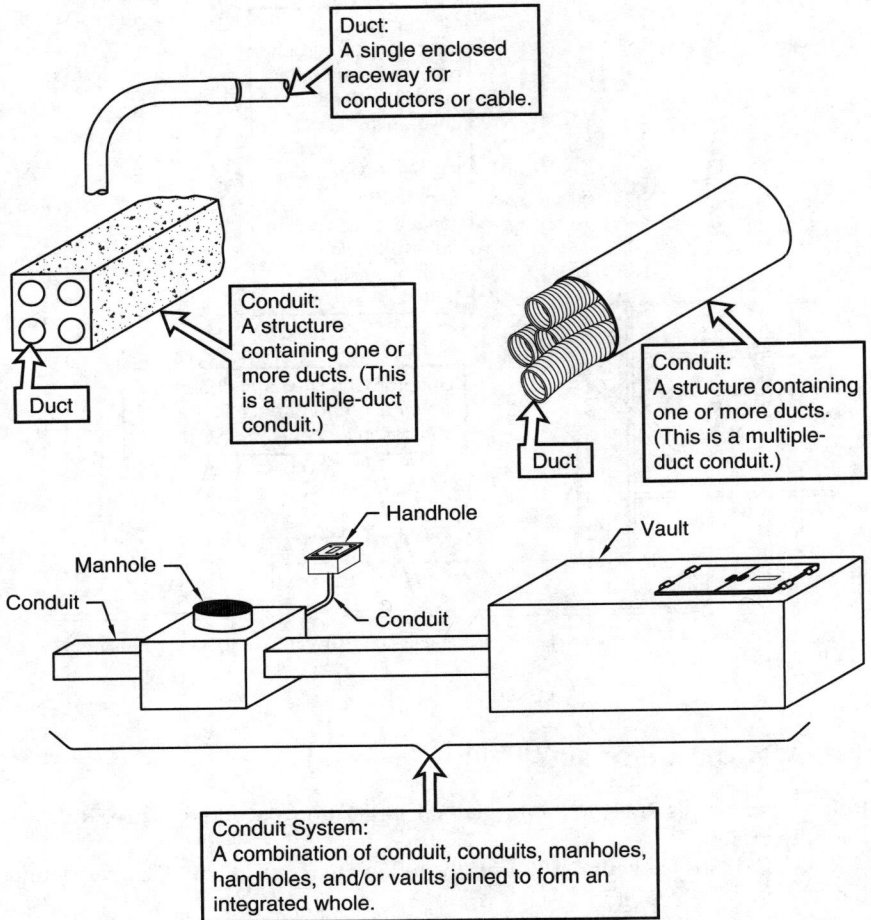

Duct:
A single enclosed raceway for conductors or cable.

Conduit:
A structure containing one or more ducts. (This is a multiple-duct conduit.)

Duct

Conduit:
A structure containing one or more ducts. (This is a multiple-duct conduit.)

Duct

Handhole

Vault

Manhole

Conduit

Conduit

Conduit System:
A combination of conduit, conduits, manholes, handholes, and/or vaults joined to form an integrated whole.

Fig. 302-1. NESC definitions of duct, conduit, and conduit system (Rule 302).

Section 31

General Requirements Applying to Underground Lines

310. REFERENCED SECTIONS

This rule references four sections related to Part 3, Underground Lines, so that rules do not have to be duplicated and the reader of the **Code** realizes that other sections are related to the information provided in Part 3. The related sections are:
- Introduction, Sec. 01
- Definitions, Sec. 02
- References, Sec. 03
- Grounding Methods, Sec. 09

The rules in Part 3, predominantly Rules 314, 342, 374, and 384, will provide the requirements for grounding underground lines and equipment. The grounding methods are provided in Sec. 09.

311. INSTALLATION AND MAINTENANCE

This rule requires that supply and communication utilities be able to locate their underground facilities. Locates are typically done as shown in the example in Fig. 311-1.

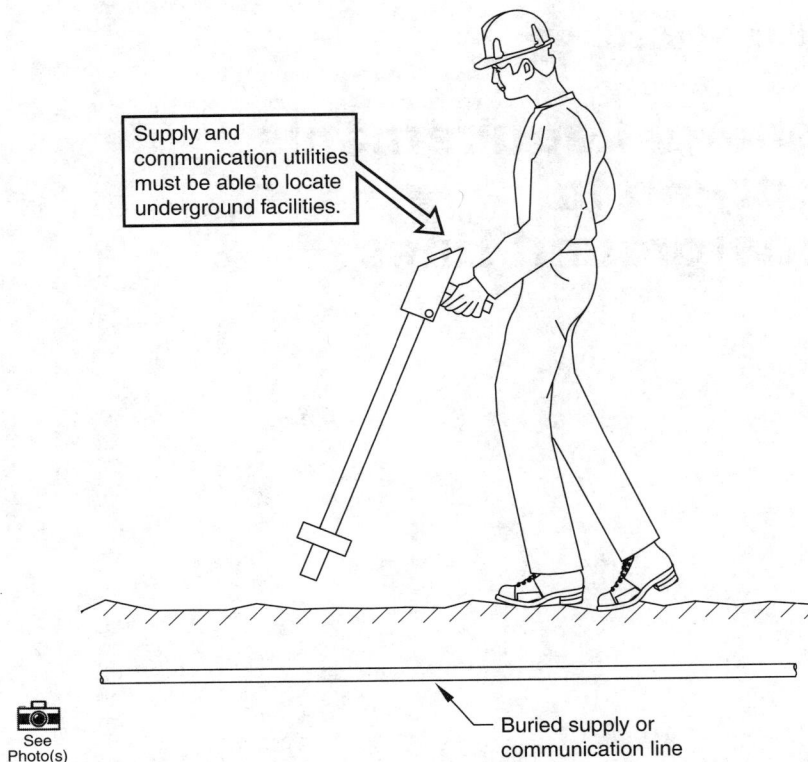

Supply and communication utilities must be able to locate underground facilities.

See Photo(s)

Buried supply or communication line

Fig. 311-1. Example of worker locating underground lines (Rule 311).

Color-coded spray paint markings are typically used to identify power, communication, and other utilities. Most localities have a "call before you dig" system in place to aid location of underground facilities. The **NESC** does not limit locates to supply and communication cables. Any underground facility (e.g., equipment, conduit systems, duct banks, vaults, etc.) must be located. Proper mapping of facilities can aid the location of facilities and is required in Part 4, Work Rules, Rule 411A2.

The **NESC** requires advance notice (a time is not specified) to owners or operators of nearby facilities that may be adversely affected by underground digging or construction by supply or communication utilities.

Rule 311C addresses emergency installations and permits laying certain supply (power) and communication cables directly on the ground. This same permission can be found in Rule 230A2d. See the discussion and figure in Rule 230A. Rule 311C is also referenced in Rule 014A2.

312. ACCESSIBILITY

Rule 312 is the underground equivalent to overhead Rule 213.

Rule 312 generally states the requirement that underground parts needing examination or adjustment during operation must be accessible to authorized supply or communication workers. This rule requires the following, all of which are discussed in more detail throughout Part 3.

- Working spaces
- Working facilities
- Clearances

Working spaces are specifically discussed in Rule 323B. This rule dimensions the working space in manholes. Various rules throughout Part 3 discuss clearances and separations between supply and communication conductors and other facilities.

The **NESC** does not specify a clear working space dimension in front of a pad-mounted transformer or primary junction box. The general requirements of this rule apply, which are that adequate working space must exist. Working space is a challenge for utilities when it comes to service transformers or other equipment on private property. Some utilities apply a notification sign on the front of pad-mounted equipment that asks the landowner to keep the area in front of the equipment clear (this sign is not a **Code** requirement). An example of the need for adequate working space is shown in Fig. 312-1.

An example of a notification sign used on pad-mounted equipment is shown in Fig. 312-2.

Additional requirements for equipment, including pad mounted equipment, can be found in Sec. 38.

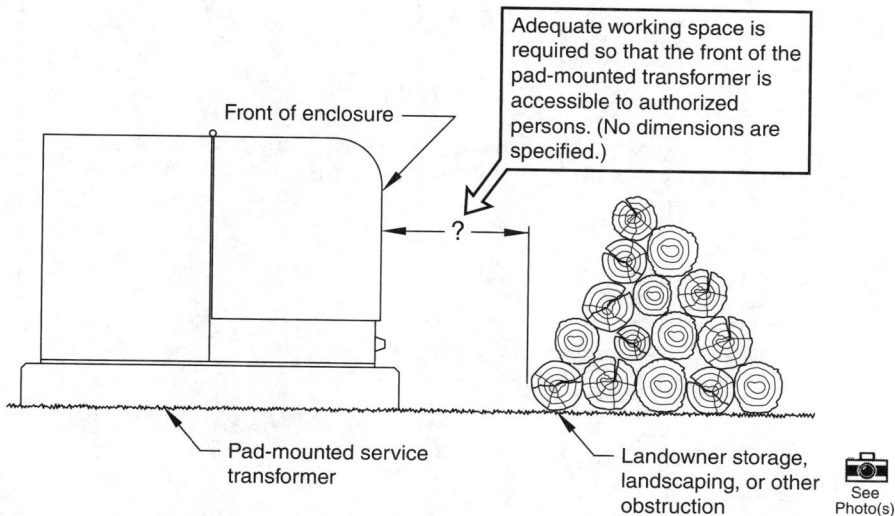

Fig. 312-1. Example of the need for adequate working space (Rule 312).

NOTICE

Adequate working space is required. Please keep shrubs, structures, stored materials, fences, etc. at least 10 ft away from this side and 2 ft away from other sides.

For information call USA Power Co. at
1-800-555-1212.

See
Photo(s)

Fig. 312-2. Example of a notification sign used on pad-mounted equipment (Rule 312).

313. INSPECTION AND TESTS
OF LINES AND EQUIPMENT

This rule is the underground equivalent of overhead Rule 214. There are slight wording differences between the overhead and underground inspection and testing rules, but the intent is similar.

Cutouts, switches, reclosers, circuit breakers, etc., are used to switch overhead lines and determine if a line is in or out of service. These same overhead devices can be used for underground switching in addition to pad-mounted switches and the termination or standoff of URD elbows.

Rule 313, like its overhead counterpart Rule 214, does not contain a specific inspection or testing checklist, it does not contain any inspection intervals (e.g., every year, 2 years, 5 years, 10 years, etc.), and it does not contain specific time periods for making corrections to NESC violations (e.g., 1 day, 1 week, 1 year, etc.). When the Code is not specific, accepted good practice must be applied per Rule 012C. See Rule 214 for a discussion and additional information on inspections.

314. GROUNDING OF CIRCUITS
AND EQUIPMENT

Rule 314 is the underground equivalent of overhead Rule 215.

Rule 314 is broken into three main paragraphs. Rule 314A reminds us that the methods of grounding are specified in Sec. 09 and the requirements of grounding are provided in this rule for circuits and conductive parts to be grounded. Rule 314B focuses on the requirements for grounding conductive parts. Rule 314C focuses on the requirements for grounding circuits. See Sec. 02, "Definitions," for a discussion of the term "effectively grounded." Not all of the grounding requirements for Part 3 are listed in this rule. Various other grounding requirements appear throughout Part 3. One important example is the grounding and bonding requirement for aboveground apparatus (pad-mounted equipment) found in Rule 384. Other Part 3 grounding rules include Rules 342 and 374.

Rule 314B provides a list of conductive parts to be grounded and includes an exception for special cases. The list does not contain conductive manhole or handhole lids that are flush with the surface of a roadway, walkway, lawn, etc. Therefore Rule 012C, which requires accepted good practice, must be used. Methods to effectively ground manhole and handhole lids do exist in the industry. The conductive parts grounding requirements of Rule 314B are outlined in Fig. 314-1.

The circuit grounding requirements of Rule 314C which are broken down into neutrals, other conductors, surge arresters, and use of earth, are outlined in Fig. 314-2.

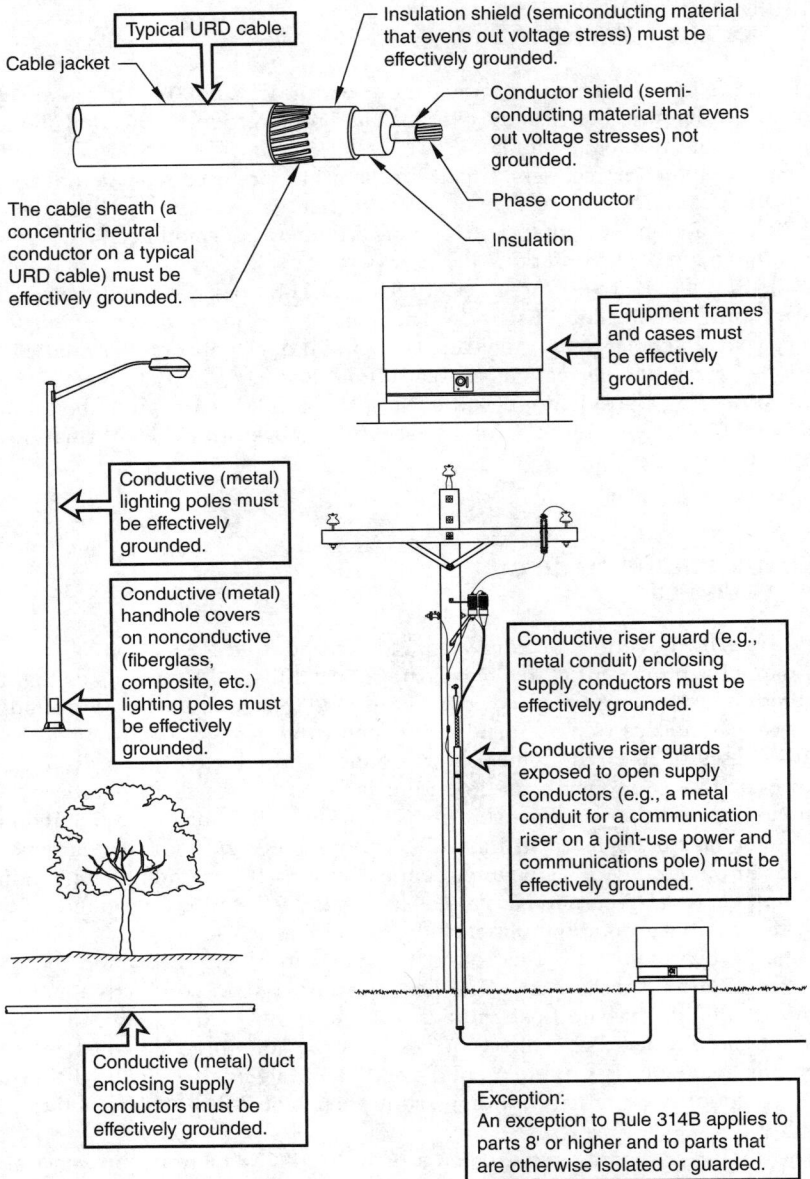

Typical URD cable.

Cable jacket

Insulation shield (semiconducting material that evens out voltage stress) must be effectively grounded.

Conductor shield (semi-conducting material that evens out voltage stresses) not grounded.

Phase conductor

Insulation

The cable sheath (a concentric neutral conductor on a typical URD cable) must be effectively grounded.

Equipment frames and cases must be effectively grounded.

Conductive (metal) lighting poles must be effectively grounded.

Conductive (metal) handhole covers on nonconductive (fiberglass, composite, etc.) lighting poles must be effectively grounded.

Conductive riser guard (e.g., metal conduit) enclosing supply conductors must be effectively grounded.

Conductive riser guards exposed to open supply conductors (e.g., a metal conduit for a communication riser on a joint-use power and communications pole) must be effectively grounded.

Conductive (metal) duct enclosing supply conductors must be effectively grounded.

Exception:
An exception to Rule 314B applies to parts 8' or higher and to parts that are otherwise isolated or guarded.

See Photo(s)

Fig. 314-1. Conductive parts to be grounded (Rule 314B).

Neutrals:
Primary neutrals, secondary neutrals, service neutrals, and common neutrals exposed to personnel contact must be effectively grounded per Sec. 09.
Exception:
Circuits designed for ground-fault detection and impedance current-limiting devices.

Surge arresters:
Surge arresters must be effectively grounded per Sec. 09.

Pad-mounted transformer

Elbow surge arrester

Primary neutral — — Secondary neutral

Other conductors:
Conductors that are not neutral conductors that are intentionally grounded must be effectively grounded where exposed to personnel contact per Sec. 09.

Phase conductor
240 V, 3Ø, 3W corner grounded delta
Phase conductor
Grounded phase conductor
Grounding connection

Use of earth as part of circuit:
This is acceptable. The earth is parallel with the neutral conductor.

VIOLATION!
Use of earth as part of circuit:
Supply circuits must not use the earth as the sole conductor (exceptions apply to HVDC systems).

— Current flow to load (on phase wire).
— Current flow from load (on concentric neutral).
Current flow from load (through earth in parallel with concentric neutral).

— Current flow to load (on phase wire).
Current flow from load (through earth due to no neutral or corroded neutral).

See Photo(s)

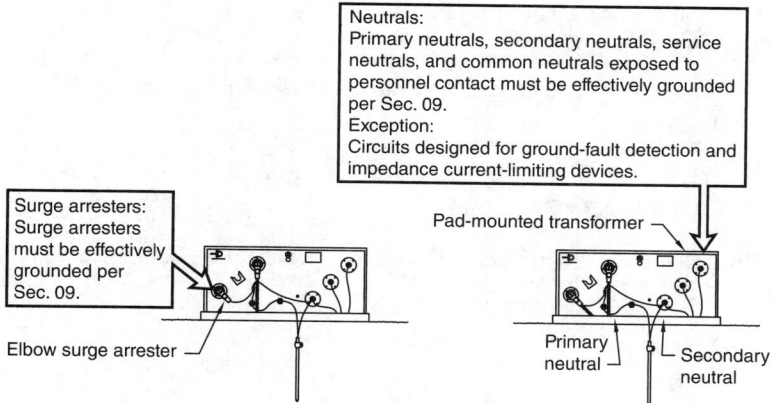

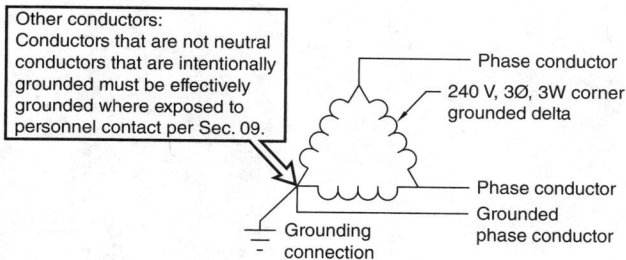

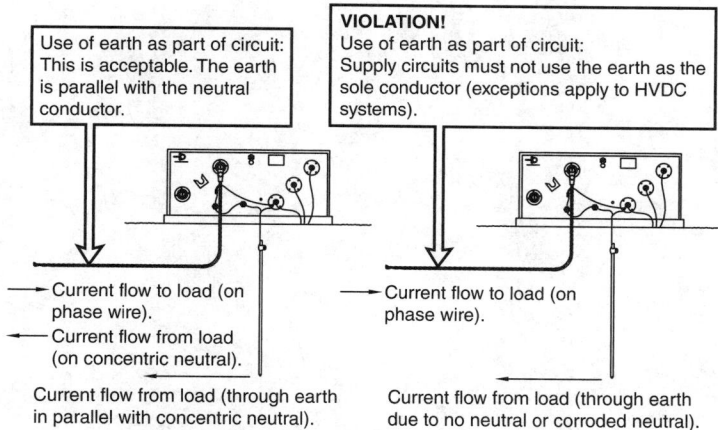

Fig. 314-2. Grounding of circuits (Rule 314C).

315. COMMUNICATIONS PROTECTIVE REQUIREMENTS

Rule 315 is the underground equivalent to overhead Rule 223. See Rule 223 for a discussion.

316. INDUCED VOLTAGE

Rule 316 is the underground equivalent to overhead Rule 212. See Rule 212 for a discussion.

Section 32

Underground Conduit Systems

The note under the title of Sec. 32 references Rule 350G for supply and communication cables installed in ducts that are not part of a conduit system. Rule 350G states that Sec. 35 (Direct-Buried Cable and Cable in Duct Not Part of a Conduit System) applies to supply and communication cables that are installed in duct that is not part of a conduit system. A typical conduit system (combination of conduit, conduits, manholes, handholes, and/or vaults joined to form an integrated whole) can be found under the streets and sidewalks of a large city. A typical duct that is not part of a conduit system can be found in a residential subdivision. See Figs. 302-1 and 32-1.

320. LOCATION

Section 32 applies to the underground structure only, not the cables in the structure. Section 33 applies to supply cables. Section 34 applies to the installation of cables in an underground structure. Rule 320 deals with the location of underground conduit systems, where they can be routed, and how far they must be separated from other underground installations.

320A. Routing. The general requirements for conduit system routing are outlined below:

- Subject conduit system to the least disturbance practical.
- When conduit system is parallel to another structure, do not locate directly over or under the other structure if practical.
- Align conduit without protrusions that would harm the cable.
- Provide sufficient bending radius to limit cable damage.

The **NESC** does not contain specific burial depths for conduit systems. Direct-buried supply cables and cables in duct not part of a conduit system covered in

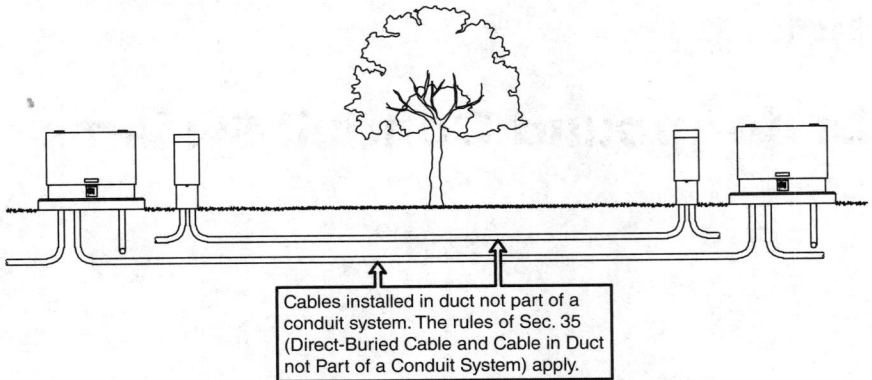

Cables installed in duct not part of a
conduit system. The rules of Sec. 35
(Direct-Buried Cable and Cable in Duct
not Part of a Conduit System) apply.

Fig. 32-1. Application of cables installed in duct not part of a conduit system (Secs. 32 and 35).

Sec. 35 do have to meet specified burial depths (covered in Rule 352 and **NESC** Table 352-1); however, no such burial requirements exist for cables installed in conduit systems. The conduit system design requires a structural review using the surface loads specified in Rule 322A3 (for the ducts and joints) and Rule 323A (for the manholes, handholes, and vaults) to determine proper burial depths. The only specific burial depth called out in Sec. 32 is for a conduit system crossing under railroad tracks. The fact that the **NESC** does not contain specific burial depths for conduit systems and the relationship between several Sec. 32 and Sec. 35 rules is shown in Fig. 320-1.

The **NESC** is not specific on conduit bending radius requirements. Rule 320A simply states that a sufficient bending radius is needed. Rule 012C, which requires accepted good practice, must be used. The National Electrical Code (NEC) provides much more detail than the **NESC** on conduit bending radius, and the NEC can be used as a reference for accepted good practice. Typically, field bends require a larger bending radius than factory bends. A bend with too tight a radius will deform or damage the conduit material. In addition to bending radius requirements, the NEC can be referenced as accepted good practice for conduit support, conduit fill (i.e., the number of conductors in any one conduit), and various other conduit application methods. The details of how conduit bending radius is typically measured are shown in Fig. 320-2.

The total bending radius of a conduit run can be determined by adding the angles of each bend. To determine how many bends are acceptable in a conduit run, a cable pulling calculation should be done per Rule 341. See Fig. 320-3.

Natural hazards due to unstable or corrosive soil should be avoided or proper construction methods should be used to minimize the hazard.

If a conduit can be installed outside a roadway, no interference with the road would occur. However, if the conduit is installed longitudinally under the road, using the shoulder or just one lane of traffic will make both installation and maintenance of the conduit conflict less with roadway traffic. The **NESC** does not specify wa conduit system burial depth under a road for a longitudinal run or for a road crossing. The road department or highway department having jurisdiction may require

Surface

Conduit systems covered in Sec. 32; general routing and strength rules apply but no burial depths are specified (except under railroad tracks). A structural review is required.

Supply cables installed in duct not part of a conduit system. The rules of Sec. 35 (Direct-Buried Cable and Cable in Duct not Part of a Conduit System) apply. **NESC** Table 352-1 specifies burial depths for supply lines.

Direct-buried supply cables covered in Sec. 35; **NESC** Table 352-1 specifies burial depths.

Direct-buried supply cable

Cables in duct not part of a conduit system

Conduit system

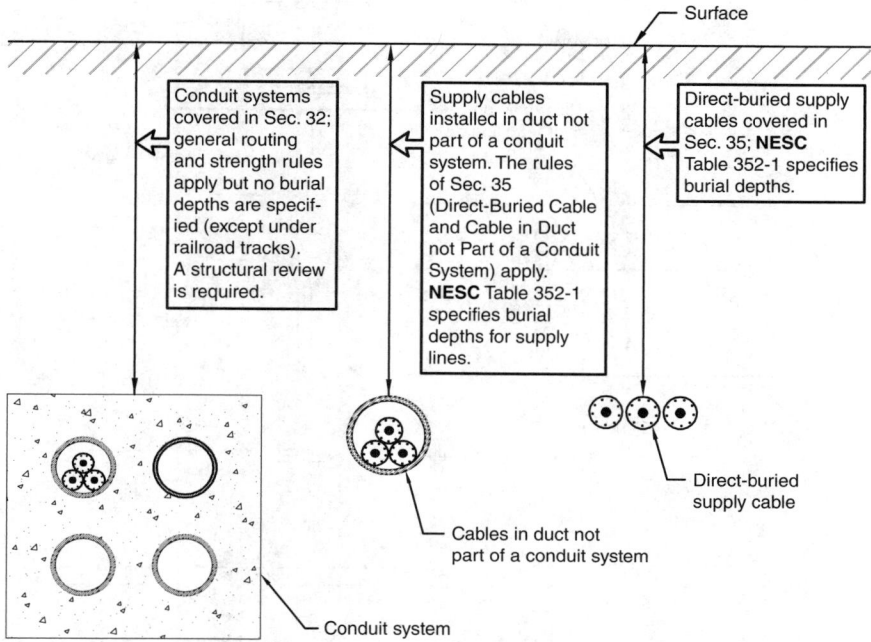

Fig. 320-1. Burial depths for conduit systems, supply cable in duct not part of a conduit system, and direct-buried supply cables (Rules 320, 322A3, 323, 350G, and 352).

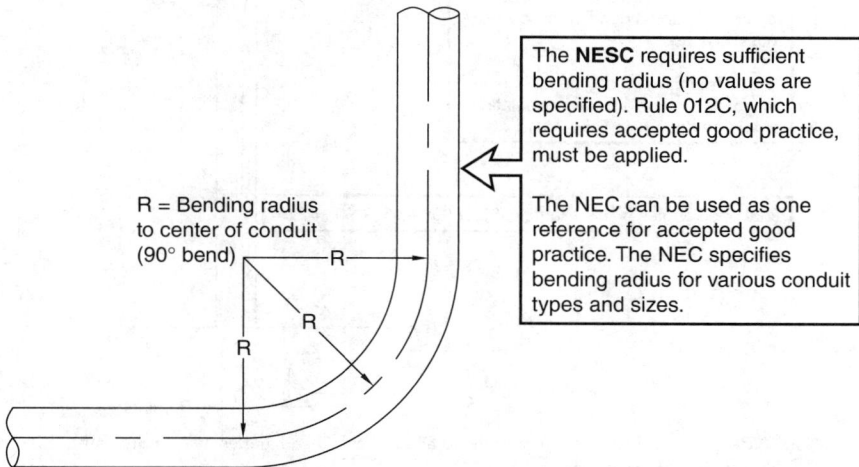

The **NESC** requires sufficient bending radius (no values are specified). Rule 012C, which requires accepted good practice, must be applied.

The NEC can be used as one reference for accepted good practice. The NEC specifies bending radius for various conduit types and sizes.

R = Bending radius to center of conduit (90° bend)

R

R

R

See Photo(s)

Fig. 320-2. Conduit bending radius (Rule 320A1c).

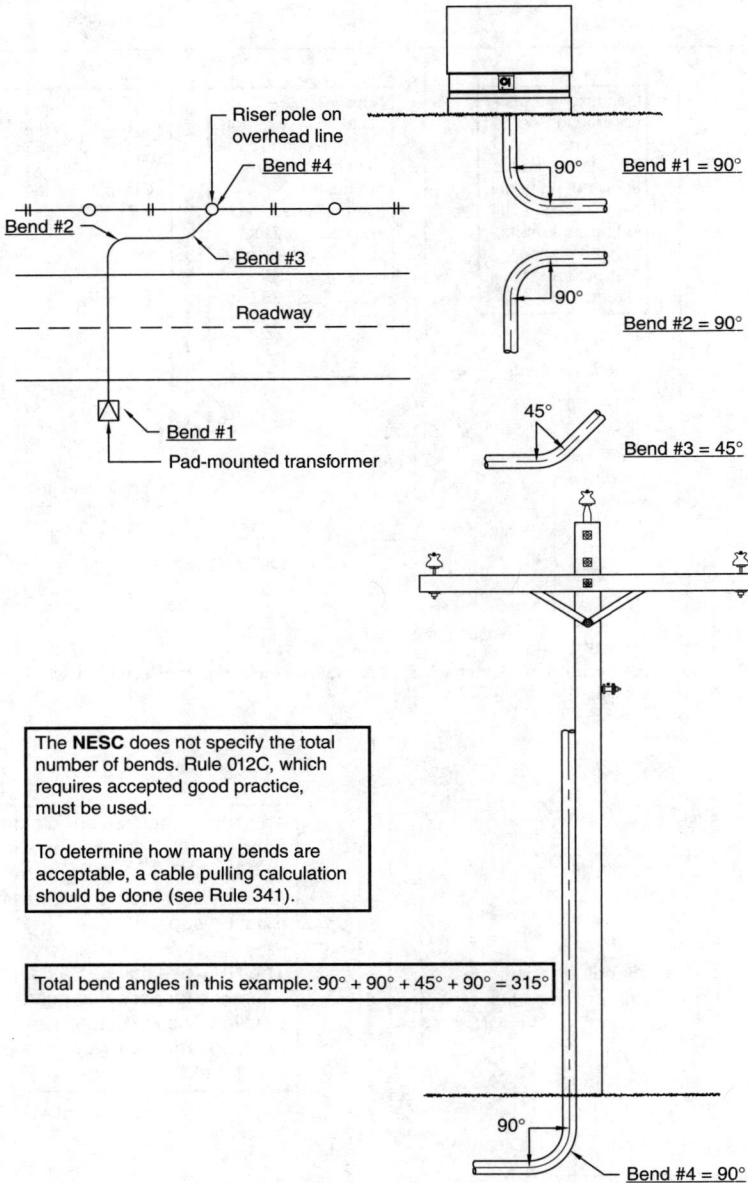

Riser pole on overhead line

Bend #4

Bend #2

Bend #3

Roadway

Bend #1

Pad-mounted transformer

90° Bend #1 = 90°

90° Bend #2 = 90°

45° Bend #3 = 45°

The **NESC** does not specify the total number of bends. Rule 012C, which requires accepted good practice, must be used.

To determine how many bends are acceptable, a cable pulling calculation should be done (see Rule 341).

Total bend angles in this example: 90° + 90° + 45° + 90° = 315°

90° Bend #4 = 90°

Fig. 320-3. Example of determining the total number of bend angles of a conduit run (Rules 320A1c and 341).

specific depths or locations to obtain a permit to install utilities under the road. The rules related to conduit routing under highways and streets are outlined in Fig. 320-4.

Conduit system routing in or on bridges and tunnels should not be damaged by traffic and should be located to provide safe access.

Conduit system routing near railroad tracks have a specific burial requirement. This is the only specific burial requirement provided in Sec. 32. The burial rules for a conduit system under a railroad track are outlined in Fig. 320-5.

Conduit systems crossing water are discussed in Rule 320A6 and in Sec. 35, "Direct Buried Cable and Cable in Duct Not Part of a Conduit System," Rule 351C5. Water crossings are sometimes referred to as submarine crossings. An underwater crossing can typically be found on lake bottoms, river bottoms, or ocean bottoms. A special cable called a "submarine cable" is commonly used for underwater crossings. Conduit can be used for the entire crossing but it is typically used near the shorelines. An example of a three-phase, 15-kV submarine cable is shown in Fig. 320-6.

Higher-voltage submarine cables are also in use and can consist of oil-filled cable with a lead outer jacket. There are no specific conduit requirements in this rule other than protecting the submarine crossing from erosion by tides or currents and locating the crossing away from where ships normally anchor. Submarine crossings frequently use conduit for some portion of the distance into the water and then the submarine cable is exposed and lies on the bottom of the body of water. An example of a submarine crossing is shown in Fig. 320-7.

Rule 96C, which requires at least four grounds in each mile, has an exception that states the rule does not apply to underwater crossings if other conditions are met.

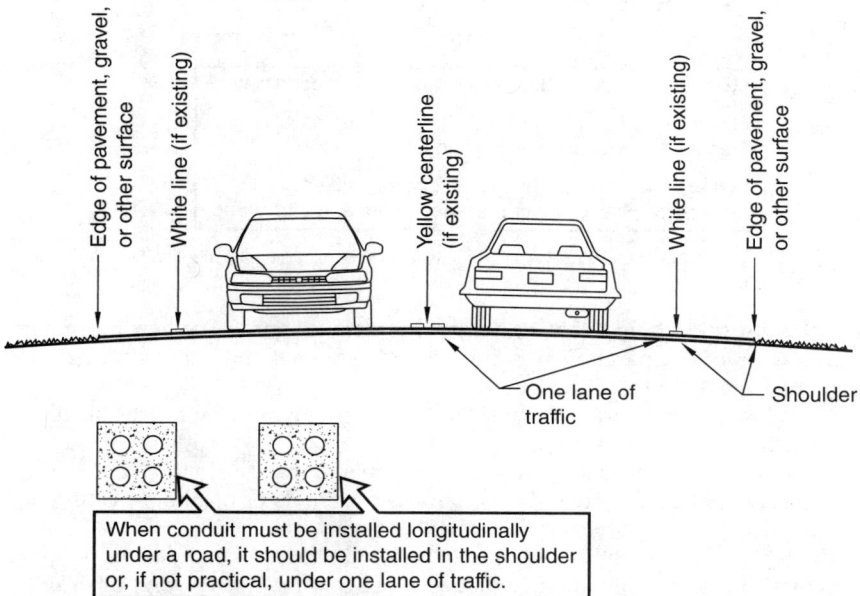

Fig. 320-4. Conduit routing under highways and streets (Rule 320A3).

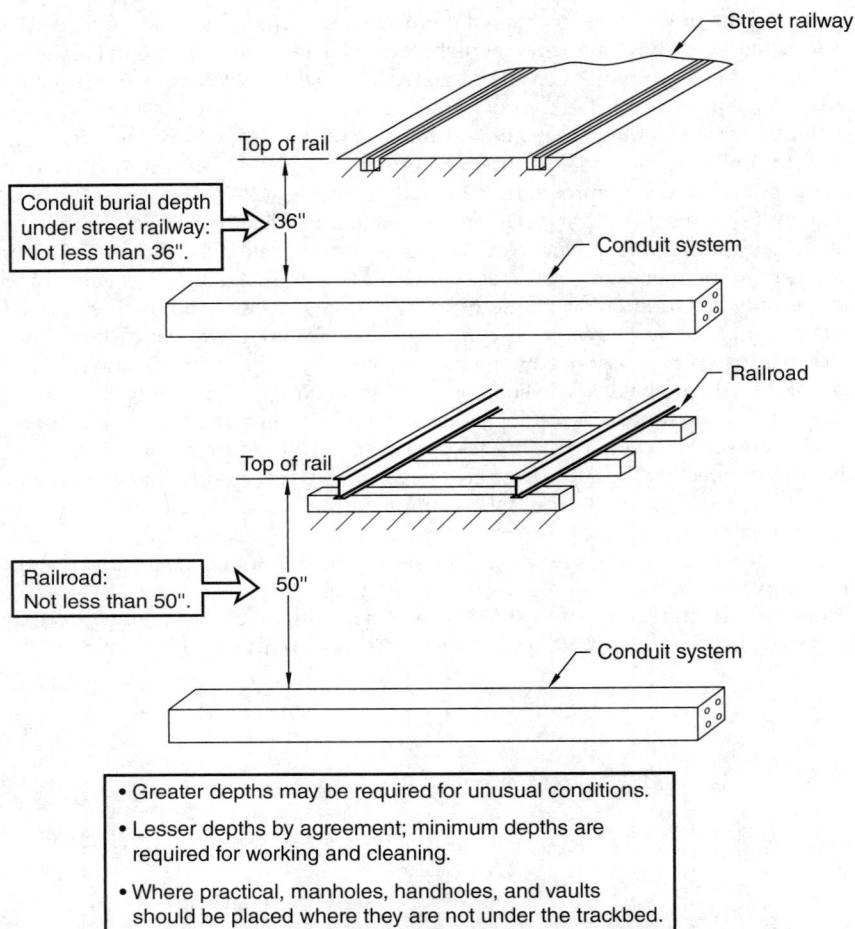

Fig. 320-5. Conduit system crossings under railroad tracks (Rule 320A5).

320B. Separation from Other Underground Installations.   The general requirements for radial separation between conduit systems and other underground structures are outlined below:

- Provide radial separation to permit maintenance while limiting the likelihood of damage to either structure.
- The parties involved should determine separations.
- Exception: Conduits may be supported from roofs of manholes, vaults, and subway tunnels with concurrence of the parties involved.

The term structure in this rule is used loosely to apply to many types of buried structures including building foundations, other conduit systems, and other utility lines such as sewer lines, water lines, gas and other lines that transport flammable material, steam lines, cryogenic lines, etc.

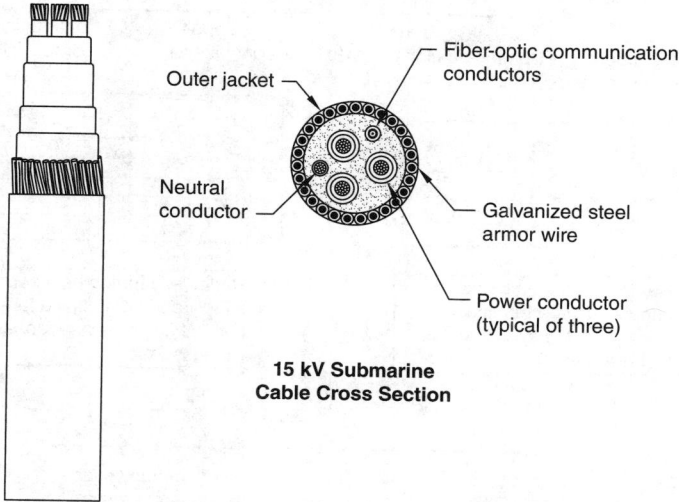

Outer jacket

Fiber-optic communication conductors

Neutral conductor

Galvanized steel armor wire

Power conductor (typical of three)

15 kV Submarine Cable Cross Section

15 kV Submarine Cable

Fig. 320-6. Example of a three-phase, 15-kV "submarine cable" (Rule 320A6).

The **Code** provides specific values for separations between supply and communication conduit systems. The separations in Rule 320B2 are between supply and communication conduit systems, not the ducts in the conduit systems. The duct material and separations of the ducts is covered in Rule 322A. The radial separation between supply and communication conduit systems are outlined in Fig. 320-8.

The requirements for conduits crossing and paralleling sanitary and storm sewers, water lines, steam lines (hot), and cryogenic lines (cold) are very general.

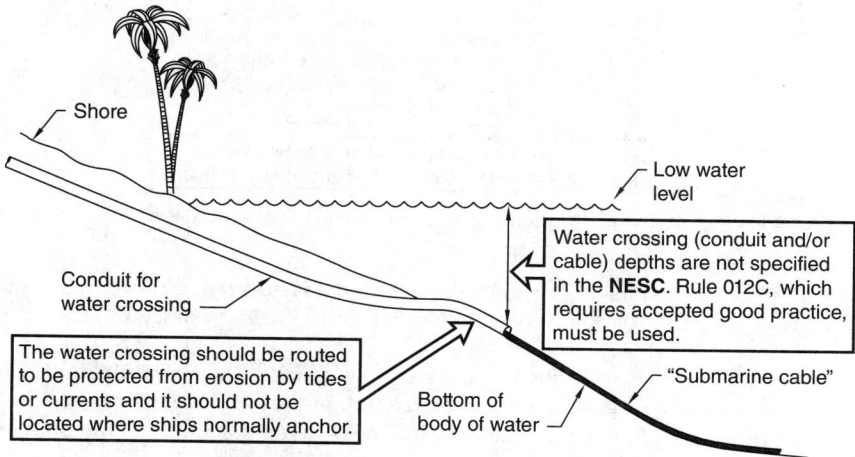

Shore

Low water level

Conduit for water crossing

Water crossing (conduit and/or cable) depths are not specified in the **NESC**. Rule 012C, which requires accepted good practice, must be used.

The water crossing should be routed to be protected from erosion by tides or currents and it should not be located where ships normally anchor.

Bottom of body of water

"Submarine cable"

Fig. 320-7. Example of a water crossing (Rule 320A6).

Supply conduit system   Communications conduit system

Radial separation between supply and communications conduit systems in concrete: Not less than 3".

Supply conduit system   Communications conduit system

Radial separation between supply and communications conduit systems in masonary: Not less than 4".

Supply conduit system   Communications conduit system

Radial separation between supply and communications conduit systems in well-tamped earth: Not less than 12".

Exception: Lesser radial separation may be used where the supply and communications utilities concur.

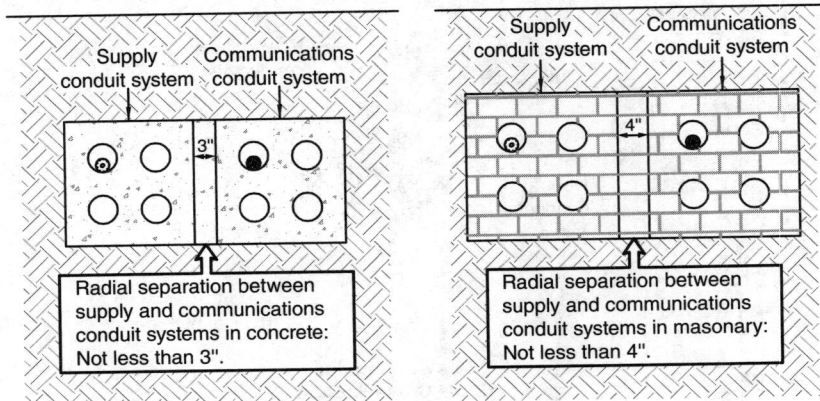

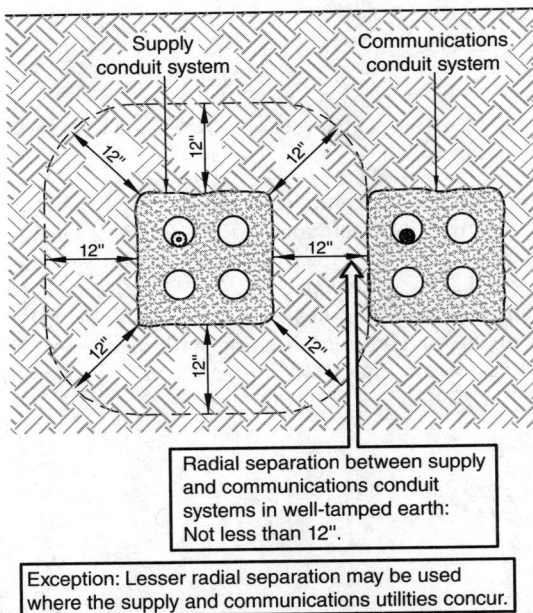

Fig. 320-8. Radial separation between supply and communication conduit systems (Rule 320B2).

No specific dimensions are provided. The **Code** is primarily concerned with providing enough separation for maintenance and avoiding damage to the lines involved. See Fig. 320-9.

The **Code** provides a specific value for separation between a conduit and a gas or other line that transports flammable material. In the case of gas lines or other lines that transport flammable material, the measurement is made to the duct in the conduit. See Fig. 320-10.

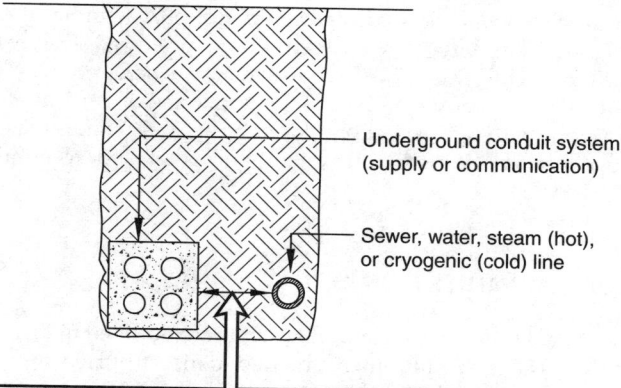

No dimensions are provided for separating a conduit system from sewer, water, steam (hot), or cryogenic (cold) lines. General separation rules apply. The separations should be determined by the parties involved.

Fig. 320-9. Radial separation from sewer, water, and steam or cryogenic lines (Rules 320B3, 320B4, and 320B6).

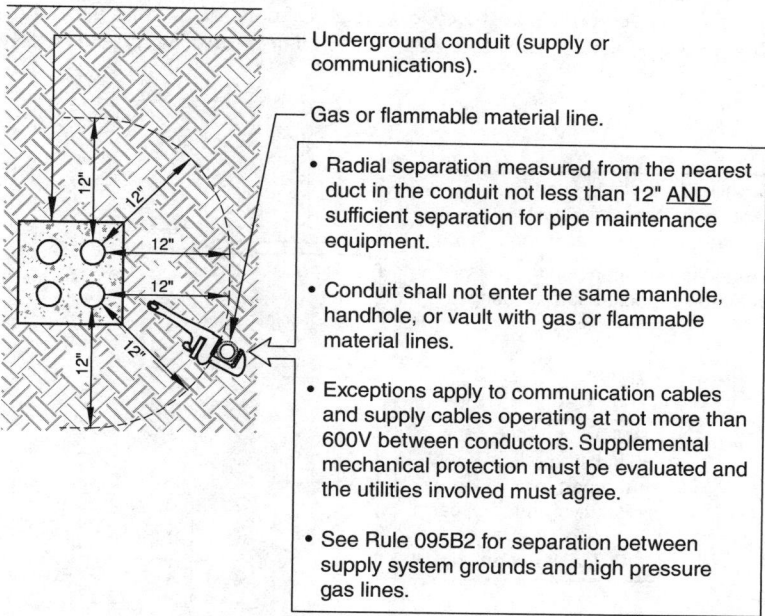

Underground conduit (supply or communications).

Gas or flammable material line.

• Radial separation measured from the nearest duct in the conduit not less than 12" <u>AND</u> sufficient separation for pipe maintenance equipment.

• Conduit shall not enter the same manhole, handhole, or vault with gas or flammable material lines.

• Exceptions apply to communication cables and supply cables operating at not more than 600V between conductors. Supplemental mechanical protection must be evaluated and the utilities involved must agree.

• See Rule 095B2 for separation between supply system grounds and high pressure gas lines.

See Photo(s)

Fig. 320-10. Radial separation from gas or other line that transports flammable material (Rule 320B5).

The conduit system locations discussed in Rule 320A indicate that vertical placement of parallel underground structures is not preferred. The conduit system separations discussed in Rule 320B1 require adequate separation between paralleling structures without any reference to the horizontal or vertical placement.

For separation between direct-buried supply and communication cables (or duct that is not part of a conduit system) and other underground structures, see Rules 353 and 354.

321. EXCAVATION AND BACKFILL

The trench and backfill rules for conduit systems are outlined in Fig. 321-1.

For trench and backfill requirements related to direct-buried supply and communication cables and cable in duct not part of a conduit system, see Rule 352.

322. CONDUIT, DUCTS, AND JOINTS

322A. General.   The general rules for conduit and ducts require the duct material and the construction of the conduit (e.g., a concrete duct bank) be designed so that a cable fault in one duct will not damage the conduit (i.e., duct bank) to an extent that would cause damage to cables in an adjacent duct. Specific separation distances between ducts in the conduit (e.g., concrete duct bank) are not specified. Rule 012C, which requires accepted good practice, must be used. The general

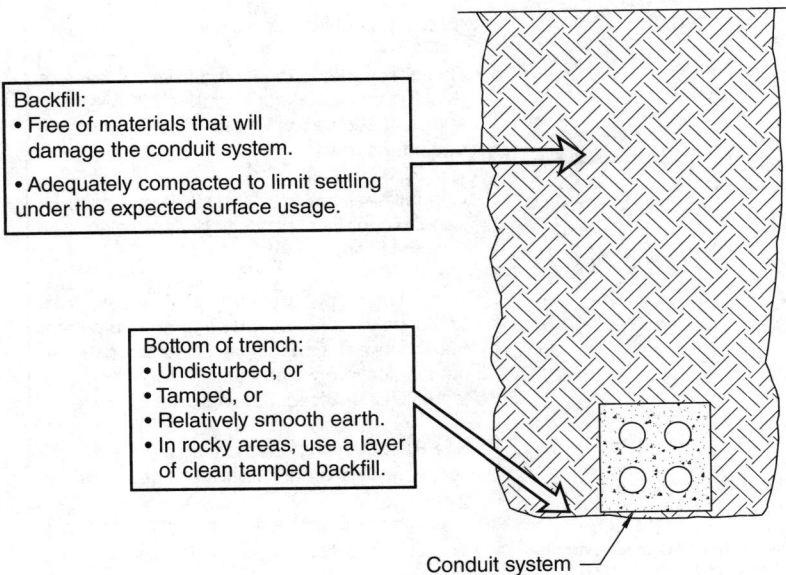

Backfill:
• Free of materials that will damage the conduit system.

• Adequately compacted to limit settling under the expected surface usage.

Bottom of trench:
• Undisturbed, or
• Tamped, or
• Relatively smooth earth.
• In rocky areas, use a layer of clean tamped backfill.

Conduit system

Fig. 321-1.   Trench and backfill requirements for conduit systems (Rule 321).

rules for conduit include surface loading requirements which are also used to determine the conduit type, as well as the burial depth. The surface loads for manholes, handholes, and vaults must also be analyzed per Rule 323A. The general rules for conduit, ducts, and joints are outlined in Fig. 322-1.

322B. Installation.   The installation rules for conduit, ducts, and joints are outlined in Fig. 322-2.

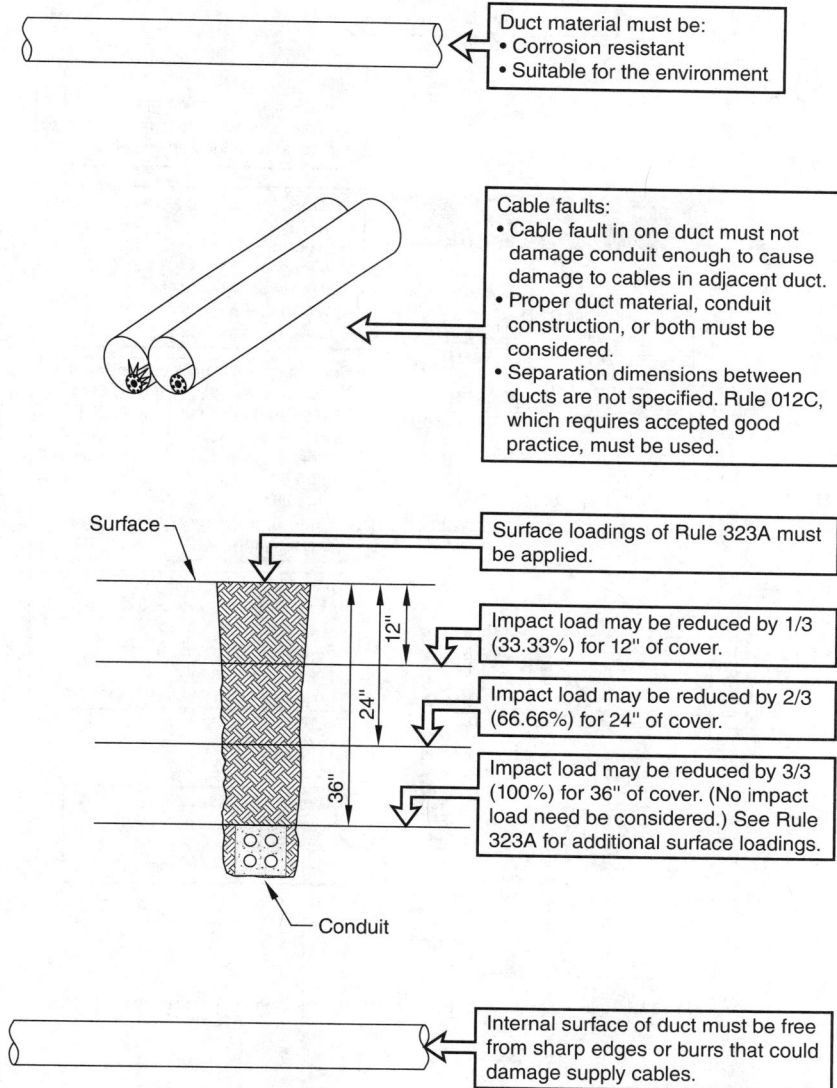

Duct material must be:
• Corrosion resistant
• Suitable for the environment

Cable faults:
• Cable fault in one duct must not damage conduit enough to cause damage to cables in adjacent duct.
• Proper duct material, conduit construction, or both must be considered.
• Separation dimensions between ducts are not specified. Rule 012C, which requires accepted good practice, must be used.

Surface

Surface loadings of Rule 323A must be applied.

12"
Impact load may be reduced by 1/3 (33.33%) for 12" of cover.

24"
Impact load may be reduced by 2/3 (66.66%) for 24" of cover.

36"
Impact load may be reduced by 3/3 (100%) for 36" of cover. (No impact load need be considered.) See Rule 323A for additional surface loadings.

Conduit

Internal surface of duct must be free from sharp edges or burrs that could damage supply cables.

Fig. 322-1. General requirements for conduit, ducts, and joints (Rule 322A).

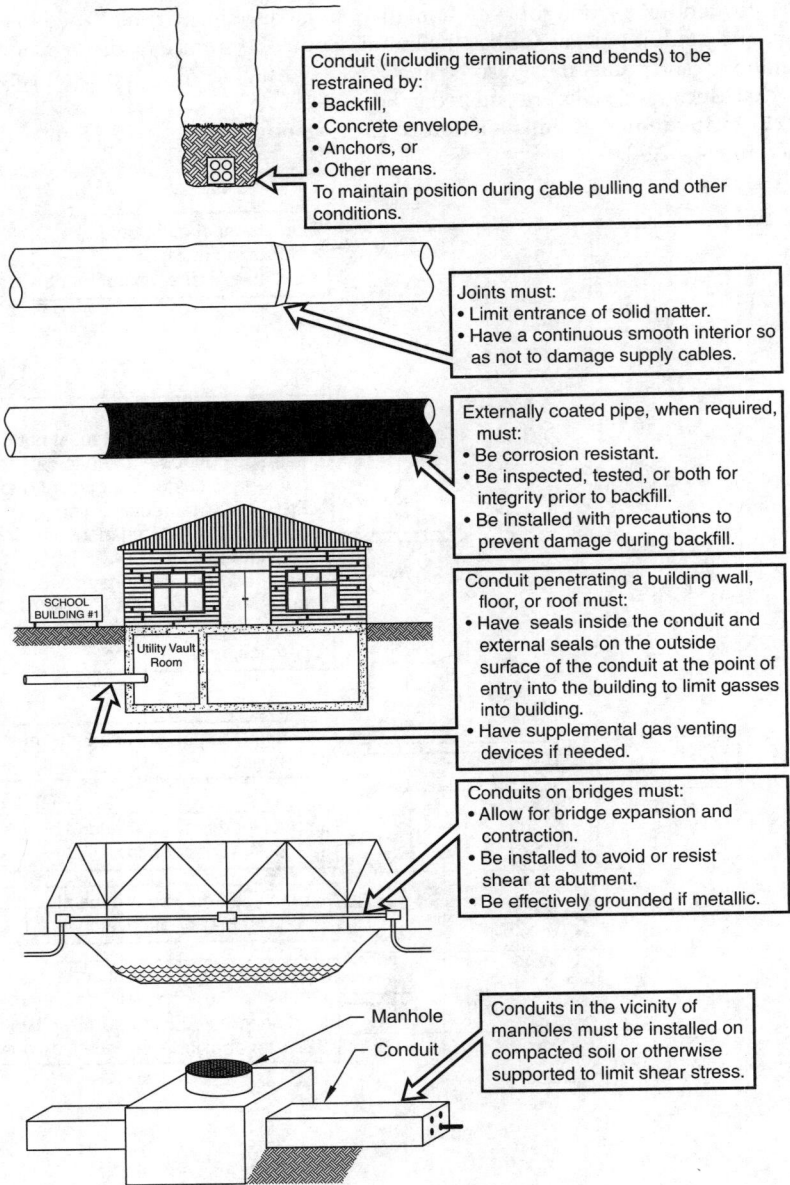

Conduit (including terminations and bends) to be restrained by:
• Backfill,
• Concrete envelope,
• Anchors, or
• Other means.
To maintain position during cable pulling and other conditions.

Joints must:
• Limit entrance of solid matter.
• Have a continuous smooth interior so as not to damage supply cables.

Externally coated pipe, when required, must:
• Be corrosion resistant.
• Be inspected, tested, or both for integrity prior to backfill.
• Be installed with precautions to prevent damage during backfill.

SCHOOL BUILDING #1

Utility Vault Room

Conduit penetrating a building wall, floor, or roof must:
• Have seals inside the conduit and external seals on the outside surface of the conduit at the point of entry into the building to limit gasses into building.
• Have supplemental gas venting devices if needed.

Conduits on bridges must:
• Allow for bridge expansion and contraction.
• Be installed to avoid or resist shear at abutment.
• Be effectively grounded if metallic.

Manhole
Conduit

Conduits in the vicinity of manholes must be installed on compacted soil or otherwise supported to limit shear stress.

Fig. 322-2. Installation requirements for ducts and joints (Rule 322B).

323. MANHOLES, HANDHOLES, AND VAULTS

323A. Strength. Manholes, handholes, and vaults must be designed to with-stand a variety of loads as outlined in Fig. 323-1.

In general, dead loads are the weight of the structure, constant weight on the structure, and permanent attachments to the structure. In general, live loads are loads that are not permanent, like a truck passing over the top of a manhole. Impact loads are generally applied in a similar fashion as live loads and are expressed as a percentage of live loads.

For roadway areas the Code provides NESC Fig. 323-1 and NESC Fig. 323-2 of a typical 10-wheel semitractor trailer truck with associated loads to use for live-load calculations. A 10-wheel semi is used as opposed to an 18-wheeler, as a 10-wheel semi will have a larger force (weight) allocated to each wheel. A note in Rule 323A1 is also provided, reminding the designer that road construction equipment may exceed the typical truck loads. For areas not subject to vehicle traffic, the Code requires that the design live load be not less than 300 lb/ft^2.

The Code requires that the live loads be increased 30 percent for impact. Impact loads can be reduced for specific burial depths (see Rule 322A).

The weight of the manhole, handhole, or vault must be sufficient to withstand hydraulic, frost, or other uplift forces. If the structure weight is not sufficient, some type of restraint or anchorage must be provided. The weight of equipment inside the structure cannot be considered, as the equipment may be removed or modified.

Examples of live loads in roadways and live loads in areas not subject to vehicular loading are shown in Fig. 323-2.

When a pulling iron is installed in a manhole, handhold, or vault, it must be rated to withstand twice the expected load. See Fig. 323-3.

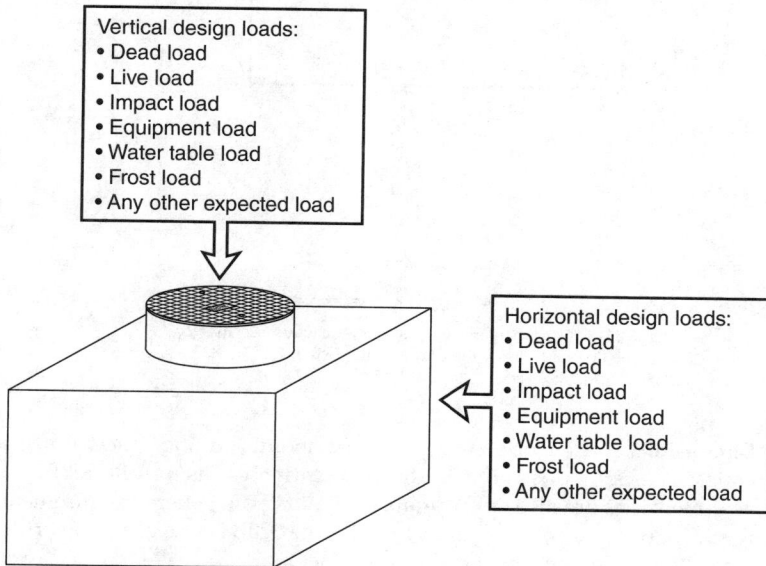

Fig. 323-1. Loads on manholes, handholes, and vaults (Rule 323A).

Live load in roadway areas must consider moving tractor-semitrailer truck with vehicle wheel loads per **NESC** Figs. 323-1 and 323-2.

Road construction loads may exceed the completed roadway loads.

Live loads must be increased by 30% for impact.

Consider hydraulic, frost, or other uplift.

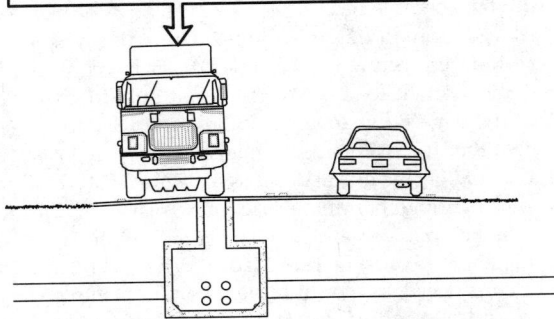

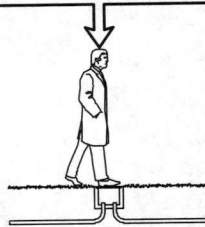

Live load in areas not subject to vehicular loading must be not less than 300 lb/ft^2.

Live loads must be increased by 30% for impact.

Consider hydraulic, frost, or other uplift.

See Photo(s)

Fig. 323-2. Examples of live loads in roadways and live loads in areas not subject to vehicular loading (Rules 323A1, 323A2, 323A3, and 323A4).

323B. Dimensions. The Code is very specific about working space dimensions in manholes. The Code does not specifically mention vaults in Rule 323B but the manhole dimensions are also appropriate for vaults. The general requirement for working space in Part 3 is found in Rule 312. Rule 323B is one of the few rules in Part 3 that actually dimensions the working space.

The rules for working space in manholes are outlined in Figs. 323-4 and 323-5.

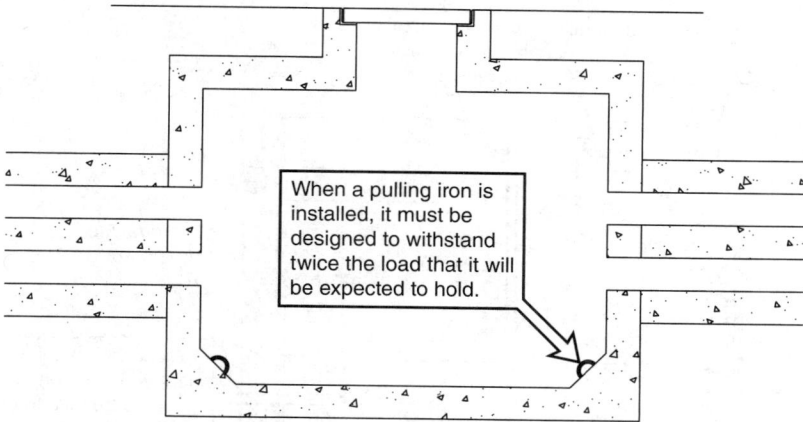

> When a pulling iron is installed, it must be designed to withstand twice the load that it will be expected to hold.

Fig. 323-3. Pulling irons in manholes, handholes, and vaults (Rule 323A5).

It is important to note that the clear working space dimensions are measured after cables and equipment are placed in the manhole. They are not the dimensions to the bare manhole wall.

323C. Manhole Access. The rules for manhole access are outlined in Figs. 323-6 through 323-11.

The access dimensions required in Rule 323C are for the manhole opening. They are not the dimensions of the lid or cover. The lid or cover could be larger than the manhole opening.

Additional rules related to manhole access require the openings to be free of protrusions that will injure workers or prevent quick egress. Where practical, they should be located outside of highways, intersections, and crosswalks to reduce traffic hazards to the workers.

The manhole access openings should not be located directly over cables or equipment. If this requirement is not practical, the manhole opening located over cables (not equipment) must have a safety sign or a protective barrier over the cables or a fixed ladder.

A manhole greater than 4 ft deep must be designed so it can be entered by means of a ladder or other suitable climbing device. If proper working space exists in the manhole per Rule 323B and the manhole opening is above the working space, ladder access should not be difficult. The ladder is not required to be permanent. Ladder requirements are also covered in Rule 323F. The Code does not consider the cables and equipment in the manhole to be suitable climbing devices.

323D. Covers. The rules for manhole and handhole covers are outlined in Fig. 323-12.

323E. Vault and Utility Tunnel Access. The general access rules for vaults and utility tunnels are similar to the general manhole access rules discussed in Rule 323C. No dimensions are provided for vault access openings. Accepted good practice must be applied. The manhole access dimensions are certainly one of

Horizontal dimension of clear working space must not be less than 3'. (See exceptions.)

Cable rack or equipment

3'

6'

Vertical dimension of clear working space must not be less than 6'. (Except as shown below.)

The 6' vertical dimension may be reduced (no value specified) if the manhole opening is within 1' horizontally of the adjacent interior side wall.

1'

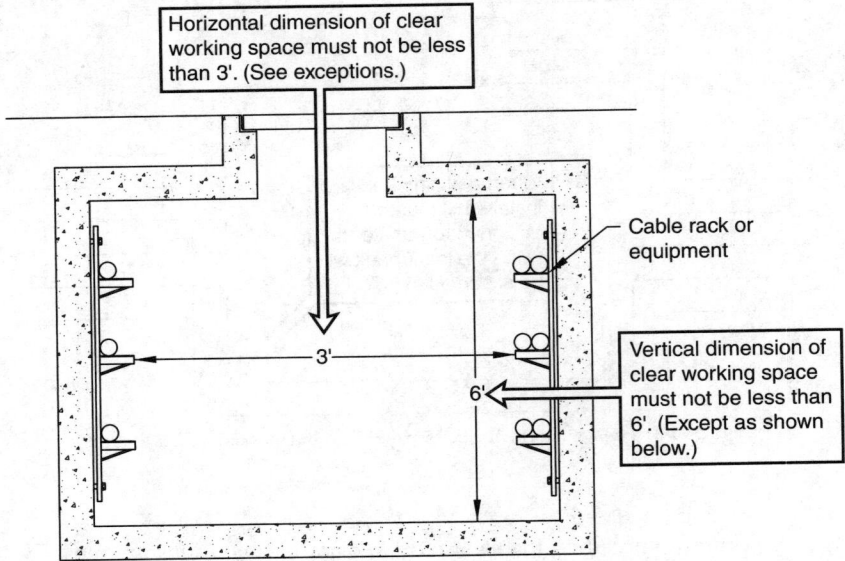

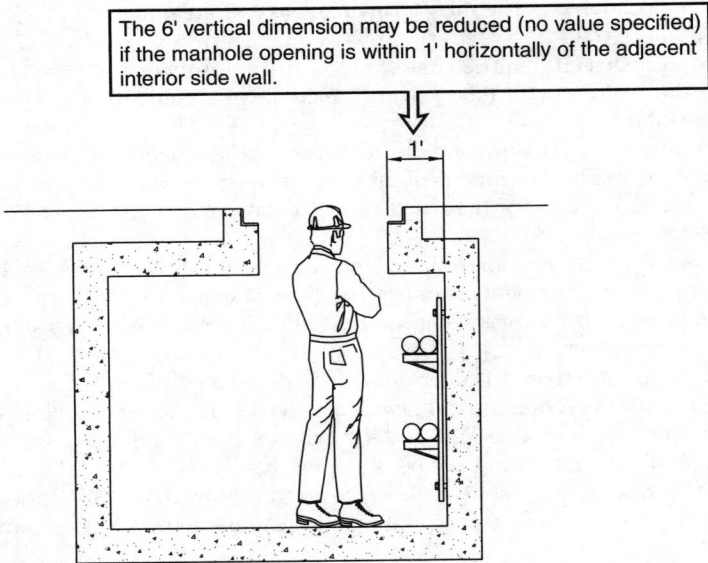

Fig. 323-4. Manhole working space dimensions (Rule 323B).

Exception 1:
If cables only (no equipment) are located on one wall and the opposite wall is unoccupied, then the 3' horizontal dimension of the clear working space can be reduced to 30".

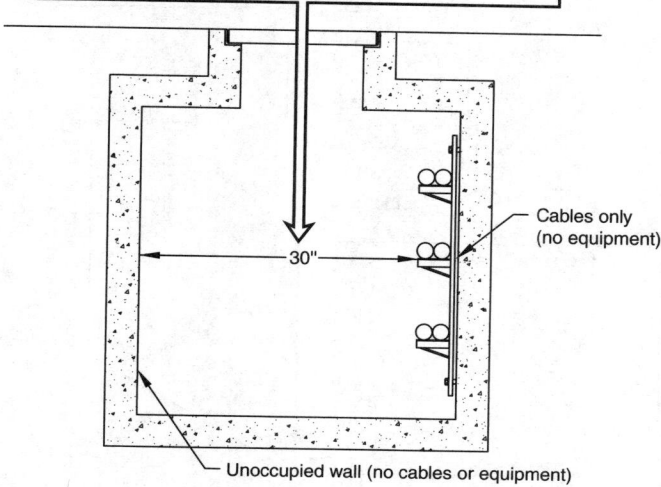

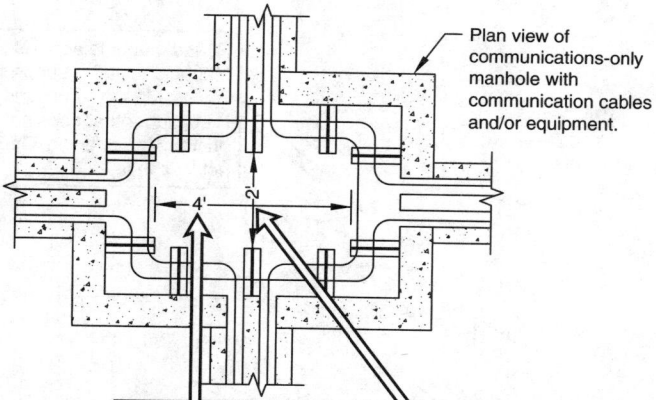

30"

Cables only
(no equipment)

Unoccupied wall (no cables or equipment)

Plan view of
communications-only
manhole with
communication cables
and/or equipment.

4'

2'

Exception 2:
In a communications-only manhole, horizontal dimensions of the clear working space may be reduced to 2' if the other horizontal dimension is increased so that the sum of the two dimensions is at least 6'.

Fig. 323-5. Exceptions to manhole working space dimensions (Rule 323B).

Round manhole access. Manhole contains supply cables. Opening not less than 26".

26"

See Photo(s)

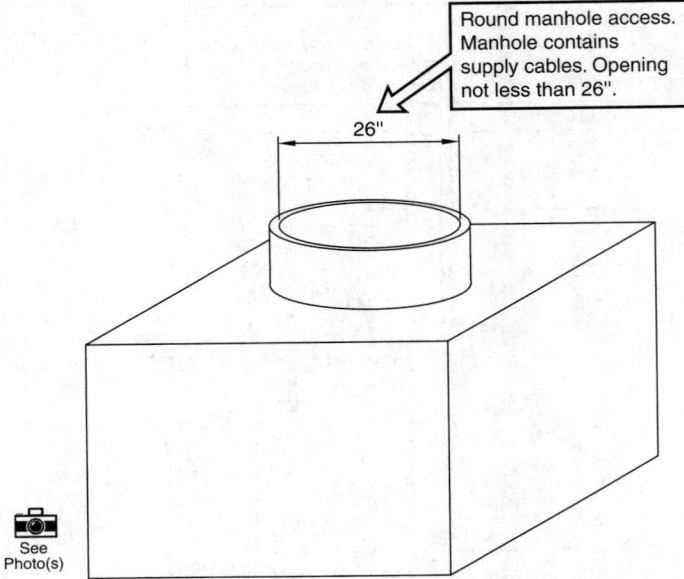

Fig. 323-6. Manhole access diameter for a round manhole opening in a manhole containing supply cables (Rule 323C1).

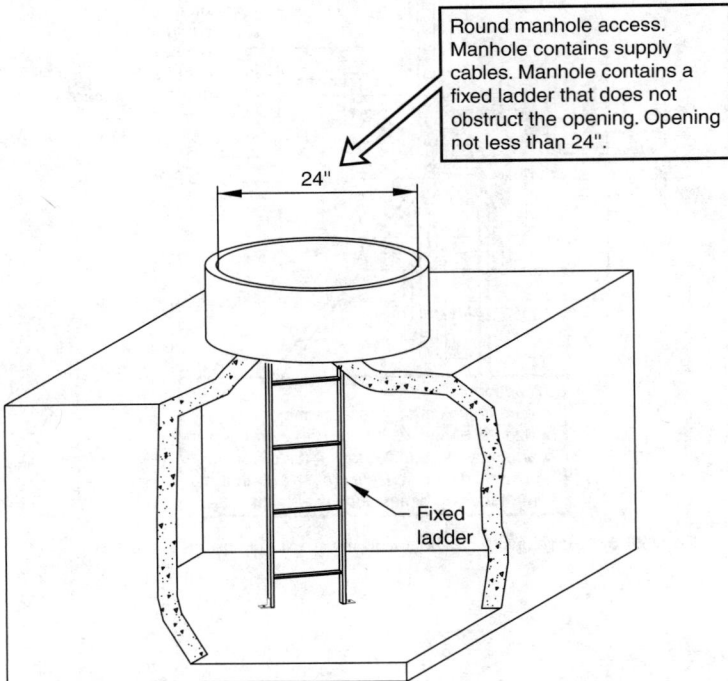

Round manhole access. Manhole contains supply cables. Manhole contains a fixed ladder that does not obstruct the opening. Opening not less than 24".

24"

Fixed ladder

Fig. 323-7. Manhole access diameter for a round manhole opening in a manhole containing supply cables (Rule 323C1).

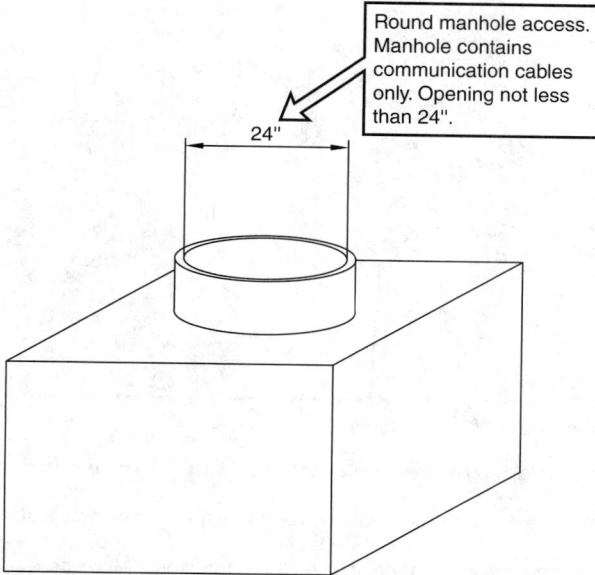

Round manhole access. Manhole contains communication cables only. Opening not less than 24".

24"

Fig. 323-8. Manhole access diameter for round manhole opening in a manhole containing communication cables only (Rule 323C1).

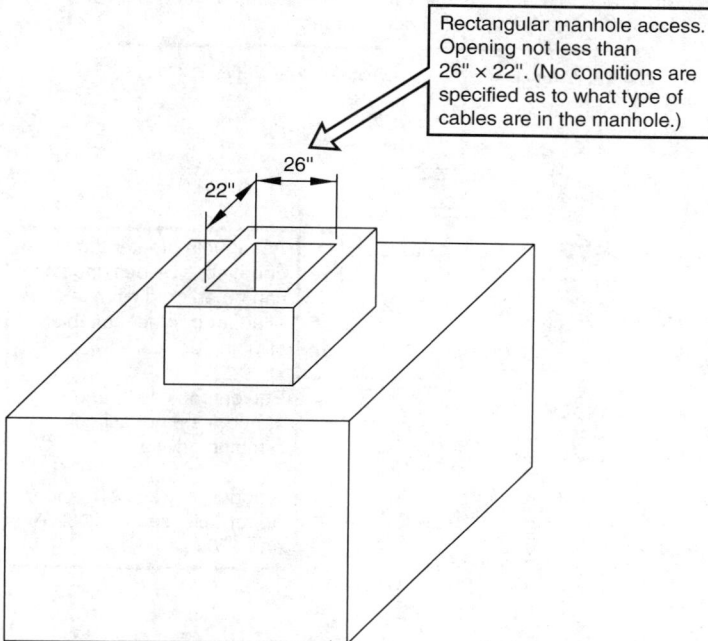

Rectangular manhole access. Opening not less than 26" × 22". (No conditions are specified as to what type of cables are in the manhole.)

26"

22"

Fig. 323-9. Manhole access size for a rectangular manhole opening (Rule 323C1).

Openings free from protrusions.

Manhole opening located so that safe access can be provided.

In highway, located outside the paved roadway when practical.

Located outside of street intersections and crosswalks where practical.

Personnel access not located over the cable or equipment. If not practical, openings can be over the cable if a safety sign or protective barrier or fixed ladder is used.

Fig. 323-10. Manhole access opening requirements (Rules 323C2, 323C3, and 323C4).

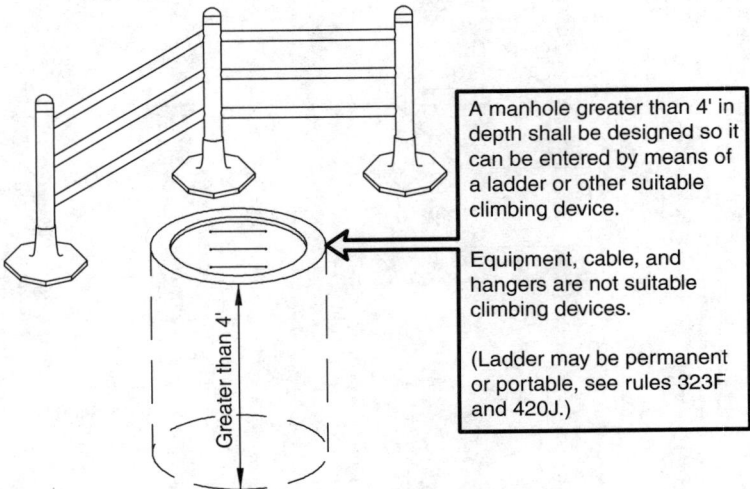

A manhole greater than 4' in depth shall be designed so it can be entered by means of a ladder or other suitable climbing device.

Equipment, cable, and hangers are not suitable climbing devices.

(Ladder may be permanent or portable, see rules 323F and 420J.)

Greater than 4'

Fig. 323-11. Manhole entry means (Rule 323C5).

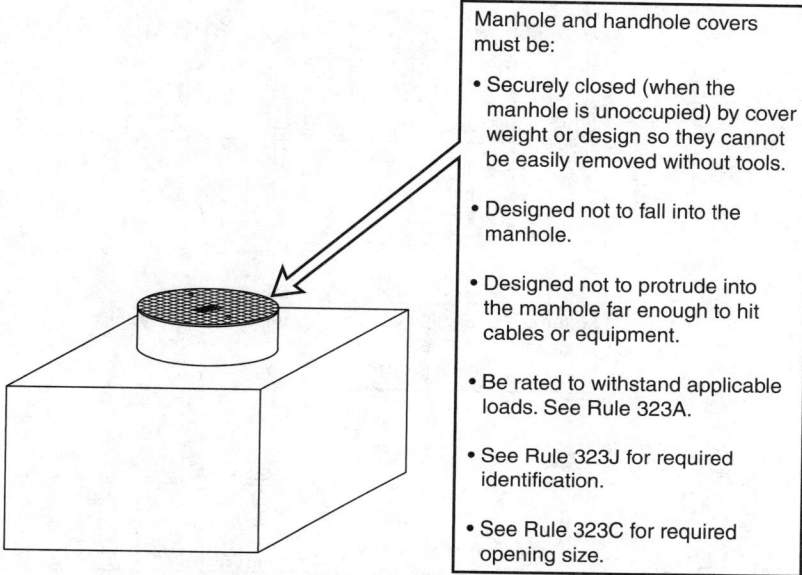

Manhole and handhole covers must be:

• Securely closed (when the manhole is unoccupied) by cover weight or design so they cannot be easily removed without tools.

• Designed not to fall into the manhole.

• Designed not to protrude into the manhole far enough to hit cables or equipment.

• Be rated to withstand applicable loads. See Rule 323A.

• See Rule 323J for required identification.

• See Rule 323C for required opening size.

Fig. 323-12. Requirements for manhole and handhole covers (Rule 323D).

the best references for accepted good practice in this case. Typically, manholes are thought to have openings or lids, and vaults and tunnels are thought to have openings, lids, or doors. Utility vaults can be found in basements and on the ground floor of buildings and even on other floors in high-rise buildings. Utility tunnels can be found under cities and college campuses. The vault and utility tunnel access rules provide additional requirements for doors accessible to the public including locking of the doors and posting of safety signs. Section 39, Rules 390 and 391, are dedicated to installations in tunnels. If an above ground vault has a ventilation opening that is not protected, unguarded energized parts and controls must be located using the safety clearance zone requirements of Part 1, Electric Supply Stations, Rule 110A and **NESC** Table 110-1. Essentially a vault with exposed energized parts is treated like a substation and a ventilation opening in the vault that is not protected is treated like a chain-link fence and requires application of Rule 110A2 and **NESC** Table 110-1. Examples of vaults and utility tunnels are shown in Fig. 323-13.

323F. Ladder Requirements. Fixed ladders in manholes and vaults are required to be corrosion-resistant. Portable ladders must meet the work rules in Part 4, Rule 420J. Additional ladder requirements related to manhole access can be found in Rule 323C. The work rules in Part 4, Rule 423B require testing for flammable gases and oxygen deficiency before entering a manhole or an unventilated vault.

323G. Drainage. Drainage for manholes, handholes, and vaults is not required to drain into sewers. Many times, natural drainage is provided by utilizing a dry

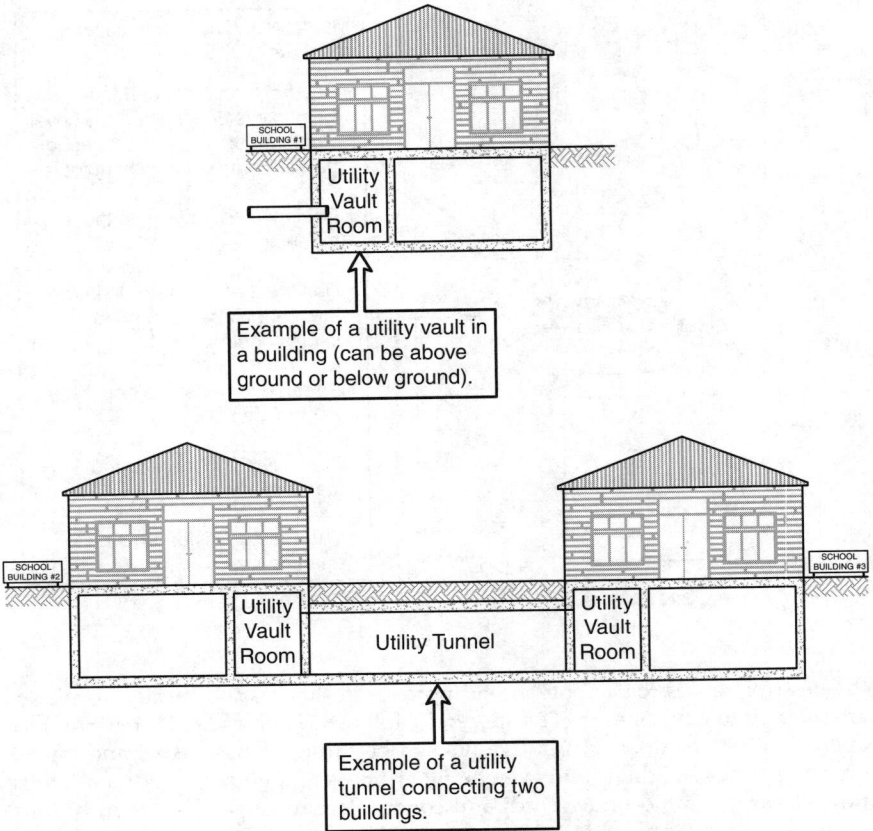

Fig. 323-13. Examples of vaults and utility tunnels (Rule 323E).

well or gravel base. When drainage into sewers is desired, a trap is a common method that can be used to keep sewer gas out of the manhole, vault, or tunnel. See Fig. 323-14.

323H. Ventilation. Ventilation must be provided if the public can somehow be affected by the gases that collect in manholes, vaults, and tunnels. An exception applies to areas under water or other impractical locations. If ventilation is not needed (typical of most manholes in city streets), the work rules in Part 4, Rule 423B require an adequate air supply for the employee while the employee is working in the manhole or unventilated vault.

323I. Mechanical Protection. If a manhole, handhole, or vault has a grated cover that allows objects to fall on the supply cables and equipment, the supply cables and equipment must be located (positioned) or guarded (physically protected) to prevent damage. This rule does not mention communication cables or

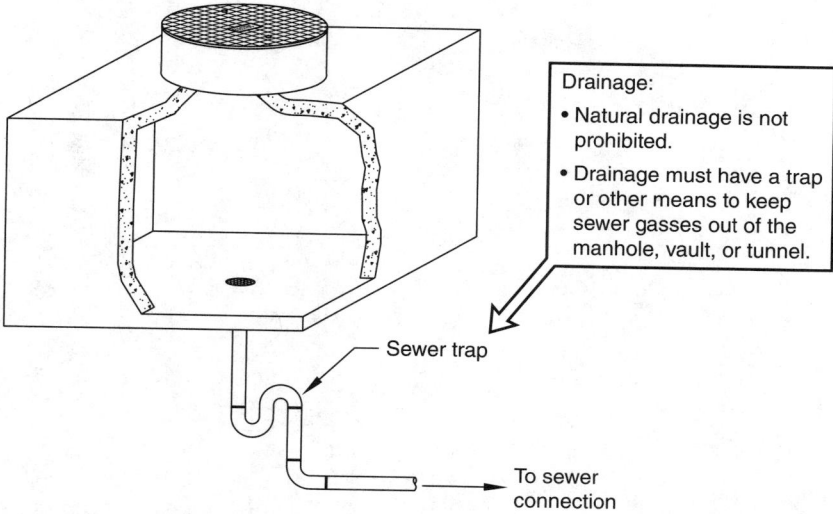

Drainage:
- Natural drainage is not prohibited.
- Drainage must have a trap or other means to keep sewer gasses out of the manhole, vault, or tunnel.

Sewer trap

To sewer connection

Fig. 323-14. Drainage requirements for manholes, vaults, and tunnels (Rule 323G).

equipment. Grated covers are typically used for venting. Cables and equipment must not be located directly under any type of manhole opening for personnel access per Rule 323C.

323J. Identification. The rules for identifying manhole and handhole covers are outlined in Fig. 323-15.

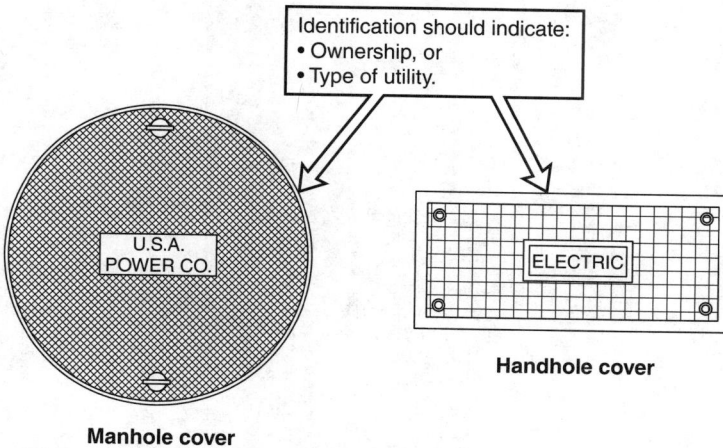

Identification should indicate:
- Ownership, or
- Type of utility.

U.S.A.
POWER CO.

ELECTRIC

Handhole cover

See Photo(s)

Manhole cover

Fig. 323-15. Requirements for identification of manhole and handhole covers (Rule 323J).

Section 33

Supply Cable

330. GENERAL

Section 33 is only applicable to supply cable. It does not address communication cables. Rule 330 starts out with a requirement that supply cables be tested in accordance with applicable standards. The rule also provides additional general requirements that are applicable to both the supply cable specifier and the supply cable manufacturer. The supply cable should be rated for the particular application and environment. The cable must also be able to withstand the fault current to which it will be subjected. Common underground supply cable ratings for distribution lines are 15 kV or 25 kV or 35 kV, 100% insulation level or 133% insulation level, full concentric neutral or 1/3 concentric neutral, ethylene-propylene rubber (EPR) insulation or cross-linked polyethylene (XLP) insulation or tree-retardant cross-linked polyethylene (TRXLP) insulation, jacketed in various aluminum and copper conductor sizes. Underground supply transmission line cables (above 35 kV) are also available and sometimes used in the industry. Rule 350 also addresses supply cable requirements.

331. SHEATHS AND JACKETS

Cable jackets protect cables from adverse environmental conditions. Older underground residential distribution (URD) cables did not have a cable jacket, and concentric-neutral corrosion became a widespread problem in some soils. Newer URD cables can be purchased with an outer jacket to protect the concentric neutral conductors. URD cables with a jacket do not have the concentric neutral in contact with the earth. The jacket may have to be stripped back to ground the concentric neutral. See Rule 096 for a grounding discussion. An example of a sheath, jacket, and shield on a typical URD supply cable is shown in Fig. 331-1.

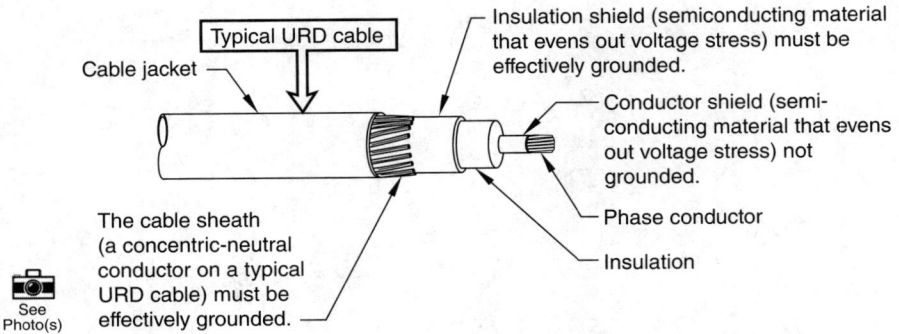

Figure 331-1. Example of a sheath, jacket, and shield on a typical URD supply cable (Rule 331).

332. SHIELDING

General shielding guidelines are provided in this rule. Organizations that specialize in developing cable standards are noted for a reference. See the figure in Rule 331 for a description of the individual components that make up a typical underground residential distribution (URD) supply cable.

333. CABLE ACCESSORIES AND JOINTS

Cable accessories and joints are the elbows, splices, and terminations used on supply cables. These accessories must be able to withstand stresses similar to the cable to which they are connected. The requirements for cable accessories in this rule focus on material specifications. Cable accessories are also discussed in Sec. 37, "Supply Cable Terminations." Section 37 focuses on the installation of the cable terminations. Examples of cable accessories and joints are shown in Fig. 333-1.

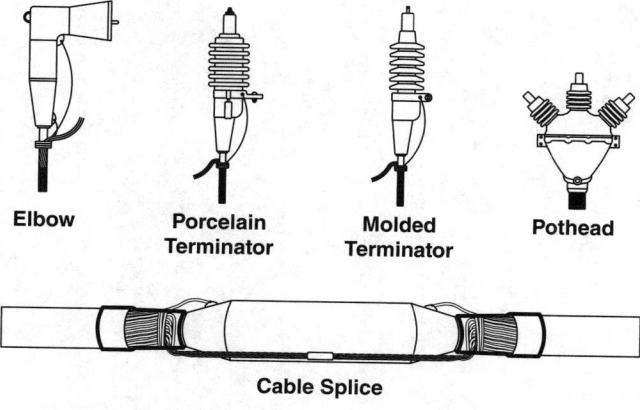

Elbow   Porcelain   Molded   Pothead
             Terminator   Terminator

Cable Splice

See Photo(s)

Figure 333-1. Examples of cable accessories and joints (Rule 333).

Section 34

Cable in
Underground Structures

340. GENERAL

Section 34 provides rules for underground cables when they are installed in underground structures (e.g., conduit systems). Section 33 provides rules for the components of a supply cable. Section 32 provides rules for the conduit system. Section 34 does not apply to direct-buried cables. See Fig. 340-1.

Per Rule 340B, a supply cable over 2 kV to ground in a conduit system consisting of nonmetallic conduit should consider the need for an effectively grounded shield, sheath, or both. When a supply cable is direct-buried or installed in duct not part of a conduit system, Rule 350B requires an effectively grounded shield, sheath, or concentric neutral with more force (via the word *shall* instead of *should*), and the requirement is for above 600 V, not 2 kV.

341. INSTALLATION

341A. General. Supply cables are mechanically stressed when they are bent. The NESC does not specify a cable bending radius. Only a general statement requiring bending to be controlled to avoid damage is provided. The cable manufacturer's recommendations should be used. A bending radius is usually expressed as a multiple of the cable diameter. When the supply cable is installed in conduit, the cable bending radius should always be compatible with the conduit bending radius. See Fig. 341-1.

Conduit system covered in Sec. 32.

Supply cable covered in Sec. 33.

Installation of underground cables in underground structures (e.g., conduit systems) covered in Sec. 34.

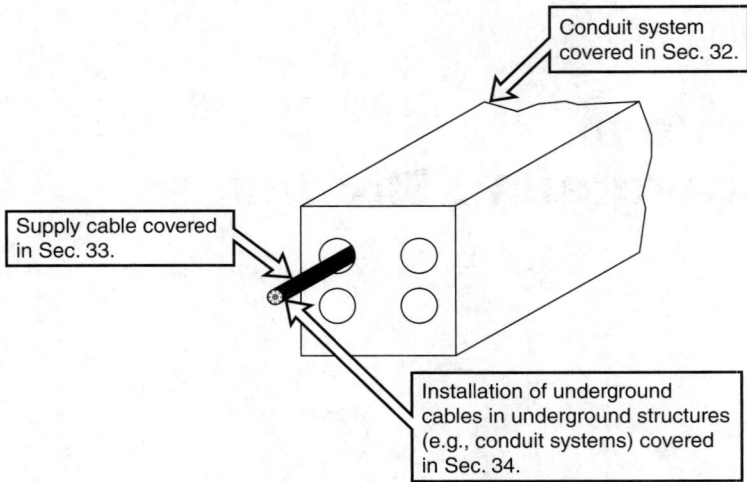

Fig. 340-1. Application of Secs. 32, 33, and 34 (Rule 340).

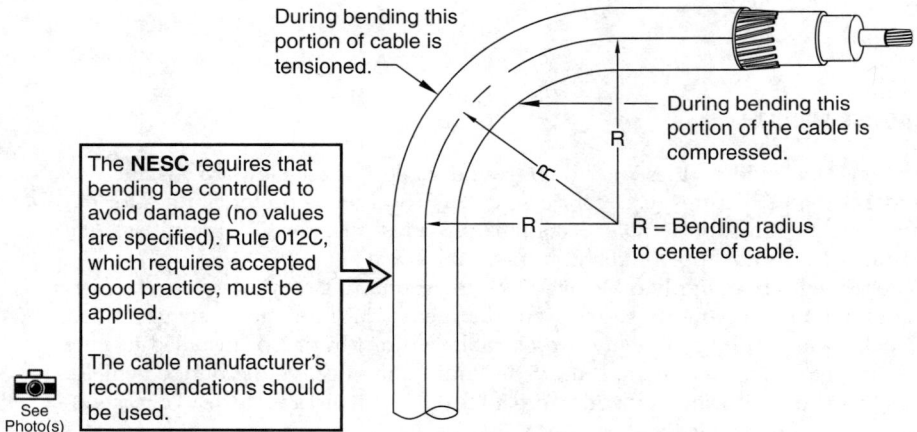

During bending this portion of cable is tensioned.

During bending this portion of the cable is compressed.

R

The **NESC** requires that bending be controlled to avoid damage (no values are specified). Rule 012C, which requires accepted good practice, must be applied.

The cable manufacturer's recommendations should be used.

R

R = Bending radius to center of cable.

See Photo(s)

Fig. 341-1. Cable bending radius (Rule 341A1).

The **NESC** does not specify conduit fill requirements. Conduit fill is the number of conductors or cables that fit in a specific conduit size. Conduit fill is normally limited to some percentage of the conduit cross-sectional area. The National Electrical Code (NEC) is a reference for acceptable good practice in this case. Many utilities develop conduit fill tables for the conduits and cables they normally stock. Information regarding conduit fill calculations is shown in Fig. 341-2.

Internal area of
the conduit (duct)

Area of
cables

Conduit fill is a comparison of the area of the
cables to the area inside the conduit (duct).
Conduit fill is typically expressed as a percent fill
(e.g., 40%). Jam ratio calculations may also be
appropriate. The **NESC** does not specify a
conduit fill percentage or jam ratio. Rule 012C,
which requires accepted good practice, must be
used. Conduit and cable manufacture data and
the National Electrical Code (NEC) are common
references for accepted good practice.

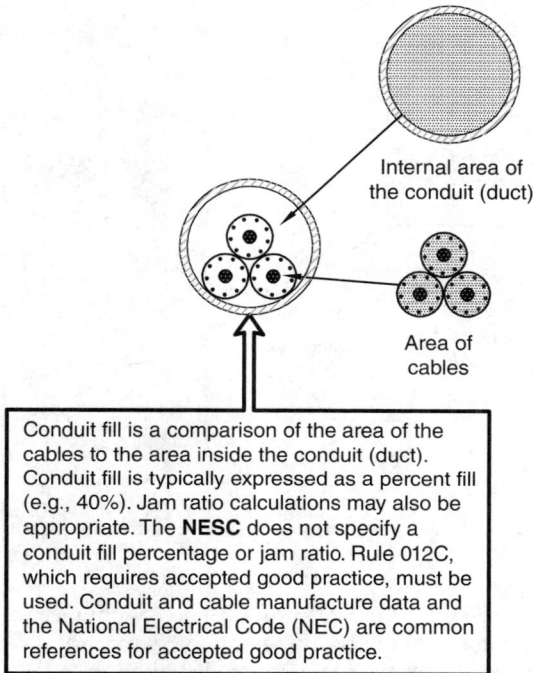

Fig. 341-2. Conduit fill calculations (Rule N/A).

The **Code** requires limiting pulling tensions and sidewall pressures to avoid cable damage. A note suggests using the cable manufacturer's recommendations. Cable manufacturers and cable lubricant suppliers can also supply engineering guidelines and software for making pulling calculations. Cable lubricants are used to make cable pulling easier, therefore reducing the pulling tension. Additional ways to reduce pulling tensions include pulling from an uphill point to a downhill point and starting the pull near a conduit bend instead of away from a conduit bend. Sidewall pressures are the forces that push on the side of the cable around a conduit bend. The same methods used to reduce pulling tension can be used to reduce sidewall pressure. Sidewall pressure is a function of the cable tension at the bend and the radius of the bend, not the diameter of the conduit. See Fig. 341-3.

The **Code** specifically discusses pulling tensions and sidewall pressures relative to supply cables. The **Code** does not discuss pulling tensions for communication cables, but communication cables require the same considerations.

The **Code** requires ducts to be cleaned of foreign material. This is typically referred to as "swabbing" the duct. The cable lubricants must be safe for the type of conduit and conduit materials. Too little friction can also be a problem. If a cable is installed on a downhill slope or vertical run, the cable may need to be restrained to prevent a cable elbow or termination from being stressed or pulled out.

Cable let-off
reel

Pulling tensions and
sidewall pressures should
be limited to avoid supply
cable damage.

Lubricant

Sidewall pressure
exists here

Pulling tensions

Pulling eye
or basket grip

Winch

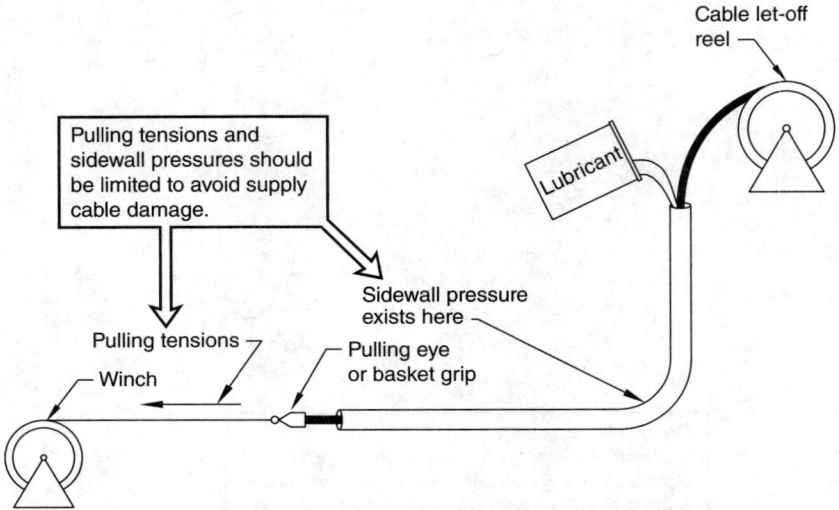

Fig. 341-3. Supply cable pulling tension and sidewall pressure (Rule 341A2).

Rule 320B requires separation between supply and communication conduit systems. Rule 341A6 reinforces that idea by saying that supply and communication cables must not be installed in the same duct unless they are operated and maintained (not necessarily owned) by the same utility. Rule 341A7 permits the installation of communication cables of different communication utilities in the same duct provided the utilities are in agreement. See the beginning of Sec. 32 for definitions of duct, conduit, and conduit system. The requirements of Rules 341A6 and 341A7 related to supply and communication cables in duct are outlined in Fig. 341-4.

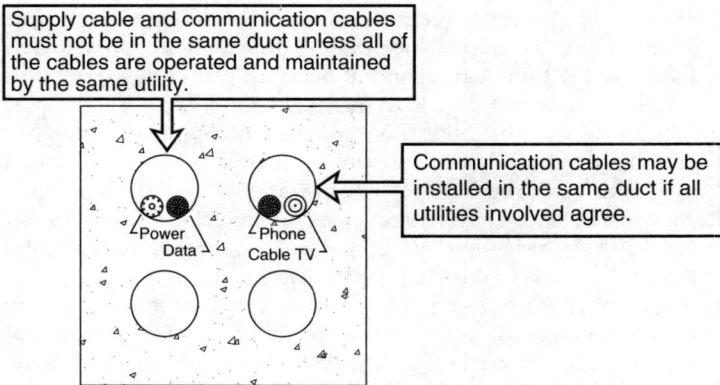

Supply cable and communication cables
must not be in the same duct unless all of
the cables are operated and maintained
by the same utility.

Communication cables may be
installed in the same duct if all
utilities involved agree.

Power
Data

Phone
Cable TV

Fig. 341-4. Supply and communication cables in duct (Rules 341A6 and 341A7).

341B. Cable in Manholes and Vaults. Rule 341B covers the installation of cable in manholes and vaults. The rule is divided into three main parts, supports, clearance, and identification. Rule 341B1 covers supports for cable in manholes and vaults. The rules for supporting cable in manholes and vaults are outlined in Fig. 341-5.

Rule 341B2 covers clearance for cables in manholes and vaults. The rules for clearance of cables in manholes and vaults focus on clearance between joint-use (supply and communication) facilities. Joint-use (supply and communication) manholes and vaults are usually not preferred by either utility. If a joint-use (supply and communication) duct bank is used, the communication conduits can branch off of the duct bank before the duct bank enters the supply manhole. The communication conduits can then enter a separate communication manhole adjacent to the supply manhole. If this system is not used and a joint-use (supply and communication) manhole is desired, it must be done with the concurrence of all the parties involved. The work rules in Part 4, Rule 433, must be applied when a communications worker

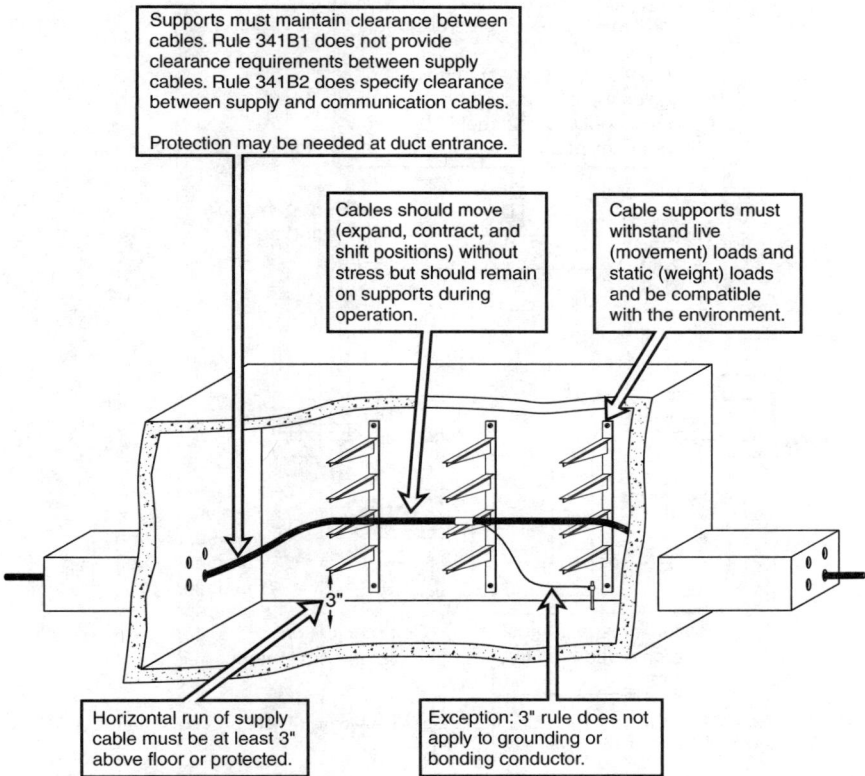

Supports must maintain clearance between cables. Rule 341B1 does not provide clearance requirements between supply cables. Rule 341B2 does specify clearance between supply and communication cables.

Protection may be needed at duct entrance.

Cables should move (expand, contract, and shift positions) without stress but should remain on supports during operation.

Cable supports must withstand live (movement) loads and static (weight) loads and be compatible with the environment.

Horizontal run of supply cable must be at least 3" above floor or protected.

Exception: 3" rule does not apply to grounding or bonding conductor.

Fig. 341-5. Supporting cable in manholes and vaults (Rule 341B1).

is in a joint-use (supply and communication) manhole. The preferred location of supply cables in a joint-use (supply and communication) manhole is below communication cables. This is the opposite position of overhead lines as supply conductors are preferred at the higher level on an overhead line per Rule 220B. The rules for clearance of cables (and equipment per Rule 341B2b) in joint-use (supply and communication) manholes and vaults are outlined in Figs. 341-6 and 341-7.

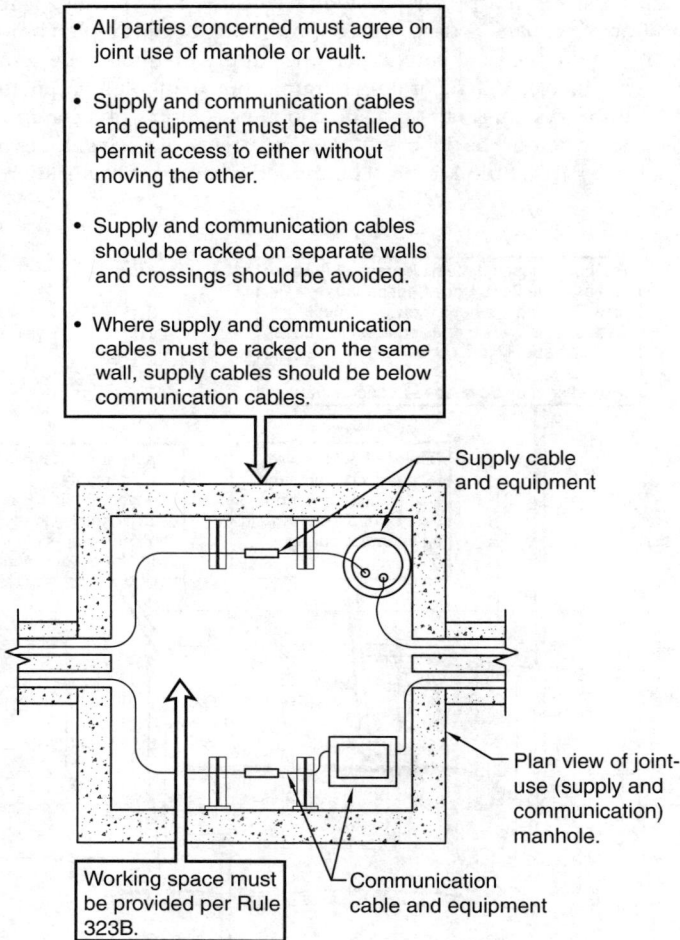

- All parties concerned must agree on joint use of manhole or vault.

- Supply and communication cables and equipment must be installed to permit access to either without moving the other.

- Supply and communication cables should be racked on separate walls and crossings should be avoided.

- Where supply and communication cables must be racked on the same wall, supply cables should be below communication cables.

Supply cable and equipment

Plan view of joint-use (supply and communication) manhole.

Working space must be provided per Rule 323B.

Communication cable and equipment

Fig. 341-6. Clearance of cables and equipment in joint-use (supply and communication) manholes and vaults (Rule 341B2).

Clearance of communication cable to:
• Supply cable of 0–15,000 V Ø-Ø: Not less than 6".
• Supply cable of 15,001–50,000 V Ø-Ø: Not less than 9".
• Supply cable of 50,001–120,000 V Ø-Ø: Not less than 12".
• Supply cable of 120,001 V Ø-Ø and above: Not less than 24".

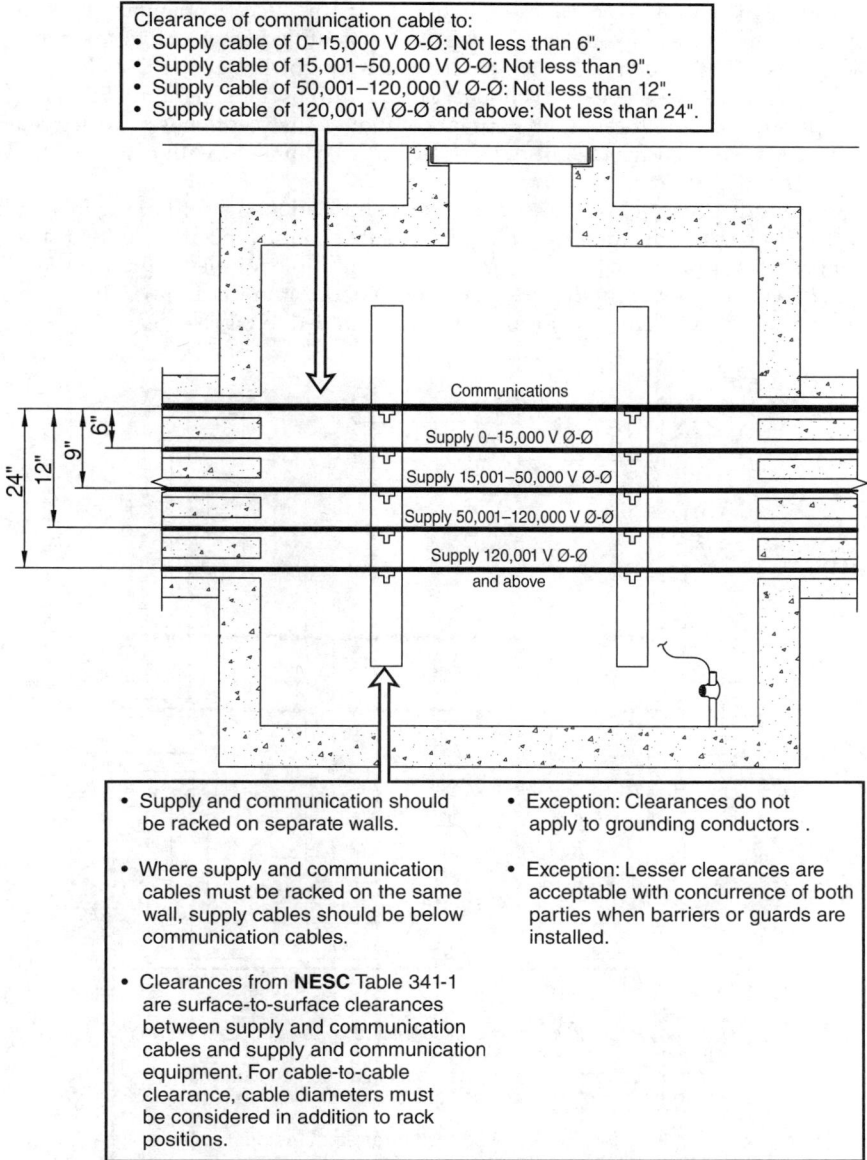

Communications

Supply 0–15,000 V Ø-Ø

Supply 15,001–50,000 V Ø-Ø

Supply 50,001–120,000 V Ø-Ø

Supply 120,001 V Ø-Ø
and above

• Supply and communication should
  be racked on separate walls.

• Where supply and communication
  cables must be racked on the same
  wall, supply cables should be below
  communication cables.

• Clearances from **NESC** Table 341-1
  are surface-to-surface clearances
  between supply and communication
  cables and supply and communication
  equipment. For cable-to-cable
  clearance, cable diameters must
  be considered in addition to rack
  positions.

• Exception: Clearances do not
  apply to grounding conductors .

• Exception: Lesser clearances are
  acceptable with concurrence of both
  parties when barriers or guards are
  installed.

Fig. 341-7. Clearance of cables and equipment in joint-use (supply and communication) manholes and vaults (Rule 341B2).

Rule 341B3 requires the cables in manholes or other access openings of a conduit system (i.e., vaults and handholes) to be identified by tags that are corrosion-resistant and suitable for the environment in which they are installed. The quality of the tag must be good enough to read with portable lighting. Brass tags and plastic compounds are both popular solutions. An exception is provided that eliminates the need for tags if maps or diagrams combined with cable position provide sufficient identification.

For joint-use manholes or vaults, a cable identification tag or marking of some type must be provided denoting the utility name and the type of cable used. This requirement applies to a manhole or vault with multiple utilities of the same kind (i.e., power and power) or joint-use (i.e., power and communication) utilities. The identification requirements of Rule 341B3 are outlined in Fig. 341-8.

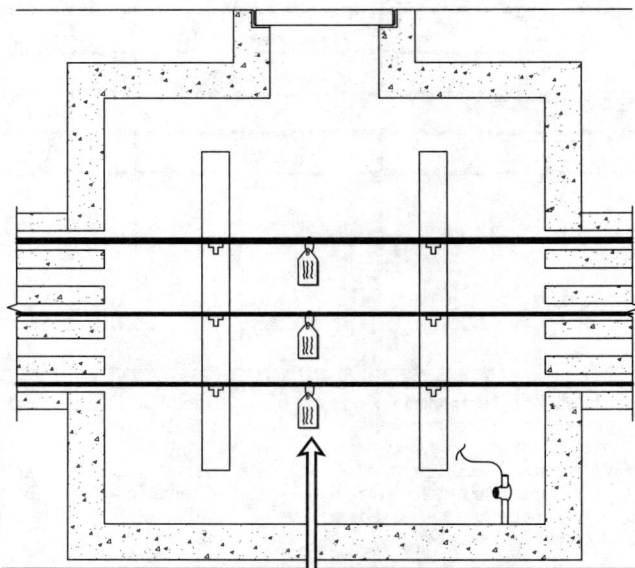

Identification:

• Cables must be permanently identified by tags or other means at manhole or other opening.
(An exception applies where position, in conjunction with diagrams or maps, provides sufficient identification.)

• Identification must be of corrosion-resistant material suitable for the environment.

• Identification must be readable with auxiliary lighting.

• Joint-use locations must identify the utility name and type of cable use.

Fig. 341-8. Identification of cables in manholes and vaults (Rule 341B3).

342. GROUNDING AND BONDING

Rule 342 provides grounding requirements in addition to the general grounding requirements in Rule 314. Cable or joints (splices) with bare metallic shields, sheaths, or concentric neutrals that are exposed to personnel contact must be effectively grounded. See the figure in Rule 331 for a description of the individual components that make up a typical underground residential distribution (URD) supply cable. Cable sheaths or shields must have a common ground to keep different cables at equal ground potential, and the grounding materials must be corrosion-resistant or suitably protected.

343. NUMBER 343 NOT USED IN THIS EDITION

344. COMMUNICATION CABLES CONTAINING SPECIAL SUPPLY CIRCUITS

Rule 344 is the underground equivalent to the overhead Rule 224B. Both this rule and Rule 224B provide special conditions for embedding a supply circuit in a communication cable. Rule 224B lists five special conditions (Rule 224B2a through 224B2e) for an overhead cable construction. Rule 344 lists six special conditions (Rules 344A1 through 344A6) for an underground cable construction. The extra condition for Rule 344 is for identification of the underground cable. Both Rule 344A and Rule 224B provide an exception when the power within the communication cable is limited.

Section 35

Direct-Buried Cable and Cable in Duct Not Part of a Conduit System

The note under the title of Sec. 35 defines how the terms duct or ducts are used in Sec. 35. This note, plus the note under the title of Sec. 32 clarifies that Sec. 32 applies to conduit systems and Sec. 35 applies to direct-buried cable and cable in duct not part of a conduit system. A typical conduit system (combination of conduit, conduits, manholes, handholes, and/or vaults joined to form an integrated whole) can be found under the streets and sidewalks of a large city. A typical duct that is not part of a conduit system can be found in a residential subdivision. One of the biggest differences between conduit systems (covered in Sec. 32) and direct-buried cable and cable in duct that is not part of a conduit system (covered in Sec. 35) is that Sec. 35 specifies burial depths and Sec. 32 does not. Section 32 requires a structural review to determine burial depths. See the discussion and Fig. 302-1, in Rule 302 and see the discussion and Figs. 32-1, and 320-1 at the beginning of Sec. 32.

350. GENERAL

Rule 350A references Sec. 33 (supply cable) and states that Sec. 33 applies to direct-buried supply cable. Additional rules in Sec. 35 supplement the supply cable rules in Sec. 33.

Per Rule 350B, a supply cable over 600 V to ground must have an effectively grounded shield, sheath, or concentric neutral that is continuous. See the figure in Rule 331 for a description of the individual components that make up an underground residential distribution (URD) supply cable.

The exception to Rule 350B recognizes that the standard splicing methods used on concentric-neutral cable do not allow the concentric neutral to remain in a concentric pattern across the splice. See Fig. 350-1.

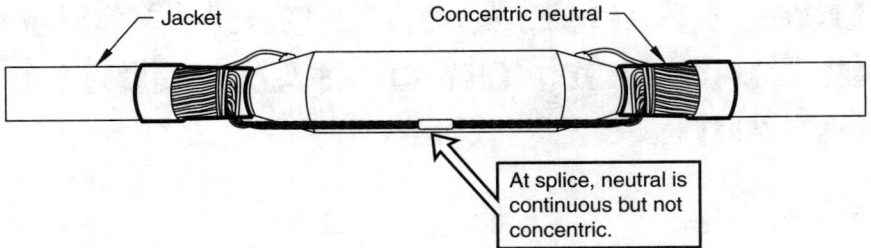

Fig. 350-1. Exception related to splices on concentric-neutral cables (Rule 350B).

Rule 350C states that supply cables meeting Rule 350B of the same supply circuit (e.g., the same three-phase feeder) may be direct-buried or installed in duct with no deliberate separation. See Fig. 350-2.

Rule 350D states that underground direct-buried cables or cables installed in duct below 600 V to ground without an effectively grounded shield or sheath must be placed in close proximity to each other. Underground direct-buried secondary service conductors typically have an insulated (not bare) neutral. The requirement to place the conductors in close proximity with no intentional separation helps reduce the path that a fault current would need to follow if a cable fault occurred. The cables that meet Rule 350B are permitted to be buried with deliberate separation but they are not required to be installed in that manner per Rule 350C. Rule 350C differs from Rule 350D as the cables in 350D are required to be installed in close proximity with no intentional separation. See Fig. 350-3.

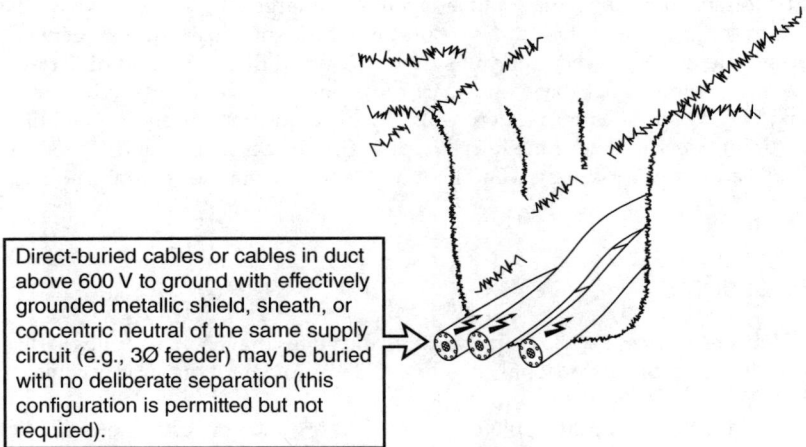

Direct-buried cables or cables in duct above 600 V to ground with effectively grounded metallic shield, sheath, or concentric neutral of the same supply circuit (e.g., 3Ø feeder) may be buried with no deliberate separation (this configuration is permitted but not required).

Fig. 350-2. Cables above 600 V to ground with no deliberate separation (Rule 350C).

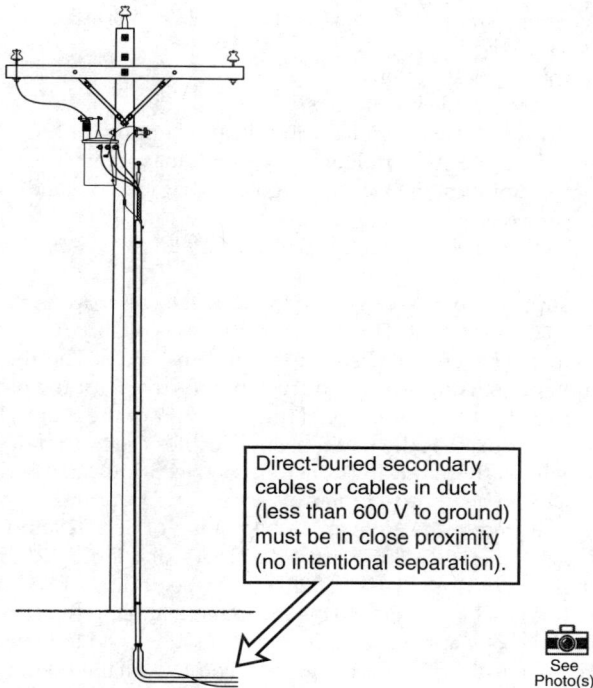

Direct-buried secondary cables or cables in duct (less than 600 V to ground) must be in close proximity (no intentional separation).

See Photo(s)

Fig. 350-3. Cables below 600 V to ground buried in close proximity (Rule 350D).

Rule 350E references Rules 344A1 through 344A5 for rules related to communication cables containing special supply circuits. These rules also apply to direct-buried communication cables and cables in duct not part of a conduit system.

Rule 350F requires marking of direct-buried supply and communication cable and cable in duct not part of a conduit system with the "lightning bolt" and the "telephone handset." NESC Fig. 350-1 provides a diagram of the requirements. The telephone handset is used for all communications, even data and cable TV. This requirement is for direct-buried cables and per Rule 350G and the title of Sec. 35, this requirement also applies to cable in duct not part of a conduit system. Cables installed in conduit systems per Sec. 32 do not need to meet this requirement, and Sec. 32, "Underground Conduit Systems," does not have any conduit labeling or color-coding requirements. A recommendation in Rule 350F states that if color coding is used as an additional method of identifying cable, the American Public Works Association Uniform Color Code for marking underground utility lines is recommended. The recommendation is not recommending color coding as an additional cable identifier, it is simply recommending that if color coding is used that the uniform color code for marking underground lines be applied. The American Public Works Association Uniform Color Code is shown below:

- White—Proposed excavation
- Pink—Temporary survey markings
- Red—Electric power lines, cables, conduit, and lighting cables
- Yellow—Gas, oil, steam, petroleum, or gaseous materials
- Orange—Communication, alarm or signal lines, cables or conduit
- Blue—Potable water
- Purple—Reclaimed water, irrigation and slurry lines
- Green—Sewers and drain lines

Sometimes supply cables are specified with three red stripes in addition to the **Code** required lightning bolt. The three red stripes typically run the length of the cable and are placed an equal distance around the cable. The recommendation does not comment on using stripes, the number of stripes, or the placement of the stripes, only the color of the additional method but not the method itself. Exception 1 to this rule addresses the fact that some cables are too small in diameter to mark or have some other marking constraints. Exception 2 states that unmarked cable from stock can be used to repair existing buried unmarked cables.

Rule 350G is related to the Note at the beginning of Sec. 32 and the Note at the beginning of this section (Sec. 35). See the discussion and Fig. 302-1, in Rule 302 and see the discussion and Figs. 32-1 and 320-1 at the beginning of Sec. 32. The recommendation in Rule 350G is similar to the recommendation in Rule 350F. The recommendation in Rule 350F applies to color coding as a method to identify cables. The recommendation in Rule 350G applies to color coding as a method to identify ducts. The recommendation is not recommending color coding as an additional duct (conduit) identifier, it is simply recommending that if color coding is used that the uniform color code for marking underground lines be applied. If concrete encasement is used around the duct, the **Code** does not contain a rule or recommendation for putting a colored dye in the concrete.

351. LOCATION AND ROUTING

The rules for locating and routing direct-buried cables and cables in duct not part of a conduit system are similar to, but slightly stricter than, the rules for locating conduit systems in Rule 320.

351A. General.   The general requirements for direct-buried cable and duct location and routing are outlined below:

- Subject cables to the least disturbance practical.
- When paralleling and directly over or under other buried facilities, the separation requirements in Rule 353 or 354 must be met.
- Install cables as straight as practical.
- Where required, cable bends must be large enough to avoid cable damage (see Rule 341 for a discussion).
- Route for safe access during construction, inspection, and maintenance.
- Plan route and structure conflicts before trenching, plowing, or boring.

351B. Natural Hazards.   Natural hazards are discussed here for direct-buried cables and in Rule 320 for conduit systems. The cable must be constructed and installed to be protected from damage. If the installation method requires switching from a direct-buried cable or duct installation to a full conduit system installation, Rule 320 would apply.

351C. Other Conditions.   Swimming pools get a lot of attention in the NESC. Part 2, Overhead Lines, and Part 3, Underground Lines, both have specific rules for swimming pool areas. Rule 351C1 addresses direct-buried supply cable or duct (not communication) in the vicinity of in-ground swimming pools. This installation is very common in residential subdivisions in warmer parts of the country. No distances are provided for communications cable or duct adjacent to in-ground swimming pools, the general locating and routing requirements in Rule 351A must be applied. The rules for locating and routing direct-buried supply cable or duct in the vicinity of in-ground swimming pools are outlined in Fig. 351-1.

Aboveground swimming pools are addressed in Rule 351C2. Buildings and other structures (including aboveground swimming pools, tanks, tool sheds, etc.) should not have direct-buried cables or duct installed under them. If they must, the structure must be properly supported so it will not damage the cables or duct. If a conduit system is chosen to protect the cables installed under or near a building, Rule 320B1 would apply.

Rule 351C3 addresses railroad tracks. Direct-buried cable or duct installed longitudinally under the ballast section of railroad tracks is to be avoided. Where it must be done, the direct-buried cable or duct is to be 50 in below the top rail. An exception to Rule 351C3 states that this clearance may be reduced by agreement of the parties concerned. A permit is normally required to cross or parallel a railroad right-of-way, and to obtain the permit, the railroad agency usually requires

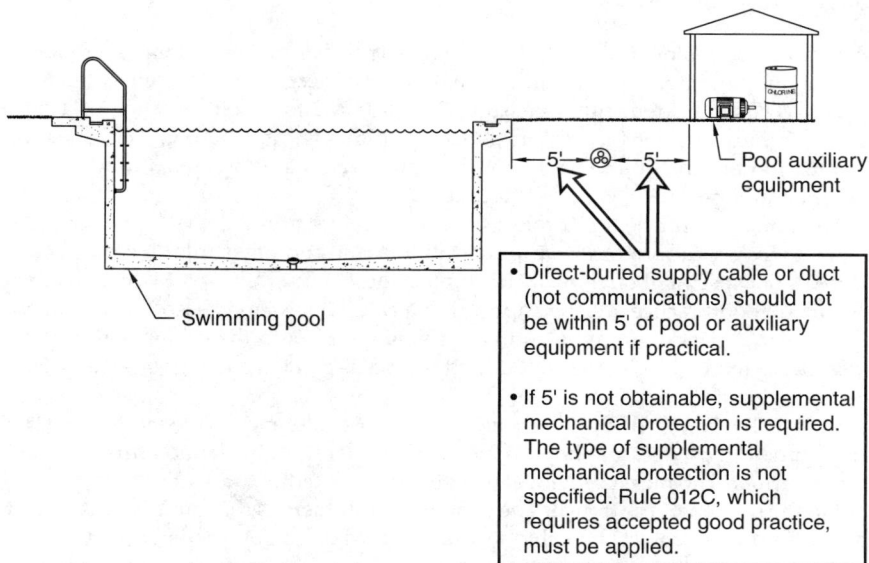

Fig. 351-1. Locating and routing direct-buried supply cable or duct near in-ground swimming pools (Rule 351C1).

the utilities to be located as far away from the tracks or as deep as possible. If a direct-buried cable or duct is installed under railroad tracks, the same burial depths required for conduit systems apply. See the figure in Rule 320A.

The Code states in Rule 351C4 that the installation of direct-buried cable or duct longitudinally under highways and streets should be avoided. The equivalent rule in the underground conduit systems section (Rule 320A3) does not have the "should be avoided" language as conduit systems are commonly found under roadways in large cities. If a conduit system is not used, cables under roadways are commonly installed in duct, not direct buried. Both the direct-buried cable or duct (Sec. 35) and underground conduit systems section (Sec. 32) define where to put the direct-buried cable or duct or the conduit system. See the figure in Rule 320A.

Water crossings are addressed in Rule 351C5 in Sec. 35, "Direct-Buried Cable and Cable in Duct Not Part of a Conduit System," and in Sec. 32, "Underground Conduit Systems." The wording is nearly identical and neither rule specifically addresses the terms "conduit" or "cable." Only the term "crossing" is used. See the water crossing discussion and the figures in Rule 320A.

Rule 351C6 addresses cables in duct attached to a bridge where permitted by the bridge owner. Rule 351C6 contains requirements for location and safe access for inspection or maintenance. Rule 322B also contains requirements for bridge attachments which are different than the requirements in Rule 351C6. It may be appropriate to apply both Rules 322B5 and 351C6 to installations on bridges. See the figure in Rule 322B for the Rule 322B5 requirements.

352. INSTALLATION

Direct-buried cable or duct installation can be done using three primary methods—trenching, plowing, and boring. Installation of underground conduit systems in Sec. 32 focuses on trenching (see Rule 321). Rule 352 for the installation of direct-buried cables or duct focuses on trenching, plowing, and boring. The rules for installation of direct-buried cable or duct by trenching, plowing, and boring are outlined in Figs. 352-1, 352-2, 352-3, and 352-4.

The phrases "sufficient to protect the cable or duct from damage" and "supplemental mechanical protection" are used in this rule and other rules in Sec. 35. The Code is not specific about how to accomplish these tasks. Rule 012C, which requires accepted good practice, must be applied. If a conduit system is used for protection, then Sec. 32, "Underground Conduit Systems," applies to the installation. See the discussion and Fig. 302-1, in Rule 302 and see the discussion and Figs. 32-1 and 320-1 at the beginning of Sec. 32.

Rule 352D, Depth of Burial, is one of the most referenced rules in Part 3. The rule starts out by defining how to measure the depth of burial of a direct-buried cable or duct. The measurement is from the surface to the top of the cable or duct, not to the bottom of the trench. The general requirement is that burial depths must be sufficient to protect the cable or duct from damage by the expected surface usage. The expected surface usage may be truck traffic or pedestrian traffic, it may involve lawn mowers or rototillers. The expected surface usage is not typically

Backfill:
- Free of materials within 4" of the cable that may damage the cable.
- Adequately compacted to limit settling under the expected surface usage.
- Do not use machine compaction within 6" of cable.

Trench dug by backhole, trenching equipment, or by hand.

Bottom of trench:
- Relatively smooth undisturbed earth, or
- Relatively smooth well-tamped earth, or
- Sand.
- In rocky areas, use a protective layer of well-tamped backfill.

See Photo(s)

Direct-buried cable

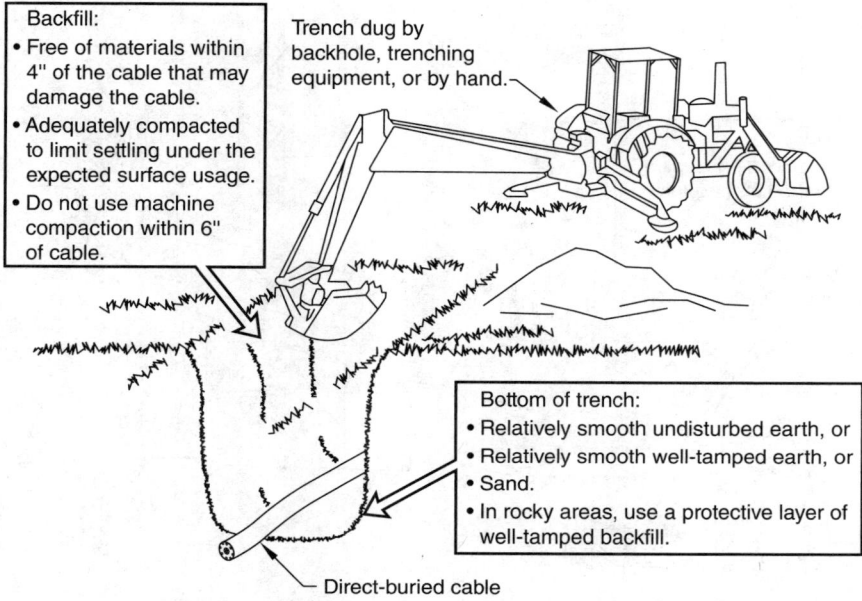

Fig. 352-1. Trenching requirements for direct-buried cable (Rule 352A1).

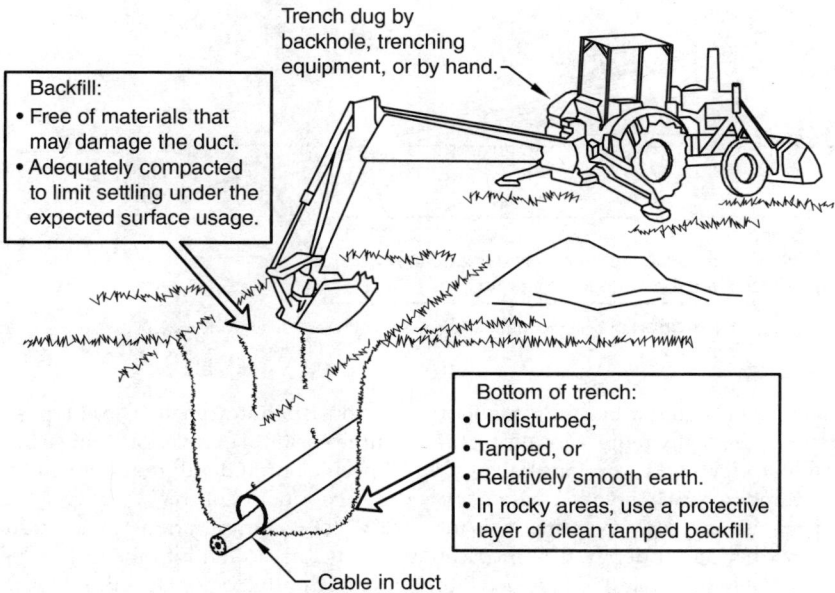

Trench dug by backhole, trenching equipment, or by hand.

Backfill:
- Free of materials that may damage the duct.
- Adequately compacted to limit settling under the expected surface usage.

Bottom of trench:
- Undisturbed,
- Tamped, or
- Relatively smooth earth.
- In rocky areas, use a protective layer of clean tamped backfill.

Cable in duct

Fig. 352-2. Trenching requirements for cable in duct (Rule 352A2).

Cable plowing using plowing equipment.

Plowing:
- In rocky areas, use a method that avoids cable damage.
- Do not damage cable by bending, side-wall pressure, or excessive cable tension.

See Photo(s)

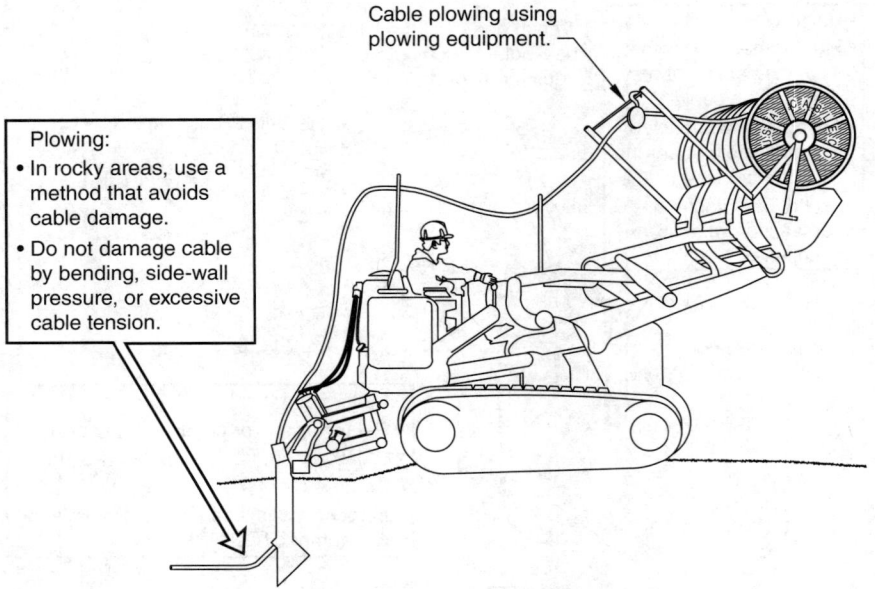

Fig. 352-3. Plowing requirements for direct-buried cables (Rule 352B).

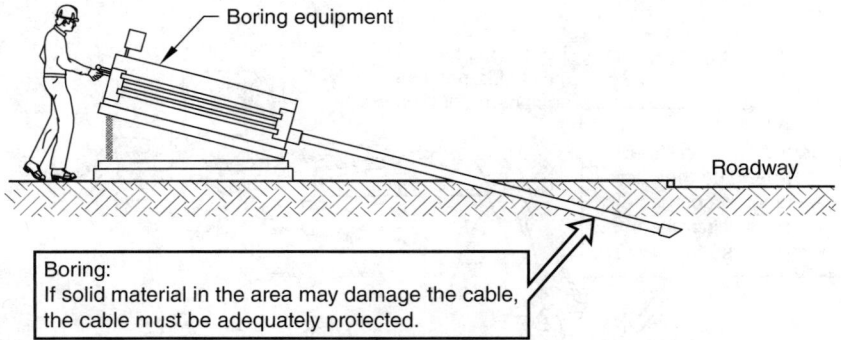

Boring equipment

Roadway

Boring:
If solid material in the area may damage the cable, the cable must be adequately protected.

See Photo(s)

Fig. 352-4. Boring requirements for direct-buried cables (Rule 352C).

a backhoe digging a building foundation or construction trench. These types of activities typically require locating buried utilities prior to excavation using a local "call before you dig" or "one-call system." Specific burial depth requirements are presented in **NESC** Table 352-1 for supply cable, conductor, or duct.

There is not a corresponding table in the **NESC** for communication cable, conductor, or duct burial depth. Communication conductors do fall into the 0- to 600-V range, but **NESC** Table 352-1 is titled Supply Cable, Conductor, or Duct Burial Depth and the voltage column specifies phase-to-phase voltage which implies supply cables. Depths for communication cable, conductor, or duct must meet the general requirements for burial depth outlined in Rule 352D1. In addition to Rule 352D1, a communication utility must also use Rule 012C, which requires accepted good practice, for communication cable burial depths. Communication cables laid on the ground do not meet the **Code** requirements for burial depth or overhead clearances.

Supply cables laid on the ground do not meet the **Code** requirements for burial depth or overhead clearances. Supply and communication cables may be laid on the ground for emergency (not temporary) installations. See Rules 230A2d and 311C. The requirements in **NESC** Table 352-1 for supply cable, conductor, or duct are graphically outlined in Fig. 352-5.

Some utilities establish supply cable burial depth values by using the **Code** clearance plus an adder. The adder (e.g., 6 in) can be thought of as a design or construction tolerance adder to maintain the required **Code** burial depth over time. There are several factors that could jeopardize a supply cable burial depth. One factor could be soil erosion. Installing a 15-kV phase-to-phase direct-buried supply cable or duct at 36 in deep can help maintain the **Code**-required 30-in depth over the life of the installation. See Rule 010 for a discussion of clearance adders.

The exception to Rule 352D2 allows lesser burial depths if the burial depths in **NESC** Table 352-1 cannot be met if supplemental mechanical protection is used. One of the most common reasons for using lesser depths than those specified in **NESC** Table 352-1 is when heavy or solid rock is encountered during trenching. The decision of what to use for supplemental mechanical protection is up to the designer. Rule 012C, which requires accepted good practice, must be applied. Conduit (duct) is a form of supplemental mechanical protection. Concrete encasement of a duct, concrete cover, concrete slabs, and wood planks are also forms of supplemental mechanical protection. The type of duct material (e.g., reinforced thermosetting resin conduit, RTRC, polyvinyl chloride conduit, PVC; high-density polyethylene conduit, HDPE; rigid metal conduit, RMC; intermediate metal conduit, IMC, etc.) and the grade of the conduit (e.g., PVC Type DB, EB, B, C, D, Schedule 40, Schedule 80, 5DR 13.5, SDR 11, SDR 9, etc.) will affect the amount of supplemental mechanical protection. The **Code** does not specify what type of duct **NESC** Table 352-1 applies to or what type of duct can be considered supplemental mechanical protection and be buried less than the depths in **NESC** Table 352-1. One reference for accepted good practice for a single duct that is not part of a conduit system is the National Electrical Code (NEC). The NEC specifies burial depths for various types of conduits (ducts) containing circuits at various voltage levels. The NEC also provides methods for reducing conduit (duct) burial depth when rock is encountered. If a conduit system is used then the rules of Sec. 32 apply. See the discussion and Fig. 302-1, in Rule 302 and see the discussion and Figs. 32-1 and 320-1 at the beginning of Sec. 32.

Rule 352E states that supply and communication cables must not be installed in the same duct unless they are operated and maintained (not necessarily owned) by the same utility. Rule 352F permits the installation of communication cables of different communication utilities in the same duct provided the utilities are in agreement. Rules 352E and 352F are similar in nature to Rules 341A6 and 341A7 (Cable in Underground Structures) and to Rules 239A2d and 239A2e (Vertical Risers). The requirements of Rules 352E and 352F are outlined in Fig. 352-6.

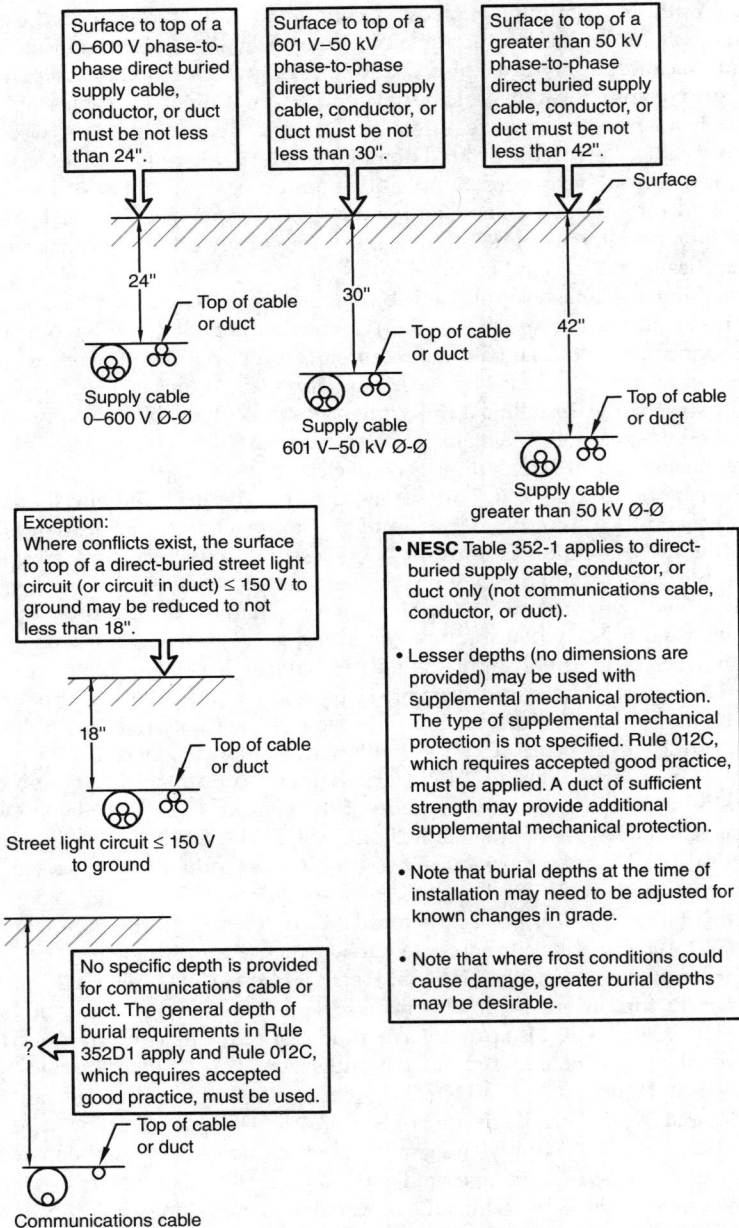

Surface to top of a 0–600 V phase-to-phase direct buried supply cable, conductor, or duct must be not less than 24".

Surface to top of a 601 V–50 kV phase-to-phase direct buried supply cable, conductor, or duct must be not less than 30".

Surface to top of a greater than 50 kV phase-to-phase direct buried supply cable, conductor, or duct must be not less than 42".

Surface

24" — Top of cable or duct

30"

42"

Top of cable or duct

Supply cable 0–600 V Ø-Ø

Supply cable 601 V–50 kV Ø-Ø

Top of cable or duct

Supply cable greater than 50 kV Ø-Ø

Exception:
Where conflicts exist, the surface to top of a direct-buried street light circuit (or circuit in duct) ≤ 150 V to ground may be reduced to not less than 18".

18" — Top of cable or duct

Street light circuit ≤ 150 V to ground

• **NESC** Table 352-1 applies to direct-buried supply cable, conductor, or duct only (not communications cable, conductor, or duct).

• Lesser depths (no dimensions are provided) may be used with supplemental mechanical protection. The type of supplemental mechanical protection is not specified. Rule 012C, which requires accepted good practice, must be applied. A duct of sufficient strength may provide additional supplemental mechanical protection.

• Note that burial depths at the time of installation may need to be adjusted for known changes in grade.

• Note that where frost conditions could cause damage, greater burial depths may be desirable.

No specific depth is provided for communications cable or duct. The general depth of burial requirements in Rule 352D1 apply and Rule 012C, which requires accepted good practice, must be used.

? — Top of cable or duct

Communications cable

Fig. 352-5. Supply cable, conductor, or duct burial depth (Rule 352D2).

| Supply cables and communication cables must not be in the same duct unless all of the cables are operated and maintained by the same utility. | Communication cables may be installed in the same duct if all utilities involved agree. |

Data
Power

Cable TV
Phone

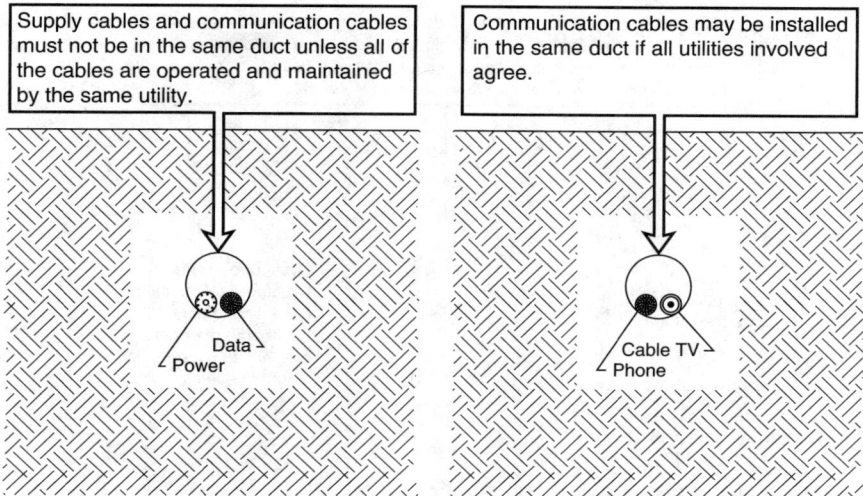

Fig. 352-6. Supply and communication cables in duct (Rules 352E and 352F).

353. DELIBERATE SEPARATIONS—EQUAL TO OR GREATER THAN 300 MM (12 IN) FROM UNDER-GROUND STRUCTURES OR OTHER CABLES

Rules 353 and 354 are closely related. As their titles indicate, Rule 353 applies to 12 in or greater separation and Rule 354 applies to less than 12 in of separation. Rule 353 for direct-buried cable or duct is similar to Rule 320B for underground conduit systems. In Rule 353 the NESC uses the term underground structures to mean sewers, water lines, gas, and other lines that transport flammable material, building foundations, steam lines, etc.

The requirement for radial (any direction) separation between a direct-buried cable or duct and another underground structure or cable in Rule 353 is 12 in or more. This value is assumed to permit maintenance on both facilities without damaging the other. If 12 in of separation is not obtainable, Rule 354 applies. When a direct-buried cable or duct is parallel and directly over or under another underground structure or cable, the parties involved must be in agreement to the method used. No such agreement is specified if a side-by-side configuration is used. Crossings require support or sufficient vertical separation to limit the transfer of detrimental loads from one system to the other. The rules related to the radial separation of direct-buried cables or duct from another underground structure or cable are outlined in Fig. 353-1.

Vertical separation between supply and communication lines is typically used by utilities when plowing in direct-buried cables or cable-in-duct. The separations in Rule 353 are from surface to surface, not center to center. Therefore, a 12-in center-to-center spacing of the plow chutes may not provide a 12-in separation between the cables or ducts depending on the cable or duct diameters.

Rule 353 applies to a direct-buried cable or duct radial separation of 12" or more. The radial separation should be adequate to permit access and maintenance without damage to either facility. For radial separation of less than 12", see Rule 354.

Radial separation (surface-to-surface)

Supply or communications direct-buried cable or duct

12" or greater

Other underground structure (sewer line, water line, gas line, flammable material line, building foundation, steam line, etc.) or cable.

Supply or communications direct-buried cable or duct

Other underground structure (sewer line, water line, gas line, flammable material line, building foundation, steam line, etc.) or cable.

12" or greater

Parallel facilities in a vertical configuration require agreement between the parties involved. (No such agreement is specified for a horizontal configuration.)

At crossings, support or sufficient vertical separation is required to limit the transfer of detrimental loads.

12" or greater

Supply or communications direct-buried cable or duct

Other underground structure (sewer line, water line, gas line, flammable material line, building foundation, steam line, etc.) or cable.

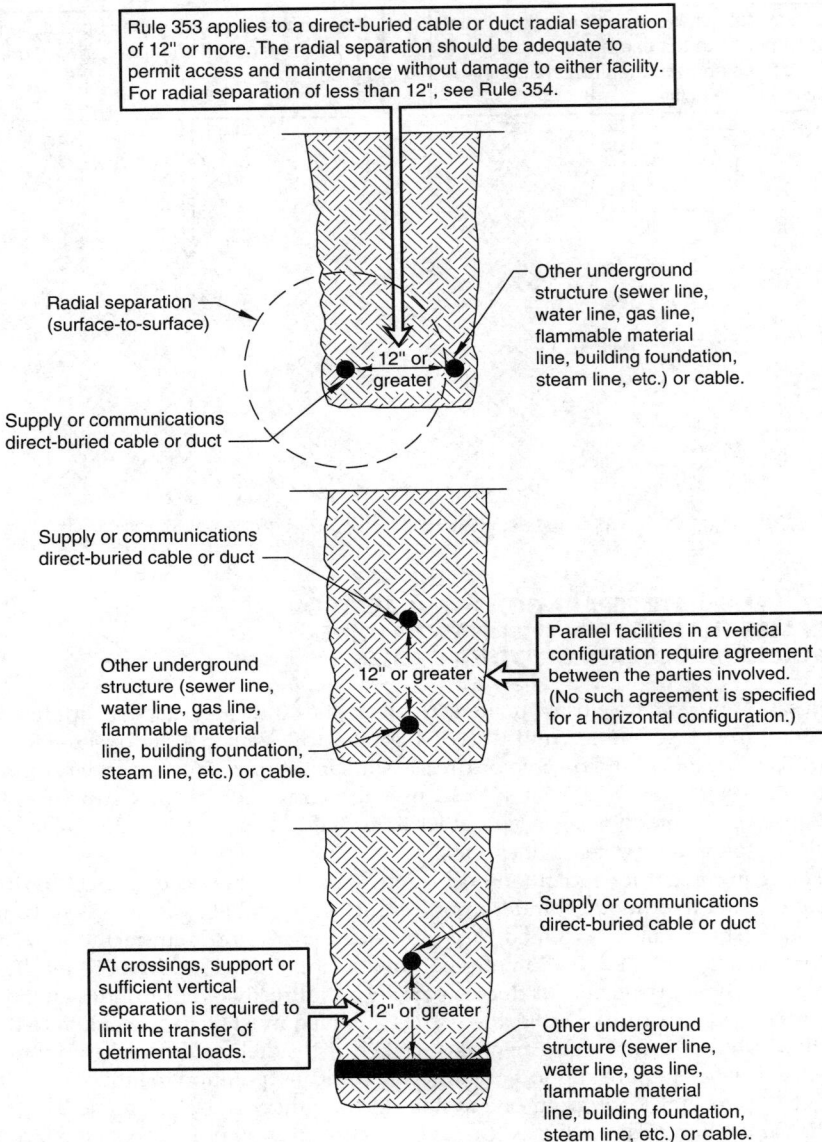

Fig. 353-1. Deliberate separation of 12 in or more between a direct-buried cable or duct and another underground structure or cable (Rule 353).

When paralleling or crossing a line that involves hot or cold temperatures like a steam (hot) line or a cryogenic (cold) line, consideration must be given to the heat or cold so it does not damage the crossing or paralleling line. Adequate separation is to be used. If separation is not obtainable, a thermal barrier is required.

354. RANDOM SEPARATION—SEPARATION LESS THAN 300 MM (12 IN) FROM UNDERGROUND STRUCTURES OR OTHER CABLES

Rules 354 and 353 are closely related. As their titles indicate, Rule 354 applies to less than 12 in of separation and Rule 353 applies to 12 in or greater separation. Random separation is sometimes referred to as random lay. The focus of the random separation rules is to prevent damage to direct-buried communication lines when a fault occurs on direct-buried supply lines. The intent is not to reduce or minimize conductive interference or "noise."

The general rules for random separation are outlined below:

- The rules apply to a direct-buried cable or duct and another underground structure or cable with a radial (any direction) separation less than 12 in.
- The radial separation between a direct-buried supply or communication cable or duct and a steam line, gas or flammable material line, or other line that transports flammable material must not be less than 12 in. (Rule 353 must be used in this case but an exception does apply between steam lines, gas or flammable material lines and supply cables operating at 300 V or below and between gas lines and communication cables.)
- Supply circuits (above 300 V to ground or 600 V phase to phase), when faulted, need to be de-energized by a protective device (i.e., fuse, recloser, circuit breaker, etc.).
- Supply and communication cables and conductors in random separation are treated as one system when considering separation from other conductors. or structures.

The exception to Rule 354A2 permits a 120/240 V, single-phase, three-wire secondary service to be less than 12 in from a gas or flammable material line provided certain conditions are met and provided the utilities agree. A similar exception permits a communication cable to be less than 12" from a steam line or gas or flammable material line provided the utilities agree. The **NESC** does not specify clearances between gas and electric meters or equipment on the building, only in the trench. See Figs. 354-1 and 354-2.

Supply cables above 600 V to ground with an effectively grounded continuous metallic shield, sheath, or concentric neutral of the same supply circuit may be buried in random separation (no deliberate separation), see Rule 350C. Supply cables of the same circuit operating below 600 V to ground without an effectively grounded shield or sheath must be buried in random separation (in close proximity with no deliberate separation), see Rule 350D.

Per Rule 354B, supply cables of multiple supply circuits may be buried in random separation (no deliberate separation) if all parties involved are in agreement, see Fig. 354-3.

Per Rule 354C, communication cables of multiple communication circuits may be buried in random separation (no deliberate separation) if all parties involved are in agreement. See Fig. 354-4.

Rule 354D focuses on supply and communication cables or conductors that are buried less than 12 in apart. If supply and communication cables are direct-buried less than 12 in apart, a variety of rules must be met. The rules include special consideration of voltage limitations, grounding and bonding requirements, protection

Electrical
Meter/Main

The **NESC** does not specify a
clearance between electrical service
equipment (or communications service
equipment) and a gas meter. This
clearance is typically based on
National Fire Protection Association
(NFPA) gas codes.

Gas Meter

Steam,
Gas, or
Flammable
Material Line

12"

Supply (Electrical) Cables
at not more than 300 V
between conductors.

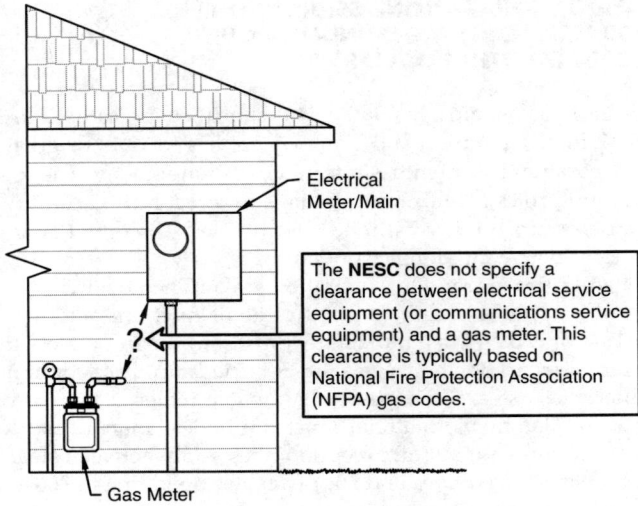

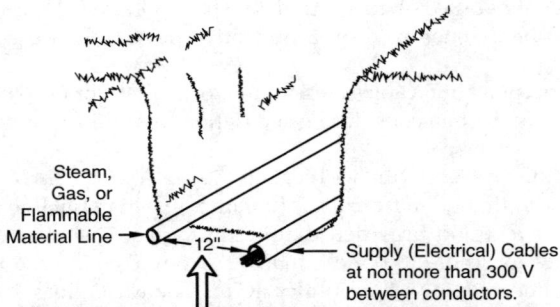

• Radial separation of supply and communications cables or
  conductors from steam, gas, or flammable material line not less
  than 12".

• Exception: For supply cables at not more than 300 V between
  conductors (e.g., a 120/240 V, 1Ø, 3W service), less than 12" may
  be used to a gas or flammable material line if:

  - Supplemental mechanical protection is used to limit the likely
    hood of detrimental heat transfer to the gas or flammable
    material line due to a cable fault. This exception does not apply
    to steam lines. (No specifics are provided as to the type of
    supplemental mechanical protection, Rule 012C, which requires
    accepted good practice, must be used.)

  - The Utilities involved agree to the reduced separation.

See
Photo(s)

Fig. 354-1. Supply line and steam, gas, or flammable material line in random
separation (Rule 354A2).

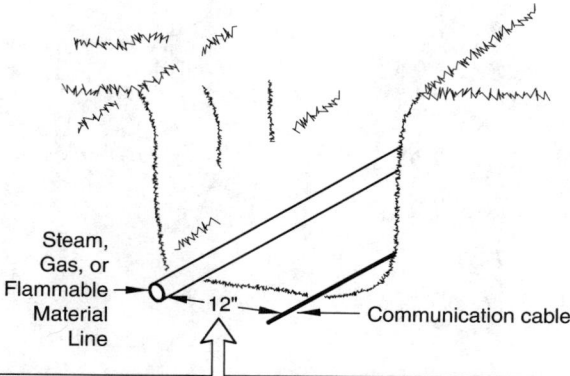

Steam, Gas, or Flammable Material Line

12"

Communication cable

- Radial separation of supply and communications cables or conductors from steam, gas, or flammable material line not less than 12".

- Exception: For communication cables, less than 12" may be used if the utilities involved agree to the reduced separation. This exception applies to steam, gas, or flammable material lines.

Fig. 354-2. Communications line and steam, gas, or flammable material line in random separation (Rule 354A2).

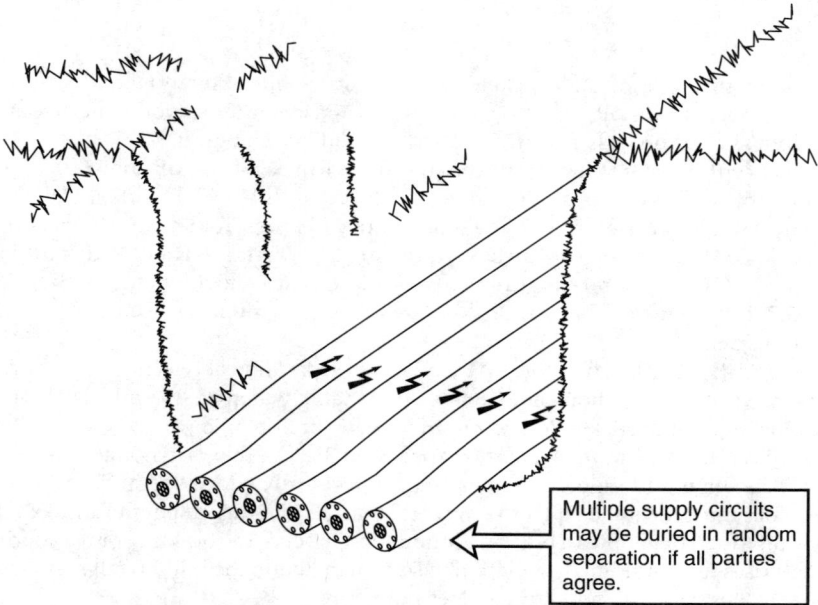

Multiple supply circuits may be buried in random separation if all parties agree.

Fig. 354-3. Multiple supply circuits in random separation (Rule 354B).

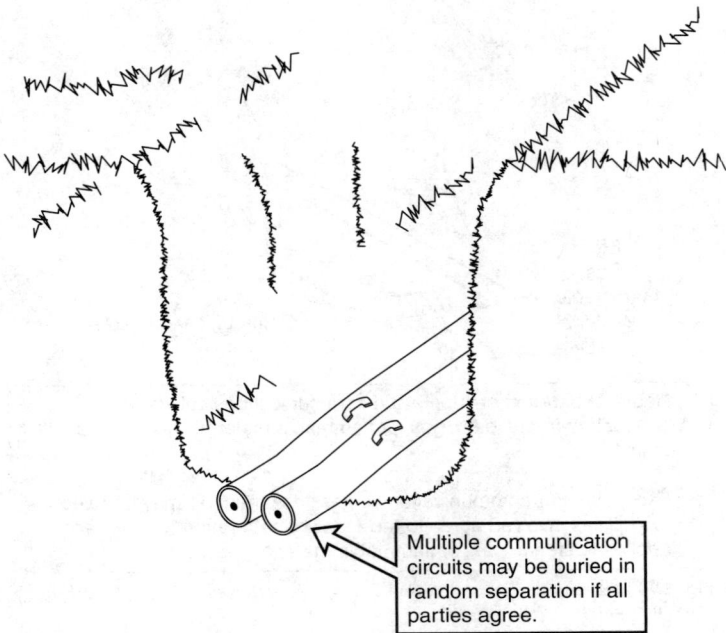

Multiple communication circuits may be buried in random separation if all parties agree.

Fig. 354-4. Multiple communication circuits in random separation (Rule 354C).

requirements, type of cable jacket, and concentric-neutral material and size. The steps needed to install direct-buried supply and communication conductors in random separation (less than 12 in apart) are outlined in Fig. 354-5.

An exception is provided at the beginning of Rule 354D for all-dielectric fiber-optic cables. This exception is similar in nature to Rules 352E and 341A6. Each rule has slightly different voltage limits and cable requirements. The exception to Rule 354D must meet Rules 354D1a through 354D1d. Rule 352E requires the use of the general requirements in Rule 350 (which includes Sec. 33). Rule 341A6 requires the use of the general requirements in Rule 340 (which includes Sec. 33).

Rule 354E provides rules for less than 12 in of separation between a direct-buried supply or communication cable and a nonmetallic water or sewer line. The rule requires that all parties involved are in agreement. Metallic water or sewer lines must have 12 in or more separation (per Rule 353) as they are not addressed in Rule 354. Building foundations are also not specifically addressed in Rule 354 and therefore must meet the equal to or greater than 12 in requirement in Rule 353. If a conduit (duct) penetrates a building foundation to serve the building or if a conduit (duct) rises up a building foundation to serve a meter on the building, it seems reasonable that the 12 in requirement does not apply.

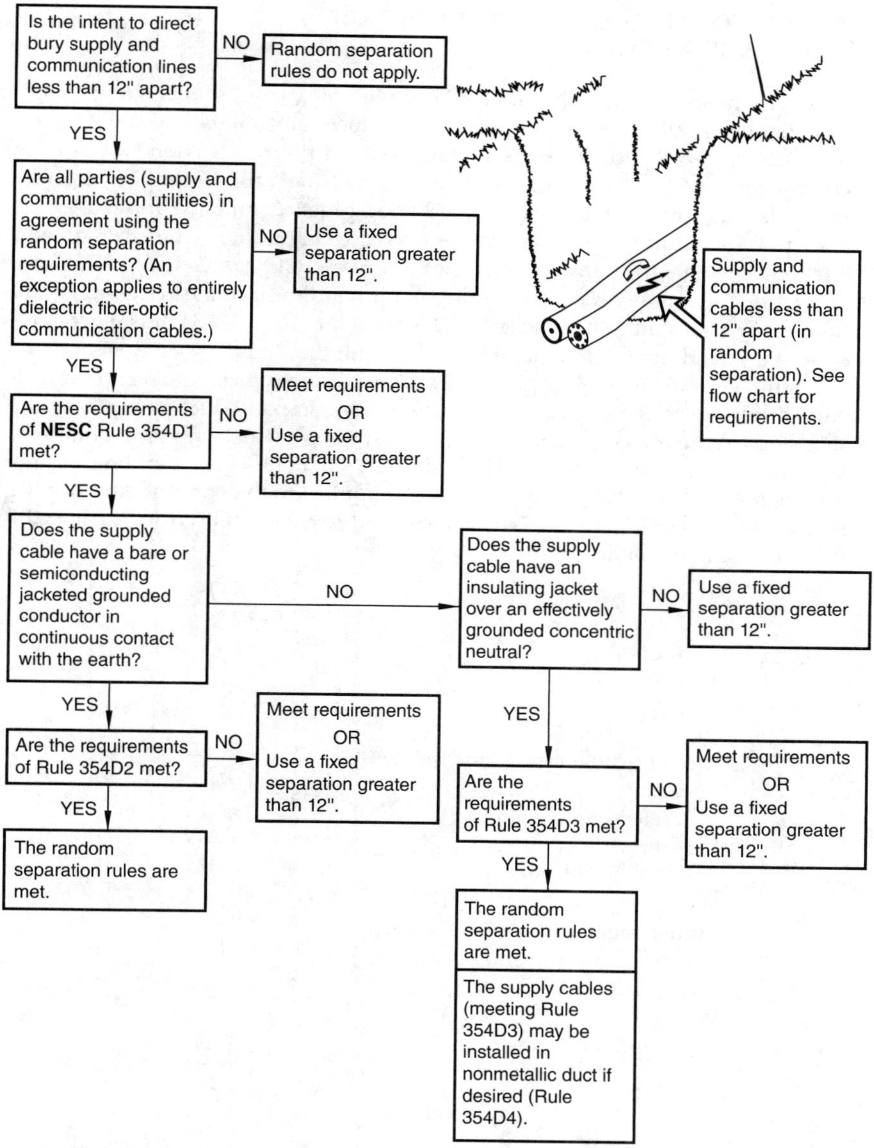

Fig. 354-5. Direct-buried supply and communication random separation requirements (Rule 354D).

355. ADDITIONAL RULES FOR DUCT NOT PART OF A CONDUIT SYSTEM

The requirements in this rule for duct not part of a conduit system are similar to some (not all) of the requirements for ducts and joints in Rule 322. See Rules 322A and 322B for a discussion and related figures. Section 35 does not address conduit (duct) bending radius, conduit (duct) fill, or pulling calculations. Rule 012C, which requires accepted good practice, must be applied. Ducts not part of a conduit system may be connected to handholes. The strength and cover requirements for handholes are not addressed in Sec. 35. Rules 323A and 323D can be used as accepted good practice for handholes connected to duct that is not part of a conduit system. Rule 355D is similar to Rule 322B4. It is more common to see a conduit system enter a building vault than it is to see a duct not part of a conduit system enter a building. For example, for a typical underground residential or commercial electrical service, the utility company duct (covered by the **NESC**) stops at the meter or meter/main enclosure on the exterior of the building and a building wiring conduit (covered by the NEC) enters the building. In this example, the National Electrical Code (NEC) applies to the conduit entering the building, not the **NESC**, and the NEC rules need to be referenced for sealing the conduit entering the building.

Section 36

Risers

360. GENERAL

The rules for risers overlap in Part 3, Underground Lines, and Part 2, Overhead Lines. Rule 239D is referenced as it focuses on mechanical protection of a riser on an overhead pole. See Rule 239 for a discussion. Rule 360 extends the protection required in Rule 239D at least 1 ft below grade. See Fig. 360-1.

Cable bending must be considered when transitioning from a buried horizontal position to a vertical riser position. See Rule 341 for a discussion of cable bending.

Risers in metallic duct or under metallic guards (i.e., U-Guard) need to have the metallic duct or metallic guard effectively grounded in accordance with Rule 314B. Rule 360C does not require risers containing supply conductors to be metal, but if they are metal, they must be effectively grounded.

361. INSTALLATION

Rule 361 provides general installation rules for risers. Rule 362 provides additional rules for pole risers and Rule 363 provides additional rules for risers entering pad-mounted equipment. The general rules for the installation of risers are outlined in Fig. 361-1.

362. POLE RISERS—ADDITIONAL REQUIREMENTS

Rules 360 and 361 apply to risers in general. Rule 362 applies specifically to pole risers. The rules for pole risers are outlined in Fig. 362-1.

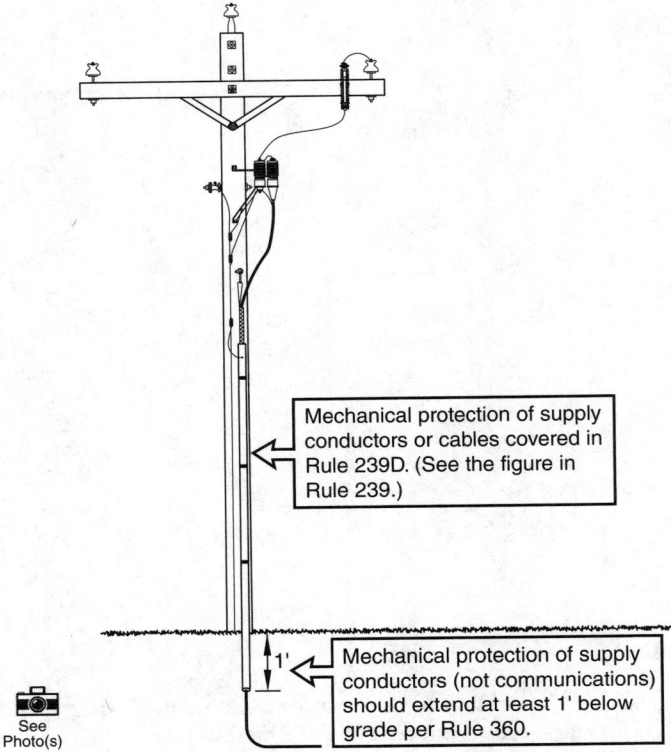

Mechanical protection of supply conductors or cables covered in Rule 239D. (See the figure in Rule 239.)

Mechanical protection of supply conductors (not communications) should extend at least 1' below grade per Rule 360.

1'

See Photo(s)

Fig. 360-1. Mechanical protection of supply risers below grade (Rule 360A).

Standoff brackets are used by some utilities to make climbing easier. If standoff brackets are used on risers, Rule 217A2c must be met. See the figure in Rule 217 and a discussion of riser conduits and climbing space in Rule 236.

363. PAD-MOUNTED INSTALLATIONS

Rules 360 and 361 apply to risers in general. Rule 363 applies specifically to underground cables rising from a horizontal position up to pad-mounted equipment. The rules for pad-mounted installations are outlined in Fig. 363-1.

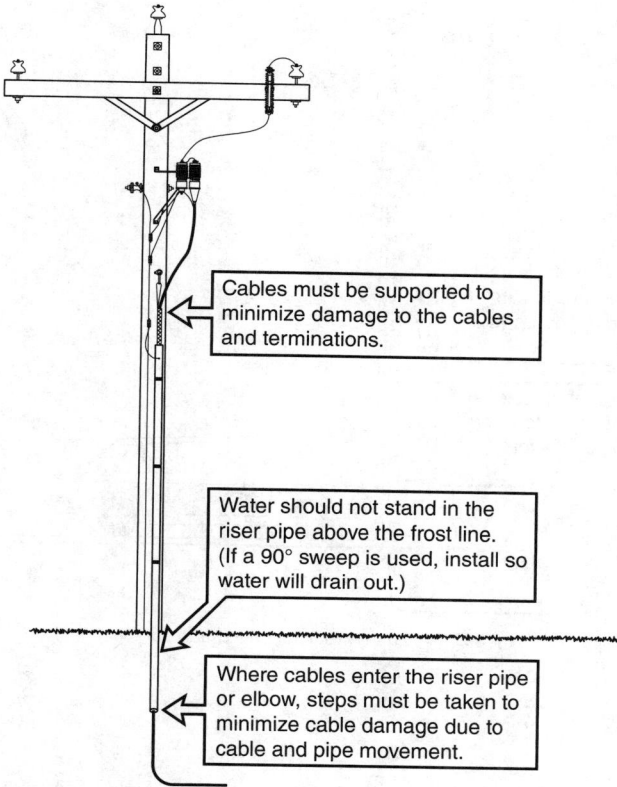

Cables must be supported to minimize damage to the cables and terminations.

Water should not stand in the riser pipe above the frost line. (If a 90° sweep is used, install so water will drain out.)

Where cables enter the riser pipe or elbow, steps must be taken to minimize cable damage due to cable and pipe movement.

Fig. 361-1. Riser general installation requirements (Rule 361).

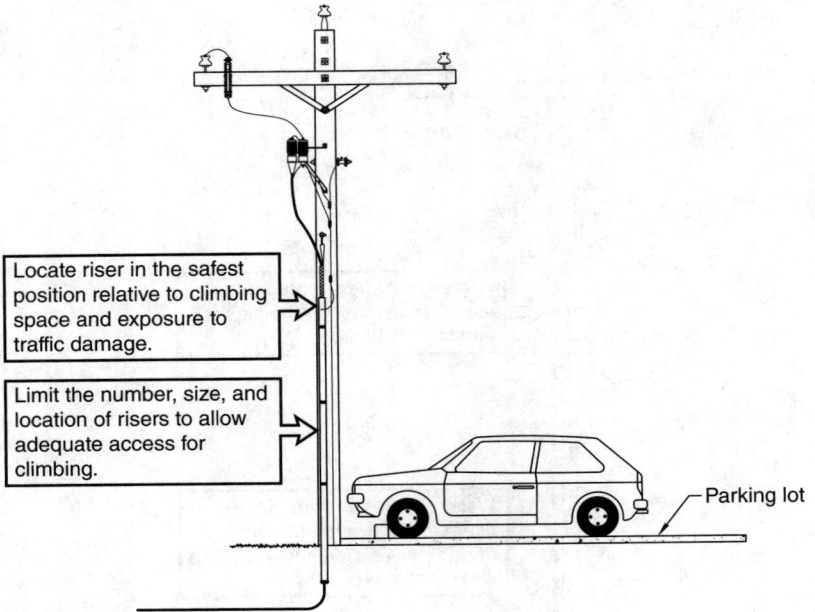

Fig. 362-1. Requirements for pole risers (Rule 362).

Locate riser in the safest position relative to climbing space and exposure to traffic damage.

Limit the number, size, and location of risers to allow adequate access for climbing.

Parking lot

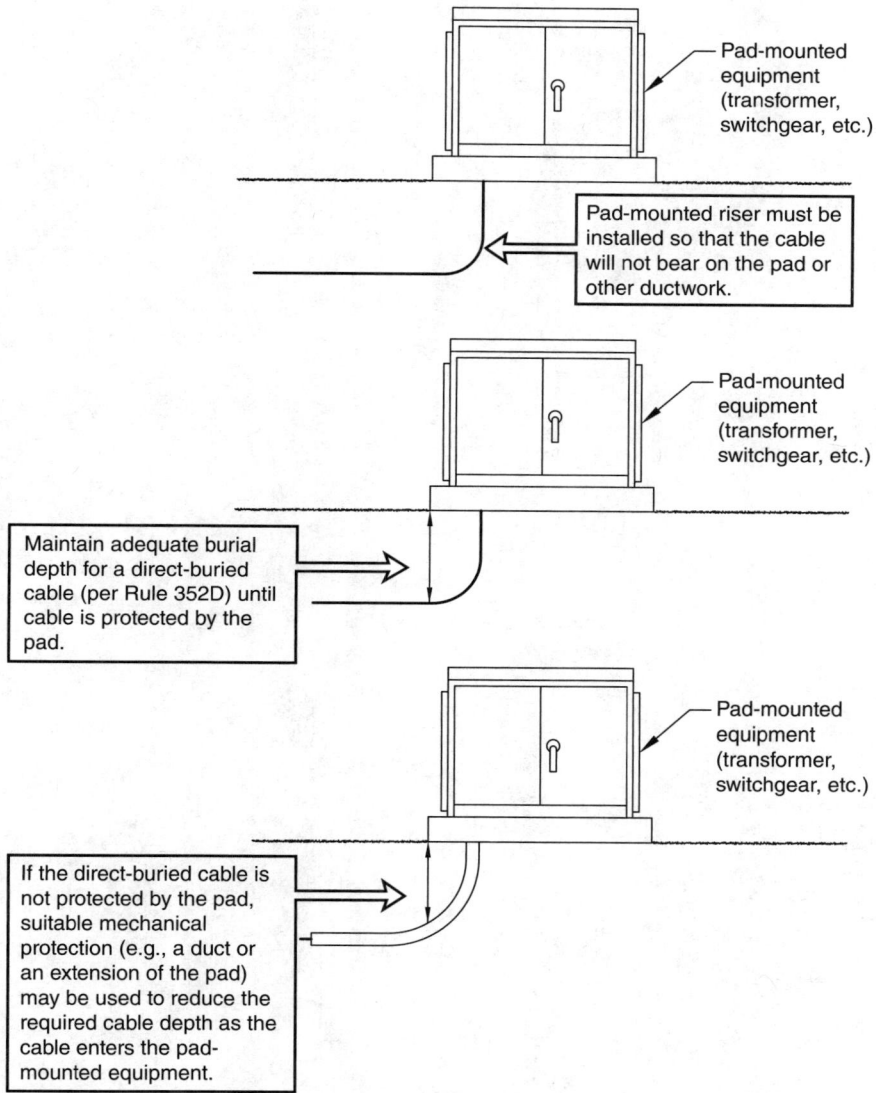

Section 37

Supply Cable Terminations

370. GENERAL

Supply cable terminations are the various items used to terminate underground cables such as elbows, cable terminators, potheads, etc. A reference to Rule 333 is made to tie Sec. 37 to Rule 333. Rule 333 focuses on the material specifications of the termination. Section 37 focuses on the installation of the termination. Cable terminators protect conductors and conductor insulation from mechanical damage, moisture, and electrical stress. Examples of supply cable terminations are shown in Fig. 370-1.

If a cable terminator is located inside a vault or pad-mounted equipment, the rules of Part 3, Underground Lines, apply. If the cable terminator is on a riser that terminates overhead inside a substation, the clearance rules found in Part 1, Electric Supply Stations, apply. If the cable terminator is on a pole outside of a substation, the clearance rules in Part 2, Overhead Lines, apply.

Equipment manufacturers determine clearance between cable terminators inside pad-mounted equipment by referencing ANSI and NEMA standards. The **NESC** provides a general statement (no specific distances) that suitable clearance must exist for the voltage and basic impulse level (BIL). Fully insulated terminations (i.e., elbows) or insulated barriers may be used to meet the necessary requirements.

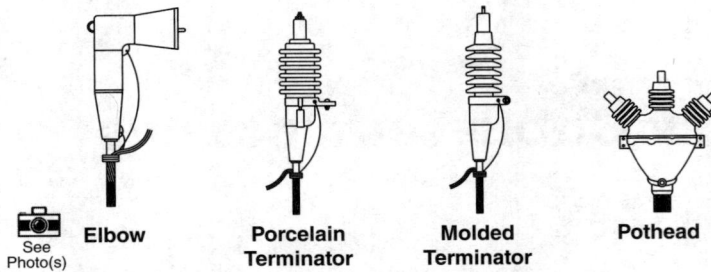

Fig. 370-1. Examples of supply cable terminations (Rule 370).

371. SUPPORT AT TERMINATIONS

Cable terminators must be installed to maintain their position. Underground residential distribution (URD) elbows must not pull out of the bushing wells they are installed in and riser pole cable terminators must not be moving around on the pole. The cable attached to the URD elbow or riser pole terminator may need to be supported to minimize stress on the cable terminator. Requirements for supporting cables are also stated in Rules 341A5 and 361B.

Supply utilities commonly use both porcelain cable terminators that are supported on a bracket or crossarm and lightweight polymer terminators that are not supported independent of the cable. The **Code** does not specify exactly how to support the cable termination, it only states that the termination must maintain its installed position.

372. IDENTIFICATION

Identification of a cable at a termination point is a simple method to maintain an organized electrical system, and it is a **Code** requirement for supply cables. Rule 372 does not apply to communication cables, as it is in Sec. 37, which is titled "Supply Cable Terminations." An exception is provided that eliminates the need for circuit identification if the position of the termination combined with maps or diagrams provides sufficient identification.

Additional identification rules for underground lines can be found in Rules 311 (underground locates), 323J (manhole and handhole covers), 341B3 (cables in manholes), 350F (marking of cables), and 385 (equipment operating in multiple). An example of supply cable identification at termination points is shown in Fig. 372-1.

The **NESC** does not contain a rule requiring the use of underground warning ribbons or tapes. The **NESC** also does not have a requirement for cable route markers. The **NESC** certainly does not prohibit the use of these items and these items are installed by some utilities. The lack of a **NESC** rule related to underground warning ribbons or tapes and cable route markers is shown in Fig. 372-2.

Suitable circuit identification must be provided for all supply cable terminations. (An exception applies where position, in conjunction with diagrams or maps, provides sufficient identification.)

See Photo(s)

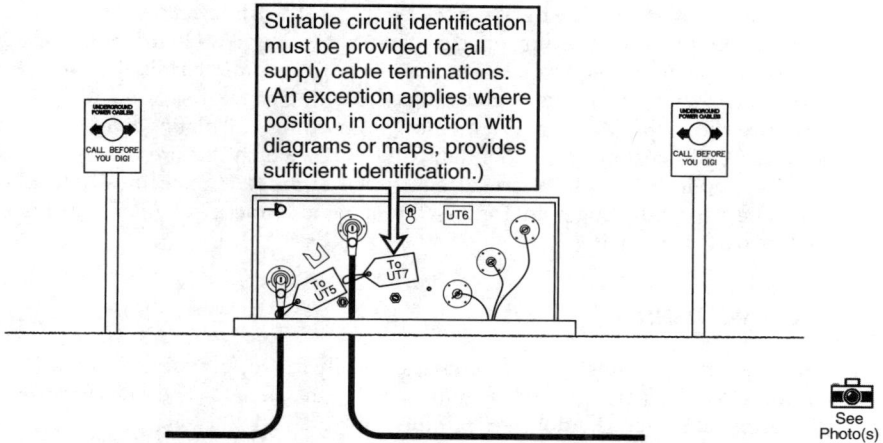

Fig. 372-1. Example of supply cable identification at termination points (Rule 372).

373. CLEARANCES IN ENCLOSURES OR VAULTS

Clearance between terminators depends on numerous factors including the type of terminator, the insulation method, and the voltage level.

This rule provides only a general statement that adequate clearance is required between supply terminations and between terminations and ground. No specific distances are provided. If terminators are in an enclosure (like a pad-mounted switch, junction box, transformer enclosure, etc.), the manufacturer will reference industry standards for the type of equipment. Adequate clearance or insulated barriers are needed for live parts in a manufactured enclosure. The burden is on the

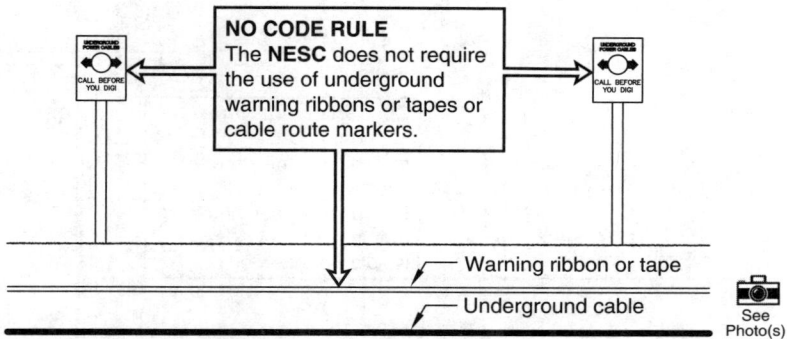

NO CODE RULE
The **NESC** does not require the use of underground warning ribbons or tapes or cable route markers.

Warning ribbon or tape

Underground cable

See Photo(s)

Fig. 372-2. Absence of Code rule related to underground warning ribbons or tapes and cable route markers (Rule N/A).

manufacturer to determine the specific clearances. The same is true for live parts in a vault, only this time the burden is on the utility designer. Guarding or isolating live parts in a vault is required. Using dead-front fittings and terminations or metal-clad switchgear are methods of guarding or isolating live parts. Adequate physical clearance above the floor is also a method of guarding or isolating live parts. Since no specific dimensions are provided, Rule 012C, which requires accepted good practice, applies. The substation rules in Part 1 of the NESC, specifically Rule 124, would be a good starting point for the accepted good practice of guarding or isolating live parts in a vault.

374. GROUNDING

Rule 374 has grounding requirements specifically for supply cable terminations in addition to the general grounding requirements in Rule 314. The rules for grounding supply cable terminations are outlined in Fig. 374-1.

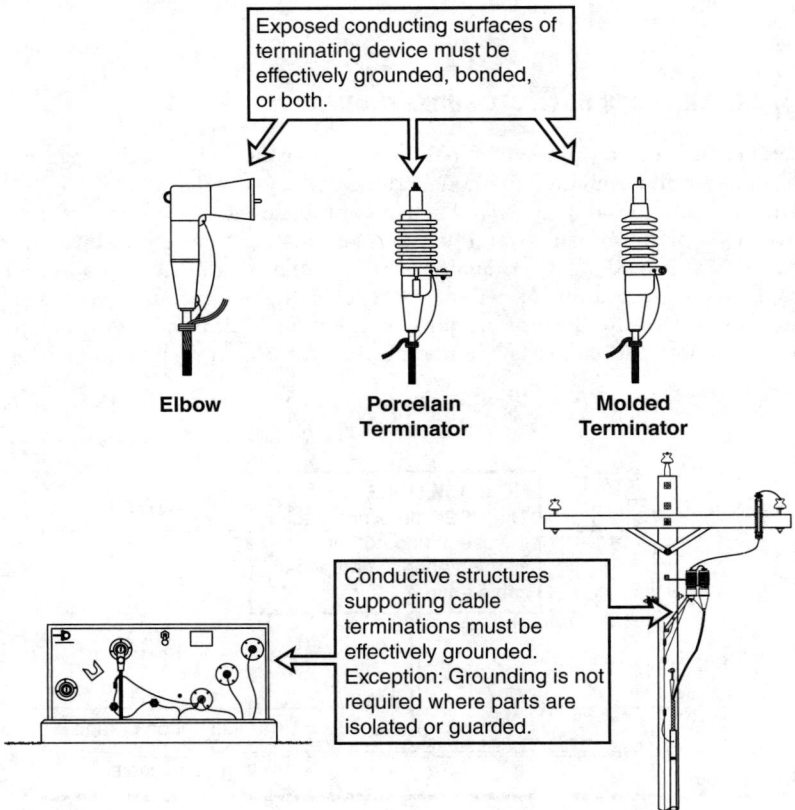

Fig. 374-1. Grounding supply cable terminations (Rule 374).

Section 38

Equipment

380. GENERAL

Rule 380 starts out by providing examples of supply and communication equipment relevant to this section. See Fig. 380-1.

Equipment located in a joint-use manhole can only be installed with the concurrence of the parties concerned. Similar wording appears in Rule 341B2b(1) for cables in joint-use (supply and communication) manholes. The term joint-use in this rule refers to simultaneous use by two or more utilities. Rule 341B2b(1) is for joint-use (supply and communication) installations. If the parties do not concur, separate conduit systems and manholes can be used or a joint conduit system can be used where the conduits leave the joint conduit system and enter separate manholes or pedestals at every termination point or pulling location.

The pads, supports, and foundations used to support equipment must be rated for the load and stress of the equipment and the equipment's operation. For example, the forces associated with installing and removing underground residential distribution (URD) elbows must be considered.

Rule 380D is similar to Rule 231A. Rule 380D specifies a distance from pad-mounted equipment to a fire hydrant. Rule 231A specifies a distance from a supporting structure (e.g., pole) to a fire hydrant. The main purpose of this rule is to allow adequate working space for the fire department to connect wrenches, hoses, and equipment to the fire hydrant. See Fig. 380-2.

The Code does not address clearance from pad-mounted equipment to a roadway or clearance of an oil-filled pad-mounted transformer to a building. Since the Code does not provide dimensions for these installation conditions, accepted good practice (Rule 012C) must be used. The absence of a Code rule related to pad-mounted equipment location adjacent to roads and buildings is outlined in Fig. 380-3.

Supply System Equipment:
Busses, transformers,
switches, etc., plus any
auxiliary equipment.

Communication System Equipment:
Repeaters, loading coils, etc., plus any
auxiliary equipment.

See Photo(s)

Supply Equipment   **Communication Equipment**

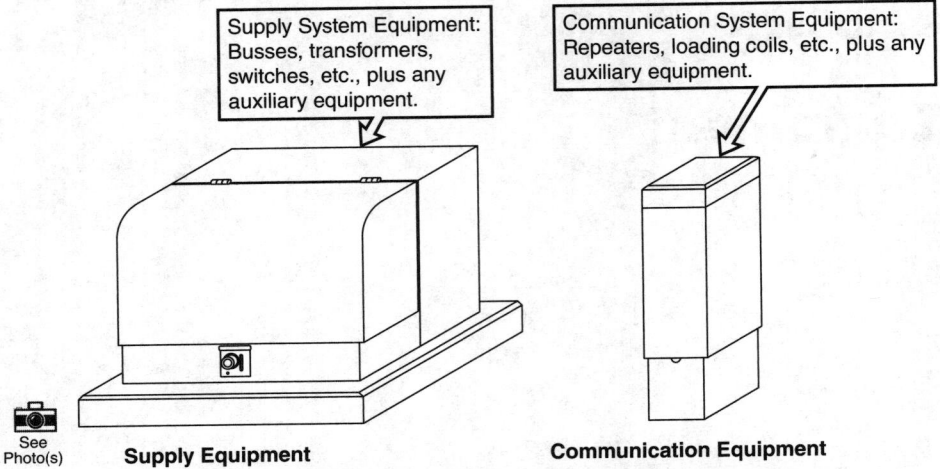

Fig. 380-1. Examples of supply and communication equipment (Rule 380).

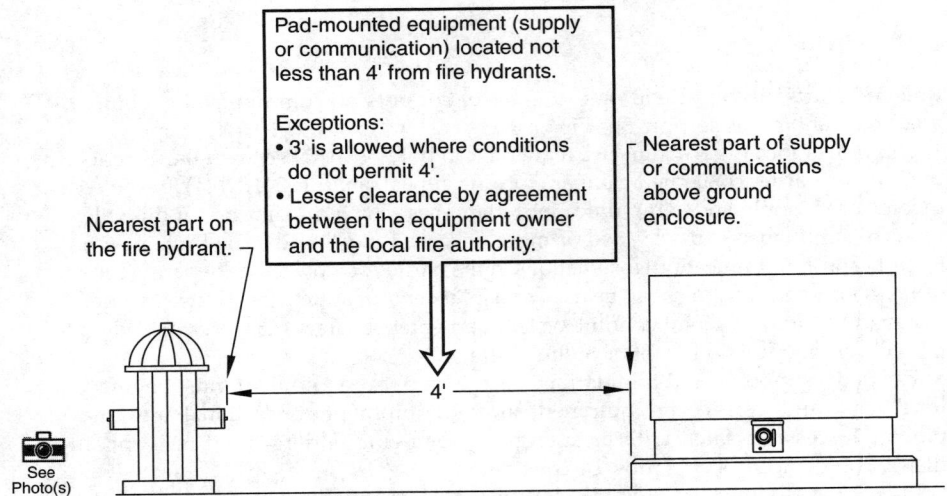

Pad-mounted equipment (supply
or communication) located not
less than 4' from fire hydrants.

Exceptions:
• 3' is allowed where conditions
  do not permit 4'.
• Lesser clearance by agreement
  between the equipment owner
  and the local fire authority.

Nearest part of supply
or communications
above ground
enclosure.

Nearest part on
the fire hydrant.

4'

See Photo(s)

Fig. 380-2. Clearance of pad-mounted equipment to fire hydrants (Rule 380D).

381. DESIGN

Equipment design and mounting must consider the following:
• Thermal conditions
• Chemical conditions
• Mechanical conditions
• Environmental conditions

NO CODE RULE
The **NESC** does not specify the clearance between pad-mounted equipment and buildings. Rule 012C, which requires accepted good practice, must be used. Rule 152A discusses liquid-filled transformer fire hazards in substations. The National Electric Code (NEC) discusses locations of various types of transformers.

NO CODE RULE
The **NESC** does not specify the clearance between pad-mounted equipment and a roadway. Rule 012C, which requires accepted good practice, must be used. Rule 231B specifies clearance of poles to roadways.

Road
Curb
?
?

See Photo(s)

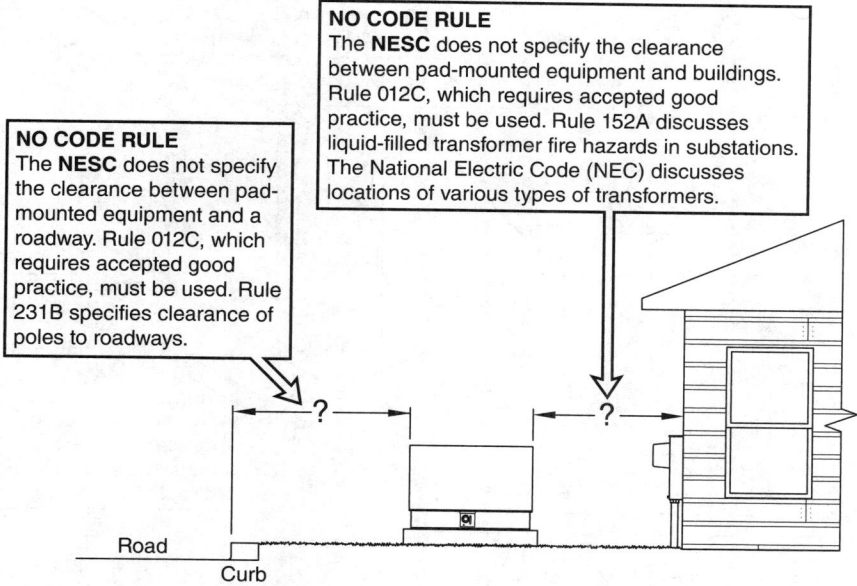

Fig. 380-3. Absence of Code rule related to underground pad-mounted equipment location adjacent to roads and buildings (Rule N/A).

Equipment and auxiliary devices must also be rated for the expected:
• Normal conditions
• Emergency conditions
• Fault conditions

The requirement for switching underground equipment is similar in nature to switching requirements for overhead switches discussed in Rule 216. Switches for underground lines must provide a clear indication of the switch contact position, and the switch handle must be marked with operating directions. The rules for underground switches are outlined in Fig. 381-1.

Remotely controlled or automatic switching devices such as pad-mounted vacuum fault interrupt switches must have provisions on the equipment to render remote or automatic controls inoperable. This feature protects workers from accidental operation or energization during maintenance.

When applying equipment containing fuses and interrupting contacts, the following must be considered during operation:
• Normal conditions
• Emergency conditions
• Fault conditions

When tools such as insulated shotgun sticks are used to handle energized devices in underground equipment, physical space or barriers must provide adequate clearance from ground or between phases.

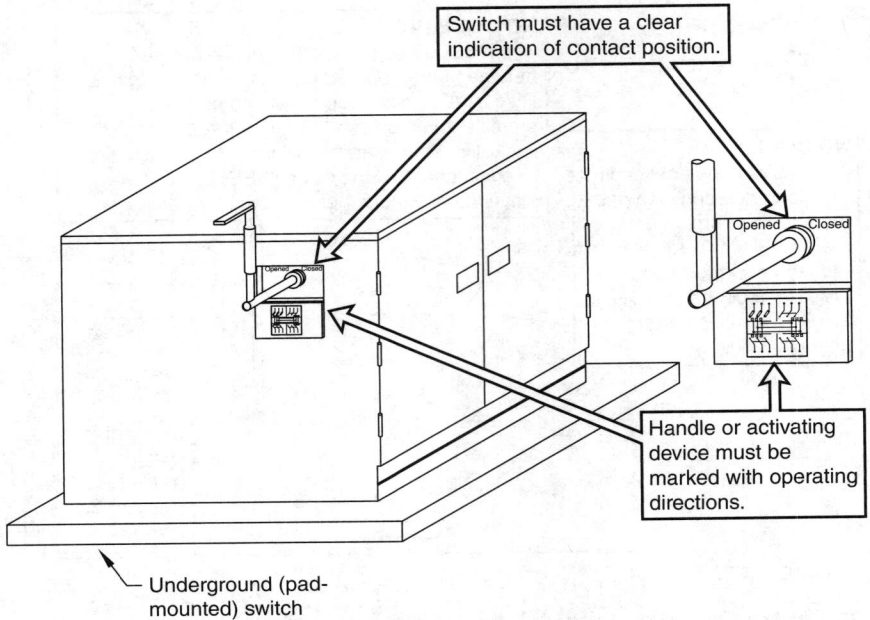

Fig. 381-1. Design of switches (Rule 381C).

Rule 381G provides the locking and access requirements for pad-mounted and other aboveground equipment. Pad-mounted and other aboveground equipment is typically not fenced in like substation equipment and it is not elevated like overhead equipment. Pad-mounted equipment has more exposure to the public than most other supply and communication facilities. Rule 381G1 applies to both supply and communication pad-mounted equipment. Pad-mounted and other aboveground equipment must have an enclosure that is locked or otherwise secured. The purpose of this is to keep out unauthorized persons (i.e., the public). Typical locking or securing methods used by both supply and communication utilities include keyed padlocks, disposable locks that can be cut with bolt cutters, penta head bolts, tamperproof bolts, etc. Rule 381G2 applies to supply pad-mounted equipment over 600 V. This requirement applies to almost all supply utility transformers, switches, junction boxes, etc., except for low-voltage secondary enclosures. Rule 381G2 requires two separate conscious acts to access exposed live parts. The first act is opening the pad-mounted enclosure that was locked or secured in Rule 381G1. It does not matter how many locks or penta head bolts are handled to open the enclosure; opening the enclosure is the first conscious act. The second conscious act must be the opening of a door (not the enclosure door in the first conscious act) or the removal of a barrier.

A pad-mounted switch with fuses is a good example of a device that requires two conscious acts to access exposed live parts. After opening the enclosure door (the first conscious act), a separate steel door or a separate fiberglass barrier exists that must be opened or removed (the second conscious act) before the live parts are exposed. A slightly different example is a typical pad-mounted service transformer that uses load break elbows for terminating the high-voltage cables. This type of transformer does not have any exposed live parts over 600 V and is usually referred to as "dead front." Opening the enclosure door is the first conscious act. No exposed live parts in excess of 600 V exist when the transformer enclosure is opened. The transformer secondary lugs (i.e., 120/240 V) may be exposed but they are under 600 V. The pulling of the insulated elbow is the second conscious act. The barrier for the second conscious act is the insulated elbow itself. See Fig. 381-2.

Rule 381G2 states that a safety sign should be visible when the first door or barrier is opened or removed. The ANSI Z535 series of signing standards are referenced in a note. Signage on pad-mounted equipment is just as critical as signage for an electric supply station. Some utilities use a warning sign on the outside of the pad-mounted enclosure and a danger sign on the inside of the pad-mounted enclosure. This approach uses the ANSI Z535 philosophy that warning is appropriate on the outer barrier (enclosure) and if that barrier is breached, a danger sign is then appropriate. Utilities should consult the ANSI documents, federal or state

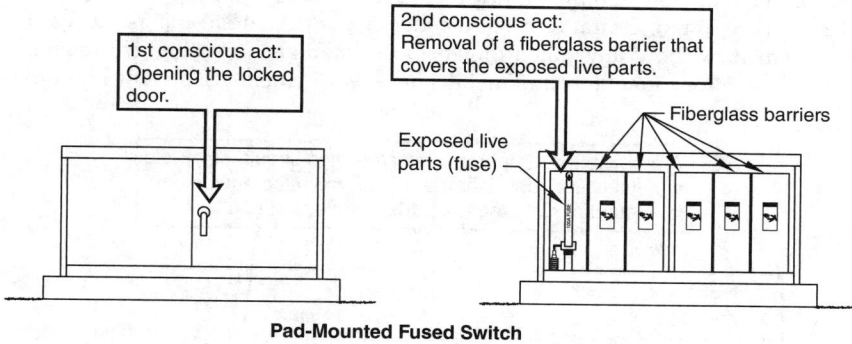

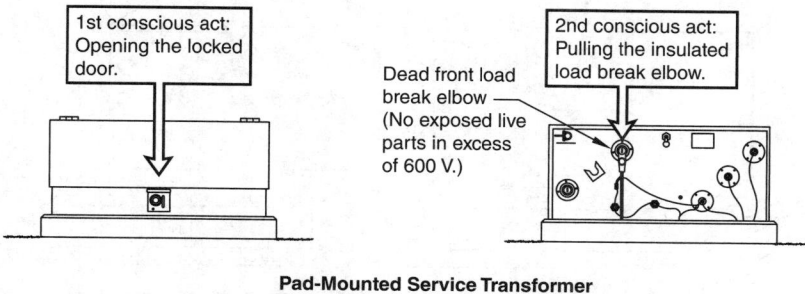

Pad-Mounted Fused Switch

Pad-Mounted Service Transformer

Fig. 381-2. Examples of access to exposed live parts in excess of 600 V (Rule 381G).

regulatory agencies, and the utility's insurance company for signing applications. It is important to note that **NESC** Rule 381G2 does not require a safety sign on the outside of pad-mounted equipment. Safety signs are only required for the inside of the equipment. See the discussion and the figure in Rule 110A for additional information on ANSI signing requirements. The safety sign requirements for pad-mounted equipment are outlined in Fig. 381-3.

382. LOCATION IN UNDERGROUND STRUCTURES

When equipment is located in the underground structures (i.e., manholes and vaults) discussed in Rule 320, the equipment must not obstruct the personnel access openings discussed in Rule 323C. The equipment must also not impede the egress of a person working in the manhole or vault per Rule 382A.

The only dimension provided in Rule 382 is the 8-in clearance from equipment to the back of a fixed ladder. Fixed ladders are not required except for the special conditions specified in Rules 323C1 and 323C4. Ladder requirements are also discussed in Rules 323C5 and 323F. No dimension is given from equipment to the front of the ladder. The clearance to the front of the ladder must meet the general requirement that the equipment must not interfere with the proper use of the ladder. The remaining clearance requirements in this rule are also general wording requirement without any stated dimensions. Equipment arrangement must consider installation, operation, and maintenance requirements. Switching equipment

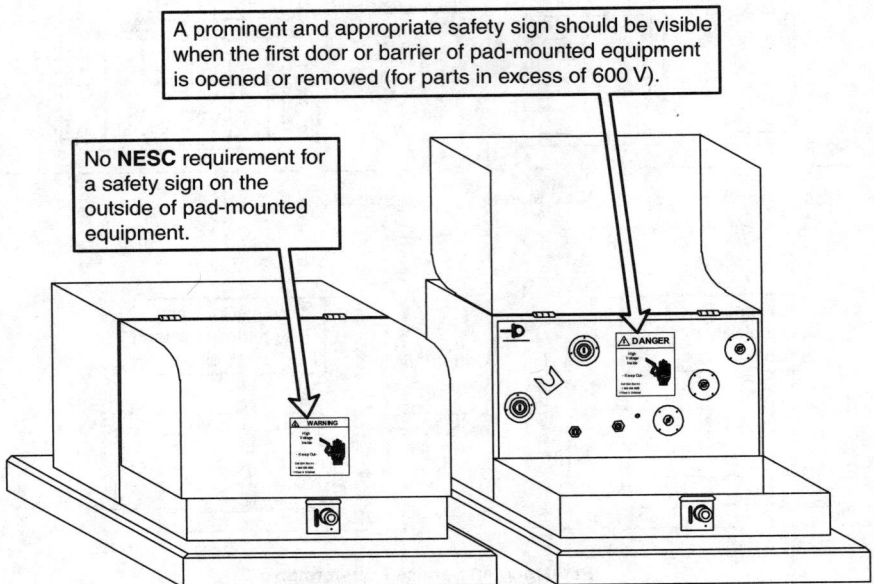

A prominent and appropriate safety sign should be visible when the first door or barrier of pad-mounted equipment is opened or removed (for parts in excess of 600 V).

No **NESC** requirement for a safety sign on the outside of pad-mounted equipment.

See Photo(s)

Fig. 381-3. Safety sign requirements for pad-mounted equipment (Rule 381G2).

must be operable from a safe position. Equipment must not interfere with the drainage or ventilation of the underground structure.

Although Rule 382 does not reference Rule 341B2 and NESC Table 341-1, the clearances in this table apply between joint-use (supply and communication) equipment.

383. INSTALLATION

The installation requirements for underground equipment (both pad-mounted equipment and equipment installed in manholes and vaults) are very general. No dimensions are provided for any of the requirements. Equipment installation must consider the following:

- Equipment weight (lifting, rolling, and mounting considerations)
- Guarding or isolating live parts from persons
- Easy access to operate, inspect, and test facilities
- Isolating or protecting live parts from conductive liquids or other materials
- Locking or securing operating controls of supply equipment

384. GROUNDING AND BONDING

Rules 384A and 384B have grounding requirements similar to the general grounding requirements in Rule 314.

Rule 384C requires bonding between aboveground metallic supply and communications enclosures that are 6 ft or less apart. The intent of this rule is to not have different potentials between adjacent metallic enclosures. Bonding between fiberglass enclosures or a metallic enclosure and a fiberglass enclosure is not required. A size is not specified for the bonding jumper. Rule 012C, which requires accepted good practice, must be used. Rule 099C, which requires an AWG No. 6 copper bond, is one option for the accepted good practice in this case. Rules 342 and 093C7 provide additional information on bonding conductors. The 6-ft spacing is an average reach for an adult person. Rule 384C does not address (or require) bonding of dozens of other metallic items (e.g., metal conduit risers, steel poles, fire hydrants, sign posts, fences, etc.) that can be adjacent to aboveground metallic supply and communications enclosures. Rule 384C clarifies that a supply pole ground is not required to be bonded to an adjacent metallic communications enclosure. A note to the rule clarifies that bonding a supply pole ground to an adjacent metallic communications pedestal is not required but it is not prohibited if the parties agree. The rules for bonding aboveground metallic supply and communications enclosures are outlined in Fig. 384-1.

385. IDENTIFICATION

Identification of equipment that operates in multiple eliminates confusion and adds safety.

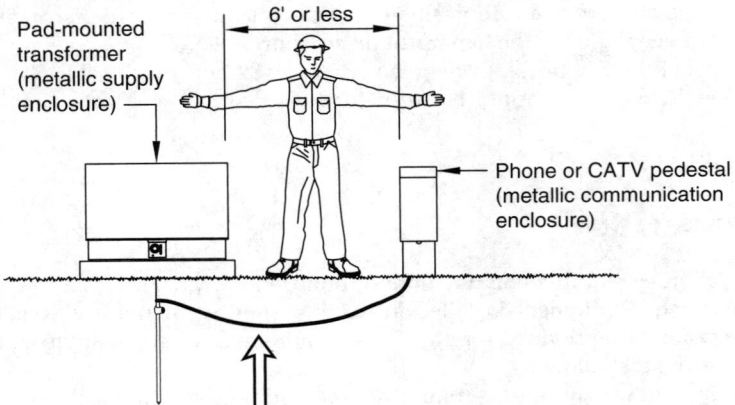

Pad-mounted transformer (metallic supply enclosure)

Phone or CATV pedestal (metallic communication enclosure)

Bond required between aboveground metallic supply and communication enclosures when separated by a distance of 6' or less. (No bond wire size is specified.) A bond is not required if one or both of the enclosures are nonmetallic. Bonding is not required, but is permitted by agreement, between a supply pole ground and a metallic communications enclosure.

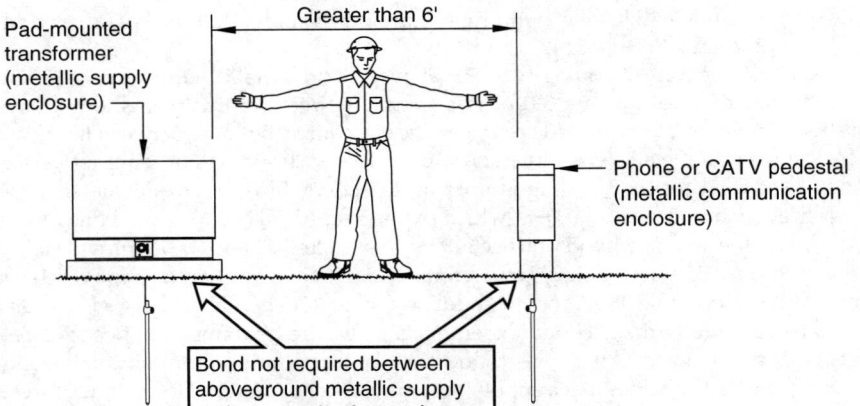

Pad-mounted transformer (metallic supply enclosure)

Phone or CATV pedestal (metallic communication enclosure)

Bond not required between aboveground metallic supply and communication enclosures when separated by a distance greater than 6'.

See Photo(s)

Fig. 384-1. Bonding aboveground metallic supply and communication enclosures (Rule 384C).

Section 39

Installation in Tunnels

390. GENERAL

Installation in tunnels must meet applicable rules found in Part 3 and the additional rules of this section. Rule 323E, Vault and Utility Tunnel Access, provides additional requirements. If unqualified persons (i.e., the general public) have access to the tunnel, the applicable requirements of Part 2 must be met. The **Code** is referring to a tunnel that could be a roadway traffic tunnel with a utility line passing through it, or the tunnel could be a dedicated utility tunnel under the surface of the ground. Utility tunnels can be found connecting buildings in urban areas or on university campuses. In some cases, the tunnels are used just for utilities accessible to qualified personnel. In other cases, the utility tunnel doubles as a public walkway. In addition to the **NESC** rules, the National Electrical Code (NEC) can be referenced for accepted good practice. A utility tunnel used as a public walkway between buildings will require adherence to the NEC. No matter what type of tunnel is involved, all parties concerned must agree on the design of the tunnel structure and the design of the utilities within the structure.

391. ENVIRONMENT

If the tunnel is accessible to the public or workers, the environment within the tunnel must be suitable for people. If the tunnel is not accessible to the public or workers, the construction would be similar to duct bank construction where cables are pulled in and out of ducts but access is only obtainable at pulling or splice locations. Rule 391A provides requirements for general environmental safety and working space. Rule 391B provides additional rules for joint-use (supply and communication) tunnels.

Part 4

Work Rules for the Operation of Electric Supply and Communications Lines and Equipment

GENERAL SECTIONS

01 INTRODUCTION   02 DEFINITIONS

03 REFERENCES   09 GROUNDING METHODS

GENERAL SECTIONS 01, 02, 03, 09

PART 1

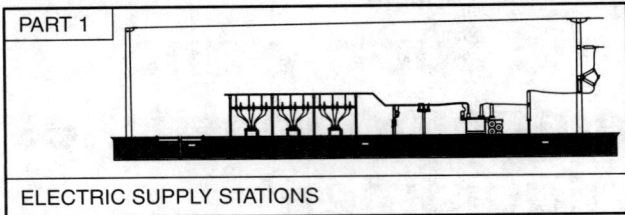

ELECTRIC SUPPLY STATIONS

PART 2

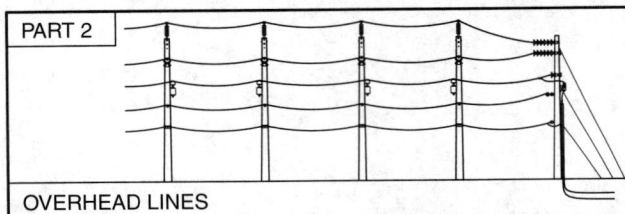

OVERHEAD LINES

PART 3

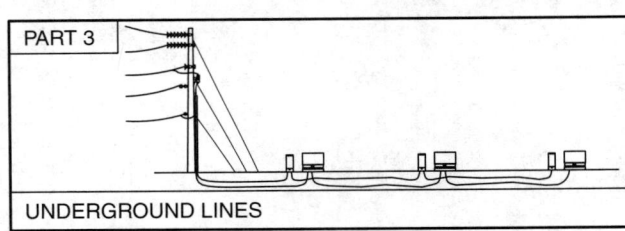

UNDERGROUND LINES

PART 4

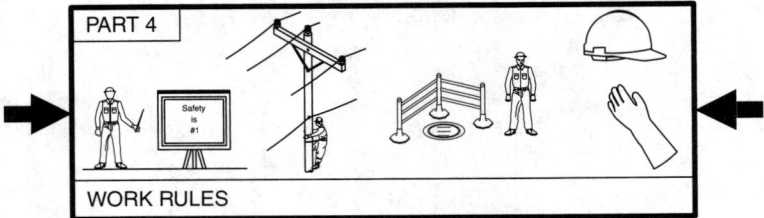

WORK RULES

Section 40

Purpose and Scope

400. PURPOSE

The purpose of Part 4, Work Rules, is similar to the purpose of the entire **NESC** outlined in Rule 010, except Rule 400 is specific to the work rules for the operation of electric supply and communication lines and equipment. Part 4 of the **NESC** focuses on practical work rules as a means of safeguarding employees and the public. The **Code** states that the intent of Part 4 is not to require unreasonable steps to comply with the rules; however, reasonable steps must be taken. A more specific statement is given in Rule 410A4. Rule 410A4 requires that the work rules be used, but if strict enforcement of the work rules seriously impedes safety, the employee in charge may temporarily modify the work rules without increasing hazards. This flexibility is needed because every conceivable situation cannot be covered in the work rules. Flexibility in applying the work rules cannot be abused in situations where the work rules apply.

401. SCOPE

The scope of Part 4, Work Rules, includes work rules to be used in the installation, operation, and maintenance of both electric supply and communications systems. Part 4 is broken down into four sections that are all interrelated. The four sections are outlined below:
- Section 41, "Supply and Communications Systems—Rules for Employers."

These rules apply to the supply and communications company.

- Section 42, "General Rules for Employ*ees*." These rules apply to the supply and communications employee working for the supply and communications company.
- Section 43, "Additional Rules for *Communications* Employ*ees*." These rules are additional rules for communications employees only. Note that Sec. 42 was for both supply and communications employees.
- Section 44, "Additional Rules for *Supply* Employ*ees*." These rules are additional rules for the supply employees only. Note that Sec. 42 was for both supply and communications employees.

The titles of **NESC** Secs. 41, 42, 43, and 44 are graphically represented in Fig. 401-1.

The format of this Handbook for Part 4, Work Rules, is different from the format of the other parts of this book. The format for Part 4 summarizes the **Code** text in an outline bulleted list. A graphic is used to visually represent which section, rule, and paragraph the outline applies to. A brief general discussion of each section is provided at the beginning of each section.

Occupational Safety and Health Administration (OSHA) regulations also apply to the operation, maintenance, and construction of electric supply and communications systems. See Rule 402 for a discussion of applicable OSHA standards.

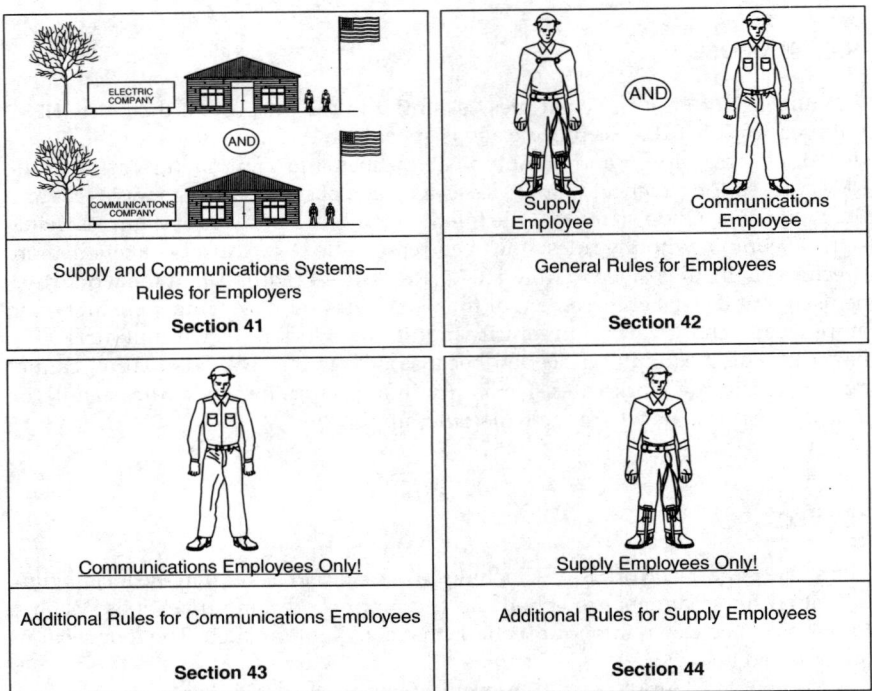

Fig. 401-1. Titles of Secs. 41, 42, 43, and 44 (Rule 401).

402. REFERENCED SECTIONS

This rule references four **NESC** sections related to Part 4, Work Rules, so that rules do not have to be duplicated and the reader of the **Code** realizes that other sections are related to the information provided in Part 4. The related sections are:
- Introduction, Sec. 01
- Definitions, Sec. 02
- References, Sec. 03
- Grounding Methods, Sec. 09

The Work Rules in Part 4 of the **NESC** are similar to (but they are not exactly the same as) the Occupational Safety and Health Administration (OSHA) regulations related to electric supply and communications systems.

The scope of this **NESC** Handbook does not include specific comments on the OSHA regulations. Many utilities tend to place more emphasis on the OSHA regulations than the work rules in Part 4 of the **NESC**, as OSHA performs both routine inspections and accident investigations. In general, utilities are required to meet the **NESC** rules and the OSHA standards; therefore, both apply, not one or the other. The Work Rules in **NESC** Part 4 are tied to the rules in other parts of the **NESC**.

The OSHA standards related to the **NESC** Work Rules are listed below and are reprinted in Appendix B of this Handbook as a reference.
- 1910.268 Telecommunications (Operation and Maintenance)
- 1910.269 (including Appendix A–G) Electric Power Generation, Transmission, and Distribution (Operation and Maintenance)
- 1926.950 through 1926.968 (Appendix A–G) 1926 Subpart V, Power Transmission and Distribution (Construction)

The OSHA 1910 series is for operation and maintenance. OSHA Standards 1910.268 and 1910.269 are listed under 1910 Subpart R—Special Industries. The OSHA 1926 series is for construction. OSHA Standards 1926.950 through 1926.968 are listed under 1926 Subpart V—Power Transmission and Distribution. OSHA uses the terms electric power and telecommunications instead of the **NESC** terms electric supply and communications. OSHA separates its standards into general industry (operation and maintenance) standards and construction standards, the **NESC** does not. OSHA has separate operations and maintenance standards and construction standards for the electric power industry. OSHA only has operation and maintenance standards for the telecommunications industry. When OSHA does not have a construction standard for a specific industry, as in the case of telecommunications, the general OSHA construction standards apply. Where specific rules do not exist in the general OSHA construction standards, applying the 1910.268 operation and maintenance standards to construction work is a prudent way of meeting OSHA's General Duty Clause (OSHA Section 5—Duties of the OSHA Act of 1970).

The OSHA 1910.268 Telecommunications (Operations and Maintenance) standard became effective in 1975. The OSHA 1910.269 Electric Power Generation, Transmission, and Distribution (Operations and Maintenance) standard became effective in 1994. The OSHA 1926.950 through 1926.968 (1926 Subpart V) Power Transmission and Distribution (Construction) standards became effective in 1972. Amendments have been made to the standards over the years. The OSHA 1926.950

through 1926.968 (1926 Subpart V) Power Transmission and Distribution (Construction) standards and the OSHA 1910.269 Electric Power Generation, Transmission, and Distribution (Operations and Maintenance) standard both received a major revision in 2014.

There are several other OSHA standards referenced throughout 1910.268, 1910.269, and 1926.950 through 1926.968 (1926 Subpart V). These related standards are referenced either without amendment or referenced and supplemented with additional information. Some of the most commonly referenced related standards are listed below:

- 1910 Subpart S—Electrical (1910.301 to 1910.399)
- 1910.151—Medical Services and First Aid (Part of 1910 Subpart K)
- 1910.147—The Control of Hazardous Energy (Lockout/Tagout) (Part of 1910 Subpart J)
- 1910 Subpart I—Personal Protective Equipment (1910.132 to 1910.138)
- 1910.137—Electrical Protective Devices (Part of 1910 Subpart I)
- 1926 Subpart M—Fall Protection (1926.500 to 1926.503)
- 1926 Subpart P—Excavations (1926.650 to 1926.652)
- 1910 Subpart D—Walking-Working Surfaces (1910.21 to 1910.30)
- 1910.67 Vehicle-Mounted Elevating and Rotating Work Platforms (Part of 1910 Subpart F)
- 1910 Subpart N—Materials Handling and Storage (1910.176 to 1910.184)
- 1926 Subpart W—Rollover Protective Structures; Overhead Protection (1926.1000 to 1926.1003 and Appendix A)

OSHA standards are very easy to access from the Internet. Each standard also has a list of requested interpretations. Each interpretation lists the question asked and the OSHA response. The phone numbers and addresses of OSHA offices are also easily accessed from the Internet. Individual states have the option to adopt the federal OSHA standards or develop their own OSHA-approved state plan. State run OSHA plans must be at least as effective as the federal OSHA program. All of this information is available at www.osha.gov.

Supply and Communications Systems—Rules for Employers

The format of this Handbook for Part 4, Work Rules, is different from the format of the other parts of this book. The format for Part 4 summarizes the **Code** text in an outline bulleted list. A graphic is used to visually represent which section, rule, and paragraph the outline applies to. A brief general discussion of each section is provided at the beginning of each section. The following brief general discussion covers selected topics in Secs. 41 and 42. This discussion is repeated at the beginning of Sec. 42.

Both employers and employees have an obligation to safety. Section 41 provides rules for supply and communications employers. Section 42 provides general rules for supply and communications employees. Section 43 provides additional rules for communications employees. Section 44 provides additional rules for supply employees.

Sections 41 and 42 are written to put the responsibility for safety on both the employer (company) and the employee (worker). The employer must designate an employee in charge to represent the company. The employee in charge is responsible for making sure employees adhere to the work rules. The employees must also assume responsibility for following safety rules. This system builds redundancy by putting the safety requirement on both the employer and employee.

Two specific examples of the relationship between Secs. 41 and 42 (employer and employee) are listed below:

- The *employer* must inform each employee of the safety rules (Rule 410A). The *employee* must read and study the safety rules (Rule 420A).
- The *employer* must have an adequate supply of protective devices and equipment (e.g., hard hats, rubber gloves, insulated tools, body belts, etc.) sufficient to enable employees to meet the requirements of the work to be undertaken (Rule 411B).

The *employee* must use personal protective equipment, the protective devices, and the special tools provided for their work and inspect these devices and tools before starting work to verify that they are in good condition (Rule 420H).

When the employee is required to perform a task, for example, inspecting personal protective equipment, the employer must have a designated employee in charge responsible for making sure the employees perform the inspection. The duties of the designated employee in charge are outlined in Rule 421A.

Section 41 (rules for employers) provides a list of typical protective devices and equipment for employees (who are covered in Sec. 42) to use. The **Code** does not dictate what protective devices and equipment must be used for a particular task. The choice of what protective devices and equipment need to be used is site-specific. The **NESC** cannot cover every conceivable site-specific situation.

Section 41 (rules for employers), Rule 411A2, requires that diagrams (i.e., maps) be maintained and on file for employees (who are covered in Sec. 42) to use. The diagrams aid the identification of structures required in Part 2, Overhead Lines, Rule 217A3, and the location of underground facilities required in Part 3, Underground Lines, Rule 311A. Accurate diagrams are useful for minimizing errors and accidents. Inaccurate maps can increase errors and accidents. Diagrams cannot be used as a substitute for applying proper safety procedures.

Section 41 (Rule 410A) requires the employer (the company) to perform an assessment to determine the potential exposure to an electric arc for employees who work on or near energized lines or equipment. The assessment considers the employees assigned tasks and work activities. For example, an engineer entering a substation to read information off of a transformer nameplate has a different potential exposure to an electric arc than a lineman replacing a pole top transformer on an energized line. In addition to the assessment of the work activity, the employer must perform an arc hazard analysis to determine the effective arc rating of clothing to be worn by the employee. The arc thermal protection value (ATPV) can be calculated in an engineering study or determined using **NESC** Tables 410-1, 410-2, and 410-3. Each table is for a different voltage range and the assumptions used to create the tables are documented in the footnotes at the bottom of the tables. Section 42 (Rule 420I) requires the employee to wear the clothing that the employer (the company) determined necessary.

Sometimes confusion exists between the application of the **NESC** Work Rules and NFPA 70E (Standard for Electrical Safety in the Workplace). A careful review of the scope of NFPA 70E indicates that NFPA 70E does not apply to supply (electric power) or communications utilities. In general, NFPA 70E applies to electrical work performed by electricians. However, even though the scope of NFPA 70E does not apply to utility companies, it can be a useful reference for accepted good practice. One practice that NFPA 70E addresses that is not commonly found in the electric utility industry is labeling equipment such as circuit breakers, fused switches, etc. with a sign or label indicating the arc hazard incident energy or required level of arc-rated clothing. Utility companies typically use training of their employees to address this issue, not signs or labels. For more information, see the NFPA 70E document in its entirety.

The Work Rules in Part 4 of the **NESC** are similar to (but they are not exactly the same as) the Occupational Safety and Health Administration (OSHA) regulations related to electric supply and communications systems. See Rule 402 for additional information on the OSHA standards that are related to the NESC Work Rules.

410. GENERAL REQUIREMENTS

410A. General

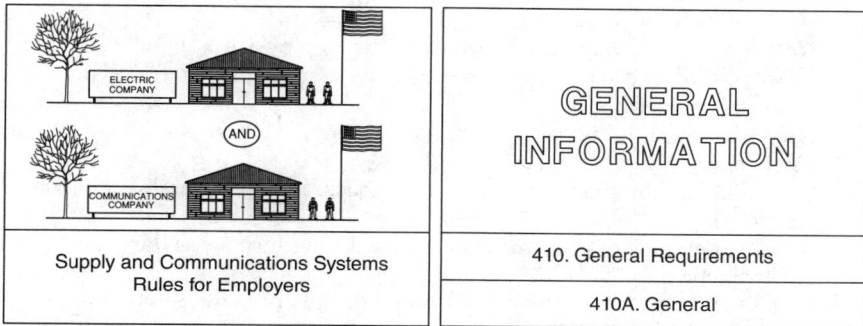

Supply and Communications Systems Rules for Employers

GENERAL INFORMATION

410. General Requirements

410A. General

- Employer must inform each employee of the safety rules.
- Employer must provide a copy of the safety rules (when necessary).
- Employer must train employees.
- Employer must ensure that each employee has demonstrated proficiency in required tasks.
- Employer must retrain employees who are not following work rules.
- Employer must ensure an assessment is performed to determine the potential exposure to electric arcs for employees who work on or near energized lines or equipment.
- If the assessment determines potential employee exposure, the following materials must not be worn (unless arc rated):
  - ✓ Acetate
  - ✓ Nylon
  - ✓ Polyester
  - ✓ Polypropylene
- If the assessment determines potential employee exposure an outer layer of clothing that could ignite and burn must not be worn.
- Assessments that determine a potential employee exposure greater than 2 cal/cm² require a detailed arc hazard analysis or use of NESC Tables 410-1 (50-1000 V), 410-2 (1.1 kV-46 kV), or 410-3 (46.1 kV-800 kV) to determine the effective arc rating of clothing to be worn by employees.
- Assessments that determine a potential employee exposure greater than 2 cal/cm² require the arc hazard analysis calculation of the estimated arc energy to be based on the available fault current, the duration of the arc in cycles, and the distance of the arc to the employee.
- Assessments that determine a potential employee exposure greater than 2 cal/cm² require the employee to cover the entire body with arc rated clothing and equipment that has an effective arc rating at least equal to the anticipated level of arc energy.

- An exception applies if the clothing required creates a hazard greater than the possible exposure to the heat energy of the electric arc, in this case clothing with a lower arc rating than required may be worn.
- Special exceptions apply to the employees hands.
- Special exceptions apply to the employees feet.
- Special exceptions apply to the employees head and face.
- Special exceptions apply to DC systems with voltages from 50 V to 250 V and 8000 A maximum fault current.
- Note that consideration can be given to the employee's assigned tasks and/or work activities when performing the assessment to determine potential exposure to electric arcs.
- Note that multiple layers of arc rated clothing have been shown by tests to block more heat than a single layer.
- Note that clothing includes shirts, pants, jackets, and coveralls in single or multiple layers.
- Note that engineering controls can be used to reduce arc energy levels and work practices can be used to reduce exposure levels.
- Employers must use procedures to assure compliance with safety rules.
- If strict enforcement of rules seriously impedes safety, the employee in charge may temporarily modify rules as long as hazards are not increased.
- If a disagreement on how to apply operating rules occurs, the employer's decision will be final; however, the decision must not be hazardous to the employee.
- Employer must provide training to employees who work around antennas to recognize and mitigate exposure to radio frequencies.
- Regulatory standards in OSHA, FCC, and IEEE can be referenced for radiation level limits.

410B. Emergency and First Aid Procedures

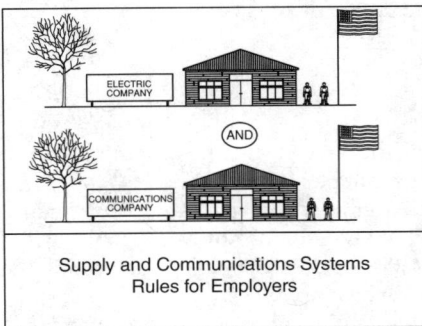

Supply and Communications Systems
Rules for Employers

410. General Requirements

410B. Emergency and First Aid Procedures

- Employer must inform employee of emergency procedures and first aid methods including resuscitation (CPR).
- Copies of emergency procedures and first aid methods must be kept where number of employees and type of work warrants their placement (e.g., in vehicles and other locations).

- Employers must regularly instruct employees on the methods of first aid and emergency procedures (if duties warrant such training).

410C. Responsibility

Supply and Communications Systems Rules for Employers	410. General Requirements
	410C. Responsibility

- Employer must select a designated person to be in charge of operations and responsible for safety.
- A crew must have only one person in charge.
- For multiple locations, one person may be in charge at each location.

411. PROTECTIVE METHODS AND DEVICES

411A. Methods

METHODS

Supply and Communications Systems Rules for Employers	411. Protective Methods and Devices
	411A. Methods

- Employer must restrict employees from access to energized or rotating equipment unless the employee is authorized.
- Diagrams (i.e., maps) of the electric system must be available to authorized employees.
- Employees are to be instructed before work starts.
- Employees are to be instructed to take additional precautions for unusual hazards.

411B. Devices and Equipment

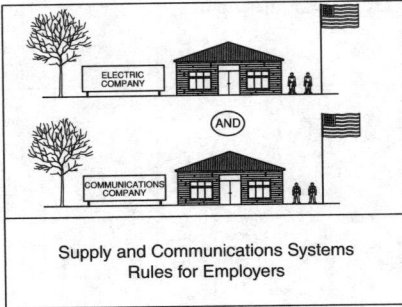

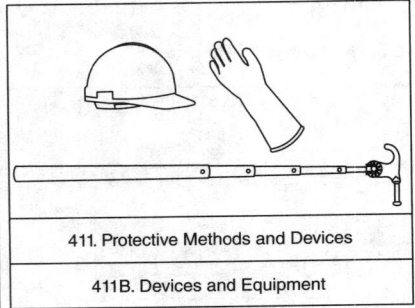

Supply and Communications Systems Rules for Employers

411. Protective Methods and Devices

411B. Devices and Equipment

- Employer must have an adequate supply of protective devices (e.g., hard hats, rubber gloves, insulated tools, body belts, etc. based on the requirements of the job).
- Employer must have an adequate supply of first aid equipment.
- Protective devices must conform to applicable standards.

411C. Inspection and Testing of Protective Devices and Equipment

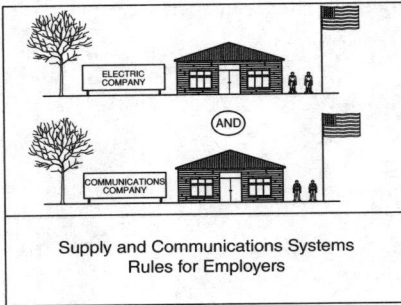

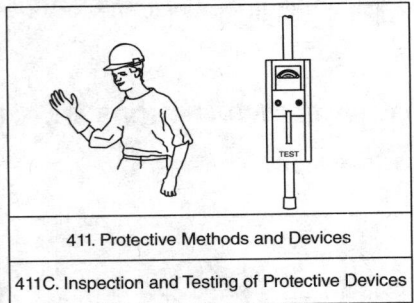

Supply and Communications Systems Rules for Employers

411. Protective Methods and Devices

411C. Inspection and Testing of Protective Devices

- Inspect or test protective devices and equipment.
- Inspect insulating gloves, sleeves, and blankets before use.
- Test insulated gloves and sleeves as required.
- Inspect climbing and fall protection equipment (e.g., line worker's body belts, lanyards, positioning straps, etc.) to ensure safety before use.

411D. ✓ Signs and Tags for Employee Safety

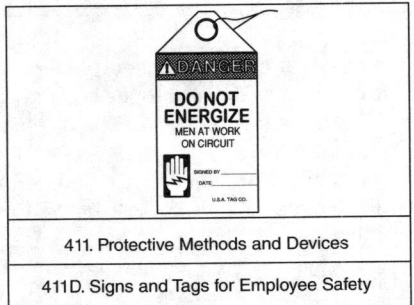

Supply and Communications Systems Rules for Employers

411. Protective Methods and Devices

411D. Signs and Tags for Employee Safety

- Safety signs and tags must meet ANSI Z535 Standards.

411E. Identification and Location

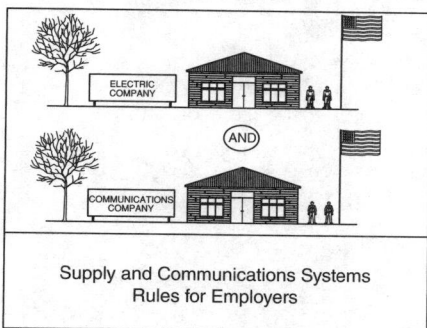

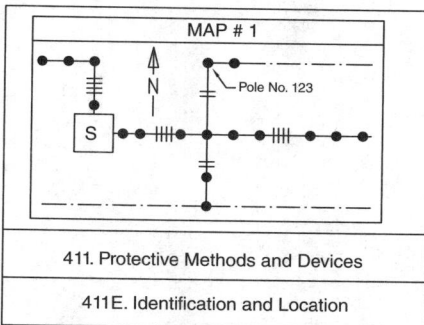

Supply and Communications Systems Rules for Employers	411. Protective Methods and Devices
	411E. Identification and Location

MAP # 1

Pole No. 123

- Provide a means to identify lines before they are worked on.
- Be able to locate underground facilities.

411F. Fall Protection

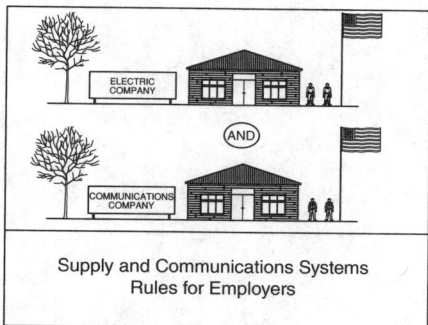

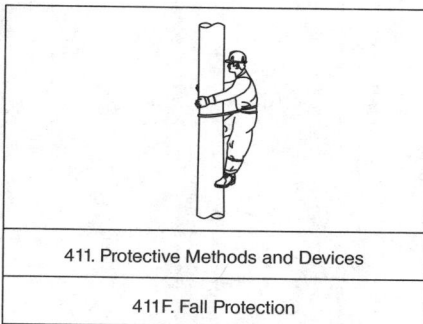

Supply and Communications Systems Rules for Employers	411. Protective Methods and Devices
	411F. Fall Protection

- Employer must develop, implement, and maintain an effective fall protection program.
- The fall protection program must include all of the following:
  - ✓ Training, retraining, and documentation
  - ✓ Guidance on equipment selection, inspection, care, and maintenance
  - ✓ Considerations concerning structural design and integrity, with particular reference to anchorages and their availability
  - ✓ Rescue plans and related training
  - ✓ Hazard recognition
- The employer must not permit the use of 100 percent leather positioning straps or nonlocking snaphooks.

Section 42

General Rules for Employees

The format of this Handbook for Part 4, Work Rules, is different from the format of the other parts of this book. The format for Part 4 summarizes the **Code** text in an outline bulleted list. A graphic is used to visually represent which section, rule, and paragraph the outline applies to. A brief general discussion of each section is provided at the beginning of each section. The following brief general discussion covers selected topics in Secs. 42 and 41. This discussion is repeated at the beginning of Sec. 41.

Both employers and employees have an obligation to safety. Section 41 provides rules for supply and communications employers. Section 42 provides general rules for supply and communications employees. Section 43 provides additional rules for communications employees. Section 44 provides additional rules for supply employees.

Sections 41 and 42 are written to put the responsibility for safety on both the employer (company) and the employee (worker). The employer must designate an employee in charge to represent the company. The employee in charge is responsible for making sure employees adhere to the work rules. The employees must also assume responsibility for following safety rules. This system builds redundancy by putting the safety requirement on both the employer and employee.

Two specific examples of the relationship between Secs. 41 and 42 (employer and employee) are listed below:

- The *employer* must inform each employee of the safety rules (Rule 410A). The *employee* must read and study the safety rules (Rule 420A).
- The *employer* must have an adequate supply of protective devices and equipment (e.g., hard hats, rubber gloves, insulated tools, body belts, etc.) sufficient

to enable employees to meet the requirements of the work to be undertaken (Rule 411B).

The *employee* must use personal protective equipment, the protective devices, and the special tools provided for their work and inspect these devices and tools before starting work to verify that they are in good condition (Rule 420H).

When the employee is required to perform a task, for example, inspecting personal protective equipment, the employer must have a designated employee in charge responsible for making sure the employees perform the inspection. The duties of the designated employee in charge are outlined in Rule 421A.

Section 41 (rules for employers) provides a list of typical protective devices and equipment for employees (who are covered in Sec. 42) to use. The **Code** does not dictate what protective devices and equipment must be used for a particular task. The choice of what protective devices and equipment need to be used is site-specific. The **NESC** cannot cover every conceivable site-specific situation.

Section 41 (rules for employers), Rule 411A2, requires that diagrams (i.e., maps) be maintained and on file for employees (who are covered in Sec. 42) to use. The diagrams aid the identification of structures required in Part 2, Overhead Lines, Rule 217A3, and the location of underground facilities required in Part 3, Underground Lines, Rule 311A. Accurate diagrams are useful for minimizing errors and accidents. Inaccurate maps can increase errors and accidents. Diagrams cannot be used as a substitute for applying proper safety procedures.

Section 41 (Rule 410A) requires the employer (the company) to perform an assessment to determine the potential exposure to an electric arc for employees who work on or near energized lines or equipment. The assessment considers the employees assigned tasks and work activities. For example, an engineer entering a substation to read information off of a transformer nameplate has a different potential exposure to an electric arc than a lineman replacing a pole top transformer on an energized line. In addition to the assessment of the work activity, the employer must perform an arc hazard analysis to determine the effective arc rating of clothing to be worn by the employee. The arc thermal protection value (ATPV) can be calculated in an engineering study or determined using **NESC** Tables 410-1, 410-2, and 410-3. Each table is for a different voltage range and the assumptions used to create the tables are documented in the footnotes at the bottom of the tables. Section 42 (Rule 420I) requires the employee to wear the clothing that the employer (the company) determined necessary.

Sometimes confusion exists between the application of the **NESC** Work Rules and NFPA 70E (Standard for Electrical Safety in the Workplace). A careful review of the scope of NFPA 70E indicates that NFPA 70E does not apply to supply (electric power) or communications utilities. In general, NFPA 70E applies to electrical work performed by electricians. However, even though the scope of NFPA 70E does not apply to utility companies, it can be a useful reference for accepted good practice. One practice that NFPA 70E addresses that is not commonly found in the electric utility industry is labeling equipment such as circuit breakers, fused switches, etc. with a sign or label indicating the arc hazard incident energy or required level of arc-rated clothing. Utility companies typically use training of their employees to address this issue, not signs or labels. For more information, see the NFPA 70E document in its entirety.

The Work Rules in Part 4 of the **NESC** are similar to (but they are not exactly the same as) the Occupational Safety and Health Administration (OSHA) regulations related to electric supply and communications systems. See Rule 402 for additional information on the OSHA standards that are related to the **NESC** Work Rules.

420. GENERAL

420A. Rules and Emergency Methods

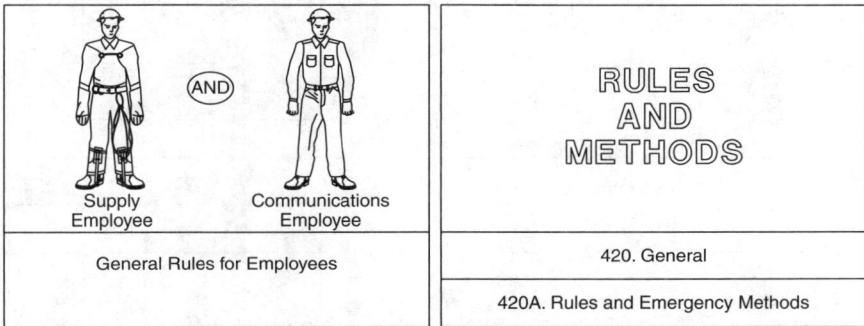

Supply Employee	AND	Communications Employee	RULES AND METHODS
General Rules for Employees			420. General
			420A. Rules and Emergency Methods

- Read and study safety rules.
- Show knowledge of safety rules.
- Be familiar with first aid, rescue techniques, and fire extinguishing.

420B. Qualification of Employees

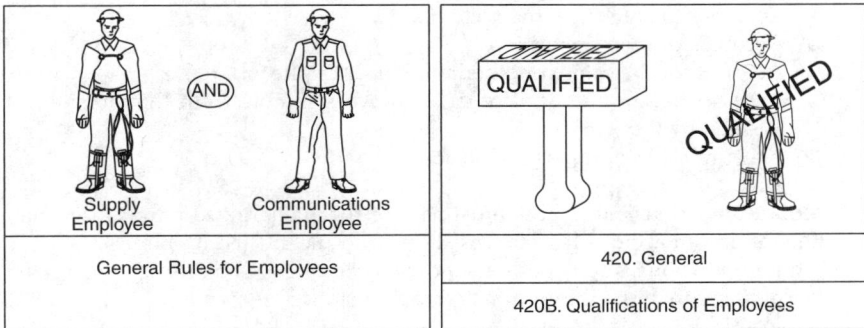

Supply Employee	AND	Communications Employee	QUALIFIED
General Rules for Employees			420. General
			420B. Qualifications of Employees

- Employees must only perform tasks for which they are trained, equipped, authorized, and directed.

- Inexperienced employees must work under experienced employees and perform only directed tasks.
- Employees operating mechanical equipment must be qualified to perform those tasks.
- If safety is in doubt, request instructions from supervisor.
- Employees who only occasionally work on electric supply lines can only do work when authorized.

420C. Safeguarding Oneself and Others

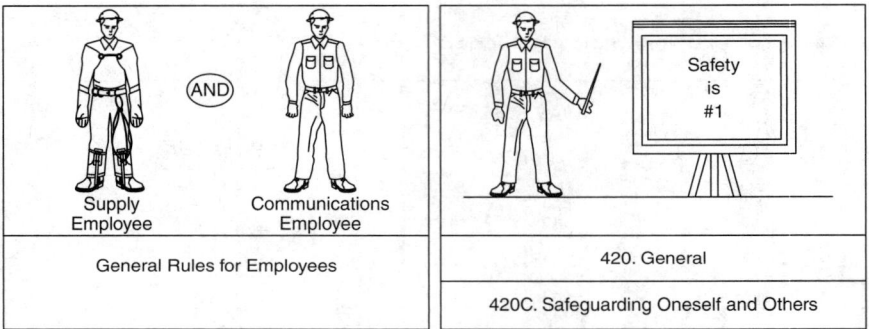

- Heed safety signs and signals.
- Warn others who are in danger or near energized lines.
- Report line or equipment defects (e.g., low clearance, broken insulators, etc.).
- Report accidentally energized items.
- Report any defect that may cause danger.
- Employees who do not work on lines and equipment must keep away from them and keep away from worksites with falling objects.
- Employees who work on energized lines must:
  - ✓ Consider the effects of their actions.
  - ✓ Account for their own safety.
  - ✓ Account for the safety of other employees on the job site.
  - ✓ Account for the safety of other employees remote from the job site but affected by the work.
  - ✓ Account for the property of others.
  - ✓ Account for the public.
- Communications employees must not approach energized parts closer than the minimum approach distances shown in Rule 431 (**NESC** Table 431-1).
- Communications employees must not take conductive objects near energized parts closer than the minimum approach distances shown in Rule 431 (**NESC** Table 431-1).
- Supply employees must not approach energized parts closer than the minimum approach distances shown in Rule 441 (**NESC** Tables 441-1 or 441-5).

- Supply employees must not take conductive objects near energized parts (without an insulating handle) closer than the minimum approach distances shown in Rule 441 (**NESC** Tables 441-1 or 441-5).
- Employees must use care when working with metal ropes, tapes, or wires in the vicinity of energized high-voltage lines due to energization and induced voltages.
- Clearance measurements from energized lines must be done with approved devices (e.g., insulated measuring sticks).

420D. Energized or Unknown Conditions

Supply Employee

Communications Employee

General Rules for Employees

420. General

420D. Energized or Unknown Conditions

- Consider equipment and lines to be energized unless they are positively known to be de-energized.
- Determine existing conditions before starting work by inspection or tests.
- Determine the operating voltage of equipment and lines before starting work.

420E. Ungrounded Metal Parts

Supply Employee

Communications Employee

General Rules for Employees

420. General

420E. Ungrounded Metal Parts

- Consider ungrounded metal parts energized at the highest voltage to which they are exposed.

420F. Arcing Conditions

Supply Employee (AND) Communications Employee

General Rules for Employees

420. General

420F. Arcing Conditions

- Keep body parts far away from devices that produce arcs during operation, such as switches.

420G. Liquid-Cell Batteries

Supply Employee (AND) Communications Employee

General Rules for Employees

420. General

420G. Liquid-Cell Batteries

- Determine that battery areas are adequately ventilated.
- Avoid smoking, open flames, or tools that produce sparks.
- Use eye and skin protection during handling.
- Take precautions to avoid short circuits and electric shocks.

420H. Tools and Protective Equipment

Supply Employee (AND) Communications Employee

General Rules for Employees

420. General

420H. Tools and Protective Equipment

- Use the personal protective equipment, devices, and tools provided for the work.
- Before starting work, carefully inspect the personal protective equipment, devices, and tools to verify they are in good condition.

420I. Clothing

Supply Employee AND Communications Employee

General Rules for Employees

420. General

420I. Clothing

- Wear clothing suitable for the assigned task and work environment.
- Employees exposed to an electric arc must wear clothing or a clothing system in accordance with Rule 410A (**NESC** Tables 410-1, 410-2, and 410-3).
- Avoid wearing exposed metal articles near energized lines.

420J. Ladders and Supports

Supply Employee AND Communications Employee

General Rules for Employees

420. General

420J. Ladders and Supports

- Verify ladders, aerial lifts, etc., are strong, in good condition, and secure before using them.
- Do not paint portable wood ladders except with a clear nonconductive coating.
- Do not reinforce portable wood ladders with metal.
- Do not use portable metal ladders near energized parts.
- Conductive portable ladders for specialized work must only be used for the work intended.

420K. Fall Protection

Supply Employee   AND   Communications Employee

General Rules for Employees

420. General

420K. Fall Protection

- Appropriate fall protection must be used when climbing, transferring, and transitioning unless doing so is not feasible or creates a greater hazard.
- Work positioning shall be rigged so the climber cannot fall more than 2 ft.
- Anchorages for work positioning must support at least twice the impact load of a fall or 3,000 lb-force, whichever is greater.
- Note that tested wood-pole fall-restriction devices are considered to meet the anchorage strength requirements when properly used.
- Note that engineering practices, design specifications, and maintenance procedures may be used to determine anchorage strength requirements, but visual inspections must be performed.
- Note that bolted attachments, step bolts, or other equipment may serve as an anchorage, but visual inspections must be performed.
- Fall protection systems must be used when working above 4 ft on poles, towers, or while working from aerial lifts, helicopters, and cable carts.
- Fall protection equipment must be inspected before use.
- Fall arrest equipment must be suitably anchored.
- Determine that the fall protection system is engaged and secure.
- Be aware of accidental disengagement of the snap hook from the D-ring by foreign objects.
- Be aware of accidental disengagement of the snap hook from the D-ring by rollout.
- Use locking snap hooks and compatible hardware.
- Do not connect snap hooks to each other.
- Do not use 100 percent leather positioning straps or nonlocking snaphooks.
- Use wire rope lanyards where the lanyard could be cut.
- Do not use wire rope lanyards near energized lines.

420L. Fire Extinguishers
- Use fire extinguishers or materials rated for energized parts or de-energize the parts first.

Supply Employee (AND) Communications Employee
General Rules for Employees

A B C
420. General
420L. Fire Extinguishers

420M. Machines or Moving Parts

Supply Employee (AND) Communications Employee
General Rules for Employees

420. General
420M. Machines or Moving Parts

- When working on moving parts, verify accidental startup will not occur by using lockout/tagout procedures.
- When working on automatic switches, stay clear of moving parts.

420N. Fuses

Supply Employee (AND) Communications Employee
General Rules for Employees

100A Fuse / 100A Fuse Link
420. General
420N. Fuses

- Use insulated gloves or tools when installing fuses on energized lines.
- Use eye protection and stand clear when installing expulsion-type fuses on energized lines.

420O. Cable Reels

Supply Employee AND Communications Employee	
General Rules for Employees	420. General
	420O. Cable Reels

- Block cable reels so they do not accidentally roll.

420P. Street and Area Lighting

Supply Employee AND Communications Employee	
General Rules for Employees	420. General
	420P. Street and Area Lighting

- Periodically examine lowering ropes or chains, supports, and fastenings.
- A device must be provided to safely disconnect each lamp on a series lighting circuit of more than 300 V before the lamp is handled.
- An exception applies when insulated devices or tools are used and the circuit is treated as a full-voltage circuit.

420Q. Communication Antennas

Supply Employee AND Communications Employee	
General Rules for Employees	420. General
	420Q. Antennas

- Use controls to mitigate radio-frequency exposure levels when working around antennas.
- Regulatory standards in OSHA, FCC, and IEEE can be referenced for radiation level limits.

421. GENERAL OPERATING ROUTINES

421A. Duties of a First-Level Supervisor or Person in Charge

Supply Employee AND Communications Employee

General Rules for Employees

421. General Operating Routines

421A. Duties of a First-Level Supervisor or Person in Charge

- Duties of the individual in charge:
  - ✓ Adopt precautions to prevent accidents.
  - ✓ See that employees are observing safety rules and operating procedures.
  - ✓ Keep records and make reports.
  - ✓ Prevent unauthorized persons from approaching the workplace.
  - ✓ Do not allow tools or devices unsuitable for the work.
  - ✓ Do not allow tools or devices to be used without testing or inspecting first.
  - ✓ Conduct a job briefing (e.g., tailgate) before each job and discuss:
    - Work procedures
    - Personal protective equipment
    - Energy source control
    - Job hazards
    - Special precautions
    - Other items as needed

421B. Area Protection

Supply Employee AND Communications Employee

General Rules for Employees

421. General Operating Routines

421B. Area Protection

UTILITY WORK AHEAD

- Areas accessible to vehicular and pedestrian traffic:
  - ✓ Prevent vehicles and pedestrians from approaching the work site.
  - ✓ Warn the public of openings or obstructions.
  - ✓ Openings or obstructions exposed at night must have warning lights and must be enclosed with protective barricades.
- Areas accessible to employees only:
  - ✓ If the work exposes energized or moving parts that are normally protected, safety signs must be displayed.
  - ✓ If the work exposes energized or moving parts that are normally protected, barricades must be erected.
  - ✓ Work on one section of a switchboard with multiple sections or work on one portion of a substation with several portions requires barriers to prevent contact with energized parts.
- Locations with crossed or fallen wires:
  - ✓ If an employee encounters crossed or fallen wires, the employee must remain on guard or use other means to prevent accidents.
  - ✓ The proper authority must be notified.
  - ✓ If qualified, and if safety rules can be met, the employee may correct the condition.

421C. Escort

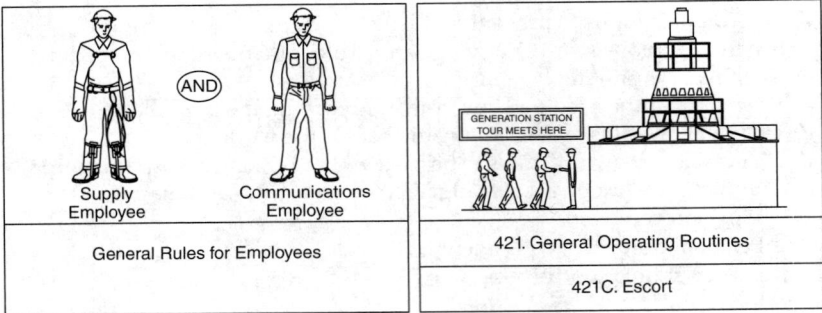

Supply Employee (AND) Communications Employee

General Rules for Employees

421. General Operating Routines

421C. Escort

- An employee responsible for safety must escort nonqualified employees or visitors near electrical lines or equipment.

422. OVERHEAD LINE OPERATING PROCEDURES

422A. Setting, Moving, or Removing Poles in or near Energized Electric Supply Lines

- Employees working on overhead lines must observe the following rules in addition to other applicable rules in Secs. 43 and 44.
- When setting, moving, or removing poles near energized lines:
  - ✓ Avoid direct contact of the pole with the energized conductors.
  - ✓ Wear insulating gloves.

Supply Employee (AND) Communications Employee	
General Rules for Employees	422. Overhead Line Operating Procedures
	422A. Setting, Moving, or Removing Poles in or in the Vicinity of Energized Electric Supply Lines

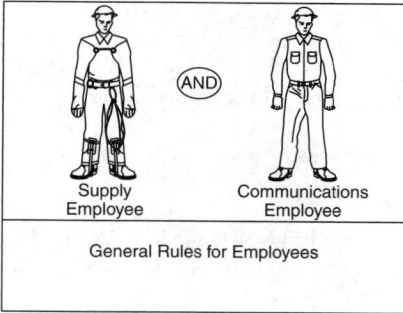

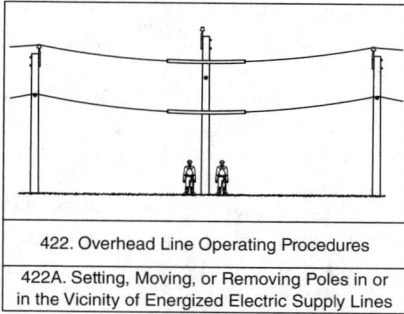

✓ Do not contact the pole with uninsulated body parts.
✓ Avoid touching trucks or other equipment unless wearing suitable protective equipment.

422B. Checking Structures Before Climbing

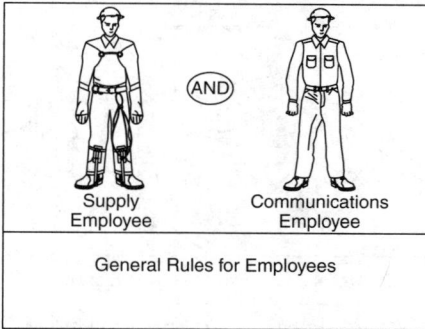

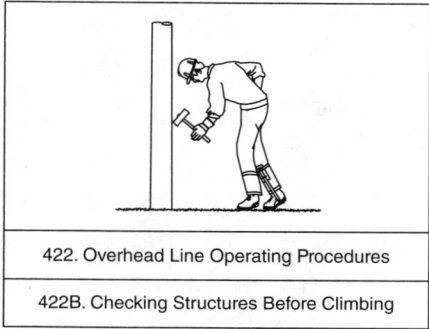

Supply Employee (AND) Communications Employee	
General Rules for Employees	422. Overhead Line Operating Procedures
	422B. Checking Structures Before Climbing

- Before climbing poles, ladders, etc., verify the structure is capable of handling the additional weight and unbalanced forces.
- Poles must not be climbed if unsafe unless guying, bracing, or other means are used to create a safe condition.

422C. Installing and Removing Wires or Cables

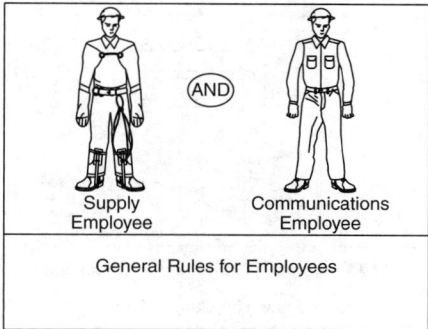

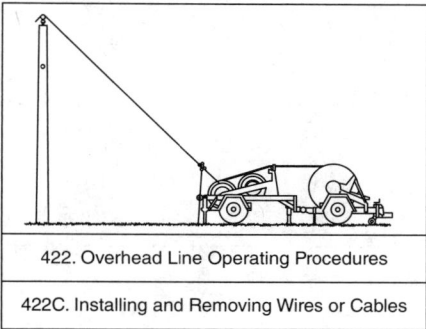

Supply Employee (AND) Communications Employee	
General Rules for Employees	422. Overhead Line Operating Procedures
	422C. Installing and Removing Wires or Cables

- Wires being installed or removed must be kept clear from energized wires.
- Wires being installed or removed that are not bonded to an effective ground must be considered energized.
- Control sag of wires being installed or removed to prevent pedestrian and vehicle traffic damage.
- Verify that structures can handle the forces associated with installing or removing wires.
- Avoid contact with moving winch lines.
- Consider the effect of a higher voltage line on a lower voltage line. Verify that the line being worked on is free from dangerous leakage and induction voltages or verify that it is effectively grounded.

423. UNDERGROUND LINE OPERATING PROCEDURES

423A. Guarding Manhole and Street Openings

Supply Employee (AND) Communications Employee

General Rules for Employees

423. Underground Line Operating Procedures

423A. Guarding Manhole and Street Openings

- Employees working on underground lines must observe the following rules in addition to other applicable rules in Secs. 43 and 44.
- Open manholes, handholes, and vaults must be protected with a barrier, temporary cover, or guard.

423B. Testing for Gas in Manholes and Unventilated Vaults

Supply Employee (AND) Communications Employee

General Rules for Employees

423. Underground Line Operating Procedures

423B. Testing for Gas in Manholes and Unventilated Vaults

- Test manhole for combustible or flammable gases before entry.
- If combustible or flammable gases exist, ventilate before entry.
- Test for oxygen deficiency.
- Make provisions for a good air supply during work.

423C. Flames

- Do not smoke in manholes.
- Use extra precaution to ensure ventilation when flames are required for work.
- Test excavation areas for combustible gases or liquids (e.g., near a gasoline service station) before using open flames.
- Provide adequate air space or a barrier to protect lines that transport flammable material when flames are required for work and the lines are exposed.

423D. Excavation

- Locate buried utilities prior to excavation.
- Existing utilities should be exposed where the bore path of guided boring or direction drilling machines crosses existing utilities.
- Hand tools used for manual excavation near supply cables must have non-conductive handles.
- Hand digging must be used when close to cables or other utilities.
- If lines that transport flammable material are broken or damaged, the employee must:

✓ Leave the excavation open
✓ Where safe, eliminate ignition sources
✓ Notify the proper authority
✓ Keep the public away
- When an employee is working in a trench or excavation in excess of 5 ft deep, or when a trench or excavation presents a cave-in hazard, shoring, sloping, or shielding methods must be used for employee protection.

423E. Identification

- Identify and protect exposed buried utilities.
- When working on one cable, protect other cables from damage.
- Before cutting a cable or opening a splice, verify its identity.
- Where multiple cables exist, the cable to be worked on must be positively identified.

423F. Operation of Power-Driven Equipment

- Keep out of manholes when power rodding.

Section 43

Additional Rules for Communications Employees

The format of this Handbook for Part 4, Work Rules, is different from the format of the other parts of this book. The format for Part 4 summarizes the **Code** text in an outline bulleted list. A graphic is used to visually represent which section, rule, and paragraph the outline applies to. A brief general discussion of each section is provided at the beginning of each section. The following brief general discussion covers selected topics in Secs. 43 and 44. This discussion is repeated at the beginning of Sec. 44.

Sections 43 and 44 provide additional rules for employees. Section 43 provides additional rules for communications employees only. Section 44 provides additional rules for supply employees only. Section 41 provides rules for supply and communications employers and Sec. 42 provides rules for supply and communications employees.

The additional rules in Sec. 43 for communications employees primarily focus on keeping the communications employee safe when the communication lines are on joint-use (supply and communication) structures or in joint-use (supply and communication) manholes with electric supply conductors. Section 43 requires that communications employees maintain minimum approach distances between the communications employee and electric supply lines and equipment. In addition to the minimum approach distances to electric supply conductors, communications employees must not position themselves above the lowest electric supply conductor exclusive of vertical runs (risers) and street lights.

The additional rules in Sec. 44 for supply employees primarily focus on minimum approach distances to energized parts, switching control procedures, work on energized lines, de-energizing lines, protective grounding, and live-line work. It is interesting to note that for a 12.47/7.2 kV line, the communication minimum worker approach distance found in **NESC** Table 431-1 and the supply worker minimum approach distance found in **NESC** Table 441-1 are both 2 ft-3 in.

The difference is that the supply worker gets additional training in Sec. 44 to go closer than the minimum approach distance. The communication worker does not get additional training to go closer than the minimum approach distance.

If a communications line is located in the supply space in accordance with the overhead line rules in Part 2 of the **NESC**, the worker who enters the supply space to work on the communications line must be trained as a supply employee (power lineman) see **NESC** Rule 224A1.

If a communications line is located below a supply line on an overhead structure and the proper communications worker safety zone clearances in Sec. 23 are met, the worker who is maintaining the communications line must be trained as a communications employee (communications lineman).

If a communications line is positioned below a supply line on an overhead structure, but the proper clearances in Sec. 23 are not met, the communications employee can correct the violation if the communications employee does not violate the minimum approach distances and other requirements in Sec. 43. If the communications employee cannot maintain the minimum approach distances in Sec. 43, the communications employee must contact a supply employee to correct the violation.

For example, the minimum approach distance rules in Sec. 43 (**NESC** Table 431-1) require the communications worker to have a 2 ft-3 in minimum approach distance to a 12.47/7.2 kV line. A common supply to communications clearance value in Rules 235 and 238 is 40 in (see Rules 235 and 238 for specific applications). This value is intended to provide enough space on the pole for the communications worker to get into position and work on the communications cable while still maintaining the 2 ft-3 in minimum approach distance to the 12.47/7.2 kV line. If a violation exists on the pole and the 40 in clearance requirement between supply and communications is only 38 in, the communications worker may be able to get into position and lower the communications cable down 2 in (from 38 in to 40 in) without violating the 2 ft-3 in minimum approach distance to the 12.47/7.2 kV line. However, if the communications cable was in violation and mounted only 1 ft below the 12.47/7.2 kV line (say right under the crossarm brace), the communications worker could not correct this violation. The communications worker would have to call on a trained supply worker (power lineman) to fix the problem.

If a communications employee is not a qualified employee, then the worker is considered an unqualified employee and the worker must maintain at least a 10-ft distance from energized lines (more than 10 ft for greater than 50 kV) per OSHA Standard 1910.333. This is commonly referred to as the OSHA 10-ft Rule. The OSHA 10-ft Rule also applies to other workers in the vicinity of power lines such as painters, roofers, etc. The **NESC** does not provide distances for unqualified workers but the OSHA standards do. The bottom line is that communications workers need the proper training or they cannot work on a communications line that is constructed jointly with a power line.

The crane and derrick equipment operation standard (OSHA 1926.1408 which is part of OSHA 1926 Subpart CC) contains minimum clearance distance requirements ranging from 10 ft to 45 ft depending on the voltage of the line with initial evaluations starting at 20 ft.

The terms qualified, qualified employee, qualified person, qualified worker, and qualified electrical worker are commonly used in the power and communication utility industry. The **NESC** and OSHA both define some of these terms. Supply (power) workers must be trained in OSHA Standard 1910.269. OSHA Standard 1910.269 states in Paragraph 1910.269(a)(2)(ii) that qualified employees shall be trained and competent in:

- The skills and techniques necessary to distinguish exposed live parts from other parts of electric equipment,
- The skills and techniques necessary to determine the nominal voltage of exposed live parts,
- The minimum approach distances specified in this section corresponding to the voltages to which the qualified employee will be exposed and the skills and techniques necessary to maintain those distances,
- The proper use of the special precautionary techniques, personal protective equipment, insulating and shielding materials, and insulated tools for working on or near exposed energized parts of electric equipment, and
- The recognition of electrical hazards to which the employee may be exposed and the skills and techniques necessary to control or avoid these hazards.

The first three bullets above outline the required training necessary for an engineer, technician, lineman, or related worker at an electric utility to go from a 10 ft (unqualified) employee down to a 2 ft-3 in employee (the minimum approach distance for a 12.47/7.2 kV line or part). The fourth bullet above outlines additional training. This additional training is much more intense for a power lineman than for an engineer or technician due to the differences in their job duties or assigned tasks. The OSHA Standards require employees to have training for the tasks that their job involves. For example, it is not necessary for an engineer for the power company to be trained in live line work procedures if the engineer does not perform live line work but the engineer does need to know minimum approach distances to energized parts to enter a substation. **NESC** Rule 421C permits nonqualified employees or visitors to be escorted in the vicinity of electric equipment or lines. In other words, a qualified employee can escort a nonqualified employee or visitor into a substation. The qualified employee is responsible for safeguarding the people in their care. The fifth bullet above is a general requirement for the safety of electrical workers.

OSHA Standards 1910.332 and 1910.333 have similar training and minimum approach distance requirements for general industry workers (including communication workers) to go from a 10 ft (unqualified) employee down to 2 ft-3 in employee (the minimum approach distance for a 12.47/7.2 kV line or part). The minimum approach distance (also referred to as MAD) is a fancy term for saying "how far to stay away." Communication workers must also be trained in OSHA Standard 1910.268. Many people do not think that communication workers need to use insulated gloves, which is true in many cases but OSHA 1910.268 does have at least three work tasks that require a communication worker to wear insulated gloves. These tasks include attaching and removing temporary bonds (a specific sequence must be followed), handling suspension strand (that is being installed on poles carrying exposed energized power conductors), and when handling poles near energized power conductors (when a possibility exists that the pole may contact a power conductor). Many people do not think that communication companies own joint-use (power and communication) poles but some do and some perform pole replacements of these poles. This work must be done by communication workers trained in specific rules for handling poles near energized power conductors, not by communication workers that normally perform cable installation and lashing work or service drop work.

The concept of minimum approach distance is outlined in Fig. 43-1.

Minimum Approach Distance:
- Employees must not approach or bring any conductive object within the minimum approach distance.
- The minimum approach distance contains an electrical component and an inadvertent (i.e., unintentional or unexpected) movement component.
- Communications worker minimum approach distances are found in **NESC** Table 431-1. (A similar table is found in OSHA 1910.268.)
- Supply worker minimum approach distances are found in **NESC** Tables 441-1 through 441-5. (Similar tables are found in OSHA 1910.269.)
- Supply workers can work inside the minimum approach distance if additional live-line rules are followed.
- Communications workers may not approach closer than the minimum approach distance.

Reach or Extended Reach:
- Reach is defined as the range of anticipated motion of an employee while performing a task.
- Extended reach is defined as the range of anticipated motion of a conductive object being held by an employee while performing a task.
- Anticipated motions include:
  ✓ Adjusting a hardhat
  ✓ Maneuvering tools
  ✓ Reaching for items being passed to the employee
  ✓ Adjusting parts
  ✓ Adjusting work positions, etc.

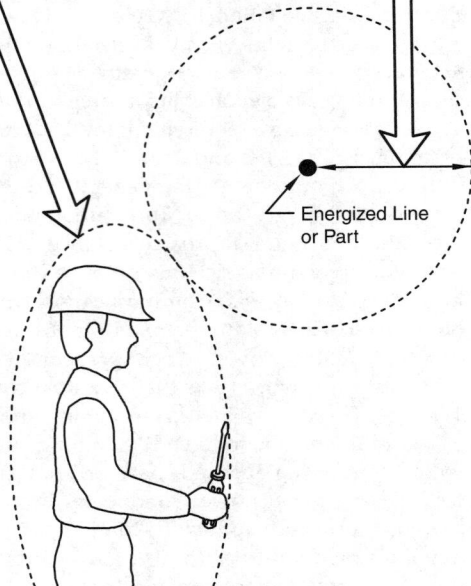

Energized Line or Part

Fig. 43-1. Minimum approach distance (Sec. 43).

The Work Rules in Part 4 of the **NESC** are similar to (but they are not exactly the same as) the Occupational Safety and Health Administration (OSHA) regulations related to electric supply and communications systems. See Rule 402 for additional information on the OSHA standards that are related to the **NESC** Work Rules.

430. GENERAL

Communications Employees Only!

Additional Rules for Communications Employees

GENERAL INFORMATION

430. General

- Section 42, "General Rules for Employees" (both supply and communications) also apply.

431. APPROACH TO ENERGIZED CONDUCTORS OR PARTS

431A. No Employee Shall Approach...

Communications Employees Only!

Additional Rules for Communications Employees

431. Approach to Energized Conductors or Parts

431A. No Employee Shall Approach...

- Communications employees must not approach energized parts closer than the minimum approach distances shown in **NESC** Table 431-1.
- Communications employees must not take conductive objects near energized parts closer than the minimum approach distances shown in **NESC** Table 431-1.
- Communications employees repairing storm damage to communications lines that are joint-use with electric supply lines must:

- ✓ Treat the supply and communications lines as energized to the highest voltage to which they are exposed, or
- ✓ Assure that the electric supply lines are de-energized and grounded per the work rules of **NESC** Sec. 44

431B. Altitude Correction

Communications Employees Only!

Additional Rules for Communications Employees

431. Approach to Energized Conductors or Parts

431B. Altitude Correction

- The minimum approach distances in **NESC** Table 431-1 are for altitudes below 12,000 ft.
- **NESC** Table 441-6 provides altitude correction factors, which must be applied to the electrical component of the minimum approach distance for altitudes above 12,000 ft.

431C. When Repairing Underground...

Communications Employees Only!

Additional Rules for Communications Employees

431. Approach to Energized Conductors or Parts

431C. When Repairing Underground...

- Communications employees repairing underground communications lines that are joint-use with damaged electric supply cables must:
  - ✓ Treat the supply and communications lines as energized to the highest voltage to which they are exposed, or
  - ✓ Assure that the electric supply lines are de-energized and grounded per the work rules of **NESC** Sec. 44

432. JOINT-USE STRUCTURES

Communications Employees Only!

Additional Rules for Communications Employees

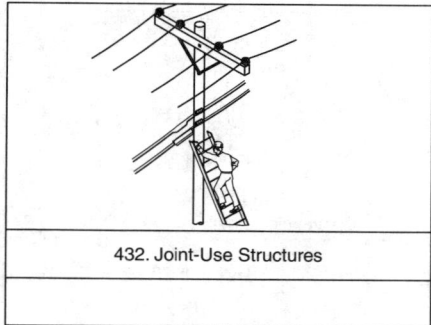

432. Joint-Use Structures

- On joint-use structures (power and communications), communications employees must not approach energized parts closer than the distances shown in **NESC** Table 431-1.
- On joint-use structures (power and communications), communications employees must not take conductive objects near energized parts closer than the distances shown in **NESC** Table 431-1.
- Communications employees must not position themselves above the lowest electric supply conductor or equipment (not including vertical risers, street lighting, and supply equipment permitted to be below the communications space).
- Note that examples of supply equipment mounted below the communications space include electronic controls, solar panels, etc.
- An exception applies to this rule when fixed rigid barriers are installed between the supply and communications facilities if the supply voltage is 140 kV or below.

433. ATTENDANT ON SURFACE AT JOINT-USE MANHOLES

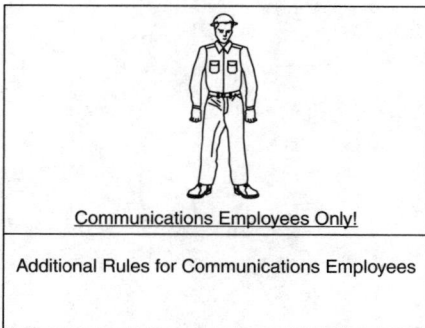

Communications Employees Only!

Additional Rules for Communications Employees

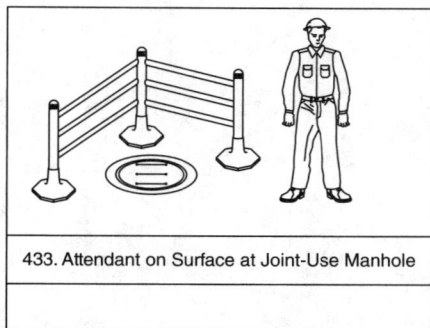

433. Attendant on Surface at Joint-Use Manhole

- Work in joint-use (power and communications) manholes requires an employee to be available on the surface to assist the worker in the manhole.

434. SHEATH CONTINUITY

Communications Employees Only!

Additional Rules for Communications Employees

434. Sheath Continuity

- Metallic or semiconductive sheath continuity must be maintained when working on underground cables.

Section 44

Additional Rules for Supply Employees

The format of this Handbook for Part 4, Work Rules, is different from the format of the other parts of this book. The format for Part 4 summarizes the **Code** text in an outline bulleted list. A graphic is used to visually represent which section, rule, and paragraph the outline applies to. A brief general discussion of each section is provided at the beginning of each section. The following brief general discussion covers selected topics in Secs. 44 and 43. This discussion is repeated at the beginning of Sec. 43.

Sections 44 and 43 provide additional rules for employees. Section 44 provides additional rules for supply employees only. Section 43 provides additional rules for communications employees only. Section 41 provides rules for supply and communications employers and Sec. 42 provides rules for supply and communications employees.

The additional rules in Sec. 44 for supply employees primarily focus on minimum approach distances to energized parts, switching control procedures, work on energized lines, de-energizing lines, protective grounding, and live-line work. It is interesting to note that for a 12.47/7.2 kV line, the communication worker minimum approach distance found in **NESC** Table 431-1 and the supply worker minimum approach distance found in **NESC** Table 441-1 are both 2 ft-3 in. The difference is that the supply worker gets additional training in Sec. 44 to go closer than the minimum approach distance. The communication worker does not get additional training to go closer than the minimum approach distance.

The additional rules in Sec. 43 for communications employees primarily focus on keeping the communications employee safe when the communications lines are on joint-use (supply and communication) structures or in joint-use (supply and communication) manholes with electric supply conductors. Section 43 requires that communications employees maintain minimum approach distances between the communications employee and electric supply lines and equipment. In addition to the minimum approach distances to electric supply conductors, communications employees must not position themselves above the lowest electric supply conductor exclusive of vertical runs (risers) and street lights.

If a communications line is located in the supply space in accordance with the overhead line rules in Part 2 of the **NESC**, the worker who enters the supply space to work on the communications line must be trained as a supply employee (power lineman) see **NESC** Rule 214A1.

If a communications line is located below a supply line on an overhead structure and the proper communications worker safety zone clearances in Sec. 23 are met, the worker who is maintaining the communications line must be trained as a communications employee (communications lineman).

If a communications line is positioned below a supply line on an overhead structure, but the proper clearances in Sec. 23 are not met, the communications employee can correct the violation if the communications employee does not violate the minimum approach distances and other requirements in Sec. 43. If the communications employee cannot maintain the minimum approach distances in Sec. 43, the communications employee must contact a supply employee to correct the violation.

For example, the minimum approach distance rules in Sec. 43 (**NESC** Table 431-1) require the communications worker to have a 2 ft-3 in minimum approach distance to a 12.47/7.2 kV line. A common supply to communications clearance value in Rules 235 and 238 is 40 in (see Rules 235 and 238 for specific applications). This value is intended to provide enough space on the pole for the communications worker to get into position and work on the communications cable while still maintaining the 2 ft-3 in minimum approach distance to the 12.47/7.2 kV line. If a violation exists on the pole and the 40 in clearance requirement between supply and communications is only 38 in, the communications worker may be able to get into position and lower the communications cable down 2 in (from 38 in to 40 in) without violating the 2 ft-3 in minimum approach distance to the 12.47/7.2 kV line. However, if the communications cable was in violation and mounted only 1 ft below the 12.47/7.2 kV line (say right under the crossarm brace) the communications worker could not correct this violation. The communications worker would have to call on a trained supply worker (power lineman) to fix the problem.

If a communications employee is not a qualified employee, then the worker is considered an unqualified employee and the worker must maintain at least a 10-ft distance from energized lines (more than 10 ft for greater than 50 kV) per OSHA Standard 1910.333. This is commonly referred to as the OSHA 10-ft Rule. The OSHA 10-ft Rule also applies to other workers in the vicinity of power lines such as painters, roofers, etc. The **NESC** does not provide distances for unqualified workers but the OSHA standards do. The bottom line is that communications workers need the proper training or they cannot work on a communications line that is constructed jointly with a power line.

The crane and derrick equipment operation standard (OSHA 1926.1408 which is part of OSHA 1926 Subpart CC) contains minimum clearance distance requirements ranging from 10 ft to 45 ft depending on the voltage of the line with initial evaluations starting at 20 ft.

The terms qualified, qualified employee, qualified person, qualified worker, and qualified electrical worker are commonly used in the power and communication utility industry. The **NESC** and OSHA both define some of these terms. Supply (power) workers must be trained in OSHA Standard 1910.269. OSHA Standard

1910.269 states in Paragraph 1910.269(a)(2)(ii) that qualified employees shall be trained and competent in:

- The skills and techniques necessary to distinguish exposed live parts from other parts of electric equipment,
- The skills and techniques necessary to determine the nominal voltage of exposed live parts,
- The minimum approach distances specified in this section corresponding to the voltages to which the qualified employee will be exposed and the skills and techniques necessary to maintain those distances,
- The proper use of the special precautionary techniques, personal protective equipment, insulating and shielding materials, and insulated tools for working on or near exposed energized parts of electric equipment, and
- The recognition of electrical hazards to which the employee may be exposed and the skills and techniques necessary to control or avoid these hazards.

The first three bullets above outline the required training necessary for an engineer, technician, lineman, or related worker at an electric utility to go from a 10 ft (unqualified) employee down to a 2 ft-3 in employee (the minimum approach distance for a 12.47/7.2 kV line or part). The fourth bullet above outlines additional training. This additional training is much more intense for a power lineman than for an engineer or technician due to the differences in their job duties or assigned tasks. The OSHA Standards require employees to have training for the tasks that their job involves. For example, it is not necessary for an engineer for the power company to be trained in live line work procedures if the engineer does not perform live line work but the engineer does need to know minimum approach distances to energized parts to enter a substation. NESC Rule 421C permits nonqualified employees or visitors to be escorted in the vicinity of electric equipment or lines. In other words, a qualified employee can escort a nonqualified employee or visitor into a substation. The qualified employee is responsible for safeguarding the people in their care. The fifth bullet above is a general requirement for the safety of electrical workers.

OSHA Standards 1910.332 and 1910.333 have similar training and minimum approach distance requirements for general industry workers (including communication workers) to go from a 10 ft (unqualified) employee down to 2 ft-3 in employee (the minimum approach distance for a 12.47/7.2 kV line or part). The minimum approach distance (also referred to as MAD) is a fancy term for saying "how far to stay away." Communication workers must also be trained in OSHA Standard 1910.268. Many people do not think that communication workers need to use insulated gloves, which is true in many cases but OSHA 1910.268 does have at least three work tasks that require insulated gloves. These tasks include attaching and removing temporary bonds (a specific sequence must be followed), handling suspension strand (that is being installed on poles carrying exposed energized power conductors), and when handling poles near energized power conductors (when a possibility exists that the pole may contact a power conductor). Many people do not think that communication companies own joint-use (power and communication) poles but some do and some perform pole replacements of these poles. This work must be done by communication workers trained in specific rules for handling poles near energized power conductors, not by communication workers that normally perform cable installation and lashing work or service drop work.

The concept of minimum approach distance is outlined in Fig. 44-1.

Minimum Approach Distance:
- Employees must not approach or bring any conductive object within the minimum approach distance.
- The minimum approach distance contains an electrical component and an inadvertent (i.e., unintentional or unexpected) movement component.
- Communications worker minimum approach distances are found in **NESC** Table 431-1. (A similar table is found in OSHA 1910.268.)
- Supply worker minimum approach distances are found in **NESC** Tables 441-1 through 441-5. (Similar tables are found in OSHA 1910.269.)
- Supply workers can work inside the minimum approach distance if additional live-line rules are followed.
- Communications workers may not approach closer than the minimum approach distance.

Reach or Extended Reach:
- Reach is defined as the range of anticipated motion of an employee while performing a task.
- Extended reach is defined as the range of anticipated motion of a conductive object being held by an employee while performing a task.
- Anticipated motions include:
  ✓ Adjusting a hardhat
  ✓ Maneuvering tools
  ✓ Reaching for items being passed to the employee
  ✓ Adjusting parts
  ✓ Adjusting work positions, etc.

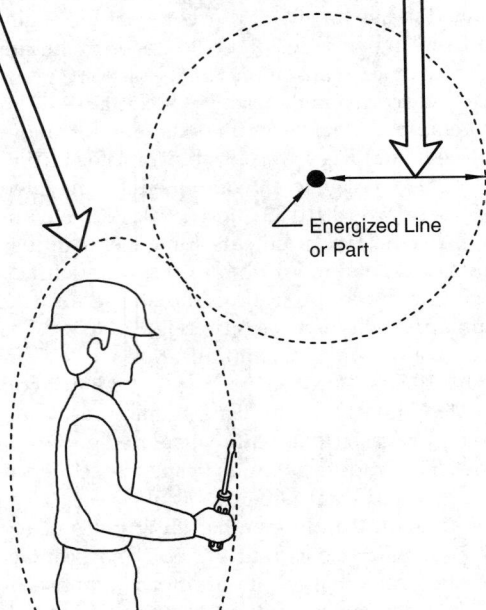

Energized Line or Part

Fig. 44-1. Minimum approach distance (Sec. 44).

The Work Rules in Part 4 of the **NESC** are similar to (but they are not exactly the same as) the Occupational Safety and Health Administration (OSHA) regulations related to electric supply and communications systems. See Rule 402 for additional information on the OSHA standards that are related to the **NESC** Work Rules.

440. GENERAL

Supply Employees Only!

Additional Rules for Supply Employees

GENERAL
INFORMATION

440. General

- Section 42, "General Rules for Employees" (both supply and communications) also apply.

441. ENERGIZED CONDUCTORS OR PARTS

441A. Minimum Approach Distance to Live Parts

Supply Employees Only!

Additional Rules for Supply Employees

441. Energized Conductors or Parts

441A. Minimum Approach Distance to Energized Lines or Parts

- Supply employees must not approach energized parts or take conductive objects near exposed energized lines or parts closer than the minimum approach distances listed in **NESC** Table 441-1 or 441-5 unless one of the following conditions is met:
  - ✓ The line or part is de-energized and grounded per Rule 444D (special exceptions apply for voltages less than 600 V).

- ✓ The employee is insulated from the energized line or part using insulated tools, rubber gloves, or rubber gloves with sleeves.
- ✓ The energized line or part is insulated from the employee and any other line or part at a different voltage.
- ✓ The employee is performing bare-hand live-line work in accordance with Rule 446.
- Note that minimum approach distances contain an electrical component and an inadvertent movement component.
  - ✓ Minimum approach distances calculated for 0.301 kV to 0.750 kV contain an electrical component plus a 1-ft inadvertent movement component.
  - ✓ Minimum approach distances calculated for 0.751 kV to 72.5 kV contain an electrical component plus a 2-ft inadvertent movement component.
  - ✓ Minimum approach distances calculated for voltages above 72.5 kV contain an electrical component plus a 1-ft inadvertent movement component.
- Note that the method used for calculating minimum approach distances was taken from OSHA 1910.269, Appendix B.
- Note that voltage ranges are contained in ANSI C84.1, Table 1.
- Note that for the purpose of Sec. 44, "reach" is defined as "the range of anticipated motion of an employee while performing a task" and "extended reach" is defined as "the range of anticipated motion of a conductive object being held by an employee while performing a task."
- Precaution for approaching voltages from 51 to 300 V:
  - ✓ Do not contact exposed energized parts unless the above conditions have been met.
- Precautions for approaching voltages from 301 V to 72.5 kV:
  - ✓ Employees must be protected from phase-to-phase and phase-to-ground differences in voltage (**NESC** Table 441-1 or 441-5 applies).
  - ✓ Exposed grounded lines, conductors, or parts in the work area must be guarded or insulated (special exceptions apply for voltages between 300 V and 750 V).
  - ✓ Rubber insulating gloves must be insulated for the maximum use voltage in **NESC** Table 441-7 and must be worn whenever the employee is within the reach or extended reach of the minimum approach distances in **NESC** Table 441-1 or 441-5 (special exceptions apply).
  - ✓ When the rubber glove method is used, it must be used with one of the two following methods:
    - Rubber insulating sleeves which are insulated for the maximum use voltage in **NESC** Table 441-7 (special exceptions apply for voltages at 750 V or less).
    - Insulating exposed energized lines or parts within the employee's reach or extended reach (this does not apply to the part being worked on under positive control).
  - ✓ When the rubber glove method is used on voltages above 15 kV phase to phase, an insulated aerial device, insulated structure-mounted platform, or other supplementary insulation must be used to support the worker.
  - ✓ Insulating cover-up shall be rated for the phase-to-phase voltage of the circuit or the phase-to-ground voltage of the circuit depending on the exposure of the circuit being worked.

✓ Determination of the phase-to-phase or phase-to-ground exposure must consider factors such as work rules, conductor spacing, worker position, tasks being performed, and any other relative factors.

✓ When insulated cover-up is used, it must be applied as the employee first approaches the energized line and it must be removed in the reverse order.

✓ Insulated cover-up must extend beyond the reach of the employee's anticipated work position or extended reach position.

• Precautions for approaching voltages above 72.5 kV:

✓ Employees must position themselves so that they are not within the reach or extended reach of the applicable minimum approach distance.

✓ The minimum approach distances in **NESC** Table 441-1 apply.

✓ In lieu of using the minimum approach distances in **NESC** Table 441-1, the minimum approach distances in **NESC** Tables 441-2 through 441-4 may be used if the per unit transient overvoltage value (T) has been determined using an engineering analysis considering system design, expected operating conditions, and control measures.

✓ Note that control measures include blocking reclosing, prohibiting switching during work, using protective air gaps, use of closing resistors and surge arresters, etc.

✓ Note that IEEE Std. 516 and OSHA 1910.269 Appendix B contain information that may be used to perform an engineering analysis to determine "T". "T" may be determined on a system basis or per-line basis.

• A temporary (transient) over voltage control device (TTOCD) which is designed and tested for installation adjacent to a work site to limit the TOV at the work site may be used to obtain a lower value of "T". (Example formulas are provided in the rule, and **NESC** Appendix D provides additional information on per-unit overvoltage factors.)

• The minimum approach distances in **NESC** Tables 441-1 through 441-5 must be increased for altitudes above 3,000 ft.

• **NESC** Table 441-6 provides altitude correction factors, which must be applied to the electrical component of the minimum approach distance.

441B. Additional Approach Requirements

Supply Employees Only!

Additional Rules for Supply Employees

441. Energized Conductors or Parts

441B. Additional Approach Requirements

- The clear insulation distance of insulators is defined as the shortest straight-line air-gap distance from the nearest energized part to the nearest grounded part.
- When working on insulators using rubber gloves or insulated tools, the clear insulation distance must not be less than the straight-line distance with tools required by Rule 441A4.
- To work on the grounded end of an open switch, all of the following conditions must be met:
  - ✓ The full air-gap distance of the switch must be maintained.
  - ✓ The air-gap distance of the switch must not be less than the electrical component of the minimum approach distance (inadvertent movement components are not required).
  - ✓ The minimum approach distance to the energized portion of the switch must be maintained.
- Special rules apply to working on insulator assemblies operating above 72.5 kV.

Supply Employees Only!

Additional Rules for Supply Employees

441. Energized Conductors or Parts

441C. Live-Line Tool Clear Insulation Length

441C. Live-Line Tool Clear Insulation Length

- For live-line tools (i.e., hot sticks) the minimum approach distances with tools required by Rule 441A4 must be maintained or exceeded between the conductive end of the tool and the employee's hands or other body parts.
- Insulated conductor support tools may be used if the clear insulation distance is at least as long as the insulator string or the distances specified in Rule 441A.
- When installing insulated conductor support tools, employee minimum approach distances must be maintained.
- Note that the conductive portion of an insulated tool can decrease the insulation value of the tool more than just the length of the conductive portion.

442. SWITCHING CONTROL PROCEDURES

442A. Designated Person

Supply Employees Only!	
Additional Rules for Supply Employees	442. Switching Control Procedures
	442A. Designated Person

- A designated person must authorize switching.
- The designated person must:
  - ✓ Keep informed of operating conditions to maintain safety
  - ✓ Maintain suitable records
  - ✓ Issue or deny authorization for switching to maintain safety

442B. Specific Work

Supply Employees Only!	
Additional Rules for Supply Employees	442. Switching Control Procedures
	442B. Specific Work

- The designated person shall give authorization before beginning work.
- The designated person shall be notified when work ends.
- Exceptions to the switching control procedures exist for emergencies and catastrophic service disruptions.

442C. Operations at Stations

Supply Employees Only!

Additional Rules for Supply Employees

442. Switching Control Procedures

442C. Operations at Stations

- Qualified employees must obtain authorization from the designated person before switching.
- Specific operating schedules may be used.
- If specific operating schedules do not exist, authorization from the designated person must be obtained for switching or starting and stopping of equipment.
- Exceptions exist for switching sections of distribution circuits and for emergency situations.

442D. Re-energizing after Work

Supply Employees Only!

Additional Rules for Supply Employees

442. Switching Control Procedures

442D. Re-energizing after Work

- Instructions to re-energize after work is complete are to be given by the designated person after the employees who requested the de-energization have reported clear.
- Employees who requested de-energization of lines for other employees or crews cannot request re-energization until the other employees or crews have reported clear.
- The above procedure must be used when more than one location is involved.

442E. Tagging Electric Supply Circuits Associated with Work Activities

Supply Employees Only!	
Additional Rules for Supply Employees	442. Switching Control Procedures
	442E. Tagging Electric Supply Circuits Associated with Work Activities

- De-energized and grounded circuits must be tagged at all points where the circuit or equipment can be energized.
- When reclosers or circuit breakers are put on one-shot, a tag must be placed at the reclosing device location.
- Exceptions exist for SCADA-controlled systems at both the SCADA operating point and the reclosing device location.
- Tags shall clearly identify the equipment or circuits being worked on.

442F. Restoration of Service after Automatic Trip

Supply Employees Only!	
Additional Rules for Supply Employees	442. Switching Control Procedures
	442F. Restoration of Service after Automatic Trip

- Tagged circuits that open automatically (i.e., a circuit with a recloser set to one-shot) must not be closed without authorization.
- When circuits open automatically, local operating rules determine how they may be closed back in with safety.

442G. Repeating Oral Messages

Supply Employees Only!

Additional Rules for Supply Employees

442. Switching Control Procedures

442G. Repeating Oral Messages

- Oral messages associated with line switching must be repeated back to the sender and the sender's identity must be obtained by the receiver.
- The sender of an oral message associated with line switching must require that the message be repeated back by the receiver and must obtain the receiver's identify.

443. WORK ON ENERGIZED LINES AND EQUIPMENT

443A. General Requirements

Supply Employees Only!

Additional Rules for Supply Employees

GENERAL INFORMATION

443. Work on Energized Lines and Equipment

443A. General Requirements

- When working on energized lines, one of the following must be done:
  - ✓ Insulate the employee from energized parts.
  - ✓ Isolate or insulate the employee from ground and other voltages other than the one being worked on.
- Treat covered conductors (i.e., tree wire that is covered with nonrated insulation) as bare energized conductors.
- Consider the effect of a higher voltage line on a lower voltage line. Verify that the line being worked on is free from dangerous leakage and induction voltages or verify that it is effectively grounded.
- Insulated supply cables that cannot be positively identified as de-energized must be pierced or severed with an appropriate tool.

- Consider the operating voltage and take appropriate precautions before cutting an insulated energized supply cable.
- When the insulating covering on an energized cable must be cut, an appropriate tool must be used.
- When the insulating coupling on an energized cable must be cut, suitable eye protection and insulating gloves with protectors must be used.
- When the insulating covering on an energized cable must be cut, extreme care must be taken to prevent short-circuiting conductors.
- Metal-measuring tapes and tapes or ropes containing metal must not be used closer to energized parts than the minimum approach distances in **NESC** Tables 441-1 or 441-5.
- Metal-measuring tapes and tapes or ropes containing metal must be used with care near energized lines due to the effect of induced voltage.
- Metallic equipment or material that is not bonded to an effective ground and could approach energized parts closer than the minimum approach distances must be treated as if it were energized at the voltage to which it is exposed.

443B. Requirement for Assisting Employee

Supply Employees Only!

Additional Rules for Supply Employees

443. Work on Energized Lines and Equipment

443B. Requirement for Assisting Employee

- Employees shall not work alone in bad weather or at night on systems of more than 750 V.
- An exception applies permitting one employee to do certain tasks.

443C. Opening and Closing Switches

Supply Employees Only!

Additional Rules for Supply Employees

443. Work on Energized Lines and Equipment

443C. Opening and Closing Switches

- A smooth continuous motion must be used to close manual switches and disconnects.
- Care must be applied to opening switches to avoid serious arcing.

443D. Working Position

Supply Employees Only!	443. Work on Energized Lines and Equipment
Additional Rules for Supply Employees	443D. Working Position

- Avoid working positions in which a shock or slip will bring the worker's body toward energized parts at a potential different from the employee's body.
- The work position should generally be from below.

443E. Protecting Employees by Switches and Disconnectors

Supply Employees Only!	443. Work on Energized Lines and Equipment
Additional Rules for Supply Employees	443E. Protecting Employees by Switches and Disconnectors

- Load break switches must be opened before non-load-break disconnects.
- Non-load-break disconnects must be closed before closing the load break switch.

443F. Making Connections

Supply Employees Only!	443. Work on Energized Lines and Equipment
Additional Rules for Supply Employees	443F. Making Connections

- When connecting de-energized lines or equipment to energized lines, the jumper wire should first be attached to the de-energized part.
- When disconnecting a line or equipment from an energized line, the source end of the jumper should be removed first.
- Loose conductors (i.e., jumpers) should be kept away from energized parts.

443G. Switchgear

Supply Employees Only!	443. Work on Energized Lines and Equipment
Additional Rules for Supply Employees	443G. Switchgear

- Switchgear must be de-energized and grounded before doing work that involves removing protective barriers unless other safety means are used.
- The switchgear safety features must be replaced after work is completed.

443H. Current Transformer Secondaries

Supply Employees Only!

Additional Rules for Supply Employees

600:5

443. Work on Energized Lines and Equipment

443H. Current Transformer Secondaries

- A current transformer secondary must be de-energized before working on it.
- If the current transformer secondary cannot be properly de-energized, the secondary circuit must be bridged (shorted) so that the secondary will not be opened.

443I. Capacitors

Supply Employees Only!

Additional Rules for Supply Employees

443. Work on Energized Lines and Equipment

443I. Capacitors

- Before working on capacitors, they must be disconnected from the energized source, shorted, and grounded.
- Any line that has capacitors must be short-circuited and grounded before it is considered de-energized.
- Before capacitors are handled, each unit must be shorted between all insulated terminals and the capacitor tank due to series-parallel operation.
- Where capacitors are installed on ungrounded racks, the racks must be grounded before working on the capacitors.
- The internal resistor of a capacitor must not be depended on to discharge the capacitor.

443J. Gas-Insulated Equipment

Supply Employees Only!	SF_6
Additional Rules for Supply Employees	443. Work on Energized Lines and Equipment
	443J. Gas-Insulated Equipment

- Employees must be instructed on the special precautions related to handling SF_6 gas.
- The by-products resulting from arcing in SF_6 gas are generally toxic and irritant.

443K. Attendant on Surface

Supply Employees Only!	
Additional Rules for Supply Employees	443. Work on Energized Lines and Equipment
	443K. Attendant on Surface

- When one employee is in a manhole, another employee must be on the surface to render assistance as required.
- The employee on the surface may enter the manhole to provide short-term assistance.
- An exception permits working alone to do certain tasks.

443L. Unintentional Grounds on Delta Circuits

Supply Employees Only!	
Additional Rules for Supply Employees	443. Work on Energized Lines and Equipment
	443L. Unintentional Grounds on Delta Circuits

- Unintentional grounds on delta circuits must be removed as soon as practical.

444. DE-ENERGIZING EQUIPMENT OR LINES TO PROTECT EMPLOYEES

444A. Application of Rule

Supply Employees Only!

Additional Rules for Supply Employees

GENERAL INFORMATION

444. De-energizing Equipment or Lines to Protect Employees

444A. Application of Rule

- When employees depend on others to operate switches to de-energize circuits or when employees must secure authorization to operate a switch, Rules 444A through 444H must be followed in order.
- If an employee is in sole charge of a circuit section and is responsible for directing the disconnecting of the section, the portions of Rules 444A through 444H that deal with a designated person may be omitted.
- Records must be kept on utility interactive systems and these systems must be capable of being visibly disconnected.

444B. Employee's Request

Supply Employees Only!

Additional Rules for Supply Employees

444. De-energizing Equipment or Lines to Protect Employees

444B. Employee's Request

- The employee in charge of the work must make a request to the designated switching person to have the line or equipment de-energized.
- The switching request must properly identify the switch or line section by position, letter, color, number, or other means.

444C. Operating Switches, Disconnectors, and Tagging

Supply Employees Only!	444. De-energizing Equipment or Lines to Protect Employees
Additional Rules for Supply Employees	444C. Operating Switches, Disconnectors, Open Points, and Tagging

- The designated switching person must direct the operation of switches and disconnects to de-energize the circuit.
- The designated switching person must request that the switches and disconnect be rendered inoperable and tagged.
- If switches that are controlled automatically or remotely can be rendered inoperable, they must be tagged at the switch location.
- If switches that are controlled automatically or remotely cannot be rendered inoperable, then they must be tagged at all points of control.
- The following information must be recorded when placing a tag:
  - ✓ Time of disconnection.
  - ✓ Name of person making the disconnection.
  - ✓ Name of person who requested the disconnection.
  - ✓ Name or title or both of the designated switching person.
- When air gaps (i.e., cut or open jumpers) created for de-energizing equipment or lines to protect employees, the air gap must be tagged and meet the minimum clearances in **NESC** Table 444-1 or separated by a properly rated insulator.

444D. Employee's Protective Grounds

Supply Employees Only!	444. De-energizing Equipment or Lines to Protect Employees
Additional Rules for Supply Employees	444D. Employee's Protective Grounds

- After switching, rendering inoperable where practical, and tagging have been completed and the employee in charge has been given permission by the designated switching person to proceed, protective grounds must be applied.

- Minimum approach distances must be maintained for making voltage tests.
- Minimum approach distances must be maintained for applying grounds.
- Temporary protective grounds must be placed and arranged to protect employees from hazardous differences in electrical potential (e.g., worksite grounds and/or other methods).
- Note that hazardous touch and step potentials may exist around grounded equipment or between separately grounded systems, additional protection measures may be needed (e.g., barriers, insulation, work practices, isolation, or ground mats).
- The minimum approach distances must be maintained from ungrounded conductors at the work location.
- Where making protective grounds is impractical or creates a hazardous condition, grounds may be omitted by special permission.
- An exception applies which allows alternative work methods under special conditions.

444E. Proceeding with Work

Supply Employees Only!

Additional Rules for Supply Employees

444. De-energizing Equipment or Lines to Protect Employees

444E. Proceeding with Work

- After lines are de-energized and grounded, the employee in charge may direct the work to be done.
- Lines may be re-energized for testing under the supervision of the employee in charge and authorization of the designated switching person.
- Additional employees in charge desiring the same equipment or lines to be de-energized and grounded must follow these same procedures.

444F. Reporting Clear—Transferring Responsibility

Supply Employees Only!

Additional Rules for Supply Employees

444. De-energizing Equipment or Lines to Protect Employees

444F. Reporting Clear—Transferring Responsibility

- Once work is completed, the employee in charge must verify that the crew assigned to the employee in charge is in the clear.
- Once work is completed, the employee in charge must remove protective grounds.
- Once work is completed, the employee in charge must report to the designated switching person that tags may be removed.
- The employee in charge may transfer responsibilities to another employee after receiving permission from the designated person and personally informing the affected persons of the transfer.

444G. Removal of Tags

Supply Employees Only!

Additional Rules for Supply Employees

444. De-energizing Equipment or Lines to Protect Employees

444G. Removal of Tags

- The designated switching person must direct the removal of tags.
- The removal of tags must be reported back to the designated switching person by the person removing them.
- Upon tag removal, the record keeping must include the following:
  ✓ Name of the person requesting removal.
  ✓ Time of removal.
  ✓ Name of the person removing the tag.
- The name of the person requesting removal must match the name of the person who requested placement of the tag unless responsibility was properly transferred.

444H. Sequence of Re-energizing

Supply Employees Only!

Additional Rules for Supply Employees

444. De-energizing Equipment or Lines to Protect Employees

444H. Sequence of Re-energizing

- The designated switching person may direct the re-energization of the line only after the removal of protective grounds and tags.

445. PROTECTIVE GROUNDS

445A. Installing Grounds

Supply Employees Only!

Additional Rules for Supply Employees

445. Protective Grounds

445A. Installing Grounds

- Extreme caution must be used to assure that the proper sequence is used when installing and removing protective grounds.
- A specific sequence must be used when installing protective grounds to a previously energized part.
- An exception to the sequence may apply to some high-voltage towers.
- 1st, grounding conductors and devices must be sized to carry anticipated fault currents.
- 2nd, one end of the grounding device must be connected to an effective ground. Grounding switches may be used.
- 3rd, a voltage test must be done utilizing appropriate (insulated) tools and minimum approach distances.
- 4th, if the line shows no voltage, grounding may be completed. If voltage is present, the source must be determined. To complete grounding, the ground device must be brought into contact with the previously energized part using insulating handles and securely attached. Where bundled conductors exist, each conductor of the bundle should be grounded. Only after completion of the fourth step can the employee approach closer than the minimum approach distances or proceed with work on the parts as grounded parts.

445B. Removing Grounds

Supply Employees Only!

Additional Rules for Supply Employees

445. Protective Grounds

445B. Removing Grounds

- Extreme caution must be used to assure that the proper sequence is used when installing and removing protective grounds.
- Grounding devices must first be removed from the de-energized parts using tools with insulated handles.
- If multiple ground cables are connected to the same grounding point, all phase connections must be removed before removing any of the ground connections.
- An exception applies to removing all of the phase connections before removing any of the ground connections if this produces a hazard such as unintentional contact of the ground with ungrounded parts. In this case the grounds may be removed individually from each phase and ground connection.
- The connection of the protective ground to the effective ground must be removed last.
- Note that a hazard may exist when de-energized lines and equipment are in proximity to energized circuits.
- Note that IEEE Std 1048 and IEEE Std 1246 contain additional information on personal protective grounding.

446. LIVE WORK

446A. Training

Supply Employees Only!

Additional Rules for Supply Employees

Safety
is
#1

446. Live Work

446A. Training

- Live work practices require adherence to the following rules in addition to other rules in Secs. 42 and 44.
- The minimum approach distances in **NESC** tables 441-1 or 441-2 shall be maintained from all grounded objects and from other conductors, lines and equipment at a different potential from the parts being worked on using the live work method.
- Employees must be trained to use rubber gloves, hot sticks, or the barehand method before using these techniques on energized lines.

446B. Equipment

Supply Employees Only!	
Additional Rules for Supply Employees	446. Live Work
	446B. Equipment

- Insulated bucket trucks, ladders, and other support equipment must be evaluated for performance at the voltages involved. Tests must be done to ensure the equipment's integrity.
- Insulated bucket trucks and other insulated aerial devices used in bare-hand work must be tested before work is started.
- IEEE Standards and ANSI Standards must be referenced for equipment operation and testing.
- Bucket trucks and other insulated aerial devices must be kept clean.
- Tools and other equipment must not be used in a manner that would reduce the insulating strength of the insulated bucket truck or other insulated aerial device.

446C. When Working on Insulators...

Supply Employees Only!	
Additional Rules for Supply Employees	446. Live Work
	446C. When Working on Insulators ...

- When insulators are worked on using live-line procedures, the clear insulation distance must not be less than the minimum approach distances of **NESC** Tables 441-1(AC) or 441-5 (DC).

446D. Bonding and Shielding for Bare-Hand Method

Supply Employees Only!	446. Live Work
Additional Rules for Supply Employees	446D. Bonding and Shielding for Bare-Hand Method

- A conductive bucket liner or other suitable conducting device must be provided to bond the insulated bucket or other insulated aerial device to the energized line.
- The employee must be bonded to the insulated bucket or other insulated aerial device by using conducting shoes, leg clips, or other means.
- Protective clothing for electrostatic shielding must be used where necessary. IEEE Std 516 provides additional information on clothing designed for electrostatic shielding.
- The aerial device must be bonded to the energized conductor before the employee contacts the energized part.

447. PROTECTION AGAINST ARCING AND OTHER DAMAGE WHILE INSTALLING AND MAINTAINING INSULATORS AND CONDUCTORS

Supply Employees Only!	447. Protection Against Arcing and Other Damage While Installing and Maintaining Insulators and Conductors
Additional Rules for Supply Employees	

- When installing and maintaining insulators and conductors, use precautions to limit damage that would cause the conductors or insulators to fall.
- Use precautions to prevent arcs from forming.
- If an arc does form, use precautions to prevent injuring or burning any parts of the supporting structure, insulators, or conductors.

Appendix A

Photographs of NESC Applications

The photos shown correspond to figures in the text with the following icon:

See
Photo(s)

Photo 011-2. Typical dividing lines between the NESC and the NEC (see Fig. 011-2).

Photo 011-3(1). Examples of NESC and NEC street lighting systems (see Fig. 011-3).

Photo 011-3(2). Examples of NESC and NEC street lighting systems (see Fig. 011-3).

Photo 011-3(3). Examples of NESC and NEC street lighting systems (see Fig. 011-3).

Photo 017-1(1). Example of nominal values (see Fig. 017-1).

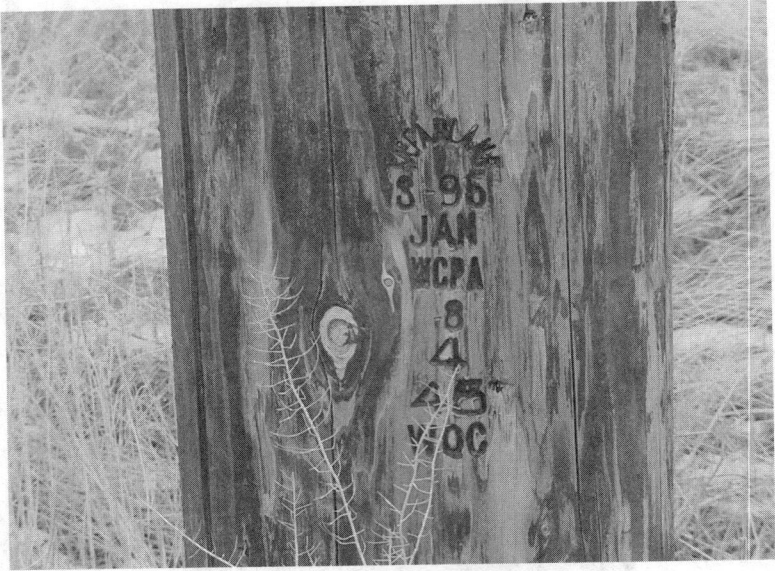

Photo 017-1(2). Example of nominal values (see Fig. 017-1).

Photo 092-4. Grounded connections for nonshielded cables over 750 V (see Fig. 092-4).

Photo 092-5. Surge arrester cable—shielding inter-connection (see Fig. 092-5).

Photo 092-6. Grounding points for a shielded cable without an insulating jacket (see Fig. 092-6).

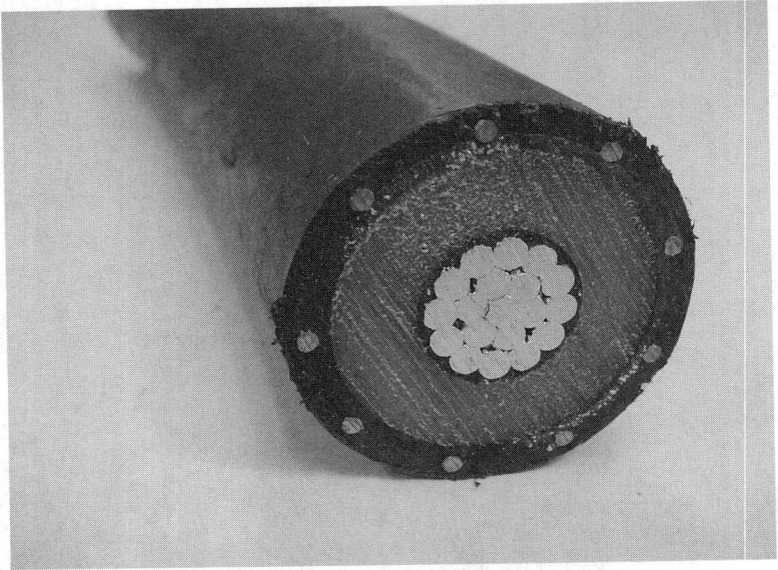

Photo 092-7. Grounding points for a shielded cable with an insulating jacket (see Fig. 092-7).

Photo 092-9. Grounding of messenger wires (see Fig. 092-9).

Photo 092-12. Example of conductive electric supply station fence grounding (see Fig. 092-12).

Photo 093-1(1). Example of a copper-grounding conductor (pole ground) and a structural metal-grounding conductor (steel pole) (see Fig. 093-1).

Photo 093-1(2). Example of a copper-grounding conductor (pole ground) and a structural metal-grounding conductor (steel pole) (see Fig. 093-1).

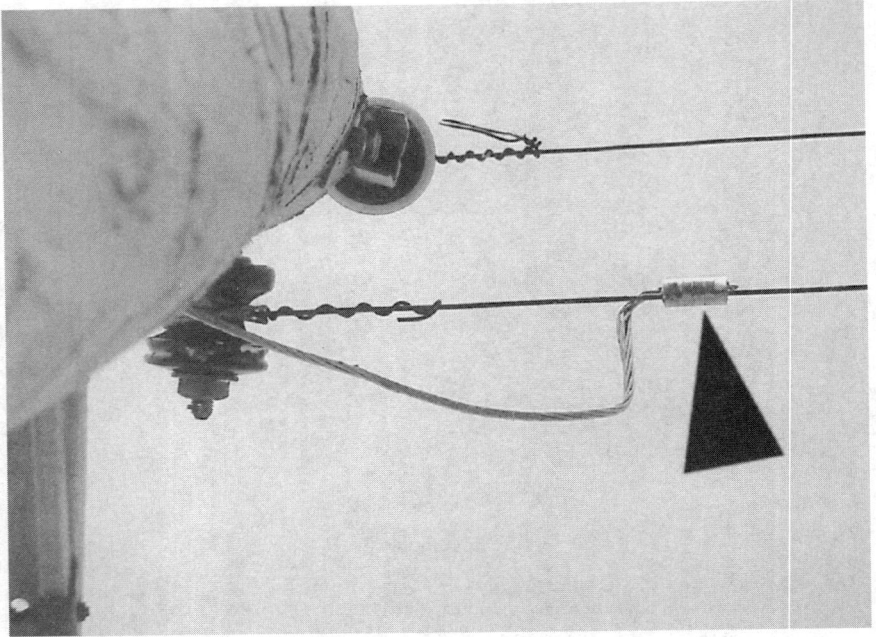

Photo 093-2. Connection of grounding conductor to grounded conductor (see Fig. 093-2).

Photo 093-6. Example of pole ground ampacity (see Fig. 093-6).

Photo 093-7(1). Requirements for grounding conductors with or without guards (see Fig. 093-7).

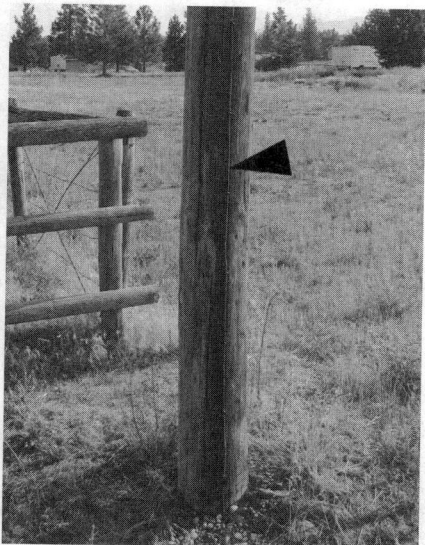

Photo 093-7(2). Requirements for grounding conductors with or without guards (see Fig. 093-7).

Photo 093-10. Example of aluminum grounding conductor transitioning to copper for burial (see Fig. 093-10).

Photo 093-12. Example of common grounding conductor for neutral and equipment (see Fig. 093-12).

Photo 094-4. Made electrodes-driven ground rods (see Fig. 094-4).

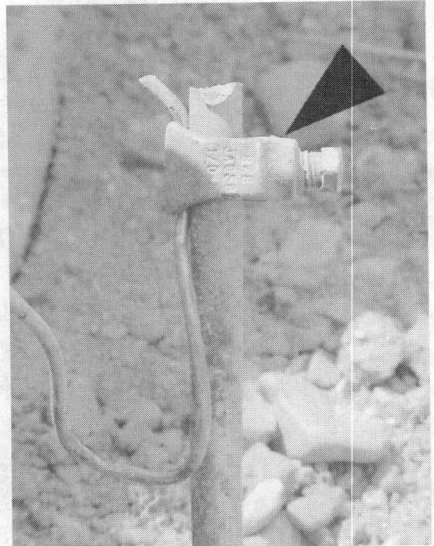

Photo 095-1. Connection of grounding conductor to grounding electrode (see Fig. 095-1).

Photo 097-3. Example of a common neutral with single grounding (see Fig. 097-3).

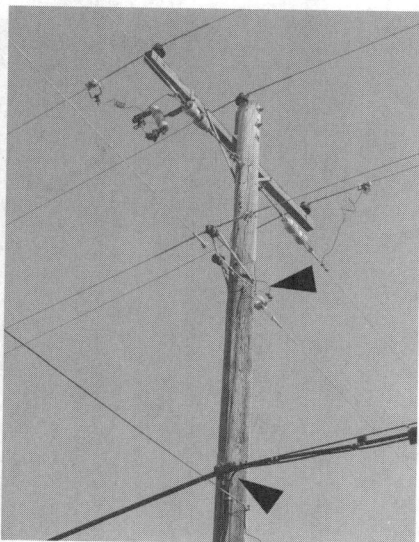

Photo 097-5. Bonding of communication systems to electric supply systems on a joint-use (power and communication) structure (see Fig. 097-5).

Photo 110-1. Example of how Rule 110A applies to Part 1, Electric Supply Stations, or Part 2, Overhead Lines (see Fig. 110-1).

Photo 110-2. Example of how Rule 110A applies to Part 1, Electric Supply Stations, or Part 3, Underground Lines (see Fig. 110-2).

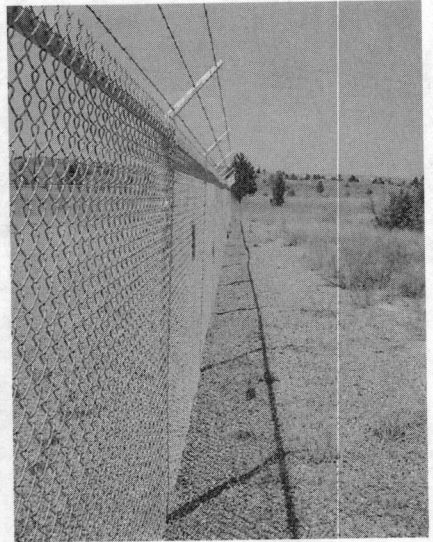

Photo 110-5. Barrier height requirements (see Fig. 110-5).

Photo 110-7. Example of how to apply the safety clearance zone to a chain-link fence per NESC Table 110-1 and NESC Fig. 110-1 (see Fig. 110-7).

Photo 110-8. Example of how to apply the safety clearance zone to an impenetrable fence per NESC Table 110-1 and NESC Fig. 110-2 (see Fig. 110-8).

Photo 110-14(1). Supporting and securing heavy equipment (see Fig. 110-14).

Photo 110-14(2). Supporting and securing heavy equipment (see Fig. 110-14).

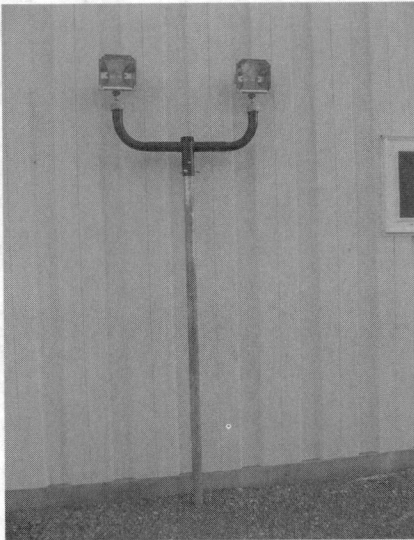

Photo 111-1. Illumination under normal conditions (see Fig. 111-1).

Photo 124-1. Example of how to apply vertical clearances per NESC Table 124-1, NESC Fig. 124-1, and Rule 124A3 (see Fig. 124-1).

Photo 124-2. Clearance measurements made to a permanent supporting surface (see Fig. 124-2).

Photo 124-4. Absence of Code rule related to bus to bus clearances (see Fig. 124-4).

Photo 124-5. Example of how to apply NESC Table 124-1 and NESC Fig. 124-2 (see Fig. 124-5).

Photo 125-3. VIOLATION! Storage materials must not be in the working space (see Fig. 125-3).

Photo 128-1. Example of identification in a substation (see Fig. 128-1).

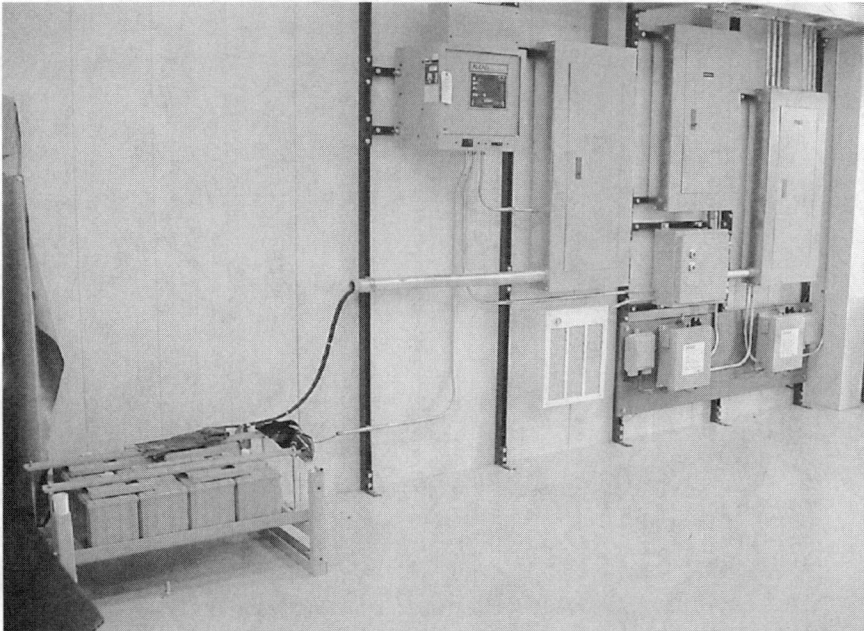

Photo 140-1. Storage batteries (see Fig. 140-1).

Photo 150-1. Example of protecting current-transformer secondary circuits (see Fig. 150-1).

Photo 153-1. Choices for short-circuit protection of power transformers (see Fig. 153-1).

Photo 162-1(1). Substation conductor mechanical protection and support (see Fig. 162-1).

Photo 162-1(2). Substation conductor mechanical protection and support (see Fig. 162-1).

Photo 163-1. Guarding a substation conductor with insulation (see Fig. 163-1).

Photo 170-1. Example of a switch position indicator (see Fig. 170-1).

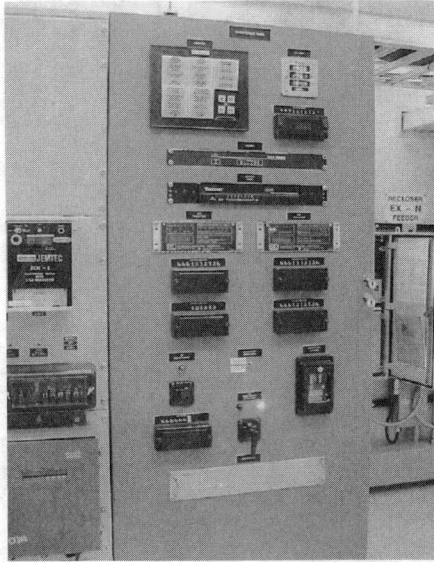

Photo 180-4. Adequate clearance for reading meters on control switchboards (see Fig. 180-4).

Photo 193-1. Arrester installation (see Fig. 193-1).

Photo 201-1. Overlap between Overhead Lines (Part 2) and Underground Lines (Part 3) (see Fig. 201-1).

Photo 201-2. Overlap between Overhead Lines (Part 2) and Electric Supply Stations (Part 1) (see Fig. 201-2).

Photo 214-1. VIOLATION! Inspection and tests of lines and equipment when in and out of service (see Fig. 214-1).

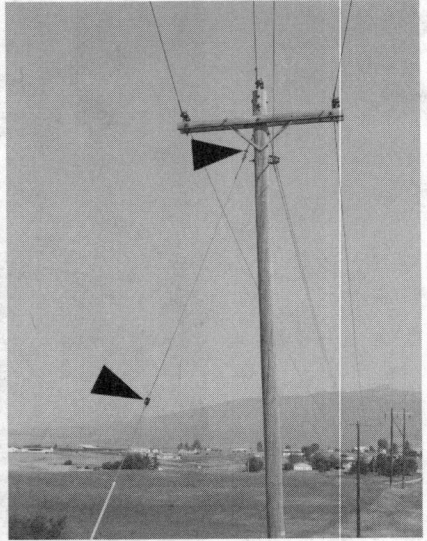

Photo 215-8. Example of a guy insulator used to limit galvanic corrosion (see Fig. 215-8).

Photo 216-1. Arrangement of overhead line switches (see Fig. 216-1).

Photo 217-2(1). Readily climbable supporting structures (see Fig. 217-2).

Photo 217-2(2). Readily climbable supporting structures (see Fig. 217-2)

Photo 217-2(3). Readily climbable supporting structures (see Fig. 217-2)

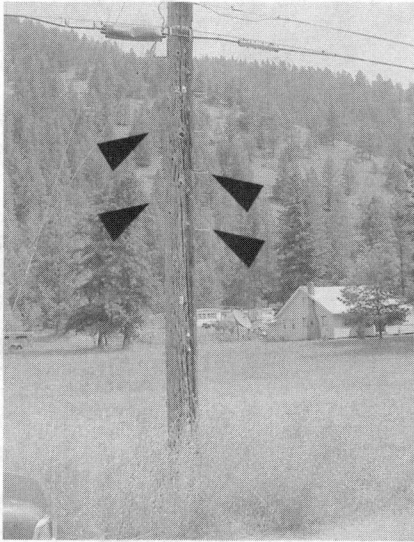

Photo 217-3. Permanently mounted and temporary steps on supporting structures (see Fig. 217-3).

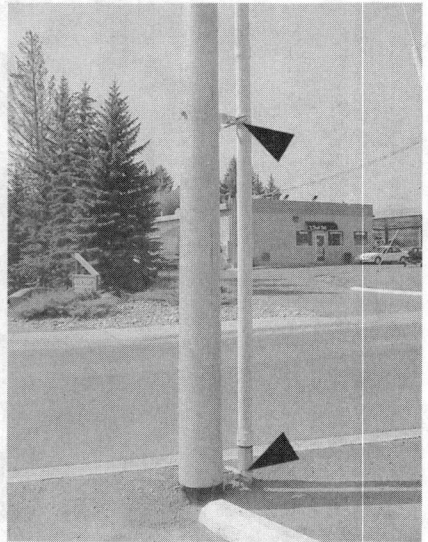

Photo 217-4. Arrangement of standoff brackets (see Fig. 217-4).

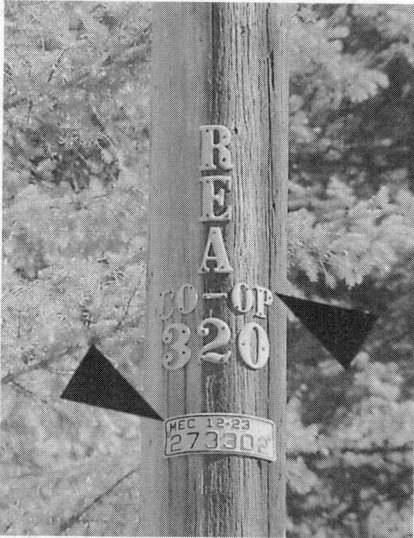

Photo 217-5. Identification of supporting structures (see Fig. 217-5).

Photo 217-6(1). VIOLATION! Attachments, decorations, and obstructions on supporting structures (see Fig. 217-6).

Photo 217-6(2). VIOLATION! Attachments, decorations, and obstructions on supporting structures (see Fig. 217-6).

Photo 217-9(1). Protection and marking of guys (see Fig. 217-9).

Photo 217-9(2). Protection and marking of guys (see Fig. 217-9).

Photo 218-1. General vegetation management requirements (see Fig. 218-1).

Photo 218-2. VIOLATION! Vegetation management at line, railroad, limited access highway, and navigable waterway crossings (see Fig. 218-2).

Photo 220-1. Standardization of levels of supply and communication conductors (see Fig. 220-1).

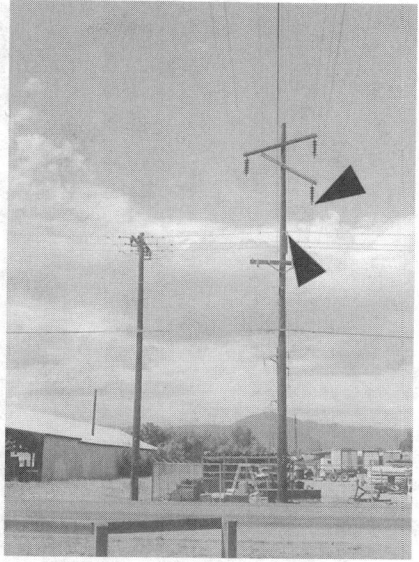

Photo 220-2. Relative levels of supply lines of different voltages at crossings (see Fig. 220-2).

Photo 220-3. Relative levels of supply lines of different voltages at structure conflict locations (see Fig. 220-3).

Photo 220-4. Relative levels of supply circuits of different voltages owned by one utility (see Fig. 220-4).

Photo 220-6(1). Example of identification of overhead conductors (see Fig. 220-6).

Photo 220-6(2). Example of identification of overhead conductors (see Fig. 220-6).

Photo 221-1(1). Avoiding conflict between two separate lines (see Fig. 221-1).

Photo 221-1(2). Avoiding conflict between two separate lines (see Fig. 221-1).

Photo 222-1. Example of joint use of structures (see Fig. 222-1).

Photo 225-1. Examples of common terms used in electric railway and trolley systems (see Fig. 225-1).

Photo 230-3. Example of cables laid on grade for emergency conditions (see Fig. 230-3).

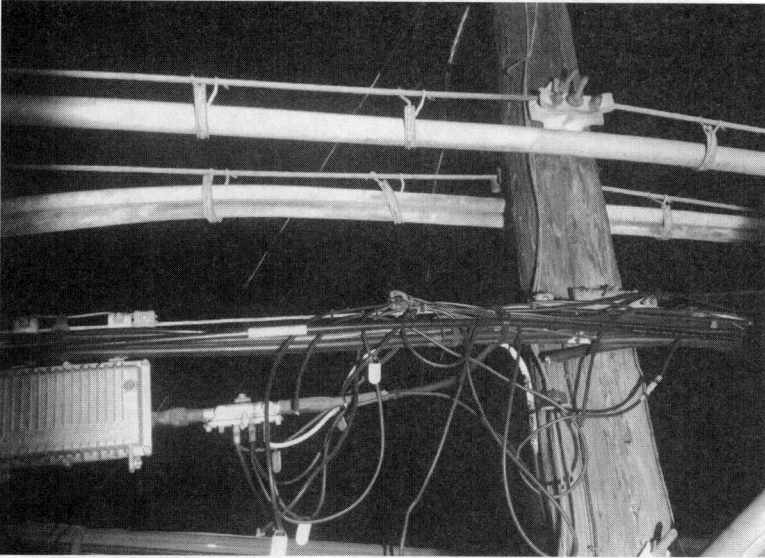

Photo 230-6(1). Construction of 230C1 supply cable (see Fig. 230-6).

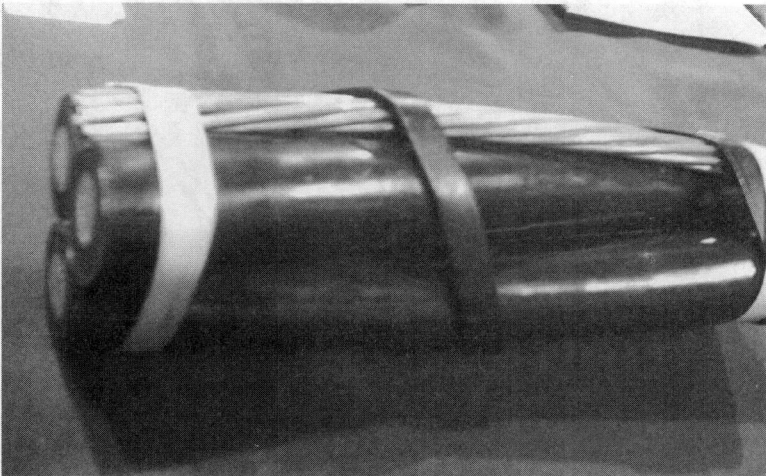

Photo 230-6(2). Construction of 230C1 supply cable (see Fig. 230-6).

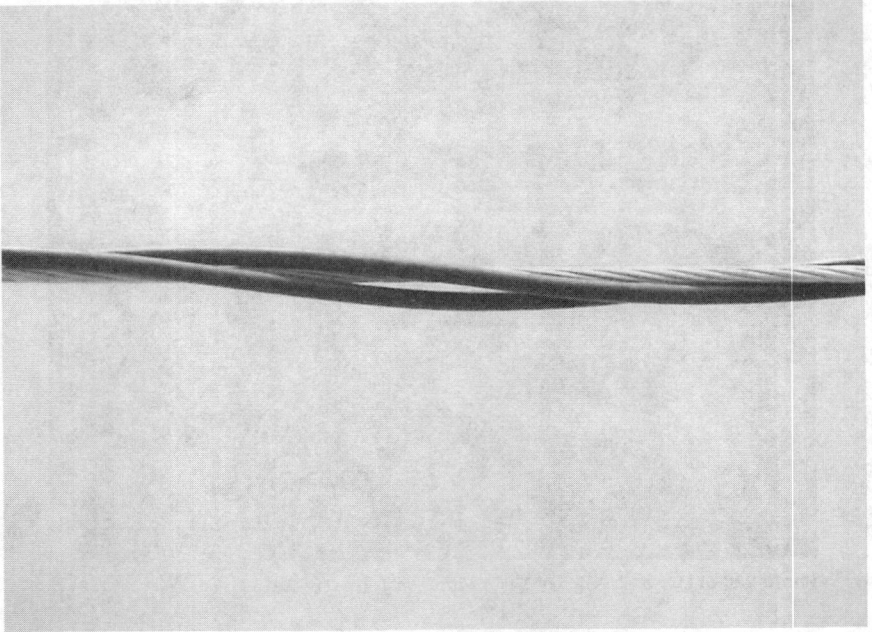

Photo 230-6(3). Construction of 230C3 supply cable (see Fig. 230-6).

Photo 230-7. Covered conductors (see Fig. 230-7).

Photo 230-8. Neutral conductors (see Fig. 230-8).

Photo 230-9. Example of an All-Dielectric Self-Supporting (ADSS) cable (see Fig. 230-9).

Photo 231-1. Clearance of a supporting structure from a fire hydrant (see Fig. 231-1).

Photo 231-2(1). VIOLATION! Clearance of a supporting structure from a curb (see Fig. 231-2).

Photo 231-2(2). Clearance of a supporting structure from a curb (see Fig. 231-2).

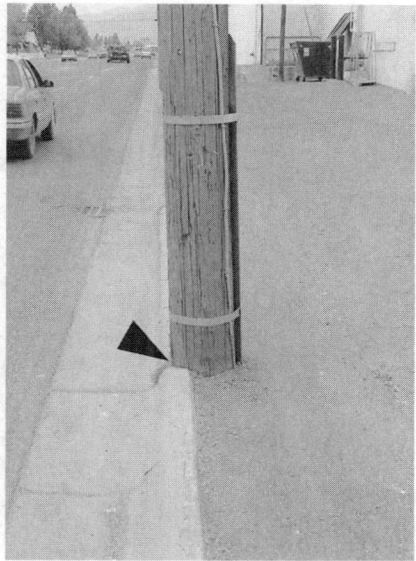

Photo 231-2(3). VIOLATION! Clearance of a supporting structure from a curb (see Fig. 231-2).

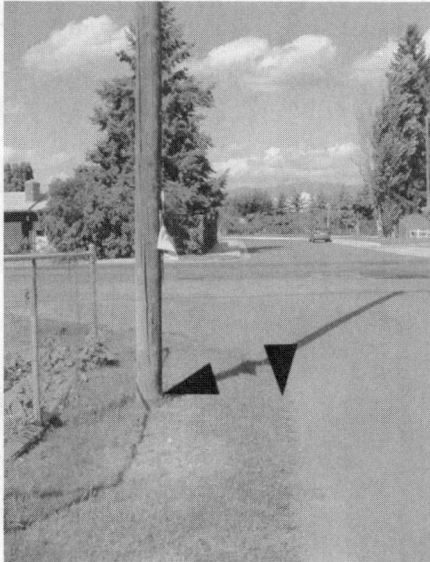

Photo 231-3. Clearance of a supporting structure from a roadway without a curb (see Fig. 231-3).

Photo 231-4. Clearance of a supporting structure from railroad tracks (see Fig. 231-4).

Photo 232-7. Common clearance values from NESC Table 232-1 (see Fig. 232-7).

Photo 232-8. Common clearance values from NESC Table 232-1 (see Fig. 232-8).

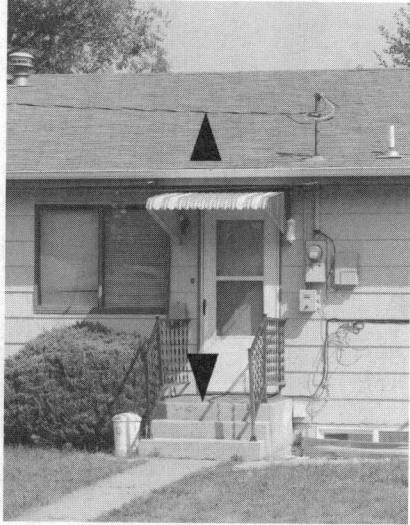

Photo 232-11. Common clearance values from NESC Table 232-1 (see Fig. 232-11).

Photo 232-17. Example of clearance to equipment cases and unguarded rigid live parts (see Fig. 232-17).

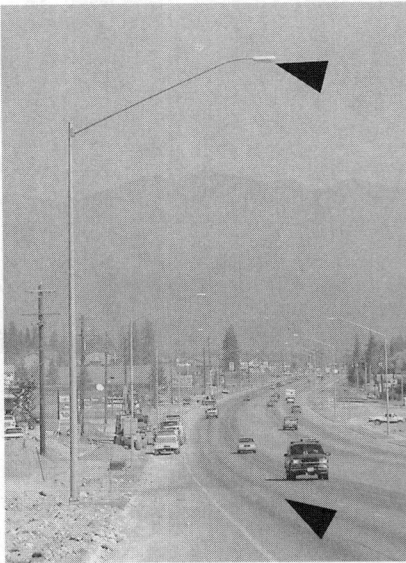

Photo 232-18. Example of clearance to street lighting (see Fig. 232-18).

Photo 232-19. Vertical clearance to land surface (see Fig. 232-19).

Photo 232-22. Federal Aviation Administration (FAA) requirements (see Fig. 232-22).

Photo 233-4(1). Example of horizontal clearance between wires carried on different supporting structures (see Fig. 233-4).

Photo 233-4(2). Example of horizontal clearance between wires carried on different supporting structures (see Fig. 233-4).

Photo 233-6. Example of vertical clearance between wires carried on different supporting structures (see Fig. 233-6).

Photo 233-7. Example of vertical clearance between wires carried on different supporting structures (see Fig. 233-7).

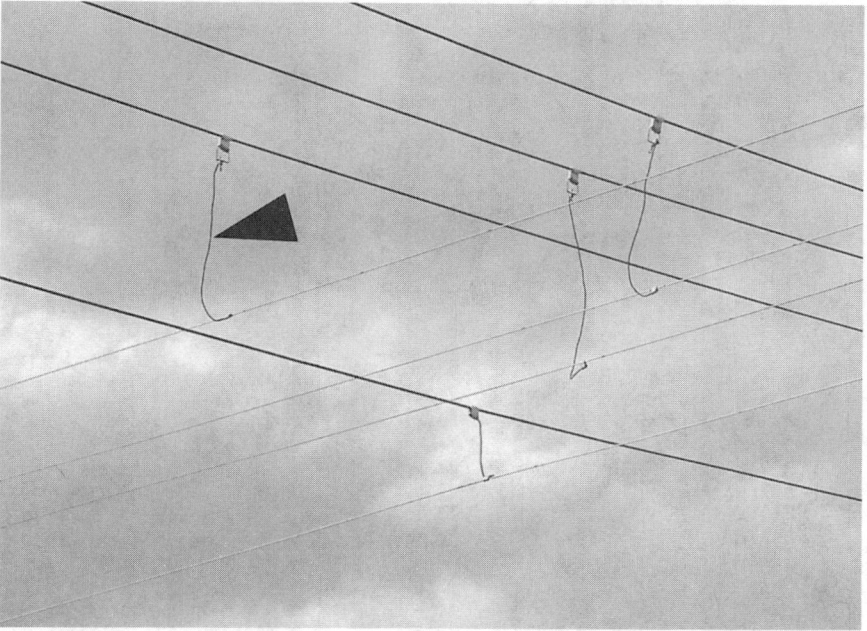

Photo 233-8. Example of the exception to vertical clearance between wires that are electrically interconnected at a crossing (see Fig. 233-8).

Photo 234-4. Example of clearance of a conductor to a street lighting pole (see Fig. 234-4).

Photo 234-5(1). Example of clearance of a conductor to a traffic signal pole (see Fig. 234-5).

Photo 234-5(2). VIOLATION! Example of clearance of a conductor to a traffic signal pole (see Fig. 234-5).

Photo 234-6(1). Application of Rules 234B, 231B, and 232B4 to bent street lighting poles (see Fig. 234-6).

Photo 234-6(2). Application of Rules 234B, 231B, and 232B4 to bent street lighting poles (see Fig. 234-6).

Photo 234-7. Example of skip-span construction (see Fig. 234-7).

Photo 234-8. Example of horizontal clearance of a conductor to a building (see Fig. 234-8).

Photo 234-10. Example of vertical clearance of a supply service drop conductor over a building but not serving it (see Fig. 234-10).

Photo 234-11. Example of clearance of a conductor to a billboard (see Fig. 234-11).

Photo 234-12. Example of horizontal clearance of a conductor to a flagpole and flag (see Fig. 234-12).

Photo 234-14. Example of clearance between a power line and irrigation equipment (see Fig. 234-14).

Photo 234-17(1). Example of clearance of supply conductors (service drops) attached to buildings (see Fig. 234-17).

Photo 234-17(2). Example of clearance of supply conductors (service drops) attached to buildings (see Fig. 234-17).

Photo 234-18. VIOLATION! Example of clearance of supply conductors (service drops) attached to buildings (see Fig. 234-18).

Photo 234-19(1). Concerns that are addressed when providing clearance to bridges (see Fig. 234-19).

Photo 234-19(2). Concerns that are addressed when providing clearance to bridges (see Fig. 234-19).

Photo 234-21(1). Concerns that are addressed when providing clearance to grain bins (see Fig. 234-21).

Photo 234-18(2). Concerns that are addressed when providing clearance to grain bins (see Fig. 234-18).

Photo 235-8. Example of vertical clearance between line conductors (see Fig. 235-8).

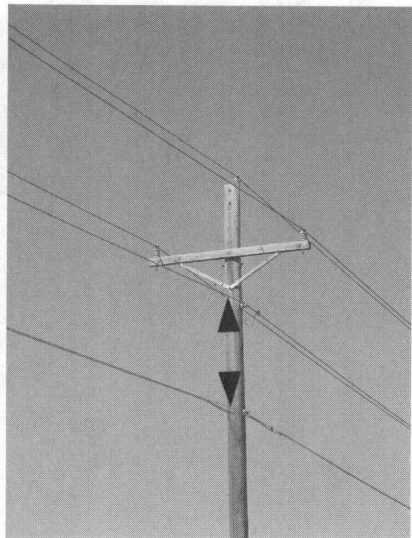

Photo 235-11. Example of vertical clearance between joint-use (supply and communication) conductors (see Fig. 235-11).

Photo 235-13. Example of vertical clearance between joint-use (supply and communication) conductors (see Fig. 235-13).

Photo 235-14. Example of vertical clearance between joint-use (supply and communication) conductors (see Fig. 235-14).

Photo 235-17. Example of clearance of a conductor to an anchor guy and a conductor to a support (see Fig. 235-17).

Photo 235-19. Examples of clearance between supply circuits of different voltages on the same support arm (see Fig. 235-19).

Photo 235-20. Conductor spacing on vertical racks or separate brackets (see Fig. 235-20).

Photo 235-22. Clearance and spacing between communication lines (see Fig. 235-22).

Photo 235-24(1). Example of clearance between supply lines and communication antennas in the supply space (see Fig. 235-24).

Photo 235-24(2). Example of clearance between supply lines and communication antennas in the supply space (see Fig. 235-24).

Photo 236-2. Example of climbing space between conductors (see Fig. 236-2).

Photo 236-4. Example of obstructions at the base of the climbing space (see Fig. 236-4).

Photo 237-2. Example of working clearances from energized equipment (see Fig. 237-2).

Photo 238-1. Definition of equipment as it applies to vertical clearance between communication and supply facilities on the same structure (see Fig. 238-1).

Photo 238-4. Example of vertical clearance between supply and communication equipment on the same structure (see Fig. 238-4).

Photo 238-5. Example of vertical clearance between supply and communication equipment on the same structure (see Fig. 238-5).

Photo 238-6. Example of vertical clearance between supply and communication equipment on the same structure (see Fig. 238-6).

Photo 238-7. Example of vertical clearance between supply and communication equipment on the same structure (see Fig. 238-7).

Photo 238-8(1). Example of vertical clearance between supply and communication equipment on the same structure (see Fig. 238-8).

Photo 238-8(2). Example of vertical clearance between supply and communication equipment on the same structure (see Fig. 238-8).

Photo 238-11. Example of vertical clearance between supply and communication equipment on the same structure (see Fig. 238-11).

Photo 238-12. Example of mounting a solar (photovoltaic) panel on a pole (see Fig. 238-12).

Photo 238-14. Example of vertical clearance between a drip loop feeding a luminaire and communication equipment (see Fig. 238-14).

Photo 238-15. Example of vertical clearance between a drip loop feeding a luminaire and communication equipment (see Fig. 238-15).

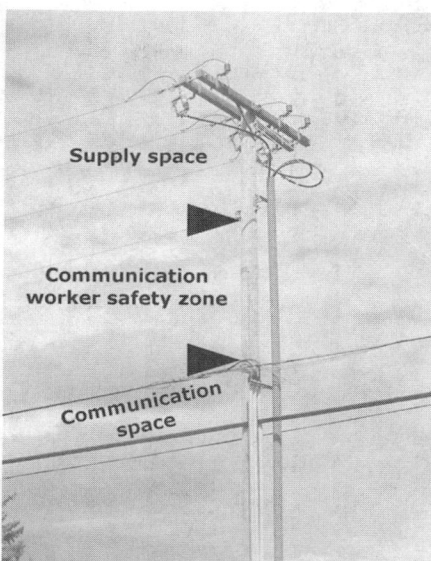

Photo 238-18. Communication worker safety zone (see Fig. 238-18).

Photo 239-1. Examples of vertical and lateral conductors (see Fig. 239-1).

Photo 239-3(1). Guarding and protection near ground (see Fig. 239-3).

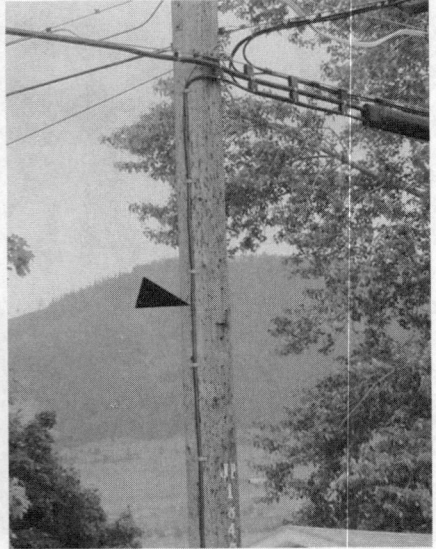

Photo 239-3(2). Guarding and protection near ground (see Fig. 239-3).

Photo 239-3(3). Guarding and protection near ground (see Fig. 239-3).

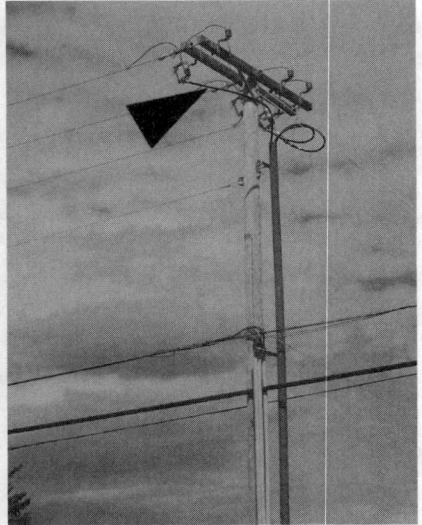

Photo 239-4. Location of vertical and lateral supply conductors on supply-line structures or within supply space on jointly used structures (see Fig. 239-4).

Photo 239-5. Location of vertical and lateral communication conductors on communication line structures (see Fig. 239-5).

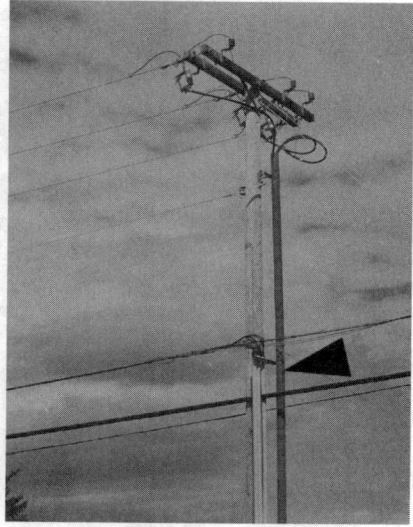

Photo 239-6. Location of vertical supply conductors and cables passing through the communication space on jointly used line structures (see Fig. 239-6).

Photo 239-7. Example of vertical supply conductors on a joint-use (supply and communication) pole (see Fig. 239-7).

Photo 239-8. VIOLATION! Example of vertical supply conductors on a joint-use (supply and communication) pole (see Fig. 239-8).

Photo 239-10. Location of vertical communication conductors passing through the supply space on jointly used structures (see Fig. 239-10).

Photo 241-3. Lines considered to be at crossings (see Fig. 241-3).

Photo 252-6. Longitudinal loads at a change in grade of construction (see Fig. 252-6).

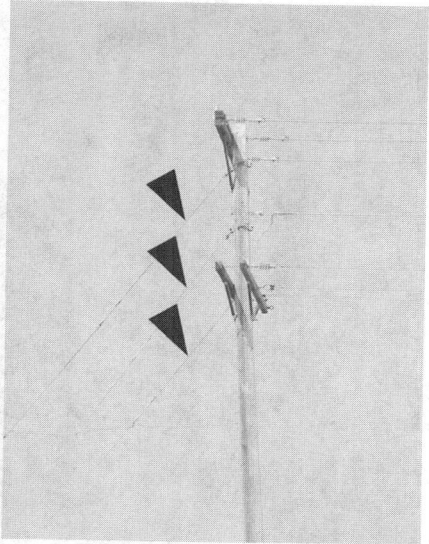

Photo 252-7(1). Longitudinal loads at deadends (see Fig. 252-7).

Photo 252-7(2). Longitudinal loads at deadends (see Fig. 252-7).

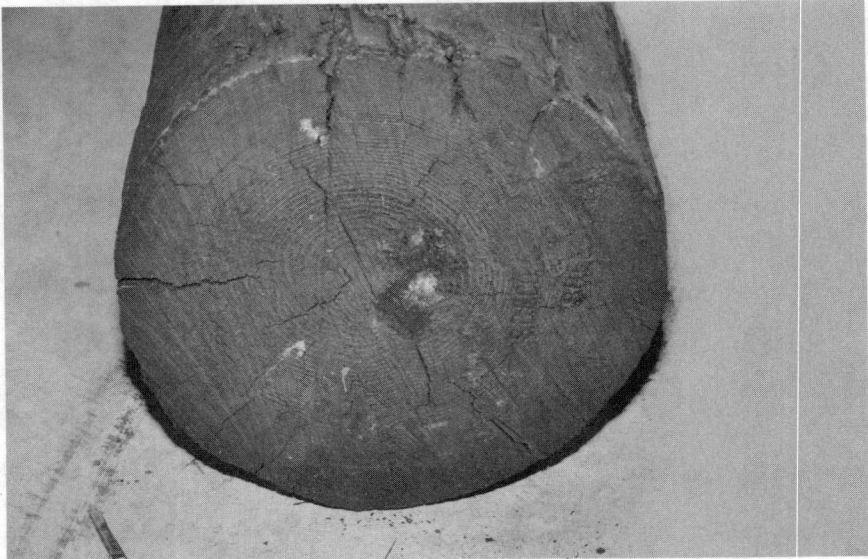

Photo 260-3. Example of how loads, material, and load and strength factors affect the size of a wood pole (see Fig. 260-3).

Photo 260-6(1). Examples of a pole tag and pole brand (see Fig. 260-6).

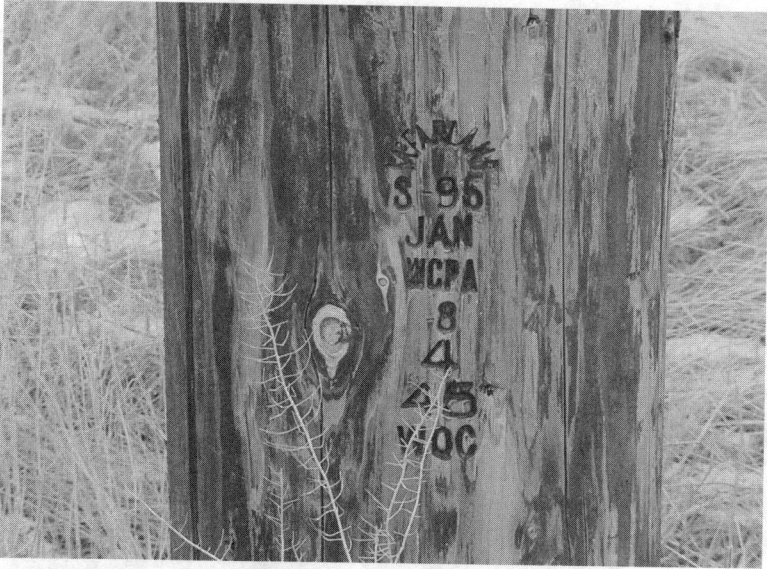

Photo 260-6(2). Examples of a pole tag and pole brand (see Fig. 260-6).

Photo 260-6(3). Examples of a pole tag and pole brand (see Fig. 260-6).

Photo 261-1(1). Strength requirements of supporting structure (see Fig. 261-1).

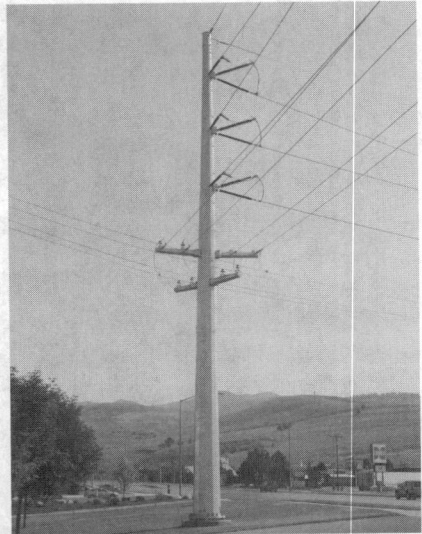

Photo 261-1(2). Strength requirements of supporting structure (see Fig. 261-1).

Photo 261-3. Strength requirements of metal, prestressed, and reinforced concrete structures (see Fig. 261-3).

Photo 261-4. Strength requirements of wood structures (see Fig. 261-4).

Photo 261-7(1). Examples of spliced and reinforced wood poles (see Fig. 261-7).

Photo 261-7(2). Examples of spliced and reinforced wood poles (see Fig. 261-7).

Photo 261-7(3). Examples of spliced and reinforced wood poles (see Fig. 261-7).

Photo 261-8. Strength requirements of fiber-reinforced polymer structures (see Fig. 261-8).

Photo 261-15. Strength of concrete and metal crossarms and braces (see Fig. 261-15).

Photo 261-16. Strength of wood crossarms and braces (see Fig. 261-16).

Photo 261-18. Strength of fiber-reinforced polymer crossarms and braces (see Fig. 261-18).

Photo 261-19(1). Example of wood and steel pins (see Fig. 261-19).

Photo 261-19(2). VIOLATION! Example of wood and steel pins (see Fig. 261-19).

Photo 261-20. Strength requirements for armless construction (see Fig. 261-20).

Photo 261-24. Example of open-wire communication conductors (see Fig. 261-24).

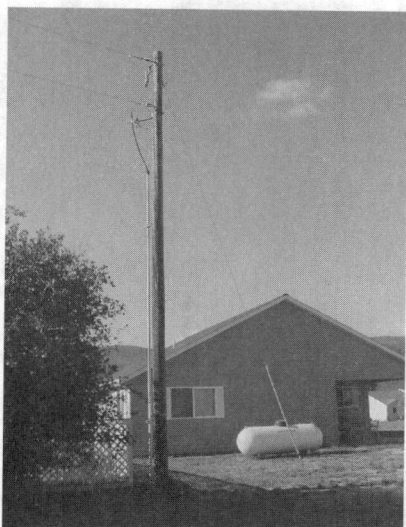

Photo 264-1(1). Examples of guying and bracing (see Fig. 264-1).

Photo 264-1(2). Examples of guying and bracing (see Fig. 264-1).

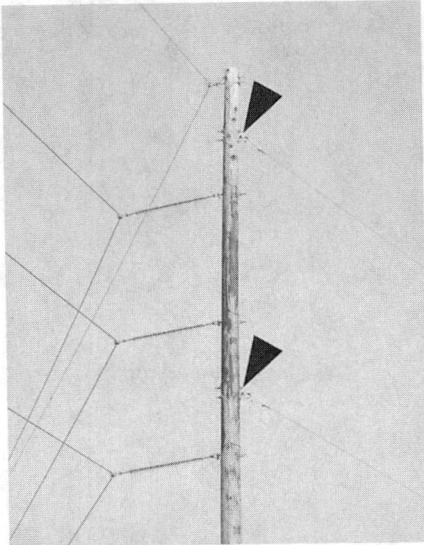

Photo 264-3(1). Point of guy attachment (see Fig. 264-3).

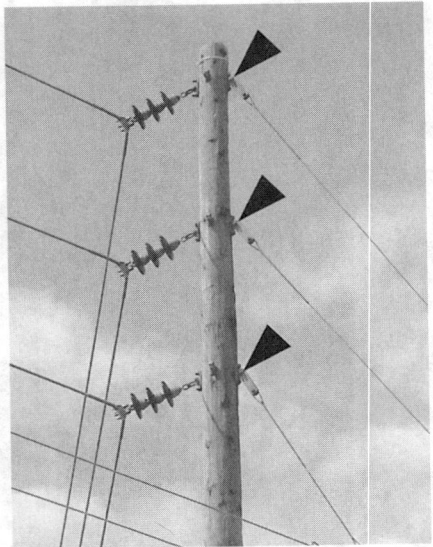

Photo 264-3(2). Point of guy attachment (see Fig. 264-3).

Photo 279-2. Properties of guy insulators (see Fig. 279-2).

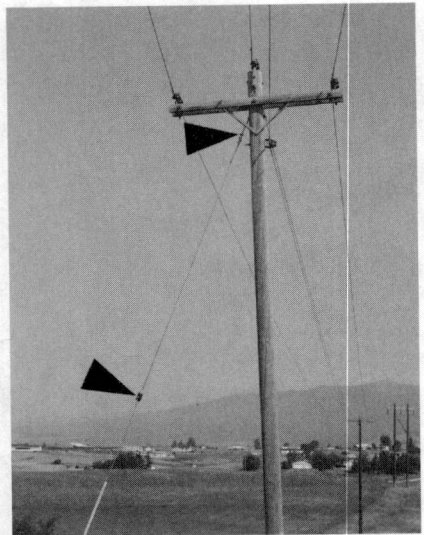

Photo 279-4. Example of a guy insulator used to limit galvanic corrosion (see Fig. 279-4).

Photo 301-1. Overlap between Underground Lines (Part 3) and Overhead Lines (Part 2) (see Fig. 301-1).

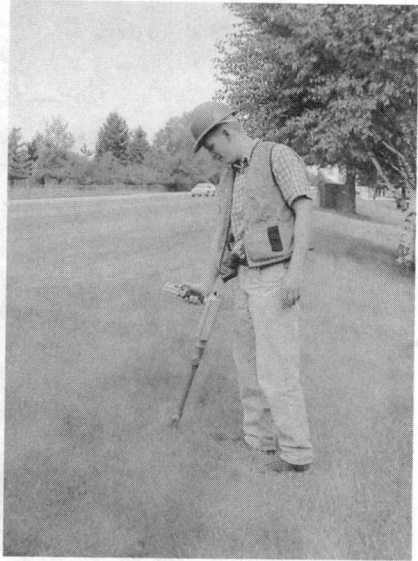

Photo 311-1. Example of worker locating underground lines (see Fig. 311-1).

Photo 312-1. Example of the need for adaquate working space (see Fig. 312-1).

Photo 312-2. Example of a notification sign used on pad-mounted equipment (see Fig. 312-2).

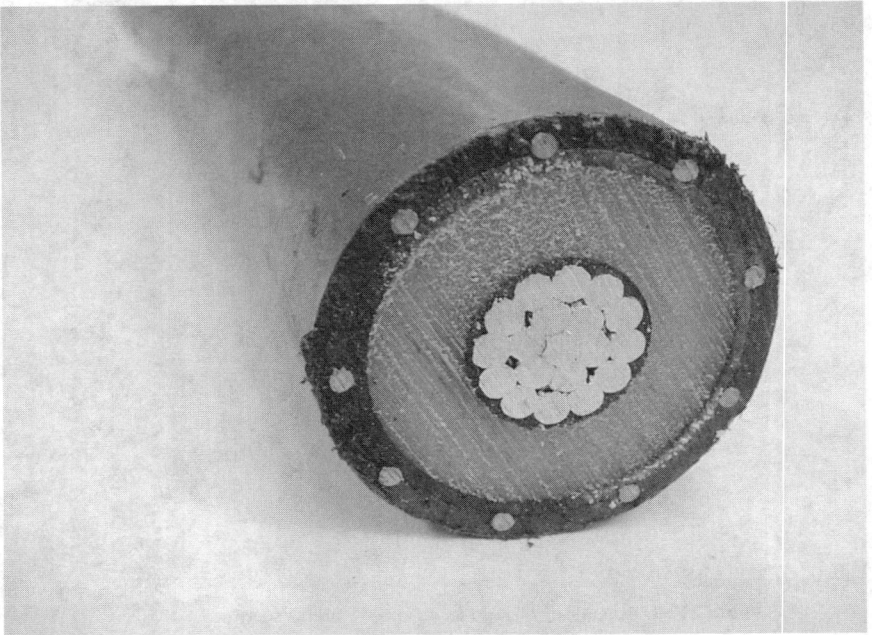

Photo 314-1. Conductive parts to be grounded (see Fig. 314-1).

Photo 314-2. Grounding of circuits (see Fig. 314-2).

Photo 320-2. Conduit bending radius (see Fig. 320-2)

Photo 320-4. Conduit routing under highways and streets (see Fig. 320-4).

Photo 320-10. VIOLATION! Radial separation from gas or other line that transports flammable material (see Fig. 320-10).

Photo 323-1. Loads on manholes, handholes, and vaults (see Fig. 323-1).

Photo 323-2(1). Examples of live loads in roadways and live loads in areas not subject to vehicular loading (see Fig. 323-2).

Photo 323-2(2). Examples of live loads in roadways and live loads in areas not subject to vehicular loading (see Fig. 323-2).

Photo 323-6. Manhole access diameter for a round manhole opening in a manhole containing supply cables (see Fig. 323-6).

Photo 323-13(1). Examples of vaults and utility tunnels (see Fig. 323-13).

Photo 323-13(2). Examples of vaults and utility tunnels (see Fig. 323-13).

Photo 323-15(1). Requirements for identification of manhole and handhole covers (see Fig. 323-15).

Photo 323-15(2). Requirements for identification of manhole and handhole covers (see Fig. 323-15).

Photo 323-15(3). Requirements for identification of manhole and handhole covers (see Fig. 323-15).

Photo 331-1. Example of a sheath, jacket, and shield on a typical URD supply cable (see Fig. 331-1).

Photo 333-1(1). Examples of cable accessories and joints (see Fig. 333-1).

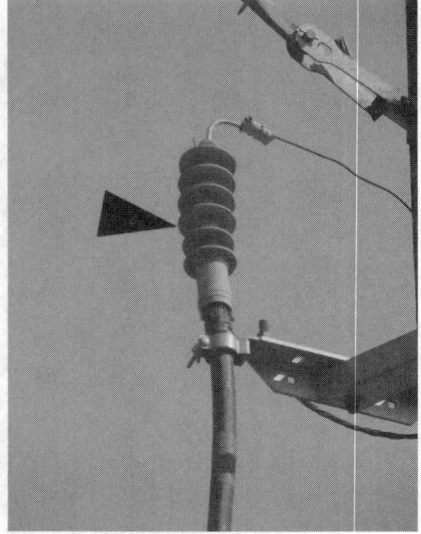

Photo 333-1(2). Examples of cable accessories and joints (see Fig. 333-1).

Photo 333-1(3). Examples of cable accessories and joints (see Fig. 333-1).

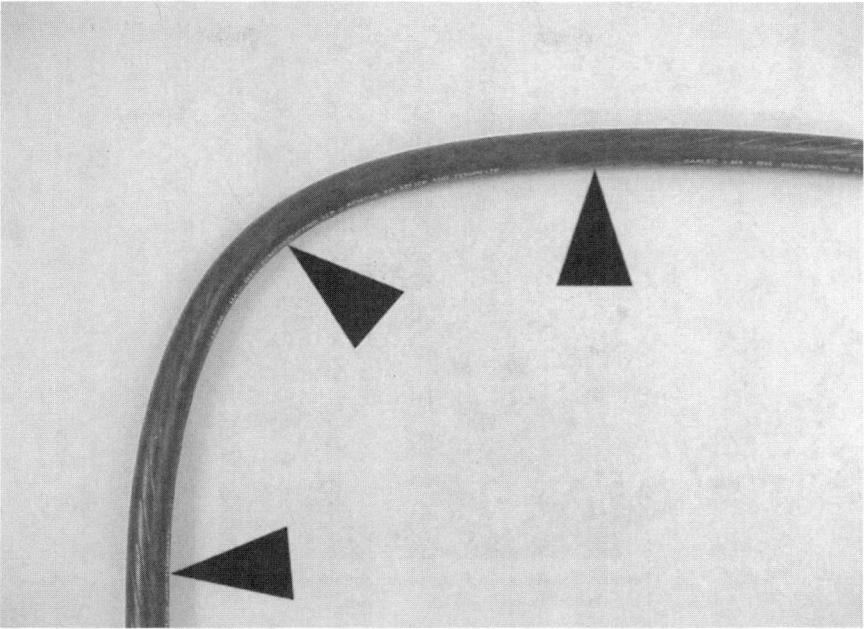

Photo 341-1. Cable bending radius (see Fig. 341-1).

Photo 350-3. Cables below 600 V to ground buried in close proximity (see Fig. 350-3).

Photo 352-1. Trenching requirements for direct-buried cable (see Fig. 352-1).

Photo 352-3. Plowing requirements for direct-buried cables (see Fig. 352-3).

Photo 352-4. Boring requirements for direct-buried cables (see Fig. 352-4).

Photo 354-1. Supply line and gas line in random separation (see Fig. 354-1).

Photo 360-1. Mechanical protection of supply risers below grade (see Fig. 360-1).

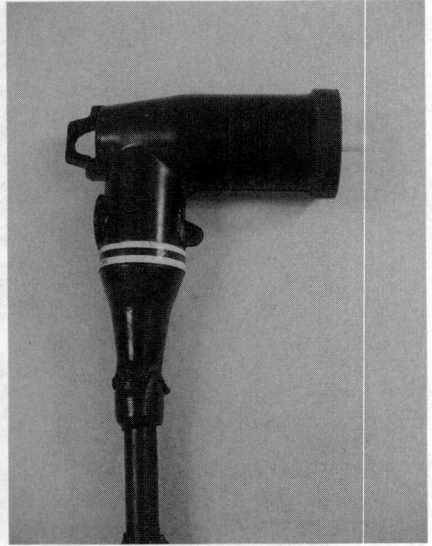

Photo 370-1(1). Examples of supply cable terminations (see Fig. 370-1).

Photo 370-1(2). Examples of supply cable terminations (see Fig. 370-1).

Photo 370-1(3). Examples of supply cable terminations (see Fig. 370-1).

Photo 372-1. Example of supply cable identification at termination points (see Fig. 372-1).

Photo 372-2. Absence of Code rule related to cable route markers (see Fig. 372-2).

Photo 380-1. Examples of supply and communication equipment (see Fig. 380-1).

Photo 380-2. Clearance of pad-mounted equipment to fire hydrants (see Fig. 380-2).

Photo 380-3(1). Absence of Code rule related to underground pad-mounted equipment location adjacent to roads and buildings (see Fig. 380-3).

Photo 380-3(2). Absence of Code rule related to underground pad-mounted equipment location adjacent to roads and buildings (see Fig. 380-3).

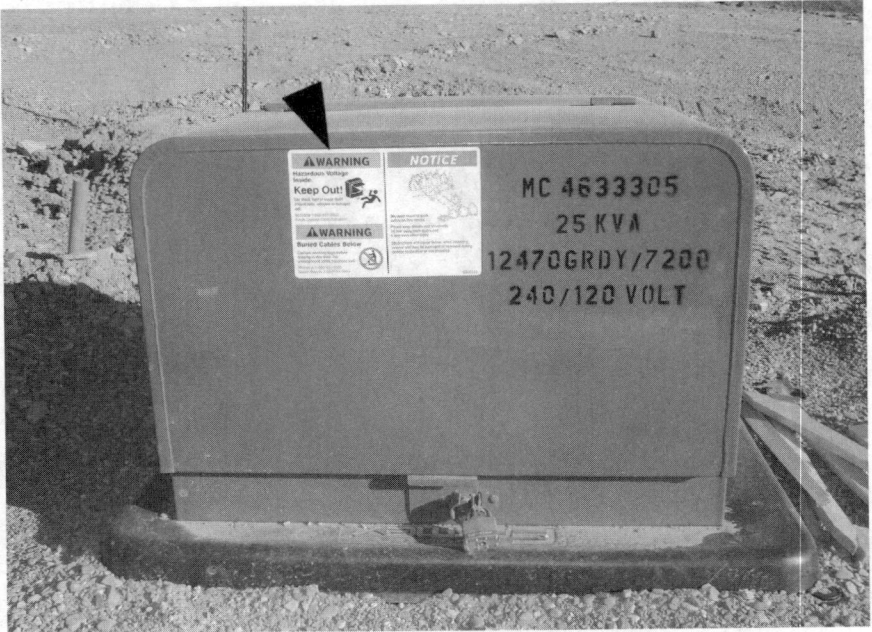

Photo 381-3(1). Safety sign requirements for pad-mounted equipment (see Fig. 381-3).

Photo 381-3(2). Safety sign requirements for pad-mounted equipment (see Fig. 381-3).

Photo 384-1. Bonding above ground metallic supply and communications enclosures (see Fig. 384-1).

Appendix B

OSHA Standards Related to the NESC Work Rules

1910.268

Telecommunications (Operation and Maintenance)
1910 Subpart R—Special Industries

1910.269 (Includes Appendix A–G)

Electric Power Generation, Transmission, and Distribution (Operation and Maintenance)
1910 Subpart R—Special Industries

1926.950 through 1926.968 (Includes Appendix A-G)

Power Transmission and Distribution (Construction)
1926 Subpart V—Power Transmission and Distribution

Note: Appendix B contains reprints of the OSHA documents listed above. The reprints were current at the time of the authoring of this Handbook. Visit www.osha.gov for updates to the standards listed above.

1910.268

Telecommunications (Operations and Maintenance) 1910 Subpart R—Special Industries

OSHA 1910.268 Paragraph Titles:

(a) Application
(b) General
(c) Training
(d) Employee protection in public work areas
(e) Tools and personal protective equipment
(f) Rubber insulating equipment
(g) Personal climbing equipment
(h) Ladders
(i) Other tools and personal protective equipment
(j) Vehicle-mounted material handling devices and other mechanical equipment
(k) Materials handling and storage
(l) Cable fault locating and testing
(m) Grounding for employee protection—pole lines
(n) Overhead lines
(o) Underground lines
(p) Microwave transmission
(q) Tree trimming—electrical hazards
(r) Buried facilities—Communications lines and power lines in the same trench
(s) Definitions

REGULATIONS (STANDARDS - 29 CFR)

TELECOMMUNICATIONS. -1910.268

- **Part Number:** 1910
- **Part Title:** Occupational Safety and Health Standards
- **Subpart:** R
- **Subpart Title:** Special Industries
- **Standard Number:** 1910.268
- **Title:** Telecommunications.

1910.268(a)

Application.

1910.268(a)(1)

This section sets forth safety and health standards that apply to the work conditions, practices, means, methods, operations, installations and processes performed at telecommunications centers and at telecommunications field installations, which are located outdoors or in building spaces used for such field installations. *Center* work includes the installation, operation, maintenance, rearrangement, and removal of communications equipment and other associated equipment in telecommunications switching centers. *Field* work includes the installation, operation, maintenance, rearrangement, and removal of conductors and other equipment used for signal or communication service, and of their supporting or containing structures, overhead or underground, on public or private rights of way, including buildings or other structures.

..1910.268(a)(2)

1910.268(a)(2)

These standards do not apply:

1910.268(a)(2)(i)

To construction work, as defined in § 1910.12, nor

1910.268(a)(2)(ii)

to installations under the exclusive control of electric utilities used for the purpose of communications or metering, or for generation, control, transformation, transmission, and distribution of electric energy, which are located in buildings used exclusively by the electric utilities for such purposes, or located outdoors on property owned or leased by the electric utilities or on public highways, streets, roads, etc., or outdoors by established rights on private property.

1910.268(a)(3)

Operations or conditions not specifically covered by this section are subject to all the applicable standards contained in this Part 1910. See § 1910.5(c). Operations which involve construction work, as defined in 1910.12 are subject to all the applicable standards contained in Part 1926 of this chapter.

1910.268(b)

General—

1910.268(b)(1)

Buildings containing telecommunications centers—

1910.268(b)(1)(i)

Illumination. Lighting in telecommunication centers shall be provided in an adequate amount such that continuing work operations, routine observations, and the passage of employees can be carried out in a safe and healthful manner. Certain specific tasks in centers, such as splicing cable and the maintenance and repair of equipment frame lineups, may require a higher level of illumination. In such cases, the employer shall install permanent lighting or portable supplemental lighting to attain a higher level of illumination shall be provided as needed to permit safe performance of the required task.

..1910.268(b)(1)(ii)

1910.268(b)(1)(ii)

Working surfaces. Guard rails and toe boards may be omitted on distribution frame mezzanine platforms to permit access to equipment. This exemption applies only on the side or sides of the platform facing the frames and only on those portions of the platform adjacent to equipped frames.

1910.268(b)(1)(iii)

Working spaces. Maintenance aisles, or wiring aisles, between equipment frame lineups are working spaces and are not an exit route for purposes of 29 CFR 1910.34.

1910.268(b)(1)(iv)

Special doors. When blastproof or power actuated doors are installed in specially designed hardsite security buildings and spaces, they shall be designed and installed so that they can be used as a means of egress in emergencies.

1910.268(b)(1)(v)

Equipment, machinery and machine guarding. When power plant machinery in telecommunications centers is operated with commutators and couplings uncovered, the adjacent housing shall be clearly marked to alert personnel to the rotating machinery.

1910.268(b)(2)

Battery handling.

1910.268(b)(2)(i)

Eye protection devices which provide side as well as frontal eye protection for employees shall be provided when measuring storage battery specific gravity or handling electrolyte, and the employer shall ensure that such devices are used by the employees. The employer shall also ensure that acid resistant gloves and aprons shall be worn for protection against spattering. Facilities for quick drenching or flushing of the eyes and body shall be provided unless the storage batteries are of the enclosed type and equipped with explosion proof vents, in which case sealed water rinse or neutralizing packs may be substituted for the quick drenching or flushing facilities. Employees assigned to work with storage batteries shall be instructed in emergency procedures such as dealing with accidental acid spills.

1910.268(b)(2)(ii)

Electrolyte (acid or base, and distilled water) for battery cells shall be mixed in a well ventilated room. Acid or base shall be poured gradually, while stirring, into the water. Water shall never be poured into concentrated (greater than 75 percent) acid solutions. Electrolyte shall never be placed in metal containers nor stirred with metal objects.

1910.268(b)(2)(iii)

When taking specific gravity readings, the open end of the hydrometer shall be covered with an acid resistant material while moving it from cell to cell to avoid splashing or throwing the electrolyte.

1910.268(b)(3)

Employers must provide employees with readily accessible, adequate, and appropriate first aid supplies. A non-mandatory example of appropriate supplies is listed in Appendix A to 29 CFR 1910.151.

..1910.268(b)(4)
1910.268(b)(4)

Hazardous materials. Highway mobile vehicles and trailers stored in garages in accordance with § 1910.110 may be equipped to carry more than one LP-gas container, but the total capacity of LP-gas containers per work vehicle stored in garages shall not exceed 100 pounds of LP-gas. All container valves shall be closed when not in use.

1910.268(b)(5)

Compressed gas. When using or transporting nitrogen cylinders in a horizontal position, special compartments, racks, or adequate blocking shall be provided to prevent cylinder movement. Regulators shall be removed or guarded before a cylinder is transported.

1910.268(b)(6)

Support structures. No employee, or any material or equipment, may be supported or permitted to be supported on any portion of a pole structure, platform, ladder, walkway or other elevated structure or aerial device unless the employer ensures that the support structure is first inspected by a competent person and it is determined to be adequately strong, in good working condition and properly secured in place.

1910.268(b)(7)

Approach distances to exposed energized overhead power lines and parts. The employer shall ensure that no employee approaches or takes any conductive object closer to any electrically energized overhead power lines and parts than prescribed in Table R-2, unless:

1910.268(b)(7)(i)

The employee is insulated or guarded from the energized parts (insulating gloves rated for the voltage involved shall be considered adequate insulation), or

1910.268(b)(7)(ii)

The energized parts are insulated or guarded from the employee and any other conductive object at a different potential, or

1910.268(b)(7)(iii)

The power conductors and equipment are deenergized and grounded.

Table R-2   Approach Distances to Exposed Energized Overhead Power Lines and Parts

Voltage range (phase to phase, RMS)	Approach distance (inches)
300 V and less	([1])
Over 300V, not over 750V	12
Over 750V not over 2 kV	18
Over 2 kV, not over 15 kV	24
Over 15 kV, not over 37 kV	36
Over 37 kV, not over 87.5 kV	42
Over 87.5 kV, not over 121 kV	48
Over 121 kV, not over 140 kV	54

[1]Avoid contact.

1910.268(b)(8)

Illumination of field work. Whenever natural light is insufficient to adequately illuminate the worksite, artificial illumination shall be provided to enable the employee to perform the work safely.

..1910.268(c)

1910.268(c)

Training. Employers shall provide training in the various precautions and safe practices described in this section and shall insure that employees do not engage in the activities to which this section applies until such employees have received proper training in the various precautions and safe practices required by this section. However, where the employer can demonstrate that an employee is already trained in the precautions and safe practices required by this section prior to his employment, training need not be provided to that employee in accordance with this section. Where training is required, it shall consist of on-the-job training or classroom-type training or a combination of both. The employer shall certify that employees have been trained by preparing a certification record which includes the identity of the person trained, the signature of the employer or the person who conducted the training, and the date the training was completed. The certification record shall be prepared at the completion of training and shall be maintained on file for the duration of the employee's employment. The certification record shall be made available upon request to the Assistant Secretary for Occupational Safety and Health. Such training shall, where appropriate, include the following subjects:

1910.268(c)(1)

Recognition and avoidance of dangers relating to encounters with harmful substances and animal, insect, or plant life;

1910.268(c)(2)

Procedures to be followed in emergency situations; and,

1910.268(c)(3)

First aid training, including instruction in artificial respiration.

1910.268(d)

Employee protection in public work areas.

1910.268(d)(1)

Before work is begun in the vicinity of vehicular or pedestrian traffic which may endanger employees, warning signs and/or flags or other traffic control devices shall be placed conspicuously to alert and channel approaching traffic. Where further protection is needed, barriers shall be utilized. At night, warning lights shall be prominently displayed, and excavated areas shall be enclosed with protective barricades.

1910.268(d)(2)

If work exposes energized or moving parts that are normally protected, danger signs shall be displayed and barricades erected, as necessary, to warn other personnel in the area.

1910.268(d)(3)

The employer shall insure that an employee finding any crossed or fallen wires which create or may create a hazardous situation at the work area:

1910.268(d)(3)(i)

Remains on guard or adopts other adequate means to warn other employees of the danger and

1910.268(d)(3)(ii)

has the proper authority notified at the earliest practical moment.

1910.268(e)

Tools and personal protective equipment—Generally. Personal protective equipment, protective devices and special tools needed for the work of employees shall be provided and the employer shall ensure that they are used by employees. Before each day's use the employer shall ensure that these personal protective devices, tools, and equipment are carefully inspected by a competent person to ascertain that they are in good condition.

..1910.268(f)

1910.268(f)

Rubber insulating equipment.

1910.268(f)(1)

Rubber insulating equipment designed for the voltage levels to be encountered shall be provided and the employer shall ensure that they are used by employees as required by this section. The requirements of § 1910.137, Electrical Protective Equipment, shall be followed except for Table I-6.

1910.268(f)(2)

The employer is responsible for the periodic retesting of all insulating gloves, blankets, and other rubber insulating equipment. This retesting shall be electrical, visual and mechanical. The following maximum retesting intervals shall apply:

Gloves, blankets, and other insulating equipment	Natural rubber	Synthetic rubber
	Months	
New	12	18
Re-issued	9	15

1910.268(f)(3)

Gloves and blankets shall be marked to indicate compliance with the retest schedule, and shall be marked with the date the next test is due. Gloves found to be defective in the field or by the tests set forth in paragraph (f)(2) of this section shall be destroyed by cutting them open from the finger to the gauntlet.

1910.268(g)

Personal climbing equipment—

1910.268(g)(1)

General. Safety belts and straps shall be provided and the employer shall ensure their use when work is performed at positions more than 4 feet above ground, on poles, and on towers, except as provided in paragraphs (n)(7) and (n)(8) of this section. No safety belts, safety straps or lanyards acquired after July 1, 1975 may be used unless they meet the tests set forth in paragraph (g)(2) of this section. The employer shall ensure that all safety belts and straps are inspected by a competent person prior to each day's use to determine that they are in safe working condition.

1910.268(g)(2)

Telecommunication lineman's body belts, safety straps, and lanyards—

..1910.268(g)(2)(i)

1910.268(g)(2)(i)

General requirements.

1910.268(g)(2)(i)(A)

Hardware for lineman's body belts, safety straps, and lanyards shall be drop forged or pressed steel and shall have a corrosion resistant finish tested to meet the requirements of the American Society for Testing and Materials B117-64, which is incorporated by reference as specified in § 1910.6 (50-hour test). Surfaces shall be smooth and free of sharp edges. Production samples of lineman's safety straps, body belts and lanyards shall be approved by a nationally recognized testing laboratory, as having been tested in accordance with and as meeting the requirements of this paragraph.

1910.268(g)(2)(i)(B)

All buckles shall withstand a 2,000-pound tensile test with a maximum permanent deformation no greater than one sixty-forth inch.

1910.268(g)(2)(i)(C)

D rings shall withstand a 5,000 pound tensile test without cracking or breaking.

1910.268(g)(2)(i)(D)

Snaphooks shall withstand a 5,000 pound tensile test, or shall withstand a 3,000-pound tensile test and a 180° bend test. Tensile failure is indicated by distortion of the snaphook sufficient to release the keeper; bend test failure is indicated by cracking of the snaphook.

1910.268(g)(2)(ii)

Specific requirements.

1910.268(g)(2)(ii)(A)(1)

All fabric used for safety straps shall be capable of withstanding an A.C. dielectric test of not less than 25,000 volts per foot "dry" for 3 minutes, without visible deterioration.

1910.268(g)(2)(ii)(A)(2)

All fabric and leather used shall be tested for leakage current. Fabric or leather may not be used if the leakage current exceeds 1 milliampere when a potential of 3,000 volts is applied to the electrodes positioned 12 inches apart.

1910.268(g)(2)(ii)(A)(3)

In lieu of alternating current tests, equivalent direct current tests may be performed.

..1910.268(g)(2)(ii)(B)

1910.268(g)(2)(ii)(B)

The cushion part of the body belt shall:

1910.268(g)(2)(ii)(B)(1)

Contain no exposed rivets on the inside. This provision does not apply to belts used by craftsmen not engaged in line work.

1910.268(g)(2)(ii)(B)(2)

Be at least three inches in width;

1910.268(g)(2)(ii)(B)(3)

Be at least five thirty-seconds (5/32) inch thick, if made of leather; and

1910.268(g)(2)(ii)(C)

[Reserved]

1910.268(g)(2)(ii)(D)

Suitable copper, steel, or equivalent liners shall be used around the bars of D rings to prevent wear between these members and the leather or fabric enclosing them.

1910.268(g)(2)(ii)(E)

All stitching shall be done with a minimum 42 pound weight nylon or equivalent thread and shall be lock stitched. Stitching parallel to an edge may not be less than three-sixteenths (3/16) inch from the edge of the narrowest member caught by the thread. The use of cross stitching on leather is prohibited.

..1910.268(g)(2)(ii)(F)

1910.268(g)(2)(ii)(F)

The keepers of snaphooks shall have a spring tension that will not allow the keeper to begin to open when a weight of 2 1/2 pounds or less is applied, but the keepers shall begin to open when a weight of four pounds is applied. In making this determination, the weight shall be supported on the keeper against the end of the nose.

1910.268(g)(2)(ii)(G)

Safety straps, lanyards, and body belts shall be tested in accordance with the following procedure:

1910.268(g)(2)(ii)(G)(1)

Attach one end of the safety strap or lanyard to a rigid support, and the other end to a 250 pound canvas bag of sand;

1910.268(g)(2)(ii)(G)(2)

Allow the 250 pound canvas bag of sand to free fall 4 feet when testing safety straps and 6 feet when testing lanyards. In each case, the strap or lanyard shall stop the fall of the 250 pound bag;

1910.268(g)(2)(ii)(G)(3)

Failure of the strap or lanyard shall be indicated by any breakage or slippage sufficient to permit the bag to fall free from the strap or lanyard.

1910.268(g)(2)(ii)(G)(4)

The entire "body belt assembly" shall be tested using on D ring. A safety strap or lanyard shall be used that is capable of passing the "impact loading test" described in paragraph (g)(2)(ii)(G)(2) of this section and attached as required in paragraph (g)(2)(ii)(G)(1) of this section. The body belt shall be secured to the 250 pound bag of sand at a point which simulates the waist of a man and shall be dropped as stated in paragraph (g)(2)(ii)(G)(2) of this section. Failure of the body belt shall be indicated by any breakage or slippage sufficient to permit the bag to fall free from the body belt.

1910.268(g)(3)

Pole climbers.

1910.268(g)(3)(i)

Pole climbers may not be used if the gaffs are less than 1 1/4 inches in length as measured on the underside of the gaff. The gaffs of pole climbers shall be covered with safety caps when not being used for their intended use.

1910.268(g)(3)(ii)

The employer shall ensure that pole climbers are inspected by a competent person for the following conditions: Fractured or cracked gaffs or leg irons, loose or dull gaffs, broken straps or buckles. If any of these conditions exist, the defect shall be corrected before the climbers are used.

1910.268(g)(3)(iii)

Pole climbers shall be inspected as required in this paragraph (g)(3) before each day's use and a gaff cut-out test performed at least weekly when in use.

..1910.268(g)(3)(iv)

1910.268(g)(3)(iv)

Pole climbers may not be worn when:

1910.268(g)(3)(iv)(A)

Working in trees (specifically designed tree climbers shall be used for tree climbing),

1910.268(g)(3)(iv)(B)

Working on ladders,

1910.268(g)(3)(iv)(C)

Working in an aerial lift,

1910.268(g)(3)(iv)(D)

Driving a vehicle, nor

1910.268(g)(3)(iv)(E)

Walking on rocky, hard, frozen, brushy or hilly terrain.

1910.268(h)

Ladders.

1910.268(h)(1)

The employer shall ensure that no employee nor any material or equipment may be supported or permitted to be supported on any portion of a ladder unless it is first determined, by inspections and checks conducted by a competent person that such ladder is adequately strong, in good condition, and properly secured in place, as required in Subpart D of this part and as required in this section.

1910.268(h)(2)

The spacing between steps or rungs permanently installed on poles and towers shall be no more than 18 inches (36 inches on any one side). This requirement also applies to fixed ladders on towers, when towers are so equipped. Spacing between steps shall be uniform above the initial unstepped section, except where working, standing, or access steps are required. Fixed ladder rungs and step rungs for poles and towers shall have a minimum diameter of 5/8". Fixed ladder rungs shall have a minimum clear width of 12 inches. Steps for poles and towers shall have a minimum clear width of 4 1/2 inches. The spacing between detachable steps may not exceed 30 inches on any one side, and these steps shall be properly secured when in use.

1910.268(h)(3)

Portable wood ladders intended for general use may not be painted but may be coated with a translucent nonconductive coating. Portable wood ladders may not be longitudinally reinforced with metal.

1910.268(h)(4)

Portable wood ladders that are not being carried on vehicles and are not in active use shall be stored where they will not be exposed to the elements and where there is good ventilation.

1910.268(h)(5)

The provisions of § 1910.25(c)(5) shall apply to rolling ladders used in telecommunications centers, except that such ladders shall have a minimum inside width, between the side rails, of at least eight inches.

..1910.268(h)(6)

1910.268(h)(6)

Climbing ladders or stairways on scaffolds used for access and egress shall be affixed or built into the scaffold by proper design and engineering, and shall be so located that their use will not disturb the stability of the scaffold. The rungs of the climbing device shall be equally spaced, but may not be less than 12 inches nominal nor more than 16 inches nominal apart. Horizontal end rungs used for platform support may also be utilized as a climbing device if such rungs meet the spacing requirement of this paragraph (h)(6), and if there is sufficient clearance between the rung and the edge of the platform to afford an adequate handhold. If a portable ladder is affixed to the scaffold, it shall be securely attached and shall have rungs meeting the spacing requirements of this paragraph (h)(6). Clearance shall be provided in the back of the ladder of not less than 6 inches from center of rung to the nearest scaffold structural member.

1910.268(h)(7)

When a ladder is supported by an aerial strand, and ladder hooks or other supports are not being used, the ladder shall be extended at least 2 feet above the strand and shall be secured to it (e.g. lashed or held by a safety strap around the strand and ladder side rail). When a ladder is supported by a pole, it shall be securely lashed to the pole unless the ladder is specifically designed to prevent movement when used in this application.

1910.268(h)(8)

The following requirements apply to metal manhole ladders.

1910.268(h)(8)(i)

Metal manhole ladders shall be free of structural defects and free of accident hazards such as sharp edges and burrs. The metal shall be protected against corrosion unless inherently corrosion-resistant.

1910.268(h)(8)(ii)

These ladders may be designed with parallel side rails, or with side rails varying uniformly in separation along the length (tapered), or with side rails flaring at the base to increase stability.

1910.268(h)(8)(iii)

The spacing of rungs or steps shall be on 12-inch centers.

1910.268(h)(8)(iv)

Connections between rungs or steps and siderails shall be constructed to insure rigidity as well as strength.

1910.268(h)(8)(v)

Rungs and steps shall be corrugated, knurled, dimpled, coated with skid-resistant material, or otherwise treated to minimize the possibility of slipping.

1910.268(h)(8)(vi)

Ladder hardware shall meet the strength requirements of the ladder's component parts and shall be of a material that is protected against corrosion unless inherently corrosion-resistant. Metals shall be so selected as to avoid excessive galvanic action.

1910.268(i)

Other tools and personal protective equipment—

..1910.268(i)(1)

1910.268(i)(1)

Head protection. Head protection meeting the requirements of ANSI Z89.2-1971, "Safety Requirements for Industrial Protective Helmets for Electrical Workers, Class B" shall be provided whenever there is exposure to possible high voltage electrical contact, and the employer shall ensure that the head protection is used by employees. ANSI Z89.2-1971 is incorporated by reference as specified in § 1910.6.

1910.268(i)(2)

Eye protection. Eye protection meeting the requirements of §1910.133 (a)(2) thru (a)(6) shall be provided and the employer shall ensure its use by employees where foreign objects may enter the eyes due to work operations such as but not limited to:

1910.268(i)(2)(i)

Drilling or chipping stone, brick or masonry, breaking concrete or pavement, etc. by hand tools (sledgehammer, etc.) or power tools such as pneumatic drills or hammers;

1910.268(i)(2)(ii)

Working on or around high speed emery or other grinding wheels unprotected by guards;

1910.268(i)(2)(iii)

Cutting or chipping terra cotta ducts, tile, etc.;

1910.268(i)(2)(iv)

Working under motor vehicles requiring hammering;

1910.268(i)(2)(v)

Cleaning operations using compressed air, steam, or sand blast;

..1910.268(i)(2)(vi)

1910.268(i)(2)(vi)

Acetylene welding or similar operations where sparks are thrown off;

1910.268(i)(2)(vii)

Using powder actuated stud drivers;

1910.268(i)(2)(viii)

Tree pruning or cutting underbrush;

1910.268(i)(2)(ix)

Handling battery cells and solutions, such as taking battery readings with a hydrometer and thermometer;

1910.268(i)(2)(x)

Removing or rearranging strand or open wire; and

1910.268(i)(2)(xi)

Performing lead sleeve wiping and while soldering.

1910.268(i)(3)

Tent heaters. Flame-type heaters may not be used within ground tents or on platforms within aerial tents unless:

..1910.268(i)(3)(i)
1910.268(i)(3)(i)

The tent covers are constructed of fire resistant materials, and

1910.268(i)(3)(ii)

Adequate ventilation is provided to maintain safe oxygen levels and avoid harmful buildup of combustion products and combustible gases.

1910.268(i)(4)

Torches.   Torches may be used on aerial splicing platforms or in buckets enclosed by tents provided the tent material is constructed of fire resistant material and the torch is turned off when not in actual use. Aerial tents shall be adequately ventilated while the torch is in operation.

1910.268(i)(5)

Portable power equipment.   Nominal 120V, or less, portable generators used for providing power at work locations do not require grounding if the output circuit is completely isolated from the frame of the unit.

1910.268(i)(6)

Vehicle-mounted utility generators.   Vehicle-mounted utility generators used for providing nominal 240V AC or less for powering portable tools and equipment need not be grounded to earth if all of the following conditions are met:

1910.268(i)(6)(i)

One side of the voltage source is solidly strapped to the metallic structure of the vehicle;

1910.268(i)(6)(ii)

Grounding-type outlets are used, with a "grounding" conductor between the outlet grounding terminal and the side of the voltage source that is strapped to the vehicle;

1910.268(i)(6)(iii)

All metallic encased tools and equipment that are powered from this system are equipped with three-wire cords and grounding-type attachment plugs, except as designated in paragraph (i)(7) of this section.

1910.268(i)(7)

Portable lights, tools, and appliances.   Portable lights, tools, and appliances having noncurrent-carrying external metal housing may be used with power equipment described in paragraph (i)(5) of this section without an equipment grounding conductor. When operated from commercial power such metal parts of these devices shall be grounded, unless these tools or appliances are protected by a system of double insulation, or its equivalent. Where such a system is employed, the equipment shall be distinctively marked to indicate double insulation.

1910.268(i)(8)

Soldering devices.   Grounding shall be omitted when using soldering irons, guns or wire-wrap tools on telecommunications circuits.

..1910.268(i)(9)

1910.268(i)(9)

Lead work.   The wiping of lead joints using melted solder, gas fueled torches, soldering irons or other appropriate heating devices, and the soldering of wires or other electrical connections do not constitute the welding, cutting and brazing described in Subpart Q of this part. When operated from commercial power the metal housing of electric solder pots shall be grounded. Electric solder pots may be used with the power equipment described in paragraph (i)(5) of this section without a grounding conductor. The employer shall ensure that wiping gloves or cloths and eye protection are used in lead wiping operations. A drip pan to catch hot lead drippings shall also be provided and used.

1910.268(j)

Vehicle-mounted material handling devices and other mechanical equipment—

1910.268(j)(1)

General.

1910.268(j)(1)(i)

The employer shall ensure that visual inspections are made of the equipment by a competent person each day the equipment is to be used to ascertain that it is in good condition.

1910.268(j)(1)(ii)

The employer shall ensure that tests shall be made at the beginning of each shift by a competent person to insure the vehicle brakes and operating systems are in proper working condition.

1910.268(j)(2)

Scrapers, loaders, dozers, graders and tractors.

1910.268(j)(2)(i)

All rubber-tired, self-propelled scrapers, rubber-tired front end loaders, rubber-tired dozers, agricultural and industrial tractors, crawler tractors, crawler-type loaders, and motor graders, with or without attachments, that are used in telecommunications work shall have rollover protective structures that meet the requirements of Subpart W of Part 1926 of this Title.

1910.268(j)(2)(ii)

Eye protection shall be provided and the employer shall ensure that it is used by employees when working in areas where flying material is generated.

1910.268(j)(3)

Vehicle-mounted elevating and rotating work platforms. These devices shall not be operated with any conductive part of the equipment closer to exposed energized power lines than the clearances set forth in Table R-2 of this section.

1910.268(j)(4)

Derrick trucks and similar equipment.

1910.268(j)(4)(i)

This equipment shall not be operated with any conductive part of the equipment closer to exposed energized power lines than the clearances set forth in Table R-2 of this section.

1910.268(j)(4)(ii)

When derricks are used to handle poles near energized power conductors, these operations shall comply with the requirements contained in paragraphs (b)(7) and (n)(11) of this section.

1910.268(j)(4)(iii)

Moving parts of equipment and machinery carried on or mounted on telecommunications line trucks shall be guarded. This may be done with barricades as specified in paragraph (d)(2) of this section.

..1910.268(j)(4)(iv)
1910.268(j)(4)(iv)

Derricks and the operation of derricks shall comply with the following requirements:

1910.268(j)(4)(iv)(A)

Manufacturer's specifications, load ratings and instructions for derrick operation shall be strictly observed.

1910.268(j)(4)(iv)(B)

Rated load capacities and instructions related to derrick operation shall be conspicuously posted on a permanent weather-resistant plate or decal in a location on the derrick that is plainly visible to the derrick operator.

1910.268(j)(4)(iv)(C)

Prior to derrick operation the parking brake must be set and the stabilizers extended if the vehicle is so equipped. When the vehicle is situated on a grade, at least two wheels must be chocked on the downgrade side.

1910.268(j)(4)(iv)(D)

Only persons trained in the operation of the derrick shall be permitted to operate the derrick.

1910.268(j)(4)(iv)(E)

Hand signals to derrick operators shall be those prescribed by ANSI B30.6-1969, "Safety Code for Derricks", which is incorporated by reference as specified in § 1910.6.

1910.268(j)(4)(iv)(F)

The employer shall ensure that the derrick and its associated equipment are inspected by a competent person at intervals set by the manufacturer but in no case less than once per year. Records shall be maintained including the dates of inspections, and necessary repairs made, if corrective action was required.

..1910.268(j)(4)(iv)(G)
1910.268(j)(4)(iv)(G)

Modifications or additions to the derrick and its associated equipment that alter its capacity or affect its safe operation shall be made only with written certification from the manufacturer, or other equivalent entity, such as a nationally recognized testing laboratory, that the modification results in the equipment being safe for its intended use. Such changes shall require the changing and posting of revised capacity and instruction decals or plates. These new ratings or limitations shall be as provided by the manufacturer or other equivalent entity.

1910.268(j)(4)(iv)(H)

Wire rope used with derricks shall be of improved plow steel or equivalent. Wire rope safety factors shall be in accordance with American National Standards Institute B30.6-1969.

1910.268(j)(4)(iv)(I)

Wire rope shall be taken out of service, or the defective portion removed, when any of the following conditions exist:

1910.268(j)(4)(iv)(I)(1)

The rope strength has been significantly reduced due to corrosion, pitting, or excessive heat, or

1910.268(j)(4)(iv)(I)(2)

The thickness of the outer wires of the rope has been reduced to two-thirds or less of the original thickness, or

1910.268(j)(4)(iv)(I)(3)

There are more than six broken wires in any one rope lay, or

1910.268(j)(4)(iv)(I)(4)

There is excessive permanent distortion caused by kinking, crushing, or severe twisting of the rope.

..1910.268(k)

1910.268(k)

Materials handling and storage—

1910.268(k)(1)

Poles.   When working with poles in piles or stacks, work shall be performed from the ends of the poles as much as possible, and precautions shall be taken for the safety of employees at the other end of the pole. During pole hauling operations, all loads shall be secured to prevent displacement. Lights, reflectors and/or flags shall be displayed on the end and sides of the load as necessary. The requirements for installation, removal, or other handling of poles in pole lines are prescribed in paragraph (n) of this section which pertains to overhead lines. In the case of hoisting machinery equipped with a positive stop loadholding device, it shall be permissible for the operator to leave his position at the controls (while a load is suspended) for the sole purpose of assisting in positioning the load prior to landing it. Prior to unloading steel, poles, crossarms, and similar material, the load shall be thoroughly examined to ascertain that the load has not shifted, that binders or stakes have not broken, and that the load is not otherwise hazardous to employees.

1910.268(k)(2)

Cable reels.   Cable reels in storage shall be checked or otherwise restrained when there is a possibility that they might accidentally roll from position.

1910.268(l)

Cable fault locating and testing.

1910.268(l)(1)

Employees involved in using high voltages to locate trouble or test cables shall be instructed in the precautions necessary for their own safety and the safety of other employees.

1910.268(l)(2)

Before the voltage is applied, cable conductors shall be isolated to the extent practicable. Employees shall be warned, by such techniques as briefing and tagging at all affected locations, to stay clear while the voltage is applied.

1910.268(m)

Grounding for employee protection—pole lines—

1910.268(m)(1)

Power conductors. Electric power conductors and equipment shall be considered as energized unless the employee can visually determine that they are bonded to one of the grounds listed in paragraph (m)(4) of this section.

1910.268(m)(2)

Nonworking open wire. Nonworking open wire communications lines shall be bonded to one of the grounds listed in paragraph (m)(4) of this section.

1910.268(m)(3)

Vertical power conduit, power ground wires and street light fixtures.

1910.268(m)(3)(i)

Metal power conduit on joint use poles, exposed vertical power ground wires, and street light fixtures which are below communications attachments or less than 20 inches above these attachments, shall be considered energized and shall be tested for voltage unless the employee can visually determine that they are bonded to the communications suspension strand or cable sheath.

1910.268(m)(3)(ii)

If no hazardous voltage is shown by the voltage test, a temporary bond shall be placed between such street light fixture, exposed vertical power grounding conductor, or metallic power conduit and the communications cable strand. Temporary bonds used for this purpose shall have sufficient conductivity to carry at least 500 amperes for a period of one second without fusing.

1910.268(m)(4)

Suitable protective grounding. Acceptable grounds for protective grounding are as follows:

1910.268(m)(4)(i)

A vertical ground wire which has been tested, found safe, and is connected to a power system multigrounded neutral or the grounded neutral of a power secondary system where there are at least three services connected;

1910.268(m)(4)(ii)

Communications cable sheath or shield and its supporting strand where the sheath or shield is:

..1910.268(m)(4)(ii)(A)

1910.268(m)(4)(ii)(A)

Bonded to an underground or buried cable which is connected to a central office ground, or

1910.268(m)(4)(ii)(B)

Bonded to an underground metallic piping system, or

1910.268(m)(4)(ii)(C)

Bonded to a power system multigrounded neutral or grounded neutral of a power secondary system which has at least three services connected;

1910.268(m)(4)(iii)

Guys which are bonded to the grounds specified in paragraphs (m)(4)(i) and (ii) of this section and which have continuity uninterrupted by an insulator; and

1910.268(m)(4)(iv)

If all of the preceding grounds are not available, arrays of driven ground rods where the resultant resistance to ground will be low enough to eliminate danger to personnel or permit prompt operation of protective devices.

..1910.268(m)(5)

1910.268(m)(5)

Attaching and removing temporary bonds. When attaching grounds (bonds), the first attachment shall be made to the protective ground. When removing bonds, the connection to the line or equipment shall be removed first. Insulating gloves shall be worn during these operations.

1910.268(m)(6)

Temporary grounding of suspension strand.

1910.268(m)(6)(i)

The suspension strand shall be grounded to the existing grounds listed in paragraph (m)(4) of this section when being placed on jointly used poles or during thunderstorm activity.

1910.268(m)(6)(ii)

Where power crossings are encountered on nonjoint lines, the strand shall be bonded to an existing ground listed in paragraph (m)(4) of this section as close as possible to the crossing. This bonding is not required where crossings are made on a common crossing pole unless there is an upward change in grade at the pole.

1910.268(m)(6)(iii)

Where roller-type bonds are used, they shall be restrained so as to avoid stressing the electrical connections.

1910.268(m)(6)(iv)

Bonds between the suspension strand and the existing ground shall be at least No. 6AWG copper.

1910.268(m)(6)(v)

Temporary bonds shall be left in place until the strand has been tensioned, dead-ended, and permanently grounded.

1910.268(m)(6)(vi)

The requirements of paragraphs (m)(6)(i) through (m)(6)(v) of this section do not apply to the installation of insulated strand.

1910.268(m)(7)

Antenna work-radio transmitting stations 3–30 MHZ.

1910.268(m)(7)(i)

Prior to grounding a radio transmitting station antenna, the employer shall insure that the rigger in charge:

..1910.268(m)(7)(i)(A)

1910.268(m)(7)(i)(A)

Prepares a danger tag signed with his signature,

1910.268(m)(7)(i)(B)

Requests the transmitting technician to shutdown the transmitter and to ground the antenna with its grounding switch,

1910.268(m)(7)(i)(C)

Is notified by the transmitting technician that the transmitter has been shutdown, and

1910.268(m)(7)(i)(D)

Tags the antenna ground switch personally in the presence of the transmitting technician after the antenna has been grounded by the transmitting technician.

1910.268(m)(7)(ii)

Power shall not be applied to the antenna, nor shall the grounding switch be opened under any circumstances while the tag is affixed.

1910.268(m)(7)(iii)(A)

Where no grounding switches are provided, grounding sticks shall be used, one on each side of line, and tags shall be placed on the grounding sticks, antenna switch, or plate power switch in a conspicuous place.

1910.268(m)(7)(iii)(B)

When necessary to further reduce excessive radio frequency pickup, ground sticks or short circuits shall be placed directly on the transmission lines near the transmitter in addition to the regular grounding switches.

1910.268(m)(7)(iii)(C)

In other cases, the antenna lines may be disconnected from ground and the transmitter to reduce pickup at the point in the field.

1910.268(m)(7)(iv)

All radio frequency line wires shall be tested for pickup with an insulated probe before they are handled either with bare hands or with metal tools.

1910.268(m)(7)(v)

The employer shall insure that the transmitting technician warn the riggers about adjacent lines which are, or may become energized.

1910.268(m)(7)(vi)

The employer shall insure that when antenna work has been completed, the rigger in charge of the job returns to the transmitter, notifies the transmitting technician in charge that work has been completed, and personally removes the tag from the antenna ground switch.

1910.268(n)

Overhead lines—

1910.268(n)(1)

Handling suspension strand.

..1910.268(n)(1)(i)
1910.268(n)(1)(i)

The employer shall insure that when handling cable suspension strand which is being installed on poles carrying exposed energized power conductors, employees shall wear insulating gloves and shall avoid body contact with the strand until after it has been tensioned, dead-ended and permanently grounded.

1910.268(n)(1)(ii)

The strand shall be restrained against upward movement during installation:

1910.268(n)(1)(ii)(A)

On joint-use poles, where there is an upward change in grade at the pole, and

1910.268(n)(1)(ii)(B)

On non-joint-use poles, where the line crosses under energized power conductors.

1910.268(n)(2)

Need for testing wood poles. Unless temporary guys or braces are attached, the following poles shall be tested in accordance with paragraph (n)(3) of this section and determined to be safe before employees are permitted to climb them:

1910.268(n)(2)(i)

Dead-end poles, except properly braced or guyed "Y" or "T" cable junction poles,

1910.268(n)(2)(ii)

Straight line poles which are not storm guyed and where adjacent span lengths exceed 165 feet,

1910.268(n)(2)(iii)

Poles at which there is a downward change in grade and which are not guyed or braced corner poles or cable junction poles,

1910.268(n)(2)(iv)

Poles which support only telephone drop wire, and

1910.268(n)(2)(v)

Poles which carry less than ten communication line wires. On joint use poles, one power line wire shall be considered as two communication wires for purposes of this paragraph (n)(2)(v).

1910.268(n)(3)

Methods for testing wood poles. One of the following methods or an equivalent method shall be used for testing wood poles:

1910.268(n)(3)(i)

Rap the pole sharply with a hammer weighing about 3 pounds, starting near the ground line and continuing upwards circumferentially around the pole to a height of approximately 6 feet. The hammer will produce a clear sound and rebound sharply when striking sound wood. Decay pockets will be indicated by a dull sound and/or a less pronounced hammer rebound. When decay pockets are indicated, the pole shall be considered unsafe. Also, prod the pole as near the ground line as possible using a pole prod or a screwdriver with a blade at least 5 inches long. If substantial decay is encountered, the pole shall be considered unsafe.

..1910.268(n)(3)(ii)

1910.268(n)(3)(ii)

Apply a horizontal force to the pole and attempt to rock it back and forth in a direction perpendicular to the line. Caution shall be exercised to avoid causing power wires to swing together. The force may be applied either by pushing with a pike pole or pulling with a rope. If the pole cracks during the test, it shall be considered unsafe.

1910.268(n)(4)

Unsafe poles or structures.   Poles or structures determined to be unsafe by test or observation may not be climbed until made safe by guying, bracing or other adequate means. Poles determined to be unsafe to climb shall, until they are made safe, be tagged in a conspicuous place to alert and warn all employees of the unsafe condition.

1910.268(n)(5)

Test requirements for cable suspension strand.

1910.268(n)(5)(i)

Before attaching a splicing platform to a cable suspension strand, the strand shall be tested and determined to have strength sufficient to support the weight of the platform and the employee. Where the strand crosses above power wires or railroad tracks it may not be tested but shall be inspected in accordance with paragraph (n)(6) of this section.

1910.268(n)(5)(ii)

The following method or an equivalent method shall be used for testing the strength of the strand: A rope, at least three-eighths inch in diameter, shall be thrown over the strand. On joint lines, the rope shall be passed over the strand using tree pruner handles or a wire raising tool. If two employees are present, both shall grip the double rope and slowly transfer their entire weight to the rope and attempt to raise themselves off the ground. If only one employee is present, one end of the rope which has been passed over the strand shall be tied o the bumper of the truck, or other equally secure anchorage. The employee then shall grasp the other end of the rope and attempt to raise himself off the ground.

..1910.268(n)(6)

1910.268(n)(6)

Inspection of strand. Where strand passes over electric power wires or railroad tracks, it shall be inspected from an elevated working position at each pole supporting the span in question. The strand may not be used to support any splicing platform, scaffold or cable car, if any of the following conditions exist:

1910.268(n)(6)(i)

Corrosion so that no galvanizing can be detected,

1910.268(n)(6)(ii)

One or more wires of the strand are broken,

1910.268(n)(6)(iii)

Worn spots, or

1910.268(n)(6)(iv)

Burn marks such as those caused by contact with electric power wires.

1910.268(n)(7)

Outside work platforms.   Unless adequate railings are provided, safety straps and body belts shall be used while working on elevated work platforms such as aerial splicing platforms, pole platforms, ladder platforms and terminal balconies.

1910.268(n)(8)

Other elevated locations.   Safety straps and body belts shall be worn when working at elevated positions on poles, towers or similar structures, which do not have adequately guarded work areas.

..1910.268(n)(9)

1910.268(n)(9)

Installing and removing wire and cable.   Before installing or removing wire or cable, the pole or structure shall be guyed, braced, or otherwise supported, as necessary, to prevent failure of the pole or structure.

1910.268(n)(10)

Avoiding contact with energized power conductors or equipment.   When cranes, derricks, or other mechanized equipment are used for setting, moving, or removing poles, all necessary precautions shall be taken to avoid contact with energized power conductors or equipment.

1910.268(n)(11)

Handling poles near energized power conductors.

1910.268(n)(11)(i)

Joint use poles may not be set, moved, or removed where the nominal voltage of open electrical power conductors exceeds 34.5kV phase to phase (20kV to ground).

1910.268(n)(11)(ii)

Poles that are to be placed, moved or removed during heavy rains, sleet or wet snow in joint lines carrying more than 8.7kV phase to phase voltage (5kV to ground) shall be guarded or otherwise prevented from direct contact with overhead energized power conductors.

1910.268(n)(11)(iii)(A)

In joint lines where the power voltage is greater than 750 volts but less than 34.5kV phase to phase (20 kV to ground), wet poles being placed, moved or removed shall be insulated with either a rubber insulating blanket, a fiberglass box guide, or equivalent protective equipment.

1910.268(n)(11)(iii)(B)

In joint lines where the power voltage is greater than 8.7 kV phase to phase (5kV to ground) but less than 34.5kV phase to phase (20 kV to ground), dry poles being placed, moved, or removed shall be insulated with either a rubber insulating blanket, a fiberglass box guide, or equivalent protective equipment.

1910.268(n)(11)(iii)(C)

Where wet or dry poles are being removed, insulation of the pole is not required if the pole is cut off 2 feet or more below the lowest power wire and also cut off near the ground line.

1910.268(n)(11)(iv)

Insulating gloves shall be worn when handling the pole with either hands or tools, when there exists a possibility that the pole may contact a power conductor. Where the voltage to ground of the power conductor exceeds 15kV to ground, Class II gloves (as defined in ANSI J6.6-1971) shall be used. For voltages not exceeding 15kV to ground, insulating gloves shall have a breakdown voltage of at least 17kV.

1910.268(n)(11)(v)

The guard or insulating material used to protect the pole shall meet the appropriate 3 minute proof test voltage requirements contained in the ANSI J6.4-1971.

..1910.268(n)(11)(vi)
1910.268(n)(11)(vi)

When there exists a possibility of contact between the pole or the vehicle-mounted equipment used to handle the pole, and an energized power conductor, the following precautions shall be observed:

1910.268(n)(11)(vi)(A)

When on the vehicle which carries the derrick, avoid all contact with the ground, with persons standing on the ground, and with all grounded objects such as guys, tree limbs, or metal sign posts. To the extent feasible, remain on the vehicle as long as the possibility of contact exists.

1910.268(n)(11)(vi)(B)

When it is necessary to leave the vehicle, step onto an insulating blanket and break all contact with the vehicle before stepping off the blanket and onto the ground. As a last resort, if a blanket is not available, the employee may jump cleanly from the vehicle.

1910.268(n)(11)(vi)(C)

When it is necessary to enter the vehicle, first step onto an insulating blanket and break all contact with the ground, grounded objects and other persons before touching the truck or derrick.

1910.268(n)(12)

Working position on poles.   Climbing and working are prohibited above the level of the lowest electric power conducter on the pole (exclusive of vertical runs and street light wiring), except:

1910.268(n)(12)(i)

Where communications facilities are attached above the electric power conductors, and a rigid fixed barrier is installed between the electric power facility and the communications facility, or

1910.268(n)(12)(ii)

Where the electric power conductors are cabled secondary service drops carrying less than 300 volts to ground and are attached 40 inches or more below the communications conductors or cables.

1910.268(n)(13)

Metal tapes and ropes.

1910.268(n)(13)(i)

Metal measuring tapes, metal measuring ropes, or tapes containing conductive strands may not be used when working near exposed energized parts.

1910.268(n)(13)(ii)

Where it is necessary to measure clearances from energized parts, only nonconductive devices shall be used.

1910.268(o)

Underground lines.   The provisions of this paragraph apply to the guarding of manholes and street openings, and to the ventilation and testing for gas in manholes and unvented vaults, where telecommunications field work is performed on or with underground lines.

1910.268(o)(1)

Guarding manholes and street openings.

1910.268(o)(1)(i)

When covers of manholes or vaults are removed, the opening shall be promptly guarded by a railing, temporary cover, or other suitable temporary barrier which is appropriate to prevent an accidental fall through the opening and to protect employees working in the manhole from foreign objects entering the manhole.

..1910.268(o)(1)(ii)

1910.268(o)(1)(ii)

While work is being performed in the manhole, a person with basic first aid training shall be immediately available to render assistance if there is cause for believing that a safety hazard exists, and if the requirements contained in paragraphs (d)(1) and (o)(1)(i) of this section do not adequately protect the employee(s). Examples of manhole worksite hazards which shall be considered to constitute a safety hazard include, but are not limited to:

1910.268(o)(1)(ii)(A)

Manhole worksites where safety hazards are created by traffic patterns that cannot be corrected by provisions of paragraph (d)(1) of this section.

1910.268(o)(1)(ii)(B)

Manhole worksites that are subject to unusual water hazards that cannot be abated by conventional means.

1910.268(o)(1)(ii)(C)

Manhole worksites that are occupied jointly with power utilities as described in paragraph (o)(3) of this section.

1910.268(o)(2)

Requirements prior to entering manholes and unvented vaults.

1910.268(o)(2)(i)

Before an employee enters a manhole, the following steps shall be taken:

1910.268(o)(2)(i)(A)

The internal atmosphere shall be tested for combustible gas and, except when continuous forced ventilation is provided, the atmosphere shall also be tested for oxygen deficiency.

1910.268(o)(2)(i)(B)

When unsafe conditions are detected by testing or other means, the work area shall be ventilated and otherwise made safe before entry.

..1910.268(o)(2)(ii)

1910.268(o)(2)(ii)

An adequate continuous supply of air shall be provided while work is performed in manholes under any of the following conditions:

1910.268(o)(2)(ii)(A)

Where combustible or explosive gas vapors have been initially detected and subsequently reduced to a safe level by ventilation,

1910.268(o)(2)(ii)(B)

Where organic solvents are used in the work procedure,

1910.268(o)(2)(ii)(C)

Where open flame torches are used in the work procedure,

1910.268(o)(2)(ii)(D)

Where the manhole is located in that portion of a public right of way open to vehicular traffic and/or exposed to a seepage of gas or gases, or

1910.268(o)(2)(ii)(E)

Where a toxic gas or oxygen deficiency is found.

1910.268(o)(2)(iii)(A)

The requirements of paragraphs (o)(2)(i) and (ii) of this section do not apply to work in central office cable vaults that are adequately ventilated.

1910.268(o)(2)(iii)(B)

The requirements of paragraphs (o)(2)(i) and (ii) of this section apply to work in unvented vaults.

1910.268(o)(3)

Joint power and telecommunication manholes. While work is being performed in a manhole occupied jointly by an electric utility and a telecommunication utility, an employee with basic first aid training shall be available in the immediate vicinity to render emergency assistance as may be required. The employee whose presence is required in the immediate vicinity for the purposes of rendering emergency assistance is not to be precluded from occasionally entering a manhole to provide assistance other than in an emergency. The requirement of this paragraph (o)(3) does not preclude a qualified employee, working alone, from entering for brief periods of time, a manhole where energized cables or equipment are in service, for the purpose of inspection, housekeeping, taking readings, or similar work if such work can be performed safely.

1910.268(o)(4)

Ladders.   Ladders shall be used to enter and exit manholes exceeding 4 feet in depth.

1910.268(o)(5)

Flames.   When open flames are used in manholes, the following precautions shall be taken to protect against the accumulation of combustible gas:

1910.268(o)(5)(i)

A test for combustible gas shall be made immediately before using the open flame device, and at least once per hour while using the device; and

1910.268(o)(5)(ii)

a fuel tank (e.g., acetylene) may not be in the manhole unless in actual use.

..*1910.268(p)*

1910.268(p)

Microwave transmission—

1910.268(p)(1)

Eye protection.   Employers shall insure that employees do not look into an open waveguide which is connected to an energized source of microwave radiation.

1910.268(p)(2)

Hazardous area.   Accessible areas associated with microwave communication systems where the electromagnetic radiation level exceeds the radiation protection guide given in § 1910.97 shall be posted as described in that section. The lower half of the warning symbol shall include the following:

Radiation in this area may exceed hazard limitations and special precautions are required. Obtain specific instruction before entering.

1910.268(p)(3)

Protective measures.   When an employee works in an area where the electromagnetic radiation exceeds the radiation protection guide, the employer shall institute measures that insure that the employee's exposure is not greater than that permitted by the radiation guide. Such measures shall include, but not be limited to those of an administrative or engineering nature or those involving personal protective equipment.

1910.268(q)

Tree trimming—electrical hazards—

..1910.268(q)(1)

1910.268(q)(1)

General.

1910.268(q)(1)(i)

Employees engaged in pruning, trimming, removing, or clearing trees from lines shall be required to consider all overhead and underground electrical power conductors to be energized with potentially fatal voltages, never to be touched (contacted) either directly or indirectly.

1910.268(q)(1)(ii)

Employees engaged in line-clearing operations shall be instructed that:

1910.268(q)(1)(ii)(A)

A direct contact is made when any part of the body touches or contacts an energized conductor, or other energized electrical fixture or apparatus.

1910.268(q)(1)(ii)(B)

An indirect contact is made when any part of the body touches any object in contact with an energized electrical conductor, or other energized fixture or apparatus.

1910.268(q)(1)(ii)(C)

An indirect contact can be made through conductive tools, tree branches, trucks, equipment, or other objects, or as a result of communications wires, cables, fences, or guy wires being accidentally energized.

1910.268(q)(1)(ii)(D)

Electric shock will occur when an employee, by either direct or indirect contact with an energized conductor, energized tree limb, tool, equipment, or other object, provides a path for the flow of electricity to a grounded object or to the ground itself. Simultaneous contact with two energized conductors will also cause electric shock which may result in serious or fatal injury.

1910.268(q)(1)(iii)

Before any work is performed in proximity to energized conductors, the system operator/owner of the energized conductors shall be contacted to ascertain if he knows of any hazards associated with the conductors which may not be readily apparent. This rule does not apply when operations are performed by or on behalf of, the system operator/owner.

1910.268(q)(2)

Working in proximity to electrical hazards.

1910.268(q)(2)(i)

Employers shall ensure that a close inspection is made by the employee and by the foremen or supervisor in charge before climbing, entering, or working around any tree, to determine whether an electrical power conductor passes through the tree, or passes within reaching distance of an employee working in the tree. If any of these conditions exist either directly or indirectly, an electrical hazard shall be considered to exist unless the system operator/owner has caused the hazard to be removed by deenergizing the lines, or installing protective equipment.

..1910.268(q)(2)(ii)

1910.268(q)(2)(ii)

Only qualified employees or trainees, familiar with the special techniques and hazards involved in line clearance, shall be permitted to perform the work if it is found that an electrical hazard exists.

1910.268(q)(2)(iii)

During all tree working operations aloft where an electrical hazard of more than 750V exists, there shall be a second employee or trainee qualified in line clearance tree trimming within normal voice communication.

1910.268(q)(2)(iv)

Where tree work is performed by employees qualified in line-clearance tree trimming and trainees qualified in line-clearance tree trimming, the clearances from energized conductors given in Table R-3 shall apply.

1910.268(q)(2)(v)

Branches hanging on an energized conductor may only be removed using appropriately insulated equipment.

Table R-3 **Minimum Working Distances from Energized Conductors for Line-Clearance Tree Trimmers and Line-Clearance Tree-Trimmer Trainees**

Voltage range (phase to phase)(kilovolts)	Minimum working distance
2.1 to 15.0	2 ft. 0 in.
15.1 to 35.0	2 ft. 4 in.
35.1 to 46.0	2 ft. 6 in.
46.1 to 72.5	3 ft. 0 in.
72.6 to 121.0	3 ft. 4 in.
138.0 to 145.0	3 ft. 6 in.
161.0 to 169.0	3 ft. 8 in.
230.0 to 242.0	5 ft. 0 in.
345.0 to 362.0	7 ft. 0 in.
500.0 to 552.0	11 ft. 0 in.
700.0 to 765.0	15 ft. 0 in.

1910.268(q)(2)(vi)

Rubber footwear, including lineman's overshoes, shall not be considered as providing any measure of safety from electrical hazards.

1910.268(q)(2)(vii)

Ladders, platforms, and aerial devices, including insulated aerial devices, may not be brought in contact with an electrical conductor. Reliance shall not be placed on their dielectric capabilities.

..1910.268(q)(2)(viii)

1910.268(q)(2)(viii)

When an aerial lift device contacts an electrical conductor, the truck supporting the aerial lift device shall be considered as energized.

1910.268(q)(3)

Storm work and emergency conditions.

1910.268(q)(3)(i)

Since storm work and emergency conditions create special hazards, only authorized representatives of the electric utility system operator/owner and not telecommunication workers may perform tree work in these situations where energized electrical power conductors are involved.

1910.268(q)(3)(ii)

When an emergency condition develops due to tree operations, work shall be suspended and the system operator/owner shall be notified immediately.

1910.268(r)

**Buried facilities—Communications lines and power
lines in the same trench** [Reserved]

1910.268(s)

Definitions—

1910.268(s)(1)

Aerial lifts. Aerial lifts include the following types of vehicle-mounted aerial devices used to elevate personnel to jobsites above ground:

1910.268(s)(1)(i)

Extensible boom platforms,

1910.268(s)(1)(ii)

Aerial ladders,

1910.268(s)(1)(iii)

Articulating boom platforms,

..1910.268(s)(1)(iv)

1910.268(s)(1)(iv)

Vertical towers,

1910.268(s)(1)(v)

A combination of any of the above defined in ANSI A92.2-1969, which is incorporated by reference as specified in § 1910.6. These devices are made of metal, wood, fiberglass reinforced plastic (FRP), or other material; are powered or manually operated; and are deemed to be aerial lifts whether or not they are capable of rotating about a substantially vertical axis.

1910.268(s)(2)

Aerial splicing platform.   This consists of a platform, approximately 3 ft. × 4 ft., used to perform aerial cable work. It is furnished with fiber or synthetic ropes for supporting the platform from aerial strand, detachable guy ropes for anchoring it, and a device for raising and lowering it with a handline.

1910.268(s)(3)

Aerial tent.   A small tent usually constructed of vinyl coated canvas which is usually supported by light metal or plastic tubing. It is designed to protect employees in inclement weather while working on ladders, aerial splicing platforms, or aerial devices.

1910.268(s)(4)

Alive or live (energized).   Electrically connected to a source of potential difference, or electrically charged so as to have a potential significantly different from that of the earth in the vicinity. The term "live" is sometimes used in the place of the term "current-carrying," where the intent is clear, to avoid repetition of the longer term.

1910.268(s)(5)

Barricade.   A physical obstruction such as tapes, cones, or "A" frame type wood and/or metal structure intended to warn and limit access to a work area.

1910.268(s)(6)

Barrier.   A physical obstruction which is intended to prevent contact with energized lines or equipment, or to prevent unauthorized access to work area.

1910.268(s)(7)

Bond.   An electrical connection from one conductive element to another for the purpose of minimizing potential differences or providing suitable conductivity for fault current or for mitigation of leakage current and electrolytic action.

1910.268(s)(8)

Cable.   A conductor with insulation, or a stranded conductor with or without insulation and other coverings (single-conductor cable), or a combination of conductors insulated from one another (multiple-conductor cable).

1910.268(s)(9)

Cable sheath.   A protective covering applied to cables.

Note: A cable sheath may consist of multiple layers of which one or more is conductive.

1910.268(s)(10)

Circuit.   A conductor or system of conductors through which an electric current is intended to flow.

..1910.268(s)(11)

1910.268(s)(11)

Communication lines.   The conductors and their supporting or containing structures for telephone, telegraph, railroad signal, data, clock, fire, police-alarm, community television antenna and other systems which are used for public or private signal or communication service, and which operate at potentials not exceeding 400 volts to ground or 750 volts between any two points of the circuit, and the transmitted power of which does not exceed 150 watts. When communications lines operate at less than 150 volts to ground, no limit is placed on the capacity of the system. Specifically designed communications cables may include communication circuits not complying with the preceding limitations, where such circuits are also used incidentally to supply power to communication equipment.

1910.268(s)(12)

Conductor.   A material, usually in the form of a wire, cable, or bus bar, suitable for carrying an electric current.

1910.268(s)(13)

Effectively grounded.   Intentionally connected to earth through a ground connection or connections of sufficiently low impedance and having sufficient current-carrying capacity to prevent the build-up of voltages which may result in undue hazard to connected equipment or to persons.

1910.268(s)(14)

Equipment.   A general term which includes materials, fittings, devices, appliances, fixtures, apparatus, and similar items used as part of, or in connection with, a supply or communications installation.

1910.268(s)(15)

Ground (reference).   That conductive body, usually earth, to which an electric potential is referenced.

1910.268(s)(16)

Ground (as a noun).   A conductive connection, whether intentional or accidental, by which an electric circuit or equipment is connected to reference ground.

1910.268(s)(17)

Ground (as a verb).   The connecting or establishment of a connection, whether by intention or accident, of an electric circuit or equipment to reference ground.

1910.268(s)(18)

Ground tent.   A small tent usually constructed of vinyl coated canvas supported by a metal or plastic frame. Its purpose is to protect employees from inclement weather while working at buried cable pedestal sites or similar locations.

1910.268(s)(19)

Grounded conductor.   A system or circuit conductor which is intentionally grounded.

..1910.268(s)(20)

1910.268(s)(20)

Grounded systems.   A system of conductors in which at least one conductor or point (usually the middle wire, or the neutral point of transformer or generator windings) is intentionally grounded, either solidly or through a current-limiting device (not a current-interrupting device).

1910.268(s)(21)

Grounding electrode conductor. (Grounding conductor). A conductor used to connect equipment or the grounded circuit of a wiring system to a grounding electrode.

1910.268(s)(22)

Insulated.   Separated from other conducting surfaces by a dielectric substance (including air space) offering a high resistance to the passage of current.

Note: When any object is said to be insulated, it is understood to be insulated in suitable manner for the conditions to which it is subjected. Otherwise, it is, within the purpose of these rules, uninsulated. Insulating coverings of conductors in one means of making the conductor insulated.

1910.268(s)(23)

Insulation (as applied to cable).   That which is relied upon to insulate the conductor from other conductors or conducting parts or from ground.

1910.268(s)(24)

Joint use.   The sharing of a common facility, such as a manhole, trench or pole, by two or more different kinds of utilities (e.g., power and telecommunications).

1910.268(s)(25)

Ladder platform.   A device designed to facilitate working aloft from an extension ladder. A typical device consists of a platform (approximately 9" × 18") hinged to a welded pipe frame. The rear edge of the platform and the bottom cross-member of the frame are equipped with latches to lock the platform to ladder rungs.

1910.268(s)(26)

Ladder seat. A removable seat used to facilitate work at an elevated position on rolling ladders in telecommunication centers.

1910.268(s)(27)

Manhole. A subsurface enclosure which personnel may enter and which is used for the purpose of installing, operating, and maintaining submersible equipment and/or cable.

1910.268(s)(28)

Manhole platform. A platform consisting of separate planks which are laid across steel platform supports. The ends of the supports are engaged in the manhole cable racks.

1910.268(s)(29)

Microwave transmission. The act of communicating or signaling utilizing a frequency between 1 GHz (gigahertz) and 300 GHz inclusively.

..1910.268(s)(30)

1910.268(s)(30)

Nominal voltage. The nominal voltage of a system or circuit is the value assigned to a system or circuit of a given voltage class for the purpose of convenient designation. The actual voltage may vary above or below this value.

1910.268(s)(31)

Pole balcony or seat. A balcony or seat used as a support for workmen at pole-mounted equipment or terminal boxes. A typical device consists of a bolted assembly of steel details and a wooden platform. Steel braces run from the pole to the underside of the balcony. A guard rail (approximately 30" high) may be provided.

1910.268(s)(32)

Pole platform. A platform intended for use by a workman in splicing and maintenance operations in an elevated position adjacent to a pole. It consists of a platform equipped at one end with a hinged chain binder for securing the platform to a pole. A brace from the pole to the underside of the platform is also provided.

1910.268(s)(33)

Qualified employee. Any worker who by reason of his training and experience has demonstrated his ability to safely perform his duties.

1910.268(s)(34)

Qualified line-clearance tree trimmer. A tree worker who through related training and on-the-job experience is familiar with the special techniques and hazards involved in line clearance.

1910.268(s)(35)

Qualified line-clearance tree-trimmer trainee. Any worker regularly assigned to a line-clearance tree-trimming crew and undergoing on-the-job training who, in the course of such training, has demonstrated his ability to perform his duties safely at his level of training.

1910.268(s)(36)

System operator/owner. The person or organization that operates or controls the electrical conductors involved.

..1910.268(s)(37)
1910.268(s)(37)

Telecommunications center. An installation of communication equipment under the exclusive control of an organization providing telecommunications service, that is located outdoors or in a vault, chamber, or a building space used primarily for such installations.

Note: Telecommunication centers are facilities established, equipped and arranged in accordance with engineered plans for the purpose of providing telecommunications service. They may be located on premises owned or leased by the organization providing telecommunication service, or on the premises owned or leased by others. This definition includes switch rooms (whether electromechanical, electronic, or computer controlled), terminal rooms, power rooms, repeater rooms, transmitter and receiver rooms, switchboard operating rooms, cable vaults, and miscellaneous communications equipment rooms. Simulation rooms of telecommunication centers for training or developmental purposes are also included.

1910.268(s)(38)

Telecommunications derricks. Rotating or nonrotating derrick structures permanently mounted on vehicles for the purpose of lifting, lowering, or positioning hardware and materials used in telecommunications work.

1910.268(s)(39)

Telecommunication line truck. A truck used to transport men, tools, and material, and to serve as a traveling workshop for telecommunication installation and maintenance work. It is sometimes equipped with a boom and auxiliary equipment for setting poles, digging holes, and elevating material or men.

1910.268(s)(40)

Telecommunication service. The furnishing of a capability to signal or communicate at a distance by means such as telephone, telegraph, police and firealarm, community antenna television, or similar system, using wire, conventional cable, coaxial cable, wave guides, microwave transmission, or other similar means.

1910.268(s)(41)

Unvented vault. An enclosed vault in which the only openings are access openings.

1910.268(s)(42)

Vault.   An enclosure above or below ground which personnel may enter, and which is used for the purpose of installing, operating, and/or maintaining equipment and/or cable which need not be of submersible design.

..1910.268(s)(43)

1910.268(s)(43)

Vented vault.   An enclosure as described in paragraph(s) (42) of this section, with provision for air changes using exhaust flue stack(s) and low level air intake(s), operating on differentials of pressure and temperature providing for air flow.

1910.268(s)(44)

Voltage of an effectively grounded circuit.   The voltage between any conductor and ground unless otherwise indicated.

1910.268(s)(45)

Voltage of a circuit not effectively grounded. The voltage between any two conductors. If one circuit is directly connected to and supplied from another circuit of higher voltage (as in the case of an autotransformer), both are considered as of the higher voltage, unless the circuit of lower voltage is effectively grounded, in which case its voltage is not determined by the circuit of higher voltage. Direct connection implies electric connection as distinguished from connection merely through electromagnetic or electrostatic induction.

[40 FR 13441, Mar. 26, 1975, as amended at 43 FR 49751, Oct. 24, 1978; 47 FR 14706, Apr. 6, 1982; 52 FR 36387, Sept. 28, 1987; 54 FR 24334, June 7, 1989; 61 FR 9227, March 7, 1996; 63 FR 33450, June 18, 1998; 67 FR 67965, Nov. 7, 2002; 69 FR 31882, June 8, 2004; 70 FR 1141, Jan. 5, 2005]

Reprinted from www.osha.gov

Occupational Safety & Health Administration
200 Constitution Avenue, NW
Washington, DC 20210

1910.269 (Includes Appendix A–G)

Electric Power Generation, Transmission, and Distribution (Operation and Maintenance) 1910 Subpart R—Special Industries

OSHA 1910.269 Paragraph Titles:

- (a) General
- (b) Medical services and first aid
- (c) Job briefing
- (d) Hazardous energy control (lockout/tagout) procedures
- (e) Enclosed spaces
- (f) Excavations
- (g) Personal protective equipment
- (h) Portable ladders and platforms
- (i) Hand and portable power equipment
- (j) Live-line tools
- (k) Materials handling and storage
- (l) Working on or near exposed energized parts
- (m) Deenergizing lines and equipment for employee protection
- (n) Grounding for the protection of employees
- (o) Testing and test facilities
- (p) Mechanical equipment
- (q) Overhead lines and live-line barehand work

(r) Line-clearance tree trimming
(s) Communication facilities
(t) Underground electrical installations
(u) Substations
(v) Power generation
(w) Special conditions
(x) Definitions

Appendix A – Flow Charts
Appendix B – Working on Exposed Energized Parts
Appendix C – Protection from Hazardous Differences in Electric Potential
Appendix D – Methods of Inspecting and Testing Wood Poles
Appendix E – Protection from Flames and Electric Arcs
Appendix F – Work-Positioning Equipment Inspection Guidelines
Appendix G – Reference Documents

ELECTRIC POWER GENERATION, TRANSMISSION, AND DISTRIBUTION – OSHA 1910.269

- **Part Number:** 1910
- **Part Title:** Occupational Safety and Health Standards
- **Subpart:** R
- **Subpart Title:** Special Industries
- **Standard Number:** 1910.269
- **Title:** Electric Power Generation, Transmission, and Distribution.
- **Appendix:** A, B, C, D, E, F, G

1910.269(a)

General

1910.269(a)(1)

Application.

1910.269(a)(1)(i)

This section covers the operation and maintenance of electric power generation, control, transformation, transmission, and distribution lines and equipment. These provisions apply to:

1910.269(a)(1)(i)(A)

Power generation, transmission, and distribution installations, including related equipment for the purpose of communication or metering that are accessible only to qualified employees;

Note to paragraph (a)(1)(i)(A):   The types of installations covered by this paragraph include the generation, transmission, and distribution installations of electric utilities, as well as equivalent installations of industrial establishments. Subpart S of this part covers supplementary electric generating equipment that is used to supply a workplace for emergency, standby, or similar purposes only. (See paragraph (a)(1)(i)(B) of this section.)

1910.269(a)(1)(i)(B)

Other installations at an electric power generating station, as follows:

1910.269(a)(1)(i)(B)(1)

Fuel and ash handling and processing installations, such as coal conveyors,

1910.269(a)(1)(i)(B)(2)

Water and steam installations, such as penstocks, pipelines, and tanks, providing a source of energy for electric generators, and

1910.269(a)(1)(i)(B)(3)

Chlorine and hydrogen systems;

1910.269(a)(1)(i)(C)

Test sites where employees perform electrical testing involving temporary measurements associated with electric power generation, transmission, and distribution in laboratories, in the field, in substations, and on lines, as opposed to metering, relaying, and routine line work;

1910.269(a)(1)(i)(D)

Work on, or directly associated with, the installations covered in paragraphs (a)(1)(i)(A) through (a)(1)(i)(C) of this section; and

1910.269(a)(1)(i)(E)

Line-clearance tree-trimming operations, as follows:

1910.269(a)(1)(i)(E)(1)

Entire § 1910.269 of this part, except paragraph (r)(1) of this section, applies to line-clearance tree-trimming operations performed by qualified employees (those who are knowledgeable in the construction and operation of the electric power generation, transmission, or distribution equipment involved, along with the associated hazards).

1910.269(a)(1)(i)(E)(2)

Paragraphs (a)(2), (a)(3), (b), (c), (g), (k), (p), and (r) of this section apply to line-clearance tree-trimming operations performed by line-clearance tree trimmers who are not qualified employees.

1910.269(a)(1)(ii)

Notwithstanding paragraph (a)(1)(i) of this section, § 1910.269 of this part does not apply:

1910.269(a)(1)(ii)(A)

To construction work, as defined in § 1910.12 of this part, except for line-clearance tree-trimming operations and work involving electric power generation installations as specified in § 1926.950(a)(3) of this chapter; or

1910.269(a)(1)(ii)(B)

To electrical installations, electrical safety-related work practices, or electrical maintenance considerations covered by Subpart S of this part.

Note 1 to paragraph (a)(1)(ii)(B): The Occupational Safety and Health Administration considers work practices conforming to §§ 1910.332 through 1910.335 as complying with the electrical safety-related work-practice requirements of § 1910.269 identified in Table 1 of Appendix A-2 to this section, provided that employers are performing the work on a generation or distribution installation meeting §§ 1910.303 through 1910.308. This table also identifies provisions in § 1910.269 that apply to work by qualified persons directly on, or associated with, installations of electric power generation, transmission, and distribution lines or equipment, regardless of compliance with §§ 1910.332 through 1910.335.

Note 2 to paragraph (a)(1)(ii)(B):   The Occupational Safety and Health Administration considers work practices performed by qualified persons and conforming to § 1910.269 as complying with §§ 1910.333(c) and 1910.335.

1910.269(a)(1)(iii)

This section applies in addition to all other applicable standards contained in this Part 1910. Employers covered under this section are not exempt from complying with other applicable provisions in Part 1910 by the operation of § 1910.5(c). Specific references in this section to other sections of Part 1910 are for emphasis only.

1910.269(a)(2)

Training.

1910.269(a)(2)(i)

All employees performing work covered by this section shall be trained as follows:

1910.269(a)(2)(i)(A)

Each employee shall be trained in, and familiar with, the safety-related work practices, safety procedures, and other safety requirements in this section that pertain to his or her job assignments.

1910.269(a)(2)(i)(B)

Each employee shall also be trained in and familiar with any other safety practices, including applicable emergency procedures (such as pole-top and manhole rescue), that are not specifically addressed by this section but that are related to his or her work and are necessary for his or her safety.

1910.269(a)(2)(i)(C)

The degree of training shall be determined by the risk to the employee for the hazard involved.

1910.269(a)(2)(ii)

Each qualified employee shall also be trained and competent in:

1910.269(a)(2)(ii)(A)

The skills and techniques necessary to distinguish exposed live parts from other parts of electric equipment,

1910.269(a)(2)(ii)(B)

The skills and techniques necessary to determine the nominal voltage of exposed live parts,

1910.269(a)(2)(ii)(C)

The minimum approach distances specified in this section corresponding to the voltages to which the qualified employee will be exposed and the skills and techniques necessary to maintain those distances,

1910.269(a)(2)(ii)(D)

The proper use of the special precautionary techniques, personal protective equipment, insulating and shielding materials, and insulated tools for working on or near exposed energized parts of electric equipment, and

1910.269(a)(2)(ii)(E)

The recognition of electrical hazards to which the employee may be exposed and the skills and techniques necessary to control or avoid these hazards.

Note to paragraph (a)(2)(ii): For the purposes of this section, a person must have the training required by paragraph (a)(2)(ii) of this section to be considered a qualified person.

1910.269(a)(2)(iii)

Each line-clearance tree trimmer who is not a qualified employee shall also be trained and competent in:

1910.269(a)(2)(iii)(A)

The skills and techniques necessary to distinguish exposed live parts from other parts of electric equipment,

1910.269(a)(2)(iii)(B)

The skills and techniques necessary to determine the nominal voltage of exposed live parts, and

1910.269(a)(2)(iii)(C)

The minimum approach distances specified in this section corresponding to the voltages to which the employee will be exposed and the skills and techniques necessary to maintain those distances.

1910.269(a)(2)(iv)

The employer shall determine, through regular supervision and through inspections conducted on at least an annual basis, that each employee is complying with the safety-related work practices required by this section.

1910.269(a)(2)(v)

An employee shall receive additional training (or retraining) under any of the following conditions:

1910.269(a)(2)(v)(A)

If the supervision or annual inspections required by paragraph (a)(2)(iv) of this section indicate that the employee is not complying with the safety-related work practices required by this section, or

1910.269(a)(2)(v)(B)

If new technology, new types of equipment, or changes in procedures necessitate the use of safety-related work practices that are different from those which the employee would normally use, or

1910.269(a)(2)(v)(C)

If he or she must employ safety-related work practices that are not normally used during his or her regular job duties.

Note to paragraph (a)(2)(v)(C): The Occupational Safety and Health Administration considers tasks that are performed less often than once per year to necessitate retraining before the performance of the work practices involved.

1910.269(a)(2)(vi)

The training required by paragraph (a)(2) of this section shall be of the classroom or on-the-job type.

1910.269(a)(2)(vii)

The training shall establish employee proficiency in the work practices required by this section and shall introduce the procedures necessary for compliance with this section.

1910.269(a)(2)(viii)

The employer shall ensure that each employee has demonstrated proficiency in the work practices involved before that employee is considered as having completed the training required by paragraph (a)(2) of this section.

Note 1 to paragraph (a)(2)(viii): Though they are not required by this paragraph, employment records that indicate that an employee has successfully completed the required training are one way of keeping track of when an employee has demonstrated proficiency.

Note 2 to paragraph (a)(2)(viii): For an employee with previous training, an employer may determine that that employee has demonstrated the proficiency required by this paragraph using the following process:

1910.269(a)(2)(viii)(1)

Confirm that the employee has the training required by paragraph (a)(2) of this section,

1910.269(a)(2)(viii)(2)

Use an examination or interview to make an initial determination that the employee understands the relevant safety-related work practices before he or she performs any work covered by this section, and

1910.269(a)(2)(viii)(3)

Supervise the employee closely until that employee has demonstrated proficiency as required by this paragraph.

1910.269(a)(3)

Information transfer.

1910.269(a)(3)(i)

Before work begins, the host employer shall inform contract employers of:

1910.269(a)(3)(i)(A)

The characteristics of the host employer's installation that are related to the safety of the work to be performed and are listed in paragraphs (a)(4)(i) through (a)(4)(v) of this section;

Note to paragraph (a)(3)(i)(A): This paragraph requires the host employer to obtain information listed in paragraphs (a)(4)(i) through (a)(4)(v) of this section if it does not have this information in existing records.

1910.269(a)(3)(i)(B)

Conditions that are related to the safety of the work to be performed, that are listed in paragraphs (a)(4)(vi) through (a)(4)(viii) of this section, and that are known to the host employer;

Note to paragraph (a)(3)(i)(B): For the purposes of this paragraph, the host employer need only provide information to contract employers that the host employer can obtain from its existing records through the exercise of reasonable diligence. This paragraph does not require the host employer to make inspections of worksite conditions to obtain this information.

1910.269(a)(3)(i)(C)

Information about the design and operation of the host employer's installation that the contract employer needs to make the assessments required by this section; and

Note to paragraph (a)(3)(i)(C): This paragraph requires the host employer to obtain information about the design and operation of its installation that contract employers need to make required assessments if it does not have this information in existing records.

1910.269(a)(3)(i)(D)

Any other information about the design and operation of the host employer's installation that is known by the host employer, that the contract employer requests, and that is related to the protection of the contract employer's employees.

Note to paragraph (a)(3)(i)(D): For the purposes of this paragraph, the host employer need only provide information to contract employers that the host employer can obtain from its existing records through the exercise of reasonable diligence. This paragraph does not require the host employer to make inspections of worksite conditions to obtain this information.

1910.269(a)(3)(ii)

Contract employers shall comply with the following requirements:

1910.269(a)(3)(ii)(A)

The contract employer shall ensure that each of its employees is instructed in the hazardous conditions relevant to the employee's work that the contract employer is aware of as a result of information communicated to the contract employer by the host employer under paragraph (a)(3)(i) of this section.

1910.269(a)(3)(ii)(B)

Before work begins, the contract employer shall advise the host employer of any unique hazardous conditions presented by the contract employer's work.

1910.269(a)(3)(ii)(C)

The contract employer shall advise the host employer of any unanticipated hazardous conditions found during the contract employer's work that the host employer did not mention under paragraph (a)(3)(i) of this section. The contract employer shall provide this information to the host employer within 2 working days after discovering the hazardous condition.

1910.269(a)(3)(iii)

The contract employer and the host employer shall coordinate their work rules and procedures so that each employee of the contract employer and the host employer is protected as required by this section.

1910.269(a)(4)

Existing characteristics and conditions. Existing characteristics and conditions of electric lines and equipment that are related to the safety of the work to be performed shall be determined before work on or near the lines or equipment is started. Such characteristics and conditions include, but are not limited to:

1910.269(a)(4)(i)

The nominal voltages of lines and equipment,

1910.269(a)(4)(ii)

The maximum switching-transient voltages,

1910.269(a)(4)(iii)

The presence of hazardous induced voltages,

1910.269(a)(4)(iv)

The presence of protective grounds and equipment grounding conductors,

1910.269(a)(4)(v)

The locations of circuits and equipment, including electric supply lines, communication lines, and fire-protective signaling circuits,

1910.269(a)(4)(vi)

The condition of protective grounds and equipment grounding conductors,

1910.269(a)(4)(vii)

The condition of poles, and

1910.269(a)(4)(viii)

Environmental conditions relating to safety.

1910.269(b)

Medical services and first aid. The employer shall provide medical services and first aid as required in § 1910.151. In addition to the requirements of § 1910.151, the following requirements also apply:

1910.269(b)(1)

First-aid training. When employees are performing work on, or associated with, exposed lines or equipment energized at 50 volts or more, persons with first-aid training shall be available as follows:

1910.269(b)(1)(i)

For field work involving two or more employees at a work location, at least two trained persons shall be available. However, for line-clearance tree trimming operations performed by line-clearance tree trimmers who are not qualified employees, only one trained person need be available if all new employees are trained in first aid within 3 months of their hiring dates.

1910.269(b)(1)(ii)

For fixed work locations such as substations, the number of trained persons available shall be sufficient to ensure that each employee exposed to electric shock can be reached within 4 minutes by a trained person. However, where the existing number of employees is insufficient to meet this requirement (at a remote substation, for example), each employee at the work location shall be a trained employee.

1910.269(b)(2)

First-aid supplies. First-aid supplies required by § 1910.151(b) shall be placed in weatherproof containers if the supplies could be exposed to the weather.

1910.269(b)(3)

First-aid kits. The employer shall maintain each first-aid kit, shall ensure that it is readily available for use, and shall inspect it frequently enough to ensure that expended items are replaced. The employer also shall inspect each first aid kit at least once per year.

1910.269(c)

Job briefing.

1910.269(c)(1)

Before each job.

1910.269(c)(1)(i)

In assigning an employee or a group of employees to perform a job, the employer shall provide the employee in charge of the job with all available information that relates to the determination of existing characteristics and conditions required by paragraph (a)(4) of this section.

1910.269(c)(1)(ii)

The employer shall ensure that the employee in charge conducts a job briefing that meets paragraphs (c)(2), (c)(3), and (c)(4) of this section with the employees involved before they start each job.

1910.269(c)(2)

Subjects to be covered. The briefing shall cover at least the following subjects: hazards associated with the job, work procedures involved, special precautions, energy-source controls, and personal protective equipment requirements.

1910.269(c)(3)

Number of briefings.

1910.269(c)(3)(i)

If the work or operations to be performed during the work day or shift are repetitive and similar, at least one job briefing shall be conducted before the start of the first job of each day or shift.

1910.269(c)(3)(ii)

Additional job briefings shall be held if significant changes, which might affect the safety of the employees, occur during the course of the work.

1910.269(c)(4)

Extent of briefing.

1910.269(c)(4)(i)

A brief discussion is satisfactory if the work involved is routine and if the employees, by virtue of training and experience, can reasonably be expected to recognize and avoid the hazards involved in the job.

1910.269(c)(4)(ii)

A more extensive discussion shall be conducted:

1910.269(c)(4)(ii)(A)

If the work is complicated or particularly hazardous, or

1910.269(c)(4)(ii)(B)

If the employee cannot be expected to recognize and avoid the hazards involved in the job.

Note to paragraph (c)(4): The briefing must address all the subjects listed in paragraph (c)(2) of this section.

1910.269(c)(5)

Working alone. An employee working alone need not conduct a job briefing. However, the employer shall ensure that the tasks to be performed are planned as if a briefing were required.

1910.269(d)

Hazardous energy control (lockout/tagout) procedures.

1910.269(d)(1)

Application. The provisions of paragraph (d) of this section apply to the use of lockout/tagout procedures for the control of energy sources in installations for the purpose of electric power generation, including related equipment for communication or metering. Locking and tagging procedures for the deenergizing of electric energy sources which are used exclusively for purposes of transmission and distribution are addressed by paragraph (m) of this section.

Note to paragraph (d)(1): Installations in electric power generation facilities that are not an integral part of, or inextricably commingled with, power generation processes or equipment are covered under § 1910.147 and Subpart S of this part.

1910.269(d)(2)

General.

1910.269(d)(2)(i)

The employer shall establish a program consisting of energy control procedures, employee training, and periodic inspections to ensure that, before any employee performs any servicing or maintenance on a machine or equipment where the unexpected energizing, start up, or release of stored energy could occur and cause injury, the machine or equipment is isolated from the energy source and rendered inoperative.

1910.269(d)(2)(ii)

The employer's energy control program under paragraph (d)(2) of this section shall meet the following requirements:

1910.269(d)(2)(ii)(A)

If an energy isolating device is not capable of being locked out, the employer's program shall use a tagout system.

1910.269(d)(2)(ii)(B)

If an energy isolating device is capable of being locked out, the employer's program shall use lockout, unless the employer can demonstrate that the use of a tagout system will provide full employee protection as follows:

1910.269(d)(2)(ii)(B)(1)

When a tagout device is used on an energy isolating device which is capable of being locked out, the tagout device shall be attached at the same location that the lockout device would have been attached, and the employer shall demonstrate that the tagout program will provide a level of safety equivalent to that obtained by the use of a lockout program.

1910.269(d)(2)(ii)(B)(2)

In demonstrating that a level of safety is achieved in the tagout program equivalent to the level of safety obtained by the use of a lockout program, the employer shall demonstrate full compliance with all tagout-related provisions of this standard together with such additional elements as are necessary to provide the equivalent safety available from the use of a lockout device. Additional means to be considered as part of the demonstration of full employee protection shall include the implementation of additional safety measures such as the removal of an isolating circuit element, blocking of a controlling switch, opening of an extra disconnecting device, or the removal of a valve handle to reduce the likelihood of inadvertent energizing.

1910.269(d)(2)(ii)(C)

After November 1, 1994, whenever replacement or major repair, renovation, or modification of a machine or equipment is performed, and whenever new machines or equipment are installed, energy isolating devices for such machines or equipment shall be designed to accept a lockout device.

1910.269(d)(2)(iii)

Procedures shall be developed, documented, and used for the control of potentially hazardous energy covered by paragraph (d) of this section.

1910.269(d)(2)(iv)

The procedure shall clearly and specifically outline the scope, purpose, responsibility, authorization, rules, and techniques to be applied to the control of hazardous energy, and the measures to enforce compliance including, but not limited to, the following:

1910.269(d)(2)(iv)(A)

A specific statement of the intended use of this procedure;

1910.269(d)(2)(iv)(B)

Specific procedural steps for shutting down, isolating, blocking and securing machines or equipment to control hazardous energy;

1910.269(d)(2)(iv)(C)

Specific procedural steps for the placement, removal, and transfer of lockout devices or tagout devices and the responsibility for them; and

1910.269(d)(2)(iv)(D)

Specific requirements for testing a machine or equipment to determine and verify the effectiveness of lockout devices, tagout devices, and other energy control measures.

1910.269(d)(2)(v)

The employer shall conduct a periodic inspection of the energy control procedure at least annually to ensure that the procedure and the provisions of paragraph (d) of this section are being followed.

1910.269(d)(2)(v)(A)

The periodic inspection shall be performed by an authorized employee who is not using the energy control procedure being inspected.

1910.269(d)(2)(v)(B)

The periodic inspection shall be designed to identify and correct any deviations or inadequacies.

1910.269(d)(2)(v)(C)

If lockout is used for energy control, the periodic inspection shall include a review, between the inspector and each authorized employee, of that employee's responsibilities under the energy control procedure being inspected.

1910.269(d)(2)(v)(D)

Where tagout is used for energy control, the periodic inspection shall include a review, between the inspector and each authorized and affected employee, of that employee's responsibilities under the energy control procedure being inspected, and the elements set forth in paragraph (d)(2)(vii) of this section.

1910.269(d)(2)(v)(E)

The employer shall certify that the inspections required by paragraph (d)(2)(v) of this section have been accomplished. The certification shall identify the machine or equipment on which the energy control procedure was being used, the date of the inspection, the employees included in the inspection, and the person performing the inspection.

Note to paragraph (d)(2)(v)(E): If normal work schedule and operation records demonstrate adequate inspection activity and contain the required information, no additional certification is required.

1910.269(d)(2)(vi)

The employer shall provide training to ensure that the purpose and function of the energy control program are understood by employees and that the knowledge and skills required for the safe application, usage, and removal of energy controls are acquired by employees. The training shall include the following:

1910.269(d)(2)(vi)(A)

Each authorized employee shall receive training in the recognition of applicable hazardous energy sources, the type and magnitude of energy available in the workplace, and in the methods and means necessary for energy isolation and control.

1910.269(d)(2)(vi)(B)

Each affected employee shall be instructed in the purpose and use of the energy control procedure.

1910.269(d)(2)(vi)(C)

All other employees whose work operations are or may be in an area where energy control procedures may be used shall be instructed about the procedures and about the prohibition relating to attempts to restart or reenergize machines or equipment that are locked out or tagged out.

1910.269(d)(2)(vii)

When tagout systems are used, employees shall also be trained in the following limitations of tags:

1910.269(d)(2)(vii)(A)

Tags are essentially warning devices affixed to energy isolating devices and do not provide the physical restraint on those devices that is provided by a lock.

1910.269(d)(2)(vii)(B)

When a tag is attached to an energy isolating means, it is not to be removed without authorization of the authorized person responsible for it, and it is never to be bypassed, ignored, or otherwise defeated.

1910.269(d)(2)(vii)(C)

Tags must be legible and understandable by all authorized employees, affected employees, and all other employees whose work operations are or may be in the area, in order to be effective.

1910.269(d)(2)(vii)(D)

Tags and their means of attachment must be made of materials which will withstand the environmental conditions encountered in the workplace.

1910.269(d)(2)(vii)(E)

Tags may evoke a false sense of security, and their meaning needs to be understood as part of the overall energy control program.

1910.269(d)(2)(vii)(F)

Tags must be securely attached to energy isolating devices so that they cannot be inadvertently or accidentally detached during use.

1910.269(d)(2)(viii)

Retraining shall be provided by the employer as follows:

1910.269(d)(2)(viii)(A)

Retraining shall be provided for all authorized and affected employees whenever there is a change in their job assignments, a change in machines, equipment, or processes that present a new hazard or whenever there is a change in the energy control procedures.

1910.269(d)(2)(viii)(B)

Retraining shall also be conducted whenever a periodic inspection under paragraph (d)(2)(v) of this section reveals, or whenever the employer has reason to believe, that there are deviations from or inadequacies in an employee's knowledge or use of the energy control procedures.

1910.269(d)(2)(viii)(C)

The retraining shall reestablish employee proficiency and shall introduce new or revised control methods and procedures, as necessary.

1910.269(d)(2)(ix)

The employer shall certify that employee training has been accomplished and is being kept up to date. The certification shall contain each employee's name and dates of training.

1910.269(d)(3)

Protective materials and hardware.

1910.269(d)(3)(i)

Locks, tags, chains, wedges, key blocks, adapter pins, self-locking fasteners, or other hardware shall be provided by the employer for isolating, securing, or blocking of machines or equipment from energy sources.

1910.269(d)(3)(ii)

Lockout devices and tagout devices shall be singularly identified; shall be the only devices used for controlling energy; may not be used for other purposes; and shall meet the following requirements:

1910.269(d)(3)(ii)(A)

Lockout devices and tagout devices shall be capable of withstanding the environment to which they are exposed for the maximum period of time that exposure is expected

1910.269(d)(3)(ii)(A)(1)

Tagout devices shall be constructed and printed so that exposure to weather conditions or wet and damp locations will not cause the tag to deteriorate or the message on the tag to become illegible.

1910.269(d)(3)(ii)(A)(2)

Tagout devices shall be so constructed as not to deteriorate when used in corrosive environments.

1910.269(d)(3)(ii)(B)

Lockout devices and tagout devices shall be standardized within the facility in at least one of the following criteria: color, shape, size. Additionally, in the case of tagout devices, print and format shall be standardized.

1910.269(d)(3)(ii)(C)

Lockout devices shall be substantial enough to prevent removal without the use of excessive force or unusual techniques, such as with the use of bolt cutters or metal cutting tools.

1910.269(d)(3)(ii)(D)

Tagout devices, including their means of attachment, shall be substantial enough to prevent inadvertent or accidental removal. Tagout device attachment means shall be of a non-reusable type, attachable by hand, self-locking, and nonreleasable with a minimum unlocking strength of no less than 50 pounds and shall have the general design and basic characteristics of being at least equivalent to a one-piece, all-environment-tolerant nylon cable tie.

1910.269(d)(3)(ii)(E)

Each lockout device or tagout device shall include provisions for the identification of the employee applying the device.

1910.269(d)(3)(ii)(F)

Tagout devices shall warn against hazardous conditions if the machine or equipment is energized and shall include a legend such as the following: Do Not Start, Do Not Open, Do Not Close, Do Not Energize, Do Not Operate.

Note to paragraph (d)(3)(ii)(F):   For specific provisions covering accident prevention tags, see § 1910.145.

1910.269(d)(4)

Energy isolation. Lockout and tagout device application and removal may only be performed by the authorized employees who are performing the servicing or maintenance.

1910.269(d)(5)

Notification. Affected employees shall be notified by the employer or authorized employee of the application and removal of lockout or tagout devices. Notification shall be given before the controls are applied and after they are removed from the machine or equipment.

Note to paragraph (d)(5):   See also paragraph (d)(7) of this section, which requires that the second notification take place before the machine or equipment is reenergized.

1910.269(d)(6)

Lockout/tagout application. The established procedures for the application of energy control (the lockout or tagout procedures) shall include the following elements and actions, and these procedures shall be performed in the following sequence:

1910.269(d)(6)(i)

Before an authorized or affected employee turns off a machine or equipment, the authorized employee shall have knowledge of the type and magnitude of the energy, the hazards of the energy to be controlled, and the method or means to control the energy.

1910.269(d)(6)(ii)

The machine or equipment shall be turned off or shut down using the procedures established for the machine or equipment. An orderly shutdown shall be used to avoid any additional or increased hazards to employees as a result of the equipment stoppage.

1910.269(d)(6)(iii)

All energy isolating devices that are needed to control the energy to the machine or equipment shall be physically located and operated in such a manner as to isolate the machine or equipment from energy sources.

1910.269(d)(6)(iv)

Lockout or tagout devices shall be affixed to each energy isolating device by authorized employees.

1910.269(d)(6)(iv)(A)

Lockout devices shall be attached in a manner that will hold the energy isolating devices in a "safe" or "off" position.

1910.269(d)(6)(iv)(B)

Tagout devices shall be affixed in such a manner as will clearly indicate that the operation or movement of energy isolating devices from the "safe" or "off" position is prohibited.

1910.269(d)(6)(iv)(B)(1)

Where tagout devices are used with energy isolating devices designed with the capability of being locked out, the tag attachment shall be fastened at the same point at which the lock would have been attached.

1910.269(d)(6)(iv)(B)(2)

Where a tag cannot be affixed directly to the energy isolating device, the tag shall be located as close as safely possible to the device, in a position that will be immediately obvious to anyone attempting to operate the device.

1910.269(d)(6)(v)

Following the application of lockout or tagout devices to energy isolating devices, all potentially hazardous stored or residual energy shall be relieved, disconnected, restrained, or otherwise rendered safe.

1910.269(d)(6)(vi)

If there is a possibility of reaccumulation of stored energy to a hazardous level, verification of isolation shall be continued until the servicing or maintenance is completed or until the possibility of such accumulation no longer exists.

1910.269(d)(6)(vii)

Before starting work on machines or equipment that have been locked out or tagged out, the authorized employee shall verify that isolation and deenergizing of the machine or equipment have been accomplished. If normally energized parts will be exposed to contact by an employee while the machine or equipment is deenergized, a test shall be performed to ensure that these parts are deenergized.

1910.269(d)(7)

Release from lockout/tagout. Before lockout or tagout devices are removed and energy is restored to the machine or equipment, procedures shall be followed and actions taken by the authorized employees to ensure the following:

1910.269(d)(7)(i)

The work area shall be inspected to ensure that nonessential items have been removed and that machine or equipment components are operationally intact.

1910.269(d)(7)(ii)

The work area shall be checked to ensure that all employees have been safely positioned or removed.

1910.269(d)(7)(iii)

After lockout or tagout devices have been removed and before a machine or equipment is started, affected employees shall be notified that the lockout or tagout devices have been removed.

1910.269(d)(7)(iv)

Each lockout or tagout device shall be removed from each energy isolating device by the authorized employee who applied the lockout or tagout device. However, if that employee is not available to remove it, the device may be removed under the direction of the employer, provided that specific procedures and training for such removal have been developed, documented, and incorporated into the employer's energy control program. The employer shall demonstrate that the specific procedure provides a degree of safety equivalent to that provided by the removal of the device by the authorized employee who applied it. The specific procedure shall include at least the following elements:

1910.269(d)(7)(iv)(A)

Verification by the employer that the authorized employee who applied the device is not at the facility;

1910.269(d)(7)(iv)(B)

Making all reasonable efforts to contact the authorized employee to inform him or her that his or her lockout or tagout device has been removed; and

1910.269(d)(7)(iv)(C)

Ensuring that the authorized employee has this knowledge before he or she resumes work at that facility.

1910.269(d)(8)

Additional requirements.

1910.269(d)(8)(i)

If the lockout or tagout devices must be temporarily removed from energy isolating devices and the machine or equipment must be energized to test or position the machine, equipment, or component thereof, the following sequence of actions shall be followed:

1910.269(d)(8)(i)(A)

Clear the machine or equipment of tools and materials in accordance with paragraph (d)(7)(i) of this section;

1910.269(d)(8)(i)(B)

Remove employees from the machine or equipment area in accordance with paragraphs (d)(7)(ii) and (d)(7)(iii) of this section;

1910.269(d)(8)(i)(C)

Remove the lockout or tagout devices as specified in paragraph (d)(7)(iv) of this section;

1910.269(d)(8)(i)(D)

Energize and proceed with the testing or positioning; and

1910.269(d)(8)(i)(E)

Deenergize all systems and reapply energy control measures in accordance with paragraph (d)(6) of this section to continue the servicing or maintenance.

1910.269(d)(8)(ii)

When servicing or maintenance is performed by a crew, craft, department, or other group, they shall use a procedure which affords the employees a level of protection equivalent to that provided by the implementation of a personal lockout or tagout device. Group lockout or tagout devices shall be used in accordance with the procedures required by paragraphs (d)(2)(iii) and (d)(2)(iv) of this section including, but not limited to, the following specific requirements:

1910.269(d)(8)(ii)(A)

Primary responsibility shall be vested in an authorized employee for a set number of employees working under the protection of a group lockout or tagout device (such as an operations lock);

1910.269(d)(8)(ii)(B)

Provision shall be made for the authorized employee to ascertain the exposure status of all individual group members with regard to the lockout or tagout of the machine or equipment;

1910.269(d)(8)(ii)(C)

When more than one crew, craft, department, or other group is involved, assignment of overall job-associated lockout or tagout control responsibility shall be given to an authorized employee designated to coordinate affected work forces and ensure continuity of protection; and

1910.269(d)(8)(ii)(D)

Each authorized employee shall affix a personal lockout or tagout device to the group lockout device, group lockbox, or comparable mechanism when he or she begins work and shall remove those devices when he or she stops working on the machine or equipment being serviced or maintained.

1910.269(d)(8)(iii)

Procedures shall be used during shift or personnel changes to ensure the continuity of lockout or tagout protection, including provision for the orderly transfer of lockout or tagout device protection between off-going and on-coming employees, to minimize their exposure to hazards from the unexpected energizing or start-up of the machine or equipment or from the release of stored energy.

1910.269(d)(8)(iv)

Whenever outside servicing personnel are to be engaged in activities covered by paragraph (d) of this section, the on-site employer and the outside employer shall inform each other of their respective lockout or tagout procedures, and each employer shall ensure that his or her personnel understand and comply with restrictions and prohibitions of the energy control procedures being used.

1910.269(d)(8)(v)

If energy isolating devices are installed in a central location and are under the exclusive control of a system operator, the following requirements apply:

1910.269(d)(8)(v)(A)

The employer shall use a procedure that affords employees a level of protection equivalent to that provided by the implementation of a personal lockout or tagout device.

1910.269(d)(8)(v)(B)

The system operator shall place and remove lockout and tagout devices in place of the authorized employee under paragraphs (d)(4), (d)(6)(iv), and (d)(7)(iv) of this section.

1910.269(d)(8)(v)(C)

Provisions shall be made to identify the authorized employee who is responsible for (that is, being protected by) the lockout or tagout device, to transfer responsibility for lockout and tagout devices, and to ensure that an authorized employee requesting removal or transfer of a lockout or tagout device is the one responsible for it before the device is removed or transferred.

Note to paragraph (d): Lockout and tagging procedures that comply with paragraphs (c) through (f) of § 1910.147 will also be deemed to comply with paragraph (d) of this section if the procedures address the hazards covered by paragraph (d) of this section.

1910.269(e)

Enclosed spaces. This paragraph covers enclosed spaces that may be entered by employees. It does not apply to vented vaults if the employer makes a determination that the ventilation system is operating to protect employees before they enter the space. This paragraph applies to routine entry into enclosed spaces in lieu of the permitspace entry requirements contained in paragraphs (d) through (k) of § 1910.146. If, after the employer takes the precautions given in paragraphs (e) and (t) of this section, the hazards remaining in the enclosed space endanger the life of an entrant or could interfere with an entrant's escape from the space, then entry into the enclosed space shall meet the permit-space entry requirements of paragraphs (d) through (k) of § 1910.146.

1910.269(e)(1)

Safe work practices. The employer shall ensure the use of safe work practices for entry into, and work in, enclosed spaces and for rescue of employees from such spaces.

1910.269(e)(2)

Training. Each employee who enters an enclosed space or who serves as an attendant shall be trained in the hazards of enclosed-space entry, in enclosed-space entry procedures, and in enclosed-space rescue procedures.

1910.269(e)(3)

Rescue equipment. Employers shall provide equipment to ensure the prompt and safe rescue of employees from the enclosed space.

1910.269(e)(4)

Evaluating potential hazards. Before any entrance cover to an enclosed space is removed, the employer shall determine whether it is safe to do so by checking for the presence of any atmospheric pressure or temperature differences and by evaluating whether there might be a hazardous atmosphere in the space. Any conditions making it unsafe to remove the cover shall be eliminated before the cover is removed.

Note to paragraph (e)(4): The determination called for in this paragraph may consist of a check of the conditions that might foreseeably be in the enclosed space. For example, the cover could be checked to see if it is hot and, if it is fastened in place, could be loosened gradually to release any residual pressure. An evaluation also needs to be made of whether conditions at the site could cause a hazardous atmosphere, such as an oxygen-deficient or flammable atmosphere, to develop within the space.

1910.269(e)(5)

Removing covers. When covers are removed from enclosed spaces, the opening shall be promptly guarded by a railing, temporary cover, or other barrier designed to prevent an accidental fall through the opening and to protect employees working in the space from objects entering the space.

1910.269(e)(6)

Hazardous atmosphere. Employees may not enter any enclosed space while it contains a hazardous atmosphere, unless the entry conforms to the permit-required confined spaces standard in § 1910.146.

1910.269(e)(7)

Attendants. While work is being performed in the enclosed space, an attendant with first-aid training shall be immediately available outside the enclosed space to provide assistance if a hazard exists because of traffic patterns in the area of the opening used for entry. The attendant is not precluded from performing other duties outside the enclosed space if these duties do not distract the attendant from: monitoring employees within the space or ensuring that it is safe for employees to enter and exit the space.

Note to paragraph (e)(7):   See paragraph (t) of this section for additional requirements on attendants for work in manholes and vaults.

1910.269(e)(8)

Calibration of test instruments. Test instruments used to monitor atmospheres in enclosed spaces shall be kept in calibration and shall have a minimum accuracy of ±10 percent.

1910.269(e)(9)

Testing for oxygen deficiency. Before an employee enters an enclosed space, the atmosphere in the enclosed space shall be tested for oxygen deficiency with a direct-reading meter or similar instrument, capable of collection and immediate analysis of data samples without the need for offsite evaluation. If continuous forced-air ventilation is provided, testing is not required provided that the procedures used ensure that employees are not exposed to the hazards posed by oxygen deficiency.

1910.269(e)(10)

Testing for flammable gases and vapors. Before an employee enters an enclosed space, the internal atmosphere shall be tested for flammable gases and vapors with a direct-reading meter or similar instrument capable of collection and immediate analysis of data samples without the need for off-site evaluation. This test shall be performed after the oxygen testing and ventilation required by paragraph (e)(9) of this section demonstrate that there is sufficient oxygen to ensure the accuracy of the test for flammability.

1910.269(e)(11)

Ventilation, and monitoring for flammable gases or vapors. If flammable gases or vapors are detected or if an oxygen deficiency is found, forced-air ventilation shall be used to maintain oxygen at a safe level and to prevent a hazardous concentration of flammable gases and vapors from accumulating. A continuous monitoring program to ensure that no increase in flammable gas or vapor concentration above safe levels occurs may be followed in lieu of ventilation if flammable gases or vapors are initially detected at safe levels.

Note to paragraph (e)(11): See the definition of "hazardous atmosphere" for guidance in determining whether a specific concentration of a substance is hazardous.

1910.269(e)(12)

Specific ventilation requirements. If continuous forced-air ventilation is used, it shall begin before entry is made and shall be maintained long enough for the employer to be able to demonstrate that a safe atmosphere exists before employees are allowed to enter the work area. The forced-air ventilation shall be so directed as to ventilate the immediate area where employees are present within the enclosed space and shall continue until all employees leave the enclosed space.

1910.269(e)(13)

Air supply. The air supply for the continuous forced-air ventilation shall be from a clean source and may not increase the hazards in the enclosed space.

1910.269(e)(14)

Open flames. If open flames are used in enclosed spaces, a test for flammable gases and vapors shall be made immediately before the open flame device is used and at least once per hour while the device is used in the space. Testing shall be conducted more frequently if conditions present in the enclosed space indicate that once per hour is insufficient to detect hazardous accumulations of flammable gases or vapors.

Note to paragraph (e)(14): See the definition of "hazardous atmosphere" for guidance in determining whether a specific concentration of a substance is hazardous.

Note to paragraph (e): Entries into enclosed spaces conducted in accordance with the permit-space entry requirements of paragraphs (d) through (k) of § 1910.146 are considered as complying with paragraph (e) of this section.

1910.269(f)

Excavations. Excavation operations shall comply with Subpart P of Part 1926 of this chapter.

1910.269(g)

Personal protective equipment.

1910.269(g)(1)

General. Personal protective equipment shall meet the requirements of Subpart I of this part.

Note to paragraph (g)(1) of this section: Paragraph (h) of § 1910.132 sets employer payment obligations for the personal protective equipment required by this section, including, but not limited to, the fall protection equipment required by paragraph (g)(2) of this section, the electrical protective equipment required by paragraph (l)(3) of this section, and the flame-resistant and arcrated clothing and other protective equipment required by paragraph (l)(8) of this section.

1910.269(g)(2)

Fall protection.

1910.269(g)(2)(i)

Personal fall arrest systems shall meet the requirements of Subpart M of Part 1926 of this chapter.

1910.269(g)(2)(ii)

Personal fall arrest equipment used by employees who are exposed to hazards from flames or electric arcs, as determined by the employer under paragraph (l)(8)(i) of this section, shall be capable of passing a drop test equivalent to that required by paragraph (g)(2)(iii)(L) of this section after exposure to an electric arc with a heat energy of 40 ± 5 cal/cm^2.

1910.269(g)(2)(iii)

Body belts and positioning straps for work-positioning equipment shall meet the following requirements:

1910.269(g)(2)(iii)(A)

Hardware for body belts and positioning straps shall meet the following requirements:

1910.269(g)(2)(iii)(A)(1)

Hardware shall be made of drop-forged steel, pressed steel, formed steel, or equivalent material.

1910.269(g)(2)(iii)(A)(2)

Hardware shall have a corrosion-resistant finish.

1910.269(g)(2)(iii)(A)(3)

Hardware surfaces shall be smooth and free of sharp edges.

1910.269(g)(2)(iii)(B)

Buckles shall be capable of withstanding an 8.9-kilonewton (2,000-pound-force) tension test with a maximum permanent deformation no greater than 0.4 millimeters (0.0156 inches).

1910.269(g)(2)(iii)(C)

D rings shall be capable of withstanding a 22-kilonewton (5,000-pound-force) tensile test without cracking or breaking.

1910.269(g)(2)(iii)(D)

Snaphooks shall be capable of withstanding a 22-kilonewton (5,000-pound-force) tension test without failure.

Note to paragraph (g)(2)(iii)(D): Distortion of the snaphook sufficient to release the keeper is considered to be tensile failure of a snaphook.

1910.269(g)(2)(iii)(E)

Top grain leather or leather substitute may be used in the manufacture of body belts and positioning straps; however, leather and leather substitutes may not be used alone as a load-bearing component of the assembly.

1910.269(g)(2)(iii)(F)

Plied fabric used in positioning straps and in load-bearing parts of body belts shall be constructed in such a way that no raw edges are exposed and the plies do not separate.

1910.269(g)(2)(iii)(G)

Positioning straps shall be capable of withstanding the following tests:

1910.269(g)(2)(iii)(G)(1)

A dielectric test of 819.7 volts, AC, per centimeter (25,000 volts per foot) for 3 minutes without visible deterioration;

1910.269(g)(2)(iii)(G)(2)

A leakage test of 98.4 volts, AC, per centimeter (3,000 volts per foot) with a leakage current of no more than 1 mA;

Note to paragraphs (g)(2)(iii)(G)(1) and (g)(2)(iii)(G)(2): Positioning straps that pass direct-current tests at equivalent voltages are considered as meeting this requirement.

1910.269(g)(2)(iii)(G)(3)

Tension tests of 20 kilonewtons (4,500 pounds-force) for sections free of buckle holes and of 15 kilonewtons (3,500 pounds-force) for sections with buckle holes;

1910.269(g)(2)(iii)(G)(4)

A buckle-tear test with a load of 4.4 kilonewtons (1,000 pounds-force); and

1910.269(g)(2)(iii)(G)(5)

A flammability test in accordance with Table R-2.

Table R-2- Flammability Test

Test method	Criteria for passing the test
Vertically suspend a 500-mm (19.7-inch) length of strapping supporting a 100-kg (220.5-lb) weight.	Any flames on the positioning strap shall self extinguish.
Use a butane or propane burner with a 76-mm (3-inch) flame.	The positioning strap shall continue to support the 100-kg (220.5-lb) mass.
Direct the flame to an edge of the strapping at a distance of 25 mm (1 inch).	
Remove the flame after 5 seconds.	
Wait for any flames on the positioning strap to stop burning.	

1910.269(g)(2)(iii)(H)

The cushion part of the body belt shall contain no exposed rivets on the inside and shall be at least 76 millimeters (3 inches) in width.

1910.269(g)(2)(iii)(I)

Tool loops shall be situated on the body of a body belt so that the 100 millimeters (4 inches) of the body belt that is in the center of the back, measuring from D ring to D ring, is free of tool loops and any other attachments.

1910.269(g)(2)(iii)(J)

Copper, steel, or equivalent liners shall be used around the bars of D rings to prevent wear between these members and the leather or fabric enclosing them.

1910.269(g)(2)(iii)(K)

Snaphooks shall be of the locking type meeting the following requirements:

1910.269(g)(2)(iii)(K)(1)

The locking mechanism shall first be released, or a destructive force shall be placed on the keeper, before the keeper will open.

1910.269(g)(2)(iii)(K)(2)

A force in the range of 6.7 N (1.5 lbf) to 17.8 N (4 lbf) shall be required to release the locking mechanism.

1910.269(g)(2)(iii)(K)(3)

With the locking mechanism released and with a force applied on the keeper against the face of the nose, the keeper may not begin to open with a force of 11.2 N (2.5 lbf) or less and shall begin to open with a maximum force of 17.8 N (4 lbf).

1910.269(g)(2)(iii)(L)

Body belts and positioning straps shall be capable of withstanding a drop test as follows:

1910.269(g)(2)(iii)(L)(1)

The test mass shall be rigidly constructed of steel or equivalent material with a mass of 100 kg (220.5 lbm). For work-positioning equipment used by employees weighing more than 140 kg (310 lbm) fully equipped, the test mass shall be increased proportionately (that is, the test mass must equal the mass of the equipped worker divided by 1.4).

1910.269(g)(2)(iii)(L)(2)

For body belts, the body belt shall be fitted snugly around the test mass and shall be attached to the teststructure anchorage point by means of a wire rope.

1910.269(g)(2)(iii)(L)(3)

For positioning straps, the strap shall be adjusted to its shortest length possible to accommodate the test and connected to the test-structure anchorage point at one end and to the test mass on the other end.

1910.269(g)(2)(iii)(L)(4)

The test mass shall be dropped an unobstructed distance of 1 meter (39.4 inches) from a supporting structure that will sustain minimal deflection during the test.

1910.269(g)(2)(iii)(L)(5)

Body belts shall successfully arrest the fall of the test mass and shall be capable of supporting the mass after the test.

1910.269(g)(2)(iii)(L)(6)

Positioning straps shall successfully arrest the fall of the test mass without breaking, and the arrest force may not exceed 17.8 kilonewtons (4,000 pounds-force). Additionally, snaphooks on positioning straps may not distort to such an extent that the keeper would release.

Note to paragraph (g)(2)(iii) of this section: When used by employees weighing no more than 140 kg (310 lbm) fully equipped, body belts and positioning straps that conform to American Society of Testing and Materials *Standard Specifications for Personal Climbing Equipment*, ASTM F887-12^{e1}, are deemed to be in compliance with paragraph (g)(2)(iii) of this section.

1910.269(g)(2)(iv)

The following requirements apply to the care and use of personal fall protection equipment.

1910.269(g)(2)(iv)(A)

Work-positioning equipment shall be inspected before use each day to determine that the equipment is in safe working condition. Work-positioning equipment that is not in safe working condition may not be used.

Note to paragraph (g)(2)(iv)(A): Appendix F to this section contains guidelines for inspecting work-positioning equipment.

1910.269(g)(2)(iv)(B)

Personal fall arrest systems shall be used in accordance with § 1926.502(d).

Note to paragraph (g)(2)(iv)(B): Fall protection equipment rigged to arrest falls is considered a fall arrest system and must meet the applicable requirements for the design and use of those systems. Fall protection equipment rigged for work positioning is considered work-positioning equipment and must meet the applicable requirements for the design and use of that equipment.

1910.269(g)(2)(iv)(C)

The employer shall ensure that employees use fall protection systems as follows:

1910.269(g)(2)(iv)(C)(1)

Each employee working from an aerial lift shall use a fall restraint system or a personal fall arrest system. Paragraph (c)(2)(v) of § 1910.67 does not apply.

1910.269(g)(2)(iv)(C)(2)

Except as provided in paragraph (g)(2)(iv)(C)(3) of this section, each employee in elevated locations more than 1.2 meters (4 feet) above the ground on poles, towers, or similar structures shall use a personal fall arrest system, work-positioning equipment, or fall restraint system, as appropriate, if the employer has not provided other fall protection meeting Subpart D of this part.

1910.269(g)(2)(iv)(C)(3)

Until March 31, 2015, a qualified employee climbing or changing location on poles, towers, or similar structures need not use fall protection equipment, unless conditions, such as, but not limited to, ice, high winds, the design of the structure (for example, no provision for holding on with hands), or the presence of contaminants on the structure, could cause the employee to lose his or her grip or footing. On and after April 1, 2015, each qualified employee climbing or changing location on poles, towers, or similar structures must use fall protection equipment unless the employer can demonstrate that climbing or changing location with fall protection is infeasible or creates a greater hazard than climbing or changing location without it.

Note 1 to paragraphs (g)(2)(iv)(C)(2) and (g)(2)(iv)(C)(3): These paragraphs apply to structures that support overhead electric power transmission and distribution lines and equipment. They do not apply to portions of buildings, such as loading docks, or to electric equipment, such as transformers and capacitors. Subpart D of this part contains the duty to provide fall protection associated with walking and working surfaces.

Note 2 to paragraphs (g)(2)(iv)(C)(2) and (g)(2)(iv)(C)(3): Until the employer ensures that employees are proficient in climbing and the use of fall protection under paragraph (a)(2)(viii) of this section, the employees are not considered "qualified employees" for the purposes of paragraphs (g)(2)(iv)(C)(2) and (g)(2)(iv)(C)(3) of this section. These paragraphs require unqualified employees (including trainees) to use fall protection any time they are more than 1.2 meters (4 feet) above the ground.

1910.269(g)(2)(iv)(D)

On and after April 1, 2015, workpositioning systems shall be rigged so that an employee can free fall no more than 0.6 meters (2 feet).

1910.269(g)(2)(iv)(E)

Anchorages for work-positioning equipment shall be capable of supporting at least twice the potential impact load of an employee's fall, or 13.3 kilonewtons (3,000 pounds-force), whichever is greater.

Note to paragraph (g)(2)(iv)(E): Wood-pole fall-restriction devices meeting American Society of Testing and Materials *Standard Specifications for Personal Climbing Equipment*, ASTM F887-12[e1], are deemed to meet the anchorage-strength requirement when they are used in accordance with manufacturers' instructions.

1910.269(g)(2)(iv)(F)

Unless the snaphook is a locking type and designed specifically for the following connections, snaphooks on work-positioning equipment may not be engaged:

1910.269(g)(2)(iv)(F)(1)

Directly to webbing, rope, or wire rope;

1910.269(g)(2)(iv)(F)(2)

To each other;

1910.269(g)(2)(iv)(F)(3)

To a D ring to which another snaphook or other connector is attached;

1910.269(g)(2)(iv)(F)(4)

To a horizontal lifeline; or

1910.269(g)(2)(iv)(F)(5)

To any object that is incompatibly shaped or dimensioned in relation to the snaphook such that accidental disengagement could occur should the connected object sufficiently depress the snaphook keeper to allow release of the object.

1910.269(h)

Portable ladders and platforms.

1910.269(h)(1)

General. Requirements for portable ladders contained in Subpart D of this part apply in addition to the requirements of paragraph (h) of this section, except as specifically noted in paragraph (h)(2) of this section.

1910.269(h)(2)

Special ladders and platforms. Portable ladders used on structures or conductors in conjunction with overhead line work need not meet § 1910.25(d)(2)(i) and (d)(2)(iii) or § 1910.26(c)(3)(iii). Portable ladders and platforms used on structures or conductors in conjunction with overhead line work shall meet the following requirements:

1910.269(h)(2)(i)

In the configurations in which they are used, portable ladders and platforms shall be capable of supporting without failure at least 2.5 times the maximum intended load.

1910.269(h)(2)(ii)

Portable ladders and platforms may not be loaded in excess of the working loads for which they are designed.

1910.269(h)(2)(iii)

Portable ladders and platforms shall be secured to prevent them from becoming dislodged.

1910.269(h)(2)(iv)

Portable ladders and platforms may be used only in applications for which they are designed.

1910.269(h)(3)

Conductive ladders. Portable metal ladders and other portable conductive ladders may not be used near exposed energized lines or equipment. However, in specialized high-voltage work, conductive ladders shall be used when the employer demonstrates that nonconductive ladders would present a greater hazard to employees than conductive ladders.

1910.269(i)

Hand and portable power equipment.

1910.269(i)(1)

General. Paragraph (i)(2) of this section applies to electric equipment connected by cord and plug. Paragraph (i)(3) of this section applies to portable and vehicle-mounted generators used to supply cord- and plug-connected equipment. Paragraph (i)(4) of this section applies to hydraulic and pneumatic tools.

1910.269(i)(2)

Cord- and plug-connected equipment. Cord- and plug-connected equipment not covered by Subpart S of this part shall comply with one of the following instead of § 1910.243(a)(5):

1910.269(i)(2)(i)

The equipment shall be equipped with a cord containing an equipment grounding conductor connected to the equipment frame and to a means for grounding the other end of the conductor (however, this option may not be used where the introduction of the ground into the work environment increases the hazard to an employee); or

1910.269(i)(2)(ii)

The equipment shall be of the double-insulated type conforming to Subpart S of this part; or

1910.269(i)(2)(iii)

The equipment shall be connected to the power supply through an isolating transformer with an ungrounded secondary of not more than 50 volts.

1910.269(i)(3)

Portable and vehicle-mounted generators. Portable and vehiclemounted generators used to supply cord- and plug-connected equipment covered by paragraph (i)(2) of this section shall meet the following requirements:

1910.269(i)(3)(i)

The generator may only supply equipment located on the generator or the vehicle and cord- and plugconnected equipment through receptacles mounted on the generator or the vehicle.

1910.269(i)(3)(ii)

The non-current-carrying metal parts of equipment and the equipment grounding conductor terminals of the receptacles shall be bonded to the generator frame.

1910.269(i)(3)(iii)

For vehicle-mounted generators, the frame of the generator shall be bonded to the vehicle frame.

1910.269(i)(3)(iv)

Any neutral conductor shall be bonded to the generator frame.

1910.269(i)(4)

Hydraulic and pneumatic tools.

1910.269(i)(4)(i)

Safe operating pressures for hydraulic and pneumatic tools, hoses, valves, pipes, filters, and fittings may not be exceeded.

Note to paragraph (i)(4)(i): If any hazardous defects are present, no operating pressure is safe, and the hydraulic or pneumatic equipment involved may not be used. In the absence of defects, the maximum rated operating pressure is the maximum safe pressure.

1910.269(i)(4)(ii)

A hydraulic or pneumatic tool used where it may contact exposed energized parts shall be designed and maintained for such use.

1910.269(i)(4)(iii)

The hydraulic system supplying a hydraulic tool used where it may contact exposed live parts shall provide protection against loss of insulating value, for the voltage involved, due to the formation of a partial vacuum in the hydraulic line.

Note to paragraph (i)(4)(iii): Use of hydraulic lines that do not have check valves and that have a separation of more than 10.7 meters (35 feet) between the oil reservoir and the upper end of the hydraulic system promotes the formation of a partial vacuum.

1910.269(i)(4)(iv)

A pneumatic tool used on energized electric lines or equipment, or used where it may contact exposed live parts, shall provide protection against the accumulation of moisture in the air supply.

1910.269(i)(4)(v)

Pressure shall be released before connections are broken, unless quickacting, self-closing connectors are used.

1910.269(i)(4)(vi)

Employers must ensure that employees do not use any part of their bodies to locate, or attempt to stop, a hydraulic leak.

1910.269(i)(4)(vii)

Hoses may not be kinked.

1910.269(j)

Live-line tools.

1910.269(j)(1)

Design of tools. Live-line tool rods, tubes, and poles shall be designed and constructed to withstand the following minimum tests:

1910.269(j)(1)(i)

If the tool is made of fiberglass-reinforced plastic (FRP), it shall withstand 328,100 volts per meter (100,000 volts per foot) of length for 5 minutes, or

Note to paragraph (j)(1)(i): Live-line tools using rod and tube that meet ASTM F711-02 (2007), *Standard Specification for Fiberglass-Reinforced Plastic (FRP) Rod and Tube Used in Live Line Tools,* are deemed to comply with paragraph (j)(1) of this section.

1910.269(j)(1)(ii)

If the tool is made of wood, it shall withstand 246,100 volts per meter (75,000 volts per foot) of length for 3 minutes, or

1910.269(j)(1)(iii)

The tool shall withstand other tests that the employer can demonstrate are equivalent.

1910.269(j)(2)

Condition of tools.

1910.269(j)(2)(i)

Each liveline tool shall be wiped clean and visually inspected for defects before use each day.

1910.269(j)(2)(ii)

If any defect or contamination that could adversely affect the insulating qualities or mechanical integrity of the live-line tool is present after wiping, the tool shall be removed from service and examined and tested according to paragraph (j)(2)(iii) of this section before being returned to service.

1910.269(j)(2)(iii)

Live-line tools used for primary employee protection shall be removed from service every 2 years, and whenever required under paragraph (j)(2)(ii) of this section, for examination, cleaning, repair, and testing as follows:

1910.269(j)(2)(iii)(A)

Each tool shall be thoroughly examined for defects.

1910.269(j)(2)(iii)(B)

If a defect or contamination that could adversely affect the insulating qualities or mechanical integrity of the live-line tool is found, the tool shall be repaired and refinished or shall be permanently removed from service. If no such defect or contamination is found, the tool shall be cleaned and waxed.

1910.269(j)(2)(iii)(C)

The tool shall be tested in accordance with paragraphs (j)(2)(iii)(D) and (j)(2)(iii)(E) of this section under the following conditions:

1910.269(j)(2)(iii)(C)(1)

After the tool has been repaired or refinished; and

1910.269(j)(2)(iii)(C)(2)

After the examination if repair or refinishing is not performed, unless the tool is made of FRP rod or foam-filled FRP tube and the employer can demonstrate that the tool has no defects that could cause it to fail during use.

1910.269(j)(2)(iii)(D)

The test method used shall be designed to verify the tool's integrity along its entire working length and, if the tool is made of fiberglass-reinforced plastic, its integrity under wet conditions.

1910.269(j)(2)(iii)(E)

The voltage applied during the tests shall be as follows:

1910.269(j)(2)(iii)(E)(1)

246,100 volts per meter (75,000 volts per foot) of length for 1 minute if the tool is made of fiberglass, or

1910.269(j)(2)(iii)(E)(2)

164,000 volts per meter (50,000 volts per foot) of length for 1 minute if the tool is made of wood, or

1910.269(j)(2)(iii)(E)(3)

Other tests that the employer can demonstrate are equivalent.

Note to paragraph (j)(2): Guidelines for the examination, cleaning, repairing, and inservice testing of live-line tools are specified in the Institute of Electrical and Electronics Engineers' *IEEE Guide for Maintenance Methods on Energized Power Lines*, IEEE Std 516-2009.

1910.269(k)

Materials handling and storage.

1910.269(k)(1)

General. Materials handling and storage shall comply with applicable materialhandling and material-storage requirements in this part, including those in Subpart N of this part.

1910.269(k)(2)

Materials storage near energized lines or equipment.

1910.269(k)(2)(i)

In areas to which access is not restricted to qualified persons only, materials or equipment may not be stored closer to energized lines or exposed energized parts of equipment than the following distances, plus a distance that provides for the maximum sag and side swing of all conductors and for the height and movement of material-handling equipment:

1910.269(k)(2)(i)(A)

For lines and equipment energized at 50 kilovolts or less, the distance is 3.05 meters (10 feet).

1910.269(k)(2)(i)(B)

For lines and equipment energized at more than 50 kilovolts, the distance is 3.05 meters (10 feet) plus 0.10 meter (4 inches) for every 10 kilovolts over 50 kilovolts.

1910.269(k)(2)(ii)

In areas restricted to qualified employees, materials may not be stored within the working space about energized lines or equipment.

Note to paragraph (k)(2)(ii): Paragraphs (u)(1) and (v)(3) of this section specify the size of the working space.

1910.269(l)

Working on or near exposed energized parts. This paragraph applies to work on exposed live parts, or near enough to them to expose the employee to any hazard they present.

1910.269(l)(1)

General.

1910.269(l)(1)(i)

Only qualified employees may work on or with exposed energized lines or parts of equipment.

1910.269(l)(1)(ii)

Only qualified employees may work in areas containing unguarded, uninsulated energized lines or parts of equipment operating at 50 volts or more.

1910.269(l)(1)(iii)

Electric lines and equipment shall be considered and treated as energized unless they have been deenergized in accordance with paragraph (d) or (m) of this section.

1910.269(l)(2)

At least two employees.

1910.269(l)(2)(i)

Except as provided in paragraph (l)(2)(ii) of this section, at least two employees shall be present while any employees perform the following types of work:

1910.269(l)(2)(i)(A)

Installation, removal, or repair of lines energized at more than 600 volts,

1910.269(l)(2)(i)(B)

Installation, removal, or repair of deenergized lines if an employee is exposed to contact with other parts energized at more than 600 volts,

1910.269(l)(2)(i)(C)

Installation, removal, or repair of equipment, such as transformers, capacitors, and regulators, if an employee is exposed to contact with parts energized at more than 600 volts,

1910.269(l)(2)(i)(D)

Work involving the use of mechanical equipment, other than insulated aerial lifts, near parts energized at more than 600 volts, and

1910.269(l)(2)(i)(E)

Other work that exposes an employee to electrical hazards greater than, or equal to, the electrical hazards posed by operations listed specifically in paragraphs (l)(2)(i)(A) through (l)(2)(i)(D) of this section.

1910.269(l)(2)(ii)

Paragraph (l)(2)(i) of this section does not apply to the following operations:

1910.269(l)(2)(ii)(A)

Routine circuit switching, when the employer can demonstrate that conditions at the site allow safe performance of this work,

1910.269(l)(2)(ii)(B)

Work performed with live-line tools when the position of the employee is such that he or she is neither within reach of, nor otherwise exposed to contact with, energized parts, and

1910.269(l)(2)(ii)(C)

Emergency repairs to the extent necessary to safeguard the general public.

<u>1910.269(l)(3)</u>

Minimum approach distances.

1910.269(l)(3)(i)

The employer shall establish minimum approach distances no less than the distances computed by Table R-3 for ac systems or Table R-8 for dc systems.

1910.269(l)(3)(ii)

No later than April 1, 2015, for voltages over 72.5 kilovolts, the employer shall determine the maximum anticipated per-unit transient overvoltage, phase-to-ground, through an engineering analysis or assume a maximum anticipated per-unit transient overvoltage, phase-to-ground, in accordance with Table R-9. When the employer uses portable protective gaps to control the maximum transient overvoltage, the value of the maximum anticipated per-unit transient overvoltage, phase-to-ground, must provide for five standard deviations between the statistical sparkover voltage of the gap and the statistical withstand voltage corresponding to the electrical component of the minimum approach distance. The employer shall make any engineering analysis conducted to determine maximum anticipated perunit transient overvoltage available upon request to employees and to the Assistant Secretary or designee for examination and copying.

Note to paragraph (l)(3)(ii):   See Appendix B to this section for information on how to calculate the maximum anticipated per-unit transient overvoltage, phase-to-ground, when the employer uses portable protective gaps to reduce maximum transient overvoltages.

1910.269(l)(3)(iii)

The employer shall ensure that no employee approaches or takes any conductive object closer to exposed energized parts than the employer's established minimum approach distance, unless:

1910.269(l)(3)(iii)(A)

The employee is insulated from the energized part (rubber insulating gloves or rubber insulating gloves and sleeves worn in accordance with paragraph (l)(4) of this section constitutes insulation of the employee from the energized part upon which the employee is working provided that the employee has control of the part in a manner sufficient to prevent exposure to uninsulated portions of the employee's body), or

1910.269(l)(3)(iii)(B)

The energized part is insulated from the employee and from any other conductive object at a different potential, or

1910.269(l)(3)(iii)(C)

The employee is insulated from any other exposed conductive object in accordance with the requirements for live-line barehand work in paragraph (q)(3) of this section.

1910.269(l)(4)

Type of insulation

1910.269(l)(4)(i)

When an employee uses rubber insulating gloves as insulation from energized parts (under paragraph (l)(3)(iii)(A) of this section), the employer shall ensure that the employee also uses rubber insulating sleeves. However, an employee need not use rubber insulating sleeves if:

1910.269(l)(4)(i)(A)

Exposed energized parts on which the employee is not working are insulated from the employee; and

1910.269(l)(4)(i)(B)

When installing insulation for purposes of paragraph (l)(4)(i)(A) of this section, the employee installs the insulation from a position that does not expose his or her upper arm to contact with other energized parts.

1910.269(l)(4)(ii)

When an employee uses rubber insulating gloves or rubber insulating gloves and sleeves as insulation from energized parts (under paragraph (l)(3)(iii)(A) of this section), the employer shall ensure that the employee:

1910.269(l)(4)(ii)(A)

Puts on the rubber insulating gloves and sleeves in a position where he or she cannot reach into the minimum approach distance, established by the employer under paragraph (l)(3)(i) of this section; and

1910.269(l)(4)(ii)(B)

Does not remove the rubber insulating gloves and sleeves until he or she is in a position where he or she cannot reach into the minimum approach distance, established by the employer under paragraph (l)(3)(i) of this section.

1910.269(l)(5)

Working position.

1910.269(l)(5)(i)

The employer shall ensure that each employee, to the extent that other safety-related conditions at the worksite permit, works in a position from which a slip or shock will not bring the employee's body into contact with exposed, uninsulated parts energized at a potential different from the employee's.

1910.269(l)(5)(ii)

When an employee performs work near exposed parts energized at more than 600 volts, but not more than 72.5 kilovolts, and is not wearing rubber insulating gloves, being protected by insulating equipment covering the energized parts, performing work using live-line tools, or performing live-line barehand work under paragraph (q)(3) of this section, the employee shall work from a position where he or she cannot reach into the minimum approach distance, established by the employer under paragraph (l)(3)(i) of this section.

1910.269(l)(6)

Making connections. The employer shall ensure that employees make connections as follows:

1910.269(l)(6)(i)

In connecting deenergized equipment or lines to an energized circuit by means of a conducting wire or device, an employee shall first attach the wire to the deenergized part;

1910.269(l)(6)(ii)

When disconnecting equipment or lines from an energized circuit by means of a conducting wire or device, an employee shall remove the source end first; and

<u>1910.269(l)(6)(iii)</u>

When lines or equipment are connected to or disconnected from energized circuits, an employee shall keep loose conductors away from exposed energized parts.

1910.269(l)(7)

Conductive articles. When an employee performs work within reaching distance of exposed energized parts of equipment, the employer shall ensure that the employee removes or renders nonconductive all exposed conductive articles, such as keychains or watch chains, rings, or wrist watches or bands, unless such articles do not increase the hazards associated with contact with the energized parts.

1910.269(l)(8)

Protection from flames and electric arcs.

1910.269(l)(8)(i)

The employer shall assess the workplace to identify employees exposed to hazards from flames or from electric arcs.

1910.269(l)(8)(ii)

For each employee exposed to hazards from electric arcs, the employer shall make a reasonable estimate of the incident heat energy to which the employee would be exposed.

Note 1 to paragraph (l)(8)(ii):   Appendix E to this section provides guidance on estimating available heat energy. The Occupational Safety and Health Administration will deem employers following the guidance in Appendix E to this section to be in compliance with paragraph (l)(8)(ii) of this section. An employer may choose a method of calculating incident heat energy not included in Appendix E to this section if the chosen method reasonably predicts the incident energy to which the employee would be exposed.

Note 2 to paragraph (l)(8)(ii):   This paragraph does not require the employer to estimate the incident heat energy exposure for every job task performed by each employee. The employer may make broad estimates that cover multiple system areas provided the employer uses reasonable assumptions about the energy-exposure distribution throughout the system and provided the estimates represent the maximum employee exposure for those areas. For example, the employer could estimate the heat energy just outside a substation feeding a radial distribution system and use that estimate for all jobs performed on that radial system.

1910.269(l)(8)(iii)

The employer shall ensure that each employee who is exposed to hazards from flames or electric arcs does not wear clothing that could melt onto his or her skin or that could ignite and continue to burn when exposed to flames or the heat energy estimated under paragraph (l)(8)(ii) of this section.

Note to paragraph (l)(8)(iii) of this section: This paragraph prohibits clothing made from acetate, nylon, polyester, rayon and polypropylene, either alone or in blends, unless the employer demonstrates that the fabric has been treated to withstand the conditions that may be encountered by the employee or that the employee wears the clothing in such a manner as to eliminate the hazard involved.

1910.269(l)(8)(iv)

The employer shall ensure that the outer layer of clothing worn by an employee, except for clothing not required to be arc rated under paragraphs (l)(8)(v)(A) through (l)(8)(v)(E) of this section, is flame resistant under any of the following conditions:

1910.269(l)(8)(iv)(A)

The employee is exposed to contact with energized circuit parts operating at more than 600 volts,

1910.269(l)(8)(iv)(B)

An electric arc could ignite flammable material in the work area that, in turn, could ignite the employee's clothing,

1910.269(l)(8)(iv)(C)

Molten metal or electric arcs from faulted conductors in the work area could ignite the employee's clothing, or

Note to paragraph (l)(8)(iv)(C): This paragraph does not apply to conductors that are capable of carrying, without failure, the maximum available fault current for the time the circuit protective devices take to interrupt the fault.

1910.269(l)(8)(iv)(D)

The incident heat energy estimated under paragraph (l)(8)(ii) of this section exceeds 2.0 cal/cm^2.

1910.269(l)(8)(v)

The employer shall ensure that each employee exposed to hazards from electric arcs wears protective clothing and other protective equipment with an arc rating greater than or equal to the heat energy estimated under paragraph (l)(8)(ii) of this section whenever that estimate exceeds 2.0 cal/cm^2. This protective equipment shall cover the employee's entire body, except as follows:

1910.269(l)(8)(v)(A)

Arc-rated protection is not necessary for the employee's hands when the employee is wearing rubber insulating gloves with protectors or, if the estimated incident energy is no more than 14 cal/cm², heavy-duty leather work gloves with a weight of at least 407 gm/m² (12 oz/yd²),

1910.269(l)(8)(v)(B)

Arc-rated protection is not necessary for the employee's feet when the employee is wearing heavy-duty work shoes or boots,

1910.269(l)(8)(v)(C)

Arc-rated protection is not necessary for the employee's head when the employee is wearing head protection meeting § 1910.135 if the estimated incident energy is less than 9 cal/cm² for exposures involving single-phase arcs in open air or 5 cal/cm² for other exposures,

1910.269(l)(8)(v)(D)

The protection for the employee's head may consist of head protection meeting § 1910.135 and a faceshield with a minimum arc rating of 8 cal/cm² if the estimated incident-energy exposure is less than 13 cal/cm² for exposures involving single-phase arcs in open air or 9 cal/cm² for other exposures, and

1910.269(l)(8)(v)(E)

For exposures involving singlephase arcs in open air, the arc rating for the employee's head and face protection may be 4 cal/cm² less than the estimated incident energy.

Note to paragraph (l)(8):   See Appendix E to this section for further information on the selection of appropriate protection.

1910.269(l)(8)(vi)

Dates.

1910.269(l)(8)(vi)(A)

The obligation in paragraph (l)(8)(ii) of this section for the employer to make reasonable estimates of incident energy commences January 1, 2015.

1910.269(l)(8)(vi)(B)

The obligation in paragraph (l)(8)(iv)(D) of this section for the employer to ensure that the outer layer of clothing worn by an employee is flame-resistant when the estimated incident heat energy exceeds 2.0 cal/cm² commences April 1, 2015.

1910.269(l)(8)(vi)(C)

The obligation in paragraph (l)(8)(v) of this section for the employer to ensure that each employee exposed to hazards from electric arcs wears the required arc-rated protective equipment commences April 1, 2015.

1910.269(l)(9)

Fuse handling. When an employee must install or remove fuses with one or both terminals energized at more than 300 volts, or with exposed parts energized at more than 50 volts, the employer shall ensure that the employee uses tools or gloves rated for the voltage. When an employee installs or removes expulsion-type fuses with one or both terminals energized at more than 300 volts, the employer shall ensure that the employee wears eye protection meeting the requirements of Subpart I of this part, uses a tool rated for the voltage, and is clear of the exhaust path of the fuse barrel.

1910.269(l)(10)

Covered (noninsulated) conductors. The requirements of this section that pertain to the hazards of exposed live parts also apply when an employee performs work in proximity to covered (noninsulated) wires.

1910.269(l)(11)

Non-current-carrying metal parts. Non-current-carrying metal parts of equipment or devices, such as transformer cases and circuit-breaker housings, shall be treated as energized at the highest voltage to which these parts are exposed, unless the employer inspects the installation and determines that these parts are grounded before employees begin performing the work.

1910.269(l)(12)

Opening and closing circuits under load.

1910.269(l)(12)(i)

The employer shall ensure that devices used by employees to open circuits under load conditions are designed to interrupt the current involved.

1910.269(l)(12)(ii)

The employer shall ensure that devices used by employees to close circuits under load conditions are designed to safely carry the current involved.

Table R-3   AC Live-Line Work Minimum Approach Distance

[The minimum approach distance (MAD; in meters) shall conform to the following equations.]

For phase-to-phase system voltages of 50 V to 300 V:[1]

MAD = avoid contact

For phase-to-phase system voltages of 301 V to 5 kV:[1]
MAD = M + D, where
   D = 0.02 m .... the electrical component of the minimum approach distance.
   M = 0.31 m for voltages up to 750 V and 0.61 m otherwise.. the inadvertent movement factor.

For phase-to-phase system voltages of 5.1 kV to 72.5 kV:[1] [4]
MAD = M + AD, where
   M = 0.61 m .... the inadvertent movement factor.
   A = the applicable value from Table R-5 .... the altitude correction factor.
   D = the value from Table R-4 corresponding to the voltage and exposure or the value of the electrical component of the minimum approach distance calculated using the method provided in Appendix B to this section. the electrical component of the minimum approach distance.

For phase-to-phase system voltages of more than 72.5 kV, nominal:[2] [4]

$$MAD = 0.3048(C + a)V_{L-G}TA + M, \text{ where}$$

C = 0.01 for phase-to-ground exposures that the employer can demonstrate consist only of air across the approach distance (gap),
0.01 for phase-to-phase exposures if the employer can demonstrate that no insulated tool spans the gap and that no large conductive object is in the gap, or
0.011 otherwise

V_{L-G} = phase-to-ground rms voltage, in kV

T = maximum anticipated per-unit transient overvoltage; for phase-to-ground exposures, T equals T_{L-G}, the maximum per-unit transient overvoltage, phase-to-ground, determined by the employer under paragraph (l)(3)(ii) of this section; for phase-to-phase exposures, T equals $1.35T_{L-G} + 0.45$

A = altitude correction factor from Table R-5

M = 0.31 m, the inadvertent movement factor

a = saturation factor, as follows:

Phase-to-Ground Exposures

$V_{Peak} = T_{L-G}V_{L-G}\sqrt{2}$	635 kV	635.1 to	915.1 to	More than
a	or less 0	915 kV	1,050 kV	1,050 kV
		$(V_{Peak} - 635)/140,000$	$(V_{Peak} - 645)/135,000$	$(V_{Peak} - 675)/125,000$

Phase-to-Phase Exposures [3]

$V_{Peak} = (1.35T_{L-G} + 0.45)$ $V_{L-G}\sqrt{2}$	630 kV	630.1 to	848.1 to	1,131.1 to	More than
a	or less 0	848 kV	1,131 kV	1,485 kV	1,485 kV
		$(V_{Peak} - 630)/155,000$	$(V_{Peak} - 633.6)/152,207$	$(V_{Peak} - 628)/153,846$	$(V_{Peak} - 350.5)/203,666$

[1]Employers may use the minimum approach distances in Table R-6. If the worksite is at an elevation of more than 900 meters (3,000 feet), see footnote 1 to Table R-6.

[2]Employers may use the minimum approach distances in Table R-7, except that the employer may not use the minimum approach distances in Table R-7 for phase-to-phase exposures if an insulated tool spans the gap or if any large conductive object is in the gap. If the worksite is at an elevation of more than 900 meters (3,000 feet), see footnote 1 to Table R-7. Employers may use the minimum approach distances in Table 14 through Table 21 in Appendix B to this section, which calculated MAD for various values of T, provided the employer follows the notes to those tables.

[3]Use the equations for phase-to-ground exposures (with VPeak for phase-to-phase exposures) unless the employer can demonstrate that no insulated tool spans the gap and that no large conductive object is in the gap.

[4]Until March 31, 2015, employers may use the minimum approach distances in Table 6 through Table 13 in Appendix B to this section.

Table R-4 Electrical Component of the Minimum Approach Distance at 5.1 to 72.5 KV [D; In meters]

Nominal voltage (kV) phase-to-phase	Phase-to-ground exposure D (m)	Phase-to-phase exposure D (m)
5.1 to 15.0	0.04	0.07
15.1 to 36.0	0.16	0.28
36.1 to 46.0	0.23	0.37
46.1 to 72.5	0.39	0.59

Table R-5 Altitude Correction Factor

Altitude above sea level (m)	A
0 to 900	1.00
901 to 1,200	1.02
1,201 to 1,500	1.05
1,501 to 1,800	1.08
1,801 to 2,100	1.11
2,101 to 2,400	1.14
2,401 to 2,700	1.17
2,701 to 3,000	1.20
3,001 to 3,600	1.25
3,601 to 4,200	1.30
4,201 to 4,800	1.35
4,801 to 5,400	1.39
5,401 to 6,000	1.44

Table R-6 Alternative Minimum Approach Distances for Voltages of 72.5 KV and Less[1]

	Distance			
	Phase-to-ground exposure		Phase-to-phase exposure	
Nominal voltage (kV) phase-to-phase	m	ft	m	ft
0.50 to 0.300[2]	Avoid Contact		Avoid Contact	
0.301 to 0.750[2]	0.33	1.09	0.33	1.09
0.751 to 5.0	0.63	2.07	0.63	2.07
5.1 to 15.0	0.65	2.14	0.68	2.24
15.1 to 36.0	0.77	2.53	0.89	2.92
36.1 to 46.0	0.84	2.76	0.98	3.22
46.1 to 72.5	1.00	3.29	1.20	3.94

[1]Employers may use the minimum approach distances in this table provided the worksite is at an elevation of 900 meters (3,000 feet) or less. If employees will be working at elevations greater than 900 meters (3,000 feet) above mean sea level, the employer shall determine minimum approach distances by multiplying the distances in this table by the correction factor in Table R-5 corresponding to the altitude of the work.

[2]For single-phase systems, use voltage-to-ground.

Table R-7   Alternative Minimum Approach Distances for Voltages of More Than 72.5 Kv[1] [2] [3]

Voltage range phase to phase (kV)	Phase-to-ground exposure		Phase-to-phase exposure	
	m	ft	m	ft
72.6 to 121.0	1.13	3.71	1.42	4.66
121.1 to 145.0	1.30	4.27	1.64	5.38
145.1 to 169.0	1.46	4.79	1.94	6.36
169.1 to 242.0	2.01	6.59	3.08	10.10
242.1 to 362.0	3.41	11.19	5.52	18.11
362.1 to 420.0	4.25	13.94	6.81	22.34
420.1 to 550.0	5.07	16.63	8.24	27.03
550.1 to 800.0	6.88	22.57	11.38	37.34

[1]Employers may use the minimum approach distances in this table provided the worksite is at an elevation of 900 meters (3,000 feet) or less. If employees will be working at elevations greater than 900 meters (3,000 feet) above mean sea level, the employer shall determine minimum approach distances by multiplying the distances in this table by the correction factor in Table R-5 corresponding to the altitude of the work.

[2]Employers may use the phase-to-phase minimum approach distances in this table provided that no insulated tool spans the gap and no large conductive object is in the gap.

[3]The clear live-line tool distance shall equal or exceed the values for the indicated voltage ranges.

Table R-8   DC Live-Line Minimum Approach Distance with Overvoltage Factor[1]
[In Meters]

Maximum anticipated per-unit transient overvoltage	Distance (m) maximum line-to-ground voltage (kV)				
	250	400	500	600	750
1.5 or less	1.12	1.60	2.06	2.62	3.61
1.6	1.17	1.69	2.24	2.86	3.98
1.7	1.23	1.82	2.42	3.12	4.37
1.8	1.28	1.95	2.62	3.39	4.79

[1]The distances specified in this table are for air, bare-hand, and live-line tool conditions. If employees will be working at elevations greater than 900 meters (3,000 feet) above mean sea level, the employer shall determine minimum approach distances by multiplying the distances in this table by the correction factor in Table R-5 corresponding to the altitude of the work.

Table R-9   Assumed Maximum Per-Unit Transient Overvoltage

Voltage range (kV)	Type of current (ac or dc)	Assumed maximum per-unit transient overvoltage
72.6 to 420.0	ac	3.5
420.1 to 550.0	ac	3.0
550.1 to 800.0	ac	2.5
250 to 750	dc	1.8

1910.269(m)

Deenergizing lines and equipment for employee protection.

1910.269(m)(1)

Application. Paragraph (m) of this section applies to the deenergizing of transmission and distribution lines and equipment for the purpose of protecting employees. See paragraph (d) of this section for requirements on the control of hazardous energy sources used in the generation of electric energy. Conductors and parts of electric equipment that have been deenergized under procedures other than those required by paragraph (d) or (m) of this section, as applicable, shall be treated as energized.

1910.269(m)(2)

General.

1910.269(m)(2)(i)

If a system operator is in charge of the lines or equipment and their means of disconnection, the employer shall designate one employee in the crew to be in charge of the clearance and shall comply with all of the requirements of paragraph (m)(3) of this section in the order specified.

1910.269(m)(2)(ii)

If no system operator is in charge of the lines or equipment and their means of disconnection, the employer shall designate one employee in the crew to be in charge of the clearance and to perform the functions that the system operator would otherwise perform under paragraph (m) of this section. All of the requirements of paragraph (m)(3) of this section apply, in the order specified, except as provided in paragraph (m)(2)(iii) of this section.

1910.269(m)(2)(iii)

If only one crew will be working on the lines or equipment and if the means of disconnection is accessible and visible to, and under the sole control of, the employee in charge of the clearance, paragraphs (m)(3)(i), (m)(3)(iii), and (m)(3)(v) of this section do not apply. Additionally, the employer does not need to use the tags required by the remaining provisions of paragraph (m)(3) of this section.

1910.269(m)(2)(iv)

If two or more crews will be working on the same lines or equipment, then:

1910.269(m)(2)(iv)(A)

The crews shall coordinate their activities under paragraph (m) of this section with a single employee in charge of the clearance for all of the crews and follow the requirements of paragraph (m) of this section as if all of the employees formed a single crew, or

1910.269(m)(2)(iv)(B)

Each crew shall independently comply with paragraph (m) of this section and, if there is no system operator in charge of the lines or equipment, shall have separate tags and coordinate deenergizing and reenergizing the lines and equipment with the other crews.

1910.269(m)(2)(v)

The employer shall render any disconnecting means that are accessible to individuals outside the employer's control (for example, the general public) inoperable while the disconnecting means are open for the purpose of protecting employees.

1910.269(m)(3)

Deenergizing lines and equipment.

1910.269(m)(3)(i)

The employee that the employer designates pursuant to paragraph (m)(2) of this section as being in charge of the clearance shall make a request of the system operator to deenergize the particular section of line or equipment. The designated employee becomes the employee in charge (as this term is used in paragraph (m)(3) of this section) and is responsible for the clearance.

1910.269(m)(3)(ii)

The employer shall ensure that all switches, disconnectors, jumpers, taps, and other means through which known sources of electric energy may be supplied to the particular lines and equipment to be deenergized are open. The employer shall render such means inoperable, unless its design does not so permit, and then ensure that such means are tagged to indicate that employees are at work.

1910.269(m)(3)(iii)

The employer shall ensure that automatically and remotely controlled switches that could cause the opened disconnecting means to close are also tagged at the points of control. The employer shall render the automatic or remote control feature inoperable, unless its design does not so permit.

1910.269(m)(3)(iv)

The employer need not use the tags mentioned in paragraphs (m)(3)(ii) and (m)(3)(iii) of this section on a network protector for work on the primary feeder for the network protector's associated network transformer when the employer can demonstrate all of the following conditions:

1910.269(m)(3)(iv)(A)

Every network protector is maintained so that it will immediately trip open if closed when a primary conductor is deenergized;

1910.269(m)(3)(iv)(B)

Employees cannot manually place any network protector in a closed position without the use of tools, and any manual override position is blocked, locked, or otherwise disabled; and

1910.269(m)(3)(iv)(C)

The employer has procedures for manually overriding any network protector that incorporate provisions for determining, before anyone places a network protector in a closed position, that: The line connected to the network protector is not deenergized for the protection of any employee working on the line; and (if the line connected to the network protector is not deenergized for the protection of any employee working on the line) the primary conductors for the network protector are energized.

1910.269(m)(3)(v)

Tags shall prohibit operation of the disconnecting means and shall indicate that employees are at work.

1910.269(m)(3)(vi)

After the applicable requirements in paragraphs (m)(3)(i) through (m)(3)(v) of this section have been followed and the system operator gives a clearance to the employee in charge, the employer shall ensure that the lines and equipment are deenergized by testing the lines and equipment to be worked with a device designed to detect voltage.

1910.269(m)(3)(vii)

The employer shall ensure the installation of protective grounds as required by paragraph (n) of this section.

<u>1910.269(m)(3)(viii)</u>

After the applicable requirements of paragraphs (m)(3)(i) through (m)(3)(vii) of this section have been followed, the lines and equipment involved may be considered deenergized.

1910.269(m)(3)(ix)

To transfer the clearance, the employee in charge (or the employee's supervisor if the employee in charge must leave the worksite due to illness or other emergency) shall inform the system operator and employees in the crew; and the new employee in charge shall be responsible for the clearance.

1910.269(m)(3)(x)

To release a clearance, the employee in charge shall:

1910.269(m)(3)(x)(A)

Notify each employee under that clearance of the pending release of the clearance;

1910.269(m)(3)(x)(B)

Ensure that all employees under that clearance are clear of the lines and equipment;

1910.269(m)(3)(x)(C)

Ensure that all protective grounds protecting employees under that clearance have been removed; and

1910.269(m)(3)(x)(D)

Report this information to the system operator and then release the clearance.

1910.269(m)(3)(xi)

Only the employee in charge who requested the clearance may release the clearance, unless the employer transfers responsibility under paragraph (m)(3)(ix) of this section.

1910.269(m)(3)(xii)

No one may remove tags without the release of the associated clearance as specified under paragraphs (m)(3)(x) and (m)(3)(xi) of this section.

1910.269(m)(3)(xiii)

The employer shall ensure that no one initiates action to reenergize the lines or equipment at a point of disconnection until all protective grounds have been removed, all crews working on the lines or equipment release their clearances, all employees are clear of the lines and equipment, and all protective tags are removed from that point of disconnection.

1910.269(n)

Grounding for the protection of employees.

1910.269(n)(1)

Application. Paragraph (n) of this section applies to grounding of generation, transmission, and distribution lines and equipment for the purpose of protecting employees. Paragraph (n)(4) of this section also applies to protective grounding of other equipment as required elsewhere in this section.

Note to paragraph (n)(1): This paragraph covers grounding of generation, transmission, and distribution lines and equipment when this section requires protective grounding and whenever the employer chooses to ground such lines and equipment for the protection of employees.

1910.269(n)(2)

General. For any employee to work transmission and distribution lines or equipment as deenergized, the employer shall ensure that the lines or equipment are deenergized under the provisions of paragraph (m) of this section and shall ensure proper grounding of the lines or equipment as specified in paragraphs (n)(3) through (n)(8) of this section. However, if the employer can demonstrate that installation of a ground is impracticable or that the conditions resulting from the installation of a ground would present greater hazards to employees than working without grounds, the lines and equipment may be treated as deenergized provided that the employer establishes that all of the following conditions apply:

1910.269(n)(2)(i)

The employer ensures that the lines and equipment are deenergized under the provisions of paragraph (m) of this section.

1910.269(n)(2)(ii)

There is no possibility of contact with another energized source.

1910.269(n)(2)(iii)

The hazard of induced voltage is not present.

1910.269(n)(3)

Equipotential zone. Temporary protective grounds shall be placed at such locations and arranged in such a manner that the employer can demonstrate will prevent each employee from being exposed to hazardous differences in electric potential.

Note to paragraph (n)(3): Appendix C to this section contains guidelines for establishing the equipotential zone required by this paragraph. The Occupational Safety and Health Administration will deem grounding practices meeting these guidelines as complying with paragraph (n)(3) of this section.

1910.269(n)(4)

Protective grounding equipment.

1910.269(n)(4)(i)

Protective grounding equipment shall be capable of conducting the maximum fault current that could flow at the point of grounding for the time necessary to clear the fault.

1910.269(n)(4)(ii)

Protective grounding equipment shall have an ampacity greater than or equal to that of No. 2 AWG copper.

1910.269(n)(4)(iii)

Protective grounds shall have an impedance low enough so that they do not delay the operation of protective devices in case of accidental energizing of the lines or equipment.

Note to paragraph (n)(4): American Society for Testing and Materials *Standard Specifications for Temporary Protective Grounds to Be Used on De-Energized Electric Power Lines and Equipment*, ASTM F855-09, contains guidelines for protective grounding equipment. The Institute of Electrical Engineers*Guide for Protective Grounding of Power Lines*, IEEE Std 1048-2003, contains guidelines for selecting and installing protective grounding equipment.

1910.269(n)(5)

Testing. The employer shall ensure that, unless a previously installed ground is present, employees test lines and equipment and verify the absence of nominal voltage before employees install any ground on those lines or that equipment.

1910.269(n)(6)

Connecting and removing grounds.

1910.269(n)(6)(i)

The employer shall ensure that, when an employee attaches a ground to a line or to equipment, the employee attaches the ground-end connection first and then attaches the other end by means of a live-line tool. For lines or equipment operating at 600 volts or less, the employer may permit the employee to use insulating equipment other than a live-line tool if the employer ensures that the line or equipment is not energized at the time the ground is connected or if the employer can demonstrate that each employee is protected from hazards that may develop if the line or equipment is energized.

1910.269(n)(6)(ii)

The employer shall ensure that, when an employee removes a ground, the employee removes the grounding device from the line or equipment using a live-line tool before he or she removes the ground-end connection. For lines or equipment operating at 600 volts or less, the employer may permit the employee to use insulating equipment other than a live-line tool if the employer ensures that the line or equipment is not energized at the time the ground is disconnected or if the employer can demonstrate that each employee is protected from hazards that may develop if the line or equipment is energized.

1910.269(n)(7)

Additional precautions. The employer shall ensure that, when an employee performs work on a cable at a location remote from the cable terminal, the cable is not grounded at the cable terminal if there is a possibility of hazardous transfer of potential should a fault occur.

1910.269(n)(8)

Removal of grounds for test. The employer may permit employees to remove grounds temporarily during tests. During the test procedure, the employer shall ensure that each employee uses insulating equipment, shall isolate each employee from any hazards involved, and shall implement any additional measures necessary to protect each exposed employee in case the previously grounded lines and equipment become energized.

1910.269(o)

Testing and test facilities.

1910.269(o)(1)

Application. Paragraph (o) of this section provides for safe work practices for high-voltage and high-power testing performed in laboratories, shops, and substations, and in the field and on electric transmission and distribution lines and equipment. It applies only to testing involving interim measurements using high voltage, high power, or combinations of high voltage and high power, and not to testing involving continuous measurements as in routine metering, relaying, and normal line work.

Note to paragraph (o)(1): OSHA considers routine inspection and maintenance measurements made by qualified employees to be routine line work not included in the scope of paragraph (o) of this section, provided that the hazards related to the use of intrinsic high-voltage or high-power sources require only the normal precautions associated with routine work specified in the other paragraphs of this section. Two typical examples of such excluded test work procedures are "phasing-out" testing and testing for a "no-voltage" condition.

1910.269(o)(2)

General requirements.

1910.269(o)(2)(i)

The employer shall establish and enforce work practices for the protection of each worker from the hazards of high-voltage or high-power testing at all test areas, temporary and permanent. Such work practices shall include, as a minimum, test area safeguarding, grounding, the safe use of measuring and control circuits, and a means providing for periodic safety checks of field test areas.

1910.269(o)(2)(ii)

The employer shall ensure that each employee, upon initial assignment to the test area, receives training in safe work practices, with retraining provided as required by paragraph (a)(2) of this section.

1910.269(o)(3)

Safeguarding of test areas.

1910.269(o)(3)(i)

The employer shall provide safeguarding within test areas to control access to test equipment or to apparatus under test that could become energized as part of the testing by either direct or inductive coupling and to prevent accidental employee contact with energized parts.

1910.269(o)(3)(ii)

The employer shall guard permanent test areas with walls, fences, or other barriers designed to keep employees out of the test areas.

1910.269(o)(3)(iii)

In field testing, or at a temporary test site not guarded by permanent fences and gates, the employer shall ensure the use of one of the following means to prevent employees without authorization from entering:

1910.269(o)(3)(iii)(A)

Distinctively colored safety tape supported approximately waist high with safety signs attached to it,

1910.269(o)(3)(iii)(B)

A barrier or barricade that limits access to the test area to a degree equivalent, physically and visually, to the barricade specified in paragraph (o)(3)(iii)(A) of this section, or

1910.269(o)(3)(iii)(C)

One or more test observers stationed so that they can monitor the entire area.

1910.269(o)(3)(iv)

The employer shall ensure the removal of the safeguards required by paragraph (o)(3)(iii) of this section when employees no longer need the protection afforded by the safeguards.

1910.269(o)(4)

Grounding practices.

1910.269(o)(4)(i)

The employer shall establish and implement safe grounding practices for the test facility.

1910.269(o)(4)(i)(A)

The employer shall maintain at ground potential all conductive parts accessible to the test operator while the equipment is operating at high voltage.

1910.269(o)(4)(i)(B)

Wherever ungrounded terminals of test equipment or apparatus under test may be present, they shall be treated as energized until tests demonstrate that they are deenergized.

1910.269(o)(4)(ii)

The employer shall ensure either that visible grounds are applied automatically, or that employees using properly insulated tools manually apply visible grounds, to the high-voltage circuits after they are deenergized and before any employee performs work on the circuit or on the item or apparatus under test. Common ground connections shall be solidly connected to the test equipment and the apparatus under test.

1910.269(o)(4)(iii)

In high-power testing, the employer shall provide an isolated ground-return conductor system designed to prevent the intentional passage of current, with its attendant voltage rise, from occurring in the ground grid or in the earth. However, the employer need not provide an isolated ground-return conductor if the employer can demonstrate that both of the following conditions exist:

1910.269(o)(4)(iii)(A)

The employer cannot provide an isolated ground-return conductor due to the distance of the test site from the electric energy source, and

1910.269(o)(4)(iii)(B)

The employer protects employees from any hazardous step and touch potentials that may develop during the test

Note to paragraph (o)(4)(iii)(B):   See Appendix C to this section for information on measures that employers can take to protect employees from hazardous step and touch potentials.

1910.269(o)(4)(iv)

For tests in which using the equipment grounding conductor in the equipment power cord to ground the test equipment would result in greater hazards to test personnel or prevent the taking of satisfactory measurements, the employer may use a ground clearly indicated in the test set-up if the employer can demonstrate that this ground affords protection for employees equivalent to the protection afforded by an equipment grounding conductor in the power supply cord.

1910.269(o)(4)(v)

The employer shall ensure that, when any employee enters the test area after equipment is deenergized, a ground is placed on the high-voltage terminal and any other exposed terminals.

1910.269(o)(4)(v)(A)

Before any employee applies a direct ground, the employer shall discharge high capacitance equipment through a resistor rated for the available energy.

1910.269(o)(4)(v)(B)

A direct ground shall be applied to the exposed terminals after the stored energy drops to a level at which it is safe to do so.

1910.269(o)(4)(vi)

If the employer uses a test trailer or test vehicle in field testing, its chassis shall be grounded. The employer shall protect each employee against hazardous touch potentials with respect to the vehicle, instrument panels, and other conductive parts accessible to employees with bonding, insulation, or isolation.

1910.269(o)(5)

Control and measuring circuits.

1910.269(o)(5)(i)

The employer may not run control wiring, meter connections, test leads, or cables from a test area unless contained in a grounded metallic sheath and terminated in a grounded metallic enclosure or unless the employer takes other precautions that it can demonstrate will provide employees with equivalent safety.

1910.269(o)(5)(ii)

The employer shall isolate meters and other instruments with accessible terminals or parts from test personnel to protect against hazards that could arise should such terminals and parts become energized during testing. If the employer provides this isolation by locating test equipment in metal compartments with viewing windows, the employer shall provide interlocks to interrupt the power supply when someone opens the compartment cover.

1910.269(o)(5)(iii)

The employer shall protect temporary wiring and its connections against damage, accidental interruptions, and other hazards. To the maximum extent possible, the employer shall keep signal, control, ground, and power cables separate from each other.

1910.269(o)(5)(iv)

If any employee will be present in the test area during testing, a test observer shall be present. The test observer shall be capable of implementing the immediate deenergizing of test circuits for safety purposes.

1910.269(o)(6)

Safety check.

1910.269(o)(6)(i)

Safety practices governing employee work at temporary or field test areas shall provide, at the beginning of each series of tests, for a routine safety check of such test areas.

1910.269(o)(6)(ii)

The test operator in charge shall conduct these routine safety checks before each series of tests and shall verify at least the following conditions:

1910.269(o)(6)(ii)(A)

Barriers and safeguards are in workable condition and placed properly to isolate hazardous areas;

1910.269(o)(6)(ii)(B)

System test status signals, if used, are in operable condition;

1910.269(o)(6)(ii)(C)

Clearly marked test-power disconnects are readily available in an emergency;

1910.269(o)(6)(ii)(D)

Ground connections are clearly identifiable;

1910.269(o)(6)(ii)(E)

Personal protective equipment is provided and used as required by Subpart I of this part and by this section; and

1910.269(o)(6)(ii)(F)

Proper separation between signal, ground, and power cables.

1910.269(p)

Mechanical equipment.

1910.269(p)(1)

General requirements.

1910.269(p)(1)(i)

The critical safety components of mechanical elevating and rotating equipment shall receive a thorough visual inspection before use on each shift.

Note to paragraph (p)(1)(i): Critical safety components of mechanical elevating and rotating equipment are components for which failure would result in free fall or free rotation of the boom.

1910.269(p)(1)(ii)

No motor vehicle or earthmoving or compacting equipment having an obstructed view to the rear may be operated on off-highway jobsites where any employee is exposed to the hazards created by the moving vehicle, unless:

1910.269(p)(1)(ii)(A)

The vehicle has a reverse signal alarm audible above the surrounding noise level, or

1910.269(p)(1)(ii)(B)

The vehicle is backed up only when a designated employee signals that it is safe to do so.

1910.269(p)(1)(iii)

Rubber-tired self-propelled scrapers, rubber-tired front-end loaders, rubber-tired dozers, wheel-type agricultural and industrial tractors, crawler-type tractors, crawler-type loaders, and motor graders, with or without attachments, shall have rollover protective structures that meet the requirements of Subpart W of Part 1926 of this chapter.

1910.269(p)(1)(iv)

The operator of an electric line truck may not leave his or her position at the controls while a load is suspended, unless the employer can demonstrate that no employee (including the operator) is endangered.

1910.269(p)(2)

Outriggers.

1910.269(p)(2)(i)

Mobile equipment, if provided with outriggers, shall be operated with the outriggers extended and firmly set, except as provided in paragraph (p)(2)(iii) of this section.

1910.269(p)(2)(ii)

Outriggers may not be extended or retracted outside of the clear view of the operator unless all employees are outside the range of possible equipment motion.

1910.269(p)(2)(iii)

If the work area or the terrain precludes the use of outriggers, the equipment may be operated only within its maximum load ratings specified by the equipment manufacturer for the particular configuration of the equipment without outriggers.

1910.269(p)(3)

Applied loads. Mechanical equipment used to lift or move lines or other material shall be used within its maximum load rating and other design limitations for the conditions under which the mechanical equipment is being used

1910.269(p)(4)

Operations near energized lines or equipment.

1910.269(p)(4)(i)

Mechanical equipment shall be operated so that the minimum approach distances, established by the employer under paragraph (l)(3)(i) of this section, are maintained from exposed energized lines and equipment. However, the insulated portion of an aerial lift operated by a qualified employee in the lift is exempt from this requirement if the applicable minimum approach distance is maintained between the uninsulated portions of the aerial lift and exposed objects having a different electrical potential.

1910.269(p)(4)(ii)

A designated employee other than the equipment operator shall observe the approach distance to exposed lines and equipment and provide timely warnings before the minimum approach distance required by paragraph (p)(4)(i) of this section is reached, unless the employer can demonstrate that the operator can accurately determine that the minimum approach distance is being maintained.

1910.269(p)(4)(iii)

If, during operation of the mechanical equipment, that equipment could become energized, the operation also shall comply with at least one of paragraphs (p)(4)(iii)(A) through (p)(4)(iii)(C) of this section.

1910.269(p)(4)(iii)(A)

The energized lines or equipment exposed to contact shall be covered with insulating protective material that will withstand the type of contact that could be made during the operation.

1910.269(p)(4)(iii)(B)

The mechanical equipment shall be insulated for the voltage involved. The mechanical equipment shall be positioned so that its uninsulated portions cannot approach the energized lines or equipment any closer than the minimum approach distances, established by the employer under paragraph (l)(3)(i) of this section.

1910.269(p)(4)(iii)(C)

Each employee shall be protected from hazards that could arise from mechanical equipment contact with energized lines or equipment. The measures used shall ensure that employees will not be exposed to hazardous differences in electric potential. Unless the employer can demonstrate that the methods in use protect each employee from the hazards that could arise if the mechanical equipment contacts the energized line or equipment, the measures used shall include all of the following techniques:

1910.269(p)(4)(iii)(C)(1)

Using the best available ground to minimize the time the lines or electric equipment remain energized,

1910.269(p)(4)(iii)(C)(2)

Bonding mechanical equipment together to minimize potential differences,

1910.269(p)(4)(iii)(C)(3)

Providing ground mats to extend areas of equipotential, and

1910.269(p)(4)(iii)(C)(4)

Employing insulating protective equipment or barricades to guard against any remaining hazardous electrical potential differences.

Note to paragraph (p)(4)(iii)(C):   Appendix C to this section contains information on hazardous step and touch potentials and on methods of protecting employees from hazards resulting from such potentials.

1910.269(q)

Overhead lines and live-line barehand work. This paragraph provides additional requirements for work performed on or near overhead lines and equipment and for live-line barehand work.

1910.269(q)(1)

General.

1910.269(q)(1)(i)

Before allowing employees to subject elevated structures, such as poles or towers, to such stresses as climbing or the installation or removal of equipment may impose, the employer shall ascertain that the structures are capable of sustaining the additional or unbalanced stresses. If the pole or other structure cannot withstand the expected loads, the employer shall brace or otherwise support the pole or structure so as to prevent failure.

Note to paragraph (q)(1)(i):   Appendix D to this section contains test methods that employers can use in ascertaining whether a wood pole is capable of sustaining the forces imposed by an employee climbing the pole. This paragraph also requires the employer to ascertain that the pole can sustain all other forces imposed by the work employees will perform.

1910.269(q)(1)(ii)

When a pole is set, moved, or removed near an exposed energized overhead conductor, the pole may not contact the conductor.

1910.269(q)(1)(iii)

When a pole is set, moved, or removed near an exposed energized overhead conductor, the employer shall ensure that each employee wears electrical protective equipment or uses insulated devices when handling the pole and that no employee contacts the pole with uninsulated parts of his or her body.

1910.269(q)(1)(iv)

To protect employees from falling into holes used for placing poles, the employer shall physically guard the holes, or ensure that employees attend the holes, whenever anyone is working nearby.

1910.269(q)(2)

Installing and removing overhead lines. The following provisions apply to the installation and removal of overhead conductors or cable (overhead lines).

1910.269(q)(2)(i)

When lines that employees are installing or removing can contact energized parts, the employer shall use the tension-stringing method, barriers, or other equivalent measures to minimize the possibility that conductors and cables the employees are installing or removing will contact energized power lines or equipment.

1910.269(q)(2)(ii)

For conductors, cables, and pulling and tensioning equipment, the employer shall provide the protective measures required by paragraph (p)(4)(iii) of this section when employees are installing or removing a conductor or cable close enough to energized conductors that any of the following failures could energize the pulling or tensioning equipment or the conductor or cable being installed or removed:

1910.269(q)(2)(ii)(A)

Failure of the pulling or tensioning equipment,

1910.269(q)(2)(ii)(B)

Failure of the conductor or cable being pulled, or

1910.269(q)(2)(ii)(C)

Failure of the previously installed lines or equipment.

1910.269(q)(2)(iii)

If the conductors that employees are installing or removing cross over energized conductors in excess of 600 volts and if the design of the circuitinterrupting devices protecting the lines so permits, the employer shall render inoperable the automatic-reclosing feature of these devices.

1910.269(q)(2)(iv)

Before employees install lines parallel to existing energized lines, the employer shall make a determination of the approximate voltage to be induced in the new lines, or work shall proceed on the assumption that the induced voltage is hazardous. Unless the employer can demonstrate that the lines that employees are installing are not subject to the induction of a hazardous voltage or unless the lines are treated as energized, temporary protective grounds shall be placed at such locations and arranged in such a manner that the employer can demonstrate will prevent exposure of each employee to hazardous differences in electric potential.

Note 1 to paragraph (q)(2)(iv): If the employer takes no precautions to protect employees from hazards associated with involuntary reactions from electric shock, a hazard exists if the induced voltage is sufficient to pass a current of 1 milliampere through a 500-ohm resistor. If the employer protects employees from injury due to involuntary reactions from electric shock, a hazard exists if the resultant current would be more than 6 milliamperes.

Note 2 to paragraph (q)(2)(iv): Appendix C to this section contains guidelines for protecting employees from hazardous differences in electric potential as required by this paragraph.

1910.269(q)(2)(v)

Reel-handling equipment, including pulling and tensioning devices, shall be in safe operating condition and shall be leveled and aligned.

1910.269(q)(2)(vi)

The employer shall ensure that employees do not exceed load ratings of stringing lines, pulling lines, conductor grips, load-bearing hardware and accessories, rigging, and hoists.

1910.269(q)(2)(vii)

The employer shall repair or replace defective pulling lines and accessories.

1910.269(q)(2)(viii)

The employer shall ensure that employees do not use conductor grips on wire rope unless the manufacturer specifically designed the grip for this application.

1910.269(q)(2)(ix)

The employer shall ensure that employees maintain reliable communications, through two-way radios or other equivalent means, between the reel tender and the pullingrig operator.

1910.269(q)(2)(x)

Employees may operate the pulling rig only when it is safe to do so.

Note to paragraph (q)(2)(x): Examples of unsafe conditions include: employees in locations prohibited by paragraph (q)(2)(xi) of this section, conductor and pulling line hang-ups, and slipping of the conductor grip.

1910.269(q)(2)(xi)

While a power-driven device is pulling the conductor or pulling line and the conductor or pulling line is in motion, the employer shall ensure that employees are not directly under overhead operations or on the crossarm, except as necessary for the employees to guide the stringing sock or board over or through the stringing sheave.

1910.269(q)(3)

Live-line barehand work. In addition to other applicable provisions contained in this section, the following requirements apply to live-line barehand work:

1910.269(q)(3)(i)

Before an employee uses or supervises the use of the live-line barehand technique on energized circuits, the employer shall ensure that the employee completes training conforming to paragraph (a)(2) of this section in the technique and in the safety requirements of paragraph (q)(3) of this section.

1910.269(q)(3)(ii)

Before any employee uses the liveline barehand technique on energized high-voltage conductors or parts, the employer shall ascertain the following information in addition to information about other existing conditions required by paragraph (a)(4) of this section:

1910.269(q)(3)(ii)(A)

The nominal voltage rating of the circuit on which employees will perform the work,

1910.269(q)(3)(ii)(B)

The clearances to ground of lines and other energized parts on which employees will perform the work, and

1910.269(q)(3)(ii)(C)

The voltage limitations of equipment employees will use.

1910.269(q)(3)(iii)

The employer shall ensure that the insulated equipment, insulated tools, and aerial devices and platforms used by employees are designed, tested, and made for live-line barehand work.

1910.269(q)(3)(iv)

The employer shall ensure that employees keep tools and equipment clean and dry while they are in use.

1910.269(q)(3)(v)

The employer shall render inoperable the automatic-reclosing feature of circuit-interrupting devices protecting the lines if the design of the devices permits.

1910.269(q)(3)(vi)

The employer shall ensure that employees do not perform work when adverse weather conditions would make the work hazardous even after the employer implements the work practices required by this section. Additionally, employees may not perform work when winds reduce the phase-to-phase or phase-to-ground clearances at the work location below the minimum approach distances specified in paragraph (q)(3)(xiv) of this section, unless insulating guards cover the grounded objects and other lines and equipment.

Note to paragraph (q)(3)(vi): Thunderstorms in the vicinity, high winds, snow storms, and ice storms are examples of adverse weather conditions that make liveline barehand work too hazardous to perform safely even after the employer implements the work practices required by this section.

1910.269(q)(3)(vii)

The employer shall provide and ensure that employees use a conductive bucket liner or other conductive device for bonding the insulated aerial device to the energized line or equipment.

1910.269(q)(3)(vii)(A)

The employee shall be connected to the bucket liner or other conductive device by the use of conductive shoes, leg clips, or other means.

1910.269(q)(3)(vii)(B)

Where differences in potentials at the worksite pose a hazard to employees, the employer shall provide electrostatic shielding designed for the voltage being worked.

1910.269(q)(3)(viii)

The employer shall ensure that, before the employee contacts the energized part, the employee bonds the conductive bucket liner or other conductive device to the energized conductor by means of a positive connection. This connection shall remain attached to the energized conductor until the employee completes the work on the energized circuit.

1910.269(q)(3)(ix)

Aerial lifts used for live-line barehand work shall have dual controls (lower and upper) as follows:

1910.269(q)(3)(ix)(A)

The upper controls shall be within easy reach of the employee in the bucket. On a two-bucket-type lift, access to the controls shall be within easy reach of both buckets.

1910.269(q)(3)(ix)(B)

The lower set of controls shall be near the base of the boom and shall be designed so that they can override operation of the equipment at any time.

1910.269(q)(3)(x)

Lower (ground-level) lift controls may not be operated with an employee in the lift except in case of emergency.

1910.269(q)(3)(xi)

The employer shall ensure that, before employees elevate an aerial lift into the work position, the employees check all controls (ground level and bucket) to determine that they are in proper working condition.

1910.269(q)(3)(xii)

The employer shall ensure that, before employees elevate the boom of an aerial lift, the employees ground the body of the truck or barricade the body of the truck and treat it as energized.

1910.269(q)(3)(xiii)

The employer shall ensure that employees perform a boom-current test before starting work each day, each time during the day when they encounter a higher voltage, and when changed conditions indicate a need for an additional test.

1910.269(q)(3)(xiii)(A)

This test shall consist of placing the bucket in contact with an energized source equal to the voltage to be encountered for a minimum of 3 minutes.

1910.269(q)(3)(xiii)(B)

The leakage current may not exceed 1 microampere per kilovolt of nominal phase-to-ground voltage.

1910.269(q)(3)(xiii)(C)

The employer shall immediately suspend work from the aerial lift when there is any indication of a malfunction in the equipment.

1910.269(q)(3)(xiv)

The employer shall ensure that employees maintain the minimum approach distances, established by the employer under paragraph (l)(3)(i) of this section, from all grounded objects and from lines and equipment at a potential different from that to which the live-line barehand equipment is bonded, unless insulating guards cover such grounded objects and other lines and equipment.

1910.269(q)(3)(xv)

The employer shall ensure that, while an employee is approaching, leaving, or bonding to an energized circuit, the employee maintains the minimum approach distances, established by the employer under paragraph (l)(3)(i) of this section, between the employee and any grounded parts, including the lower boom and portions of the truck and between the employee and conductive objects energized at different potentials.

1910.269(q)(3)(xvi)

While the bucket is alongside an energized bushing or insulator string, the employer shall ensure that employees maintain the phase-to ground minimum approach distances, established by the employer under paragraph (l)(3)(i) of this section, between all parts of the bucket and the grounded end of the bushing or insulator string or any other grounded surface.

1910.269(q)(3)(xvii)

The employer shall ensure that employees do not use handlines between the bucket and the boom or between the bucket and the ground. However, employees may use nonconductive-type handlines from conductor to ground if not supported from the bucket. The employer shall ensure that no one uses ropes used for live-line barehand work for other purposes.

1910.269(q)(3)(xviii)

The employer shall ensure that employees do not pass uninsulated equipment or material between a pole or structure and an aerial lift while an employee working from the bucket is bonded to an energized part.

1910.269(q)(3)(xix)

A nonconductive measuring device shall be readily accessible to employees performing live-line barehand work to assist them in maintaining the required minimum approach distance.

1910.269(q)(4)

Towers and structures. The following requirements apply to work performed on towers or other structures that support overhead lines.

1910.269(q)(4)(i)

The employer shall ensure that no employee is under a tower or structure while work is in progress, except when the employer can demonstrate that such a working position is necessary to assist employees working above.

1910.269(q)(4)(ii)

The employer shall ensure that employees use tag lines or other similar devices to maintain control of tower sections being raised or positioned, unless the employer can demonstrate that the use of such devices would create a greater hazard to employees.

1910.269(q)(4)(iii)

The employer shall ensure that employees do not detach the loadline from a member or section until they safely secure the load.

1910.269(q)(4)(iv)

The employer shall ensure that, except during emergency restoration procedures, employees discontinue work when adverse weather conditions would make the work hazardous in spite of the work practices required by this section.

Note to paragraph (q)(4)(iv): Thunderstorms in the vicinity, high winds, snow storms, and ice storms are examples of adverse weather conditions that make this work too hazardous to perform even after the employer implements the work practices required by this section.

1910.269(r)

Line-clearance tree trimming operations. This paragraph provides additional requirements for line-clearance tree-trimming operations and for equipment used in these operations.

1910.269(r)(1)

Electrical hazards. This paragraph does not apply to qualified employees.

1910.269(r)(1)(i)

Before an employee climbs, enters, or works around any tree, a determination shall be made of the nominal voltage of electric power lines posing a hazard to employees. However, a determination of the maximum nominal voltage to which an employee will be exposed may be made instead, if all lines are considered as energized at this maximum voltage.

1910.269(r)(1)(ii)

There shall be a second line-clearance tree trimmer within normal (that is, unassisted) voice communication under any of the following conditions:

1910.269(r)(1)(ii)(A)

If a line-clearance tree trimmer is to approach more closely than 3.05 meters (10 feet) to any conductor or electric apparatus energized at more than 750 volts or

1910.269(r)(1)(ii)(B)

If branches or limbs being removed are closer to lines energized at more than 750 volts than the distances listed in Table R-5, Table R-6, Table R-7, and Table R-8 or

1910.269(r)(1)(ii)(C)

If roping is necessary to remove branches or limbs from such conductors or apparatus.

1910.269(r)(1)(iii)

Line-clearance tree trimmers shall maintain the minimum approach distances from energized conductors given in Table R-5, Table R-6, Table R-7, and Table R-8.

1910.269(r)(1)(iv)

Branches that are contacting exposed energized conductors or equipment or that are within the distances specified in Table R-5, Table R-6, Table R-7, and Table R-8 may be removed only through the use of insulating equipment.

Note to paragraph (r)(1)(iv): A tool constructed of a material that the employer can demonstrate has insulating qualities meeting paragraph (j)(1) of this section is considered as insulated under paragraph (r)(1)(iv) of this section if the tool is clean and dry.

1910.269(r)(1)(v)

Ladders, platforms, and aerial devices may not be brought closer to an energized part than the distances listed in Table R-5, Table R-6, Table R-7, and Table R-8.

1910.269(r)(1)(vi)

Line-clearance tree-trimming work may not be performed when adverse weather conditions make the work hazardous in spite of the work practices required by this section. Each employee performing line-clearance tree trimming work in the aftermath of a storm or under similar emergency conditions shall be trained in the special hazards related to this type of work.

Note to paragraph (r)(1)(vi): Thunderstorms in the immediate vicinity, high winds, snow storms, and ice storms are examples of adverse weather conditions that are presumed to make line-clearance tree trimming work too hazardous to perform safely.

1910.269(r)(2)

Brush chippers.

1910.269(r)(2)(i)

Brush chippers shall be equipped with a locking device in the ignition system.

1910.269(r)(2)(ii)

Access panels for maintenance and adjustment of the chipper blades and associated drive train shall be in place and secure during operation of the equipment.

1910.269(r)(2)(iii)

Brush chippers not equipped with a mechanical infeed system shall be equipped with an infeed hopper of length sufficient to prevent employees from contacting the blades or knives of the machine during operation.

1910.269(r)(2)(iv)

Trailer chippers detached from trucks shall be chocked or otherwise secured.

1910.269(r)(2)(v)

Each employee in the immediate area of an operating chipper feed table shall wear personal protective equipment as required by Subpart I of this part.

1910.269(r)(3)

Sprayers and related equipment.

1910.269(r)(3)(i)

Walking and working surfaces of sprayers and related equipment shall be covered with slip-resistant material. If slipping hazards cannot be eliminated, slip-resistant footwear or handrails and stair rails meeting the requirements of Subpart D of this part may be used instead of slip-resistant material.

1910.269(r)(3)(ii)

Equipment on which employees stand to spray while the vehicle is in motion shall be equipped with guardrails around the working area. The guardrail shall be constructed in accordance with Subpart D of this part.

1910.269(r)(4)

Stump cutters.

1910.269(r)(4)(i)

Stump cutters shall be equipped with enclosures or guards to protect employees.

1910.269(r)(4)(ii)

Each employee in the immediate area of stump grinding operations including the stump cutter operator) shall wear personal protective equipment as required by Subpart I of this part.

1910.269(r)(5)

Gasoline-engine power saws. Gasoline-engine power saw operations shall meet the requirements of § 1910.266(e) and the following:

1910.269(r)(5)(i)

Each power saw weighing more than 6.8 kilograms (15 pounds, service weight) that is used in trees shall be supported by a separate line, except when work is performed from an aerial lift and except during topping or removing operations where no supporting limb will be available.

1910.269(r)(5)(ii)

Each power saw shall be equipped with a control that will return the saw to idling speed when released.

1910.269(r)(5)(iii)

Each power saw shall be equipped with a clutch and shall be so adjusted that the clutch will not engage the chain drive at idling speed.

1910.269(r)(5)(iv)

A power saw shall be started on the ground or where it is otherwise firmly supported. Drop starting of saws over 6.8 kilograms (15 pounds), other than chain saws, is permitted outside of the bucket of an aerial lift only if the area below the lift is clear of personnel.

Note to paragraph (r)(5)(iv): Paragraph (e)(2)(vi) of § 1910.266 prohibits drop starting of chain saws.

1910.269(r)(5)(v)

A power saw engine may be started and operated only when all employees other than the operator are clear of the saw.

1910.269(r)(5)(vi)

A power saw may not be running when the saw is being carried up into a tree by an employee.

1910.269(r)(5)(vii)

Power saw engines shall be stopped for all cleaning, refueling, adjustments, and repairs to the saw or motor, except as the manufacturer's servicing procedures require otherwise.

1910.269(r)(6)

Backpack power units for use in pruning and clearing.

1910.269(r)(6)(i)

While a backpack power unit is running, no one other than the operator may be within 3.05 meters (10 feet) of the cutting head of a brush saw.

1910.269(r)(6)(ii)

A backpack power unit shall be equipped with a quick shutoff switch readily accessible to the operator.

1910.269(r)(6)(iii)

Backpack power unit engines shall be stopped for all cleaning, refueling, adjustments, and repairs to the saw or motor, except as the manufacturer's servicing procedures require otherwise.

1910.269(r)(7)

Rope.

1910.269(r)(7)(i)

Climbing ropes shall be used by employees working aloft in trees. These ropes shall have a minimum diameter of 12 millimeters (0.5 inch) with a minimum breaking strength of 10.2 kilonewtons (2,300 pounds). Synthetic rope shall have elasticity of not more than 7 percent.

1910.269(r)(7)(ii)

Rope shall be inspected before each use and, if unsafe (for example, because of damage or defect), may not be used.

1910.269(r)(7)(iii)

Rope shall be stored away from cutting edges and sharp tools. Rope contact with corrosive chemicals, gas, and oil shall be avoided.

1910.269(r)(7)(iv)

When stored, rope shall be coiled and piled, or shall be suspended, so that air can circulate through the coils.

1910.269(r)(7)(v)

Rope ends shall be secured to prevent their unraveling.

1910.269(r)(7)(vi)

Climbing rope may not be spliced to effect repair.

1910.269(r)(7)(vii)

A rope that is wet, that is contaminated to the extent that its insulating capacity is impaired, or that is otherwise not considered to be insulated for the voltage involved may not be used near exposed energized lines.

1910.269(r)(8)

Fall protection. Each employee shall be tied in with a climbing rope and safety saddle when the employee is working above the ground in a tree, unless he or she is ascending into the tree.

1910.269(s)

Communication facilities.

1910.269(s)(1)

Microwave transmission.

1910.269(s)(1)(i)

The employer shall ensure that no employee looks into an open waveguide or antenna connected to an energized microwave source.

1910.269(s)(1)(ii)

If the electromagnetic-radiation level within an accessible area associated with microwave communications systems exceeds the radiation-protection guide specified by § 1910.97(a)(2), the employer shall post the area with warning signs containing the warning symbol described in § 1910.97(a)(3). The lower half of the warning symbol shall include the following statements, or ones that the employer can demonstrate are equivalent: "Radiation in this area may exceed hazard limitations and special precautions are required. Obtain specific instruction before entering."

1910.269(s)(1)(iii)

When an employee works in an area where the electromagnetic radiation could exceed the radiationprotection guide, the employer shall institute measures that ensure that the employee's exposure is not greater than that permitted by that guide. Such measures may include administrative and engineering controls and personal protective equipment.

1910.269(s)(2)

Power-line carrier. The employer shall ensure that employees perform power-line carrier work, including work on equipment used for coupling carrier current to power line conductors, in accordance with the requirements of this section pertaining to work on energized lines.

1910.269(t)

Underground electrical installations. This paragraph provides additional requirements for work on underground electrical installations.

1910.269(t)(1)

Access. The employer shall ensure that employees use a ladder or other climbing device to enter and exit a manhole or subsurface vault exceeding 1.22 meters (4 feet) in depth. No employee may climb into or out of a manhole or vault by stepping on cables or hangers.

1910.269(t)(2)

Lowering equipment into manholes.

1910.269(t)(2)(i)

Equipment used to lower materials and tools into manholes or vaults shall be capable of supporting the weight to be lowered and shall be checked for defects before use.

1910.269(t)(2)(ii)

Before anyone lowers tools or material into the opening for a manhole or vault, each employee working in the manhole or vault shall be clear of the area directly under the opening.

1910.269(t)(3)

Attendants for manholes and vaults.

1910.269(t)(3)(i)

While work is being performed in a manhole or vault containing energized electric equipment, an employee with first-aid training shall be available on the surface in the immediate vicinity of the manhole or vault entrance to render emergency assistance.

1910.269(t)(3)(ii)

Occasionally, the employee on the surface may briefly enter a manhole or vault to provide nonemergency assistance.

Note 1 to paragraph (t)(3)(ii): Paragraph (e)(7) of this section may also require an attendant and does not permit this attendant to enter the manhole or vault.

Note 2 to paragraph (t)(3)(ii): Paragraph (l)(1)(ii) of this section requires employees entering manholes or vaults containing unguarded, uninsulated energized lines or parts of electric equipment operating at 50 volts or more to be qualified.

1910.269(t)(3)(iii)

For the purpose of inspection, housekeeping, taking readings, or similar work, an employee working alone may enter, for brief periods of time, a manhole or vault where energized cables or equipment are in service if the employer can demonstrate that the employee will be protected from all electrical hazards.

1910.269(t)(3)(iv)

The employer shall ensure that employees maintain reliable communications, through two-way radios or other equivalent means, among all employees involved in the job.

1910.269(t)(4)

Duct rods. The employer shall ensure that, if employees use duct rods, the employees install the duct rods in the direction presenting the least hazard to employees. The employer shall station an employee at the far end of the duct line being rodded to ensure that the employees maintain the required minimum approach distances.

1910.269(t)(5)

Multiple cables. When multiple cables are present in a work area, the employer shall identify the cable to be worked by electrical means, unless its identity is obvious by reason of distinctive appearance or location or by other readily apparent means of identification. The employer shall protect cables other than the one being worked from damage.

1910.269(t)(6)

Moving cables. Except when paragraph (t)(7)(ii) of this section permits employees to perform work that could cause a fault in an energized cable in a manhole or vault, the employer shall ensure that employees inspect energized cables to be moved for abnormalities.

1910.269(t)(7)

Protection against faults.

1910.269(t)(7)(i)

Where a cable in a manhole or vault has one or more abnormalities that could lead to a fault or be an indication of an impending fault, the employer shall deenergize the cable with the abnormality before any employee may work in the manhole or vault, except when service-load conditions and a lack of feasible alternatives require that the cable remain energized. In that case, employees may enter the manhole or vault provided the employer protects them from the possible effects of a failure using shields or other devices that are capable of containing the adverse effects of a fault. The employer shall treat the following abnormalities as indications of impending faults unless the employer can demonstrate that the conditions could not lead to a fault: Oil or compound leaking from cable or joints, broken cable sheaths or joint sleeves, hot localized surface temperatures of cables or joints, or joints swollen beyond normal tolerance.

1910.269(t)(7)(ii)

If the work employees will perform in a manhole or vault could cause a fault in a cable, the employer shall deenergize that cable before any employee works in the manhole or vault, except when service-load conditions and a lack of feasible alternatives require that the cable remain energized. In that case, employees may enter the manhole or vault provided the employer protects them from the possible effects of a failure using shields or other devices that are capable of containing the adverse effects of a fault.

1910.269(t)(8)

Sheath continuity. When employees perform work on buried cable or on cable in a manhole or vault, the employer shall maintain metallic-sheath continuity, or the cable sheath shall be treated as energized.

1910.269(u)

Substations. This paragraph provides additional requirements for substations and for work performed in them.

1910.269(u)(1)

Access and working space. The employer shall provide and maintain sufficient access and working space about electric equipment to permit ready and safe operation and maintenance of such equipment by employees.

Note to paragraph (u)(1): American National Standard *National Electrical Safety Code,* ANSI/IEEE C2-2012 contains guidelines for the dimensions of access and working space about electric equipment in substations. Installations meeting the ANSI provisions comply with paragraph (u)(1) of this section. The Occupational Safety and Health Administration will determine whether an installation that does not conform to this ANSI standard complies with paragraph (u)(1) of this section based on the following criteria:

(1) Whether the installation conforms to the edition of ANSI C2 that was in effect when the installation was made,
(2) Whether the configuration of the installation enables employees to maintain the minimum approach distances, established by the employer under paragraph (l)(3)(i) of this section, while the employees are working on exposed, energized parts, and
(3) Whether the precautions taken when employees perform work on the installation provide protection equivalent to the protection provided by access and working space meeting ANSI/IEEE C2-2012.

1910.269(u)(2)

Draw-out-type circuit breakers. The employer shall ensure that, when employees remove or insert draw-out-type circuit breakers, the breaker is in the open position. The employer shall also render the control circuit inoperable if the design of the equipment permits.

1910.269(u)(3)

Substation fences. Conductive fences around substations shall be grounded. When a substation fence is expanded or a section is removed, fence sections shall be isolated, grounded, or bonded as necessary to protect employees from hazardous differences in electric potential.

Note to paragraph (u)(3): IEEE Std 80-2000, *IEEE Guide for Safety in AC Substation Grounding,* contains guidelines for protection against hazardous differences in electric potential.

1910.269(u)(4)

Guarding of rooms and other spaces containing electric supply equipment.

1910.269(u)(4)(i)

Rooms and other spaces in which electric supply lines or equipment are installed shall meet the requirements of paragraphs (u)(4)(ii) through (u)(4)(v) of this section under the following conditions:

1910.269(u)(4)(i)(A)

If exposed live parts operating at 50 to 150 volts to ground are within 2.4 meters (8 feet) of the ground or other working surface inside the room or other space,

1910.269(u)(4)(i)(B)

If live parts operating at 151 to 600 volts to ground and located within 2.4 meters (8 feet) of the ground or other working surface inside the room or other space are guarded only by location, as permitted under paragraph (u)(5)(i) of this section, or

1910.269(u)(4)(i)(C)

If live parts operating at more than 600 volts to ground are within the room or other space, unless:

1910.269(u)(4)(i)(C)(1)

The live parts are enclosed within grounded, metal-enclosed equipment whose only openings are designed so that foreign objects inserted in these openings will be deflected from energized parts, or

1910.269(u)(4)(i)(C)(2)

The live parts are installed at a height, above ground and any other working surface, that provides protection at the voltage on the live parts corresponding to the protection provided by a 2.4-meter (8-foot) height at 50 volts.

1910.269(u)(4)(ii)

Fences, screens, partitions, or walls shall enclose the rooms and other spaces so as to minimize the possibility that unqualified persons will enter.

1910.269(u)(4)(iii)

Unqualified persons may not enter the rooms or other spaces while the electric supply lines or equipment are energized.

1910.269(u)(4)(iv)

The employer shall display signs at entrances to the rooms and other spaces warning unqualified persons to keep out.

1910.269(u)(4)(v)

The employer shall keep each entrance to a room or other space locked, unless the entrance is under the observation of a person who is attending the room or other space for the purpose of preventing unqualified employees from entering.

<u>1910.269(u)(5)</u>

Guarding of energized parts.

1910.269(u)(5)(i)

The employer shall provide guards around all live parts operating at more than 150 volts to ground without an insulating covering unless the location of the live parts gives sufficient clearance (horizontal, vertical, or both) to minimize the possibility of accidental employee contact.

Note to paragraph (u)(5)(i): American National Standard *National Electrical Safety Code*, ANSI/IEEE C2-2002 contains guidelines for the dimensions of clearance distances about electric equipment in substations. Installations meeting the ANSI provisions comply with paragraph (u)(5)(i) of this section. The Occupational Safety and Health Administration will determine whether an installation that does not conform to this ANSI standard complies with paragraph (u)(5)(i) of this section based on the following criteria:

(1) Whether the installation conforms to the edition of ANSI C2 that was in effect when the installation was made,
(2) Whether each employee is isolated from energized parts at the point of closest approach; and
(3) Whether the precautions taken when employees perform work on the installation provide protection equivalent to the protection provided by horizontal and vertical clearances meeting ANSI/IEEE C2-2002.

1910.269(u)(5)(ii)

Except for fuse replacement and other necessary access by qualified persons, the employer shall maintain guarding of energized parts within a compartment during operation and maintenance functions to prevent accidental contact with energized parts and to prevent dropped tools or other equipment from contacting energized parts.

1910.269(u)(5)(iii)

Before guards are removed from energized equipment, the employer shall install barriers around the work area to prevent employees who are not working on the equipment, but who are in the area, from contacting the exposed live parts.

1910.269(u)(6)

Substation entry.

1910.269(u)(6)(i)

Upon entering an attended substation, each employee, other than employees regularly working in the station, shall report his or her presence to the employee in charge of substation activities to receive information on special system conditions affecting employee safety.

1910.269(u)(6)(ii)

The job briefing required by paragraph (c) of this section shall cover information on special system conditions affecting employee safety, including the location of energized equipment in or adjacent to the work area and the limits of any deenergized work area.

1910.269(v)

Power generation. This paragraph provides additional requirements and related work practices for power generating plants.

1910.269(v)(1)

Interlocks and other safety devices.

1910.269(v)(1)(i)

Interlocks and other safety devices shall be maintained in a safe, operable condition.

1910.269(v)(1)(ii)

No interlock or other safety device may be modified to defeat its function, except for test, repair, or adjustment of the device.

1910.269(v)(2)

Changing brushes. Before exciter or generator brushes are changed while the generator is in service, the exciter or generator field shall be checked to determine whether a ground condition exists. The brushes may not be changed while the generator is energized if a ground condition exists.

1910.269(v)(3)

Access and working space. The employer shall provide and maintain sufficient access and working space about electric equipment to permit ready and safe operation and maintenance of such equipment by employees.

Note to paragraph (v)(3) of this section: American National Standard *National Electrical Safety Code*, ANSI/IEEE C2-2012 contains guidelines for the dimensions of access and working space about electric equipment in substations. Installations meeting the ANSI provisions comply with paragraph (v)(3) of this section. The Occupational Safety and Health Administration will determine whether an installation that does not conform to this ANSI standard complies with paragraph (v)(3) of this section based on the following criteria:

(1) Whether the installation conforms to the edition of ANSI C2 that was in effect when the installation was made;

(2) Whether the configuration of the installation enables employees to maintain the minimum approach distances, established by the employer under paragraph (l)(3)(i) of this section, while the employees are working on exposed, energized parts, and;

(3) Whether the precautions taken when employees perform work on the installation provide protection equivalent to the protection provided by access and working space meeting ANSI/IEEE C2-2012.

1910.269(v)(4)

Guarding of rooms and other spaces containing electric supply equipment.

1910.269(v)(4)(i)

Rooms and other spaces in which electric supply lines or equipment are installed shall meet the requirements of paragraphs (v)(4)(ii) through (v)(4)(v) of this section under the following conditions:

1910.269(v)(4)(i)(A)

If exposed live parts operating at 50 to 150 volts to ground are within 2.4 meters (8 feet) of the ground or other working surface inside the room or other space,

1910.269(v)(4)(i)(B)

If live parts operating at 151 to 600 volts to ground and located within 2.4 meters (8 feet) of the ground or other working surface inside the room or other space are guarded only by location, as permitted under paragraph (v)(5)(i) of this section, or

1910.269(v)(4)(i)(C)

If live parts operating at more than 600 volts to ground are within the room or other space, unless:

1910.269(v)(4)(i)(C)(1)

The live parts are enclosed within grounded, metal-enclosed equipment whose only openings are designed so that foreign objects inserted in these openings will be deflected from energized parts, or

1910.269(v)(4)(i)(C)(2)

The live parts are installed at a height, above ground and any other working surface, that provides protection at the voltage on the live parts corresponding to the protection provided by a 2.4-meter (8-foot) height at 50 volts.

1910.269(v)(4)(ii)

Fences, screens, partitions, or walls shall enclose the rooms and other spaces so as to minimize the possibility that unqualified persons will enter.

1910.269(v)(4)(iii)

Unqualified persons may not enter the rooms or other spaces while the electric supply lines or equipment are energized.

1910.269(v)(4)(iv)

The employer shall display signs at entrances to the rooms and other spaces warning unqualified persons to keep out.

1910.269(v)(4)(v)

The employer shall keep each entrance to a room or other space locked, unless the entrance is under the observation of a person who is attending the room or other space for the purpose of preventing unqualified employees from entering.

<u>1910.269(v)(5)</u>

Guarding of energized parts.

1910.269(v)(5)(i)

The employer shall provide guards around all live parts operating at more than 150 volts to ground without an insulating covering unless the location of the live parts gives sufficient clearance (horizontal, vertical, or both) to minimize the possibility of accidental employee contact.

Note to paragraph (v)(5)(i): American National Standard *National Electrical Safety Code*, ANSI/IEEE C2-2002 contains guidelines for the dimensions of clearance distances about electric equipment in substations. Installations meeting the ANSI provisions comply with paragraph (v)(5)(i) of this section. The Occupational Safety and Health Administration will determine whether an installation that does not conform to this ANSI standard complies with paragraph (v)(5)(i) of this section based on the following criteria:

(1) Whether the installation conforms to the edition of ANSI C2 that was in effect when the installation was made;
(2) Whether each employee is isolated from energized parts at the point of closest approach; and
(3) Whether the precautions taken when employees perform work on the installation provide protection equivalent to the protection provided by horizontal and vertical clearances meeting ANSI/IEEE C2-2002.

1910.269(v)(5)(ii)

Except for fuse replacement and other necessary access by qualified persons, the employer shall maintain guarding of energized parts within a compartment during operation and maintenance functions to prevent accidental contact with energized parts and to prevent dropped tools or other equipment from contacting energized parts.

1910.269(v)(5)(iii)

Before guards are removed from energized equipment, the employer shall install barriers around the work area to prevent employees who are not working on the equipment, but who are in the area, from contacting the exposed live parts.

1910.269(v)(6)

Water or steam spaces. The following requirements apply to work in water and steam spaces associated with boilers:

1910.269(v)(6)(i)

A designated employee shall inspect conditions before work is permitted and after its completion. Eye protection, or full face protection if necessary, shall be worn at all times when condenser, heater, or boiler tubes are being cleaned.

1910.269(v)(6)(ii)

Where it is necessary for employees to work near tube ends during cleaning, shielding shall be installed at the tube ends.

1910.269(v)(7)

Chemical cleaning of boilers and pressure vessels. The following requirements apply to chemical cleaning of boilers and pressure vessels:

1910.269(v)(7)(i)

Areas where chemical cleaning is in progress shall be cordoned off to restrict access during cleaning. If flammable liquids, gases, or vapors or combustible materials will be used or might be produced during the cleaning process, the following requirements also apply:

1910.269(v)(7)(i)(A)

The area shall be posted with signs restricting entry and warning of the hazards of fire and explosion; and

1910.269(v)(7)(i)(B)

Smoking, welding, and other possible ignition sources are prohibited in these restricted areas.

1910.269(v)(7)(ii)

The number of personnel in the restricted area shall be limited to those necessary to accomplish the task safely.

1910.269(v)(7)(iii)

There shall be ready access to water or showers for emergency use.

Note to paragraph (v)(7)(iii): See § 1910.141 for requirements that apply to the water supply and to washing facilities.

1910.269(v)(7)(iv)

Employees in restricted areas shall wear protective equipment meeting the requirements of Subpart I of this part and including, but not limited to, protective clothing, boots, goggles, and gloves.

1910.269(v)(8)

Chlorine systems.

1910.269(v)(8)(i)

Chlorine system enclosures shall be posted with signs restricting entry and warning of the hazard to health and the hazards of fire and explosion.

Note to paragraph (v)(8)(i): See Subpart Z of this part for requirements necessary to protect the health of employees from the effects of chlorine.

1910.269(v)(8)(ii)

Only designated employees may enter the restricted area. Additionally, the number of personnel shall be limited to those necessary to accomplish the task safely.

1910.269(v)(8)(iii)

Emergency repair kits shall be available near the shelter or enclosure to allow for the prompt repair of leaks in chlorine lines, equipment, or containers.

1910.269(v)(8)(iv)

Before repair procedures are started, chlorine tanks, pipes, and equipment shall be purged with dry air and isolated from other sources of chlorine.

1910.269(v)(8)(v)

The employer shall ensure that chlorine is not mixed with materials that would react with the chlorine in a dangerously exothermic or other hazardous manner.

1910.269(v)(9)

Boilers.

1910.269(v)(9)(i)

Before internal furnace or ash hopper repair work is started, overhead areas shall be inspected for possible falling objects. If the hazard of falling objects exists, overhead protection such as planking or nets shall be provided.

1910.269(v)(9)(ii)

When opening an operating boiler door, employees shall stand clear of the opening of the door to avoid the heat blast and gases which may escape from the boiler.

1910.269(v)(10)

Turbine generators.

1910.269(v)(10)(i)

Smoking and other ignition sources are prohibited near hydrogen or hydrogen sealing systems, and signs warning of the danger of explosion and fire shall be posted.

1910.269(v)(10)(ii)

Excessive hydrogen makeup or abnormal loss of pressure shall be considered as an emergency and shall be corrected immediately.

1910.269(v)(10)(iii)

A sufficient quantity of inert gas shall be available to purge the hydrogen from the largest generator.

1910.269(v)(11)

Coal and ash handling.

1910.269(v)(11)(i)

Only designated persons may operate railroad equipment.

1910.269(v)(11)(ii)

Before a locomotive or locomotive crane is moved, a warning shall be given to employees in the area.

1910.269(v)(11)(iii)

Employees engaged in switching or dumping cars may not use their feet to line up drawheads.

1910.269(v)(11)(iv)

Drawheads and knuckles may not be shifted while locomotives or cars are in motion.

1910.269(v)(11)(v)

When a railroad car is stopped for unloading, the car shall be secured from displacement that could endanger employees.

1910.269(v)(11)(vi)

An emergency means of stopping dump operations shall be provided at railcar dumps.

1910.269(v)(11)(vii)

The employer shall ensure that employees who work in coal- or ashhandling conveyor areas are trained and knowledgeable in conveyor operation and in the requirements of paragraphs (v)(11)(viii) through (v)(11)(xii) of this section.

1910.269(v)(11)(viii)

Employees may not ride a coalor ash-handling conveyor belt at any time. Employees may not cross over the conveyor belt, except at walkways, unless the conveyor's energy source has been deenergized and has been locked out or tagged in accordance with paragraph (d) of this section

1910.269(v)(11)(ix)

A conveyor that could cause injury when started may not be started until personnel in the area are alerted by a signal or by a designated person that the conveyor is about to start.

1910.269(v)(11)(x)

If a conveyor that could cause injury when started is automatically controlled or is controlled from a remote location, an audible device shall be provided that sounds an alarm that will be recognized by each employee as a warning that the conveyor will start and that can be clearly heard at all points along the conveyor where personnel may be present. The warning device shall be actuated by the device starting the conveyor and shall continue for a period of time before the conveyor starts that is long enough to allow employees to move clear of the conveyor system. A visual warning may be used in place of the audible device if the employer can demonstrate that it will provide an equally effective warning in the particular circumstances involved. However if the employer can demonstrate that the system's function would be seriously hindered by the required time delay, warning signs may be provided in place of the audible warning device. If the system was installed before January 31, 1995, warning signs may be provided in place of the audible warning device until such time as the conveyor or its control system is rebuilt or rewired. These warning signs shall be clear, concise, and legible and shall indicate that conveyors and allied equipment may be started at any time, that danger exists, and that personnel must keep clear. These warning signs shall be provided along the conveyor at areas not guarded by position or location.

1910.269(v)(11)(xi)

Remotely and automatically controlled conveyors, and conveyors that have operating stations which are not manned or which are beyond voice and visual contact from drive areas, loading areas, transfer points, and other locations on the conveyor path not guarded by location, position, or guards shall be furnished with emergency stop buttons, pull cords, limit switches, or similar emergency stop devices. However, if the employer can demonstrate that the design, function, and operation of the conveyor do not expose an employee to hazards, an emergency stop device is not required.

1910.269(v)(11)(xi)(A)

Emergency stop devices shall be easily identifiable in the immediate vicinity of such locations.

1910.269(v)(11)(xi)(B)

An emergency stop device shall act directly on the control of the conveyor involved and may not depend on the stopping of any other equipment.

1910.269(v)(11)(xi)(C)

Emergency stop devices shall be installed so that they cannot be overridden from other locations.

1910.269(v)(11)(xii)

Where coal-handling operations may produce a combustible atmosphere from fuel sources or from flammable gases or dust, sources of ignition shall be eliminated or safely controlled to prevent ignition of the combustible atmosphere.

Note to paragraph (v)(11)(xii): Locations that are hazardous because of the presence of combustible dust are classified as Class II hazardous locations. See § 1910.307.

1910.269(v)(11)(xiii)

An employee may not work on or beneath overhanging coal in coal bunkers, coal silos, or coal storage areas, unless the employee is protected from all hazards posed by shifting coal.

1910.269(v)(11)(xiv)

An employee entering a bunker or silo to dislodge the contents shall wear a body harness with lifeline attached. The lifeline shall be secured to a fixed support outside the bunker and shall be attended at all times by an employee located outside the bunker or facility.

1910.269(v)(12)

Hydroplants and equipment. Employees working on or close to water gates, valves, intakes, forebays, flumes, or other locations where increased or decreased water flow or levels may pose a significant hazard shall be warned and shall vacate such dangerous areas before water flow changes are made.

1910.269(w)

Special conditions.

1910.269(w)(1)

Capacitors. The following additional requirements apply to work on capacitors and on lines connected to capacitors.

Note to paragraph (w)(1):   See paragraphs (m) and (n) of this section for requirements pertaining to the deenergizing and grounding of capacitor installations.

1910.269(w)(1)(i)

Before employees work on capacitors, the employer shall disconnect the capacitors from energized sources and short circuit the capacitors. The employer shall ensure that the employee short circuiting the capacitors waits at least 5 minutes from the time of disconnection before applying the short circuit,

1910.269(w)(1)(ii)

Before employees handle the units, the employer shall short circuit each unit in series-parallel capacitor banks between all terminals and the capacitor case or its rack. If the cases of capacitors are on ungrounded substation racks, the employer shall bond the racks to ground.

1910.269(w)(1)(iii)

The employer shall short circuit any line connected to capacitors before the line is treated as deenergized.

1910.269(w)(2)

Current transformer secondaries. The employer shall ensure that employees do not open the secondary of a current transformer while the transformer is energized. If the employer cannot deenergize the primary of the current transformer before employees perform work on an instrument, a relay, or other section of a current transformer secondary circuit, the employer shall bridge the circuit so that the current transformer secondary does not experience an open-circuit condition.

1910.269(w)(3)

Series streetlighting.

1910.269(w)(3)(i)

If the open-circuit voltage exceeds 600 volts, the employer shall ensure that employees work on series streetlighting circuits in accordance with paragraph (q) or (t) of this section, as appropriate.

1910.269(w)(3)(ii)

Before any employee opens a series loop, the employer shall deenergize the street-lighting transformer and isolate it from the source of supply or shall bridge the loop to avoid an open-circuit condition.

1910.269(w)(4)

Illumination. The employer shall provide sufficient illumination to enable the employee to perform the work safely.

1910.269(w)(5)

Protection against drowning.

1910.269(w)(5)(i)

Whenever an employee may be pulled or pushed, or might fall, into water where the danger of drowning exists, the employer shall provide the employee with, and shall ensure that the employee uses, a U.S. Coast Guardapproved personal flotation device.

1910.269(w)(5)(ii)

The employer shall maintain each personal flotation device in safe condition and shall inspect each personal flotation device frequently enough to ensure that it does not have rot, mildew, water saturation, or any other condition that could render the device unsuitable for use.

1910.269(w)(5)(iii)

An employee may cross streams or other bodies of water only if a safe means of passage, such as a bridge, is available.

1910.269(w)(6)

Employee protection in public work areas.

1910.269(w)(6)(i)

Traffic-control signs and traffic-control devices used for the protection of employees shall meet § 1926.200(g)(2) of this chapter.

1910.269(w)(6)(ii)

Before employees begin work in the vicinity of vehicular or pedestrian traffic that may endanger them, the employer shall place warning signs or flags and other traffic-control devices in conspicuous locations to alert and channel approaching traffic.

1910.269(w)(6)(iii)

The employer shall use barricades where additional employee protection is necessary.

1910.269(w)(6)(iv)

The employer shall protect excavated areas with barricades.

1910.269(w)(6)(v)

The employer shall display warning lights prominently at night.

1910.269(w)(7)

Backfeed. When there is a possibility of voltage backfeed from sources of cogeneration or from the secondary system (for example, backfeed from more than one energized phase feeding a common load), the requirements of paragraph (l) of this section apply if employees will work the lines or equipment as energized, and the requirements of paragraphs (m) and (n) of this section apply if employees will work the lines or equipment as deenergized.

1910.269(w)(8)

Lasers. The employer shall install, adjust, and operate laser equipment in accordance with § 1926.54 of this chapter.

1910.269(w)(9)

Hydraulic fluids. Hydraulic fluids used for the insulated sections of equipment shall provide insulation for the voltage involved.

<u>1910.269(x)</u>

Definitions.

Affected employee. An employee whose job requires him or her to operate or use a machine or equipment on which servicing or maintenance is being performed under lockout or tagout, or whose job requires him or her to work in an area in which such servicing or maintenance is being performed.

Attendant. An employee assigned to remain immediately outside the entrance to an enclosed or other space to render assistance as needed to employees inside the space.

Authorized employee. An employee who locks out or tags out machines or equipment in order to perform servicing or maintenance on that machine or equipment. An affected employee becomes an authorized employee when that employee's duties include performing servicing or maintenance covered under this section.

Automatic circuit recloser. A self-controlled device for automatically interrupting and reclosing an alternating-current circuit, with a predetermined sequence of opening and reclosing followed by resetting, hold closed, or lockout.

Barricade. A physical obstruction such as tapes, cones, or A-frame type wood or metal structures that provides a warning about, and limits access to, a hazardous area.

Barrier. A physical obstruction that prevents contact with energized lines or equipment or prevents unauthorized access to a work area.

Bond. The electrical interconnection of conductive parts designed to maintain a common electric potential.

Bus. A conductor or a group of conductors that serve as a common connection for two or more circuits.

Bushing. An insulating structure that includes a through conductor or that provides a passageway for such a conductor, and that, when mounted on a barrier, insulates the conductor from the barrier for the purpose of conducting current from one side of the barrier to the other.

Cable. A conductor with insulation, or a stranded conductor with or without insulation and other coverings (singleconductor cable), or a combination of conductors insulated from one another (multiple-conductor cable).

Cable sheath. A conductive protective covering applied to cables.

Note to the definition of "cable sheath": A cable sheath may consist of multiple layers one or more of which is conductive.

Circuit. A conductor or system of conductors through which an electric current is intended to flow.

Clearance (between objects). The clear distance between two objects measured surface to surface.

Clearance (for work). Authorization to perform specified work or permission to enter a restricted area.

Communication lines. (See *Lines*; (1) *Communication lines.*)

Conductor. A material, usually in the form of a wire, cable, or bus bar, used for carrying an electric current.

Contract employer. An employer, other than a host employer, that performs work covered by this section under contract.

Covered conductor. A conductor covered with a dielectric having no rated insulating strength or having a rated insulating strength less than the voltage of the circuit in which the conductor is used.

Current-carrying part. A conducting part intended to be connected in an electric circuit to a source of voltage. Non-current-carrying parts are those not intended to be so connected.

Deenergized. Free from any electrical connection to a source of potential difference and from electric charge; not having a potential that is different from the potential of the earth.

Note to the definition of "deenergized": The term applies only to current-carrying parts, which are sometimes energized (alive).

Designated employee (designated person). An employee (or person) who is assigned by the employer to perform specific duties under the terms of this section and who has sufficient knowledge of the construction and operation of the equipment, and the hazards involved, to perform his or her duties safely.

Electric line truck. A truck used to transport personnel, tools, and material for electric supply line work.

Electric supply equipment. Equipment that produces, modifies, regulates, controls, or safeguards a supply of electric energy.

Electric supply lines. (See *Lines*; (2) *Electric supply lines.*)

Electric utility. An organization responsible for the installation, operation, or maintenance of an electric supply system.

Enclosed space. A working space, such as a manhole, vault, tunnel, or shaft, that has a limited means of egress or entry, that is designed for periodic employee entry under normal operating conditions, and that, under normal conditions, does not contain a hazardous atmosphere, but may contain a hazardous atmosphere under abnormal conditions.

Note to the definition of "enclosed space": The Occupational Safety and Health Administration does not consider spaces that are enclosed but not designed for employee entry under normal operating conditions to be enclosed spaces for the purposes of this section. Similarly, the Occupational Safety and Health Administration does not consider spaces that are enclosed and that are expected to contain a hazardous atmosphere to be enclosed spaces for the purposes of this section. Such spaces meet the definition of permit spaces in § 1910.146, and entry into them must conform to that standard.

Energized (alive, live). Electrically connected to a source of potential difference, or electrically charged so as to have a potential significantly different from that of earth in the vicinity.

Energy isolating device. A physical device that prevents the transmission or release of energy, including, but not limited to, the following: a manually operated electric circuit breaker, a disconnect switch, a manually operated switch, a slide gate, a slip blind, a line valve, blocks, and any similar device with a visible indication of the position of the device. (Push buttons, selector switches, and other control-circuit-type devices are not energy isolating devices.)

Energy source. Any electrical, mechanical, hydraulic, pneumatic, chemical, nuclear, thermal, or other energy source that could cause injury to employees.

Entry (as used in paragraph (e) of this section). The action by which a person passes through an opening into an enclosed space. Entry includes ensuing work activities in that space and is considered to have occurred as soon as any part of the entrant's body breaks the plane of an opening into the space.

Equipment (electric),. A general term including material, fittings, devices, appliances, fixtures, apparatus, and the like used as part of or in connection with an electrical installation.

Exposed, Exposed to contact (as applied to energized parts). Not isolated or guarded.

Fall restraint system. A fall protection system that prevents the user from falling any distance.

First-aid training. Training in the initial care, including cardiopulmonary resuscitation (which includes chest compressions, rescue breathing, and, as appropriate, other heart and lung resuscitation techniques), performed by a person who is not a medical practitioner, of a sick or injured person until definitive medical treatment can be administered.

Ground. A conducting connection, whether planned or unplanned, between an electric circuit or equipment and the earth, or to some conducting body that serves in place of the earth.

Grounded. Connected to earth or to some conducting body that serves in place of the earth.

Guarded. Covered, fenced, enclosed, or otherwise protected, by means of suitable covers or casings, barrier rails or screens, mats, or platforms, designed to minimize the possibility, under normal conditions, of dangerous approach or inadvertent contact by persons or objects.

Note to the definition of "guarded": Wires that are insulated, but not otherwise protected, are not guarded.

Hazardous atmosphere. An atmosphere that may expose employees to the risk of death, incapacitation, impairment of ability to self-rescue (that is, escape unaided from an enclosed space), injury, or acute illness from one or more of the following causes:

(1) Flammable gas, vapor, or mist in excess of 10 percent of its lower flammable limit (LFL);
(2) Airborne combustible dust at a concentration that meets or exceeds its LFL;

Note to the definition of "hazardous atmosphere" (2):   This concentration may be approximated as a condition in which the dust obscures vision at a distance of 1.52 meters (5 feet) or less.
(3) Atmospheric oxygen concentration below 19.5 percent or above 23.5 percent;
(4) Atmospheric concentration of any substance for which a dose or a permissible exposure limit is published in Subpart G, *Occupational Health and Environmental Control*, or in Subpart Z, *Toxic and Hazardous Substances*, of this part and which could result in employee exposure in excess of its dose or permissible exposure limit;

Note to the definition of "hazardous atmosphere" (4):   An atmospheric concentration of any substance that is not capable of causing death, incapacitation, impairment of ability to self-rescue, injury, or acute illness due to its health effects is not covered by this provision.
(5) Any other atmospheric condition that is immediately dangerous to life or health.

Note to the definition of "hazardous atmosphere" (5):   For air contaminants for which the Occupational Safety and Health Administration has not determined a dose or permissible exposure limit, other sources of information, such as Material Safety Data Sheets that comply with the Hazard Communication Standard, § 1910.1200, published information, and internal documents can provide guidance in establishing acceptable atmospheric conditions.

High-power tests. Tests in which the employer uses fault currents, load currents, magnetizing currents, and linedropping currents to test equipment, either at the equipment's rated voltage or at lower voltages.

High-voltage tests. Tests in which the employer uses voltages of approximately 1,000 volts as a practical minimum and in which the voltage source has sufficient energy to cause injury.

High wind. A wind of such velocity that one or more of the following hazards would be present:

(1) The wind could blow an employee from an elevated location,
(2) The wind could cause an employee or equipment handling material to lose control of the material, or
(3) The wind would expose an employee to other hazards not controlled by the standard involved.

Note to the definition of "high wind":   The Occupational Safety and Health Administration normally considers winds exceeding 64.4 kilometers per hour (40 miles per hour), or 48.3 kilometers per hour (30 miles per hour) if the work involves material handling, as meeting this criteria, unless the employer takes precautions to protect employees from the hazardous effects of the wind.

Host employer. An employer that operates, or that controls the operating procedures for, an electric power generation, transmission, or distribution installation on which a contract employer is performing work covered by this section.

Note to the definition of "host employer": The Occupational Safety and Health Administration will treat the electric utility or the owner of the installation as the host employer if it operates or controls operating procedures for the installation. If the electric utility or installation owner neither operates nor controls operating procedures for the installation, the Occupational Safety and Health Administration will treat the employer that the utility or owner has contracted with to operate or control the operating procedures for the installation as the host employer. In no case will there be more than one host employer.

Immediately dangerous to life or health (IDLH). Any condition that poses an immediate or delayed threat to life or that would cause irreversible adverse health effects or that would interfere with an individual's ability to escape unaided from a permit space.

Note to the definition of "immediately dangerous to life or health": Some materials-hydrogen fluoride gas and cadmium vapor, for example-may produce immediate transient effects that, even if severe, may pass without medical attention, but are followed by sudden, possibly fatal collapse 12-72 hours after exposure. The victim "feels normal" from recovery from transient effects until collapse. Such materials in hazardous quantities are considered to be "immediately" dangerous to life or health

Insulated. Separated from other conducting surfaces by a dielectric (including air space) offering a high resistance to the passage of current.

Note to the definition of "insulated": When any object is said to be insulated, it is understood to be insulated for the conditions to which it normally is subjected. Otherwise, it is, for the purpose of this section, uninsulated.

Insulation (cable). Material relied upon to insulate the conductor from other conductors or conducting parts or from ground.

Isolated. Not readily accessible to persons unless special means for access are used.

Line-clearance tree trimmer. An employee who, through related training or on-the-job experience or both, is familiar with the special techniques and hazards involved in line-clearance tree trimming.

Note 1 to the definition of "line-clearance tree trimmer": An employee who is regularly assigned to a line-clearance tree-trimming crew and who is undergoing on-the-job training and who, in the course of such training, has demonstrated an ability to perform duties safely at his or her level of training and who is under the direct supervision of a line-clearance tree trimmer is considered to be a line-clearance tree trimmer for the performance of those duties.

Note 2 to the definition of "line-clearance tree trimmer": A line-clearance tree trimmer is not considered to be a "qualified employee" under this section unless he or she has the training required for a qualified employee under paragraph (a)(2)(ii) of this section. However, under the electrical safety-related work practices standard in Subpart S of this part, a line-clearance tree trimmer is considered to be a "qualified employee". Tree trimming performed by such "qualified employees" is not subject to the electrical safety-related work practice requirements contained in §§ 1910.331 through 1910.335 of this part. (See also the note following § 1910.332(b)(3) of this part for information regarding the training an employee must have to be considered a qualified employee under §§ 1910.331 through 1910.335 of this part.)

Line-clearance tree trimming. The pruning, trimming, repairing, maintaining, removing, or clearing of trees, or the cutting of brush, that is within the following distance of electric supply lines and equipment:

(1) For voltages to ground of 50 kilovolts or less-3.05 meters (10 feet);

(2) For voltages to ground of more than 50 kilovolts-3.05 meters (10 feet) plus 0.10 meters (4 inches) for every 10 kilovolts over 50 kilovolts.

Lines. (1) *Communication lines.* The conductors and their supporting or containing structures which are used for public or private signal or communication service, and which operate at potentials not exceeding 400 volts to ground or 750 volts between any two points of the circuit, and the transmitted power of which does not exceed 150 watts. If the lines are operating at less than 150 volts, no limit is placed on the transmitted power of the system. Under certain conditions, communication cables may include communication circuits exceeding these limitations where such circuits are also used to supply power solely to communication equipment.

Note to the definition of "communication lines": Telephone, telegraph, railroad signal, data, clock, fire, police alarm, cable television, and other systems conforming to this definition are included. Lines used for signaling purposes, but not included under this definition, are considered as electric supply lines of the same voltage.

(2) *Electric supply lines.* Conductors used to transmit electric energy and their necessary supporting or containing structures. Signal lines of more than 400 volts are always supply lines within this section, and those of less than 400 volts are considered as supply lines, if so run and operated throughout.

Manhole. A subsurface enclosure that personnel may enter and that is used for installing, operating, and maintaining submersible equipment or cable.

Minimum approach distance. The closest distance an employee may approach an energized or a grounded object.

Note to the definition of "minimum approach distance": Paragraph (l)(3)(i) of this section requires employers to establish minimum approach distances.

Personal fall arrest system. A system used to arrest an employee in a fall from a working level.

Qualified employee (qualified person). An employee (person) knowledgeable in the construction and operation of the electric power generation, transmission, and distribution equipment involved, along with the associated hazards.

Note 1 to the definition of "qualified employee (qualified person)": An employee must have the training required by (a)(2)(ii) of this section to be a qualified employee.

Note 2 to the definition of "qualified employee (qualified person)": Except under (g)(2)(iv)(C)(2) and (g)(2)(iv)(C)(3) of this section, an employee who is undergoing on-the-job training and who has demonstrated, in the course of such training, an ability to perform duties safely at his or her level of training and who is under the direct supervision of a qualified person is a qualified person for the performance of those duties.

Statistical sparkover voltage. A transient overvoltage level that produces a 97.72-percent probability of sparkover (that is, two standard deviations above the voltage at which there is a 50-percent probability of sparkover).

Statistical withstand voltage. A transient overvoltage level that produces a 0.14-percent probability of sparkover (that is, three standard deviations below the voltage at which there is a 50-percent probability of sparkover).

Switch. A device for opening and closing or for changing the connection of a circuit. In this section, a switch is manually operable, unless otherwise stated.

System operator. A qualified person designated to operate the system or its parts.

Vault. An enclosure, above or below ground, that personnel may enter and that is used for installing, operating, or maintaining equipment or cable.

Vented vault. A vault that has provision for air changes using exhaustflue stacks and low-level air intakes operating on pressure and temperature differentials that provide for airflow that precludes a hazardous atmosphere from developing.

Voltage. The effective (root mean square, or rms) potential difference between any two conductors or between a conductor and ground. This section expresses voltages in nominal values, unless otherwise indicated. The nominal voltage of a system or circuit is the value assigned to a system or circuit of a given voltage class for the purpose of convenient designation. The operating voltage of the system may vary above or below this value.

Work-positioning equipment. A body belt or body harness system rigged to allow an employee to be supported on an elevated vertical surface, such as a utility pole or tower leg, and work with both hands free while leaning.

[59 FR 40672, Aug. 9, 1994; 59 FR 51672, Oct. 12, 1994; 79 FR 20633-20659, July 10, 2014; 79 FR 56960, September 24, 2014]

Reprinted from www.osha.gov

Occupational Safety & Health Administration
200 Constitution Avenue, NW
Washington, DC 20210

Appendix A to §1910.269 – Flow Charts

- **Part Number:** 1910
- **Part Title:** Occupational Safety and Health Standards
- **Subpart:** R
- **Subpart Title:** Special Industries
- **Standard Number:** 1910.269 App A
- **Title:** Flow Charts.

This appendix presents information, in the form of flow charts, that illustrates the scope and application of § 1910.269. This appendix addresses the interface between § 1910.269 and Subpart S of this Part (Electrical), between § 1910.269 and § 1910.146 (Permitrequired confined spaces), and between § 1910.269 and § 1910.147 (The control of hazardous energy (lockout/tagout)). These flow charts provide guidance for employers trying to implement the requirements of § 1910.269 in combination with other General Industry Standards contained in Part 1910. Employers should always consult the relevant standards, in conjunction with this appendix, to ensure compliance with all applicable requirements.

**Appendix A-1 to §1910.269 – Application of §1910.269
and Subpart S of this Part to Electrical Installations.**

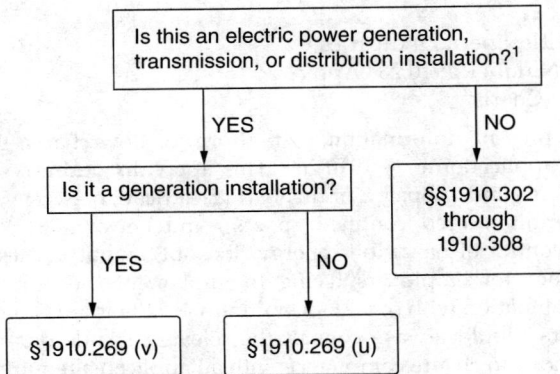

Is this an electric power generation,
transmission, or distribution installation?[1]

YES   NO

Is it a generation installation?

§§1910.302
through
1910.308

YES   NO

§1910.269 (v)   §1910.269 (u)

[1]This chart applies to electrical installation design
requirements only. See Appendix A-2 for electrical
safety-related work practices. Supplementary electric
generating equipment that is used to supply a workplace
for emergency, standby, or similar purposed only is not
considered an electric power generation installation.

Appendix A-2 to §1910.269 – Application of §1910.269 and Subpart S of this Part to Electrical Safety-Related Work Practices[1].

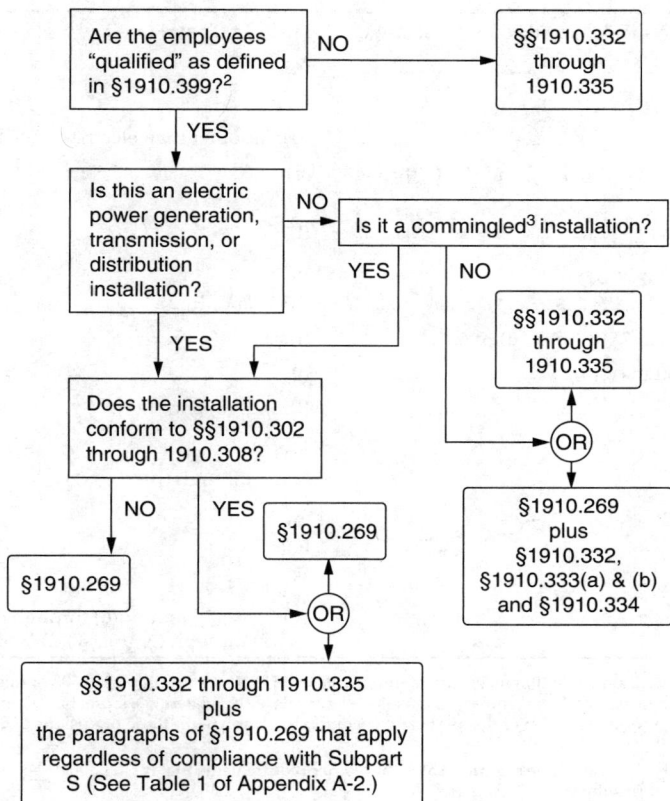

```
┌─────────────────────┐   NO    ┌──────────────┐
│ Are the employees   │────────▶│ §§1910.332   │
│ "qualified" as      │         │  through     │
│ defined in          │         │  1910.335    │
│ §1910.399?[2]       │         └──────────────┘
└─────────────────────┘
          │ YES
          ▼
┌─────────────────────┐   NO    ┌─────────────────────────────────┐
│ Is this an electric │────────▶│ Is it a commingled[3] installation? │
│ power generation,   │         └─────────────────────────────────┘
│ transmission, or    │           YES          NO
│ distribution        │                          │
│ installation?       │                          ▼
└─────────────────────┘                  ┌──────────────┐
          │ YES                           │ §§1910.332   │
          ▼                               │  through     │
┌─────────────────────┐                   │  1910.335    │
│ Does the            │                   └──────────────┘
│ installation        │                          ▲
│ conform to          │                        (OR)
│ §§1910.302          │                          │
│ through 1910.308?   │                  ┌──────────────────┐
└─────────────────────┘                  │   §1910.269      │
   NO        YES                         │     plus         │
    │     ┌─────────┐                    │  §1910.332,      │
    │     │§1910.269│                    │ §1910.333(a) & (b)│
    ▼     └─────────┘                    │  and §1910.334   │
┌─────────┐    ▲                         └──────────────────┘
│§1910.269│  (OR)
└─────────┘    │
┌────────────────────────────────────┐
│ §§1910.332 through 1910.335         │
│            plus                     │
│ the paragraphs of §1910.269 that    │
│ apply regardless of compliance with │
│ Subpart S (See Table 1 of           │
│ Appendix A-2.)                      │
└────────────────────────────────────┘
```

[1]This flowchart applies only to the electrical safety-related work practice and training requirements in § 1910.269 and §§ 1910.332 through 1910.335.

[2]See §§ 1910.269(a)(1)(ii)(B) and 1910.331(b) and (c)(1)

[3]This means commingled to the extent that the electric power generation, transmission, or distribution installation poses the greater hazard.

Table 1. Electrical Safety-Related Work Practices in §1910.269

Complaince with Subpart S Will comply with These Paragraphs of §1910.269[1]	Paragraphs that Apply Regardless of Compliance with Subpart S[2]
(d), electric-shock hazards only	(a)(2), (a)(3) and (a)(4).
(h)(3)	(b)
(i)(2) and (i)(3)	(c)
(k)	(d), for other than electric-shock hazards.
(l)(1) through (l)(5), (l)(7), and (l)(10) through (l)(12)	(e)
(m)	(f)
(p)(4)	(g)
(s)(2)	(h)(1) and (h)(2).
(u)(1) and (u)(3) through (u)(5)	(i)(4)
(v)(3) through (v)(5)	(j)
(w)(1) and (w)(7)	(l)(6), (l)(8) and (l)(9).
	(n)
	(o)
	(p)(1) through (p)(3).
	(q)
	(r)
	(s)(1)
	(t)
	(u)(2) and (u)(6)
	(v)(1), (v)(2), and (v)(6) through (v)(12).
	(w)(2) through (w)(6), (w)(8), and (w)(9).

[1]If the electrical installation meets the requirements of §§ 1910.302 through 1910.308 of this part, then the electrical installation and any associated electrical safety-related work practices conforming to §§ 1910.332 through 1910.335 of this part are considered to comply with these provisions of § 1910.269 of this part.

[2]These provisions include electrical safety and other requirements that must be met regardless of compliance with subpart S of this part.

Appendix A-3 to §1910.269 – Application of §1910.269 and Subpart S of this Part to Tree-Trimming Operations.

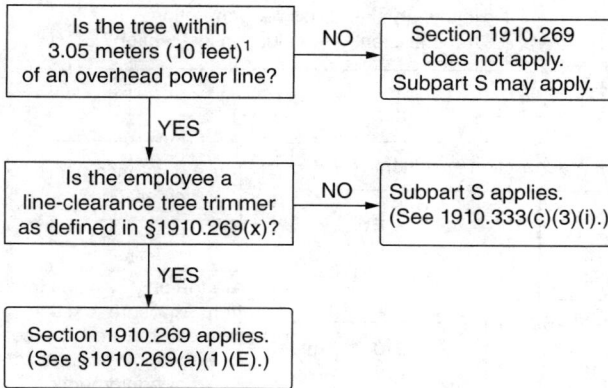

```
┌─────────────────────────┐
│     Is the tree within  │      NO    ┌──────────────────────┐
│  3.05 meters (10 feet)¹ │  ────────▶ │   Section 1910.269    │
│  of an overhead power   │            │    does not apply.    │
│         line?           │            │  Subpart S may apply. │
└─────────────────────────┘            └──────────────────────┘
             │ YES
             ▼
┌─────────────────────────┐
│   Is the employee a     │      NO    ┌──────────────────────────┐
│ line-clearance tree     │  ────────▶ │   Subpart S applies.      │
│ trimmer as defined in   │            │ (See 1910.333(c)(3)(i).)  │
│   §1910.269(x)?         │            └──────────────────────────┘
└─────────────────────────┘
             │ YES
             ▼
┌─────────────────────────┐
│ Section 1910.269 applies.│
│ (See §1910.269(a)(1)(E).)│
└─────────────────────────┘
```

[1]3.05 meters (10 feet) plus 0.10 meters (4 inches for every 10 kilovolts over 50 kilovolts.

Note: Paragraph (t) of § 1910.269 contains additional requirements for work in manholes and underground vaults.

Appendix A-4 to §1910.269 – Application of §§1910.147, 1910.269 and 1910.333 to Hazardous Energy Control Procedures (Lockout/Tagout).

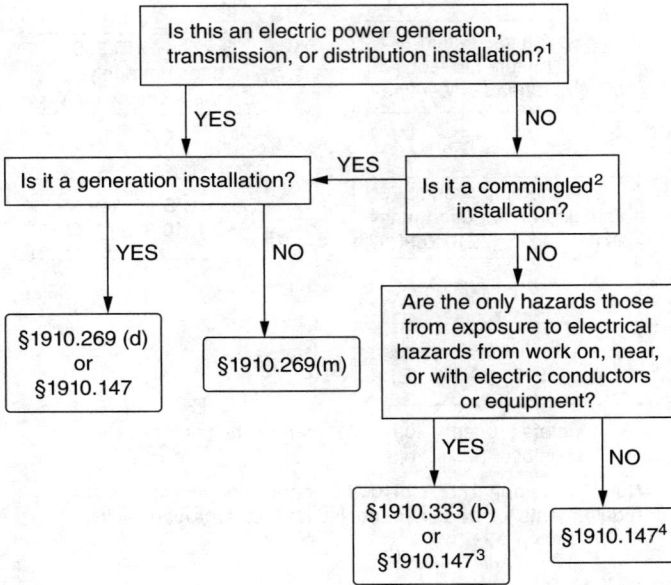

Is this an electric power generation, transmission, or distribution installation?[1]

YES → Is it a generation installation?

NO → Is it a commingled[2] installation?

Is it a commingled[2] installation? → **YES** → Is it a generation installation?

Is it a generation installation?
- **YES** → §1910.269 (d) or §1910.147
- **NO** → §1910.269(m)

Is it a commingled[2] installation?
- **NO** → Are the only hazards those from exposure to electrical hazards from work on, near, or with electric conductors or equipment?

Are the only hazards those from exposure to electrical hazards from work on, near, or with electric conductors or equipment?
- **YES** → §1910.333 (b) or §1910.147[3]
- **NO** → §1910.147[4]

[1]If a generation, transmission, or distribution installation conforms to §§ 1910.302 through 1910.308, the lockout and tagging procedures of § 1910.333(b) may be followed for electrickshock hazards.

[2]This means commingled to the extent that the electric power generation, transmission, or distribution installation poses the greater hazard.

[3]Paragraphs (b)(2)(iii)(D) and (b)(2)(iv)(B) of § 1910.333 still apply.

[4]Paragraph (b) of § 1910.333 applies to any electrical hazards from work on, near, or with electric conductiors and equipment.

Appendix A-5 to §1910.269 – Application of §§1910.146 and 1910.269 to Permit-Required Confined Spaces.

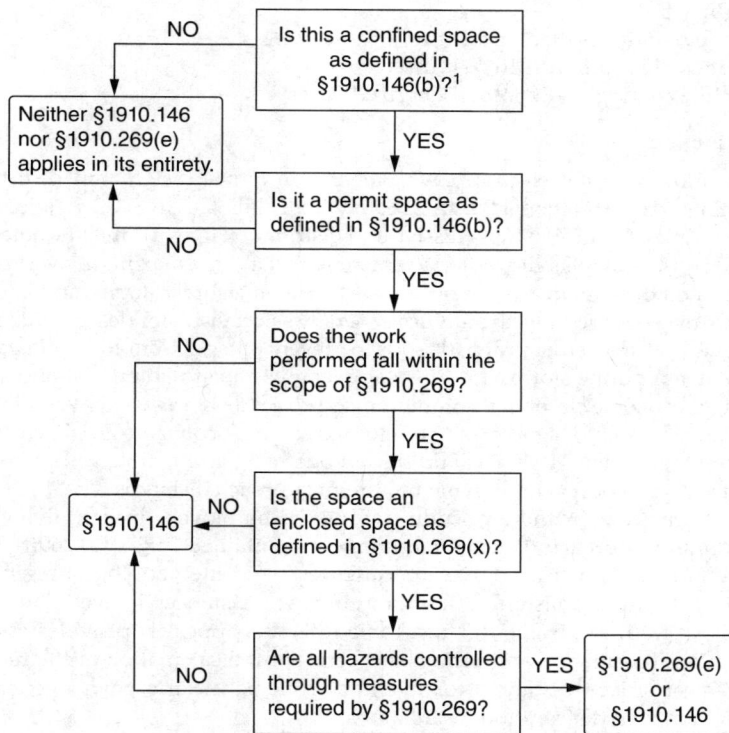

Is this a confined space as defined in §1910.146(b)?[1]

→ NO → Neither §1910.146 nor §1910.269(e) applies in its entirety.

↓ YES

Is it a permit space as defined in §1910.146(b)?

NO → Neither §1910.146 nor §1910.269(e) applies in its entirety.

↓ YES

Does the work performed fall within the scope of §1910.269?

NO → §1910.146

↓ YES

Is the space an enclosed space as defined in §1910.269(x)?

NO → §1910.146

↓ YES

Are all hazards controlled through measures required by §1910.269?

NO → §1910.146

YES → §1910.269(e) or §1910.146

[1]See §1910.146(c) for general nonentry requirements that apply to all confined spaces.

Reprinted from www.osha.gov

Occupational Safety & Health Administration
200 Constitution Avenue, NW
Washington, DC 20210

Appendix B to §1910.269 – Working on Exposed Energized Parts

- **Part Number:** 1910
- **Part Title:** Occupational Safety and Health Standards
- **Subpart:** R
- **Subpart Title:** Special Industries
- **Standard Number:** 1910.269 App B
- **Title:** Working on Exposed Energized Parts.

I. Introduction

Electric utilities design electric power generation, transmission, and distribution installations to meet National Electrical Safety Code (NESC), ANSI C2, requirements. Electric utilities also design transmission and distribution lines to limit line outages as required by system reliability criteria [1] and to withstand the maximum overvoltages impressed on the system. Conditions such as switching surges, faults, and lightning can cause overvoltages. Electric utilities generally select insulator design and lengths and the clearances to structural parts so as to prevent outages from contaminated line insulation and during storms. Line insulator lengths and structural clearances have, over the years, come closer to the minimum approach distances used by workers. As minimum approach distances and structural clearances converge, it is increasingly important that system designers and system operating and maintenance personnel understand the concepts underlying minimum approach distances.

The information in this appendix will assist employers in complying with the minimum approach-distance requirements contained in § 1910.269(l)(3) and (q)(3). Employers must use the technical criteria and methodology presented in this appendix in establishing minimum approach distances in accordance with § 1910.269(l)(3)(i) and Table R-3 and Table R-8. This appendix provides essential background information and technical criteria for the calculation of the required minimum approach distances for live-line work on electric power generation, transmission, and distribution installations.

Unless an employer is using the maximum transient overvoltages specified in Table R-9 for voltages over 72.5 kilovolts, the employer must use persons knowledgeable in the techniques discussed in this appendix, and competent in the field of electric transmission and distribution system design, to determine the maximum transient overvoltage.

II. General

A. *Definitions.* The following definitions from § 1910.269(x) relate to work on or near electric power generation, transmission, and distribution lines and equipment and the electrical hazards they present.

Exposed. . . . Not isolated or guarded.

Guarded. Covered, fenced, enclosed, or otherwise protected, by means of suitable covers or casings, barrier rails or screens, mats, or platforms, designed to minimize the possibility, under normal conditions, of dangerous approach or inadvertent contact by persons or objects.

---

[1]Federal, State, and local regulatory bodies and electric utilities set reliability requirements that limit the number and duration of system outages.

Note to the definition of "guarded": Wires that are insulated, but not otherwise protected, are not guarded.
Insulated. Separated from other conducting surfaces by a dielectric (including air space) offering a high resistance to the passage of current.

Note to the definition of "insulated": When any object is said to be insulated, it is understood to be insulated for the conditions to which it normally is subjected. Otherwise, it is, for the purpose of this section, uninsulated.
Isolated. Not readily accessible to persons unless special means for access are used.
Statistical sparkover voltage. A transient overvoltage level that produces a 97.72-percent probability of sparkover (that is, two standard deviations above the voltage at which there is a 50-percent probability of sparkover).
Statistical withstand voltage. A transient overvoltage level that produces a 0.14-percent probability of sparkover (that is, three standard deviations below the voltage at which there is a 50-percent probability of sparkover).
B. *Installations energized at 50 to 300 volts.* The hazards posed by installations energized at 50 to 300 volts are the same as those found in many other workplaces. That is not to say that there is no hazard, but the complexity of electrical protection required does not compare to that required for highvoltage systems. The employee must avoid contact with the exposed parts, and the protective equipment used (such as rubber insulating gloves) must provide insulation for the voltages involved.
C. *Exposed energized parts over 300 volts AC.* Paragraph (l)(3)(i) of § 1910.269 requires the employer to establish minimum approach distances no less than the distances computed by Table R-3 for ac systems so that employees can work safely without risk of sparkover.[2]
Unless the employee is using electrical protective equipment, air is the insulating medium between the employee and energized parts. The distance between the employee and an energized part must be sufficient for the air to withstand the maximum transient overvoltage that can reach the worksite under the working conditions and practices the employee is using. This distance is the minimum air insulation distance, and it is equal to the electrical component of the minimum approach distance. Normal system design may provide or include a means (such as lightning arrestors) to control maximum anticipated transient overvoltages, or the employer may use temporary devices (portable protective gaps) or measures (such as preventing automatic circuit breaker reclosing) to achieve the same result. Paragraph (l)(3)(ii) of § 1910.269 requires the employer to determine the maximum anticipated per-unit transient overvoltage, phase-to-ground, through an engineering analysis or assume a maximum anticipated per-unit transient overvoltage, phase-to-ground, in accordance with Table R-9, which specifies the following maximums for ac systems:
72.6 to 420.0 kilovolts-3.5 per unit
420.1 to 550.0 kilovolts-3.0 per unit
550.1 to 800.0 kilovolts-2.5 per unit
See paragraph IV.A.2, later in this appendix, for additional discussion of maximum transient overvoltages.
D. *Types of exposures.* Employees working on or near energized electric power generation, transmission, and distribution systems face two kinds of exposures: Phaseto-ground and phase-to-phase. The exposure is phase-to-ground: (1) With

[2]Sparkover is a disruptive electric discharge in which an electric arc forms and electric current passes through air.

respect to an energized part, when the employee is at ground potential or (2) with respect to ground, when an employee is at the potential of the energized part during live-line barehand work. The exposure is phase-tophase, with respect to an energized part, when an employee is at the potential of another energized part (at a different potential) during live-line barehand work.

III. Determination of Minimum Approach Distances for AC Voltages Greater Than 300 Volts

A. *Voltages of 301 to 5,000 volts*. Test data generally forms the basis of minimum air insulation distances. The lowest voltage for which sufficient test data exists is 5,000 volts, and these data indicate that the minimum air insulation distance at that voltage is 20 millimeters (1 inch). Because the minimum air insulation distance increases with increasing voltage, and, conversely, decreases with decreasing voltage, an assumed minimum air insulation distance of 20 millimeters will protect against sparkover at voltages of 301 to 5,000 volts. Thus, 20 millimeters is the electrical component of the minimum approach distance for these voltages.

B. *Voltages of 5.1 to 72.5 kilovolts*. For voltages from 5.1 to 72.5 kilovolts, the Occupational Safety and Health Administration bases the methodology for calculating the electrical component of the minimum approach distance on Institute of Electrical and Electronic Engineers (IEEE) Standard 4-1995, *Standard Techniques for High-Voltage Testing*. Table 1 lists the critical sparkover distances from that standard as listed in IEEE Std 516-2009, *IEEE Guide for Maintenance Methods on Energized Power Lines*.

Table 1   Sparkover Distance for Rod-To-Rod Gap

60 Hz Rod-to-Rod sparkover (kV peak)	Gap spacing from IEEE Std 4-1995 (cm)
25	2
36	3
46	4
53	5
60	6
70	8
79	10
86	12
95	14
104	16
112	18
120	20
143	25
167	30
192	35
218	40
243	45
270	50
322	60

Source: IEEE Std 516-2009.

To use this table to determine the electrical component of the minimum approach distance, the employer must determine the peak phase-to-ground transient overvoltage and select a gap from the table that corresponds to that voltage as a withstand voltage rather than a critical sparkover voltage. To calculate the electrical component of the minimum approach distance for voltages between 5 and 72.5 kilovolts, use the following procedure:

1. Divide the phase-to-phase voltage by the square root of 3 to convert it to a phase-to-ground voltage.
2. Multiply the phase-to-ground voltage by the square root of 2 to convert the rms value of the voltage to the peak phase-to-ground voltage.
3. Multiply the peak phase-to-ground voltage by the maximum per-unit transient overvoltage, which, for this voltage range, is 3.0, as discussed later in this appendix. This is the maximum phase-to-ground transient overvoltage, which corresponds to the withstand voltage for the relevant exposure.[3]
4. Divide the maximum phase-to-ground transient overvoltage by 0.85 to determine the corresponding critical sparkover voltage. (The critical sparkover voltage is 3 standard deviations (or 15 percent) greater than the withstand voltage.)
5. Determine the electrical component of the minimum approach distance from Table 1 through interpolation.

Table 2 illustrates how to derive the electrical component of the minimum approach distance for voltages from 5.1 to 72.5 kilovolts, before the application of any altitude correction factor, as explained later.

Table 2   Calculating the Electrical Component of Mad 751 V to 72.5 kV

Step	Maximum system phase-to-phase voltage (kV)			
	15	36	46	72.5
1. Divide by $\sqrt{3}$......................	8.7	20.8	26.6	41.9
2. Multiply by $\sqrt{2}$.....................	12.2	29.4	37.6	59.2
3. Multiply by 3.0....................	36.7	88.2	112.7	177.6
4. Divide by 0.85.....................	43.2	103.7	132.6	208.9
5. Interpolate from Table 1..	3+(7.2/10)*1	14+(8.7/9)*2	20+(12.6/23)*5	35+(16.9/26)*5
Electrical component of MAD (cm)...............................	3.72	15.93	22.74	38.25

C. *Voltages of 72.6 to 800 kilovolts.* For voltages of 72.6 kilovolts to 800 kilovolts, this section bases the electrical component of minimum approach distances, before the application of any altitude correction factor, on the following formula:

Equation 1-For Voltages of 72.6 kV to 800 kV

$$D = 0.3048(C + a) V_{L-G}T$$

Where:

D = Electrical component of the minimum approach distance in air in meters;
C = a correction factor associated with the variation of gap sparkover with voltage;

---

[3]The withstand voltage is the voltage at which sparkover is not likely to occur across a specified distance. It is the voltage taken at the 3s point below the sparkover voltage, assuming that the sparkover curve follows a normal distribution.

a = A factor relating to the saturation of air at system voltages of 345 kilovolts or higher;[4]

V_{L-G} = Maximum system line-to-ground rms voltage in kilovolts-it should be the "actual" maximum, or the normal highest voltage for the range (for example, 10 percent above the nominal voltage); and

T = Maximum transient overvoltage factor in per unit.

In Equation 1, C is 0.01: (1) For phase-to-ground exposures that the employer can demonstrate consist only of air across the approach distance (gap) and (2) for phase-tophase exposures if the employer can demonstrate that no insulated tool spans the gap and that no large conductive object is in the gap. Otherwise,C is 0.011.

In Equation 1, the term a varies depending on whether the employee's exposure is phase-to-ground or phase-to-phase and on whether objects are in the gap. The employer must use the equations in Table 3 to calculate a. Sparkover test data with insulation spanning the gap form the basis for the equations for phase-to-ground exposures, and sparkover test data with only air in the gap form the basis for the equations for phase-tophase exposures. The phase-to-ground equations result in slightly higher values of a, and, consequently, produce larger minimum approach distances, than the phase-to-phase equations for the same value of V_{Peak}.

Table 3 Equations for Calculating the Surge Factor, a

Phase-to-ground exposures

$\dfrac{V_{Peak} = T_{L-G}V_{L-G}\sqrt{2}}{a}$	635 kV or less 0	635.1 to 915 kV $(V_{Peak}- 635)/140{,}000$	915.1 to 1,050 kV $(V_{Peak}-645)/135{,}000$
$\dfrac{V_{Peak} = T_{L-G}V_{L-G}\sqrt{2}}{a}$	More than 1,050 kV $(V_{Peak}-675)/125{,}000$		

Phase-to-phase exposures 1

$\dfrac{V_{Peak} = (1.35T_{L-G} + 0.45)V_{L-G}\sqrt{2}}{a}$	2. 630 kV or less 0	630.1 to 848 kV $(V_{Peak}-630)/155{,}000$	848.1 to 1,131 kV $(V_{Peak}-633.6)/152{,}207$
$\dfrac{V_{Peak} = (1.35TL\text{-}G + 0.45)VL\text{-}G\sqrt{2}}{a}$	1,131.1 to 1,485 kV $(V_{Peak}-628)/153{,}846$	More than 1,485 kV $(V_{Peak}-350.5)/203{,}666$	

[1]Use the equations for phase-to-ground exposures (with V_{Peak} for phase-to-phase exposures) unless the employer can demonstrate that no insulated tool spans the gap and that no large conductive object is in the gap.

In Equation 1, T is the maximum transient overvoltage factor in per unit. As noted earlier, § 1910.269(l)(3)(ii) requires the employer to determine the maximum anticipated per-unit transient overvoltage, phase-to-ground, through an engineering analysis or assume a maximum anticipated per-unit transient overvoltage, phase-to-ground, in accordance with Table R-9. For phase-to-ground exposures, the employer uses this value, called T_{L-G}, as T in Equation 1. IEEE Std 516-2009 provides the following formula to calculate the phase-to-phase maximum transient overvoltage, T_{L-L}, from T_{L-G}:

$$T_{L-L} = 1.35T_{L-G} + 0.45$$

[4]Test data demonstrates that the saturation factor is greater than 0 at peak voltages of about 630 kilovolts. Systems operating at 345 kilovolts (or maximum system voltages of 362 kilovolts) can have peak maximum transient overvoltages exceeding 630 kilovolts. Table R-3 sets equations for calculating a based on peak voltage.

For phase-to-phase exposures, the employer uses this value as T in Equation 1.

D. *Provisions for inadvertent movement.* The minimum approach distance must include an "adder" to compensate for the inadvertent movement of the worker relative to an energized part or the movement of the part relative to the worker. This "adder" must account for this possible inadvertent movement and provide the worker with a comfortable and safe zone in which to work. Employers must add the distance for inadvertent movement (called the "ergonomic component of the minimum approach distance") to the electrical component to determine the total safe minimum approach distances used in live-line work.

The Occupational Safety and Health Administration based the ergonomic component of the minimum approach distance on response time-distance analysis. This technique uses an estimate of the total response time to a hazardous incident and converts that time to the distance traveled. For example, the driver of a car takes a given amount of time to respond to a "stimulus" and stop the vehicle. The elapsed time involved results in the car's traveling some distance before coming to a complete stop. This distance depends on the speed of the car at the time the stimulus appears and the reaction time of the driver.

In the case of live-line work, the employee must first perceive that he or she is approaching the danger zone. Then, the worker responds to the danger and must decelerate and stop all motion toward the energized part. During the time it takes to stop, the employee will travel some distance. This is the distance the employer must add to the electrical component of the minimum approach distance to obtain the total safe minimum approach distance.

At voltages from 751 volts to 72.5 kilovolts,[5] the electrical component of the minimum approach distance is smaller than the ergonomic component. At 72.5 kilovolts, the electrical component is only a little more than 0.3 meters (1 foot). An ergonomic component of the minimum approach distance must provide for all the worker's unanticipated movements. At these voltages, workers generally use rubber insulating gloves; however, these gloves protect only a worker's hands and arms. Therefore, the energized object must be at a safe approach distance to protect the worker's face. In this case, 0.61 meters (2 feet) is a sufficient and practical ergonomic component of the minimum approach distance.

For voltages between 72.6 and 800 kilovolts, employees must use different work practices during energized line work. Generally, employees use live-line tools (hot sticks) to perform work on energized equipment. These tools, by design, keep the energized part at a constant distance from the employee and, thus, maintain the appropriate minimum approach distance automatically.

The location of the worker and the type of work methods the worker is using also influence the length of the ergonomic component of the minimum approach distance. In this higher voltage range, the employees use work methods that more tightly control their movements than when the workers perform work using rubber insulating gloves. The worker, therefore, is farther from the energized line or equipment and must be more precise in his or her movements just to perform the work. For these reasons, this section adopts an ergonomic component of the minimum approach distance of 0.31 m (1 foot) for voltages between 72.6 and 800 kilovolts.

[5]For voltages of 50 to 300 volts, Table R-3 specifies a minimum approach distance of "avoid contact." The minimum approach distance for this voltage range contains neither an electrical component nor an ergonomic component.

Table 4 summarizes the ergonomic component of the minimum approach distance for various voltage ranges.

Table 4 Ergonomic Component of Minimum Approach Distance

Voltage range (kV)	Distance	
	m	ft
0.301 to 0.750.	0.31	1.0
0.751 to 72.5	0.61	2.0
72.6 to 800	0.31	1.0

Note: The employer must add this distance to the electrical component of the minimum approach distance to obtain the full minimum approach distance.

The ergonomic component of the minimum approach distance accounts for errors in maintaining the minimum approach distance (which might occur, for example, if an employee misjudges the length of a conductive object he or she is holding), and for errors in judging the minimum approach distance. The ergonomic component also accounts for inadvertent movements by the employee, such as slipping. In contrast, the working position selected to properly maintain the minimum approach distance must account for all of an employee's reasonably likely movements and still permit the employee to adhere to the applicable minimum approach distance. (See Figure 1.) Reasonably likely movements include an employee's adjustments to tools, equipment, and working positions and all movements needed to perform the work. For example, the employee should be able to perform all of the following actions without straying into the minimum approach distance:

Adjust his or her hardhat,
maneuver a tool onto an energized part with a reasonable amount of overreaching or underreaching,
reach for and handle tools, material, and equipment passed to him or her, and
adjust tools, and replace components on them, when necessary during the work procedure.

The training of qualified employees required under § 1910.269(a)(2), and the job planning and briefing required under § 1910.269(c), must address selection of a proper working position.

BILLING CODE 4510-26-P

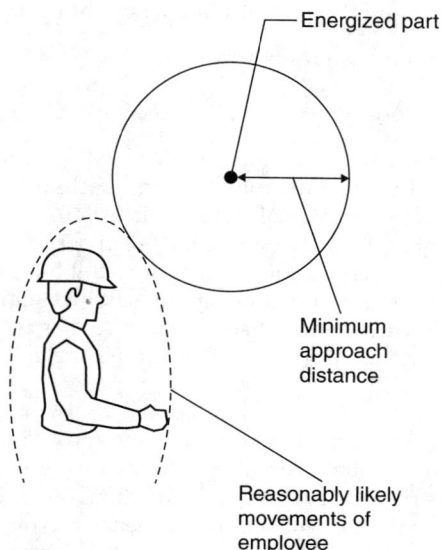

Figure 1. Maintaining the Minimum Approach Distance

BILLING CODE 4510-26-C

E. *Miscellaneous correction factors.* Changes in the air medium that forms the insulation influences the strength of an air gap. A brief discussion of each factor follows.
1. *Dielectric strength of air.* The dielectric strength of air in a uniform electric field at standard atmospheric conditions is approximately 3 kilovolts per millimeter.[6]
The pressure, temperature, and humidity of the air, the shape, dimensions, and separation of the electrodes, and the characteristics of the applied voltage (wave shape) affect the disruptive gradient.
2. *Atmospheric effect.* The empirically determined electrical strength of a given gap is normally applicable at standard atmospheric conditions (20 °C, 101.3 kilopascals, 11 grams/cubic centimeter humidity). An increase in the density (humidity) of the air inhibits sparkover for a given air gap. The combination of temperature and air pressure that results in the lowest gap sparkover voltage is high temperature and low pressure. This combination of conditions is not likely to occur. Low air pressure, generally associated with high humidity, causes increased electrical strength. An average air pressure generally correlates with low humidity. Hot and dry working conditions normally result in reduced electrical strength. The equations for minimum approach distances in Table R-3 assume standard atmospheric conditions.
3. *Altitude.* The reduced air pressure at high altitudes causes a reduction in the electrical strength of an air gap. An employer must increase the minimum approach

[6]For the purposes of estimating arc length, § 1910.269 generally assumes a more conservative dielectric strength of 10 kilovolts per 25.4 millimeters, consistent with assumptions made in consensus standards such as the National Electrical Safety Code (IEEE C2-2012). The more conservative value accounts for variables such as electrode shape, wave shape, and a certain amount of overvoltage.

distance by about 3 percent per 300 meters (1,000 feet) of increased altitude for altitudes above 900 meters (3,000 feet). Table R-5 specifies the altitude correction factor that the employer must use in calculating minimum approach distances.

IV. Determining Minimum Approach Distances

A. Factors Affecting Voltage Stress at the Worksite

1. *System voltage (nominal).* The nominal system voltage range determines the voltage for purposes of calculating minimum approach distances. The employer selects the range in which the nominal system voltage falls, as given in the relevant table, and uses the highest value within that range in perunit calculations.

2. *Transient overvoltages.* Operation of switches or circuit breakers, a fault on a line or circuit or on an adjacent circuit, and similar activities may generate transient overvoltages on an electrical system. Each overvoltage has an associated transient voltage wave shape. The wave shape arriving at the site and its magnitude vary considerably.

In developing requirements for minimum approach distances, the Occupational Safety and Health Administration considered the most common wave shapes and the magnitude of transient overvoltages found on electric power generation, transmission, and distribution systems. The equations in Table R-3 for minimum approach distances use per-unit maximum transient overvoltages, which are relative to the nominal maximum voltage of the system. For example, a maximum transient overvoltage value of 3.0 per unit indicates that the highest transient overvoltage is 3.0 times the nominal maximum system voltage.

Table 5   Magnitude of Typical Transient Overvoltages

Cause	Magnitude (per unit)
Energized 200-mile line without closing resistors	3.5
Energized 200-mile line with one-step closing resistor	2.1
Energized 200-mile line with multistep resistor	2.5
Reclosing with trapped charge one-step resistor	2.2
Opening surge with single restrike	3.0
Fault initiation unfaulted phase	2.1
Fault initiation adjacent circuit	2.5
Fault clearing	1.7 to 1.9

3. *Typical magnitude of overvoltages.* Table 5 lists the magnitude of typical transient overvoltages.

4. *Standard deviation-air-gap withstand.* For each air gap length under the same atmospheric conditions, there is a statistical variation in the breakdown voltage. The probability of breakdown against voltage has a normal (Gaussian) distribution. The standard deviation of this distribution varies with the wave shape, gap geometry, and atmospheric conditions. The withstand voltage of the air gap is

three standard deviations (3s) below the critical sparkover voltage. (The critical sparkover voltage is the crest value of the impulse wave that, under specified conditions, causes sparkover 50 percent of the time. An impulse wave of three standard deviations below this value, that is, the withstand voltage, has a probability of sparkover of approximately 1 in 1,000.)

5. *Broken Insulators.* Tests show reductions in the insulation strength of insulator strings with broken skirts. Broken units may lose up to 70 percent of their withstand capacity. Because an employer cannot determine the insulating capability of a broken unit without testing it, the employer must consider damaged units in an insulator to have no insulating value. Additionally, the presence of a live-line tool alongside an insulator string with broken units may further reduce the overall insulating strength. The number of good units that must be present in a string for it to be "insulated" as defined by § 1910.269(x) depends on the maximum overvoltage possible at the worksite.

B. Minimum Approach Distances Based on Known, Maximum-Anticipated Per-Unit Transient Overvoltages

1. *Determining the minimum approach distance for AC systems.* Under § 1910.269(l)(3)(ii), the employer must determine the maximum anticipated per-unit transient overvoltage, phase-to-ground, through an engineering analysis or must assume a maximum anticipated per-unit transient overvoltage, phase-to-ground, in accordance with Table R-9. When the employer conducts an engineering analysis of the system and determines that the maximum transient overvoltage is lower than specified by Table R-9, the employer must ensure that any conditions assumed in the analysis, for example, that employees block reclosing on a circuit or install portable protective gaps, are present during energized work. To ensure that these conditions are present, the employer may need to institute new livework procedures reflecting the conditions and limitations set by the engineering analysis.

2. *Calculation of reduced approach distance values.* An employer may take the following steps to reduce minimum approach distances when the maximum transient overvoltage on the system (that is, the maximum transient overvoltage without additional steps to control overvoltages) produces unacceptably large minimum approach distances:

Step 1. Determine the maximum voltage (with respect to a given nominal voltage range) for the energized part.

Step 2. Determine the technique to use to control the maximum transient overvoltage. (See paragraphs IV.C and IV.D of this appendix.) Determine the maximum transient overvoltage that can exist at the worksite with that form of control in place and with a confidence level of 3s. This voltage is the withstand voltage for the purpose of calculating the appropriate minimum approach distance.

Step 3. Direct employees to implement procedures to ensure that the control technique is in effect during the course of the work.

Step 4. Using the new value of transient overvoltage in per unit, calculate the required minimum approach distance from Table R-3.

C. Methods of Controlling Possible Transient Overvoltage Stress Found on a System

1. *Introduction.* There are several means of controlling overvoltages that occur on transmission systems. For example, the employer can modify the operation of circuit breakers or other switching devices to reduce switching transient overvoltages. Alternatively, the employer can hold the overvoltage to an acceptable level by installing surge arresters or portable protective gaps on the system. In addition, the employer can change the transmission system to minimize the effect of switching operations. Section 4.8 of IEEE Std 516-2009 describes various ways of controlling, and thereby reducing, maximum transient overvoltages.

2. *Operation of circuit breakers.*[7] The maximum transient overvoltage that can reach the worksite is often the result of switching on the line on which employees are working. Disabling automatic reclosing during energized line work, so that the line will not be reenergized after being opened for any reason, limits the maximum switching surge overvoltage to the larger of the opening surge or the greatest possible fault-generated surge, provided that the devices (for example, insertion resistors) are operable and will function to limit the transient overvoltage and that circuit breaker restrikes do not occur. The employer must ensure the proper functioning of insertion resistors and other overvoltage-limiting devices when the employer's engineering analysis assumes their proper operation to limit the overvoltage level. If the employer cannot disable the reclosing feature (because of system operating conditions), other methods of controlling the switching surge level may be necessary.

Transient surges on an adjacent line, particularly for double circuit construction, may cause a significant overvoltage on the line on which employees are working. The employer›s engineering analysis must account for coupling to adjacent lines.

3. *Surge arresters.* The use of modern surge arresters allows a reduction in the basic impulse-insulation levels of much transmission system equipment. The primary function of early arresters was to protect the system insulation from the effects of lightning. Modern arresters not only dissipate lightning-caused transients, but may also control many other system transients caused by switching or faults.

The employer may use properly designed arresters to control transient overvoltages along a transmission line and thereby reduce the requisite length of the insulator string and possibly the maximum transient overvoltage on the line.[8]

4. *Switching Restrictions.* Another form of overvoltage control involves establishing switching restrictions, whereby the employer prohibits the operation of circuit breakers until certain system conditions are present. The employer restricts switching by using a tagging system, similar to that used for a permit, except that the common term used for this activity is a "hold-off" or "restriction." These terms indicate that the restriction does not prevent operation, but only modifies the operation during the livework activity.

---

[7]The detailed design of a circuit interrupter, such as the design of the contacts, resistor insertion, and breaker timing control, are beyond the scope of this appendix. The design of the system generally accounts for these features. This appendix only discusses features that can limit the maximum switching transient overvoltage on a system.

[8]Surge arrester application is beyond the scope of this appendix. However, if the employer installs the arrester near the work site, the application would be similar to the protective gaps discussed in paragraph IV.D of this appendix.

D. Minimum Approach Distance Based on Control of Maximum Transient Overvoltage at the Worksite

When the employer institutes control of maximum transient overvoltage at the worksite by installing portable protective gaps, the employer may calculate the minimum approach distance as follows:

Step 1. Select the appropriate withstand voltage for the protective gap based on system requirements and an acceptable probability of gap sparkover.[9]

Step 2. Determine a gap distance that provides a withstand voltage [10] greater than or equal to the one selected in the first step.[11]

Step 3. Use 110 percent of the gap's critical sparkover voltage to determine the phase-to-ground peak voltage at gap sparkover ($V_{PPG Peak}$).

Step 4. Determine the maximum transient overvoltage, phase-to-ground, at the worksite from the following formula:

$$T = \frac{V_{PPG Peak}}{V_{L-G}\sqrt{2}}$$

Step 5. Use this value of T[12] in the equation in Table R-3 to obtain the minimum approach distance. If the worksite is no more than 900 meters (3,000 feet) above sea level, the employer may use this value of T to determine the minimum approach distance from Table 14 through Table 21.

Note: All rounding must be to the next higher value (that is, always round up).

Sample protective gap calculations.

Problem: Employees are to perform work on a 500-kilovolt transmission line at sea level that is subject to transient overvoltages of 2.4 p.u. The maximum operating voltage of the line is 550 kilovolts. Determine the length of the protective gap that will provide the minimum practical safe approach distance. Also, determine what that minimum approach distance is

Step 1. Calculate the smallest practical maximum transient overvoltage (1.25 times the crest phase-to-ground voltage):[13]

$$550 \, kV \times \frac{\sqrt{2}}{\sqrt{3}} \times 1.25 = 561 \, kV$$

This value equals the withstand voltage of the protective gap.

Step 2. Using test data for a particular protective gap, select a gap that has a critical sparkover voltage greater than or equal to:

$$561 \, kV \div 0.85 = 660 \, kV$$

[9]The employer should check the withstand voltage to ensure that it results in a probability of gap flashover that is acceptable from a system outage perspective. (In other words, a gap sparkover will produce a system outage. The employer should determine whether such an outage will impact overall system performance to an acceptable degree.) In general, the withstand voltage should be at least 1.25 times the maximum crest operating voltage.

[10]The manufacturer of the gap provides, based on test data, the critical sparkover voltage for each gap spacing (for example, a critical sparkover voltage of 665 kilovolts for a gap spacing of 1.2 meters). The withstand voltage for the gap is equal to 85 percent of its critical sparkover voltage.

[11]Switch steps 1 and 2 if the length of the protective gap is known.

[12]IEEE Std 516-2009 states that most employers add 0.2 to the calculated value of T as an additional safety factor.

[13]To eliminate sparkovers due to minor system disturbances, the employer should use a withstand voltage no lower than 1.25 p.u. Note that this is a practical, or operational, consideration only. It may be feasible for the employer to use lower values of withstand voltage.

For example, if a protective gap with a 1.22-m (4.0-foot) spacing tested to a critical sparkover voltage of 665 kilovolts (crest), select this gap spacing.

Step 3. The phase-to-ground peak voltage at gap sparkover (VPPG Peak) is 110 percent of the value from the previous step:

$$665 \, kV \times 1.10 = 732 \, kV$$

This value corresponds to the withstand voltage of the electrical component of the minimum approach distance.

Step 4. Use this voltage to determine the worksite value of *T*:

$$T = \frac{732}{564} = 1.7 \, p.u.$$

Step 5. Use this value of *T* in the equation in Table R-3 to obtain the minimum approach distance, or look up the minimum approach distance in Table 14 through Table 21:

$$MAD = 2.29\text{m} \ (7.6 \ \text{ft}).$$

E. Location of Protective Gaps

1. *Adjacent structures.* The employer may install the protective gap on a structure adjacent to the worksite, as this practice does not significantly reduce the protection afforded by the gap.

2. *Terminal stations.* Gaps installed at terminal stations of lines or circuits provide a level of protection; however, that level of protection may not extend throughout the length of the line to the worksite. The use of substation terminal gaps raises the possibility that separate surges could enter the line at opposite ends, each with low enough magnitude to pass the terminal gaps without sparkover. When voltage surges occur simultaneously at each end of a line and travel toward each other, the total voltage on the line at the point where they meet is the arithmetic sum of the two surges. A gap installed within 0.8 km (0.5 mile) of the worksite will protect against such intersecting waves. Engineering studies of a particular line or system may indicate that employers can adequately protect employees by installing gaps at even more distant locations. In any event, unless using the default values for *T* from Table R-9, the employer must determine *T* at the worksite.

3. *Worksite.* If the employer installs protective gaps at the worksite, the gap setting establishes the worksite impulse insulation strength. Lightning strikes as far as 6 miles from the worksite can cause a voltage surge greater than the gap withstand voltage, and a gap sparkover can occur. In addition, the gap can sparkover from overvoltages on the line that exceed the withstand voltage of the gap. Consequently, the employer must protect employees from hazards resulting from any sparkover that could occur.

F. *Disabling automatic reclosing.* There are two reasons to disable the automatic-reclosing feature of circuit-interrupting devices while employees are performing live-line work:

- To prevent reenergization of a circuit faulted during the work, which could create a hazard or result in more serious injuries or damage than the injuries or damage produced by the original fault;

Table 6- Minimum Approach Distances Until December 31, 2014

Voltage range phase to phase (kV)	Phase-to-ground exposure		Phase-to-phase exposure	
	m	ft	m	ft
0.05 to 1.0	Avoid Contact		Avoid Contact	
1.1 to 15.0	0.64	2.10	0.66	2.20
15.1 to 36.0	0.72	2.30	0.77	2.60
36.1 to 46.0	0.77	2.60	0.85	2.80
46.1 to 72.5	0.90	3.00	1.05	3.50
72.6 to 121	0.95	3.20	1.29	4.30
138 to 145	1.09	3.60	1.50	4.90
161 to 169	1.22	4.00	1.71	5.70
230 to 242	1.59	5.30	2.27	7.50
345 to 362	2.59	8.50	3.80	12.50
500 to 550	3.42	11.30	5.50	18.10
765 to 800	4.53	14.90	7.91	26.00

Note: The clear live-line tool distance must equal or exceed the values for the indicated voltage ranges.

Table 7- Minimum Approach Distances Until March 31, 2015-72.6 To 121.0 kV with Overvoltage Factor

T (p.u.)	Phase-to-ground exposure		Phase-to-phase exposure	
	m	ft	m	ft
2.0	0.74	2.42	1.09	3.58
2.1	0.76	2.50	1.09	3.58
2.2	0.79	2.58	1.12	3.67
2.3	0.81	2.67	1.14	3.75
2.4	0.84	2.75	1.17	3.83
2.5	0.84	2.75	1.19	3.92
2.6	0.86	2.83	1.22	4.00
2.7	0.89	2.92	1.24	4.08
2.8	0.91	3.00	1.24	4.08
2.9	0.94	3.08	1.27	4.17
3.0	0.97	3.17	1.30	4.25

Note 1: The employer may apply the distance specified in this table only where the employer determines the maximum anticipated per-unit transient overvoltage by engineering analysis. (Table 6 applies otherwise.)
Note 2: The distances specified in this table are the air, bare-hand, and live-line tool distances.

Table 8-   Minimum Approach Distances Until March 31, 2015-121.1 To 145.0 kV with Overvoltage Factor

	Phase-to-ground exposure		Phase-to-ground exposure	
T (p.u.)	m	ft	m	ft
2.0	0.84	2.75	1.24	4.08
2.1	0.86	2.83	1.27	4.17
2.2	0.89	2.92	1.30	4.25
2.3	0.91	3.00	1.32	4.33
2.4	0.94	3.08	1.35	4.42
2.5	0.97	3.17	1.37	4.50
2.6	0.99	3.25	1.40	4.58
2.7	1.02	3.33	1.42	4.67
2.8	1.04	3.42	1.45	4.75
2.9	1.07	3.50	1.47	4.83
3.0	1.09	3.58	1.50	4.92

Note 1: The employer may apply the distance specified in this table only where the employer determines the maximum anticipated per-unit transient overvoltage by engineering analysis. (Table 6 applies otherwise.)

Note 2: The distances specified in this table are the air, bare-hand, and live-line tool distances.

Table 9-   Minimum Approach Distances Until March 31, 2015-145.1 To 169.0 kV with Overvoltage Factor

	Phase-to-ground exposure		Phase-to-phase exposure	
T (p.u.)	m	ft	m	ft
2.0	0.91	3.00	1.42	4.67
2.1	0.97	3.17	1.45	4.75
2.2	0.99	3.25	1.47	4.83
2.3	1.02	3.33	1.50	4.92
2.4	1.04	3.42	1.52	5.00
2.5	1.07	3.50	1.57	5.17
2.6	1.12	3.67	1.60	5.25
2.7	1.14	3.75	1.63	5.33
2.8	1.17	3.83	1.65	5.42
2.9	1.19	3.92	1.68	5.50
3.0	1.22	4.00	1.73	5.67

Note 1: The employer may apply the distance specified in this table only where the employer determines the maximum anticipated per-unit transient overvoltage by engineering analysis. (Table 6 applies otherwise.)

Note 2: The distances specified in this table are the air, bare-hand, and live-line tool distances.

Table 10- Minimum Approach Distances Until March 31, 2015-169.1 To 242.0 kV with Overvoltage Factor

T (p.u.)	Phase-to-ground exposure		Phase-to-ground exposure	
	m	ft	m	ft
2.0	1.17	3.83	1.85	6.08
2.1	1.22	4.00	1.91	6.25
2.2	1.24	4.08	1.93	6.33
2.3	1.30	4.25	1.98	6.50
2.4	1.35	4.42	2.01	6.58
2.5	1.37	4.50	2.06	6.75
2.6	1.42	4.67	2.11	6.92
2.7	1.47	4.83	2.13	7.00
2.8	1.50	4.92	2.18	7.17
2.9	1.55	5.08	2.24	7.33
3.0	1.60	5.25	2.29	7.50

Note 1: The employer may apply the distance specified in this table only where the employer determines the maximum anticipated per-unit transient overvoltage by engineering analysis. (Table 6 applies otherwise.)
Note 2: The distances specified in this table are the air, bare-hand, and live-line tool distances.

Table 11- Minimum Approach Distances Until March 31, 2015-242.1 To 362.0 kV with Overvoltage Factor

T (p.u.)	Phase-to-ground exposure		Phase-to-ground exposure	
	m	ft	m	ft
2.0	1.60	5.25	2.62	8.58
2.1	1.65	5.42	2.69	8.83
2.2	1.75	5.75	2.79	9.17
2.3	1.85	6.08	2.90	9.50
2.4	1.93	6.33	3.02	9.92
2.5	2.03	6.67	3.15	10.33
2.6	2.16	7.08	3.28	10.75
2.7	2.26	7.42	3.40	11.17
2.8	2.36	7.75	3.53	11.58
2.9	2.49	8.17	3.68	12.08
3.0	2.59	8.50	3.81	12.50

Note 1: The employer may apply the distance specified in this table only where the employer determines the maximum anticipated per-unit transient overvoltage by engineering analysis. (Table 6 applies otherwise.)
Note 2: The distances specified in this table are the air, bare-hand, and live-line tool distances.

Table 12- Minimum Approach Distances Until March 31, 2015-362.1 To 552.0 kV with Overvoltage Factor

T (p.u.)	Phase-to-ground exposure		Phase-to-ground exposure	
	m	ft	m	ft
1.5	1.83	6.00	2.24	7.33
1.6	1.98	6.50	2.67	8.75
1.7	2.13	7.00	3.10	10.17
1.8	2.31	7.58	3.53	11.58
1.9	2.46	8.08	4.01	13.17
2.0	2.67	8.75	4.52	14.83
2.1	2.84	9.33	4.75	15.58
2.2	3.02	9.92	4.98	16.33
2.3	3.20	10.50	5.23	17.17
2.4	3.43	11.25	5.51	18.08

Note 1: The employer may apply the distance specified in this table only where the employer determines the maximum anticipated per-unit transient overvoltage by engineering analysis. (Table 6 applies otherwise.)
Note 2: The distances specified in this table are the air, bare-hand, and live-line tool distances.

Table 13- Minimum Approach Distances Until March 31, 2015-552.1 To 800.0 kV with Overvoltage Factor

T (p.u.)	Phase-to-ground exposure		Phase-to-ground exposure	
	m	ft	m	ft
1.5	2.95	9.67	3.68	12.08
1.6	3.25	10.67	4.42	14.50
1.7	3.56	11.67	5.23	17.17
1.8	3.86	12.67	6.07	19.92
1.9	4.19	13.75	6.99	22.92
2.0	4.55	14.92	7.92	26.00

Note 1: The employer may apply the distance specified in this table only where the employer determines the maximum anticipated per-unit transient overvoltage by engineering analysis. (Table 6 applies otherwise.)
Note 2: The distances specified in this table are the air, bare-hand, and live-line tool distances.

B. Alternative minimum approach distances. Employers may use the minimum approach distances in Table 14 through Table 21 provided that the employer follows the notes to those tables.

Table 14- AC Minimum Approach Distances-72.6 To 121.0 kV

T (p.u.)	Phase-to-ground exposure		Phase-to-ground exposure	
	m	ft	m	ft
1.5	0.67	2.2	0.84	2.8
1.6	0.69	2.3	0.87	2.9
1.7	0.71	2.3	0.90	3.0
1.8	0.74	2.4	0.93	3.1
1.9	0.76	2.5	0.96	3.1
2.0	0.78	2.6	0.99	3.2
2.1	0.81	2.7	1.01	3.3
2.2	0.83	2.7	1.04	3.4
2.3	0.85	2.8	1.07	3.5
2.4	0.88	2.9	1.10	3.6
2.5	0.90	3.0	1.13	3.7
2.6	0.92	3.0	1.16	3.8
2.7	0.95	3.1	1.19	3.9
2.8	0.97	3.2	1.22	4.0
2.9	0.99	3.2	1.24	4.1
3.0	1.02	3.3	1.27	4.2
3.1	1.04	3.4	1.30	4.3
3.2	1.06	3.5	1.33	4.4
3.3	1.09	3.6	1.36	4.5
3.4	1.11	3.6	1.39	4.6
3.5	1.13	3.7	1.42	4.7

Table 15-   **AC Minimum Approach Distances-121.1 To 145.0 kV**

T (p.u.)	Phase-to-ground exposure		Phase-to-ground exposure	
	m	ft	m	ft
1.5	0.74	2.4	0.95	3.1
1.6	0.76	2.5	0.98	3.2
1.7	0.79	2.6	1.02	3.3
1.8	0.82	2.7	1.05	3.4
1.9	0.85	2.8	1.08	3.5
2.0	0.88	2.9	1.12	3.7
2.1	0.90	3.0	1.15	3.8
2.2	0.93	3.1	1.19	3.9
2.3	0.96	3.1	1.22	4.0
2.4	0.99	3.2	1.26	4.1
2.5	1.02	3.3	1.29	4.2
2.6	1.04	3.4	1.33	4.4
2.7	1.07	3.5	1.36	4.5
2.8	1.10	3.6	1.39	4.6
2.9	1.13	3.7	1.43	4.7
3.0	1.16	3.8	1.46	4.8
3.1	1.19	3.9	1.50	4.9
3.2	1.21	4.0	1.53	5.0
3.3	1.24	4.1	1.57	5.2
3.4	1.27	4.2	1.60	5.2
3.5	1.30	4.3	1.64	5.4

Table 16- AC Minimum Approach Distances-145.1 To 169.0 kV

T (p.u.)	Phase-to-ground exposure		Phase-to-ground exposure	
	m	ft	m	ft
1.5	0.81	2.7	1.05	3.4
1.6	0.84	2.8	1.09	3.6
1.7	0.87	2.9	1.13	3.7
1.8	0.90	3.0	1.17	3.8
1.9	0.94	3.1	1.21	4.0
2.0	0.97	3.2	1.25	4.1
2.1	1.00	3.3	1.29	4.2
2.2	1.03	3.4	1.33	4.4
2.3	1.07	3.5	1.37	4.5
2.4	1.10	3.6	1.41	4.6
2.5	1.13	3.7	1.45	4.8
2.6	1.17	3.8	1.49	4.9
2.7	1.20	3.9	1.53	5.0
2.8	1.23	4.0	1.57	5.2
2.9	1.26	4.1	1.61	5.3
3.0	1.30	4.3	1.65	5.4
3.1	1.33	4.4	1.70	5.6
3.2	1.36	4.5	1.76	5.8
3.3	1.39	4.6	1.82	6.0
3.4	1.43	4.7	1.88	6.2
3.5	1.46	4.8	1.94	6.4

Table 17- AC Minimum Approach Distances-169.1 To 242.0 kV

T (p.u.)	Phase-to-ground exposure		Phase-to-ground exposure	
	m	ft	m	ft
1.5	1.02	3.3	1.37	4.5
1.6	1.06	3.5	1.43	4.7
1.7	1.11	3.6	1.48	4.9
1.8	1.16	3.8	1.54	5.1
1.9	1.21	4.0	1.60	5.2
2.0	1.25	4.1	1.66	5.4
2.1	1.30	4.3	1.73	5.7
2.2	1.35	4.4	1.81	5.9
2.3	1.39	4.6	1.90	6.2
2.4	1.44	4.7	1.99	6.5
2.5	1.49	4.9	2.08	6.8
2.6	1.53	5.0	2.17	7.1
2.7	1.58	5.2	2.26	7.4
2.8	1.63	5.3	2.36	7.7
2.9	1.67	5.5	2.45	8.0
3.0	1.72	5.6	2.55	8.4
3.1	1.77	5.8	2.65	8.7
3.2	1.81	5.9	2.76	9.1
3.3	1.88	6.2	2.86	9.4
3.4	1.95	6.4	2.97	9.7
3.5	2.01	6.6	3.08	10.1

Table 18-   AC Minimum Approach Distances-242.1 To 362.0 kV

T (p.u.)	Phase-to-ground exposure		Phase-to-ground exposure	
	m	ft	m	ft
1.5	1.37	4.5	1.99	6.5
1.6	1.44	4.7	2.13	7.0
1.7	1.51	5.0	2.27	7.4
1.8	1.58	5.2	2.41	7.9
1.9	1.65	5.4	2.56	8.4
2.0	1.72	5.6	2.71	8.9
2.1	1.79	5.9	2.87	9.4
2.2	1.87	6.1	3.03	9.9
2.3	1.97	6.5	3.20	10.5
2.4	2.08	6.8	3.37	11.1
2.5	2.19	7.2	3.55	11.6
2.6	2.29	7.5	3.73	12.2
2.7	2.41	7.9	3.91	12.8
2.8	2.52	8.3	4.10	13.5
2.9	2.64	8.7	4.29	14.1
3.0	2.76	9.1	4.49	14.7
3.1	2.88	9.4	4.69	15.4
3.2	3.01	9.9	4.90	16.1
3.3	3.14	10.3	5.11	16.8
3.4	3.27	10.7	5.32	17.5
3.5	3.41	11.2	5.52	18.1

Table 19- AC Minimum Approach Distances-362.1 To 420.0 kV

T (p.u.)	Phase-to-ground exposure		Phase-to-ground exposure	
	m	ft	m	ft
1.5	1.53	5.0	2.40	7.9
1.6	1.62	5.3	2.58	8.5
1.7	1.70	5.6	2.75	9.0
1.8	1.78	5.8	2.94	9.6
1.9	1.88	6.2	3.13	10.3
2.0	1.99	6.5	3.33	10.9
2.1	2.12	7.0	3.53	11.6
2.2	2.24	7.3	3.74	12.3
2.3	2.37	7.8	3.95	13.0
2.4	2.50	8.2	4.17	13.7
2.5	2.64	8.7	4.40	14.4
2.6	2.78	9.1	4.63	15.2
2.7	2.93	9.6	4.87	16.0
2.8	3.07	10.1	5.11	16.8
2.9	3.23	10.6	5.36	17.6
3.0	3.38	11.1	5.59	18.3
3.1	3.55	11.6	5.82	19.1
3.2	3.72	12.2	6.07	19.9
3.3	3.89	12.8	6.31	20.7
3.4	4.07	13.4	6.56	21.5
3.5	4.25	13.9	6.81	22.3

Table 20-   AC Minimum Approach Distances-420.1 To 550.0 kV

T (p.u.)	Phase-to-ground exposure		Phase-to-ground exposure	
	m	ft	m	ft
1.5	1.95	6.4	3.46	11.4
1.6	2.11	6.9	3.73	12.2
1.7	2.28	7.5	4.02	13.2
1.8	2.45	8.0	4.31	14.1
1.9	2.62	8.6	4.61	15.1
2.0	2.81	9.2	4.92	16.1
2.1	3.00	9.8	5.25	17.2
2.2	3.20	10.5	5.55	18.2
2.3	3.40	11.2	5.86	19.2
2.4	3.62	11.9	6.18	20.3
2.5	3.84	12.6	6.50	21.3
2.6	4.07	13.4	6.83	22.4
2.7	4.31	14.1	7.18	23.6
2.8	4.56	15.0	7.52	24.7
2.9	4.81	15.8	7.88	25.9
3.0	5.07	16.6	8.24	27.0

Table 21- AC Minimum Approach Distances-550.1 To 800.0 kV

T (p.u.)	Phase-to-ground exposure		Phase-to-ground exposure	
	m	ft	m	ft
1.5	3.16	10.4	5.97	19.6
1.6	3.46	11.4	6.43	21.1
1.7	3.78	12.4	6.92	22.7
1.8	4.12	13.5	7.42	24.3
1.9	4.47	14.7	7.93	26.0
2.0	4.83	15.8	8.47	27.8
2.1	5.21	17.1	9.02	29.6
2.2	5.61	18.4	9.58	31.4
2.3	6.02	19.8	10.16	33.3
2.4	6.44	21.1	10.76	35.3
2.5	6.88	22.6	11.38	37.3

- To prevent any transient overvoltage caused by the switching surge that would result if the circuit were reenergized.

However, due to system stability considerations, it may not always be feasible to disable the automatic-reclosing feature.

V. Minimum Approach-Distance Tables

A. *Legacy tables.* Employers may use the minimum approach distances in Table 6 through Table 13 until March 31, 2015.

Notes to Table 14 through Table 21:

1. The employer must determine the maximum anticipated per-unit transient overvoltage, phase-to-ground, through an engineering analysis, as required by § 1910.269(l)(3)(ii), or assume a maximum anticipated per-unit transient over-voltage, phase-to-ground, in accordance with Table R-9.
2. For phase-to-phase exposures, the employer must demonstrate that no insulated tool spans the gap and that no large conductive object is in the gap.
3. The worksite must be at an elevation of 900 meters (3,000 feet) or less above sea level.

[79 FR 20665-20677, July 10, 2014; 79 FR 56962, September 24, 2014]

Reprinted from www.osha.gov

Occupational Safety & Health Administration
200 Constitution Avenue, NW
Washington, DC 20210

Appendix C to §1910.269 – Protection From Hazardous Differences in Electric Potential

- **Part Number:** 1910
- **Part Title:** Occupational Safety and Health Standards
- **Subpart:** R
- **Subpart Title:** Special Industries
- **Standard Number:** 1910.269 App C
- **Title:** Protection from Hazardous Differences in Electric Potential

I. Introduction

Current passing through an impedance impresses voltage across that impedance. Even conductors have some, albeit low, value of impedance. Therefore, if a "grounded"[1] object, such as a crane or deenergized and grounded power line, results in a ground fault on a power line, voltage is impressed on that grounded object. The voltage impressed on the grounded object depends largely on the voltage on the line, on the impedance of the faulted conductor, and on the impedance to "true," or "absolute," ground represented by the object. If the impedance of the object causing the fault is relatively large, the voltage impressed on the object is essentially the phase-to-ground system voltage. However, even faults to grounded power lines or to well grounded transmission towers or substation structures (which have relatively low values of impedance to ground) can result in hazardous voltages.[2] In all cases, the degree of the hazard depends on the magnitude of the current through the employee and the time of exposure. This appendix discusses methods of protecting workers against the possibility that grounded objects, such as cranes and other mechanical equipment, will contact energized power lines and that deenergized and grounded power lines will become accidentally energized.

II. Voltage-Gradient Distribution

A. *Voltage-gradient distribution curve.* Absolute, or true, ground serves as a reference and always has a voltage of 0 volts above ground potential. Because there is an impedance between a grounding electrode and absolute ground, there will be a voltage difference between the grounding electrode and absolute ground under ground-fault conditions. Voltage dissipates from the grounding electrode (or from the grounding point) and creates a ground potential gradient. The voltage decreases rapidly with increasing distance from the grounding electrode. A voltage drop associated with this dissipation of voltage is a ground potential. Figure 1 is a typical voltage-gradient distribution curve (assuming a uniform soil texture).

---

[1]This appendix generally uses the term "grounded" only with respect to grounding that the employer intentionally installs, for example, the grounding an employer installs on a deenergized conductor. However, in this case, the term "grounded" means connected to earth, regardless of whether or not that connection is intentional.

[2]Thus, grounding systems for transmission towers and substation structures should be designed to minimize the step and touch potentials involved.

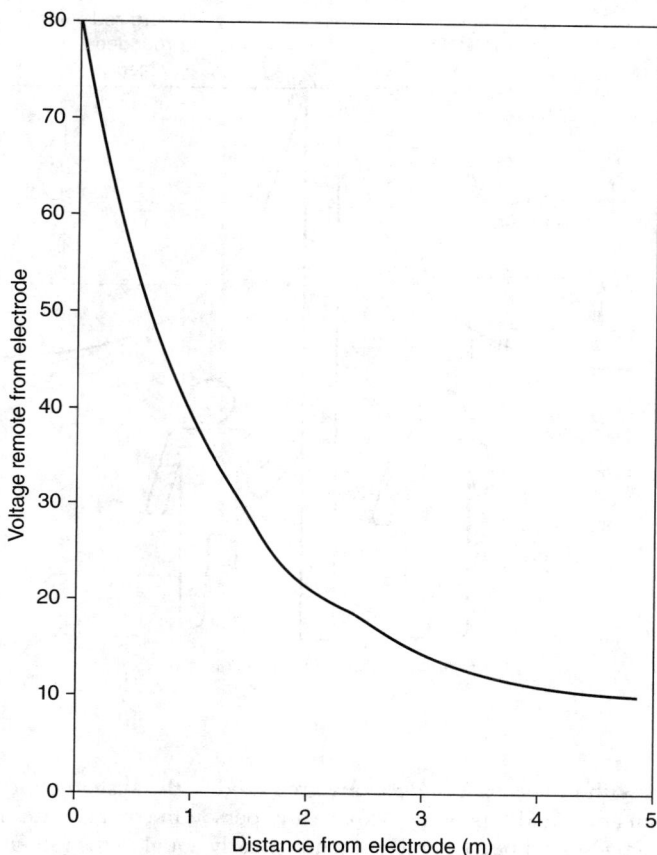

Figure 1. Typical Voltage - Grandiet Distribution Curve

B. *Step and touch potentials*. Figure 1 also shows that workers are at risk from step and touch potentials. Step potential is the voltage between the feet of a person standing near an energized grounded object (the electrode). In Figure 1, the step potential is equal to the difference in voltage between two points at different distances from the electrode (where the points represent the location of each foot in relation to the electrode). A person could be at risk of injury during a fault simply by standing near the object.

Touch potential is the voltage between the energized grounded object (again, the electrode) and the feet of a person in contact with the object. In Figure 1, the touch potential is equal to the difference in voltage between the electrode (which is at a distance of 0 meters) and a point some distance away from the electrode (where the point represents the location of the feet of the person in contact with the object). The touch potential could be nearly the full voltage across the grounded object if that object is grounded at a point remote from the place where the person is

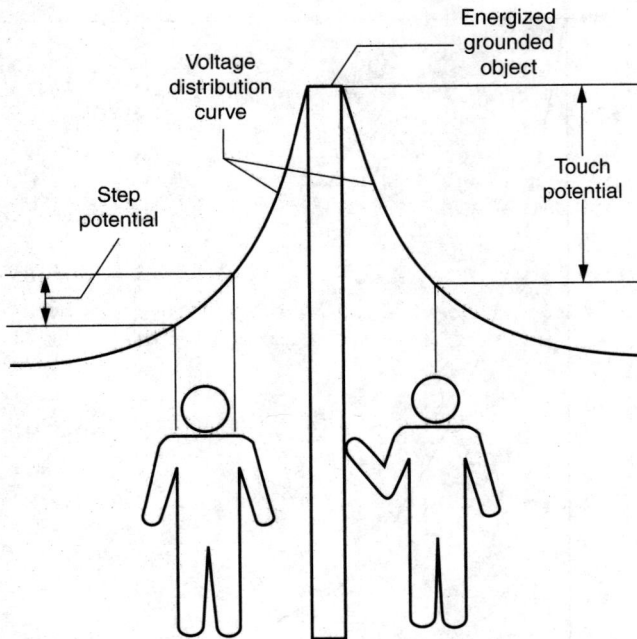

Figure 2.   Step and Touch Potentials

in contact with it. For example, a crane grounded to the system neutral and that contacts an energized line would expose any person in contact with the crane or its uninsulated load line to a touch potential nearly equal to the full fault voltage. Figure 2 illustrates step and touch potentials.

III. Protecting Workers From Hazardous Differences in Electrical Potential

A. *Definitions*. The following definitions apply to section III of this appendix:

Bond. The electrical interconnection of conductive parts designed to maintain a common electric potential.

Bonding cable (bonding jumper). A cable connected to two conductive parts to bond the parts together.

Cluster bar. A terminal temporarily attached to a structure that provides a means for the attachment and bonding of grounding and bonding cables to the structure.

Ground. A conducting connection between an electric circuit or equipment and the earth, or to some conducting body that serves in place of the earth.

Grounding cable (grounding jumper). A cable connected between a deenergized part and ground. Note that grounding cables carry fault current and bonding cables generally do not. A cable that bonds two conductive parts but carries substantial fault current (for example, a jumper connected between one phase and a grounded phase) is a grounding cable.

Ground mat (grounding grid). A temporarily or permanently installed metallic mat or grating that establishes an equipotential surface and provides connection points for attaching grounds.

B. *Analyzing the hazard.* The employer can use an engineering analysis of the power system under fault conditions to determine whether hazardous step and touch voltages will develop. The analysis should determine the voltage on all conductive objects in the work area and the amount of time the voltage will be present. Based on the this analysis, the employer can select appropriate measures and protective equipment, including the measures and protective equipment outlined in Section III of this appendix, to protect each employee from hazardous differences in electric potential. For example, from the analysis, the employer will know the voltage remaining on conductive objects after employees install bonding and grounding equipment and will be able to select insulating equipment with an appropriate rating, as described in paragraph III.C.2 of this appendix.

C. *Protecting workers on the ground.* The employer may use several methods, including equipotential zones, insulating equipment, and restricted work areas, to protect employees on the ground from hazardous differences in electrical potential.

1. An equipotential zone will protect workers within it from hazardous step and touch potentials. (See Figure 3.) Equipotential zones will not, however, protect employees located either wholly or partially outside the protected area. The employer can establish an equipotential zone for workers on the ground, with respect to a grounded object, through the use of a metal mat connected to the grounded object. The employer can use a grounding grid to equalize the voltage within the grid or bond conductive objects in the immediate work area to minimize the potential between the objects and between each object and ground. (Bonding an object outside the work area can increase the touch potential to that object, however.) Section III.D of this appendix discusses equipotential zones for employees working on deenergized and grounded power lines.

2. Insulating equipment, such as rubber gloves, can protect employees handling grounded equipment and conductors from hazardous touch potentials. The insulating equipment must be rated for the highest voltage that can be impressed on the grounded objects under fault conditions (rather than for the full system voltage).

3. Restricting employees from areas where hazardous step or touch potentials could arise can protect employees not directly involved in performing the operation. The employer must ensure that employees on the ground in the vicinity of transmission structures are at a distance where step voltages would be insufficient to cause injury. Employees must not handle grounded conductors or equipment likely to become energized to hazardous voltages unless the employees are within an equipotential zone or protected by insulating equipment.

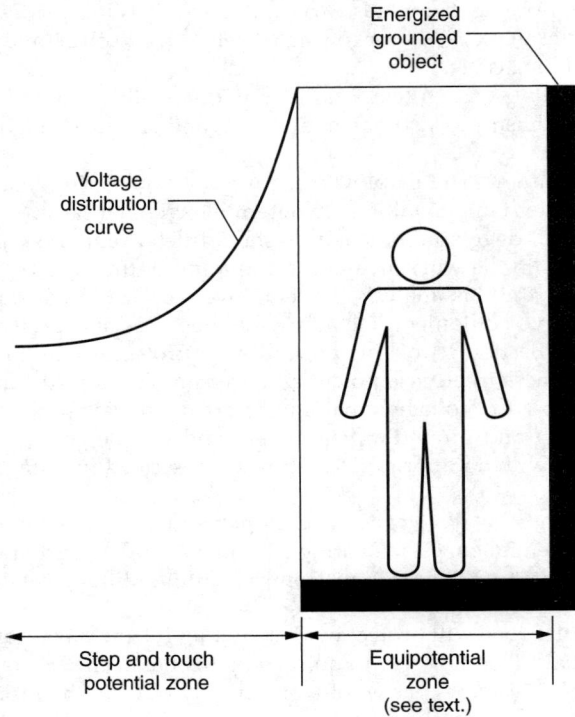

Figure 3. Protection from Ground-Potential Gradients

D. *Protecting employees working on deenergized and grounded power lines.* This Section III.D of Appendix C establishes guidelines to help employers comply with requirements in § 1910.269(n) for using protective grounding to protect employees working on deenergized power lines. Paragraph (n) of § 1910.269 applies to grounding of transmission and distribution lines and equipment for the purpose of protecting workers. Paragraph (n)(3) of § 1910.269 requires temporary protective grounds to be placed at such locations and arranged in such a manner that the employer can demonstrate will prevent exposure of each employee to hazardous differences in electric potential.[3] Sections III.D.1 and III.D.2 of this appendix provide guidelines that employers can use in making the demonstration required by § 1910.269(n)(3). Section III.D.1 of this appendix provides guidelines on how the employer can determine whether particular grounding practices expose employees to hazardous differences in electric potential. Section III.D.2 of this appendix describes grounding methods that the employer can use in lieu of an engineering analysis to make the demonstration required by § 1910.269(n)(3). The Occupational Safety and Health Administration will consider employers that comply with the criteria in this appendix as meeting § 1910.269(n)(3).

[3] The protective grounding required by § 1910.269(n) limits to safe values the potential differences between accessible objects in each employee's work environment. Ideally, a protective grounding system would create a true equipotential zone in which every point is at the same electric potential. In practice, current passing through the grounding and bonding elements creates potential differences. If these potential differences are hazardous, the employer may not treat the zone as an equipotential zone.

Finally, Section III.D.3 of this appendix discusses other safety considerations that will help the employer comply with other requirements in § 1910.269(n). Following these guidelines will protect workers from hazards that can occur when a deenergized and grounded line becomes energized.

1. *Determining safe body current limits.* This Section III.D.1 of Appendix C provides guidelines on how an employer can determine whether any differences in electric potential to which workers could be exposed are hazardous as part of the demonstration required by § 1910.269(n)(3).

Institute of Electrical and Electronic Engineers (IEEE) Standard 1048-2003, IEEE Guide for Protective Grounding of Power Lines, provides the following equation for determining the threshold of ventricular fibrillation when the duration of the electric shock is limited:

$$I = \frac{116}{\sqrt{t}}$$

where *I* is the current through the worker's body, and t is the duration of the current in seconds. This equation represents the ventricular fibrillation threshold for 95.5 percent of the adult population with a mass of 50 kilograms (110 pounds) or more. The equation is valid for current durations between 0.0083 to 3.0 seconds.

To use this equation to set safe voltage limits in an equipotential zone around the worker, the employer will need to assume a value for the resistance of the worker's body. IEEE Std 1048-2003 states that «total body resistance is usually taken as 1000 Ω for determining . . . body current limits.» However, employers should be aware that the impedance of a worker's body can be substantially less than that value. For instance, IEEE Std 1048-2003 reports a minimum hand-to-hand resistance of 610 ohms and an internal body resistance of 500 ohms. The internal resistance of the body better represents the minimum resistance of a worker's body when the skin resistance drops near zero, which occurs, for example, when there are breaks in the worker's skin, for instance, from cuts or from blisters formed as a result of the current from an electric shock, or when the worker is wet at the points of contact.

Employers may use the IEEE Std 1048-2003 equation to determine safe body current limits only if the employer protects workers from hazards associated with involuntary muscle reactions from electric shock (for example, the hazard to a worker from falling as a result of an electric shock). Moreover, the equation applies only when the duration of the electric shock is limited. If the precautions the employer takes, including those required by applicable standards, do not adequately protect employees from hazards associated with involuntary reactions from electric shock, a hazard exists if the induced voltage is sufficient to pass a current of 1 milliampere through a 500-ohm resistor. (The 500-ohm resistor represents the resistance of an employee. The 1-milliampere current is the threshold of perception.) Finally, if the employer protects employees from injury due to involuntary reactions from electric shock, but the duration of the electric shock is unlimited (that is, when the fault current at the work location will be insufficient to trip the devices protecting the circuit), a hazard exists if the resultant current would be more than 6 milliamperes (the recognized let-go threshold for workers[4]).

---

[4]Electric current passing through the body has varying effects depending on the amount of the current. At the let-go threshold, the current overrides a person's control over his or her muscles. At that level, an employee grasping an object will not be able to let go of the object. The let-go threshold varies from person to person; however, the recognized value for workers is 6 milliamperes.

2. *Acceptable methods of grounding for employers that do not perform an engineering determination.* The grounding methods presented in this section of this appendix ensure that differences in electric potential are as low as possible and, therefore, meet § 1910.269(n)(3) without an engineering determination of the potential differences. These methods follow two principles: (i) The grounding method must ensure that the circuit opens in the fastest available clearing time, and (ii) the grounding method must ensure that the potential differences between conductive objects in the employee's work area are as low as possible.

Paragraph (n)(3) of § 1910.269 does not require grounding methods to meet the criteria embodied in these principles. Instead, the paragraph requires that protective grounds be "placed at such locations and arranged in such a manner that the employer can demonstrate will prevent exposure of each employee to hazardous differences in electric potential." However, when the employer's grounding practices do not follow these two principles, the employer will need to perform an engineering analysis to make the demonstration required by § 1910.269(n)(3).

i. *Ensuring that the circuit opens in the fastest available clearing time.* Generally, the higher the fault current, the shorter the clearing times for the same type of fault. Therefore, to ensure the fastest available clearing time, the grounding method must maximize the fault current with a low impedance connection to ground. The employer accomplishes this objective by grounding the circuit conductors to the best ground available at the worksite. Thus, the employer must ground to a grounded system neutral conductor, if one is present. A grounded system neutral has a direct connection to the system ground at the source, resulting in an extremely low impedance to ground. In a substation, the employer may instead ground to the substation grid, which also has an extremely low impedance to the system ground and, typically, is connected to a grounded system neutral when one is present. Remote system grounds, such as pole and tower grounds, have a higher impedance to the system ground than grounded system neutrals and substation grounding grids; however, the employer may use a remote ground when lower impedance grounds are not available. In the absence of a grounded system neutral, substation grid, and remote ground, the employer may use a temporary driven ground at the worksite.

In addition, if employees are working on a three-phase system, the grounding method must short circuit all three phases. Short circuiting all phases will ensure faster clearing and lower the current through the grounding cable connecting the deenergized line to ground, thereby lowering the voltage across that cable. The short circuit need not be at the worksite; however, the employer must treat any conductor that is not grounded at the worksite as energized because the ungrounded conductors will be energized at fault voltage during a fault.

ii. *Ensuring that the potential differences between conductive objects in the employee's work area are as low as possible.* To achieve as low a voltage as possible across any two conductive objects in the work area, the employer must bond all conductive objects in the work area. This section of this appendix discusses how to create a zone that minimizes differences in electric potential between conductive objects in the work area.

The employer must use bonding cables to bond conductive objects, except for metallic objects bonded through metal-to-metal contact. The employer must ensure that metal-to-metal contacts are tight and free of contamination, such as oxidation, that can increase the impedance across the connection. For example, a bolted connection between metal lattice tower members is acceptable if the connection is tight and free of corrosion and other contamination. Figure 4 shows how to create an equipotential zone for metal lattice towers.

Wood poles are conductive objects. The poles can absorb moisture and conduct electricity, particularly at distribution and transmission voltages. Consequently, the employer must either: (1) Provide a conductive platform, bonded to a grounding cable, on which the worker stands or (2) use cluster bars to bond wood poles to the grounding cable. The employer must ensure that employees install the cluster bar below, and close to, the worker's feet. The inner portion of the wood pole is more conductive than the outer shell, so it is important that the cluster bar be in conductive contact with a metal spike or nail that penetrates the wood to a depth greater than or equal to the depth the worker's climbing gaffs will penetrate the wood. For example, the employer could mount the cluster bar on a bare pole ground wire fastened to

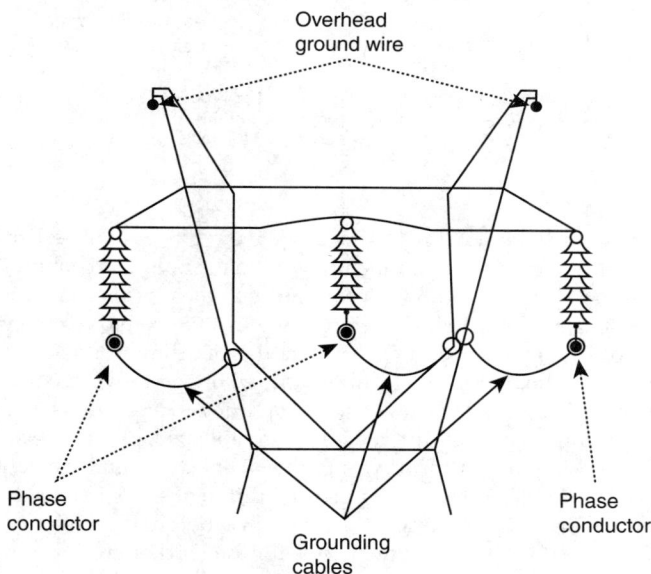

Overhead
ground wire

Phase
conductor

Phase
conductor

Grounding
cables

Notes:
  1. Employers must ground overhead ground wires that are within reach of the employee.
  2. The grounding cable must be as short as practicable; therefore, the attachment points between the grounding cable and the tower may be different from that shown in the figure.

Figure 4.   Equipotential Zone for Metal Lattice Tower

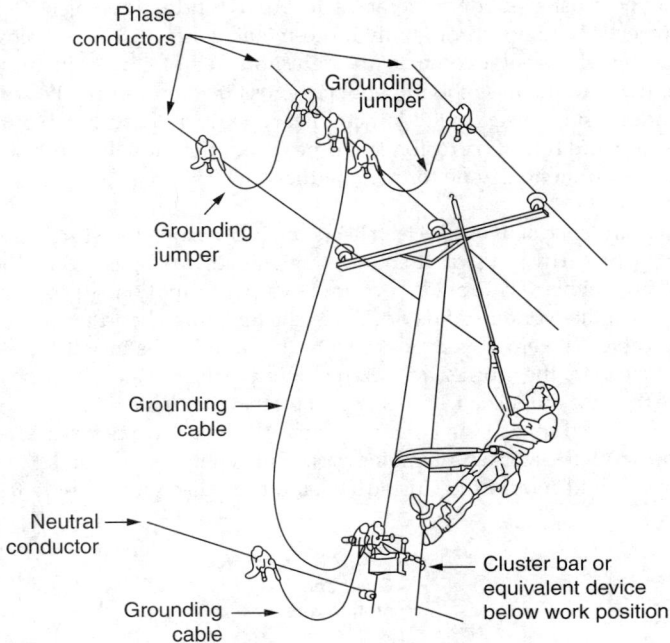

Figure 5.   Equipotential Grounding for Wood Poles
Figure reprinted with permission from Hubbell Power Systems, Inc.
(Hubbell) OSHA revised the figure from Hubbell's original.

the pole with nails or staples that penetrate to the required depth. Alternatively, the employer may temporarily nail a conductive strap to the pole and connect the strap to the cluster bar. Figure 5 shows how to create an equipotential zone for wood poles. For underground systems, employers commonly install grounds at the points of disconnection of the underground cables. These grounding points are typically remote from the manhole or underground vault where employees will be working on the cable. Workers in contact with a cable grounded at a remote location can experience hazardous potential differences if the cable becomes energized or if a fault occurs on a different, but nearby, energized cable. The fault current causes potential gradients in the earth, and a potential difference will exist between the earth where the worker is standing and the earth where the cable is grounded. Consequently, to create an equipotential zone for the worker, the employer must provide a means of connecting the deenergized cable to ground at the worksite by having the worker stand on a conductive mat bonded to the deenergized cable. If the cable is cut, the employer must install a bond across the opening in the cable or install one bond on each side of the opening to ensure that the separate cable ends are at the same potential. The employer must protect the worker from any hazardous differences in potential any time there is no bond between the mat and the cable (for example, before the worker installs the bonds).

3. *Other safety-related considerations.* To ensure that the grounding system is safe and effective, the employer should also consider the following factors:[5]
i. *Maintenance of grounding equipment.* It is essential that the employer properly maintain grounding equipment. Corrosion in the connections between grounding cables and clamps and on the clamp surface can increase the resistance of the cable, thereby increasing potential differences. In addition, the surface to which a clamp attaches, such as a conductor or tower member, must be clean and free of corrosion and oxidation to ensure a low-resistance connection. Cables must be free of damage that could reduce their current-carrying capacity so that they can carry the full fault current without failure. Each clamp must have a tight connection to the cable to ensure a low resistance and to ensure that the clamp does not separate from the cable during a fault.

ii. *Grounding cable length and movement.* The electromagnetic forces on grounding cables during a fault increase with increasing cable length. These forces can cause the cable to move violently during a fault and can be high enough to damage the cable or clamps and cause the cable to fail. In addition, flying cables can injure workers. Consequently, cable lengths should be as short as possible, and grounding cables that might carry high fault current should be in positions where the cables will not injure workers during a fault.

Reprinted from www.osha.gov

Occupational Safety & Health Administration
200 Constitution Avenue, NW
Washington, DC 20210

---

[5]This appendix only discusses factors that relate to ensuring an equipotential zone for employees. The employer must consider other factors in selecting a grounding system that is capable of conducting the maximum fault current that could flow at the point of grounding for the time necessary to clear the fault, as required by § 1910.269(n)(4)(i). IEEE Std 1048-2003 contains guidelines for selecting and installing grounding equipment that will meet § 1910.269(n)(4)(i).

Appendix D to § 1910.269-Methods of Inspecting and Testing Wood Poles

- **Part Number:** 1910
- **Part Title:** Occupational Safety and Health Standards
- **Subpart:** R
- **Subpart Title:** Special Industries
- **Standard Number:** 1910.269 App D
- **Title:** Methods of Inspecting and Testing Wood Poles.

I. Introduction

When employees are to perform work on a wood pole, it is important to determine the condition of the pole before employees climb it. The weight of the employee, the weight of equipment to be installed, and other working stresses (such as the removal or retensioning of conductors) can lead to the failure of a defective pole or a pole that is not designed to handle the additional stresses.[1] For these reasons, it is essential that, before an employee climbs a wood pole, the employer ascertain that the pole is capable of sustaining the stresses of the work. The determination that the pole is capable of sustaining these stresses includes an inspection of the condition of the pole.

If the employer finds the pole to be unsafe to climb or to work from, the employer must secure the pole so that it does not fail while an employee is on it. The employer can secure the pole by a line truck boom, by ropes or guys, or by lashing a new pole alongside it. If a new one is lashed alongside the defective pole, employees should work from the new one.

II. Inspecting Wood Poles

A qualified employee should inspect wood poles for the following conditions:[2]
A. *General condition.* Buckling at the ground line or an unusual angle with respect to the ground may indicate that the pole has rotted or is broken.
B. *Cracks.* Horizontal cracks perpendicular to the grain of the wood may weaken the pole. Vertical cracks, although not normally considered to be a sign of a defective pole, can pose a hazard to the climber, and the employee should keep his or her gaffs away from them while climbing.
C. *Holes.* Hollow spots and woodpecker holes can reduce the strength of a wood pole.
D. *Shell rot and decay.* Rotting and decay are cutout hazards and possible indications of the age and internal condition of the pole.
E. *Knots.* One large knot or several smaller ones at the same height on the pole may be evidence of a weak point on the pole.
F. *Depth of setting.* Evidence of the existence of a former ground line substantially above the existing ground level may be an indication that the pole is no longer buried to a sufficient depth.

---

[1] A properly guyed pole in good condition should, at a minimum, be able to handle the weight of an employee climbing it.

[2] The presence of any of these conditions is an indication that the pole may not be safe to climb or to work from. The employee performing the inspection must be qualified to make a determination as to whether it is safe to perform the work without taking additional precautions.

G. *Soil conditions.* Soft, wet, or loose soil around the base of the pole may indicate that the pole will not support any change in stress.

H. *Burn marks.* Burning from transformer failures or conductor faults could damage the pole so that it cannot withstand changes in mechanical stress.

III. Testing Wood Poles

The following tests, which are from § 1910.268(n)(3), are acceptable methods of testing wood poles:

A. *Hammer test.* Rap the pole sharply with a hammer weighing about 1.4 kg (3 pounds), starting near the ground line and continuing upwards circumferentially around the pole to a height of approximately 1.8 meters (6 feet). The hammer will produce a clear sound and rebound sharply when striking sound wood. Decay pockets will be indicated by a dull sound or a less pronounced hammer rebound. Also, prod the pole as near the ground line as possible using a pole prod or a screwdriver with a blade at least 127 millimeters (5 inches) long. If substantial decay is present, the pole is unsafe.

B. *Rocking test.* Apply a horizontal force to the pole and attempt to rock it back and forth in a direction perpendicular to the line. Exercise caution to avoid causing power lines to swing together. Apply the force to the pole either by pushing it with a pike pole or pulling the pole with a rope. If the pole cracks during the test, it is unsafe.

Reprinted from www.osha.gov

Occupational Safety & Health Administration
200 Constitution Avenue, NW
Washington, DC 20210

Appendix E to § 1910.269-Protection From Flames and Electric Arcs

- **Part Number:** 1910
- **Part Title:** Occupational Safety and Health Standards
- **Subpart:** R
- **Subpart Title:** Special Industries
- **Standard Number:** 1910.269 App E
- **Title:** Protection From Flames and Electric Arcs.

I. Introduction

Paragraph (l)(8) of § 1910.269 addresses protecting employees from flames and electric arcs. This paragraph requires employers to: (1) Assess the workplace for flame and electric-arc hazards (paragraph (l)(8)(i)); (2) estimate the available heat energy from electric arcs to which employees would be exposed (paragraph (l)(8)(ii)); (3) ensure that employees wear clothing that will not melt, or ignite and continue to burn, when exposed to flames or the estimated heat energy (paragraph (l)(8)(iii)); and (4) ensure that employees wear flame-resistant clothing 1 and protective clothing and other protective equipment that has an arc rating greater than or equal to the available heat energy under certain conditions (paragraphs (l)(8)(iv) and (l)(8) (v)). This appendix contains information to help employers estimate available heat energy as required by § 1910.269(l)(8)(ii), select protective clothing and other protective equipment with an arc rating suitable for the available heat energy as required by § 1910.269(l)(8)(v), and ensure that employees do not wear flammable clothing that could lead to burn injury as addressed by §§ 1910.269(l)(8)(iii) and (l)(8)(iv).

II. Assessing the Workplace for Flame and Electric-Arc Hazards

Paragraph (l)(8)(i) of § 1910.269 requires the employer to assess the workplace to identify employees exposed to hazards from flames or from electric arcs. This provision ensures that the employer evaluates employee exposure to flames and electric arcs so that employees who face such exposures receive the required protection. The employer must conduct an assessment for each employee who performs work on or near exposed, energized parts of electric circuits.

A. Assessment Guidelines

Sources electric arcs. Consider possible sources of electric arcs, including:

- Energized circuit parts not guarded or insulated,
- Switching devices that produce electric arcs in normal operation,
- Sliding parts that could fault during operation (for example, rack-mounted circuit breakers), and
- Energized electric equipment that could fail (for example, electric equipment with damaged insulation or with evidence of arcing or overheating).

Exposure to flames. Identify employees exposed to hazards from flames. Factors to consider include:

- The proximity of employees to open flames, and

- For flammable material in the work area, whether there is a reasonable likelihood that an electric arc or an open flame can ignite the material.

Probability that an electric arc will occur. Identify employees exposed to electric-arc hazards. The Occupational Safety and Health Administration will consider an employee exposed to electric-arc hazards if there is a reasonable likelihood that an electric arc will occur in the employee's work area, in other words, if the probability of such an event is higher than it is for the normal operation of enclosed equipment. Factors to consider include:

- For energized circuit parts not guarded or insulated, whether conductive objects can come too close to or fall onto the energized parts,
- For exposed, energized circuit parts, whether the employee is closer to the part than the minimum approach distance established by the employer (as permitted by § 1910.269(l)(3)(iii)).
- Whether the operation of electric equipment with sliding parts that could fault during operation is part of the normal operation of the equipment or occurs during servicing or maintenance, and
- For energized electric equipment, whether there is evidence of impending failure, such as evidence of arcing or overheating.

B. Examples

Table 1 provides task-based examples of exposure assessments.

Table 1-   Example Assessments for Various Tasks

Task		Is employee exposed to flame or electricarc hazard?
Normal operation of enclosed equipment, such as closing or opening a switch.	The employer properly installs and maintains enclosed equipment, and there is no evidence of impending failure.	No.
	There is evidence of arcing or overheating	Yes.
	Parts of the equipment are loose or sticking, or the equipment otherwise exhibits signs of lack of maintenance.	Yes.
Servicing electric equipment, such as racking in a circuit breaker or replacing a switch		Yes.
Inspection of electric equipment with exposed energized parts.	The employee is not holding conductive objects and remains outside the minimum approach distance established by the employer.	No.
	The employee is holding a conductive object, such as a flashlight, that could fall or otherwise contact energized parts (irrespective of whether the employee maintains the minimum approach distance).	Yes.
	The employee is closer than the minimum approach distance established by the employer (for example, when wearing rubber insulating gloves or rubber insulating gloves and sleeves).	Yes.
Using open flames, for example, in wiping cable splice sleeves		Yes.

III. Protection Against Burn Injury

A. Estimating Available Heat Energy

Calculation methods. Paragraph (l)(8)(ii) of § 1910.269 provides that, for each employee exposed to an electric-arc hazard, the employer must make a reasonable estimate of the heat energy to which the employee would be exposed if an arc occurs. Table 2 lists various methods of calculating values of available heat energy from an electric circuit. The Occupational Safety and Health Administration does not endorse any of these specific methods. Each method requires the input of various parameters, such as fault current, the expected length of the electric arc, the distance from the arc to the employee, and the clearing time for the fault (that is, the time the circuit protective devices take to open the circuit and clear the fault). The employer can precisely determine some of these parameters, such as the fault current and the clearing time, for a given system. The employer will need to estimate other parameters, such as the length of the arc and the distance between the arc and the employee, because such parameters vary widely.

Table 2- Methods of Calculating Incident Heat Energy from an Electric Arc

1. *Standard for Electrical Safety Requirements for Employee Workplaces*, NFPA 70E-2012, Annex D, "Sample Calculation of Flash Protection Boundary."
2. Doughty, T.E., Neal, T.E., and Floyd II, H.L., "Predicting Incident Energy to Better Manage the Electric Arc Hazard on 600 V Power Distribution Systems,"*Record of Conference Papers IEEE IAS 45th Annual Petroleum and Chemical Industry Conference,* September 28-30, 1998.
3. *Guide for Performing Arc-Flash Hazard Calculations*, IEEE Std 1584-2002, 1584a-2004 (Amendment 1 to IEEE Std 1584-2002), and 1584b-2011 (Amendment 2: Changes to Clause 4 of IEEE Std 1584-2002).*
4. ARCPRO, a commercially available software program developed by Kinectrics, Toronto, ON, CA.

*This appendix refers to IEEE Std 1584-2002 with both amendments as IEEE Std 1584b-2011.

The amount of heat energy calculated by any of the methods is approximately inversely proportional to the square of the distance between the employee and the arc. In other words, if the employee is very close to the arc, the heat energy is very high; but if the employee is just a few more centimeters away, the heat energy drops substantially. Thus, estimating the distance from the arc to the employee is key to protecting employees.

The employer must select a method of estimating incident heat energy that provides a reasonable estimate of incident heat energy for the exposure involved. Table 3 shows which methods provide reasonable estimates for various exposures.

Table 3- Selecting a Reasonable Incident-Energy Calculation Method[1]

Incident-energy calculation method	600 V and Less[2]			601 V to 15 kV[2]			More than 15 kV		
	1Φ	3Φa	3Φb	1Φ	3Φa	3Φb	1Φ	3Φa	3Φb
NFPA 70E-2012 Annex D (Lee equation)	Y-C	Y	N	Y-C	Y-C	N	N[3]	N[3]	N[3]
Doughty, Neal, and Floyd	Y-C	Y	Y	N	N	N	N	N	N
IEEE Std 1584b-2011	Y	Y	Y	Y	Y	Y	N	N	N
ARCPRO	y	N	N	Y	N	N	Y	Y[4]	Y[4]

Key:

1Φ: Single-phase arc in open air.

3Φa: Three-phase arc in open air.

3Φb: Three-phase arc in an enclosure (box).

Y: Acceptable; produces a reasonable estimate of incident heat energy from this type of electric arc.

N: Not acceptable; does not produce a reasonable estimate of incident heat energy from this type of electric arc.

Y-C: Acceptable; produces a reasonable, but conservative, estimate of incident heat energy from this type of electric arc.

Notes:

[1]Although the Occupational Safety and Health Administration will consider these methods reasonable for enforcement purposes when employers use the methods in accordance with this table, employers should be aware that the listed methods do not necessarily result in estimates that will provide full protection from internal faults in transformers and similar equipment or from arcs in underground manholes or vaults.

[2]At these voltages, the presumption is that the arc is three-phase unless the employer can demonstrate that only one phase is present or that the spacing of the phases is sufficient to prevent a multi-phase arc from occurring.

[3]Although the Occupational Safety and Health Administration will consider this method acceptable for purposes of assessing whether incident energy exceeds 2.0 cal/cm², the results at voltages of more than 15 kilovolts are extremely conservative and unrealistic.

[4]The Occupational Safety and Health Administration will deem the results of this method reasonable when the employer adjusts them using the conversion factors for three-phase arcs in open air or in an enclosure, as indicated in the program's instructions.

Selecting a reasonable distance from the employee to the arc. In estimating available heat energy, the employer must make some reasonable assumptions about how far the employee will be from the electric arc. Table 4 lists reasonable distances from the employee to the electric arc. The distances in Table 4 are consistent with national consensus standards, such as the Institute of Electrical and Electronic Engineers' *National Electrical Safety Code,* ANSI/IEEE C2-2012, and *IEEE Guide for Performing Arc-Flash Hazard Calculations,* IEEE Std 1584b-2011. The employer is free to use other reasonable distances, but must consider equipment enclosure size and the working distance to the employee in selecting a distance from the employee to the arc. The Occupational Safety and Health Administration will consider a distance reasonable when the employer bases it on equipment size and working distance.

Table 4- Selecting a Reasonable Distance from the Employee to the Electric Arc

Class of equipment	Single-phase arc mm (inches)	Three-phase arc mm (inches)
Cable	*NA	455 (18)
Low voltage MCCs and panelboards	NA	455 (18)
Low-voltage switchgear	NA	610 (24)
5-kV switchgear	NA	910 (36)
15-kV switchgear	NA	910 (36)
Single conductors in air (up to 46 kilovolts), work with rubber insulating gloves	380 (15)	NA
Single conductors in air, work with live-line tools and live-line barehand work	$MAD - (2 \times kV \times 2.54)$ $(MAD - (2 \times kV/10))$[†]	NA

*NA = not applicable.
[†]The terms in this equation are:
MAD = The applicable minimum approach distance, and
kV = The system voltage in kilovolts.

Selecting a reasonable arc gap. For a single-phase arc in air, the electric arc will almost always occur when an energized conductor approaches too close to ground. Thus, an employer can determine the arc gap, or arc length, for these exposures by the dielectric strength of air and the voltage on the line. The dielectric strength of air is approximately 10 kilovolts for every 25.4 millimeters (1 inch). For example, at 50 kilovolts, the arc gap would be $50 \div 10 \times 25.4$ (or 50×2.54), which equals 127 millimeters (5 inches).

For three-phase arcs in open air and in enclosures, the arc gap will generally be dependent on the spacing between parts energized at different electrical potentials. Documents such as IEEE Std 1584b-2011 provide information on these distances. Employers may select a reasonable arc gap from Table 5, or they may select any other reasonable arc gap based on sparkover distance or on the spacing between (1) live parts at different potentials or (2) live parts and grounded parts (for example, bus or conductor spacings in equipment). In any event, the employer must use an estimate that reasonably resembles the actual exposures faced by the employee.

Table 5-   Selecting a Reasonable Arc Gap

Class of equipment	Single-phase arc mm (inches)	Three-phase arc mm [1](inches)
Cable	NA[2]	13 (0.5).
Low voltage MCCs and panelboards	NA	25 (1.0).
Low-voltage switchgear	NA	32 (1.25).
5-kV switchgear	NA	104 (4.0).
15-kV switchgear	NA	152 (6.0).
Single conductors in air, 15 kV and less......................	51 (2.0)	Phase conductor spacing.
Single conductor in air, more than 15 kV	Voltage in $kV \times 2.54$ (Voltage in $kV \times 0.1$), but no less than 51 mm (2 inches).	Phase conductor spacing.

[1]Source: IEEE Std 1584b-2011.
[2]NA = not applicable.

Making estimates over multiple system areas. The employer need not estimate the heat-energy exposure for every job task performed by each employee. Paragraph (l)(8)(ii) of § 1910.269 permits the employer to make broad estimates that cover multiple system areas provided that: (1) The employer uses reasonable assumptions about the energy-exposure distribution throughout the system, and (2) the estimates represent the maximum exposure for those areas. For example, the employer can use the maximum fault current and clearing time to cover several system areas at once.

Incident heat energy for single-phase-toground exposures. Table 6 and Table 7 provide incident heat energy levels for openair, phase-to-ground electric-arc exposures typical for overhead systems.[2] Table 6 presents estimates of available energy for employees using rubber insulating gloves to perform work on overhead systems operating at 4 to 46 kilovolts. The table assumes that the employee will be 380 millimeters (15 inches) from the electric arc, which is a reasonable estimate for rubber insulating glove work. Table 6 also assumes that the arc length equals the sparkover distance for the maximum transient overvoltage of each voltage range.[3] To use the table, an employer would use the voltage, maximum fault current, and maximum clearing time for a system area and, using the appropriate

[1]Flame-resistant clothing includes clothing that is inherently flame resistant and clothing chemically treated with a flame retardant. (See ASTM F1506-10a,*Standard Performance Specification for Flame Resistant Textile Materials for Wearing Apparel for Use by Electrical Workers Exposed to Momentary Electric Arc and Related Thermal Hazards*, and ASTM F1891-12 *Standard Specification for Arc and Flame Resistant Rainwear*.)
[2]The Occupational Safety and Health Administration used metric values to calculate the clearing times in Table 6 and Table 7. An employer may use English units to calculate clearing times instead even though the results will differ slightly.
[3]The Occupational Safety and Health Administration based this assumption, which is more conservative than the arc length specified in Table 5, on Table 410-2 of the 2012 NESC.

voltage range and fault-current and clearingtime values corresponding to the next higher values listed in the table, select the appropriate heat energy (4, 5, 8, or 12 cal/cm^2) from the table. For example, an employer might have a 12,470-volt power line supplying a system area. The power line can supply a maximum fault current of 8 kiloamperes with a maximum clearing time of 10 cycles. For rubber glove work, this system falls in the 4.0-to-15.0-kilovolt range; the next-higher fault current is 10 kA (the second row in that voltage range); and the clearing time is under 18 cycles (the first column to the right of the fault current column). Thus, the available heat energy for this part of the system will be 4 cal/cm^2 or less (from the column heading), and the employer could select protection with a 5-cal/cm^2 rating to meet § 1910.269(l)(8)(v). Alternatively, an employer could select a base incident-energy value and ensure that the clearing times for each voltage range and fault current listed in the table do not exceed the corresponding clearing time specified in the table. For example, an employer that provides employees with arc-flash protective equipment rated at 8 cal/cm^2 can use the table to determine if any system area exceeds 8 cal/cm^2 by checking the clearing time for the highest fault current for each voltage range and ensuring that the clearing times do not exceed the values specified in the 8-cal/cm^2 column in the table.

Table 7 presents similar estimates for employees using live-line tools to perform work on overhead systems operating at voltages of 4 to 800 kilovolts. The table assumes that the arc length will be equal to the sparkover distance[4] and that the employee will be a distance from the arc equal to the minimum approach distance minus twice the sparkover distance.

The employer will need to use other methods for estimating available heat energy in situations not addressed by Table 6 or Table 7. The calculation methods listed in Table 2 and the guidance provided in Table 3 will help employers do this. For example, employers can use IEEE Std 1584b-2011 to estimate the available heat energy (and to select appropriate protective equipment) for many specific conditions, including lowervoltage, phase-to-phase arc, and enclosed arc exposures.

---

4 The dielectric strength of air is about 10 kilovolts for every 25.4 millimeters (1 inch). Thus, the employer can estimate the arc length in millimeters to be the phase-to-ground voltage in kilovolts multiplied by 2.54 (or voltage (in kilovolts) × 2.54).

Table 6- Incident Heat Energy for Various Fault Currents, Clearing Times, and Voltages of 4.0 to 46.0 Kv: Rubber Insulating Glove Exposures Involving Phase-to-Ground Arcs in Open Air Only ˙†‡

Voltage range (kV) **	Fault current (kA)	Maximum clearing time (cycles)			
		4 cal/cm²	5 cal/cm²	8 cal/cm²	12 cal/cm²
4.0 to 15.0	5	46	58	92	138
	10	18	22	36	54
	15	10	12	20	30
	20	6	8	13	19
15.1 to 25.0	5	28	34	55	83
	10	11	14	23	34
	15	7	8	13	20
	20	4	5	9	13
25.1 to 36.0	5	21	26	42	62
	10	9	11	18	26
	15	5	6	10	16
	20	4	4	7	11
36.1 to 46.0	5	16	20	32	48
	10	7	9	14	21
	15	4	5	8	13
	20	3	4	6	9

Notes:* This table is for open-air, phase-to-ground electric-arc exposures. It is not for phase-to-phase arcs or enclosed arcs (arc in a box).
†The table assumes that the employee will be 380 mm (15 in.) from the electric arc. The table also assumes the arc length to be the sparkover distance for the maximum transient overvoltage of each voltage range (see Appendix B to § 1910.269), as follows:
  4.0 to 15.0 kV 51 mm (2 in.)
  15.1 to 25.0 kV 102 mm (4 in.)
  25.1 to 36.0 kV 152 mm (6 in.)
  36.1 to 46.0 kV 229 mm (9 in.)
‡The Occupational Safety and Health Administration calculated the values in this table using the ARCPRO method listed in Table 2.
**The voltage range is the phase-to-phase system voltage.

Table 7- Incident Heat Energy for Various Fault Currents, Clearing Times, and Voltages: Live-Line Tool Exposures Involving Phase-to-Ground Arcs in Open Air Only[*†‡•]

Voltage range (kV)**	Fault current (kA)	Maximum clearing time (cycles)			
		4 cal/cm²	5 cal/cm²	8 cal/cm²	12 cal/cm²
4.0 to 15.0	5	197	246	394	591
	10	73	92	147	220
	15	39	49	78	117
	20	24	31	49	73
15.1 to 25.0	5	197	246	394	591
	10	75	94	150	225
	15	41	51	82	122
	20	26	33	52	78
25.1 to 36.0	5	138	172	275	413
	10	53	66	106	159
	15	30	37	59	89
	20	19	24	38	58
36.1 to 46.0	5	129	161	257	386
	10	51	64	102	154
	15	29	36	58	87
	20	19	24	38	57
46.1 to 72.5	20	18	23	36	55
	30	10	13	20	30
	40	6	8	13	19
	50	4	6	9	13
72.6 to 121.0	20	10	12	20	30
	30	6	7	11	17
	40	4	5	7	11
	50	3	3	5	8
121.1 to 145.0	20	12	15	24	35
	30	7	9	15	22
	40	5	6	10	15
	50	4	5	8	11

145.1 to 169.0	20	12	15	24	36
	30	7	9	15	22
	40	5	7	10	16
	50	4	5	8	12
169.1 to 242.0	20	13	17	27	40
	30	8	10	17	25
	40	6	7	12	17
	50	4	5	9	13
242.1 to 362.0	20	25	32	51	76
	30	16	19	31	47
	40	11	14	22	33
	50	8	10	16	25
362.1 to 420.0	20	12	15	25	37
	30	8	10	15	23
	40	5	7	11	16
	50	4	5	8	12
420.1 to 550.0	20	23	29	47	70
	30	14	18	29	43
	40	10	13	20	30
	50	8	9	15	23
550.1 to 800.0	20	25	31	50	75
	30	15	19	31	46
	40	11	13	21	32
	50	8	10	16	24

Notes:

*This table is for open-air, phase-to-ground electric-arc exposures. It is not for phase-to-phase arcs or enclosed arcs (arc in a box).

†The table assumes the arc length to be the sparkover distance for the maximum phase-to-ground voltage of each voltage range (see Appendix B to this section). The table also assumes that the employee will be the minimum approach distance minus twice the arc length from the electric arc.

‡The Occupational Safety and Health Administration calculated the values in this table using the ARCPRO method listed in Table 2.

*For voltages of more than 72.6 kV, employers may use this table only when the minimum approach distance established under § 1910.269(l)(3)(i) is greater than or equal to the following values:

   72.6 to 121.0 kV 1.02 m.
   121.1 to 145.0 kV 1.16 m.
   145.1 to 169.0 kV 1.30 m.
   169.1 to 242.0 kV 1.72 m.
   242.1 to 362.0 kV 2.76 m.
   362.1 to 420.0 kV 2.50 m.
   420.1 to 550.0 kV 3.62 m.
   550.1 to 800.0 kV 4.83 m.

**The voltage range is the phase-to-phase system voltage.

B. Selecting Protective Clothing and Other Protective Equipment

Paragraph (l)(8)(v) of § 1910.269 requires employers, in certain situations, to select protective clothing and other protective equipment with an arc rating that is greater than or equal to the incident heat energy estimated under § 1910.269(l)(8)(ii). Based on laboratory testing required by ASTM F1506-10a, the expectation is that protective clothing with an arc rating equal to the estimated incident heat energy will be capable of preventing second-degree burn injury to an employee exposed to that incident heat energy from an electric arc. Note that actual electric-arc exposures may be more or less severe than the estimated value because of factors such as arc movement, arc length, arcing from reclosing of the system, secondary fires or explosions, and weather conditions. Additionally, for arc rating based on the fabric's arc thermal performance value [5](ATPV), a worker exposed to incident energy at the arc rating has a 50-percent chance of just barely receiving a second-degree burn. Therefore, it is possible (although not likely) that an employee will sustain a second-degree (or worse) burn wearing clothing conforming to § 1910.269(l)(8)(v) under certain circumstances. However, reasonable employer estimates and maintaining appropriate minimum approach distances for employees should limit burns to relatively small burns that just barely extend beyond the epidermis (that is, just barely a seconddegree burn). Consequently, protective clothing and other protective equipment meeting § 1910.269(l)(8)(v) will provide an appropriate degree of protection for an employee exposed to electric-arc hazards.

Paragraph (l)(8)(v) of § 1910.269 does not require arc-rated protection for exposures of 2 cal/cm^2 or less. Untreated cotton clothing will reduce a 2-cal/cm^2 exposure below the 1.2- to 1.5-cal/cm^2 level necessary to cause burn injury, and this material should not ignite at such low heat energy levels. Although § 1910.269(l)(8)(v) does not require clothing to have an arc rating when exposures are 2 cal/cm^2 or less, § 1910.269(l)(8)(iv) requires the outer layer of clothing to be flame resistant under certain conditions, even when the estimated incident heat energy is less than 2 cal/cm^2, as discussed later in this appendix.

Additionally, it is especially important to ensure that employees do not wear undergarments made from fabrics listed in the note to § 1910.269(l)(8)(iii) even when the outer layer is flame resistant or arc rated. These fabrics can melt or ignite easily when an electric arc occurs. Logos and name tags made from non-flame-resistant material can adversely affect the arc rating or the flameresistant characteristics of arc-rated or flameresistant clothing. Such logos and name tags may violate § 1910.269(l)(8)(iii), (l)(8)(iv), or (l)(8)(v).

Paragraph (l)(8)(v) of § 1910.269 requires that arc-rated protection cover the employee's entire body, with limited exceptions for the employee's hands, feet,

[5]ASTM F1506-10a defines "arc thermal performance value" as "the incident energy on a material or a multilayer system of materials that results in a 50% probability that sufficient heat transfer through the tested specimen is predicted to cause the onset of a second-degree skin burn injury based on the Stoll [footnote] curve, cal/cm2." The footnote to this definition reads: "Derived from: Stoll, A. M., and Chianta, M. A., 'Method and Rating System for Evaluations of Thermal Protection,' Aerospace Medicine, Vol 40, 1969, pp. 1232-1238 and Stoll, A. M., and Chianta, M. A., 'Heat Transfer through Fabrics as Related to Thermal Injury,' Transactions-New York Academy of Sciences, Vol 33(7), Nov. 1971, pp. 649-670."

| For any estimated incident heat energy .... | When the employee is wearing rubber insulating gloves with protectors. |
| If the estimated incident heat energy does not exceed 14 cal/cm² ..... | When the employee is wearing heavy-duty leather work gloves with a weight of at least 407 gm/m²(12 oz/yd²). |

face, and head. Paragraph (l)(8)(v)(A) of § 1910.269 provides that arc-rated protection is not necessary for the employee's hands under the following conditions: Paragraph (l)(8)(v)(B) of § 1910.269 provides that arc-rated protection is not necessary for the employee's feet when the employee is wearing heavy-duty work

Minimum head and face protection

Exposure	None *	Arc-rated faceshield with a minimum rating of 8 cal/cm²*	Arc-rated hood or faceshield with balaclava
Single-phase, open air	2-8 cal/cm²	9-12 cal/cm²	13 cal/cm² or higher†.
Three-phase	2-4 cal/cm²	5-8 cal/cm²	9 cal/cm² or higher‡.

*These ranges assume that employees are wearing hardhats meeting the specifications in § 1910.135 or § 1926.100(b)(2), as applicable.
†The arc rating must be a minimum of 4 cal/cm2 less than the estimated incident energy. Note that § 1910.269(l)(8)(v)(E) permits this type of head and face protection, with a minimum arc rating of 4 cal/cm² less than the estimated incident energy, at any incident energy level.
‡Note that § 1910.269(l)(8)(v) permits this type of head and face protection at any incident energy level.

shoes or boots. Finally, § 1910.269(l)(8)(v)(C), (l)(8)(v)(D), and (l)(8)(v)(E) require arc-rated head and face protection as follows:

IV. Protection Against Ignition

Paragraph (l)(8)(iii) of § 1910.269 prohibits clothing that could melt onto an employee's skin or that could ignite and continue to burn when exposed to flames or to the available heat energy estimated by the employer under § 1910.269(l)(8)(ii). Meltable fabrics, such as acetate, nylon, polyester, and polypropylene, even in blends, must be avoided. When these fibers melt, they can adhere to the skin, thereby transferring heat rapidly, exacerbating burns, and complicating treatment. These outcomes can result even if the meltable fabric is not directly next to the skin. The remainder of this section focuses on the prevention of ignition.

Paragraph (l)(8)(v) of § 1910.269 generally requires protective clothing and other protective equipment with an arc rating greater than or equal to the employer's estimate of available heat energy. As explained earlier in this appendix, untreated cotton is usually acceptable for exposures of 2 cal/cm² or less. [6]If the exposure is greater than that, the employee generally must wear flame-resistant clothing with a suitable arc rating in accordance with § 1910.269(l)(8)(iv) and (l)(8)(v). However,

[6]See § 1910.269(l)(8)(iv)(A), (l)(8)(iv)(B), and (l)(8)(iv)(C) for conditions under which employees must wear flame-resistant clothing as the outer layer of clothing even when the incident heat energy does not exceed 2 cal/cm².

even if an employee is wearing a layer of flame-resistant clothing, there are circumstances under which flammable layers of clothing would be uncovered, and an electric arc could ignite them. For example, clothing ignition is possible if the employee is wearing flammable clothing under the flame-resistant clothing and the underlayer is uncovered because of an opening in the flame-resistant clothing. Thus, for purposes of § 1910.269(l)(8)(iii), it is important for the employer to consider the possibility of clothing ignition even when an employee is wearing flame-resistant clothing with a suitable arc rating.

Under § 1910.269(l)(8)(iii), employees may not wear flammable clothing in conjunction with flame-resistant clothing if the flammable clothing poses an ignition hazard.[7] Although outer flame-resistant layers may not have openings that expose flammable inner layers, when an outer flame-resistant layer would be unable to resist breakopen,[8] the next (inner) layer must be flame-resistant if it could ignite.

Non-flame-resistant clothing can ignite even when the heat energy from an electric arc is insufficient to ignite the clothing. For example, nearby flames can ignite an employee›s clothing; and, even in the absence of flames, electric arcs pose ignition hazards beyond the hazard of ignition from incident energy under certain conditions. In addition to requiring flame-resistant clothing when the estimated incident energy exceeds 2.0 cal/cm², § 1910.269(l)(8)(iv) requires flame-resistant clothing when: The employee is exposed to contact with energized circuit parts operating at more than 600 volts (§ 1910.269(l)(8)(iv)(A)), an electric arc could ignite flammable material in the work area that, in turn, could ignite the employee's clothing (§ 1910.269(l)(8)(iv)(B)), and molten metal or electric arcs from faulted conductors in the work area could ignite the employee's clothing (§ 1910.269(l)(8)(iv)(C)). For example, grounding conductors can become a source of heat energy if they cannot carry fault current without failure. The employer must consider these possible sources of electric arcs[9] in determining whether the employee's clothing could ignite under § 1910.269(l)(8)(iv)(C).

Reprinted from www.osha.gov

Occupational Safety & Health Administration
200 Constitution Avenue, NW
Washington, DC 20210

---

[7]Paragraph (l)(8)(iii) of § 1910.269 prohibits clothing that could ignite and continue to burn when exposed to the heat energy estimated under paragraph (l)(8)(ii) of that section.

[8]Breakopen occurs when a hole, tear, or crack develops in the exposed fabric such that the fabric no longer effectively blocks incident heat energy.

[9]Static wires and pole grounds are examples of grounding conductors that might not be capable of carrying fault current without failure. Grounds that can carry the maximum available fault current are not a concern, and employers need not consider such grounds a possible electric arc source.

Appendix F to § 1910.269-Work-Positioning Equipment Inspection Guidelines

Part Number: 1910
Part Title: Occupational Safety and Health Standards
Subpart: R
Subpart Title: Special Industries
Standard Number: 1910.269 App F
Title: Work-Positioning Equipment Inspection Guidelines.

I. Body Belts

Inspect body belts to ensure that:
A. The hardware has no cracks, nicks, distortion, or corrosion;
B. No loose or worn rivets are present;
C. The waist strap has no loose grommets;
D. The fastening straps are not 100-percent leather; and
E. No worn materials that could affect the safety of the user are present.

II. Positioning Straps

Inspect positioning straps to ensure that:
A. The warning center of the strap material is not exposed;
B. No cuts, burns, extra holes, or fraying of strap material is present;
C. Rivets are properly secured;
D. Straps are not 100-percent leather; and
E. Snaphooks do not have cracks, burns, or corrosion.

III. Climbers

Inspect pole and tree climbers to ensure that:
A. Gaffs are at least as long as the manufacturer's recommended minimums (generally 32 and 51 millimeters (1.25 and 2.0 inches) for pole and tree climbers, respectively, measured on the underside of the gaff);

Note: Gauges are available to assist in determining whether gaffs are long enough and shaped to easily penetrate poles or trees.

B. Gaffs and leg irons are not fractured or cracked;
C. Stirrups and leg irons are free of excessive wear;
D. Gaffs are not loose;
E. Gaffs are free of deformation that could adversely affect use;
F. Gaffs are properly sharpened; and
G. There are no broken straps or buckles.
[79 FR 20691, July 10, 2014]

Reprinted from www.osha.gov

Occupational Safety & Health Administration
200 Constitution Avenue, NW
Washington, DC 20210

Appendix G to § 1910.269-Reference Documents

- **Part Number:** 1910
- **Part Title:** Occupational Safety and Health Standards
- **Subpart:** R
- **Subpart Title:** Special Industries
- **Standard Number:** 1910.269 App G
- **Title:** Reference Documents.

The references contained in this appendix provide information that can be helpful in understanding and complying with the requirements contained in § 1910.269. The national consensus standards referenced in this appendix contain detailed specifications that employers may follow in complying with the more performance-based requirements of § 1910.269. Except as specifically noted in § 1910.269, however, the Occupational Safety and Health Administration will not necessarily deem compliance with the national consensus standards to be compliance with the provisions of § 1910.269.

ANSI/SIA A92.2-2009,*American National Standard for Vehicle-Mounted Elevating and Rotating Aerial Devices.*

ANSI Z133-2012,*American National Standard Safety Requirements for Arboricultural Operations-Pruning, Trimming, Repairing, Maintaining, and Removing Trees, and Cutting Brush.*

ANSI/IEEE Std 935-1989,*IEEE Guide on Terminology for Tools and Equipment to Be Used in Live Line Working.*

ASME B20.1-2012,*Safety Standard for Conveyors and Related Equipment.*

ASTM D120-09,*Standard Specification for Rubber Insulating Gloves.*

ASTM D149-09 (2013),*Standard Test Method for Dielectric Breakdown Voltage and Dielectric Strength of Solid Electrical Insulating Materials at Commercial Power Frequencies.*

ASTM D178-01 (2010),*Standard Specification for Rubber Insulating Matting.*

ASTM D1048-12,*Standard Specification for Rubber Insulating Blankets.*

ASTM D1049-98 (2010),*Standard Specification for Rubber Insulating Covers.*

ASTM D1050-05 (2011),*Standard Specification for Rubber Insulating Line Hose.*

ASTM D1051-08,*Standard Specification for Rubber Insulating Sleeves.*

ASTM F478-09,*Standard Specification for In-Service Care of Insulating Line Hose and Covers.*

ASTM F479-06 (2011),*Standard Specification for In-Service Care of Insulating Blankets.*

ASTM F496-08,*Standard Specification for In-Service Care of Insulating Gloves and Sleeves.*

ASTM F711-02 (2007),*Standard Specification for Fiberglass-Reinforced Plastic (FRP) Rod and Tube Used in Live Line Tools.*

ASTM F712-06 (2011),*Standard Test Methods and Specifications for Electrically Insulating Plastic Guard Equipment for Protection of Workers.*

ASTM F819-10,*Standard Terminology Relating to Electrical Protective Equipment for Workers.*

ASTM F855-09,*Standard Specifications for Temporary Protective Grounds to Be Used on De-energized Electric Power Lines and Equipment.*

ASTM F887-12[e1],*Standard Specifications for Personal Climbing Equipment.*

ASTM F914/F914M-10,*Standard Test Method for Acoustic Emission for Aerial Personnel Devices Without Supplemental Load Handling Attachments.*

ASTM F1116-03 (2008),*Standard Test Method for Determining Dielectric Strength of Dielectric Footwear.*

ASTM F1117-03 (2008),*Standard Specification for Dielectric Footwear.*

ASTM F1236-96 (2012),*Standard Guide for Visual Inspection of Electrical Protective Rubber Products.*

ASTM F1430/F1430M-10,*Standard Test Method for Acoustic Emission Testing of Insulated and Non-Insulated Aerial Personnel Devices with Supplemental Load Handling Attachments.*

ASTM F1505-10,*Standard Specification for Insulated and Insulating Hand Tools.*

ASTM F1506-10a,*Standard Performance Specification for Flame Resistant and Arc Rated Textile Materials for Wearing Apparel for Use by Electrical Workers Exposed to Momentary Electric Arc and Related Thermal Hazards.*

ASTM F1564-13,*Standard Specification for Structure-Mounted Insulating Work Platforms for Electrical Workers.*

ASTM F1701-12,*Standard Specification for Unused Polypropylene Rope with Special Electrical Properties.*

ASTM F1742-03 (2011),*Standard Specification for PVC Insulating Sheeting.*

ASTM F1796-09,*Standard Specification for High Voltage Detectors-Part 1 Capacitive Type to be Used for Voltages Exceeding 600 Volts AC.*

ASTM F1797-09[e1],*Standard Test Method for Acoustic Emission Testing of Insulated and Non-Insulated Digger Derricks.*

ASTM F1825-03 (2007),*Standard Specification for Clampstick Type Live Line Tools.*

ASTM F1826-00 (2011),*Standard Specification for Live Line and Measuring Telescoping Tools.*

ASTM F1891-12,*Standard Specification for Arc and Flame Resistant Rainwear.*

ASTM F1958/F1958M-12,*Standard Test Method for Determining the Ignitability of Non-flame-Resistant Materials for Clothing by Electric Arc Exposure Method Using Mannequins.*

ASTM F1959/F1959M-12,*Standard Test Method for Determining the Arc Rating of Materials for Clothing.*

IEEE Stds 4-1995,4a-2001(Amendment to IEEE Standard Techniques for High-Voltage Testing), *IEEE Standard Techniques for High-Voltage Testing.*

IEEE Std 62-1995,*IEEE Guide for Diagnostic Field Testing of Electric Power Apparatus-Part 1: Oil Filled Power Transformers, Regulators, and Reactors.*

IEEE Std 80-2000,*Guide for Safety in AC Substation Grounding.*

IEEE Std 100-2000,*The Authoritative Dictionary of IEEE Standards Terms Seventh Edition.*

IEEE Std 516-2009,*IEEE Guide for Maintenance Methods on Energized Power Lines.*

IEEE Std 524-2003,*IEEE Guide to the Installation of Overhead Transmission Line Conductors*

IEEE Std 957-2005,*IEEE Guide for Cleaning Insulators.*

IEEE Std 1048-2003,*IEEE Guide for Protective Grounding of Power Lines.*

IEEE Std 1067-2005,*IEEE Guide for In-Service Use, Care, Maintenance, and Testing of Conductive Clothing for Use on Voltages up to 765 kV AC and ±750 kV DC.*

IEEE Std 1307-2004,*IEEE Standard for Fall Protection for Utility Work.*

IEEE Stds 1584-2002,1584a-2004 Amendment 1 to IEEE Std 1584-2002), and 1584b-2011 (Amendment 2: Changes to Clause 4 of IEEE Std 1584-2002), *IEEE Guide for Performing Arc-Flash Hazard Calculations.*
IEEE C2-2012,*National Electrical Safety Code.*
NFPA 70E-2012,*Standard for Electrical Safety in the Workplace.*
[79 FR 20691-20692, July 10, 2014]

Reprinted from www.osha.gov

Occupational Safety & Health Administration
200 Constitution Avenue, NW
Washington, DC 20210

Power Transmission and Distribution (Construction) 1926 Subpart V—Power Transmission and Distribution

OSHA 1926.950–1926.968 Paragraph Titles:

OSHA 1926 Subpart V—Authority for 1926 Subpart V
Authority

OSHA 1926.950—General
(a) Application
(b) Training
(c) Information transfer
(d) Existing characteristics and conditions

OSHA 1926.951—Medical Services and First Aid
(a) General
(b) First-Aid training

OSHA 1926.952—Job Briefing
(a) Before each job
(b) Subjects to be covered
(c) Number of briefings
(d) Extent of briefing
(e) Working alone

OSHA 1926.953—Enclosed spaces
(a) General
(b) Safe work practices
(c) Training
(d) Rescue equipment
(e) Evaluating potential hazards
(f) Removing covers
(g) Hazardous atmosphere
(h) Attendants
(i) Calibration of test instruments
(j) Testing for oxygen deficiency
(k) Testing for flammable gases and vapors
(l) Ventilation, and monitoring for flammable gases or vapors
(m) Specific ventilation requirements
(n) Air supply
(o) Open flames

OSHA 1926.954—Personal Protective Equipment
(a) General
(b) Fall protection

OSHA 1926.955—Portable Ladders and Platforms
(a) General
(b) Special ladders and platforms
(c) Conductive ladders

OSHA 1926.956—Hand and Portable Power Equipment
(a) General
(b) Cord- and plug-connected equipment
(c) Portable and vehicle-mounted generators
(d) Hydraulic and pneumatic tools

OSHA 1926.957—Live-line Tools
(a) Design of tools
(b) Condition of tools

OSHA 1926.958—Materials Handling and Storage
(a) General
(b) Materials storage near energized lines or equipment

OSHA 1926.959—Mechanical Equipment
(a) General requirements
(b) Outriggers
(c) Applied loads
(d) Operations near energized lines or equipment

OSHA 1926.960—Working On or Near Exposed Energized Parts
(a) Application
(b) General
(c) Live work
(d) Working position
(e) Making connections

(f) Conductive articles
(g) Protection from flames and electric arcs
(h) Fuse handling
(i) Covered (noninsulated) conductors
(j) Non-current-carrying metal parts
(k) Opening and closing circuits under load

OSHA 1926.961—Deenergizing Lines and Equipment for Employee Protection
(a) Application
(b) General
(c) Deenergizing lines and equipment

OSHA 1926.962—Grounding for the Protection of Employees
(a) Application
(b) General
(c) Equipotential zone
(d) Protective grounding equipment
(e) Testing
(f) Connecting and removing grounds
(g) Additional precautions
(h) Removal of grounds for test

OSHA 1926.963—Testing and Test Facilities
(a) Application
(b) General requirements
(c) Safeguarding of test areas
(d) Grounding practices
(e) Control and measuring circuits
(f) Safety check

OSHA 1926.964—Overhead Lines and Live-Line Barehand Work
(a) General
(b) Installing and removing overhead lines
(c) Live-line barehand work
(d) Towers and structures

OSHA 1926.965—Underground Electrical Installations
(a) Application
(b) Access
(c) Lowering equipment into manholes
(d) Attendants for manholes and vaults
(e) Duct rods
(f) Multiple cables
(g) Moving cables
(h) Protection against faults

OSHA 1926.966—Substations
(a) Application
(b) Access and working space
(c) Draw-out-type circuit breakers
(d) Substation fences
(e) Guarding of rooms and other spaces containing electric supply equipment

(f) Guarding of energized parts
(g) Substation entry

OSHA 1926.967—Special Conditions
(a) Capacitors
(b) Current transformer secondaries
(c) Series streetlighting
(d) Illumination
(e) Protection against drowning
(f) Excavations
(g) Employee protection in public work areas
(h) Backfeed
(i) Lasers
(j) Hydraulic fluids
(k) Communication facilities

OSHA 1926.968—Definitions

Appendix A – Reserve
Appendix B – Working on Exposed Energized Parts
Appendix C – Protection from Hazardous Differences in Electric Potential
Appendix D – Methods of Inspecting and Testing Wood Poles
Appendix E – Protection from Flames and Electric Arcs
Appendix F – Work-Positioning Equipment Inspection Guidelines
Appendix G – Reference Documents

REGULATIONS (STANDARDS - 29 CFR)

ELECTRIC POWER TRANSMISSION AND DISTRIBUTION - 1926 SUBPART V

- **Part Number:** 1926
- **Part Title:** Safety and Health Regulations for Construction
- **Subpart:** V
- **Subpart Title:** Electric Power Transmission and Distribution
- **Standard Number:** 1926 Subpart V
- **Title:** Authority for 1926 Subpart V
- **Appendix:** A, B, C, D, E, F, G
- **GPO Source:** e-CFR

Authority: 40 U.S.C. 3701 et seq.; 29 U.S.C. 653, 655, 657; Secretary of Labor's Order No. 1-2012 (77 FR 3912); and 29 CFR part 1911.

[59 FR 40730, Aug. 9, 1994; 75 FR 48135, Aug. 9, 2010; 78 FR 32116, May 29, 2013; 79 FR 20696, July 10, 2014; 79 FR 56962, September 24, 2014; 80 FR 25518, May 4, 2015; 80 FR 60040, October 5, 2015]

Reprinted from www.osha.gov

Occupational Safety & Health Administration
200 Constitution Avenue, NW
Washington, DC 20210

REGULATIONS (STANDARDS - 29 CFR)

GENERAL - 1926.950

- **Part Number:** 1926
- **Part Title:** Safety and Health Regulations for Construction
- **Subpart:** V
- **Subpart Title:** Electric Power Transmission and Distribution
- **Standard Number:** 1926.950
- **Title:** General.
- **GPO Source:** e-CFR

1926.950(a)

Application.

1926.950(a)(1)

Scope.

1926.950(a)(1)(i)

This subpart, except for paragraph (a)(3) of this section, covers the construction of electric power transmission and distribution lines and equipment. As used in this subpart, the term "construction" includes the erection of new electric transmission and distribution lines and equipment, and the alteration, conversion, and improvement of existing electric transmission and distribution lines and equipment.

Note to paragraph (a)(1)(i): An employer that complies with § 1910.269 of this chapter will be considered in compliance with requirements in this subpart that do not reference other subparts of this part. Compliance with § 1910.269 of this chapter will not excuse an employer from compliance obligations under other subparts of this part.

1926.950(a)(1)(ii)

Notwithstanding paragraph (a)(1)(i) of this section, this subpart does not apply to electrical safety-related work practices for unqualified employees.

1926.950(a)(2)

Other Part 1926 standards. This subpart applies in addition to all other applicable standards contained in this Part 1926. Employers covered under this subpart are not exempt from complying with other applicable provisions in Part 1926 by the operation of § 1910.5(c) of this chapter. Specific references in this subpart to other sections of Part 1926 are provided for emphasis only.

1926.950(a)(3)

Applicable Part 1910 requirements. Line-clearance treetrimming operations and work involving electric power generation installations shall comply with § 1910.269 of this chapter.

1926.950(b)

Training.

1926.950(b)(1)

All employees.

1926.950(b)(1)(i)

Each employee shall be trained in, and familiar with, the safety-related work practices, safety procedures, and other safety requirements in this subpart that pertain to his or her job assignments.

1926.950(b)(1)(ii)

Each employee shall also be trained in and familiar with any other safety practices, including applicable emergency procedures (such as pole-top and manhole rescue), that are not specifically addressed by this subpart but that are related to his or her work and are necessary for his or her safety.

1926.950(b)(1)(iii)

The degree of training shall be determined by the risk to the employee for the hazard involved.

1926.950(b)(2)

Qualified employees. Each qualified employee shall also be trained and competent in:

1926.950(b)(2)(i)

The skills and techniques necessary to distinguish exposed live parts from other parts of electric equipment,

1926.950(b)(2)(ii)

The skills and techniques necessary to determine the nominal voltage of exposed live parts,

1926.950(b)(2)(iii)

The minimum approach distances specified in this subpart corresponding to the voltages to which the qualified employee will be exposed and the skills and techniques necessary to maintain those distances,

1926.950(b)(2)(iv)

The proper use of the special precautionary techniques, personal protective equipment, insulating and shielding materials, and insulated tools for working on or near exposed energized parts of electric equipment, and

1926.950(b)(2)(v)

The recognition of electrical hazards to which the employee may be exposed and the skills and techniques necessary to control or avoid these hazards.

Note to paragraph (b)(2): For the purposes of this subpart, a person must have the training required by paragraph (b)(2) of this section to be considered a qualified person.

1926.950(b)(3)

Supervision and annual inspection. The employer shall determine, through regular supervision and through inspections conducted on at least an annual basis, that each employee is complying with the safetyrelated work practices required by this subpart.

1926.950(b)(4)

Additional training. An employee shall receive additional training (or retraining) under any of the following conditions:

1926.950(b)(4)(i)

If the supervision or annual inspections required by paragraph (b)(3) of this section indicate that the employee is not complying with the safety-related work practices required by this subpart, or

1926.950(b)(4)(ii)

If new technology, new types of equipment, or changes in procedures necessitate the use of safety-related work practices that are different from those which the employee would normally use, or

1926.950(b)(4)(iii)

If he or she must employ safetyrelated work practices that are not normally used during his or her regular job duties.

Note to paragraph (b)(4)(iii): The Occupational Safety and Health Administration considers tasks that are performed less often than once per year to necessitate retraining before the performance of the work practices involved.

1926.950(b)(5)

Type of training. The training required by paragraph (b) of this section shall be of the classroom or on-the-job type.

1926.950(b)(6)

Training goals. The training shall establish employee proficiency in the work practices required by this subpart and shall introduce the procedures necessary for compliance with this subpart.

1926.950(b)(7)

Demonstration of proficiency. The employer shall ensure that each employee has demonstrated proficiency in the work practices involved before that employee is considered as having completed the training required by paragraph (b) of this section.

Note 1 to paragraph (b)(7): Though they are not required by this paragraph, employment records that indicate that an employee has successfully completed the required training are one way of keeping track of when an employee has demonstrated proficiency.

Note 2 to paragraph (b)(7): For an employee with previous training, an employer may determine that that employee has demonstrated the proficiency required by this paragraph using the following process: (1) Confirm that the employee has the training required by paragraph (b) of this section, (2) use an examination or interview to make an initial determination that the employee understands the relevant safetyrelated work practices before he or she performs any work covered by this subpart, and (3) supervise the employee closely until that employee has demonstrated proficiency as required by this paragraph.

1926.950(c)

Information transfer.

1926.950(c)(1)

Host employer responsibilities. Before work begins, the host employer shall inform contract employers of:

1926.950(c)(1)(i)

The characteristics of the host employer's installation that are related to the safety of the work to be performed and are listed in paragraphs (d)(1) through (d)(5) of this section;

Note to paragraph (c)(1)(i): This paragraph requires the host employer to obtain information listed in paragraphs (d)(1) through (d)(5) of this section if it does not have this information in existing records.

1926.950(c)(1)(ii)

Conditions that are related to the safety of the work to be performed, that are listed in paragraphs (d)(6) through (d)(8) of this section, and that are known to the host employer;

Note to paragraph (c)(1)(ii): For the purposes of this paragraph, the host employer need only provide information to contract employers that the host employer can obtain from its existing records through the exercise of reasonable diligence. This paragraph does not require the host employer to make inspections of worksite conditions to obtain this information.

1926.950(c)(1)(iii)

Information about the design and operation of the host employer's installation that the contract employer needs to make the assessments required by this subpart; and

Note to paragraph (c)(1)(iii): This paragraph requires the host employer to obtain information about the design and operation of its installation that contract employers need to make required assessments if it does not have this information in existing records.

1926.950(c)(1)(iv)

Any other information about the design and operation of the host employer's installation that is known by the host employer, that the contract employer requests, and that is related to the protection of the contract employer's employees.

Note to paragraph (c)(1)(iv): For the purposes of this paragraph, the host employer need only provide information to contract employers that the host employer can obtain from its existing records through the exercise of reasonable diligence. This paragraph does not require the host employer to make inspections of worksite conditions to obtain this information.

1926.950(c)(2)

Contract employer responsibilities.

1926.950(c)(2)(i)

The contract employer shall ensure that each of its employees is instructed in the hazardous conditions relevant to the employee's work that the contract employer is aware of as a result of information communicated to the contract employer by the host employer under paragraph (c)(1) of this section.

1926.950(c)(2)(ii)

Before work begins, the contract employer shall advise the host employer of any unique hazardous conditions presented by the contract employer's work

1926.950(c)(2)(iii)

The contract employer shall advise the host employer of any unanticipated hazardous conditions found during the contract employer's work that the host employer did not mention under paragraph (c)(1) of this section. The contract employer shall provide this information to the host employer within 2 working days after discovering the hazardous condition.

1926.950(c)(3)

Joint host- and contract-employer responsibilities. The contract employer and the host employer shall coordinate their work rules and procedures so that each employee of the contract employer and the host employer is protected as required by this subpart.

1926.950(d)

Existing characteristics and conditions. Existing characteristics and conditions of electric lines and equipment that are related to the safety of the work to be performed shall be determined before work on or near the lines or equipment is started. Such characteristics and conditions include, but are not limited to:

1926.950(d)(1)

The nominal voltages of lines and equipment,

1926.950(d)(2)

The maximum switching-transient voltages,

1926.950(d)(3)

The presence of hazardous induced voltages,

1926.950(d)(4)

The presence of protective grounds and equipment grounding conductors,

1926.950(d)(5)

The locations of circuits and equipment, including electric supply lines, communication lines, and fire protective signaling circuits,

1926.950(d)(6)

The condition of protective grounds and equipment grounding conductors,

1926.950(d)(7)

The condition of poles, and

1926.950(d)(8)

Environmental conditions relating to safety.

[79 FR 20696-20697, July 10, 2014]

Reprinted from www.osha.gov

Occupational Safety & Health Administration
200 Constitution Avenue, NW
Washington, DC 20210

REGULATIONS (STANDARDS - 29 CFR)

MEDICAL SERVICES AND FIRST AID - 1926.951
- **Part Number:** 1926
- **Part Title:** Safety and Health Regulations for Construction
- **Subpart:** V
- **Subpart Title:** Electric Power Transmission and Distribution
- **Standard Number:** 1926.951
- **Title:** Medical services and first aid.
- **GPO Source:** e-CFR

1926.951(a)

General. The employer shall provide medical services and first aid as required in § 1926.50.

1926.951(b)

First-aid training. In addition to the requirements of § 1926.50, when employees are performing work on, or associated with, exposed lines or equipment energized at 50 volts or more, persons with first-aid training shall be available as follows:

1926.951(b)(1)

Field work. For field work involving two or more employees at a work location, at least two trained persons shall be available.

1926.951(b)(2)

Fixed work locations. For fixed work locations such as substations, the number of trained persons available shall be sufficient to ensure that each employee exposed to electric shock can be reached within 4 minutes by a trained person. However, where the existing number of employees is insufficient to meet this requirement (at a remote substation, for example), each employee at the work location shall be a trained employee.

[59 FR 40730, Aug. 9, 1994; 79 FR 20698, July 10, 2014]

Reprinted from www.osha.gov

Occupational Safety & Health Administration
200 Constitution Avenue, NW
Washington, DC 20210

REGULATIONS (STANDARDS - 29 CFR)

JOB BRIEFING - 1926.952
- **Part Number:** 1926
- **Part Title:** Safety and Health Regulations for Construction
- **Subpart:** V
- **Subpart Title:** Electric Power Transmission and Distribution
- **Standard Number:** 1926.952
- **Title:** Job briefing.
- **GPO Source:** e-CFR

1926.952(a)

Before each job.

1926.952(a)(1)

Information provided by the employer. In assigning an employee or a group of employees to perform a job, the employer shall provide the employee in charge of the job with all available information that relates to the determination of existing characteristics and conditions required by § 1926.950(d).

1926.952(a)(2)

Briefing by the employee in charge. The employer shall ensure that the employee in charge conducts a job briefing that meets paragraphs (b), (c), and (d) of this section with the employees involved before they start each job.

1926.952(b)

Subjects to be covered. The briefing shall cover at least the following subjects: Hazards associated with the job, work procedures involved, special precautions, energy-source controls, and personal protective equipment requirements.

1926.952(c)

Number of briefings.

1926.952(c)(1)

At least one before each day or shift. If the work or operations to be performed during the work day or shift are repetitive and similar, at least one job briefing shall be conducted before the start of the first job of each day or shift.

1926.952(c)(2)

Additional briefings. Additional job briefings shall be held if significant changes, which might affect the safety of the employees, occur during the course of the work.

1926.952(d)

Extent of briefing.

1926.952(d)(1)

Short discussion. A brief discussion is satisfactory if the work involved is routine and if the employees, by virtue of training and experience, can reasonably be expected to recognize and avoid the hazards involved in the job.

1926.952(d)(2)

Detailed discussion. A more extensive discussion shall be conducted:

1926.952(d)(2)(i)

If the work is complicated or particularly hazardous, or

1926.952(d)(2)(ii)

If the employee cannot be expected to recognize and avoid the hazards involved in the job.

Note to paragraph (d): The briefing must address all the subjects listed in paragraph (b) of this section.

1926.952(e)

Working alone. An employee working alone need not conduct a job briefing. However, the employer shall ensure that the tasks to be performed are planned as if a briefing were required.

[75 FR 48135, Aug. 9, 2010; 78 FR 32116, May 29, 2013; 79 FR 20698, July 10, 2014]

Reprinted from www.osha.gov

Occupational Safety & Health Administration
200 Constitution Avenue, NW
Washington, DC 20210

REGULATIONS (STANDARDS - 29 CFR)

ENCLOSED SPACES - 1926.953
- **Part Number:** 1926
- **Part Title:** Safety and Health Regulations for Construction
- **Subpart:** V
- **Subpart Title:** Electric Power Transmission and Distribution
- **Standard Number:** 1926.953
- **Title:** Enclosed spaces.
- **GPO Source:** e-CFR

1926.953(a)

General. This section covers enclosed spaces that may be entered by employees. It does not apply to vented vaults if the employer makes a determination that the ventilation system is operating to protect employees before they enter the space. This section applies to routine entry into enclosed spaces. If, after the employer takes the precautions given in this section and in § 1926.965, the hazards remaining in the enclosed space endanger the life of an entrant or could interfere with an entrant's escape from the space, then entry into the enclosed space must meet the permit space entry requirements of subpart AA of this part. For routine entries where the hazards remaining in the enclosed space do not endanger the life of an entrant or interfere with an entrant's escape from the space, this section applies in lieu of the permitspace entry requirements contained in §§ 1926.1204 through 926.1211.

1926.953(b)

Safe work practices. The employer shall ensure the use of safe work practices for entry into, and work in, enclosed spaces and for rescue of employees from such spaces.

1926.953(c)

Training. Each employee who enters an enclosed space or who serves as an attendant shall be trained in the hazards of enclosed-space entry, in enclosed-space entry procedures, and in enclosed-space rescue procedures.

1926.953(d)

Rescue equipment. Employers shall provide equipment to ensure the prompt and safe rescue of employees from the enclosed space.

1926.953(e)

Evaluating potential hazards. Before any entrance cover to an enclosed space is removed, the employer shall determine whether it is safe to do so by checking for the presence of any atmospheric pressure or temperature differences and by evaluating whether there might be a hazardous atmosphere in the space. Any conditions making it unsafe to remove the cover shall be eliminated before the cover is removed.

Note to paragraph (e): The determination called for in this paragraph may consist of a check of the conditions that might foreseeably be in the enclosed space. For example, the cover could be checked to see if it is hot and, if it is fastened in place, could be loosened gradually to release any residual pressure. An evaluation also needs to be made of whether conditions at the site could cause a hazardous atmosphere, such as an oxygen-deficient or flammable atmosphere, to develop within the space.

1926.953(f)

Removing covers. When covers are removed from enclosed spaces, the opening shall be promptly guarded by a railing, temporary cover, or other barrier designed to prevent an accidental fall through the opening and to protect employees working in the space from objects entering the space.

1926.953(g)

Hazardous atmosphere. Employees may not enter any enclosed space while it contains a hazardous atmosphere, unless the entry conforms to the confined spaces in construction standard in subpart AA of this part.

1926.953(h)

Attendants. While work is being performed in the enclosed space, an attendant with first-aid training shall be immediately available outside the enclosed space to provide assistance if a hazard exists because of traffic patterns in the area of the opening used for entry. The attendant is not precluded from performing other duties outside the enclosed space if these duties do not distract the attendant from: Monitoring employees within the space or ensuring that it is safe for employees to enter and exit the space.

Note to paragraph (h): See § 1926.965 for additional requirements on attendants for work in manholes and vaults.

1926.953(i)

Calibration of test instruments. Test instruments used to monitor atmospheres in enclosed spaces shall be kept in calibration and shall have a minimum accuracy of ±10 percent.

1926.953(j)

Testing for oxygen deficiency. Before an employee enters an enclosed space, the atmosphere in the enclosed space shall be tested for oxygen deficiency with a direct-reading meter or similar instrument, capable of collection and immediate analysis of data samples without the need for offsite evaluation. If continuous forced-air ventilation is provided, testing is not required provided that the procedures used ensure that employees are not exposed to the hazards posed by oxygen deficiency.

1926.953(k)

Testing for flammable gases and vapors. Before an employee enters an enclosed space, the internal atmosphere shall be tested for flammable gases and vapors with a direct-reading meter or similar instrument capable of collection and immediate analysis of data samples without the need for off-site evaluation. This test shall be performed after the oxygen testing and ventilation required by paragraph (j) of this section demonstrate that there is sufficient oxygen to ensure the accuracy of the test for flammability.

1926.953(l)

Ventilation, and monitoring for flammable gases or vapors. If flammable gases or vapors are detected or if an oxygen deficiency is found, forced-air ventilation shall be used to maintain oxygen at a safe level and to prevent a hazardous concentration of flammable gases and vapors from accumulating. A continuous monitoring program to ensure that no increase in flammable gas or vapor concentration above safe levels occurs may be followed in lieu of ventilation if flammable gases or vapors are initially detected at safe levels.

Note to paragraph (l): See the definition of "hazardous atmosphere" for guidance in determining whether a specific concentration of a substance is hazardous.

1926.953(m)

Specific ventilation requirements. If continuous forced-air ventilation is used, it shall begin before entry is made and shall be maintained long enough for the employer to be able to demonstrate that a safe atmosphere exists before employees are allowed to enter the work area. The forced-air ventilation shall be so directed as to ventilate the immediate area where employees are present within the enclosed space and shall continue until all employees leave the enclosed space.

1926.953(n)

Air supply. The air supply for the continuous forced-air ventilation shall be from a clean source and may not increase the hazards in the enclosed space.

1926.953(o)

Open flames. If open flames are used in enclosed spaces, a test for flammable gases and vapors shall be made immediately before the open flame device is used and at least once per hour while the device is used in the space. Testing shall be conducted more frequently if conditions present in the enclosed space indicate that once per hour is insufficient to detect hazardous accumulations of flammable gases or vapors.

Note to paragraph (o): See the definition of "hazardous atmosphere" for guidance in determining whether a specific concentration of a substance is hazardous.

Note to § 1926.953: Entries into enclosed spaces conducted in accordance with the permit space entry requirements of subpart AA of this part are considered as complying with this section.

[75 FR 48135, Aug. 9, 2010; 78 FR 32116, May 29, 2013; 79 FR 20698-20699, July 10, 2014; 80 FR 25518, May 4, 2015]

Reprinted from www.osha.gov

Occupational Safety & Health Administration
200 Constitution Avenue, NW
Washington, DC 20210

REGULATIONS (STANDARDS - 29 CFR)

PERSONAL PROTECTIVE EQUIPMENT - 1926.954
- **Part Number:** 1926
- **Part Title:** Safety and Health Regulations for Construction
- **Subpart:** V
- **Subpart Title:** Electric Power Transmission and Distribution
- **Standard Number:** 1926.954
- **Title:** Personal protective equipment.
- **GPO Source:** e-CFR

1926.954(a)

General. Personal protective equipment shall meet the requirements of Subpart E of this part.

Note to paragraph (a): Paragraph (d) of § 1926.95 sets employer payment obligations for the personal protective equipment required by this subpart, including, but not limited to, the fall protection equipment required by paragraph (b) of this section, the electrical protective equipment required by § 1926.960(c), and the flame-resistant and arc-rated clothing and other protective equipment required by § 1926.960(g).

1926.954(b)

Fall protection.

1926.954(b)(1)

Personal fall arrest systems.

1926.954(b)(1)(i)

Personal fall arrest systems shall meet the requirements of Subpart M of this part.

1926.954(b)(1)(ii)

Personal fall arrest equipment used by employees who are exposed to hazards from flames or electric arcs, as determined by the employer under § 1926.960(g)(1), shall be capable of passing a drop test equivalent to that required by paragraph (b) (2)(xii) of this section after exposure to an electric arc with a heat energy of 40±5 cal/cm².

1926.954(b)(2)

Work-positioning equipment. Body belts and positioning straps for Work-positioning equipment shall meet the following requirements:

1926.954(b)(2)(i)

Hardware for body belts and positioning straps shall meet the following requirements:

1926.954(b)(2)(i)(A)

Hardware shall be made of dropforged steel, pressed steel, formed steel, or equivalent material.

1926.954(b)(2)(i)(B)

Hardware shall have a corrosion-resistant finish.

1926.954(b)(2)(i)(C)

Hardware surfaces shall be smooth and free of sharp edges.

1926.954(b)(2)(ii)

Buckles shall be capable of withstanding an 8.9-kilonewton (2,000-pound-force) tension test with a maximum permanent deformation no greater than 0.4 millimeters (0.0156 inches).

1926.954(b)(2)(iii)

D rings shall be capable of withstanding a 22-kilonewton (5,000-pound-force) tensile test without cracking or breaking.

1926.954(b)(2)(iv)

Snaphooks shall be capable of withstanding a 22-kilonewton (5,000-pound-force) tension test without failure.

Note to paragraph (b)(2)(iv): Distortion of the snaphook sufficient to release the keeper is considered to be tensile failure of a snaphook.

1926.954(b)(2)(v)

Top grain leather or leather substitute may be used in the manufacture of body belts and positioning straps; however, leather and leather substitutes may not be used alone as a load-bearing component of the assembly.

1926.954(b)(2)(vi)

Plied fabric used in positioning straps and in load-bearing parts of body belts shall be constructed in such a way that no raw edges are exposed and the plies do not separate.

1926.954(b)(2)(vii)

Positioning straps shall be capable of withstanding the following tests:

1926.954(b)(2)(vii)(A)

A dielectric test of 819.7 volts, AC, per centimeter (25,000 volts per foot) for 3 minutes without visible deterioration;

1926.954(b)(2)(vii)(B)

A leakage test of 98.4 volts, AC, per centimeter (3,000 volts per foot) with a leakage current of no more than 1 mA;

Note to paragraphs (b)(2)(vii)(A) and (b)(2)(vii)(B): Positioning straps that pass direct-current tests at equivalent voltages are considered as meeting this requirement.

1926.954(b)(2)(vii)(C)

Tension tests of 20 kilonewtons (4,500 pounds-force) for sections free of buckle holes and of 15 kilonewtons (3,500 pounds-force) for sections with buckle holes;

1926.954(b)(2)(vii)(D)

A buckle-tear test with a load of 4.4 kilonewtons (1,000 pounds-force); and

1926.954(b)(2)(vii)(E)

A flammability test in accordance with Table V-1.

Table V-1   Flammability Test

Test method	Criteria for passing the test
Vertically suspend a 500-mm (19.7-inch) length of strapping supporting a 100-kg (220.5-lb) weight.	Any flames on the positioning strap shall self extinguish.
Use a butane or propane burner with a 76-mm (3-inch) flame	The positioning strap shall continue to support the 100-kg (220.5-lb) mass.
Direct the flame to an edge of the strapping at a distance of 25 mm (1 inch).	
Remove the flame after 5 seconds.	
Wait for any flames on the positioning strap to stop burning.	

1926.954(b)(2)(viii)

The cushion part of the body belt shall contain no exposed rivets on the inside and shall be at least 76 millimeters (3 inches) in width.

1926.954(b)(2)(ix)

Tool loops shall be situated on the body of a body belt so that the 100 millimeters (4 inches) of the body belt that is in the center of the back, measuring from D ring to D ring, is free of tool loops and any other attachments.

1926.954(b)(2)(x)

Copper, steel, or equivalent liners shall be used around the bars of D rings to prevent wear between these members and the leather or fabric enclosing them.

1926.954(b)(2)(xi)

Snaphooks shall be of the locking type meeting the following requirements:

1926.954(b)(2)(xi)(A)

The locking mechanism shall first be released, or a destructive force shall be placed on the keeper, before the keeper will open.

1926.954(b)(2)(xi)(B)

A force in the range of 6.7 N (1.5 lbf) to 17.8 N (4 lbf) shall be required to release the locking mechanism.

1926.954(b)(2)(xi)(C)

With the locking mechanism released and with a force applied on the keeper against the face of the nose, the keeper may not begin to open with a force of 11.2 N (2.5 lbf) or less and shall begin to open with a maximum force of 17.8 N (4 lbf).

1926.954(b)(2)(xii)

Body belts and positioning straps shall be capable of withstanding a drop test as follows:

1926.954(b)(2)(xii)(A)

The test mass shall be rigidly constructed of steel or equivalent material with a mass of 100 kg (220.5 lbm). For work-positioning equipment used by employees weighing more than 140 kg (310 lbm) fully equipped, the test mass shall be increased proportionately (that is, the test mass must equal the mass of the equipped worker divided by 1.4).

1926.954(b)(2)(xii)(B)

For body belts, the body belt shall be fitted snugly around the test mass and shall be attached to the test-structure anchorage point by means of a wire rope.

1926.954(b)(2)(xii)(C)

For positioning straps, the strap shall be adjusted to its shortest length possible to accommodate the test and connected to the test-structure anchorage point at one end and to the test mass on the other end.

1926.954(b)(2)(xii)(D)

The test mass shall be dropped an unobstructed distance of 1 meter (39.4 inches) from a supporting structure that will sustain minimal deflection during the test.

1926.954(b)(2)(xii)(E)

Body belts shall successfully arrest the fall of the test mass and shall be capable of supporting the mass after the test.

1926.954(b)(2)(xii)(F)

Positioning straps shall successfully arrest the fall of the test mass without breaking, and the arrest force may not exceed 17.8 kilonewtons (4,000 pounds-force). Additionally, snaphooks on positioning straps may not distort to such an extent that the keeper would release.

Note to paragraph (b)(2): When used by employees weighing no more than 140 kg (310 lbm) fully equipped, body belts and positioning straps that conform to American Society of Testing and Materials *Standard Specifications for Personal Climbing Equipment*, ASTM F887-12^{e1}, are deemed to be in compliance with paragraph (b)(2) of this section.

1926.954(b)(3)

Care and use of personal fall protection equipment.

1926.954(b)(3)(i)

Work-positioning equipment shall be inspected before use each day to determine that the equipment is in safe working condition. Work-positioning equipment that is not in safe working condition may not be used.

Note to paragraph (b)(3)(i): Appendix F to this subpart contains guidelines for inspecting work-positioning equipment.

1926.954(b)(3)(ii)

Personal fall arrest systems shall be used in accordance with § 1926.502(d).

Note to paragraph (b)(3)(ii): Fall protection equipment rigged to arrest falls is considered a fall arrest system and must meet the applicable requirements for the design and use of those systems. Fall protection equipment rigged for work positioning is considered work-positioning equipment and must meet the applicable requirements for the design and use of that equipment.

1926.954(b)(3)(iii)

The employer shall ensure that employees use fall protection systems as follows:

1926.954(b)(3)(iii)(A)

Each employee working from an aerial lift shall use a fall restraint system or a personal fall arrest system. Paragraph (b)(2)(v) of § 1926.453 does not apply.

1926.954(b)(3)(iii)(B)

Except as provided in paragraph (b)(3)(iii)(C) of this section, each employee in elevated locations more than 1.2 meters (4 feet) above the ground on poles, towers, or similar structures shall use a personal fall arrest system, work-positioning equipment, or fall restraint system, as appropriate, if the employer has not provided other fall protection meeting Subpart M of this part.

1926.954(b)(3)(iii)(C)

Until March 31, 2015, a qualified employee climbing or changing location on poles, towers, or similar structures need not use fall protection equipment, unless conditions, such as, but not limited to, ice, high winds, the design of the structure (for example, no provision for holding on with hands), or the presence of contaminants on the structure, could cause the employee to lose his or her grip or footing. On and after April 1, 2015, each qualified employee climbing or changing location on poles, towers, or similar structures unless the employer can demonstrate that climbing or changing location with fall protection is infeasible or creates a greater hazard than climbing or changing location without it.

Note 1 to paragraphs (b)(3)(iii)(B) and (b)(3)(iii)(C): These paragraphs apply to structures that support overhead electric power transmission and distribution lines and equipment. They do not apply to portions of buildings, such as loading docks, or to electric equipment, such as transformers and capacitors. Subpart M of this part contains the duty to provide fall protection associated with walking and working surfaces.

Note 2 to paragraphs (b)(3)(iii)(B) and (b)(3)(iii)(C): Until the employer ensures that employees are proficient in climbing and the use of fall protection under §1926.950(b)(7), the employees are not considered "qualified employees" for the purposes of paragraphs (b)(3)(iii)(B) and (b)(3)(iii)(C) of this section. These paragraphs require unqualified employees (including trainees) to use fall protection any time they are more than 1.2 meters (4 feet) above the ground.

1926.954(b)(3)(iv)

On and after April 1, 2015, Work-positioning systems shall be rigged so that an employee can free fall no more than 0.6 meters (2 feet).

1926.954(b)(3)(v)

Anchorages for work-positioning equipment shall be capable of supporting at least twice the potential impact load of an employee's fall, or 13.3 kilonewtons (3,000 pounds-force), whichever is greater.

Note to paragraph (b)(3)(v): Wood-pole fall-restriction devices meeting American Society of Testing and Materials *Standard Specifications for Personal Climbing Equipment*, ASTM F887-12[e1], are deemed to meet the anchorage-strength requirement when they are used in accordance with manufacturers' instructions.

1926.954(b)(3)(vi)

Unless the snaphook is a locking type and designed specifically for the following connections, snaphooks on work-positioning equipment may not be engaged:

1926.954(b)(3)(vi)(A)

Directly to webbing, rope, or wire rope;

1926.954(b)(3)(vi)(B)

To each other;

1926.954(b)(3)(vi)(C)

To a D ring to which another snaphook or other connector is attached;

1926.954(b)(3)(vi)(D)

To a horizontal lifeline; or

1926.954(b)(3)(vi)(E)

To any object that is incompatibly shaped or dimensioned in relation to the snaphook such that accidental disengagement could occur should the connected object sufficiently depress the snaphook keeper to allow release of the object.

[79 FR 20699-20700, July 10, 2014]

Reprinted from www.osha.gov

Occupational Safety & Health Administration
200 Constitution Avenue, NW
Washington, DC 20210

REGULATIONS (STANDARDS - 29 CFR)

PORTABLE LADDERS AND PLATFORMS - 1926.955
- **Part Number:** 1926
- **Part Title:** Safety and Health Regulations for Construction
- **Subpart:** V
- **Subpart Title:** Electric Power Transmission and Distribution
- **Standard Number**: 1926.955
- **Title:** Portable ladders and platforms.
- **GPO Source:** e-CFR

1926.955(a)

General. Requirements for portable ladders contained in Subpart X of this part apply in addition to the requirements of this section, except as specifically noted in paragraph (b) of this section.

1926.955(b)

Special ladders and platforms. Portable ladders used on structures or conductors in conjunction with overhead line work need not meet § 1926.1053(b)(5)(i) and (b)(12). Portable ladders and platforms used on structures or conductors in conjunction with overhead line work shall meet the following requirements:

1926.955(b)(1)

Design load. In the configurations in which they are used, portable platforms shall be capable of supporting without failure at least 2.5 times the maximum intended load.

1926.955(b)(2)

Maximum load. Portable ladders and platforms may not be loaded in excess of the working loads for which they are designed.

1926.955(b)(3)

Securing in place. Portable ladders and platforms shall be secured to prevent them from becoming dislodged.

1926.955(b)(4)

Intended use. Portable ladders and platforms may be used only in applications for which they are designed.

1926.955(c)

Conductive ladders. Portable metal ladders and other portable conductive ladders may not be used near exposed energized lines or equipment. However, in specialized high-voltage work, conductive ladders shall be used when the employer demonstrates that nonconductive ladders would present a greater hazard to employees than conductive ladders.

[79 FR 20700-20701, July 10, 2014]

Reprinted from www.osha.gov

Occupational Safety & Health Administration
200 Constitution Avenue, NW
Washington, DC 20210

REGULATIONS (STANDARDS - 29 CFR)

HAND AND PORTABLE POWER EQUIPMENT - 1926.956
- **Part Number:** 1926
- **Part Title:** Safety and Health Regulations for Construction
- **Subpart:** V
- **Subpart Title:** Electric Power Transmission and Distribution
- **Standard Number:** 1926.956
- **Title:** Hand and portable power equipment.
- **GPO Source:** e-CFR

1926.956(a)

General. Paragraph (b) of this section applies to electric equipment connected by cord and plug. Paragraph (c) of this section applies to portable and vehicle-mounted generators used to supply cord- and plug-connected equipment. Paragraph (d) of this section applies to hydraulic and pneumatic tools.

1926.956(b)

Cord- and plug-connected equipment. Cord- and plug-connected equipment not covered by Subpart K of this part shall comply with one of the following instead of § 1926.302(a)(1):

1926.956(b)(1)

The equipment shall be equipped with a cord containing an equipment grounding conductor connected to the equipment frame and to a means for grounding the other end of the conductor (however, this option may not be used where the introduction of the ground into the work environment increases the hazard to an employee); or

1926.956(b)(2)

The equipment shall be of the double-insulated type conforming to Subpart K of this part; or

1926.956(b)(3)

The equipment shall be connected to the power supply through an isolating transformer with an ungrounded secondary of not more than 50 volts.

1926.956(c)

Portable and vehicle-mounted generators. Portable and vehicle-mounted generators used to supply cord- and plug-connected equipment covered by paragraph (b) of this section shall meet the following requirements:

1926.956(c)(1)

Equipment to be supplied. The generator may only supply equipment located on the generator or the vehicle and cord- and plug-connected equipment through receptacles mounted on the generator or the vehicle.

1926.956(c)(2)

Equipment grounding. The non-current-carrying metal parts of equipment and the equipment grounding conductor terminals of the receptacles shall be bonded to the generator frame.

1926.956(c)(3)

Bonding the frame. For vehicle-mounted generators, the frame of the generator shall be bonded to the vehicle frame.

1926.956(c)(4)

Bonding the neutral conductor. Any neutral conductor shall be bonded to the generator frame.

1926.956(d)

Hydraulic and pneumatic tools.

1926.956(d)(1)

Hydraulic fluid in insulating tools. Paragraph (d)(1) of § 1926.302 does not apply to hydraulic fluid used in insulating sections of hydraulic tools.

1926.956(d)(2)

Operating pressure. Safe operating pressures for hydraulic and pneumatic tools, hoses, valves, pipes, filters, and fittings may not be exceeded.

Note to paragraph (d)(2): If any hazardous defects are present, no operating pressure is safe, and the hydraulic or pneumatic equipment involved may not be used. In the absence of defects, the maximum rated operating pressure is the maximum safe pressure.

1926.956(d)(3)

Work near energized parts. A hydraulic or pneumatic tool used where it may contact exposed energized parts shall be designed and maintained for such use.

1926.956(d)(4)

Protection against vacuum formation. The hydraulic system supplying a hydraulic tool used where it may contact exposed live parts shall provide protection against loss of insulating value, for the voltage involved, due to the formation of a partial vacuum in the hydraulic line.

Note to paragraph (d)(4): Use of hydraulic lines that do not have check valves and that have a separation of more than 10.7 meters (35 feet) between the oil reservoir and the upper end of the hydraulic system promotes the formation of a partial vacuum.

1926.956(d)(5)

Protection against the accumulation of moisture. A pneumatic tool used on energized electric lines or equipment, or used where it may contact exposed live parts, shall provide protection against the accumulation of moisture in the air supply.

1926.956(d)(6)

Breaking connections. Pressure shall be released before connections are broken, unless quick-acting, self-closing connectors are used.

1926.956(d)(7)

Leaks. Employers must ensure that employees do not use any part of their bodies to locate, or attempt to stop, a hydraulic leak.

1926.956(d)(8)

Hoses. Hoses may not be kinked.

[79 FR 20701, July 10, 2014]

Reprinted from www.osha.gov

Occupational Safety & Health Administration
200 Constitution Avenue, NW
Washington, DC 20210

REGULATIONS (STANDARDS - 29 CFR)

LIVE-LINE TOOLS - 1926.957
- **Part Number:** 1926
- **Part Title:** Safety and Health Regulations for Construction
- **Subpart:** V
- **Subpart Title:** Electric Power Transmission and Distribution
- **Standard Number:** 1926.957
- **Title:** Live-line tools.
- **GPO Source:** e-CFR

1926.957(a)

Design of tools. Live-line tool rods, tubes, and poles shall be designed and constructed to withstand the following minimum tests:

1926.957(a)(1)

Fiberglass-reinforced plastic. If the tool is made of fiberglass-reinforced plastic (FRP), it shall withstand 328,100 volts per meter (100,000 volts per foot) of length for 5 minutes, or

Note to paragraph (a)(1): Live-line tools using rod and tube that meet ASTM F711-02 (2007), *Standard Specification for Fiberglass-Reinforced Plastic (FRP) Rod and Tube Used in Live Line Tools,* are deemed to comply with paragraph (a)(1) of this section.

1926.957(a)(2)

Wood. If the tool is made of wood, it shall withstand 246,100 volts per meter (75,000 volts per foot) of length for 3 minutes, or

1926.957(a)(3)

Equivalent tests. The tool shall withstand other tests that the employer can demonstrate are equivalent.

1926.957(b)

Condition of tools.

1926.957(b)(1)

Daily inspection. Each live-line tool shall be wiped clean and visually inspected for defects before use each day.

1926.957(b)(2)

Defects. If any defect or contamination that could adversely affect the insulating qualities or mechanical integrity of the live-line tool is present after wiping, the tool shall be removed from service and examined and tested according to paragraph (b)(3) of this section before being returned to service.

1926.957(b)(3)

Biennial inspection and testing. Live-line tools used for primary employee protection shall be removed from service every 2 years, and whenever required under paragraph (b)(2) of this section, for examination, cleaning, repair, and testing as follows:

1926.957(b)(3)(i)

Each tool shall be thoroughly examined for defects.

1926.957(b)(3)(ii)

If a defect or contamination that could adversely affect the insulating qualities or mechanical integrity of the live-line tool is found, the tool shall be repaired and refinished or shall be permanently removed from service. If no such defect or contamination is found, the tool shall be cleaned and waxed.

1926.957(b)(3)(iii)

The tool shall be tested in accordance with paragraphs (b)(3)(iv) and (b)(3)(v) of this section under the following conditions:

1926.957(b)(3)(iii)(A)

After the tool has been repaired or refinished; and

1926.957(b)(3)(iii)(B)

After the examination if repair or refinishing is not performed, unless the tool is made of FRP rod or foam-filled FRP tube and the employer can demonstrate that the tool has no defects that could cause it to fail during use.

1926.957(b)(3)(iv)

The test method used shall be designed to verify the tool's integrity along its entire working length and, if the tool is made of fiberglass-reinforced plastic, its integrity under wet conditions.

1926.957(b)(3)(v)

The voltage applied during the tests shall be as follows:

1926.957(b)(3)(v)(A)

246,100 volts per meter (75,000 volts per foot) of length for 1 minute if the tool is made of fiberglass, or

1926.957(b)(3)(v)(B)

164,000 volts per meter (50,000 volts per foot) of length for 1 minute if the tool is made of wood, or

1926.957(b)(3)(v)(C)

Other tests that the employer can demonstrate are equivalent.

Note to paragraph (b): Guidelines for the examination, cleaning, repairing, and inservice testing of live-line tools are specified in the Institute of Electrical and Electronics Engineers› *IEEE Guide for Maintenance Methods on Energized Power Lines*, IEEE Std 516-2009.

[79 FR 20701-20702, July 10, 2014]

Reprinted from www.osha.gov

Occupational Safety & Health Administration
200 Constitution Avenue, NW
Washington, DC 20210

REGULATIONS (STANDARDS - 29 CFR)

MATERIALS HANDLING AND STORAGE - 1926.958
- **Part Number:** 1926
- **Part Title:** Safety and Health Regulations for Construction
- **Subpart:** V
- **Subpart Title:** Electric Power Transmission and Distribution
- **Standard Number:** 1926.958
- **Title:** Materials handling and storage.
- **GPO Source:** e-CFR

1926.958(a)

General. Materials handling and storage shall comply with applicable material-handling and material-storage requirements in this part, including those in Subparts N and CC of this part.

1926.958(b)

Materials storage near energized lines or equipment.

1926.958(b)(1)

Unrestricted areas. In areas to which access is not restricted to qualified persons only, materials or equipment may not be stored closer to energized lines or exposed energized parts of equipment than the following distances, plus a distance that provides for the maximum sag and side swing of all conductors and for the height and movement of material-handling equipment:

1926.958(b)(1)(i)

For lines and equipment energized at 50 kilovolts or less, the distance is 3.05 meters (10 feet).

1926.958(b)(1)(ii)

For lines and equipment energized at more than 50 kilovolts, the distance is 3.05 meters (10 feet) plus 0.10 meter (4 inches) for every 10 kilovolts over 50 kilovolts.

1926.958(b)(2)

Restricted areas. In areas restricted to qualified employees, materials may not be stored within the working space about energized lines or equipment.

Note to paragraph (b)(2): Paragraph (b) of § 1926.966 specifies the size of the working space.

[79 FR 20702, July 10, 2014]

Reprinted from www.osha.gov

Occupational Safety & Health Administration
200 Constitution Avenue, NW
Washington, DC 20210

REGULATIONS (STANDARDS - 29 CFR)

MECHANICAL EQUIPMENT - 1926.959
- **Part Number:** 1926
- **Part Title:** Safety and Health Regulations for Construction
- **Subpart:** V
- **Subpart Title:** Electric Power Transmission and Distribution
- **Standard Number:** 1926.959
- **Title:** Mechanical equipment.
- **Applicable Standards:** 1926.959; 1926.960
- **GPO Source:** e-CFR

1926.959(a)

General requirements.

1926.959(a)(1)

Other applicable requirements. Mechanical equipment shall be operated in accordance with applicable requirements in this part, including Subparts N, O, and CC of this part, except that § 1926.600(a)(6) does not apply to operations performed by qualified employees.

1926.959(a)(2)

Inspection before use. The critical safety components of mechanical elevating and rotating equipment shall receive a thorough visual inspection before use on each shift.

Note to paragraph (a)(2): Critical safety components of mechanical elevating and rotating equipment are components for which failure would result in free fall or free rotation of the boom.

1926.959(a)(3)

Operator. The operator of an electric line truck may not leave his or her position at the controls while a load is suspended, unless the employer can demonstrate that no employee (including the operator) is endangered.

1926.959(b)

Outriggers.

1926.959(b)(1)

Extend outriggers. Mobile equipment, if provided with outriggers, shall be operated with the outriggers extended and firmly set, except as provided in paragraph (b)(3) of this section.

1926.959(b)(2)

Clear view. Outriggers may not be extended or retracted outside of the clear view of the operator unless all employees are outside the range of possible equipment motion.

1926.959(b)(3)

Operation without outriggers. If the work area or the terrain precludes the use of outriggers, the equipment may be operated only within its maximum load ratings specified by the equipment manufacturer for the particular configuration of the equipment without outriggers.

1926.959(c)

Applied loads. Mechanical equipment used to lift or move lines or other material shall be used within its maximum load rating and other design limitations for the conditions under which the mechanical equipment is being used.

1926.959(d)

Operations near energized lines or equipment.

1926.959(d)(1)

Minimum approach distance. Mechanical equipment shall be operated so that the minimum approach distances, established by the employer under § 1926.960(c)(1)(i), are maintained from exposed energized lines and equipment. However, the insulated portion of an aerial lift operated by a qualified employee in the lift is exempt from this requirement if the applicable minimum approach distance is maintained between the uninsulated portions of the aerial lift and exposed objects having a different electrical potential.

1926.959(d)(2)

Observer. A designated employee other than the equipment operator shall observe the approach distance to exposed lines and equipment and provide timely warnings before the minimum approach distance required by paragraph (d)(1) of this section is reached, unless the employer can demonstrate that the operator can accurately determine that the minimum approach distance is being maintained.

1926.959(d)(3)

Extra precautions. If, during operation of the mechanical equipment, that equipment could become energized, the operation also shall comply with at least one of paragraphs (d)(3)(i) through (d)(3)(iii) of this section.

1926.959(d)(3)(i)

The energized lines or equipment exposed to contact shall be covered with insulating protective material that will withstand the type of contact that could be made during the operation.

1926.959(d)(3)(ii)

The mechanical equipment shall be insulated for the voltage involved. The mechanical equipment shall be positioned so that its uninsulated portions cannot approach the energized lines or equipment any closer than the minimum approach distances, established by the employer under § 1926.960(c)(1)(i).

1926.959(d)(3)(iii)

Each employee shall be protected from hazards that could arise from mechanical equipment contact with energized lines or equipment. The measures used shall ensure that employees will not be exposed to hazardous differences in electric potential. Unless the employer can demonstrate that the methods in use protect each employee from the hazards that could arise if the mechanical equipment contacts the energized line or equipment, the measures used shall include all of the following techniques:

1926.959(d)(3)(iii)(A)

Using the best available ground to minimize the time the lines or electric equipment remain energized,

1926.959(d)(3)(iii)(B)

Bonding mechanical equipment together to minimize potential differences,

1926.959(d)(3)(iii)(C)

Providing ground mats to extend areas of equipotential, and

1926.959(d)(3)(iii)(D)

Employing insulating protective equipment or barricades to guard against any remaining hazardous electrical potential differences.

Note to paragraph (d)(3)(iii): Appendix C to this subpart contains information on hazardous step and touch potentials and on methods of protecting employees from hazards resulting from such potentials.

[79 FR 20702, July 10, 2014]

Reprinted from www.osha.gov

Occupational Safety & Health Administration
200 Constitution Avenue, NW
Washington, DC 20210

REGULATIONS (STANDARDS - 29 CFR)

WORKING ON OR NEAR EXPOSED ENERGIZED PARTS - 1926.960

- **Part Number:** 1926
- **Part Title:** Safety and Health Regulations for Construction
- **Subpart:** V
- **Subpart Title:** Electric Power Transmission and Distribution
- **Standard Number:** 1926.960
- **Title:** Working on or near exposed energized parts.
- **GPO Source:** e-CFR

1926.960(a)

Application. This section applies to work on exposed live parts, or near enough to them to expose the employee to any hazard they present.

1926.960(b)

General.

1926.960(b)(1)

Qualified employees only.

1926.960(b)(1)(i)

Only qualified employees may work on or with exposed energized lines or parts of equipment.

1926.960(b)(1)(ii)

Only qualified employees may work in areas containing unguarded, uninsulated energized lines or parts of equipment operating at 50 volts or more.

1926.960(b)(2)

Treat as energized. Electric lines and equipment shall be considered and treated as energized unless they have been deenergized in accordance with § 1926.961.

1926.960(b)(3)

At least two employees.

1926.960(b)(3)(i)

Except as provided in paragraph (b)(3)(ii) of this section, at least two employees shall be present while any employees perform the following types of work:

1926.960(b)(3)(i)(A)

Installation, removal, or repair of lines energized at more than 600 volts,

1926.960(b)(3)(i)(B)

Installation, removal, or repair of deenergized lines if an employee is exposed to contact with other parts energized at more than 600 volts,

1926.960(b)(3)(i)(C)

Installation, removal, or repair of equipment, such as transformers, capacitors, and regulators, if an employee is exposed to contact with parts energized at more than 600 volts,

1926.960(b)(3)(i)(D)

Work involving the use of mechanical equipment, other than insulated aerial lifts, near parts energized at more than 600 volts, and

1926.960(b)(3)(i)(E)

Other work that exposes an employee to electrical hazards greater than, or equal to, the electrical hazards posed by operations listed specifically in paragraphs (b)(3)(i)(A) through (b)(3)(i)(D) of this section.

1926.960(b)(3)(ii)

Paragraph (b)(3)(i) of this section does not apply to the following operations:

1926.960(b)(3)(ii)(A)

Routine circuit switching, when the employer can demonstrate that conditions at the site allow safe performance of this work,

1926.960(b)(3)(ii)(B)

Work performed with live-line tools when the position of the employee is such that he or she is neither within reach of, nor otherwise exposed to contact with, energized parts, and

1926.960(b)(3)(ii)(C)

Emergency repairs to the extent necessary to safeguard the general public.

1926.960(c)

Live work.

1926.960(c)(1)

Minimum approach distances.

1926.960(c)(1)(i)

The employer shall establish minimum approach distances no less than the distances computed by Table V-2 for ac systems or Table V-7 for dc systems.

1926.960(c)(1)(ii)

No later than April 1, 2015, for voltages over 72.5 kilovolts, the employer shall determine the maximum anticipated per-unit transient overvoltage, phase-to-ground, through an engineering analysis or assume a maximum anticipated per-unit transient overvoltage, phase-to-ground, in accordance with Table V-8. When the employer uses portable protective gaps to control the maximum transient overvoltage, the value of the maximum anticipated per-unit transient overvoltage, phase-to-ground, must provide for five standard deviations between the statistical sparkover voltage of the gap and the statistical withstand voltage corresponding to the electrical component of the minimum approach distance. The employer shall make any engineering analysis conducted to determine maximum anticipated perunit transient overvoltage available upon request to employees and to the Assistant Secretary or designee for examination and copying.

Note to paragraph (c)(1)(ii): See Appendix B to this subpart for information on how to calculate the maximum anticipated per-unit transient overvoltage, phase-to-ground, when the employer uses portable protective gaps to reduce maximum transient overvoltages.

1926.960(c)(1)(iii)

The employer shall ensure that no employee approaches or takes any conductive object closer to exposed energized parts than the employer's established minimum approach distance, unless:

1926.960(c)(1)(iii)(A)

The employee is insulated from the energized part (rubber insulating gloves or rubber insulating gloves and sleeves worn in accordance with paragraph (c)(2) of this section constitutes insulation of the employee from the energized part upon which the employee is working provided that the employee has control of the part in a manner sufficient to prevent exposure to uninsulated portions of the employee's body), or

1926.960(c)(1)(iii)(B)

The energized part is insulated from the employee and from any other conductive object at a different potential, or

1926.960(c)(1)(iii)(C)

The employee is insulated from any other exposed conductive object in accordance with the requirements for live-line barehand work in § 1926.964(c).

1926.960(c)(2)

Type of insulation.

1926.960(c)(2)(i)

When an employee uses rubber insulating gloves as insulation from energized parts (under paragraph (c)(1)(iii)(A) of this section), the employer shall ensure that the employee also uses rubber insulating sleeves. However, an employee need not use rubber insulating sleeves if:

1926.960(c)(2)(i)(A)

Exposed energized parts on which the employee is not working are insulated from the employee; and

1926.960(c)(2)(i)(B)

When installing insulation for purposes of paragraph (c)(2)(i)(A) of this section, the employee installs the insulation from a position that does not expose his or her upper arm to contact with other energized parts.

1926.960(c)(2)(ii)

When an employee uses rubber insulating gloves or rubber insulating gloves and sleeves as insulation from energized parts (under paragraph (c)(1)(iii)(A) of this section), the employer shall ensure that the employee:

1926.960(c)(2)(ii)(A)

Puts on the rubber insulating gloves and sleeves in a position where he or she cannot reach into the minimum approach distance, established by the employer under paragraph (c)(1) of this section; and

1926.960(c)(2)(ii)(B)

Does not remove the rubber insulating gloves and sleeves until he or she is in a position where he or she cannot reach into the minimum approach distance, established by the employer under paragraph (c)(1) of this section.

1926.960(d)

Working position.

1926.960(d)(1)

Working from below. The employer shall ensure that each employee, to the extent that other safety-related conditions at the worksite permit, works in a position from which a slip or shock will not bring the employee's body into contact with exposed, uninsulated parts energized at a potential different from the employee's.

1926.960(d)(2)

Requirements for working without electrical protective equipment. When an employee performs work near exposed parts energized at more than 600 volts, but not more than 72.5 kilovolts, and is not wearing rubber insulating gloves, being protected by insulating equipment covering the energized parts, performing work using live-line tools, or performing live-line barehand work under § 1926.964(c), the employee shall work from a position where he or she cannot reach into the minimum approach distance, established by the employer under paragraph (c)(1) of this section.

1926.960(e)

Making connections. The employer shall ensure that employees make connections as follows:

1926.960(e)(1)

Connecting. In connecting deenergized equipment or lines to an energized circuit by means of a conducting wire or device, an employee shall first attach the wire to the deenergized part;

1926.960(e)(2)

Disconnecting. When disconnecting equipment or lines from an energized circuit by means of a conducting wire or device, an employee shall remove the source end first; and

1926.960(e)(3)

Loose conductors. When lines or equipment are connected to or disconnected from energized circuits, an employee shall keep loose conductors away from exposed energized parts.

1926.960(f)

Conductive articles. When an employee performs work within reaching distance of exposed energized parts of equipment, the employer shall ensure that the employee removes or renders nonconductive all exposed conductive articles, such as keychains or watch chains, rings, or wrist watches or bands, unless such articles do not increase the hazards associated with contact with the energized parts.

1926.960(g)

Protection from flames and electric arcs.

1926.960(g)(1)

Hazard assessment. The employer shall assess the workplace to identify employees exposed to hazards from flames or from electric arcs.

1926.960(g)(2)

Estimate of available heat energy. For each employee exposed to hazards from electric arcs, the employer shall make a reasonable estimate of the incident heat energy to which the employee would be exposed.

Note 1 to paragraph (g)(2): Appendix E to this subpart provides guidance on estimating available heat energy. The Occupational Safety and Health Administration will deem employers following the guidance in Appendix E to this subpart to be in compliance with paragraph (g)(2) of this section. An employer may choose a method of calculating incident heat energy not included in Appendix E to this subpart if the chosen method reasonably predicts the incident energy to which the employee would be exposed.

Note 2 to paragraph (g)(2): This paragraph does not require the employer to estimate the incident heat energy exposure for every job task performed by each employee. The employer may make broad estimates that cover multiple system areas provided the employer uses reasonable assumptions about the energy-exposure distribution throughout the system and provided the estimates represent the maximum employee exposure for those areas. For example, the employer could estimate the heat energy just outside a substation feeding a radial distribution system and use that estimate for all jobs performed on that radial system.

1926.960(g)(3)

Prohibited clothing. The employer shall ensure that each employee who is exposed to hazards from flames or electric arcs does not wear clothing that could melt onto his or her skin or that could ignite and continue to burn when exposed to flames or the heat energy estimated under paragraph (g)(2) of this section.

Note to paragraph (g)(3): This paragraph prohibits clothing made from acetate, nylon, polyester, rayon and polypropylene, either alone or in blends, unless the employer demonstrates that the fabric has been treated to withstand the conditions that may be encountered by the employee or that the employee wears the clothing in such a manner as to eliminate the hazard involved.

1926.960(g)(4)

Flame-resistant clothing. The employer shall ensure that the outer layer of clothing worn by an employee, except for clothing not required to be arc rated under paragraphs (g)(5)(i) through (g)(5)(v) of this section, is flame resistant under any of the following conditions:

1926.960(g)(4)(i)

The employee is exposed to contact with energized circuit parts operating at more than 600 volts,

1926.960(g)(4)(ii)

An electric arc could ignite flammable material in the work area that, in turn, could ignite the employee's clothing,

1926.960(g)(4)(iii)

Molten metal or electric arcs from faulted conductors in the work area could ignite the employee's clothing, or

Note to paragraph (g)(4)(iii): This paragraph does not apply to conductors that are capable of carrying, without failure, the maximum available fault current for the time the circuit protective devices take to interrupt the fault.

1926.960(g)(4)(iv)

The incident heat energy estimated under paragraph (g)(2) of this section exceeds 2.0 cal/cm².

1926.960(g)(5)

Arc rating. The employer shall ensure that each employee exposed to hazards from electric arcs wears protective clothing and other protective equipment with an arc rating greater than or equal to the heat energy estimated under paragraph (g)(2) of this section whenever that estimate exceeds 2.0 cal/cm^2. This protective equipment shall cover the employee's entire body, except as follows:

1926.960(g)(5)(i)

Arc-rated protection is not necessary for the employee's hands when the employee is wearing rubber insulating gloves with protectors or, if the estimated incident energy is no more than 14 cal/cm^2, heavy-duty leather work gloves with a weight of at least 407 gm/m^2 (12 oz/yd^2),

1926.960(g)(5)(ii)

Arc-rated protection is not necessary for the employee's feet when the employee is wearing heavy-duty work shoes or boots,

1926.960(g)(5)(iii)

Arc-rated protection is not necessary for the employee's head when the employee is wearing head protection meeting § 1926.100(b)(2) if the estimated incident energy is less than 9 cal/cm^2 for exposures involving single-phase arcs in open air or 5 cal/cm^2 for other exposures,

1926.960(g)(5)(iv)

The protection for the employee's head may consist of head protection meeting § 1926.100(b)(2) and a faceshield with a minimum arc rating of 8 cal/cm^2 if the estimated incidentenergy exposure is less than 13 cal/cm^2 for exposures involving single-phase arcs in open air or 9 cal/cm^2 for other exposures, and

1926.960(g)(5)(v)

For exposures involving singlephase arcs in open air, the arc rating for the employee's head and face protection may be 4 cal/cm^2 less than the estimated incident energy.

Note to paragraph (g): See Appendix E to this subpart for further information on the selection of appropriate protection.

1926.960(g)(6)

Dates.

1926.960(g)(6)(i)

The obligation in paragraph (g)(2) of this section for the employer to make reasonable estimates of incident energy commences January 1, 2015.

1926.960(g)(6)(ii)

The obligation in paragraph (g)(4)(iv) of this section for the employer to ensure that the outer layer of clothing worn by an employee is flame-resistant when the estimated incident heat energy exceeds 2.0 cal/cm^2 commences April 1, 2015.

1926.960(g)(6)(iii)

The obligation in paragraph (g)(5) of this section for the employer to ensure that each employee exposed to hazards from electric arcs wears the required arc-rated protective equipment commences April 1, 2015.

1926.960(h)

Fuse handling. When an employee must install or remove fuses with one or both terminals energized at more than 300 volts, or with exposed parts energized at more than 50 volts, the employer shall ensure that the employee uses tools or gloves rated for the voltage. When an employee installs or removes expulsion-type fuses with one or both terminals energized at more than 300 volts, the employer shall ensure that the employee wears eye protection meeting the requirements of Subpart E of this part, uses a tool rated for the voltage, and is clear of the exhaust path of the fuse barrel.

1926.960(i)

Covered (noninsulated) conductors. The requirements of this section that pertain to the hazards of exposed live parts also apply when an employee performs work in proximity to covered (noninsulated) wires.

1926.960(j)

Non-current-carrying metal parts. Non-current-carrying metal parts of equipment or devices, such as transformer cases and circuit-breaker housings, shall be treated as energized at the highest voltage to which these parts are exposed, unless the employer inspects the installation and determines that these parts are grounded before employees begin performing the work.

1926.960(k)

Opening and closing circuits under load.

1926.960(k)(1)

The employer shall ensure that devices used by employees to open circuits under load conditions are designed to interrupt the current involved.

1926.960(k)(2)

The employer shall ensure that devices used by employees to close circuits under load conditions are designed to safely carry the current involved.

[79 FR 20702-20708, July 10, 2014; 79 FR 56962, September 24, 2014]

Reprinted from www.osha.gov

Occupational Safety & Health Administration
200 Constitution Avenue, NW
Washington, DC 20210

Table V-2   Ac Live-Line Work Minimum Approach Distance

The minimum approach distance (MAD; in meters) shall comform to the following equations.

For phase-to-phase system voltages of 50 V to 300 V: [1]

MAD = avoid contact

For phase-to-phase system voltages of 301 V to 5kV: [1]

$MAD = M + D$, where

$D = 0.02$ m	the electrical component of the minimum approach distance
$M = 0.31$ m for voltages up to 750V and 0.61 m otherwise	the inadvertent movement factor

For phase-to-phase system voltages of 5.1 kV to 72.5V: [1,4]

$MAD = M + AD$, where

$M = 0.61$ m	the inadvertent movement factor
A = the applicable value from Table V-4	the altitude correction factor
D = the value from Table V-3 corresponding to the voltage and exposure or the value of the electrical component of the minimum approach distance calculated using the method provided in Appendix B to this subpart	the electrical component of the minimum approach distance

For phase-to-phase system voltages of more than 72.5 kV, nominal: [2,4]

$MAD = 0.3048(C+a)V_{L\text{-}G}TA+M$, where

$C=$	0.01 for phase-to-ground exposures that the employer can demonstrate consist only of air across the approach distance (gap), 0.01 for phase-to-phase exposures if the employer can demonstrate that no insulated tool spans the gap and the no large conductive object is in the gap, or 0.011 otherwise
$V_{L\text{-}G} =$	phase-to-ground rms voltage, in kV
$T =$	maximum anticipated per-unit transient overvoltage; for phase-to-ground exposures, T equals $T_{L\text{-}G}$, the maximum per-unit transient overvoltage, phase-to-ground, determined by the employer under paragraph (c)(1)(ii) of this section; for phase-to-phase exposures, T equals $1.35T_{L\text{-}G}+0.45$

(Continued)

Table V-2   Ac Live-Line Work Minimum Approach Distance

$A =$	altitude correction factor from Table V-4
$M =$	0.31 m, the inadvertent movement factor
$a =$	saturation factor, as follows:

Phase-to-Ground Exposure

$V_{Peak} =$ $T_{L\text{-}G}V_{L\text{-}G}\sqrt{2}$	635 kV or less	635.1 to 915 kV	915.1 to 1,050 kV	More than 1,050 kV
a	0	$(V_{Peak}\text{-}635)/$ 140,000	$(V_{Peak}\text{-}645)/$ 135,000	$(V_{Peak}\text{-}675)/125,000$

Phase-to-Ground Exposure[3]

$V_{Peak} = (1.35T_{L\text{-}G} +$ $0.45)V_{L\text{-}g}\sqrt{2}$	630 kV or less	630.1 to 848 kV	848.1 to 1,131 kV	1,131.1 to 1,485 kV More than 1,485 kV
a	0	$(V_{Peak}\text{-}630)/$ 155,000	$(V_{Peak}\text{-}633.6)/$ 152,207	$(V_{Peak}\text{-}628)/153,846$ $(V_{Peak}\text{-}350.5)/203,666$

[1]Employers may use the minimum approach distances in Table V-5. If the worksite is at an elevation of more than 900 metes (3,000 feet), see footnote 1 to Table V-5

[2]Employers may use the minimum approach distances in Table V-6, except that the employer may not use the minimum approach distances in Table V-6 for phase-to-phase exposures if an insulated tool spans the gap or if any large conductive object is in the gap. If the worksite is at an elevation of more than 900 meters (3,00 feet), see footnote 1 to Table V-6. Employers may use the minimum approach distance in Table 7 through Table 14 in Appendix B to this subpart, which calculated MAD for various values of T, provided the employer follows the notes to those tables.

[3]Use the equations for phase-to-ground exposures (with V_{Peak} for phase-to-phase exposures) unless the employer can demonstrate that no insulated tool spans the gap and that no large conductive objects is in the gap.

[4]Until March 31, 2015, employers may use the minimum approach distances in Table 6 in Appendix B to this subpart

Table V-3   Electrical Component of the Minimum Approach Distance (D; in Meters) at 5.1 to 72.5 KV

Nominal voltage (kV) phase-to-phase	Phase-to-ground exposure D (m)	Phase-to-phase exposure D (m)
5.1 to 15.0	0.04	0.07
15.1 to 36.0	0.16	0.28
36.1 to 46.0	0.23	0.37
46.1 to 72.5	0.39	0.59

Table V-4   Altitude Correction Factor

Altitude above sea level (m)	A
0 to 900	1.00
901 to 1,200	1.02
1,201 to 1,500	1.05
1,501 to 1,800	1.08
1,801 to 2,100	1.11
2,101 to 2,400	1.14
2,401 to 2,700	1.17
2,701 to 3,000	1.20
3,001 to 3,600	1.25
3,601 to 4,200	1.30
4,201 to 4,800	1.35
4,801 to 5,400	1.39
5,401 to 6,000	1.44

Table V-5   Alternative Minimum Approach Distances for Voltages of 72.5 Kv and Less[1]

Nominal voltage (kV) phase-to-phase	Distance			
	Phase-to-ground exposure		Phase-to-phase exposure	
	m	ft	m	ft
0.50 0.300[2]	Avoid contact		Avoid contact	
0.301 to 0.750[2]	0.33	1.09	0.33	1.09
0.751 to 5.0	0.63	2.07	0.63	2.07
5.1 to 15.0	0.65	2.14	0.68	2.24
15.1 to 36.0	0.77	2.53	0.89	2.92
36.1 to 46.0	0.84	2.76	0.98	3.22
46.1 to 72.5	1.00	3.29	1.20	3.94

[1] Employers may use the minimum approach distances in this table provided the worksite is at an elevation of 900 meters (3,000 feet) or less. If employees will be working at elevations greater than 900 meters (3,000 feet) above mean sea level, the employer shall determine minimum approach distances by multiplying the distances in this table by the correction factor in Table V-4 corresponding to the altitude of the work.

[2] For single-phase systems, use voltage-to-ground.

Table V-6   Alternative Minimum Approach Distances for Voltages of More than 72.5 KV[1, 2, 3]

Voltage range phase to phase (kV)	Phase-to-ground exposure		Phase-to-ground exposure	
	m	ft	m	ft
72.6 to 121.0..	1.13	3.71	1.42	4.66
121.1 to 145.0	1.30	4.27	1.64	5.38
145.1 to 169.0	1.46	4.79	1.94	6.36
169.1 to 242.0	2.01	6.59	3.08	10.10
242.1 to 362.0	3.41	11.19	5.52	18.11
362.1 to 420.0	4.25	13.94	6.81	22.34
420.1 to 550.0	5.07	16.63	8.24	27.03
550.1 to 800.0	6.88	22.57	11.38	37.34

[1] Employers may use the minimum approach distances in this table provided the worksite is at an elevation of 900 meters (3,000 feet) or less. If employees will be working at elevations greater than 900 meters (3,000 feet) above mean sea level, the employer shall determine minimum approach distances by multiplying the distances in this table by the correction factor in Table V-4 corresponding to the altitude of the work.

[2] Employers may use the phase-to-phase minimum approach distances in this table provided that no insulated tool spans the gap and no large conductive object is in the gap.

[3] The clear live-line tool distance shall equal or exceed the values for the indicated voltage ranges.

Table V-7   DC Live-Line Minimum Approach Distance (in Meters) with Overvoltage Factor[1]

Maximum anticipated per-unit transient overvoltage	distance (m) maximum line-to-ground voltage (kV)				
	250	400	500	600	750
1.5 or less	1.12	1.60	2.06	2.62	3.61
1.6	1.17	1.69	2.24	2.86	3.98
1.7	1.23	1.82	2.42	3.12	4.37
1.8	1.28	1.95	2.62	3.39	4.79

[1] The distances specified in this table are for air, bare-hand, and live-line tool conditions. If employees will be working at elevations greater than 900 meters (3,000 feet) above mean sea level, the employer shall determine minimum approach distances by multiplying the distances in this table by the correction factor in Table V-4 corresponding to the altitude of the work.

Table V-8   Assumed Maximum Per-Unit Transient Overvoltage

Voltage range (kV)	Type of current (ac or dc)	Assumed maximum per-unit transient overvoltage
72.6 to 420.0...	ac	3.5
420.1 to 550.0..	ac	3.0
550.1 to 800.0	ac	2.5
250 to 750	dc	1.8

REGULATIONS (STANDARDS - 29 CFR)

DEENERGIZING LINES AND EQUIPMENT FOR EMPLOYEE PROTECTION - 1926.961
- **Part Number:** 1926
- **Part Title:** Safety and Health Regulations for Construction
- **Subpart:** V
- **Subpart Title:** Electric Power Transmission and Distribution
- **Standard Number:** 1926.961
- **Title:** Deenergizing lines and equipment for employee protection.
- **GPO Source:** e-CFR

1926.961(a)

Application. This section applies to the deenergizing of transmission and distribution lines and equipment for the purpose of protecting employees. Conductors and parts of electric equipment that have been deenergized under procedures other than those required by this section shall be treated as energized.

1926.961(b)

General.

1926.961(b)(1)

System operator. If a system operator is in charge of the lines or equipment and their means of disconnection, the employer shall designate one employee in the crew to be in charge of the clearance and shall comply with all of the requirements of paragraph (c) of this section in the order specified.

1926.961(b)(2)

No system operator. If no system operator is in charge of the lines or equipment and their means of disconnection, the employer shall designate one employee in the crew to be in charge of the clearance and to perform the functions that the system operator would otherwise perform under this section. All of the requirements of paragraph (c) of this section apply, in the order specified, except as provided in paragraph (b)(3) of this section.

1926.961(b)(3)

Single crews working with the means of disconnection under the control of the employee in charge of the clearance. If only one crew will be working on the lines or equipment and if the means of disconnection is accessible and visible to, and under the sole control of, the employee in charge of the clearance, paragraphs (c)(1), (c)(3), and (c)(5) of this section do not apply. Additionally, the employer does not need to use the tags required by the remaining provisions of paragraph (c) of this section.

1926.961(b)(4)

Multiple crews. If two or more crews will be working on the same lines or equipment, then:

1926.961(b)(4)(i)

The crews shall coordinate their activities under this section with a single employee in charge of the clearance for all of the crews and follow the requirements of this section as if all of the employees formed a single crew, or

1926.961(b)(4)(ii)

Each crew shall independently comply with this section and, if there is no system operator in charge of the lines or equipment, shall have separate tags and coordinate deenergizing and reenergizing the lines and equipment with the other crews.

1926.961(b)(5)

Disconnecting means accessible to general public. The employer shall render any disconnecting means that are accessible to individuals outside the employer's control (for example, the general public) inoperable while the disconnecting means are open for the purpose of protecting employees.

1926.961(c)

Deenergizing lines and equipment.

1926.961(c)(1)

Request to deenergize. The employee that the employer designates pursuant to paragraph (b) of this section as being in charge of the clearance shall make a request of the system operator to deenergize the particular section of line or equipment. The designated employee becomes the employee in charge (as this term is used in paragraph (c) of this section) and is responsible for the clearance.

1926.961(c)(2)

Open disconnecting means. The employer shall ensure that all switches, disconnectors, jumpers, taps, and other means through which known sources of electric energy may be supplied to the particular lines and equipment to be deenergized are open. The employer shall render such means inoperable, unless its design does not so permit, and then ensure that such means are tagged to indicate that employees are at work.

1926.961(c)(3)

Automatically and remotely controlled switches. The employer shall ensure that automatically and remotely controlled switches that could cause the opened disconnecting means to close are also tagged at the points of control. The employer shall render the automatic or remote control feature inoperable, unless its design does not so permit.

1926.961(c)(4)

Network protectors. The employer need not use the tags mentioned in paragraphs (c)(2) and (c)(3) of this section on a network protector for work on the primary feeder for the network protector's associated network transformer when the employer can demonstrate all of the following conditions:

1926.961(c)(4)(i)

Every network protector is maintained so that it will immediately trip open if closed when a primary conductor is deenergized;

1926.961(c)(4)(ii)

Employees cannot manually place any network protector in a closed position without the use of tools, and any manual override position is blocked, locked, or otherwise disabled; and

1926.961(c)(4)(iii)

The employer has procedures for manually overriding any network protector that incorporate provisions for determining, before anyone places a network protector in a closed position, that: The line connected to the network protector is not deenergized for the protection of any employee working on the line; and (if the line connected to the network protector is not deenergized for the protection of any employee working on the line) the primary conductors for the network protector are energized.

1926.961(c)(5)

Tags. Tags shall prohibit operation of the disconnecting means and shall indicate that employees are at work.

1926.961(c)(6)

Test for energized condition. After the applicable requirements in paragraphs (c)(1) through (c)(5) of this section have been followed and the system operator gives a clearance to the employee in charge, the employer shall ensure that the lines and equipment are deenergized by testing the lines and equipment to be worked with a device designed to detect voltage.

1926.961(c)(7)

Install grounds. The employer shall ensure the installation of protective grounds as required by § 1926.962.

1926.961(c)(8)

Consider lines and equipment deenergized. After the applicable requirements of paragraphs (c)(1) through (c)(7) of this section have been followed, the lines and equipment involved may be considered deenergized.

1926.961(c)(9)

Transferring clearances. To transfer the clearance, the employee in charge (or the employee's supervisor if the employee in charge must leave the worksite due to illness or other emergency) shall inform the system operator and employees in the crew; and the new employee in charge shall be responsible for the clearance.

1926.961(c)(10)

Releasing clearances. To release a clearance, the employee in charge shall:

1926.961(c)(10)(i)

Notify each employee under that clearance of the pending release of the clearance;

1926.961(c)(10)(ii)

Ensure that all employees under that clearance are clear of the lines and equipment;

1926.961(c)(10)(iii)

Ensure that all protective grounds protecting employees under that clearance have been removed; and

1926.961(c)(10)(iv)

Report this information to the system operator and then release the clearance.

1926.961(c)(11)

Person releasing clearance. Only the employee in charge who requested the clearance may release the clearance, unless the employer transfers responsibility under paragraph (c)(9) of this section.

1926.961(c)(12)

Removal of tags. No one may remove tags without the release of the associated clearance as specified under paragraphs (c)(10) and (c)(11) of this section.

1926.961(c)(13)

Reenergizing lines and equipment. The employer shall ensure that no one initiates action to reenergize the lines or equipment at a point of disconnection until all protective grounds have been removed, all crews working on the lines or equipment release their clearances, all employees are clear of the lines and equipment, and all protective tags are removed from that point of disconnection.

[79 FR 20708-20709, July 10, 2014]

Reprinted from www.osha.gov

Occupational Safety & Health Administration
200 Constitution Avenue, NW
Washington, DC 20210

REGULATIONS (STANDARDS - 29 CFR)

GROUNDING FOR THE PROTECTION OF EMPLOYEES - 1926.962
- **Part Number:** 1926
- **Part Title:** Safety and Health Regulations for Construction
- **Subpart:** V
- **Subpart Title:** Electric Power Transmission and Distribution
- **Standard Number:** 1926.962
- **Title:** Grounding for the protection of employees.
- **GPO Source:** e-CFR

1926.962(a)

Application. This section applies to grounding of transmission and distribution lines and equipment for the purpose of protecting employees. Paragraph (d) of this section also applies to protective grounding of other equipment as required elsewhere in this Subpart.

Note to paragraph (a): This section covers grounding of transmission and distribution lines and equipment when this subpart requires protective grounding and whenever the employer chooses to ground such lines and equipment for the protection of employees.

1926.962(b)

General. For any employee to work transmission and distribution lines or equipment as deenergized, the employer shall ensure that the lines or equipment are deenergized under the provisions of § 1926.961 and shall ensure proper grounding of the lines or equipment as specified in paragraphs (c) through (h) of this section. However, if the employer can demonstrate that installation of a ground is impracticable or that the conditions resulting from the installation of a ground would present greater hazards to employees than working without grounds, the lines and equipment may be treated as deenergized provided that the employer establishes that all of the following conditions apply:

1926.962(b)(1)

Deenergized. The employer ensures that the lines and equipment are deenergized under the provisions of § 1926.961.

1926.962(b)(2)

No possibility of contact. There is no possibility of contact with another energized source.

1926.962(b)(3)

No induced voltage. The hazard of induced voltage is not present.

1926.962(c)

Equipotential zone. Temporary protective grounds shall be placed at such locations and arranged in such a manner that the employer can demonstrate will prevent each employee from being exposed to hazardous differences in electric potential.

Note to paragraph (c): Appendix C to this subpart contains guidelines for establishing the equipotential zone required by this paragraph. The Occupational Safety and Health Administration will deem grounding practices meeting these guidelines as complying with paragraph (c) of this section.

1926.962(d)

Protective grounding equipment.

1926.962(d)(1)

Ampacity.

1926.962(d)(1)(i)

Protective grounding equipment shall be capable of conducting the maximum fault current that could flow at the point of grounding for the time necessary to clear the fault.

1926.962(d)(1)(ii)

Protective grounding equipment shall have an ampacity greater than or equal to that of No. 2 AWG copper.

1926.962(d)(2)

Impedance. Protective grounds shall have an impedance low enough so that they do not delay the operation of protective devices in case of accidental energizing of the lines or equipment.

Note to paragraph (d): American Society for Testing and Materials *Standard Specifications for Temporary Protective Grounds to Be Used on De-Energized Electric Power Lines and Equipment*, ASTM F855-09, contains guidelines for protective grounding equipment. The Institute of Electrical Engineers *Guide for Protective Grounding of Power Lines*, IEEE Std 1048-2003, contains guidelines for selecting and installing protective grounding equipment.

1926.962(e)

Testing. The employer shall ensure that, unless a previously installed ground is present, employees test lines and equipment and verify the absence of nominal voltage before employees install any ground on those lines or that equipment.

1926.962(f)

Connecting and removing grounds.

1926.962(f)(1)

Order of connection. The employer shall ensure that, when an employee attaches a ground to a line or to equipment, the employee attaches the ground-end connection first and then attaches the other end by means of a live-line tool. For lines or equipment operating at 600 volts or less, the employer may permit the employee to use insulating equipment other than a live-line tool if the employer ensures that the line or equipment is not energized at the time the ground is connected or if the employer can demonstrate that each employee is protected from hazards that may develop if the line or equipment is energized.

1926.962(f)(2)

Order of removal. The employer shall ensure that, when an employee removes a ground, the employee removes the grounding device from the line or equipment using a live-line tool before he or she removes the groundend connection. For lines or equipment operating at 600 volts or less, the employer may permit the employee to use insulating equipment other than a live-line tool if the employer ensures that the line or equipment is not energized at the time the ground is disconnected or if the employer can demonstrate that each employee is protected from hazards that may develop if the line or equipment is energized.

1926.962(g)

Additional precautions. The employer shall ensure that, when an employee performs work on a cable at a location remote from the cable terminal, the cable is not grounded at the cable terminal if there is a possibility of hazardous transfer of potential should a fault occur.

1926.962(h)

Removal of grounds for test. The employer may permit employees to remove grounds temporarily during tests. During the test procedure, the employer shall ensure that each employee uses insulating equipment, shall isolate each employee from any hazards involved, and shall implement any additional measures necessary to protect each exposed employee in case the previously grounded lines and equipment become energized.

[79 FR 20709-20710, July 10, 2014]

Reprinted from www.osha.gov

Occupational Safety & Health Administration
200 Constitution Avenue, NW
Washington, DC 20210

REGULATIONS (STANDARDS - 29 CFR)

TESTING AND TEST FACILITIES - 1926.963
- **Part Number:** 1926
- **Part Title:** Safety and Health Regulations for Construction
- **Subpart:** V
- **Subpart Title:** Electric Power Transmission and Distribution
- **Standard Number:** 1926.963
- **Title:** Testing and test facilities.
- **GPO Source:** e-CFR

1926.963(a)

Application. This section provides for safe work practices for high-voltage and high-power testing performed in laboratories, shops, and substations, and in the field and on electric transmission and distribution lines and equipment. It applies only to testing involving interim measurements using high voltage, high power, or combinations of high voltage and high power, and not to testing involving continuous measurements as in routine metering, relaying, and normal line work.

Note to paragraph (a): OSHA considers routine inspection and maintenance measurements made by qualified employees to be routine line work not included in the scope of this section, provided that the hazards related to the use of intrinsic highvoltage or high-power sources require only the normal precautions associated with routine work specified in the other paragraphs of this subpart. Two typical examples of such excluded test work procedures are "phasing-out" testing and testing for a "no-voltage" condition.

1926.963(b)

General requirements.

1926.963(b)(1)

Safe work practices. The employer shall establish and enforce work practices for the protection of each worker from the hazards of high-voltage or high-power testing at all test areas, temporary and permanent. Such work practices shall include, as a minimum, test area safeguarding, grounding, the safe use of measuring and control circuits, and a means providing for periodic safety checks of field test areas.

1926.963(b)(2)

Training. The employer shall ensure that each employee, upon initial assignment to the test area, receives training in safe work practices, with retraining provided as required by § 1926.950(b).

1926.963(c)

Safeguarding of test areas.

1926.963(c)(1)

Safeguarding. The employer shall provide safeguarding within test areas to control access to test equipment or to apparatus under test that could become energized as part of the testing by either direct or inductive coupling and to prevent accidental employee contact with energized parts.

1926.963(c)(2)

Permanent test areas. The employer shall guard permanent test areas with walls, fences, or other barriers designed to keep employees out of the test areas.

1926.963(c)(3)

Temporary test areas. In field testing, or at a temporary test site not guarded by permanent fences and gates, the employer shall ensure the use of one of the following means to prevent employees without authorization from entering:

1926.963(c)(3)(i)

Distinctively colored safety tape supported approximately waist high with safety signs attached to it,

1926.963(c)(3)(ii)

A barrier or barricade that limits access to the test area to a degree equivalent, physically and visually, to the barricade specified in paragraph (c)(3)(i) of this section, or

1926.963(c)(3)(iii)

One or more test observers stationed so that they can monitor the entire area.

1926.963(c)(4)

Removal of safeguards. The employer shall ensure the removal of the safeguards required by paragraph (c)(3) of this section when employees no longer need the protection afforded by the safeguards.

1926.963(d)

Grounding practices.

1926.963(d)(1)

Establish and implement practices. The employer shall establish and implement safe grounding practices for the test facility.

1926.963(d)(1)(i)

The employer shall maintain at ground potential all conductive parts accessible to the test operator while the equipment is operating at high voltage.

1926.963(d)(1)(ii)

Wherever ungrounded terminals of test equipment or apparatus under test may be present, they shall be treated as energized until tests demonstrate that they are deenergized.

1926.963(d)(2)

Installation of grounds. The employer shall ensure either that visible grounds are applied automatically, or that employees using properly insulated tools manually apply visible grounds, to the high-voltage circuits after they are deenergized and before any employee performs work on the circuit or on the item or apparatus under test. Common ground connections shall be solidly connected to the test equipment and the apparatus under test.

1926.963(d)(3)

Isolated ground return. In highpower testing, the employer shall provide an isolated ground-return conductor system designed to prevent the intentional passage of current, with its attendant voltage rise, from occurring in the ground grid or in the earth. However, the employer need not provide an isolated ground-return conductor if the employer can demonstrate that both of the following conditions exist:

1926.963(d)(3)(i)

The employer cannot provide an isolated ground-return conductor due to the distance of the test site from the electric energy source, and

1926.963(d)(3)(ii)

The employer protects employees from any hazardous step and touch potentials that may develop during the test.

Note to paragraph (d)(3)(ii): See Appendix C to this subpart for information on measures that employers can take to protect employees from hazardous step and touch potentials.

1926.963(d)(4)

Equipment grounding conductors. For tests in which using the equipment grounding conductor in the equipment power cord to ground the test equipment would result in greater hazards to test personnel or prevent the taking of satisfactory measurements, the employer may use a ground clearly indicated in the test set-up if the employer can demonstrate that this ground affords protection for employees equivalent to the protection afforded by an equipment grounding conductor in the power supply cord.

1926.963(d)(5)

Grounding after tests. The employer shall ensure that, when any employee enters the test area after equipment is deenergized, a ground is placed on the high-voltage terminal and any other exposed terminals.

1926.963(d)(5)(i)

Before any employee applies a direct ground, the employer shall discharge high capacitance equipment or apparatus through a resistor rated for the available energy.

1926.963(d)(5)(ii)

A direct ground shall be applied to the exposed terminals after the stored energy drops to a level at which it is safe to do so.

1926.963(d)(6)

Grounding test vehicles. If the employer uses a test trailer or test vehicle in field testing, its chassis shall be grounded. The employer shall protect each employee against hazardous touch potentials with respect to the vehicle, instrument panels, and other conductive parts accessible to employees with bonding, insulation, or isolation.

1926.963(e)

Control and measuring circuits.

1926.963(e)(1)

Control wiring. The employer may not run control wiring, meter connections, test leads, or cables from a test area unless contained in a grounded metallic sheath and terminated in a grounded metallic enclosure or unless the employer takes other precautions that it can demonstrate will provide employees with equivalent safety.

1926.963(e)(2)

Instruments. The employer shall isolate meters and other instruments with accessible terminals or parts from test personnel to protect against hazards that could arise should such terminals and parts become energized during testing. If the employer provides this isolation by locating test equipment in metal compartments with viewing windows, the employer shall provide interlocks to interrupt the power supply when someone opens the compartment cover.

1926.963(e)(3)

Routing temporary wiring. The employer shall protect temporary wiring and its connections against damage, accidental interruptions, and other hazards. To the maximum extent possible, the employer shall keep signal, control, ground, and power cables separate from each other.

1926.963(e)(4)

Test observer. If any employee will be present in the test area during testing, a test observer shall be present. The test observer shall be capable of implementing the immediate deenergizing of test circuits for safety purposes.

1926.963(f)

Safety check.

1926.963(f)(1)

Before each test. Safety practices governing employee work at temporary or field test areas shall provide, at the beginning of each series of tests, for a routine safety check of such test areas.

1926.963(f)(2)

Conditions to be checked. The test operator in charge shall conduct these routine safety checks before each series of tests and shall verify at least the following conditions:

1926.963(f)(2)(i)

Barriers and safeguards are in workable condition and placed properly to isolate hazardous areas;

1926.963(f)(2)(ii)

System test status signals, if used, are in operable condition;

1926.963(f)(2)(iii)

Clearly marked test-power disconnects are readily available in an emergency;

1926.963(f)(2)(iv)

Ground connections are clearly identifiable;

1926.963(f)(2)(v)

Personal protective equipment is provided and used as required by Subpart E of this part and by this subpart; and

1926.963(f)(2)(vi)

Proper separation between signal, ground, and power cables.

[79 FR 20710-20711, July 10, 2014]

Reprinted from www.osha.gov

Occupational Safety & Health Administration
200 Constitution Avenue, NW
Washington, DC 20210

REGULATIONS (STANDARDS - 29 CFR)

OVERHEAD LINES AND LIVE-LINE BAREHAND WORK - 1926.964
- **Part Number:** 1926
- **Part Title:** Safety and Health Regulations for Construction
- **Subpart:** V
- **Subpart Title:** Electric Power Transmission and Distribution
- **Standard Number:** 1926.964
- **Title:** Overhead lines and live-line barehand work.
- **GPO Source:** e-CFR

1926.964(a)

General.

1926.964(a)(1)

Application. This section provides additional requirements for work performed on or near overhead lines and equipment and for live-line barehand work.

1926.964(a)(2)

Checking structure before climbing. Before allowing employees to subject elevated structures, such as poles or towers, to such stresses as climbing or the installation or removal of equipment may impose, the employer shall ascertain that the structures are capable of sustaining the additional or unbalanced stresses. If the pole or other structure cannot withstand the expected loads, the employer shall brace or otherwise support the pole or structure so as to prevent failure.

Note to paragraph (a)(2): Appendix D to this subpart contains test methods that employers can use in ascertaining whether a wood pole is capable of sustaining the forces imposed by an employee climbing the pole. This paragraph also requires the employer to ascertain that the pole can sustain all other forces imposed by the work employees will perform.

1926.964(a)(3)

Setting and moving poles.

1926.964(a)(3)(i)

When a pole is set, moved, or removed near an exposed energized overhead conductor, the pole may not contact the conductor.

1926.964(a)(3)(ii)

When a pole is set, moved, or removed near an exposed energized overhead conductor, the employer shall ensure that each employee wears electrical protective equipment or uses insulated devices when handling the pole and that no employee contacts the pole with uninsulated parts of his or her body.

1926.964(a)(3)(iii)

To protect employees from falling into holes used for placing poles, the employer shall physically guard the holes, or ensure that employees attend the holes, whenever anyone is working nearby.

1926.964(b)

Installing and removing overhead lines. The following provisions apply to the installation and removal of overhead conductors or cable (overhead lines).

1926.964(b)(1)

Tension stringing method. When lines that employees are installing or removing can contact energized parts, the employer shall use the tensionstringing method, barriers, or other equivalent measures to minimize the possibility that conductors and cables the employees are installing or removing will contact energized power lines or equipment.

1926.964(b)(2)

Conductors, cables, and pulling and tensioning equipment. For conductors, cables, and pulling and tensioning equipment, the employer shall provide the protective measures required by § 1926.959(d)(3) when employees are installing or removing a conductor or cable close enough to energized conductors that any of the following failures could energize the pulling or tensioning equipment or the conductor or cable being installed or removed:

1926.964(b)(2)(i)

Failure of the pulling or tensioning equipment,

1926.964(b)(2)(ii)

Failure of the conductor or cable being pulled, or

1926.964(b)(2)(iii)

Failure of the previously installed lines or equipment.

1926.964(b)(3)

Disable automatic-reclosing feature. If the conductors that employees are installing or removing cross over energized conductors in excess of 600 volts and if the design of the circuit-interrupting devices protecting the lines so permits, the employer shall render inoperable the automatic-reclosing feature of these devices.

1926.964(b)(4)

Induced voltage.

1926.964(b)(4)(i)

Before employees install lines parallel to existing energized lines, the employer shall make a determination of the approximate voltage to be induced in the new lines, or work shall proceed on the assumption that the induced voltage is hazardous.

1926.964(b)(4)(ii)

Unless the employer can demonstrate that the lines that employees are installing are not subject to the induction of a hazardous voltage or unless the lines are treated as energized, temporary protective grounds shall be placed at such locations and arranged in such a manner that the employer can demonstrate will prevent exposure of each employee to hazardous differences in electric potential.

Note to paragraph (b)(4)(ii): Appendix C to this subpart contains guidelines for protecting employees from hazardous differences in electric potential as required by this paragraph.

Note to paragraph (b)(4): If the employer takes no precautions to protect employees from hazards associated with involuntary reactions from electric shock, a hazard exists if the induced voltage is sufficient to pass a current of 1 milliampere through a 500-ohm resistor. If the employer protects employees from injury due to involuntary reactions from electric shock, a hazard exists if the resultant current would be more than 6 milliamperes.

1926.964(b)(5)

Safe operating condition. Reelhandling equipment, including pulling and tensioning devices, shall be in safe operating condition and shall be leveled and aligned.

1926.964(b)(6)

Load ratings. The employer shall ensure that employees do not exceed load ratings of stringing lines, pulling lines, conductor grips, load-bearing hardware and accessories, rigging, and hoists.

1926.964(b)(7)

Defective pulling lines. The employer shall repair or replace defective pulling lines and accessories.

1926.964(b)(8)

Conductor grips. The employer shall ensure that employees do not use conductor grips on wire rope unless the manufacturer specifically designed the grip for this application.

1926.964(b)(9)

Communications. The employer shall ensure that employees maintain reliable communications, through twoway radios or other equivalent means, between the reel tender and the pullingrig operator.

1926.964(b)(10)

Operation of pulling rig. Employees may operate the pulling rig only when it is safe to do so.

Note to paragraph (b)(10): Examples of unsafe conditions include: employees in locations prohibited by paragraph (b)(11) of this section, conductor and pulling line hangups, and slipping of the conductor grip.

1926.964(b)(11)

Working under overhead operations. While a power-driven device is pulling the conductor or pulling line and the conductor or pulling line is in motion, the employer shall ensure that employees are not directly under overhead operations or on the crossarm, except as necessary for the employees to guide the stringing sock or board over or through the stringing sheave.

1926.964(c)

Live-line barehand work. In addition to other applicable provisions contained in this subpart, the following requirements apply to live-line barehand work:

1926.964(c)(1)

Training. Before an employee uses or supervises the use of the live-line barehand technique on energized circuits, the employer shall ensure that the employee completes training conforming to § 1926.950(b) in the technique and in the safety requirements of paragraph (c) of this section.

1926.964(c)(2)

Existing conditions. Before any employee uses the live-line barehand technique on energized high-voltage conductors or parts, the employer shall ascertain the following information in addition to information about other existing conditions required by § 1926.950(d):

1926.964(c)(2)(i)

The nominal voltage rating of the circuit on which employees will perform the work,

1926.964(c)(2)(ii)

The clearances to ground of lines and other energized parts on which employees will perform the work, and

1926.964(c)(2)(iii)

The voltage limitations of equipment employees will use.

1926.964(c)(3)

Insulated tools and equipment.

1926.964(c)(3)(i)

The employer shall ensure that the insulated equipment, insulated tools, and aerial devices and platforms used by employees are designed, tested, and made for live-line barehand work.

1926.964(c)(3)(ii)

The employer shall ensure that employees keep tools and equipment clean and dry while they are in use.

1926.964(c)(4)

Disable automatic-reclosing feature. The employer shall render inoperable the automatic-reclosing feature of circuit-interrupting devices protecting the lines if the design of the devices permits.

1926.964(c)(5)

Adverse weather conditions. The employer shall ensure that employees do not perform work when adverse weather conditions would make the work hazardous even after the employer implements the work practices required by this subpart. Additionally, employees may not perform work when winds reduce the phase-to-phase or phase-to-ground clearances at the work location below the minimum approach distances specified in paragraph (c)(13) of this section, unless insulating guards cover the grounded objects and other lines and equipment.

Note to paragraph (c)(5): Thunderstorms in the vicinity, high winds, snow storms, and ice storms are examples of adverse weather conditions that make live-line barehand work too hazardous to perform safely even after the employer implements the work practices required by this subpart.

1926.964(c)(6)

Bucket liners and electrostatic shielding. The employer shall provide and ensure that employees use a conductive bucket liner or other conductive device for bonding the insulated aerial device to the energized line or equipment.

1926.964(c)(6)(i)

The employee shall be connected to the bucket liner or other conductive device by the use of conductive shoes, leg clips, or other means.

1926.964(c)(6)(ii)

Where differences in potentials at the worksite pose a hazard to employees, the employer shall provide electrostatic shielding designed for the voltage being worked.

1926.964(c)(7)

Bonding the employee to the energized part. The employer shall ensure that, before the employee contacts the energized part, the employee bonds the conductive bucket liner or other conductive device to the energized conductor by means of a positive connection. This connection shall remain attached to the energized conductor until the employee completes the work on the energized circuit.

1926.964(c)(8)

Aerial-lift controls. Aerial lifts used for live-line barehand work shall have dual controls (lower and upper) as follows:

1926.964(c)(8)(i)

The upper controls shall be within easy reach of the employee in the bucket. On a two-bucket-type lift, access to the controls shall be within easy reach of both buckets.

1926.964(c)(8)(ii)

The lower set of controls shall be near the base of the boom and shall be designed so that they can override operation of the equipment at any time.

1926.964(c)(9)

Operation of lower controls. Lower (ground-level) lift controls may not be operated with an employee in the lift except in case of emergency.

1926.964(c)(10)

Check controls. The employer shall ensure that, before employees elevate an aerial lift into the work position, the employees check all controls (ground level and bucket) to determine that they are in proper working condition.

1926.964(c)(11)

Body of aerial lift truck. The employer shall ensure that, before employees elevate the boom of an aerial lift, the employees ground the body of the truck or barricade the body of the truck and treat it as energized.

1926.964(c)(12)

Boom-current test. The employer shall ensure that employees perform a boom-current test before starting work each day, each time during the day when they encounter a higher voltage, and when changed conditions indicate a need for an additional test.

1926.964(c)(12)(i)

This test shall consist of placing the bucket in contact with an energized source equal to the voltage to be encountered for a minimum of 3 minutes.

1926.964(c)(12)(ii)

The leakage current may not exceed 1 microampere per kilovolt of nominal phase-to-ground voltage.

1926.964(c)(12)(iii)

The employer shall immediately suspend work from the aerial lift when there is any indication of a malfunction in the equipment.

1926.964(c)(13)

Minimum approach distance. The employer shall ensure that employees maintain the minimum approach distances, established by the employer under § 1926.960(c)(1)(i), from all grounded objects and from lines and equipment at a potential different from that to which the live-line barehand equipment is bonded, unless insulating guards cover such grounded objects and other lines and equipment.

1926.964(c)(14)

Approaching, leaving, and bonding to energized part. The employer shall ensure that, while an employee is approaching, leaving, or bonding to an energized circuit, the employee maintains the minimum approach distances, established by the employer under § 1926.960(c)(1)(i), between the employee and any grounded parts, including the lower boom and portions of the truck and between the employee and conductive objects energized at different potentials.

1926.964(c)(15)

Positioning bucket near energized bushing or insulator string. While the bucket is alongside an energized bushing or insulator string, the employer shall ensure that employees maintain the phase-to-ground minimum approach distances, established by the employer under § 1926.960(c)(1)(i), between all parts of the bucket and the grounded end of the bushing or insulator string or any other grounded surface.

1926.964(c)(16)

Handlines. The employer shall ensure that employees do not use handlines between the bucket and the boom or between the bucket and the ground. However, employees may use nonconductive-type handlines from conductor to ground if not supported from the bucket. The employer shall ensure that no one uses ropes used for live-line barehand work for other purposes.

1926.964(c)(17)

Passing objects to employee. The employer shall ensure that employees do not pass uninsulated equipment or material between a pole or structure and an aerial lift while an employee working from the bucket is bonded to an energized part.

1926.964(c)(18)

Nonconductive measuring device. A nonconductive measuring device shall be readily accessible to employees performing live-line barehand work to assist them in maintaining the required minimum approach distance.

1926.964(d)

Towers and structures. The following requirements apply to work performed on towers or other structures that support overhead lines.

1926.964(d)(1)

Working beneath towers and structures. The employer shall ensure that no employee is under a tower or structure while work is in progress, except when the employer can demonstrate that such a working position is necessary to assist employees working above.

1926.964(d)(2)

Tag lines. The employer shall ensure that employees use tag lines or other similar devices to maintain control of tower sections being raised or positioned, unless the employer can demonstrate that the use of such devices would create a greater hazard to employees.

1926.964(d)(3)

Disconnecting load lines. The employer shall ensure that employees do not detach the loadline from a member or section until they safely secure the load.

1926.964(d)(4)

Adverse weather conditions. The employer shall ensure that, except during emergency restoration procedures, employees discontinue work when adverse weather conditions would make the work hazardous in spite of the work practices required by this subpart.

Note to paragraph (d)(4): Thunderstorms in the vicinity, high winds, snow storms, and ice storms are examples of adverse weather conditions that make this work too hazardous to perform even after the employer implements the work practices required by this subpart.

[79 FR 20711-20713, July 10, 2014]

Reprinted from www.osha.gov

Occupational Safety & Health Administration
200 Constitution Avenue, NW
Washington, DC 20210

REGULATIONS (STANDARDS - 29 CFR)

UNDERGROUND ELECTRICAL INSTALLATIONS - 1926.965
- **Part Number:** 1926
- **Part Title:** Safety and Health Regulations for Construction
- **Subpart:** V
- **Subpart Title:** Electric Power Transmission and Distribution
- **Standard Number:** 1926.965
- **Title:** Underground electrical installations.
- **GPO Source:** e-CFR

1926.965(a)

Application. This section provides additional requirements for work on underground electrical installations.

1926.965(b)

Access. The employer shall ensure that employees use a ladder or other climbing device to enter and exit a manhole or subsurface vault exceeding 1.22 meters (4 feet) in depth. No employee may climb into or out of a manhole or vault by stepping on cables or hangers.

1926.965(c)

Lowering equipment into manholes.

1926.965(c)(1)

Hoisting equipment. Equipment used to lower materials and tools into manholes or vaults shall be capable of supporting the weight to be lowered and shall be checked for defects before use.

1926.965(c)(2)

Clear the area of employees. Before anyone lowers tools or material into the opening for a manhole or vault, each employee working in the manhole or vault shall be clear of the area directly under the opening.

1926.965(d)

Attendants for manholes and vaults.

1926.965(d)(1)

When required. While work is being performed in a manhole or vault containing energized electric equipment, an employee with first-aid training shall be available on the surface in the immediate vicinity of the manhole or vault entrance to render emergency assistance.

1926.965(d)(2)

Brief entries allowed. Occasionally, the employee on the surface may briefly enter a manhole or vault to provide nonemergency assistance.

Note 1 to paragraph (d)(2): Paragraph (h) of 1926.953 may also require an attendant and does not permit this attendant to enter the manhole or vault.

Note 2 to paragraph (d)(2): Paragraph (b)(1)(ii) of § 1926.960 requires employees entering manholes or vaults containing unguarded, uninsulated energized lines or parts of electric equipment operating at 50 volts or more to be qualified.

1926.965(d)(3)

Entry without attendant. For the purpose of inspection, housekeeping, taking readings, or similar work, an employee working alone may enter, for brief periods of time, a manhole or vault where energized cables or equipment are in service if the employer can demonstrate that the employee will be protected from all electrical hazards.

1926.965(d)(4)

Communications. The employer shall ensure that employees maintain reliable communications, through twoway radios or other equivalent means, among all employees involved in the job.

1926.965(e)

Duct rods. The employer shall ensure that, if employees use duct rods, the employees install the duct rods in the direction presenting the least hazard to employees. The employer shall station an employee at the far end of the duct line being rodded to ensure that the employees maintain the required minimum approach distances.

1926.965(f)

Multiple cables. When multiple cables are present in a work area, the employer shall identify the cable to be worked by electrical means, unless its identity is obvious by reason of distinctive appearance or location or by other readily apparent means of identification. The employer shall protect cables other than the one being worked from damage.

1926.965(g)

Moving cables. Except when paragraph (h)(2) of this section permits employees to perform work that could cause a fault in an energized cable in a manhole or vault, the employer shall ensure that employees inspect energized cables to be moved for abnormalities.

1926.965(h)

Protection against faults.

1926.965(h)(1)

Cables with abnormalities. Where a cable in a manhole or vault has one or more abnormalities that could lead to a fault or be an indication of an impending fault, the employer shall deenergize the cable with the abnormality before any employee may work in the manhole or vault, except when service-load conditions and a lack of feasible alternatives require that the cable remain energized. In that case, employees may enter the manhole or vault provided the employer protects them from the possible effects of a failure using shields or other devices that are capable of containing the adverse effects of a fault. The employer shall treat the following abnormalities as indications of impending faults unless the employer can demonstrate that the conditions could not lead to a fault: Oil or compound leaking from cable or joints, broken cable sheaths or joint sleeves, hot localized surface temperatures of cables or joints, or joints swollen beyond normal tolerance.

1926.965(h)(2)

Work-related faults. If the work employees will perform in a manhole or vault could cause a fault in a cable, the employer shall deenergize that cable before any employee works in the manhole or vault, except when serviceload conditions and a lack of feasible alternatives require that the cable remain energized. In that case, employees may enter the manhole or vault provided the employer protects them from the possible effects of a failure using shields or other devices that are capable of containing the adverse effects of a fault.

1926.965(h)(2)(i)

Sheath continuity. When employees perform work on buried cable or on cable in a manhole or vault, the employer shall maintain metallic-sheath continuity, or the cable sheath shall be treated as energized.

[79 FR 20713, July 10, 2014]

Reprinted from www.osha.gov

Occupational Safety & Health Administration
200 Constitution Avenue, NW
Washington, DC 20210

REGULATIONS (STANDARDS - 29 CFR)

SUBSTATIONS - 1926.966
- **Part Number:** 1926
- **Part Title:** Safety and Health Regulations for Construction
- **Subpart:** V
- **Subpart Title:** Electric Power Transmission and Distribution
- **Standard Number:** 1926.966
- **Title:** Substations.
- **GPO Source:** e-CFR

1926.966(a)

Application. This section provides additional requirements for substations and for work performed in them.

1926.966(b)

Access and working space. The employer shall provide and maintain sufficient access and working space about electric equipment to permit ready and safe operation and maintenance of such equipment by employees.

Note to paragraph (b): American National Standard *National Electrical Safety Code,* ANSI/IEEE C2-2012 contains guidelines for the dimensions of access and working space about electric equipment in substations. Installations meeting the ANSI provisions comply with paragraph (b) of this section. The Occupational Safety and Health Administration will determine whether an installation that does not conform to this ANSI standard complies with paragraph (b) of this section based on the following criteria:

(1) Whether the installation conforms to the edition of ANSI C2 that was in effect when the installation was made;
(2) Whether the configuration of the installation enables employees to maintain the minimum approach distances, established by the employer under § 1926.960(c)(1)(i), while the employees are working on exposed, energized parts; and
(3) Whether the precautions taken when employees perform work on the installation provide protection equivalent to the protection provided by access and working space meeting ANSI/IEEE C2-2012.

1926.966(c)

Draw-out-type circuit breakers. The employer shall ensure that, when employees remove or insert draw-outtype circuit breakers, the breaker is in the open position. The employer shall also render the control circuit inoperable if the design of the equipment permits.

1926.966(d)

Substation fences. Conductive fences around substations shall be grounded. When a substation fence is expanded or a section is removed, fence sections shall be isolated, grounded, or bonded as necessary to protect employees from hazardous differences in electric potential.

Note to paragraph (d): IEEE Std 80-2000, *IEEE Guide for Safety in AC Substation Grounding*, contains guidelines for protection against hazardous differences in electric potential.

1926.966(e)

Guarding of rooms and other spaces containing electric supply equipment.

1926.966(e)(1)

When to guard rooms and other spaces. Rooms and other spaces in which electric supply lines or equipment are installed shall meet the requirements of paragraphs (e)(2) through (e)(5) of this section under the following conditions:

1926.966(e)(1)(i)

If exposed live parts operating at 50 to 150 volts to ground are within 2.4 meters (8 feet) of the ground or other working surface inside the room or other space,

1926.966(e)(1)(ii)

If live parts operating at 151 to 600 volts to ground and located within 2.4 meters (8 feet) of the ground or other working surface inside the room or other space are guarded only by location, as permitted under paragraph (f)(1) of this section, or

1926.966(e)(1)(iii)

If live parts operating at more than 600 volts to ground are within the room or other space, unless:

1926.966(e)(1)(iii)(A)

The live parts are enclosed within grounded, metal-enclosed equipment whose only openings are designed so that foreign objects inserted in these openings will be deflected from energized parts, or

1926.966(e)(1)(iii)(B)

The live parts are installed at a height, above ground and any other working surface, that provides protection at the voltage on the live parts corresponding to the protection provided by a 2.4-meter (8-foot) height at 50 volts.

1926.966(e)(2)

Prevent access by unqualified persons. Fences, screens, partitions, or walls shall enclose the rooms and other spaces so as to minimize the possibility that unqualified persons will enter.

1926.966(e)(3)

Restricted entry. Unqualified persons may not enter the rooms or other spaces while the electric supply lines or equipment are energized.

1926.966(e)(4)

Warning signs. The employer shall display signs at entrances to the rooms and other spaces warning unqualified persons to keep out.

1926.966(e)(5)

Entrances to rooms and other. The employer shall keep each entrance to a room or other space locked, unless the entrance is under the observation of a person who is attending the room or other space for the purpose of preventing unqualified employees from entering.

1926.966(f)

Guarding of energized parts.

1926.966(f)(1)

Type of guarding. The employer shall provide guards around all live parts operating at more than 150 volts to ground without an insulating covering unless the location of the live parts gives sufficient clearance (horizontal, vertical, or both) to minimize the possibility of accidental employee contact.

Note to paragraph (f)(1): American National Standard *National Electrical Safety Code,* ANSI/IEEE C2-2002 contains guidelines for the dimensions of clearance distances about electric equipment in substations. Installations meeting the ANSI provisions comply with paragraph (f)(1) of this section. The Occupational Safety and Health Administration will determine whether an installation that does not conform to this ANSI standard complies with paragraph (f)(1) of this section based on the following criteria:

(1) Whether the installation conforms to the edition of ANSI C2 that was in effect when the installation was made;
(2) Whether each employee is isolated from energized parts at the point of closest approach; and
(3) Whether the precautions taken when employees perform work on the installation provide protection equivalent to the protection provided by horizontal and vertical clearances meeting ANSI/IEEE C2-2002.

1926.966(f)(2)

Maintaining guards during operation. Except for fuse replacement and other necessary access by qualified persons, the employer shall maintain guarding of energized parts within a compartment during operation and maintenance functions to prevent accidental contact with energized parts and to prevent dropped tools or other equipment from contacting energized parts.

1926.966(f)(3)

Temporary removal of guards. Before guards are removed from energized equipment, the employer shall install barriers around the work area to prevent employees who are not working on the equipment, but who are in the area, from contacting the exposed live parts.

1926.966(g)

Substation entry.

1926.966(g)(1)

Report upon entering. Upon entering an attended substation, each employee, other than employees regularly working in the station, shall report his or her presence to the employee in charge of substation activities to receive information on special system conditions affecting employee safety.

1926.966(g)(2)

Job briefing. The job briefing required by § 1926.952 shall cover information on special system conditions affecting employee safety, including the location of energized equipment in or adjacent to the work area and the limits of any deenergized work area.

[79 FR 20713-20714, July 10, 2014]

Reprinted from www.osha.gov

Occupational Safety & Health Administration
200 Constitution Avenue, NW
Washington, DC 20210

REGULATIONS (STANDARDS - 29 CFR)

SPECIAL CONDITIONS - 1926.967
- **Part Number:** 1926
- **Part Title:** Safety and Health Regulations for Construction
- **Subpart:** V
- **Subpart Title:** Electric Power Transmission and Distribution
- **Standard Number:** 1926.967
- **Title:** Special conditions.
- **GPO Source:** e-CFR

1926.967(a)

Capacitors. The following additional requirements apply to work on capacitors and on lines connected to capacitors.

Note to paragraph (a): See §§ 1926.961 and 1926.962 for requirements pertaining to the deenergizing and grounding of capacitor installations.

1926.967(a)(1)

Disconnect from energized source. Before employees work on capacitors, the employer shall disconnect the capacitors from energized sources and short circuit the capacitors. The employer shall ensure that the employee short circuiting the capacitors waits at least 5 minutes from the time of disconnection before applying the short circuit,

1926.967(a)(2)

Short circuiting units. Before employees handle the units, the employer shall short circuit each unit in series-parallel capacitor banks between all terminals and the capacitor case or its rack. If the cases of capacitors are on ungrounded substation racks, the employer shall bond the racks to ground.

1926.967(a)(3)

Short circuiting connected lines. The employer shall short circuit any line connected to capacitors before the line is treated as deenergized.

1926.967(b)

Current transformer secondaries. The employer shall ensure that employees do not open the secondary of a current transformer while the transformer is energized. If the employer cannot deenergize the primary of the current transformer before employees perform work on an instrument, a relay, or other section of a current transformer secondary circuit, the employer shall bridge the circuit so that the current transformer secondary does not experience an open-circuit condition.

1926.967(c)

Series streetlighting.

1926.967(c)(1)

Applicable requirements. If the open-circuit voltage exceeds 600 volts, the employer shall ensure that employees work on series streetlighting circuits in accordance with § 1926.964 or § 1926.965, as appropriate.

1926.967(c)(2)

Opening a series loop. Before any employee opens a series loop, the employer shall deenergize the streetlighting transformer and isolate it from the source of supply or shall bridge the loop to avoid an open-circuit condition.

1926.967(d)

Illumination. The employer shall provide sufficient illumination to enable the employee to perform the work safely.

Note to paragraph (d): See § 1926.56, which requires specific levels of illumination.

1926.967(e)

Protection against drowning.

1926.967(e)(1)

Personal flotation devices. Whenever an employee may be pulled or pushed, or might fall, into water where the danger of drowning exists, the employer shall provide the employee with, and shall ensure that the employee uses, a personal flotation device meeting § 1926.106.

1926.967(e)(2)

Maintaining flotation devices in safe condition. The employer shall maintain each personal flotation device in safe condition and shall inspect each personal flotation device frequently enough to ensure that it does not have rot, mildew, water saturation, or any other condition that could render the device unsuitable for use.

1926.967(e)(3)

Crossing bodies of water. An employee may cross streams or other bodies of water only if a safe means of passage, such as a bridge, is available.

1926.967(f)

Excavations. Excavation operations shall comply with Subpart P of this part.

1926.967(g)

Employee protection in public work areas.

1926.967(g)(1)

Traffic control devices. Traffic-control signs and traffic-control devices used for the protection of employees shall meet § 1926.200(g)(2).

1926.967(g)(2)

Controlling traffic. Before employees begin work in the vicinity of vehicular or pedestrian traffic that may endanger them, the employer shall place warning signs or flags and other trafficcontrol devices in conspicuous locations to alert and channel approaching traffic.

1926.967(g)(3)

Barricades. The employer shall use barricades where additional employee protection is necessary.

1926.967(g)(4)

Excavated areas. The employer shall protect excavated areas with barricades.

1926.967(g)(5)

Warning lights. The employer shall display warning lights prominently at night.

1926.967(h)

Backfeed. When there is a possibility of voltage backfeed from sources of cogeneration or from the secondary system (for example, backfeed from more than one energized phase feeding a common load), the requirements of § 1926.967 apply if employees will work the lines or equipment as energized, and the requirements of §§ 1926.961 and 1926.962 apply if employees will work the lines or equipment as deenergized.

1926.967(i)

Lasers. The employer shall install, adjust, and operate laser equipment in accordance with § 1926.54.

1926.967(j)

Hydraulic fluids. Hydraulic fluids used for the insulated sections of equipment shall provide insulation for the voltage involved.

1926.967(k)

Communication facilities.

1926.967(k)(1)

Microwave transmission.

1926.967(k)(1)(i)

The employer shall ensure that no employee looks into an open waveguide or antenna connected to an energized microwave source.

1926.967(k)(1)(ii)

If the electromagnetic-radiation level within an accessible area associated with microwave communications systems exceeds the radiation-protection guide specified by § 1910.97(a)(2) of this chapter, the employer shall post the area with warning signs containing the warning symbol described in § 1910.97(a)(3) of this chapter. The lower half of the warning symbol shall include the following statements, or ones that the employer can demonstrate are equivalent: "Radiation in this area may exceed hazard limitations and special precautions are required. Obtain specific instruction before entering."

1926.967(k)(1)(iii)

When an employee works in an area where the electromagnetic radiation could exceed the radiationprotection guide, the employer shall institute measures that ensure that the employee's exposure is not greater than that permitted by that guide. Such measures may include administrative and engineering controls and personal protective equipment.

1926.967(k)(2)

Power-line carrier. The employer shall ensure that employees perform power-line carrier work, including work on equipment used for coupling carrier current to power line conductors, in accordance with the requirements of this subpart pertaining to work on energized lines.

[79 FR 20714-20715, July 10, 2014]

Reprinted from www.osha.gov

Occupational Safety & Health Administration
200 Constitution Avenue, NW
Washington, DC 20210

REGULATIONS (STANDARDS - 29 CFR)

DEFINITIONS - 1926.968
- **Part Number:** 1926
- **Part Title:** Safety and Health Regulations for Construction
- **Subpart:** V
- **Subpart Title:** Electric Power Transmission and Distribution
- **Standard Number:** 1926.968
- **Title:** Definitions.
- **GPO Source:** e-CFR

Attendant. An employee assigned to remain immediately outside the entrance to an enclosed or other space to render assistance as needed to employees inside the space.

Automatic circuit recloser. A selfcontrolled device for automatically interrupting and reclosing an alternating-current circuit, with a predetermined sequence of opening and reclosing followed by resetting, hold closed, or lockout.

Barricade. A physical obstruction such as tapes, cones, or A-frame type wood or metal structures that provides a warning about, and limits access to, a hazardous area.

Barrier. A physical obstruction that prevents contact with energized lines or equipment or prevents unauthorized access to a work area.

Bond. The electrical interconnection of conductive parts designed to maintain a common electric potential.

Bus. A conductor or a group of conductors that serve as a common connection for two or more circuits.

Bushing. An insulating structure that includes a through conductor or that provides a passageway for such a conductor, and that, when mounted on a barrier, insulates the conductor from the barrier for the purpose of conducting current from one side of the barrier to the other.

Cable. A conductor with insulation, or a stranded conductor with or without insulation and other coverings (single-conductor cable), or a combination of conductors insulated from one another (multiple-conductor cable).

Cable sheath. A conductive protective covering applied to cables.

Note to the definition of "cable sheath": A cable sheath may consist of multiple layers one or more of which is conductive.

Circuit. A conductor or system of conductors through which an electric current is intended to flow.

Clearance (between objects). The clear distance between two objects measured surface to surface.

Clearance (for work). Authorization to perform specified work or permission to enter a restricted area.

Communication lines. (See *Lines;* (1) *Communication lines.*)

Conductor. A material, usually in the form of a wire, cable, or bus bar, used for carrying an electric current.

Contract employer. An employer, other than a host employer, that performs work covered by Subpart V of this part under contract.

Covered conductor. A conductor covered with a dielectric having no rated insulating strength or having a rated insulating strength less than the voltage of the circuit in which the conductor is used.

Current-carrying part. A conducting part intended to be connected in an electric circuit to a source of voltage. Non-current-carrying parts are those not intended to be so connected.

Deenergized. Free from any electrical connection to a source of potential difference and from electric charge; not having a potential that is different from the potential of the earth.

Note to the definition of "deenergized": The term applies only to current-carrying parts, which are sometimes energized (alive).

Designated employee (designated person). An employee (or person) who is assigned by the employer to perform specific duties under the terms of this subpart and who has sufficient knowledge of the construction and operation of the equipment, and the hazards involved, to perform his or her duties safely.

Electric line truck. A truck used to transport personnel, tools, and material for electric supply line work.

Electric supply equipment. Equipment that produces, modifies, regulates, controls, or safeguards a supply of electric energy.

Electric supply lines. (See "Lines; (2) Electric supply lines.")

Electric utility. An organization responsible for the installation, operation, or maintenance of an electric supply system.

Enclosed space. A working space, such as a manhole, vault, tunnel, or shaft, that has a limited means of egress or entry, that is designed for periodic employee entry under normal operating conditions, and that, under normal conditions, does not contain a hazardous atmosphere, but may contain a hazardous atmosphere under abnormal conditions.

Note to the definition of "Enclosed space": The Occupational Safety and Health Administration does not consider spaces that are enclosed but not designed for employee entry under normal operating conditions to be enclosed spaces for the purposes of this subpart. Similarly, the Occupational Safety and Health Administration does not consider spaces that are enclosed and that are expected to contain a hazardous atmosphere to be enclosed spaces for the purposes of this subpart. Such spaces meet the definition of permit spaces in subpart AA of this part, and entry into them must conform to that standard.

Energized (alive, live). Electrically connected to a source of potential difference, or electrically charged so as to have a potential significantly different from that of earth in the vicinity.

Energy source. Any electrical, mechanical, hydraulic, pneumatic, chemical, nuclear, thermal, or other energy source that could cause injury to employees.

Entry (as used in § 1926.953). The action by which a person passes through an opening into an enclosed space. Entry includes ensuing work activities in that space and is considered to have occurred as soon as any part of the entrant's body breaks the plane of an opening into the space.

Equipment (electric). A general term including material, fittings, devices, appliances, fixtures, apparatus, and the like used as part of or in connection with an electrical installation.

Exposed, Exposed to contact (as applied to energized parts). Not isolated or guarded.

Fall restraint system. A fall protection system that prevents the user from falling any distance.

First-aid training. Training in the initial care, including cardiopulmonary resuscitation (which includes chest compressions, rescue breathing, and, as appropriate, other heart and lung resuscitation techniques), performed by a person who is not a medical practitioner, of a sick or injured person until definitive medical treatment can be administered.

Ground. A conducting connection, whether planned or unplanned, between an electric circuit or equipment and the earth, or to some conducting body that serves in place of the earth.

Grounded. Connected to earth or to some conducting body that serves in place of the earth.

Guarded. Covered, fenced, enclosed, or otherwise protected, by means of suitable covers or casings, barrier rails or screens, mats, or platforms, designed to minimize the possibility, under normal conditions, of dangerous approach or inadvertent contact by persons or objects.

Note to the definition of "guarded": Wires that are insulated, but not otherwise protected, are not guarded.

Hazardous atmosphere. An atmosphere that may expose employees to the risk of death, incapacitation, impairment of ability to self-rescue (that is, escape unaided from an enclosed space), injury, or acute illness from one or more of the following causes:

1. Flammable gas, vapor, or mist in excess of 10 percent of its lower flammable limit (LFL);
2. Airborne combustible dust at a concentration that meets or exceeds its LFL;

Note to the definition of "hazardous atmosphere" (2): This concentration may be approximated as a condition in which the dust obscures vision at a distance of 1. 52 meters (5 feet) or less.

3. Atmospheric oxygen concentration below 19.5 percent or above 23.5 percent;
4. Atmospheric concentration of any substance for which a dose or a permissible exposure limit is published in Subpart D, *Occupational Health and Environmental Controls*, or in Subpart Z, *Toxic and Hazardous Substances*, of this part and which could result in employee exposure in excess of its dose or permissible exposure limit;

Note to the definition of "hazardous atmosphere" (4): An atmospheric concentration of any substance that is not capable of causing death, incapacitation, impairment of ability to self-rescue, injury, or acute illness due to its health effects is not covered by this provision.

5. Any other atmospheric condition that is immediately dangerous to life or health.

Note to the definition of "hazardous atmosphere" (5): For air contaminants for which the Occupational Safety and Health Administration has not determined a dose or permissible exposure limit, other sources of information, such as Material Safety Data Sheets that comply with the Hazard Communication Standard, § 1926.59, published information, and internal documents can provide guidance in establishing acceptable atmospheric conditions.

High-power tests. Tests in which the employer uses fault currents, load currents, magnetizing currents, and linedropping currents to test equipment, either at the equipment›s rated voltage or at lower voltages.

High-voltage tests. Tests in which the employer uses voltages of approximately 1,000 volts as a practical minimum and in which the voltage source has sufficient energy to cause injury.

High wind. A wind of such velocity that one or more of the following hazards would be present:

1. The wind could blow an employee from an elevated location,
2. The wind could cause an employee or equipment handling material to lose control of the material, or
3. The wind would expose an employee to other hazards not controlled by the standard involved.

Note to the definition of "high wind": The Occupational Safety and Health Administration normally considers winds exceeding 64.4 kilometers per hour (40 miles per hour), or 48.3 kilometers per hour (30 miles per hour) if the work involves material handling, as meeting this criteria, unless the employer takes precautions to protect employees from the hazardous effects of the wind.

Host employer. An employer that operates, or that controls the operating procedures for, an electric power generation, transmission, or distribution installation on which a contract employer is performing work covered by Subpart V of this part.

Note to the definition of "host employer": The Occupational Safety and Health Administration will treat the electric utility or the owner of the installation as the host employer if it operates or controls operating procedures for the installation. If the electric utility or installation owner neither operates nor controls operating procedures for the installation, the Occupational Safety and Health Administration will treat the employer that the utility or owner has contracted with to operate or control the operating procedures for the installation as the host employer. In no case will there be more than one host employer.

Immediately dangerous to life or health (IDLH). Any condition that poses an immediate or delayed threat to life or that would cause irreversible adverse health effects or that would interfere with an individual's ability to escape unaided from a permit space.

Note to the definition of "immediately dangerous to life or health": Some materials-hydrogen fluoride gas and cadmium vapor, for example-may produce immediate transient effects that, even if severe, may pass without medical attention, but are followed by sudden, possibly fatal collapse 12-72 hours after exposure. The victim "feels normal" from recovery from transient effects until collapse. Such materials in hazardous quantities are considered to be "immediately" dangerous to life or health.

Insulated. Separated from other conducting surfaces by a dielectric (including air space) offering a high resistance to the passage of current.

Note to the definition of "insulated": When any object is said to be insulated, it is understood to be insulated for the conditions to which it normally is subjected. Otherwise, it is, for the purpose of this subpart, uninsulated.

Insulation (cable). Material relied upon to insulate the conductor from other conductors or conducting parts or from ground.

Isolated. Not readily accessible to persons unless special means for access are used.

Line-clearance tree trimming. The pruning, trimming, repairing, maintaining, removing, or clearing of trees, or the cutting of brush, that is within the following distance of electric supply lines and equipment:
1. For voltages to ground of 50 kilovolts or less-3.05 meters (10 feet);
2. For voltages to ground of more than 50 kilovolts-3.05 meters (10 feet) plus 0.10 meters (4 inches) for every 10 kilovolts over 50 kilovolts.

Lines.
1. *Communication lines.* The conductors and their supporting or containing structures which are used for public or private signal or communication service, and which operate at potentials not exceeding 400 volts to ground or 750 volts between any two points of the circuit, and the transmitted power of which does not exceed 150 watts. If the lines are operating at less than 150 volts, no limit is placed on the transmitted power of the system. Under certain conditions, communication cables may include communication circuits exceeding these limitations where such circuits are also used to supply power solely to communication equipment.

Note to the definition of "communication lines": Telephone, telegraph, railroad signal, data, clock, fire, police alarm, cable television, and other systems conforming to this definition are included. Lines used for signaling purposes, but not included under this definition, are considered as electric supply lines of the same voltage.
2. *Electric supply lines.* Conductors used to transmit electric energy and their necessary supporting or containing structures. Signal lines of more than 400 volts are always supply lines within this subpart, and those of less than 400 volts are considered as supply lines, if so run and operated throughout.

Manhole. A subsurface enclosure that personnel may enter and that is used for installing, operating, and maintaining submersible equipment or cable.

Minimum approach distance. The closest distance an employee may approach an energized or a grounded object.

Note to the definition of "minimum approach distance": Paragraph (c)(1)(i) of § 1926.960 requires employers to establish minimum approach distances.

Personal fall arrest system. A system used to arrest an employee in a fall from a working level.

Qualified employee (qualified person). An employee (person) knowledgeable in the construction and operation of the electric power generation, transmission, and distribution equipment involved, along with the associated hazards.

Note 1 to the definition of "qualified employee (qualified person)": An employee must have the training required by § 1926.950(b)(2) to be a qualified employee.

Note 2 to the definition of "qualified employee (qualified person)": Except under § 1926.954(b)(3)(iii), an employee who is undergoing on-the-job training and who has demonstrated, in the course of such training, an ability to perform duties safely at his or her level of training and who is under the direct supervision of a qualified person is a qualified person for the performance of those duties.

Statistical sparkover voltage. A transient overvoltage level that produces a 97.72-percent probability of sparkover (that is, two standard deviations above the voltage at which there is a 50-percent probability of sparkover).

Statistical withstand voltage. A transient overvoltage level that produces a 0.14-percent probability of sparkover (that is, three standard deviations below the voltage at which there is a 50-percent probability of sparkover).

Switch. A device for opening and closing or for changing the connection of a circuit. In this subpart, a switch is manually operable, unless otherwise stated.

System operator. A qualified person designated to operate the system or its parts.

Vault. An enclosure, above or below ground, that personnel may enter and that is used for installing, operating, or maintaining equipment or cable.

Vented vault. A vault that has provision for air changes using exhaustflue stacks and low-level air intakes operating on pressure and temperature differentials that provide for airflow that precludes a hazardous atmosphere from developing.

Voltage. The effective (root mean square, or rms) potential difference between any two conductors or between a conductor and ground. This subpart expresses voltages in nominal values, unless otherwise indicated. The nominal voltage of a system or circuit is the value assigned to a system or circuit of a given voltage class for the purpose of convenient designation. The operating voltage of the system may vary above or below this value.

Work-positioning equipment. A body belt or body harness system rigged to allow an employee to be supported on an elevated vertical surface, such as a utility pole or tower leg, and work with both hands free while leaning.

[79 FR 20715-20717, July 10, 2014; 79 FR 56962, September 24, 2014; 80 FR 25518, May 4, 2015]

Reprinted from www.osha.gov

Occupational Safety & Health Administration
200 Constitution Avenue, NW
Washington, DC 20210

REGULATIONS (STANDARDS - 29 CFR)

- **Part Number:** 1926
- **Part Title:** Safety and Health Regulations for Construction
- **Subpart:** V
- **Subpart Title:** Electric Power Transmission and Distribution
- **Standard Number:** 1926 Subpart V App A
- **Title:** Appendix A to Subpart V of Part 1926 - [Reserved]
- **GPO Source:** e-CFR

Appendix A to Subpart V of Part 1926-[Reserved]

[79 FR 20717, July 10, 2014]

Reprinted from www.osha.gov

Occupational Safety & Health Administration
200 Constitution Avenue, NW
Washington, DC 20210

REGULATIONS (STANDARDS - 29 CFR)

- **Part Number:** 1926
- **Part Title:** Safety and Health Regulations for Construction
- **Subpart:** V
- **Subpart Title:** Electric Power Transmission and Distribution
- **Standard Number:** 1926 Subpart V App B
- **Title:** Appendix B to Subpart V of Part 1926-Working on Exposed Energized Parts
- **GPO Source:** e-CFR

Appendix B to Subpart V of Part 1926-Working on Exposed Energized Parts

I. Introduction

Electric utilities design electric power generation, transmission, and distribution installations to meet National Electrical Safety Code (NESC), ANSI C2, requirements. Electric utilities also design transmission and distribution lines to limit line outages as required by system reliability criteria [1] and to withstand the maximum overvoltages impressed on the system. Conditions such as switching surges, faults, and lightning can cause overvoltages. Electric utilities generally select insulator design and lengths and the clearances to structural parts so as to prevent outages from contaminated line insulation and during storms. Line insulator lengths and structural clearances have, over the years, come closer to the minimum approach distances used by workers. As minimum approach distances and structural clearances converge, it is increasingly important that system designers and system operating and maintenance personnel understand the concepts underlying minimum approach distances.

The information in this appendix will assist employers in complying with the minimum approach-distance requirements contained in §§ 1926.960(c)(1) and 1926.964(c). Employers must use the technical criteria and methodology presented in this appendix in establishing minimum approach distances in accordance with § 1926.960(c)(1)(i) and Table V-2 and Table V-7. This appendix provides essential background information and technical criteria for the calculation of the required minimum approach distances for live-line work on electric power generation, transmission, and distribution installations.

Unless an employer is using the maximum transient overvoltages specified in Table V-8 for voltages over 72.5 kilovolts, the employer must use persons knowledgeable in the techniques discussed in this appendix, and competent in the field of electric transmission and distribution system design, to determine the maximum transient overvoltage.

II. General

A. *Definitions.* The following definitions from § 1926.968 relate to work on or near electric power generation, transmission, and distribution lines and equipment and the electrical hazards they present.

---

[1]Federal, State, and local regulatory bodies and electric utilities set reliability requirements that limit the number and duration of system outages.

Exposed. . . . Not isolated or guarded.

Guarded. Covered, fenced, enclosed, or otherwise protected, by means of suitable covers or casings, barrier rails or screens, mats, or platforms, designed to minimize the possibility, under normal conditions, of dangerous approach or inadvertent contact by persons or objects.

Note to the definition of "guarded": Wires that are insulated, but not otherwise protected, are not guarded.

Insulated. Separated from other conducting surfaces by a dielectric (including air space) offering a high resistance to the passage of current.

Note to the definition of " insulated": When any object is said to be insulated, it is understood to be insulated for the conditions to which it normally is subjected. Otherwise, it is, for the purpose of this subpart, uninsulated.

Isolated. Not readily accessible to persons unless special means for access are used.

Statistical sparkover voltage. A transient overvoltage level that produces a 97.72-percent probability of sparkover (that is, two standard deviations above the voltage at which there is a 50-percent probability of sparkover).

Statistical withstand voltage. A transient overvoltage level that produces a 0.14-percent probability of sparkover (that is, three standard deviations below the voltage at which there is a 50-percent probability of sparkover).

B. *Installations energized at 50 to 300 volts.* The hazards posed by installations energized at 50 to 300 volts are the same as those found in many other workplaces. That is not to say that there is no hazard, but the complexity of electrical protection required does not compare to that required for highvoltage systems. The employee must avoid contact with the exposed parts, and the protective equipment used (such as rubber insulating gloves) must provide insulation for the voltages involved.

C. *Exposed energized parts over 300 volts AC.* Paragraph (c)(1)(i) of § 1926.960 requires the employer to establish minimum approach distances no less than the distances computed by Table V-2 for ac systems so that employees can work safely without risk of sparkover.[2]

Unless the employee is using electrical protective equipment, air is the insulating medium between the employee and energized parts. The distance between the employee and an energized part must be sufficient for the air to withstand the maximum transient overvoltage that can reach the worksite under the working conditions and practices the employee is using. This distance is the minimum air insulation distance, and it is equal to the electrical component of the minimum approach distance.

Normal system design may provide or include a means (such as lightning arrestors) to control maximum anticipated transient overvoltages, or the employer may use temporary devices (portable protective gaps) or measures (such as preventing automatic circuit breaker reclosing) to achieve the same result. Paragraph (c)(1)(ii) of § 1926.960 requires the employer to determine the maximum

[2]Sparkover is a disruptive electric discharge in which an electric arc forms and electric current passes through air.

anticipated per-unit transient overvoltage, phase-to-ground, through an engineering analysis or assume a maximum anticipated per-unit transient overvoltage, phase-to-ground, in accordance with Table V-8, which specifies the following maximums for ac systems:

72.6 to 420.0 kilovolts .... 3.5 per unit.

420.1 to 550.0 kilovolts .... 3.0 per unit.

550.1 to 800.0 kilovolts .... 2.5 per unit.

See paragraph IV.A.2, later in this appendix, for additional discussion of maximum transient overvoltages.

D. *Types of exposures.* Employees working on or near energized electric power generation, transmission, and distribution systems face two kinds of exposures: Phaseto-ground and phase-to-phase. The exposure is phase-to-ground: (1) With respect to an energized part, when the employee is at ground potential or (2) with respect to ground, when an employee is at the potential of the energized part during live-line barehand work. The exposure is phase-tophase, with respect to an energized part, when an employee is at the potential of another energized part (at a different potential) during live-line barehand work.

III. Determination of Minimum Approach Distances for AC Voltages Greater Than 300 Volts

A. *Voltages of 301 to 5,000 volts.* Test data generally forms the basis of minimum air insulation distances. The lowest voltage for which sufficient test data exists is 5,000 volts, and these data indicate that the minimum air insulation distance at that voltage is 20 millimeters (1 inch). Because the minimum air insulation distance increases with increasing voltage, and, conversely, decreases with decreasing voltage, an assumed minimum air insulation distance of 20 millimeters will protect against sparkover at voltages of 301 to 5,000 volts. Thus, 20 millimeters is the electrical component of the minimum approach distance for these voltages.

B. *Voltages of 5.1 to 72.5 kilovolts.* For voltages from 5.1 to 72.5 kilovolts, the Occupational Safety and Health Administration bases the methodology for calculating the electrical component of the minimum approach distance on Institute of Electrical and Electronic Engineers (IEEE) Standard 4-1995, *Standard Techniques for High-Voltage Testing.* Table 1 lists the critical sparkover distances from that standard as listed in IEEE Std 516-2009, *IEEE Guide for Maintenance Methods on Energized Power Lines.*

To use this table to determine the electrical component of the minimum approach distance, the employer must determine the peak phase-to-ground transient overvoltage and select a gap from the table that corresponds to that voltage as a withstand voltage rather than a critical sparkover voltage. To calculate the electrical component of the minimum approach distance for voltages between 5 and 72.5 kilovolts, use the following procedure:

1. Divide the phase-to-phase voltage by the square root of 3 to convert it to a phase-to-ground voltage.

Table 1 Sparkover Distance for Rod-to-Rod Gap

60 Hz rod-to-rod sparkover (kV peak)	Gap spacing from IEEE Std 4-1995 (cm)
25	2
36	3
46	4
53	5
60	6
70	8
79	10
86	12
95	14
104	16
112	18
120	20
143	25
167	30
192	35
218	40
243	45
270	50
322	60

Source: IEEE Std 516-2009.

2. Multiply the phase-to-ground voltage by the square root of 2 to convert the rms value of the voltage to the peak phase-to-ground voltage.
3. Multiply the peak phase-to-ground voltage by the maximum per-unit transient overvoltage, which, for this voltage range, is 3.0, as discussed later in this appendix. This is the maximum phase-to-ground transient overvoltage, which corresponds to the withstand voltage for the relevant exposure.[3]
4. Divide the maximum phase-to-ground transient overvoltage by 0.85 to determine the corresponding critical sparkover voltage. (The critical sparkover voltage is 3 standard deviations (or 15 percent) greater than the withstand voltage.)
5. Determine the electrical component of the minimum approach distance from Table 1 through interpolation.

Table 2 illustrates how to derive the electrical component of the minimum approach distance for voltages from 5.1 to 72.5 kilovolts, before the application of any altitude correction factor, as explained later.

C. *Voltages of 72.6 to 800 kilovolts.* For voltages of 72.6 kilovolts to 800 kilovolts, this subpart bases the electrical component of minimum approach distances, before the application of any altitude correction factor, on the following formula:

---

[3]The withstand voltage is the voltage at which sparkover is not likely to occur across a specified distance. It is the voltage taken at the 3s point below the sparkover voltage, assuming that the sparkover curve follows a normal distribution.

Table 2 Calculating The Electrical Component of Mad-751 V to 72.5 KV

Step	Maximum system phase-to-phase voltage (kV)			
	15	36	46	72.5
1. Divide by √3..................	8.7	20.8	26.6	41.9
2. Multiply by √2...............	12.2	29.4	37.6	59.2
3. Multiply by 3.0..............	36.7	88.2	112.7	177.6
4. Divide by 0.85...............	43.2	103.7	132.6	208.9
5. Interpolate from Table 1............................	3+(7.2/10)*1	14+(8.7/9)*2	20+(12.6/23)*5	35+(16.9/26)*5
Electrical component of MAD (cm)	3.72	15.93	22.74	38.25

Equation 1-For voltages of 72.6 kV to 800 kV

$$D = 0.3048(C + a)V_{L-G}T$$

Where:

D = Electrical component of the minimum approach distance in air in meters;

C = a correction factor associated with the variation of gap sparkover with voltage;

a = A factor relating to the saturation of air at system voltages of 345 kilovolts or higher; [4]

V_{L-G} = Maximum system line-to-ground rms voltage in kilovolts-it should be the "actual" maximum, or the normal highest voltage for the range (for example, 10 percent above the nominal voltage); and

T = Maximum transient overvoltage factor in per unit.

In Equation 1, C is 0.01: (1) For phase-to-ground exposures that the employer can demonstrate consist only of air across the approach distance (gap) and (2) for phase-tophase exposures if the employer can demonstrate that no insulated tool spans the gap and that no large conductive object is in the gap. Otherwise, C is 0.011.

In Equation 1, the term a varies depending on whether the employee's exposure is phase-to-ground or phase-to-phase and on whether objects are in the gap. The employer must use the equations in Table 3 to calculate a. Sparkover test data with insulation spanning the gap form the basis for the equations for phase-to-ground exposures, and sparkover test data with only air in the gap form the basis for the equations for phase-tophase exposures. The phase-to-ground equations result in slightly higher values of a, and, consequently, produce larger minimum approach distances, than the phase-to-phase equations for the same value of V_{Peak}.

In Equation 1, T is the maximum transient overvoltage factor in per unit. As noted earlier, § 1926.960(c)(1)(ii) requires the employer to determine the maximum anticipated per-unit transient overvoltage, phase-to-ground, through an engineering analysis or assume a maximum anticipated per-unit transient overvoltage,

---

[4] Test data demonstrates that the saturation factor is greater than 0 at peak voltages of about 630 kilovolts. Systems operating at 345 kilovolts (or maximum system voltages of 362 kilovolts) can have peak maximum transient overvoltages exceeding 630 kilovolts. Table V-2 sets equations for calculating a based on peak voltage.

Table 3   Equations for Calculating the Surge Factor,[a]

Phase-to-Ground Exposures			
$V_{Peak}=T_{L-G}V_{L-G}\sqrt{2}$	635 kV or less	635.1 to 915 kV	915.1 to 1,050 kV
a	0	$(V_{Peak}-635)/140,000$	$(V_{Peak}-645)/135,000$
$V_{Peak}=T_{L-G}V_{L-G}\sqrt{2}$	More than 1,050 kV		
a	$(V_{Peak}-675)/125,000$		

Phase-to-Phase Exposures[1]			
$V_{Peak}=(1.35T_{L-G}+0.45)V_{L-G}\sqrt{2}$	630 kV or less	630.1 to 848 kV	848.1 to 1,131 kV
a	0	$(V_{Peak}-630)/155,000$	$(V_{Peak}-633.6)/152,207$
$V_{Peak}=(1.35T_{L-G}+0.45)V_{L-G}\sqrt{2}$	1,131 to 1,485 kV	More than 1,485 kV	
a	$(V_{Peak}-628)/153,846$	$(V_{Peak}-350.5)/203,666$	

[1] Use the equations for the phase-to-ground exposures (with V_{Peak} for phase-to-phase exposures) unless the employer can demonstrate that no insulated tools spans the gap and that no large conductive object is in the gap.

phase-to-ground, in accordance with Table V-8. For phase-to-ground exposures, the employer uses this value, called T_{L-G}, as T in Equation 1. IEEE Std 516-2009 provides the following formula to calculate the phase-to-phase maximum transient overvoltage, T_{L-L}, from T_{L-G}:

$$T_{L-L} = 1.35T_{L-G} + 0.45.$$

For phase-to-phase exposures, the employer uses this value as T in Equation 1.

D. *Provisions for inadvertent movement.* The minimum approach distance must include an "adder" to compensate for the inadvertent movement of the worker relative to an energized part or the movement of the part relative to the worker. This "adder" must account for this possible inadvertent movement and provide the worker with a comfortable and safe zone in which to work. Employers must add the distance for inadvertent movement (called the "ergonomic component of the minimum approach distance") to the electrical component to determine the total safe minimum approach distances used in liveline work.

The Occupational Safety and Health Administration based the ergonomic component of the minimum approach distance on response time-distance analysis. This technique uses an estimate of the total response time to a hazardous incident and converts that time to the distance traveled. For example, the driver of a car takes a given amount of time to respond to a "stimulus" and stop the vehicle. The elapsed time involved results in the car's traveling some distance before coming to a complete stop. This distance depends on the speed of the car at the time the stimulus appears and the reaction time of the driver.

In the case of live-line work, the employee must first perceive that he or she is approaching the danger zone. Then, the worker responds to the danger and must decelerate and stop all motion toward the energized part. During the time it takes to stop, the employee will travel some distance. This is the distance the employer must add to the electrical component of the minimum approach distance to obtain the total safe minimum approach distance.

At voltages from 751 volts to 72.5 kilovolts,[5] the electrical component of the minimum approach distance is smaller than the ergonomic component. At 72.5 kilovolts, the electrical component is only a little more than 0.3 meters (1 foot). An ergonomic component of the minimum approach distance must provide for all the worker's unanticipated movements. At these voltages, workers generally use rubber insulating gloves; however, these gloves protect only a worker's hands and arms. Therefore, the energized object must be at a safe approach distance to protect the worker's face. In this case, 0.61 meters (2 feet) is a sufficient and practical ergonomic component of the minimum approach distance.

For voltages between 72.6 and 800 kilovolts, employees must use different work practices during energized line work. Generally, employees use live-line tools (hot sticks) to perform work on energized equipment. These tools, by design, keep the energized part at a constant distance from the employee and, thus, maintain the appropriate minimum approach distance automatically.

The location of the worker and the type of work methods the worker is using also influence the length of the ergonomic component of the minimum approach distance. In this higher voltage range, the employees use work methods that more tightly control their movements than when the workers perform work using rubber insulating gloves. The worker, therefore, is farther from the energized line or equipment and must be more precise in his or her movements just to perform the work. For these reasons, this subpart adopts an ergonomic component of the minimum approach distance of 0.31 m (1 foot) for voltages between 72.6 and 800 kilovolts.

Table 4 summarizes the ergonomic component of the minimum approach distance for various voltage ranges.

Note: The employer must add this distance to the electrical component of the minimum approach distance to obtain the full minimum approach distance.

Table 4 Ergonomic Component Of Minimum Approach Distance

Voltage range (kV)	Distance	
	m	ft
0.301 to 0.750	0.31	1.0
0.751 to 72.5	0.61	2.0
72.6 to 800	0.31	1.0

[5] For voltages of 50 to 300 volts, Table V-2 specifies a minimum approach distance of "avoid contact." The minimum approach distance for this voltage range contains neither an electrical component nor an ergonomic component.

The ergonomic component of the minimum approach distance accounts for errors in maintaining the minimum approach distance (which might occur, for example, if an employee misjudges the length of a conductive object he or she is holding), and for errors in judging the minimum approach distance. The ergonomic component also accounts for inadvertent movements by the employee, such as slipping. In contrast, the working position selected to properly maintain the minimum approach distance must account for all of an employee's reasonably likely movements and still permit the employee to adhere to the applicable minimum approach distance. (See Figure 1.) Reasonably likely movements include an employee's adjustments to tools, equipment, and working positions and all movements needed to perform the work. For example, the employee should be able to perform all of the following actions without straying into the minimum approach distance:

- Adjust his or her hardhat,
- maneuver a tool onto an energized part with a reasonable amount of over-reaching or underreaching,
- reach for and handle tools, material, and equipment passed to him or her, and
- adjust tools, and replace components on them, when necessary during the work procedure.

The training of qualified employees required under § 1926.950, and the job planning and briefing required under § 1926.952, must address selection of a proper working position.

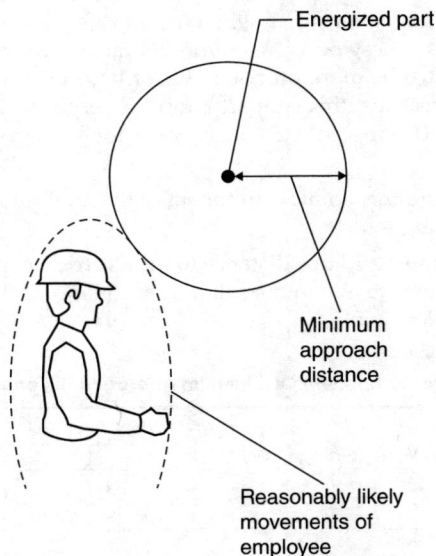

Figure 1. Maintaining the Minimum Approach Distance

E. *Miscellaneous correction factors.* Changes in the air medium that forms the insulation influences the strength of an air gap. A brief discussion of each factor follows.

1. *Dielectric strength of air.* The dielectric strength of air in a uniform electric field at standard atmospheric conditions is approximately 3 kilovolts per millimeter.[6]
The pressure, temperature, and humidity of the air, the shape, dimensions, and separation of the electrodes, and the characteristics of the applied voltage (wave shape) affect the disruptive gradient.

2. *Atmospheric effect.* The empirically determined electrical strength of a given gap is normally applicable at standard atmospheric conditions (20 °C, 101.3 kilopascals, 11 grams/cubic centimeter humidity). An increase in the density (humidity) of the air inhibits sparkover for a given air gap. The combination of temperature and air pressure that results in the lowest gap sparkover voltage is high temperature and low pressure. This combination of conditions is not likely to occur. Low air pressure, generally associated with high humidity, causes increased electrical strength. An average air pressure generally correlates with low humidity. Hot and dry working conditions normally result in reduced electrical strength. The equations for minimum approach distances in Table V-2 assume standard atmospheric conditions.

3. *Altitude.* The reduced air pressure at high altitudes causes a reduction in the electrical strength of an air gap. An employer must increase the minimum approach distance by about 3 percent per 300 meters (1,000 feet) of increased altitude for altitudes above 900 meters (3,000 feet). Table V-4 specifies the altitude correction factor that the employer must use in calculating minimum approach distances.

IV. Determining Minimum Approach Distances

A. Factors Affecting Voltage Stress at the Worksite

1. *System voltage (nominal).* The nominal system voltage range determines the voltage for purposes of calculating minimum approach distances. The employer selects the range in which the nominal system voltage falls, as given in the relevant table, and uses the highest value within that range in perunit calculations.

2. *Transient overvoltages.* Operation of switches or circuit breakers, a fault on a line or circuit or on an adjacent circuit, and similar activities may generate transient overvoltages on an electrical system. Each overvoltage has an associated transient voltage wave shape. The wave shape arriving at the site and its magnitude vary considerably.
In developing requirements for minimum approach distances, the Occupational Safety and Health Administration considered the most common wave shapes and the magnitude of transient overvoltages found on electric power generation, transmission, and distribution systems. The equations in Table V-2 for minimum approach distances use per-unit maximum transient overvoltages, which are relative to the nominal maximum voltage of the system. For example, a maximum

---

[6]For the purposes of estimating arc length, Subpart V generally assumes a more conservative dielectric strength of 10 kilovolts per 25.4 millimeters, consistent with assumptions made in consensus standards such as the National Electrical Safety Code (IEEE C2-2012). The more conservative value accounts for variables such as electrode shape, wave shape, and a certain amount of overvoltage.

Table 5   Magnitude of Typical Transient Overvoltages

Cause	Magnitude (per unit)
Energized 200-mile line without closing resistors	3.5
Energized 200-mile line with one-step closing resistor	2.1
Energized 200-mile line with multistep resistor	2.5
Reclosing with trapped charge one-step resistor	2.2
Opening surge with single restrike	3.0
Fault initiation unfaulted phase	2.1
Fault initiation adjacent circuit	2.5
Fault clearing	1.7 to 1.9

transient overvoltage value of 3.0 per unit indicates that the highest transient over-voltage is 3.0 times the nominal maximum system voltage.

3. *Typical magnitude of overvoltages.* Table 5 lists the magnitude of typical transient overvoltages.

4. *Standard deviation-air-gap withstand.* For each air gap length under the same atmospheric conditions, there is a statistical variation in the breakdown voltage. The probability of breakdown against voltage has a normal (Gaussian) distribution. The standard deviation of this distribution varies with the wave shape, gap geometry, and atmospheric conditions. The withstand voltage of the air gap is three standard deviations (3s) below the critical sparkover voltage. (The critical sparkover voltage is the crest value of the impulse wave that, under specified conditions, causes sparkover 50 percent of the time. An impulse wave of three standard deviations below this value, that is, the withstand voltage, has a probability of sparkover of approximately 1 in 1,000.)

5. *Broken Insulators.* Tests show reductions in the insulation strength of insulator strings with broken skirts. Broken units may lose up to 70 percent of their withstand capacity. Because an employer cannot determine the insulating capability of a broken unit without testing it, the employer must consider damaged units in an insulator to have no insulating value. Additionally, the presence of a live-line tool alongside an insulator string with broken units may further reduce the overall insulating strength. The number of good units that must be present in a string for it to be "insulated" as defined by § 1926.968 depends on the maximum overvoltage possible at the worksite.

B. Minimum Approach Distances Based on Known, Maximum-Anticipated Per-Unit Transient Overvoltages

1. *Determining the minimum approach distance for AC systems.* Under § 1926.960(c)(1)(ii), the employer must determine the maximum anticipated per-unit transient overvoltage, phase-to-ground, through an engineering analysis or must assume a maximum anticipated per-unit transient overvoltage, phase-to-ground, in accordance with Table V-8. When the employer conducts an engineering analysis of the system and determines that the maximum transient overvoltage is lower than specified by Table V-8, the employer must ensure that any conditions assumed in the analysis, for example, that employees block reclosing on a circuit or install

portable protective gaps, are present during energized work. To ensure that these conditions are present, the employer may need to institute new livework procedures reflecting the conditions and limitations set by the engineering analysis.

2. *Calculation of reduced approach distance values.* An employer may take the following steps to reduce minimum approach distances when the maximum transient overvoltage on the system (that is, the maximum transient overvoltage without additional steps to control overvoltages) produces unacceptably large minimum approach distances:

Step 1. Determine the maximum voltage (with respect to a given nominal voltage range) for the energized part.

Step 2. Determine the technique to use to control the maximum transient overvoltage. (See paragraphs IV.C and IV.D of this appendix.) Determine the maximum transient overvoltage that can exist at the worksite with that form of control in place and with a confidence level of 3s . This voltage is the withstand voltage for the purpose of calculating the appropriate minimum approach distance.

Step 3. Direct employees to implement procedures to ensure that the control technique is in effect during the course of the work.

Step 4. Using the new value of transient overvoltage in per unit, calculate the required minimum approach distance from Table V-2.

C. Methods of Controlling Possible Transient Overvoltage Stress Found on a System

1. *Introduction.* There are several means of controlling overvoltages that occur on transmission systems. For example, the employer can modify the operation of circuit breakers or other switching devices to reduce switching transient overvoltages. Alternatively, the employer can hold the overvoltage to an acceptable level by installing surge arresters or portable protective gaps on the system. In addition, the employer can change the transmission system to minimize the effect of switching operations. Section 4.8 of IEEE Std 516-2009 describes various ways of controlling, and thereby reducing, maximum transient overvoltages.

2. *Operation of circuit breakers.*[7] The maximum transient overvoltage that can reach the worksite is often the result of switching on the line on which employees are working. Disabling automatic reclosing during energized line work, so that the line will not be reenergized after being opened for any reason, limits the maximum switching surge overvoltage to the larger of the opening surge or the greatest possible fault-generated surge, provided that the devices (for example, insertion resistors) are operable and will function to limit the transient overvoltage and that circuit breaker restrikes do not occur. The employer must ensure the proper functioning of insertion resistors and other overvoltage-limiting devices when the employer's engineering analysis assumes their proper operation to limit the overvoltage level. If the employer cannot disable the reclosing feature (because of system operating conditions), other methods of controlling the switching surge level may be necessary.

---

[7]The detailed design of a circuit interrupter, such as the design of the contacts, resistor insertion, and breaker timing control, are beyond the scope of this appendix. The design of the system generally accounts for these features. This appendix only discusses features that can limit the maximum switching transient overvoltage on a system.

Transient surges on an adjacent line, particularly for double circuit construction, may cause a significant overvoltage on the line on which employees are working. The employer's engineering analysis must account for coupling to adjacent lines

3. *Surge arresters.* The use of modern surge arresters allows a reduction in the basic impulse-insulation levels of much transmission system equipment. The primary function of early arresters was to protect the system insulation from the effects of lightning. Modern arresters not only dissipate lightning-caused transients, but may also control many other system transients caused by switching or faults.

The employer may use properly designed arresters to control transient overvoltages along a transmission line and thereby reduce the requisite length of the insulator string and possibly the maximum transient overvoltage on the line.[8]

4. *Switching Restrictions.* Another form of overvoltage control involves establishing switching restrictions, whereby the employer prohibits the operation of circuit breakers until certain system conditions are present. The employer restricts switching by using a tagging system, similar to that used for a permit, except that the common term used for this activity is a "hold-off" or "restriction." These terms indicate that the restriction does not prevent operation, but only modifies the operation during the livework activity.

D. Minimum Approach Distance Based on Control of Maximum Transient Overvoltage at the Worksite

When the employer institutes control of maximum transient overvoltage at the worksite by installing portable protective gaps, the employer may calculate the minimum approach distance as follows:

Step 1. Select the appropriate withstand voltage for the protective gap based on system requirements and an acceptable probability of gap sparkover.[9]

Step 2. Determine a gap distance that provides a withstand voltage [10] greater than or equal to the one selected in the first step.[11]

Step 3. Use 110 percent of the gap's critical sparkover voltage to determine the phase-to-ground peak voltage at gap sparkover ($V_{PPG\ Peak}$).

Step 4. Determine the maximum transient overvoltage, phase-to-ground, at the worksite from the following formula:

$$T = \frac{V_{PPG\ Peak}}{V_{L-G}\sqrt{2}}.$$

[8]Surge arrester application is beyond the scope of this appendix. However, if the employer installs the arrester near the work site, the application would be similar to the protective gaps discussed in paragraph IV.D of this appendix.

[9]The employer should check the withstand voltage to ensure that it results in a probability of gap flashover that is acceptable from a system outage perspective. (In other words, a gap sparkover will produce a system outage. The employer should determine whether such an outage will impact overall system performance to an acceptable degree.) In general, the withstand voltage should be at least 1.25 times the maximum crest operating voltage.

[10]The manufacturer of the gap provides, based on test data, the critical sparkover voltage for each gap spacing (for example, a critical sparkover voltage of 665 kilovolts for a gap spacing of 1.2 meters). The withstand voltage for the gap is equal to 85 percent of its critical sparkover voltage.

[11]Switch steps 1 and 2 if the length of the protective gap is known.

Step 5. Use this value of T^{12} in the equation in Table V-2 to obtain the minimum approach distance. If the worksite is no more than 900 meters (3,000 feet) above sea level, the employer may use this value of T to determine the minimum approach distance from Table 7 through Table 14.

Note: All rounding must be to the next higher value (that is, always round up).

Sample protective gap calculations.

Problem: Employees are to perform work on a 500-kilovolt transmission line at sea level that is subject to transient overvoltages of 2.4 p.u. The maximum operating voltage of the line is 550 kilovolts. Determine the length of the protective gap that will provide the minimum practical safe approach distance. Also, determine what that minimum approach distance is.

Step 1. Calculate the smallest practical maximum transient overvoltage (1.25 times the crest phase-to-ground voltage): [13]

$$550\ kV \times \frac{\sqrt{2}}{\sqrt{3}} \times 1.25 = 561\ kV$$

This value equals the withstand voltage of the protective gap.

Step 2. Using test data for a particular protective gap, select a gap that has a critical sparkover voltage greater than or equal to:

$$561\ kV \div 0.85 = 660\ kV$$

For example, if a protective gap with a 1.22-m (4.0-foot) spacing tested to a critical sparkover voltage of 665 kilovolts (crest), select this gap spacing.

Step 3. The phase-to-ground peak voltage at gap sparkover (VPPG Peak) is 110 percent of the value from the previous step:

$$665\ kV \times 1.10 = 732\ kV$$

This value corresponds to the withstand voltage of the electrical component of the minimum approach distance.

Step 4. Use this voltage to determine the worksite value of T:

$$T = \frac{732}{564} = 1.7\ p.u.$$

Step 5. Use this value of T in the equation in Table V-2 to obtain the minimum approach distance, or look up the minimum approach distance in Table 7 through Table 14:

$$MAD = 2.29\ m(7.6ft)$$

---

[12]IEEE Std 516-2009 states that most employers add 0.2 to the calculated value of T as an additional safety factor.

[13] To eliminate sparkovers due to minor system disturbances, the employer should use a withstand voltage no lower than 1.25 p.u. Note that this is a practical, or operational, consideration only. It may be feasible for the employer to use lower values of withstand voltage.

E. Location of Protective Gaps

1. *Adjacent structures.* The employer may install the protective gap on a structure adjacent to the worksite, as this practice does not significantly reduce the protection afforded by the gap.
2. *Terminal stations.* Gaps installed at terminal stations of lines or circuits provide a level of protection; however, that level of protection may not extend throughout the length of the line to the worksite. The use of substation terminal gaps raises the possibility that separate surges could enter the line at opposite ends, each with low enough magnitude to pass the terminal gaps without sparkover. When voltage surges occur simultaneously at each end of a line and travel toward each other, the total voltage on the line at the point where they meet is the arithmetic sum of the two surges. A gap installed within 0.8 km (0.5 mile) of the worksite will protect against such intersecting waves. Engineering studies of a particular line or system may indicate that employers can adequately protect employees by installing gaps at even more distant locations. In any event, unless using the default values for T from Table V-8, the employer must determine T at the worksite.

Worksite. If the employer installs protective gaps at the worksite, the gap setting establishes the worksite impulse insulation strength. Lightning strikes as far as 6 miles from the worksite can cause a voltage surge greater than the gap withstand voltage, and a gap sparkover can occur. In addition, the gap can sparkover from overvoltages on the line that exceed the withstand voltage of the gap. Consequently, the employer must protect employees from hazards resulting from any sparkover that could occur.

F. Disabling automatic reclosing. There are two reasons to disable the automatic reclosing feature of circuit-interrupting devices while employees are performing liveline work:

- To prevent reenergization of a circuit faulted during the work, which could create a hazard or result in more serious injuries or damage than the injuries or damage produced by the original fault;
- To prevent any transient overvoltage caused by the switching surge that would result if the circuit were reenergized. However, due to system stability considerations, it may not always be feasible to disable the automatic-reclosing feature.

V. Minimum Approach-Distance Tables

A. *Legacy tables.* Employers may use the minimum approach distances in Table 6 until March 31, 2015.

B. *Alternative minimum approach distances.* Employers may use the minimum approach distances in Table 7 through Table 14 provided that the employer follows the notes to those tables.

Table 6   Minimum Approach Distances Until March 31, 2015

Voltage range phase to phase (kV)	Phase-to-ground exposure		Phase-to-phase exposure	
	m	ft	m	ft
2.1 to 15.0	0.64	2.1	0.61	2.0
15.1 to 35.0	0.71	2.3	0.71	2.3
35.1 to 46.0	0.76	2.5	0.76	2.5
46.1 to 72.5	0.91	3.0	0.91	3.0
72.6 to 121	1.02	3.3	1.37	4.5
138 to 145	1.07	3.5	1.52	5.0
161 to 169	1.12	3.7	1.68	5.5
230 to 242	1.52	5.0	2.54	8.3
345 to 362*	2.13	7.0	4.06	13.3
500 to 552*	3.35	11.0	6.10	20.0
700 to 765*	4.57	15.0	9.45	31.0

*The minimum approach distance may be the shortest distance between the energized part and the grounded surface.

Table 7   Ac Minimum Approach Distances-72.6 to 121.0 KV

T (p.u.)	Phase-to-ground exposure		Phase-to-phase exposure	
	m	ft	m	ft
1.5	0.67	2.2	0.84	2.8
1.6	0.69	2.3	0.87	2.9
1.7	0.71	2.3	0.90	3.0
1.8	0.74	2.4	0.93	3.1
1.9	0.76	2.5	0.96	3.1
2.0	0.78	2.6	0.99	3.2
2.1	0.81	2.7	1.01	3.3
2.2	0.83	2.7	1.04	3.4
2.3	0.85	2.8	1.07	3.5
2.4	0.88	2.9	1.10	3.6
2.5	0.90	3.0	1.13	3.7
2.6	0.92	3.0	1.16	3.8
2.7	0.95	3.1	1.19	3.9
2.8	0.97	3.2	1.22	4.0
2.9	0.99	3.2	1.24	4.1
3.0	1.02	3.3	1.27	4.2
3.1	1.04	3.4	1.30	4.3
3.2	1.06	3.5	1.33	4.4
3.3	1.09	3.6	1.36	4.5
3.4	1.11	3.6	1.39	4.6
3.5	1.13	3.7	1.42	4.7

Table 8   Ac Minimum Approach Distances-121.1 to 145.0 KV

T (p.u.)	Phase-to-ground exposure		Phase-to-phase exposure	
	m	ft	m	ft
1.5	0.74	2.4	0.95	3.1
1.6	0.76	2.5	0.98	3.2
1.7	0.79	2.6	1.02	3.3
1.8	0.82	2.7	1.05	3.4
1.9	0.85	2.8	1.08	3.5
2.0	0.88	2.9	1.12	3.7
2.1	0.90	3.0	1.15	3.8
2.2	0.93	3.1	1.19	3.9
2.3	0.96	3.1	1.22	4.0
2.4	0.99	3.2	1.26	4.1
2.5	1.02	3.3	1.29	4.2
2.6	1.04	3.4	1.33	4.4
2.7	1.07	3.5	1.36	4.5
2.8	1.10	3.6	1.39	4.6
2.9	1.13	3.7	1.43	4.7
3.0	1.16	3.8	1.46	4.8
3.1	1.19	3.9	1.50	4.9
3.2	1.21	4.0	1.53	5.0
3.3	1.24	4.1	1.57	5.2
3.4	1.27	4.2	1.60	5.2
3.5	1.30	4.3	1.64	5.4

Table 10   Ac Minimum Approach Distances-169.1 to 242.0 KV

T (p.u.)	Phase-to-ground exposure		Phase-to-phase exposure	
	m	ft	m	ft
1.5	1.02	3.3	1.37	4.5
1.6	1.06	3.5	1.43	4.7
1.7	1.11	3.6	1.48	4.9
1.8	1.16	3.8	1.54	5.1
1.9	1.21	4.0	1.60	5.2
2.0	1.25	4.1	1.66	5.4
2.1	1.30	4.3	1.73	5.7
2.2	1.35	4.4	1.81	5.9
2.3	1.39	4.6	1.90	6.2
2.4	1.44	4.7	1.99	6.5
2.5	1.49	4.9	2.08	6.8
2.6	1.53	5.0	2.17	7.1
2.7	1.58	5.2	2.26	7.4
2.8	1.63	5.3	2.36	7.7
2.9	1.67	5.5	2.45	8.0
3.0	1.72	5.6	2.55	8.4
3.1	1.77	5.8	2.65	8.7
3.2	1.81	5.9	2.76	9.1
3.3	1.88	6.2	2.86	9.4
3.4	1.95	6.4	2.97	9.7
3.5	2.01	6.6	3.08	10.1

Table 11   Ac Minimum Approach Distances-242.1 to 362.0 KV

T (p.u.)	Phase-to-ground exposure		Phase-to-phase exposure	
	m	ft	m	ft
1.5	1.37	4.5	1.99	6.5
1.6	1.44	4.7	2.13	7.0
1.7	1.51	5.0	2.27	7.4
1.8	1.58	5.2	2.41	7.9
1.9	1.65	5.4	2.56	8.4
2.0	1.72	5.6	2.71	8.9
2.1	1.79	5.9	2.87	9.4
2.2	1.87	6.1	3.03	9.9
2.3	1.97	6.5	3.20	10.5
2.4	2.08	6.8	3.37	11.1
2.5	2.19	7.2	3.55	11.6
2.6	2.29	7.5	3.73	12.2
2.7	2.41	7.9	3.91	12.8
2.8	2.52	8.3	4.10	13.5
2.9	2.64	8.7	4.29	14.1
3.0	2.76	9.1	4.49	14.7
3.1	2.88	9.4	4.69	15.4
3.2	3.01	9.9	4.90	16.1
3.3	3.14	10.3	5.11	16.8
3.4	3.27	10.7	5.32	17.5
3.5	3.41	11.2	5.52	18.1

Table 12   Ac Minimum Approach Distances-362.1 to 420.0 KV

T (p.u.)	Phase-to-ground exposure		Phase-to-phase exposure	
	m	ft	m	ft
1.5	1.53	5.0	2.40	7.9
1.6	1.62	5.3	2.58	8.5
1.7	1.70	5.6	2.75	9.0
1.8	1.78	5.8	2.94	9.6
1.9	1.88	6.2	3.13	10.3
2.0	1.99	6.5	3.33	10.9
2.1	2.12	7.0	3.53	11.6
2.2	2.24	7.3	3.74	12.3
2.3	2.37	7.8	3.95	13.0
2.4	2.50	8.2	4.17	13.7
2.5	2.64	8.7	4.40	14.4
2.6	2.78	9.1	4.63	15.2
2.7	2.93	9.6	4.87	16.0
2.8	3.07	10.1	5.11	16.8
2.9	3.23	10.6	5.36	17.6
3.0	3.38	11.1	5.59	18.3
3.1	3.55	11.6	5.82	19.1
3.2	3.72	12.2	6.07	19.9
3.3	3.89	12.8	6.31	20.7
3.4	4.07	13.4	6.56	21.5
3.5	4.25	13.9	6.81	22.3

Table 13   Ac Minimum Approach Distances-420.1 to 550.0 KV

T (p.u.)	Phase-to-ground exposure		Phase-to-phase exposure	
	m	ft	m	ft
1.5	1.95	6.4	3.46	11.4
1.6	2.11	6.9	3.73	12.2
1.7	2.28	7.5	4.02	13.2
1.8	2.45	8.0	4.31	14.1
1.9	2.62	8.6	4.61	15.1
2.0	2.81	9.2	4.92	16.1
2.1	3.00	9.8	5.25	17.2
2.2	3.20	10.5	5.55	18.2
2.3	3.40	11.2	5.86	19.2
2.4	3.62	11.9	6.18	20.3
2.5	3.84	12.6	6.50	21.3
2.6	4.07	13.4	6.83	22.4
2.7	4.31	14.1	7.18	23.6
2.8	4.56	15.0	7.52	24.7
2.9	4.81	15.8	7.88	25.9
3.0	5.07	16.6	8.24	27.0

Table 14   Ac Minimum Approach Distances-550.1 to 800.0 KV

T (p.u.)	Phase-to-ground exposure		Phase-to-phase exposure	
	m	ft	m	ft
1.5	3.16	10.4	5.97	19.6
1.6	3.46	11.4	6.43	21.1
1.7	3.78	12.4	6.92	22.7
1.8	4.12	13.5	7.42	24.3
1.9	4.47	14.7	7.93	26.0
2.0	4.83	15.8	8.47	27.8
2.1	5.21	17.1	9.02	29.6
2.2	5.61	18.4	9.58	31.4
2.3	6.02	19.8	10.16	33.3
2.4	6.44	21.1	10.76	35.3
2.5	6.88	22.6	11.38	37.3

Notes to Table 7 through Table 14:

1. The employer must determine the maximum anticipated per-unit transient overvoltage, phase-to-ground, through an engineering analysis, as required by § 1926.960(c)(1)(ii), or assume a maximum anticipated per-unit transient over-voltage, phase-to-ground, in accordance with Table V-8.
2. For phase-to-phase exposures, the employer must demonstrate that no insulated tool spans the gap and that no large conductive object is in the gap.
3. The worksite must be at an elevation of 900 meters (3,000 feet) or less above sea level.

[79 FR 20717-20728, July 10, 2014; 79 FR 56962, September 24, 2014]

Reprinted from www.osha.gov

Occupational Safety & Health Administration
200 Constitution Avenue, NW
Washington, DC 20210

REGULATIONS (STANDARDS - 29 CFR)

- **Part Number:** 1926
- **Part Title:** Safety and Health Regulations for Construction
- **Subpart:** V
- **Subpart Title:** Electric Power Transmission and Distribution
- **Standard Number:** 1926 Subpart V App C
- **Title:** Appendix C to Subpart V of Part 1926-Protection From Hazardous Differences in Electric Potential
- **GPO Source:** e-CFR

Appendix C to Subpart V of Part 1926-Protection From Hazardous Differences in Electric Potential

I. Introduction

Current passing through an impedance impresses voltage across that impedance. Even conductors have some, albeit low, value of impedance. Therefore, if a "grounded" [1] object, such as a crane or deenergized and grounded power line, results in a ground fault on a power line, voltage is impressed on that grounded object. The voltage impressed on the grounded object depends largely on the voltage on the line, on the impedance of the faulted conductor, and on the impedance to "true," or "absolute," ground represented by the object. If the impedance of the object causing the fault is relatively large, the voltage impressed on the object is essentially the phase-to-ground system voltage. However, even faults to grounded power lines or to well grounded transmission towers or substation structures (which have relatively low values of impedance to ground) can result in hazardous voltages.[2] In all cases, the degree of the hazard depends on the magnitude of the current through the employee and the time of exposure. This appendix discusses methods of protecting workers against the possibility that grounded objects, such as cranes and other mechanical equipment, will contact energized power lines and that deenergized and grounded power lines will become accidentally energized.

II. Voltage-Gradient Distribution

A. *Voltage-gradient distribution curve.* Absolute, or true, ground serves as a reference and always has a voltage of 0 volts above ground potential. Because there is an impedance between a grounding electrode and absolute ground, there will be a voltage difference between the grounding electrode and absolute ground under ground-fault conditions. Voltage dissipates from the grounding electrode (or from the grounding point) and creates a ground potential gradient. The voltage

---

[1]This appendix generally uses the term "grounded" only with respect to grounding that the employer intentionally installs, for example, the grounding an employer installs on a deenergized conductor. However, in this case, the term "grounded" means connected to earth, regardless of whether or not that connection is intentional.

[2]Thus, grounding systems for transmission towers and substation structures should be designed to minimize the step and touch potentials involved.

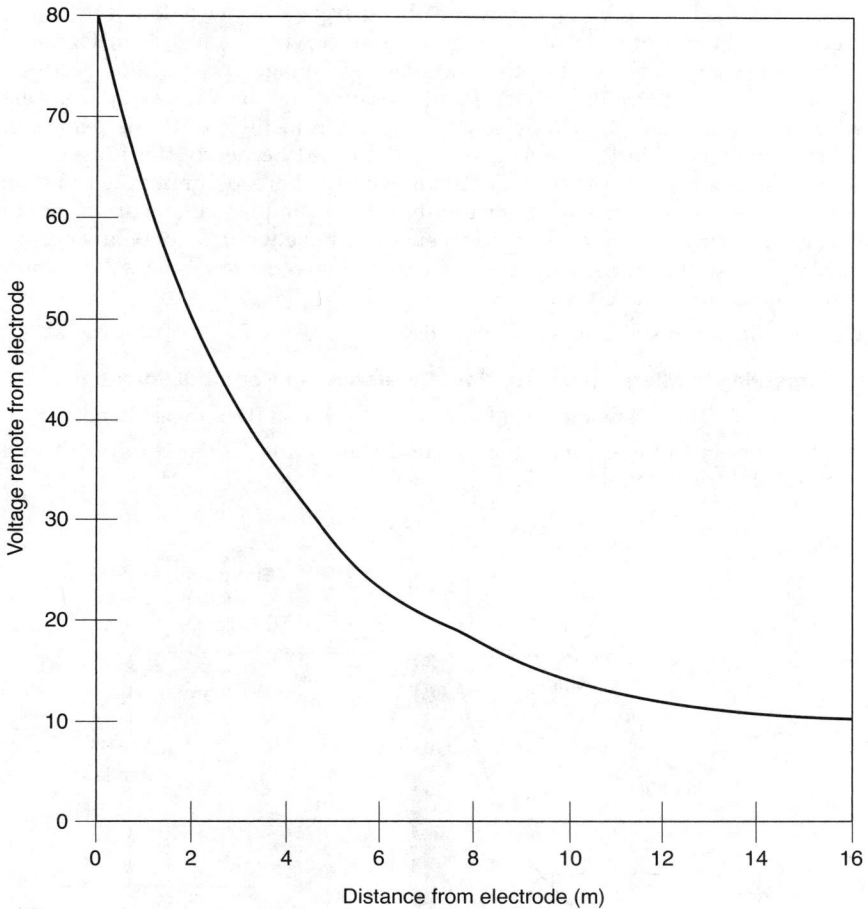

Figure 1.   Typical Voltage - Gradient Distribution Curve

decreases rapidly with increasing distance from the grounding electrode. A voltage drop associated with this dissipation of voltage is a ground potential. Figure 1 is a typical voltage-gradient distribution curve (assuming a uniform soil texture).

B. *Step and touch potentials.* Figure 1 also shows that workers are at risk from step and touch potentials. Step potential is the voltage between the feet of a person standing near an energized grounded object (the electrode). In Figure 1, the step potential is equal to the difference in voltage between two points at different distances from the electrode (where the points represent the location of each foot in relation to the electrode). A person could be at risk of injury during a fault simply by standing near the object.

Touch potential is the voltage between the energized grounded object (again, the electrode) and the feet of a person in contact with the object. In Figure 1, the touch potential is equal to the difference in voltage between the electrode (which is at a distance of 0 meters) and a point some distance away from the electrode (where the point represents the location of the feet of the person in contact with the object). The touch potential could be nearly the full voltage across the grounded object if that object is grounded at a point remote from the place where the person is in contact with it. For example, a crane grounded to the system neutral and that contacts an energized line would expose any person in contact with the crane or its uninsulated load line to a touch potential nearly equal to the full fault voltage.

Figure 2 illustrates step and touch potentials.

III. Protecting Workers From Hazardous Differences in Electrical Potential

A. *Definitions*. The following definitions apply to section III of this appendix:

Bond. The electrical interconnection of conductive parts designed to maintain a common electric potential.

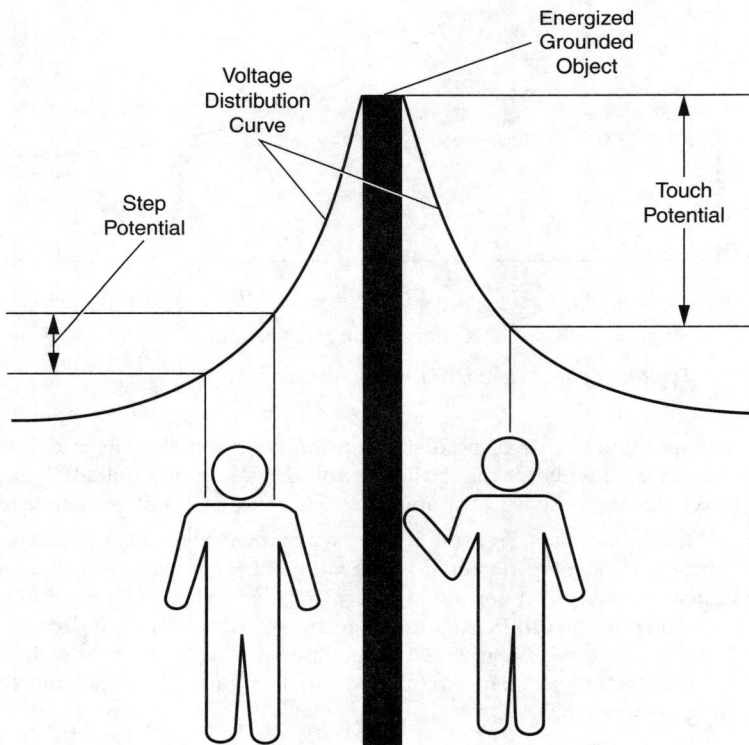

Figure 2. Step and Touch Potentials

Bonding cable (bonding jumper). A cable connected to two conductive parts to bond the parts together.

Cluster bar. A terminal temporarily attached to a structure that provides a means for the attachment and bonding of grounding and bonding cables to the structure.

Ground. A conducting connection between an electric circuit or equipment and the earth, or to some conducting body that serves in place of the earth.

Grounding cable (grounding jumper). A cable connected between a deenergized part and ground. Note that grounding cables carry fault current and bonding cables generally do not. A cable that bonds two conductive parts but carries substantial fault current (for example, a jumper connected between one phase and a grounded phase) is a grounding cable.

Ground mat (grounding grid). A temporarily or permanently installed metallic mat or grating that establishes an equipotential surface and provides connection points for attaching grounds.

B. *Analyzing the hazard.* The employer can use an engineering analysis of the power system under fault conditions to determine whether hazardous step and touch voltages will develop. The analysis should determine the voltage on all conductive objects in the work area and the amount of time the voltage will be present. Based on the this analysis, the employer can select appropriate measures and protective equipment, including the measures and protective equipment outlined in Section III of this appendix, to protect each employee from hazardous differences in electric potential. For example, from the analysis, the employer will know the voltage remaining on conductive objects after employees install bonding and grounding equipment and will be able to select insulating equipment with an appropriate rating, as described in paragraph III.C.2 of this appendix.

C. *Protecting workers on the ground.* The employer may use several methods, including equipotential zones, insulating equipment, and restricted work areas, to protect employees on the ground from hazardous differences in electrical potential.

1. An equipotential zone will protect workers within it from hazardous step and touch potentials. (See Figure 3.) Equipotential zones will not, however, protect employees located either wholly or partially outside the protected area. The employer can establish an equipotential zone for workers on the ground, with respect to a grounded object, through the use of a metal mat connected to the grounded object. The employer can use a grounding grid to equalize the voltage within the grid or bond conductive objects in the immediate work area to minimize the potential between the objects and between each object and ground. (Bonding an object outside the work area can increase the touch potential to that object, however.) Section III.D of this appendix discusses equipotential zones for employees working on deenergized and grounded power lines.
2. Insulating equipment, such as rubber gloves, can protect employees handling grounded equipment and conductors from hazardous touch potentials. The insulating equipment must be rated for the highest voltage that can be impressed on the grounded objects under fault conditions (rather than for the full system voltage).
3. Restricting employees from areas where hazardous step or touch potentials could arise can protect employees not directly involved in performing the operation.

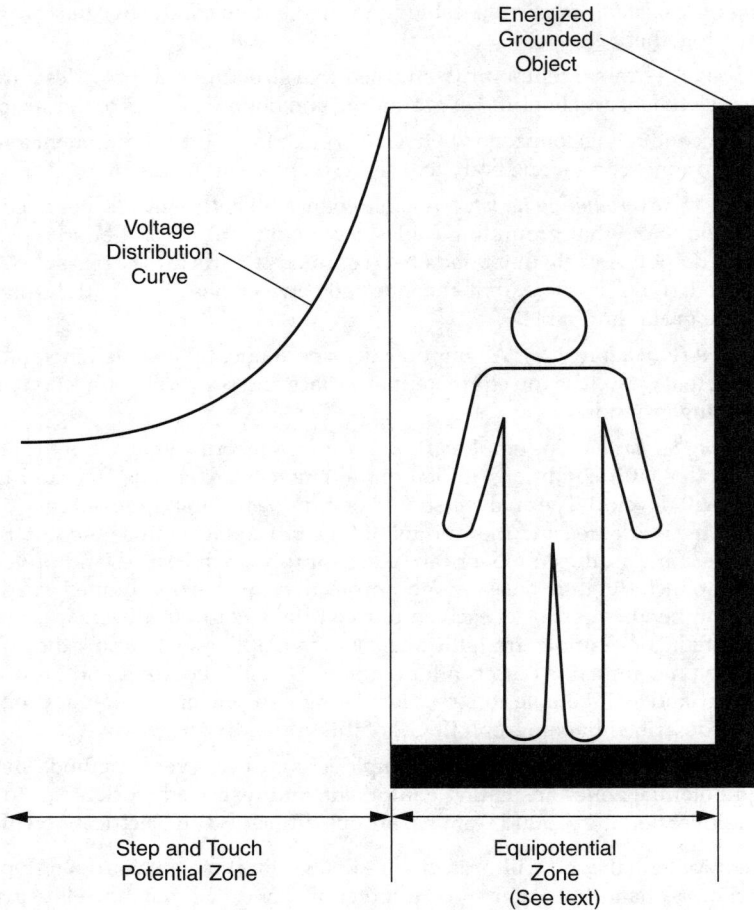

Figure 3.   Protection from Ground-Potential Gradients

The employer must ensure that employees on the ground in the vicinity of transmission structures are at a distance where step voltages would be insufficient to cause injury. Employees must not handle grounded conductors or equipment likely to become energized to hazardous voltages unless the employees are within an equipotential zone or protected by insulating equipment.

D. *Protecting employees working on deenergized and grounded power lines.* This Section III.D of Appendix C establishes guidelines to help employers comply with requirements in § 1926.962 for using protective grounding to protect employees working on deenergized power lines. Section 1926.962 applies to grounding of transmission and distribution lines and equipment for the purpose of protecting workers. Paragraph (c) of § 1926.962 requires temporary protective grounds to be placed at such locations and arranged in such a manner that the employer can demonstrate will prevent exposure of each employee to hazardous differences in electric potential.[3] Sections III.D.1 and III.D.2 of this appendix provide guidelines that employers can use in making the demonstration required by § 1926.962(c). Section III.D.1 of this appendix provides guidelines on how the employer can determine whether particular grounding practices expose employees to hazardous differences in electric potential. Section III.D.2 of this appendix describes grounding methods that the employer can use in lieu of an engineering analysis to make the demonstration required by § 1926.962(c). The Occupational Safety and Health Administration will consider employers that comply with the criteria in this appendix as meeting § 1926.962(c).

Finally, Section III.D.3 of this appendix discusses other safety considerations that will help the employer comply with other requirements in § 1926.962. Following these guidelines will protect workers from hazards that can occur when a deenergized and grounded line becomes energized.

1. *Determining safe body current limits.* This Section III.D.1 of Appendix C provides guidelines on how an employer can determine whether any differences in electric potential to which workers could be exposed are hazardous as part of the demonstration required by § 1926.962(c).

Institute of Electrical and Electronic Engineers (IEEE) Standard 1048-2003, *IEEE Guide for Protective Grounding of Power Lines*, provides the following equation for determining the threshold of ventricular fibrillation when the duration of the electric shock is limited:

$$I = \frac{116}{\sqrt{t}},$$

where *I* is the current through the worker's body, and t is the duration of the current in seconds. This equation represents the ventricular fibrillation threshold for 95.5 percent of the adult population with a mass of 50 kilograms (110 pounds) or more. The equation is valid for current durations between 0.0083 to 3.0 seconds.

To use this equation to set safe voltage limits in an equipotential zone around the worker, the employer will need to assume a value for the resistance of the worker's body. IEEE Std 1048-2003 states that "total body resistance is usually taken as 1000 Ω for determining . . . body current limits." However, employers should be aware that the impedance of a worker's body can be substantially less than that value. For instance, IEEE Std 1048-2003 reports a minimum hand-to-hand resistance of 610 ohms and an internal body resistance of 500 ohms. The internal resistance of the body better represents the minimum resistance of a worker's body when the skin resistance drops near zero, which occurs, for example, when there are breaks in the worker's skin, for instance, from cuts or from blisters formed as a result of the current from an electric shock, or when the worker is wet at the points of contact.

Employers may use the IEEE Std 1048-2003 equation to determine safe body current limits only if the employer protects workers from hazards associated with involuntary muscle reactions from electric shock (for example, the hazard to a

---

[3] The protective grounding required by § 1926.962 limits to safe values the potential differences between accessible objects in each employee's work environment. Ideally, a protective grounding system would create a true equipotential zone in which every point is at the same electric potential. In practice, current passing through the grounding and bonding elements creates potential differences. If these potential differences are hazardous, the employer may not treat the zone as an equipotential zone.

worker from falling as a result of an electric shock). Moreover, the equation applies only when the duration of the electric shock is limited. If the precautions the employer takes, including those required by applicable standards, do not adequately protect employees from hazards associated with involuntary reactions from electric shock, a hazard exists if the induced voltage is sufficient to pass a current of 1 milliampere through a 500-ohm resistor. (The 500-ohm resistor represents the resistance of an employee. The 1-milliampere current is the threshold of perception.) Finally, if the employer protects employees from injury due to involuntary reactions from electric shock, but the duration of the electric shock is unlimited (that is, when the fault current at the work location will be insufficient to trip the devices protecting the circuit), a hazard exists if the resultant current would be more than 6 milliamperes (the recognized let-go threshold for workers[4]).

2. *Acceptable methods of grounding for employers that do not perform an engineering determination.* The grounding methods presented in this section of this appendix ensure that differences in electric potential are as low as possible and, therefore, meet § 1926.962(c) without an engineering determination of the potential differences. These methods follow two principles: (i) The grounding method must ensure that the circuit opens in the fastest available clearing time, and (ii) the grounding method must ensure that the potential differences between conductive objects in the employee's work area are as low as possible.

Paragraph (c) of § 1926.962 does not require grounding methods to meet the criteria embodied in these principles. Instead, the paragraph requires that protective grounds be "placed at such locations and arranged in such a manner that the employer can demonstrate will prevent exposure of each employee to hazardous differences in electric potential." However, when the employer's grounding practices do not follow these two principles, the employer will need to perform an engineering analysis to make the demonstration required by § 1926.962(c).

i. *Ensuring that the circuit opens in the fastest available clearing time.* Generally, the higher the fault current, the shorter the clearing times for the same type of fault. Therefore, to ensure the fastest available clearing time, the grounding method must maximize the fault current with a low impedance connection to ground. The employer accomplishes this objective by grounding the circuit conductors to the best ground available at the worksite. Thus, the employer must ground to a grounded system neutral conductor, if one is present. A grounded system neutral has a direct connection to the system ground at the source, resulting in an extremely low impedance to ground. In a substation, the employer may instead ground to the substation grid, which also has an extremely low impedance to the system ground and, typically, is connected to a grounded system neutral when one is present. Remote system grounds, such as pole and tower grounds, have a higher impedance to the system ground than grounded system neutrals and substation grounding grids; however, the employer may use a remote ground when lower impedance grounds are not available. In the absence of a grounded system neutral, substation grid, and remote ground, the employer may use a temporary driven ground at the worksite.

---

[4] Electric current passing through the body has varying effects depending on the amount of the current. At the let-go threshold, the current overrides a person's control over his or her muscles. At that level, an employee grasping an object will not be able to let go of the object. The let-go threshold varies from person to person; however, the recognized value for workers is 6 milliamperes.

In addition, if employees are working on a three-phase system, the grounding method must short circuit all three phases. Short circuiting all phases will ensure faster clearing and lower the current through the grounding cable connecting the deenergized line to ground, thereby lowering the voltage across that cable. The short circuit need not be at the worksite; however, the employer must treat any conductor that is not grounded at the worksite as energized because the ungrounded conductors will be energized at fault voltage during a fault.

ii. *Ensuring that the potential differences between conductive objects in the employee's work area are as low as possible.* To achieve as low a voltage as possible across any two conductive objects in the work area, the employer must bond all conductive objects in the work area. This section of this appendix discusses how to create a zone that minimizes differences in electric potential between conductive objects in the work area.

The employer must use bonding cables to bond conductive objects, except for metallic objects bonded through metal-to-metal contact. The employer must ensure that metal-to-metal contacts are tight and free of contamination, such as oxidation, that can increase the impedance across the connection. For example, a bolted connection between metal lattice tower members is acceptable if the connection is tight and free of corrosion and other contamination. Figure 4 shows how to create an equipotential zone for metal lattice towers.

Wood poles are conductive objects. The poles can absorb moisture and conduct electricity, particularly at distribution and transmission voltages. Consequently, the employer must either: (1) Provide a conductive platform, bonded to a grounding

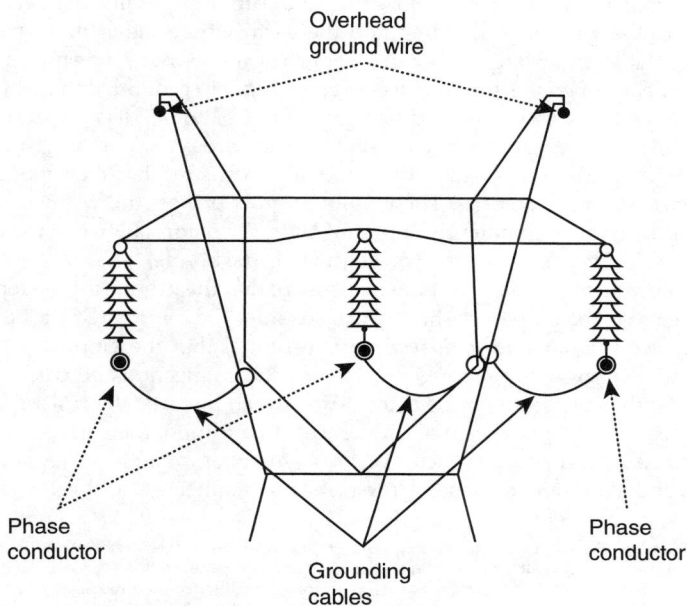

Figure 4.   Equipotential Zone for Metal Lattice Tower

cable, on which the worker stands or (2) use cluster bars to bond wood poles to the grounding cable. The employer must ensure that employees install the cluster bar below, and close to, the worker's feet. The inner portion of the wood pole is more conductive than the outer shell, so it is important that the cluster bar be in conductive contact with a metal spike or nail that penetrates the wood to a depth greater than or equal to the depth the worker's climbing gaffs will penetrate the wood. For example, the employer could mount the cluster bar on a bare pole ground wire fastened to the pole with nails or staples that penetrate to the required depth. Alternatively, the employer may temporarily nail a conductive strap to the pole and connect the strap to the cluster bar. Figure 5 shows how to create an equipotential zone for wood poles.

Notes:

1. Employers must ground overhead ground wires that are within reach of the employee.
2. The grounding cable must be as short as practicable; therefore, the attachment points between the grounding cable and the tower may be different from the shown in the figure.

For underground systems, employers commonly install grounds at the points of disconnection of the underground cables. These grounding points are typically remote from the manhole or underground vault where employees will be working on the cable. Workers in contact with a cable grounded at a remote location can experience hazardous potential differences if the cable becomes energized or if a fault occurs on a different, but nearby, energized cable. The fault current causes potential gradients in the earth, and a potential difference will exist between the earth where the worker is standing and the earth where the cable is grounded. Consequently, to create an equipotential zone for the worker, the employer must provide a means of connecting the deenergized cable to ground at the worksite by having the worker stand on a conductive mat bonded to the deenergized cable. If the cable is cut, the employer must install a bond across the opening in the cable or install one bond on each side of the opening to ensure that the separate cable ends are at the same potential. The employer must protect the worker from any hazardous differences in potential any time there is no bond between the mat and the cable (for example, before the worker installs the bonds).

3. Other safety-related considerations. To ensure that the grounding system is safe and effective, the employer should also consider the following factors:[5]

i. *Maintenance of grounding equipment.* It is essential that the employer properly maintain grounding equipment. Corrosion in the connections between grounding cables and clamps and on the clamp surface can increase the resistance of the cable, thereby increasing potential differences. In addition, the surface to which a clamp attaches, such as a conductor or tower member, must be clean and free of corrosion and oxidation to ensure a low-resistance connection. Cables must be free

---

[5] This appendix only discusses factors that relate to ensuring an equipotential zone for employees. The employer must consider other factors in selecting a grounding system that is capable of conducting the maximum fault current that could flow at the point of grounding for the time necessary to clear the fault, as required by § 1926.962(d)(1)(i). IEEE Std 1048-2003 contains guidelines for selecting and installing grounding equipment that will meet § 1926.962(d)(1)(i).

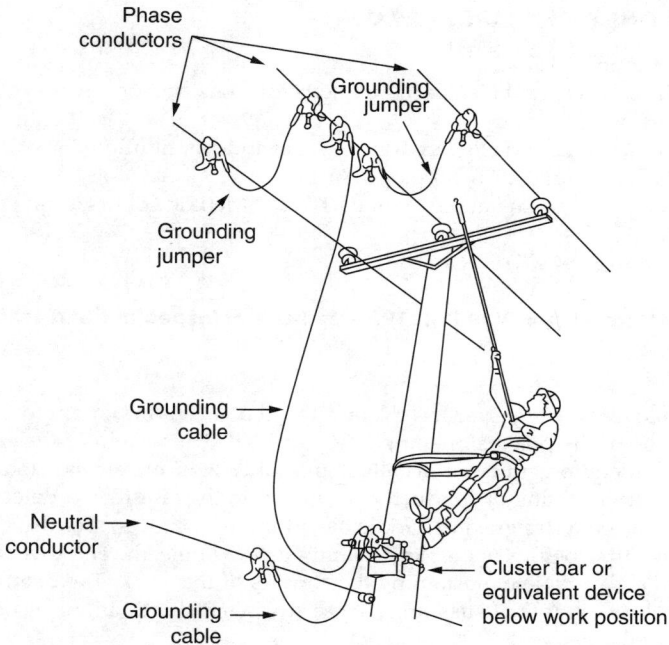

Figure 5. Equipotential Grounding for Wood Poles

of damage that could reduce their current-carrying capacity so that they can carry the full fault current without failure. Each clamp must have a tight connection to the cable to ensure a low resistance and to ensure that the clamp does not separate from the cable during a fault.

ii. *Grounding cable length and movement.* The electromagnetic forces on grounding cables during a fault increase with increasing cable length. These forces can cause the cable to move violently during a fault and can be high enough to damage the cable or clamps and cause the cable to fail. In addition, flying cables can injure workers. Consequently, cable lengths should be as short as possible, and grounding cables that might carry high fault current should be in positions where the cables will not injure workers during a fault.

[79 FR 20728-20736, July 10, 2014]

Reprinted from www.osha.gov

Occupational Safety & Health Administration
200 Constitution Avenue, NW
Washington, DC 20210

REGULATIONS (STANDARDS - 29 CFR)

- **Part Number:** 1926
- **Part Title:** Safety and Health Regulations for Construction
- **Subpart:** V
- **Subpart Title:** Electric Power Transmission and Distribution
- **Standard Number:** 1926 Subpart V App D
- **Title:** Appendix D to Subpart V of Part 1926 - Methods of Inspecting and Testing Wood Poles
- **GPO Source:** e-CFR

Appendix D to Subpart V of Part 1926-Methods of Inspecting and Testing Wood Poles

I. Introduction

When employees are to perform work on a wood pole, it is important to determine the condition of the pole before employees climb it. The weight of the employee, the weight of equipment to be installed, and other working stresses (such as the removal or retensioning of conductors) can lead to the failure of a defective pole or a pole that is not designed to handle the additional stresses.[1] For these reasons, it is essential that, before an employee climbs a wood pole, the employer ascertain that the pole is capable of sustaining the stresses of the work. The determination that the pole is capable of sustaining these stresses includes an inspection of the condition of the pole.

If the employer finds the pole to be unsafe to climb or to work from, the employer must secure the pole so that it does not fail while an employee is on it. The employer can secure the pole by a line truck boom, by ropes or guys, or by lashing a new pole alongside it. If a new one is lashed alongside the defective pole, employees should work from the new one.

II. Inspecting Wood Poles

A qualified employee should inspect wood poles for the following conditions:[2]
A. *General condition.* Buckling at the ground line or an unusual angle with respect to the ground may indicate that the pole has rotted or is broken.
B. *Cracks.* Horizontal cracks perpendicular to the grain of the wood may weaken the pole. Vertical cracks, although not normally considered to be a sign of a defective pole, can pose a hazard to the climber, and the employee should keep his or her gaffs away from them while climbing.
C. *Holes.* Hollow spots and woodpecker holes can reduce the strength of a wood pole.
D. *Shell rot and decay.* Rotting and decay are cutout hazards and possible indications of the age and internal condition of the pole.
E. *Knots.* One large knot or several smaller ones at the same height on the pole may be evidence of a weak point on the pole.

[1]A properly guyed pole in good condition should, at a minimum, be able to handle the weight of an employee climbing it.

[2]The presence of any of these conditions is an indication that the pole may not be safe to climb or to work from. The employee performing the inspection must be qualified to make a determination as to whether it is safe to perform the work without taking additional precautions.

F. *Depth of setting.* Evidence of the existence of a former ground line substantially above the existing ground level may be an indication that the pole is no longer buried to a sufficient depth.

G. *Soil conditions.* Soft, wet, or loose soil around the base of the pole may indicate that the pole will not support any change in stress.

H. *Burn marks.* Burning from transformer failures or conductor faults could damage the pole so that it cannot withstand changes in mechanical stress.

III. Testing Wood Poles

The following tests, which are from § 1910.268(n)(3) of this chapter, are acceptable methods of testing wood poles:

A. *Hammer test.* Rap the pole sharply with a hammer weighing about 1.4 kg (3 pounds), starting near the ground line and continuing upwards circumferentially around the pole to a height of approximately 1.8 meters (6 feet). The hammer will produce a clear sound and rebound sharply when striking sound wood. Decay pockets will be indicated by a dull sound or a less pronounced hammer rebound. Also, prod the pole as near the ground line as possible using a pole prod or a screwdriver with a blade at least 127 millimeters (5 inches) long. If substantial decay is present, the pole is unsafe.

B. *Rocking test.* Apply a horizontal force to the pole and attempt to rock it back and forth in a direction perpendicular to the line. Exercise caution to avoid causing power lines to swing together. Apply the force to the pole either by pushing it with a pike pole or pulling the pole with a rope. If the pole cracks during the test, it is unsafe.

[79 FR 20736, July 10, 2014]

Reprinted from www.osha.gov

Occupational Safety & Health Administration
200 Constitution Avenue, NW
Washington, DC 20210

REGULATIONS (STANDARDS - 29 CFR)

- **Part Number:** 1926
- **Part Title:** Safety and Health Regulations for Construction
- **Subpart:** V
- **Subpart Title:** Electric Power Transmission and Distribution
- **Standard Number:** 1926 Subpart V App E
- **Title:** Appendix E to Subpart V of Part 1926 - Protection From Flames and Electric Arcs
- **GPO Source:** e-CFR

Appendix E to Subpart V of Part 1926 - Protection From Flames and Electric Arcs

I. Introduction

Paragraph (g) of § 1926.960 addresses protecting employees from flames and electric arcs. This paragraph requires employers to: (1) Assess the workplace for flame and electric-arc hazards (paragraph (g)(1)); (2) estimate the available heat energy from electric arcs to which employees would be exposed (paragraph (g)(2)); (3) ensure that employees wear clothing that will not melt, or ignite and continue to burn, when exposed to flames or the estimated heat energy (paragraph (g)(3)); and (4) ensure that employees wear flame-resistant clothing [1] and protective clothing and other protective equipment that has an arc rating greater than or equal to the available heat energy under certain conditions (paragraphs (g)(4) and (g)(5)). This appendix contains information to help employers estimate available heat energy as required by § 1926.960(g)(2), select protective clothing and other protective equipment with an arc rating suitable for the available heat energy as required by § 1926.960(g)(5), and ensure that employees do not wear flammable clothing that could lead to burn injury as addressed by §§ 1926.960(g)(3) and (g)(4).

II. Assessing the Workplace for Flame and Electric-Arc Hazards

Paragraph (g)(1) of § 1926.960 requires the employer to assess the workplace to identify employees exposed to hazards from flames or from electric arcs. This provision ensures that the employer evaluates employee exposure to flames and electric arcs so that employees who face such exposures receive the required protection. The employer must conduct an assessment for each employee who performs work on or near exposed, energized parts of electric circuits.

A. Assessment Guidelines

Sources electric arcs. Consider possible sources of electric arcs, including:

- Energized circuit parts not guarded or insulated,
- Switching devices that produce electric arcs in normal operation,

---

[1]Flame-resistant clothing includes clothing that is inherently flame resistant and clothing chemically treated with a flame retardant. (See ASTM F1506-10a, *Standard Performance Specification for Flame Resistant Textile Materials for Wearing Apparel for Use by Electrical Workers Exposed to Momentary Electric Arc and Related Thermal Hazards*, and ASTM F1891-12 *Standard Specification for Arc and Flame Resistant Rainwear.*)

- Sliding parts that could fault during operation (for example, rack-mounted circuit breakers), and
- Energized electric equipment that could fail (for example, electric equipment with damaged insulation or with evidence of arcing or overheating).

Exposure to flames. Identify employees exposed to hazards from flames. Factors to consider include:

- The proximity of employees to open flames, and
- For flammable material in the work area, whether there is a reasonable likelihood that an electric arc or an open flame can ignite the material.

Probability that an electric arc will occur. Identify employees exposed to electric-arc hazards. The Occupational Safety and Health Administration will consider an employee exposed to electric-arc hazards if there is a reasonable likelihood that an electric arc will occur in the employee's work area, in other words, if the probability of such an event is higher than it is for the normal operation of enclosed equipment. Factors to consider include:

- For energized circuit parts not guarded or insulated, whether conductive objects can come too close to or fall onto the energized parts,
- For exposed, energized circuit parts, whether the employee is closer to the part than the minimum approach distance established by the employer (as permitted by § 1926.960(c)(1)(iii)).
- Whether the operation of electric equipment with sliding parts that could fault during operation is part of the normal operation of the equipment or occurs during servicing or maintenance, and
- For energized electric equipment, whether there is evidence of impending failure, such as evidence of arcing or overheating.

B. Examples

Table 1 provides task-based examples of exposure assessments.

III. Protection Against Burn Injury

A. Estimating Available Heat Energy

Calculation methods. Paragraph (g)(2) of § 1926.960 provides that, for each employee exposed to an electric-arc hazard, the employer must make a reasonable estimate of the heat energy to which the employee would be exposed if an arc occurs. Table 2 lists various methods of calculating values of available heat energy from an electric circuit. The Occupational Safety and Health Administration does not endorse any of these specific methods. Each method requires the input of various parameters, such as fault current, the expected length of the electric arc, the distance from the arc to the employee, and the clearing time for the fault (that is, the time the circuit protective devices take to open the circuit and clear the fault). The employer can precisely determine some of these parameters, such as the fault current and the clearing time, for a given system. The employer will need to estimate other parameters, such as the length of the arc and the distance between the arc and the employee, because such parameters vary widely.

Table 1   Example Assessments for Various Tasks

Task		Is employee exposed to flame or electric arc hazard?
Normal operation of enclosed equipment, such as closing or opening a switch.	The employer properly installs and maintains enclosed equipment, and there is no evidence of impending failure.	No.
	There is evidence of arcing or overheating	Yes.
	Parts of the equipment are loose or sticking, or the equipment otherwise exhibits signs of lack of maintenance.	Yes.
Servicing electric equipment, such as racking in a circuit breaker or replacing a switch		Yes.
Inspection of electric equipment with exposed energized parts.	The employee is not holding conductive objects and remains outside the minimum approach distance established by the employer.	No.
	The employee is holding a conductive object, such as a flashlight, that could fall or otherwise contact energized parts (irrespective of whether the employee maintains the minimum approach distance).	Yes.
	The employee is closer than the minimum approach distance established by the employer (for example, when wearing rubber insulating gloves or rubber insulating gloves and sleeves).	Yes.
Using open flames, for example, in wiping cable splice sleeves		Yes.

Table 2   Methods Of Calculating Incident Heat Energy from an Electric ARC

1. Standard for Electrical Safety Requirements for Employee Workplaces, NFPA 70E-2012, Annex D, "Sample Calculation of Flash Protection Boundary."
2. Doughty, T.E., Neal, T.E., and Floyd II, H.L., "Predicting Incident Energy to Better Manage the Electric Arc Hazard on 600 V Power Distribution Systems," *Record of Conference Papers IEEE IAS 45th Annual Petroleum and Chemical Industry Conference*, September 28-30, 1998.
3. *Guide for Performing Arc-Flash Hazard Calculations*, IEEE Std 1584-2002, 1584a--2004 (Amendment 1 to IEEE Std 1584-2002), and 1584b-2011 (Amendment 2: Changes to Clause 4 of IEEE Std 1584-2002).*
4. ARCPRO, a commercially available software program developed by Kinectrics, Toronto, ON, CA.

*This appendix refers to IEEE Std 1584-2002 with both amendments as IEEE Std 1584b-2011.

The amount of heat energy calculated by any of the methods is approximately inversely proportional to the square of the distance between the employee and the arc. In other words, if the employee is very close to the arc, the heat energy is very high; but if the employee is just a few more centimeters away, the heat energy drops substantially. Thus, estimating the distance from the arc to the employee is key to protecting employees. The employer must select a method of estimating incident heat energy that provides a reasonable estimate of incident heat energy for the exposure involved. Table 3 shows which methods provide reasonable estimates for various exposures.

Selecting a reasonable distance from the employee to the arc. In estimating available heat energy, the employer must make some reasonable assumptions about how far the employee will be from the electric arc. Table 4 lists reasonable distances from the employee to the electric arc. The distances in Table 4 are consistent with national consensus standards, such as the Institute of Electrical and Electronic Engineers' *National Electrical Safety Code*, ANSI/IEEE C2-2012, and *IEEE Guide for Performing Arc-Flash Hazard Calculations*, IEEE Std 1584b-2011. The employer is free to use other reasonable distances, but must consider equipment enclosure size and the working distance to the employee in selecting a distance from the employee to the arc. The Occupational Safety and Health Administration will consider a distance reasonable when the employer bases it on equipment size and working distance.

Table 3   Selecting A Reasonable Incident-Energy Calculation Method[1]

Incident-energy calculation method	600 V and Less[2]			601 V to 15 kV[2]			More than 15 kV		
	1Φ	3Φa	3Φb	1Φ	3Φa	3Φb	1Φ	3Φa	3Φb
NFPA 70E-2012 Annex D (Lee equation)	Y-C	Y	N	Y-C	Y-C	N	N[3]	N[3]	N[3]
Doughty, Neal, and Floyd	Y-C	Y	Y	N	N	N	N	N	N
IEEE Std 1584b-2011	Y	Y	Y	Y	Y	Y	N	N	N
ARCPRO	Y	N	N	Y	N	N	Y	Y[4]	Y[4]

Key:
1Φ: Single-phase arc in open air
3Φa: Three-phase arc in open air
3Φb: Three-phase arc in an enclosure (box)
 Y: Acceptable; produces a reasonable estimate of incident heat energy from this type of electric arc
 N: Not acceptable; does not produce a reasonable estimate of incident heat energy from this type of electric arc
 Y-C: Acceptable; produces a reasonable, but conservative, estimate of incident heat energy from this type of electric arc.
 Notes:[1] Although the Occupational Safety and Health Administration will consider these methods reasonable for enforcement purposes when employers use the methods in accordance with this table, employers should be aware that the listed methods do not necessarily result in estimates that will provide full protection from internal faults in transformers and similar equipment or from arcs in underground manholes or vaults.
 [2] At these voltages, the presumption is that the arc is three-phase unless the employer can demonstrate that only one phase is present or that the spacing of the phases is sufficient to prevent a multiphase arc from occurring.
 [3] Although the Occupational Safety and Health Administration will consider this method acceptable for purposes of assessing whether incident energy exceeds 2.0 cal/cm², the results at voltages of more than 15 kilovolts are extremely conservative and unrealistic.
 [4] The Occupational Safety and Health Administration will deem the results of this method reasonable when the employer adjusts them using the conversion factors for three-phase arcs in open air or in an enclosure, as indicated in the program's instructions.

Table 4   Selecting a Reasonable Distance from the Employee to the Electric ARC

Class of equipment	Single-phase arc mm (inches)	Three-phase arc mm (inches)
Cable	NA*	455 (18)
Low voltage MCCs and panelboards	NA	455 (18)
Low-voltage switchgear	NA	610 (24)
5-kV switchgear	NA	910 (36)
15-kV switchgear	NA	910 (36)
Single conductors in air (up to 46 kilovolts), work with rubber insulating gloves	380 (15)	NA
Single conductors in air, work with live-line tools and live-line barehand work	MAD$-(2{\times}kV{\times}2.54)$ MAD$-(2{\times}kV/10))^{\dagger}$	NA

* NA = not applicable.
† The terms in this equation are:
MAD = The applicable minimum approach distance, and
kV = The system voltage in kilovolts.

Selecting a reasonable arc gap.   For a single-phase arc in air, the electric arc will almost always occur when an energized conductor approaches too close to ground. Thus, an employer can determine the arc gap, or arc length, for these exposures by the dielectric strength of air and the voltage on the line. The dielectric strength of air is approximately 10 kilovolts for every 25.4 millimeters (1 inch). For example, at 50 kilovolts, the arc gap would be 50 ÷ 10 × 25.4 (or 50 × 2.54), which equals 127 millimeters (5 inches).

For three-phase arcs in open air and in enclosures, the arc gap will generally be dependent on the spacing between parts energized at different electrical potentials. Documents such as IEEE Std 1584b-2011 provide information on these distances. Employers may select a reasonable arc gap from Table 5, or they may select any other reasonable arc gap based on sparkover distance or on the spacing between (1) live parts at different potentials or (2) live parts and grounded parts (for example, bus or conductor spacings in equipment). In any event, the employer must use an estimate that reasonably resembles the actual exposures faced by the employee.

Table 5   Selecting a Reasonable ARC Gap

Class of equipment	Single-phase arc mm (inches)	Three-phase arc mm [1] (inches)
Cable	NA[2]	13 (0.5)
Low voltage MCCs and panel-boards	NA	25 (1.0)
Low-voltage switchgear	NA	32 (1.25)
5-kV switchgear	NA	104 (4.0)
15-kV switchgear	NA	152 (6.0)
Single conductors in air, 15 kV and less	51 (2.0)	Phase conductor spacings.
Single conductor in air, more than 15 kV	Voltage in kV × 2.54 (Voltage in kV × 0.1), but no less than 51 mm (2 inches)	Phase conductor spacings.

[1] Source: IEEE Std 1584b-2011.
[2] NA = not applicable.

Making estimates over multiple system areas. The employer need not estimate the heat-energy exposure for every job task performed by each employee. Paragraph (g)(2) of § 1926.960 permits the employer to make broad estimates that cover multiple system areas provided that: (1) The employer uses reasonable assumptions about the energy-exposure distribution throughout the system, and (2) the estimates represent the maximum exposure for those areas. For example, the employer can use the maximum fault current and clearing time to cover several system areas at once.

Incident heat energy for single-phase-toground exposures. Table 6 and Table 7 provide incident heat energy levels for openair, phase-to-ground electric-arc exposures typical for overhead systems.[2] Table 6 presents estimates of available energy for employees using rubber insulating gloves to perform work on overhead systems operating at 4 to 46 kilovolts. The table assumes that the employee will be 380 millimeters (15 inches) from the electric arc, which is a reasonable estimate for rubber insulating glove work. Table 6 also assumes that the arc length equals the sparkover distance for the maximum transient overvoltage of each voltage range.[3] To use the table, an employer would use the voltage, maximum fault current, and maximum clearing time for a system area and, using the appropriate voltage range and fault-current and clearingtime values corresponding to the next higher values listed in the table, select the appropriate heat energy (4, 5, 8, or 12 cal/cm²) from the table. For example, an employer might have a 12,470-volt power line supplying a system area. The power line can supply a maximum fault current of

[2]The Occupational Safety and Health Administration used metric values to calculate the clearing times in Table 6 and Table 7. An employer may use English units to calculate clearing times instead even though the results will differ slightly.
[3]The Occupational Safety and Health Administration based this assumption, which is more conservative than the arc length specified in Table 5, on Table 410-2 of the 2012 NESC.

8 kiloamperes with a maximum clearing time of 10 cycles. For rubber glove work, this system falls in the 4.0-to-15.0-kilovolt range; the next-higher fault current is 10 kA (the second row in that voltage range); and the clearing time is under 18 cycles (the first column to the right of the fault current column). Thus, the available heat energy for this part of the system will be 4 cal/cm² or less (from the column heading), and the employer could select protection with a 5-cal/cm² rating to meet § 1926.960(g)(5). Alternatively, an employer could select a base incident-energy value and ensure that the clearing times for each voltage range and fault current listed in the table do not exceed the corresponding clearing time specified in the table. For example, an employer that provides employees with arc-flash protective equipment rated at 8 cal/cm² can use the table to determine if any system area exceeds 8 cal/cm² by checking the clearing time for the highest fault current for each voltage range and ensuring that the clearing times do not exceed the values specified in the 8-cal/cm2 column in the table.

Table 7 presents similar estimates for employees using live-line tools to perform work on overhead systems operating at voltages of 4 to 800 kilovolts. The table assumes that the arc length will be equal to the sparkover distance [4] and that the employee will be a distance from the arc equal to the minimum approach distance minus twice the sparkover distance.

The employer will need to use other methods for estimating available heat energy in situations not addressed by Table 6 or Table 7. The calculation methods listed in Table 2 and the guidance provided in Table 3 will help employers do this. For example, employers can use IEEE Std 1584b-2011 to estimate the available heat energy (and to select appropriate protective equipment) for many specific conditions, including lowervoltage, phase-to-phase arc, and enclosed arc exposures.

---

[4] The dielectric strength of air is about 10 kilovolts for every 25.4 millimeters (1 inch). Thus, the employer can estimate the arc length in millimeters to be the phase-to-ground voltage in kilovolts multiplied by 2.54 (or voltage (in kilovolts) × 2.54).

Table 6   Incident Heat Energy for Various Fault Currents, Clearing Times, and Voltages of 4.0 to 46.0 KV: Rubber Insulating Glove Exposures Involving Phase-to-Ground Arcs in Open Air Only*†‡

Voltage range (kV) **	Fault current (kA)	Maximum clearing time (cycles)			
		4 cal/cm²	5 cal/cm²	8 cal/cm²	12 cal/cm²
4.0 to 15.0…….	5	46	58	92	138
	10	18	22	36	54
	15	10	12	20	30
	20	6	8	13	19
15.1 to 25.0……	5	28	34	55	83
	10	11	14	23	34
	15	7	8	13	20
	20	4	5	9	13
25.1 to 36.0……	5	21	26	42	62
	10	9	11	18	26
	15	5	6	10	16
	20	4	4	7	11
36.1 to 46.0……	5	16	20	32	48
	10	7	9	14	21
	15	4	5	8	13
	20	3	4	6	9

Notes:
    * This table is for open-air, phase-to-ground electric-arc exposures. It is not for phase-to-phase arcs or enclosed arcs (arc in a box).
    † The table assumes that the employee will be 380 mm (15 in.) from the electric arc. The table also assumes the arc length to be the sparkover distance for the maximum transient overvoltage of each voltage range (see Appendix B to this subpart), as follows:
    4.0 to 15.0 kV 51 mm (2 in.)
    15.1 to 25.0 kV 102 mm (4 in.)
    25.1 to 36.0 kV 152 mm (6 in.)
    36.1 to 46.0 kV 229 mm (9 in.)
    ‡ The Occupational Safety and Health Administration calculated the values in this table using the ARCPRO method listed in Table 2.
    ** The voltage range is the phase-to-phase system voltage.

Table 7   Incident Heat Energy for Various Fault Currents, Clearing Times, and Voltages: Live-Line Tool Exposures Involving Phase-to-Ground Arcs in Open Air Only [*][†][‡][#]

Voltage range (kV) **	Fault current (kA)	Maximum clearing time (cycles)			
		4 cal/cm²	5 cal/cm²	8 cal/cm²	12 cal/cm²
4.0 to 15.0 ……..	5	197	246	394	591
	10	73	92	147	220
	15	39	49	78	117
	20	24	31	49	73
15.1 to 25.0 ……..	5	197	246	394	591
	10	75	94	150	225
	15	41	51	82	122
	20	26	33	52	78
25.1 to 36.0 ……..	5	138	172	275	413
	10	53	66	106	159
	15	30	37	59	89
	20	19	24	38	58
36.1 to 46.0 ……..	5	129	161	257	386
	10	51	64	102	154
	15	29	36	58	87
	20	19	24	38	57
46.1 to 72.5 ……..	20	18	23	36	55
	30	10	13	20	30
	40	6	8	13	19
	50	4	6	9	13
72.6 to 121.0 ……	20	10	12	20	30
	30	6	7	11	17
	40	4	5	7	11
	50	3	3	5	8
121.1 to 145.0…	20	12	15	24	35
	30	7	9	15	22
	40	5	6	10	15
	50	4	5	8	11
145.1 to 169.0…	20	12	15	24	36
	30	7	9	15	22
	40	5	7	10	16
	50	4	5	8	12
169.1 to 242.0…	20	13	17	27	40
	30	8	10	17	25
	40	6	7	12	17
	50	4	5	9	13

(Continued)

Table 7   Incident Heat Energy for Various Fault Currents, Clearing Times, and Voltages: Live-Line Tool Exposures Involving Phase-to-Ground Arcs in Open Air Only * † ‡ #

Voltage range (kV) **	Fault current (kA)	Maximum clearing time (cycles)			
		4 cal/cm²	5 cal/cm²	8 cal/cm²	12 cal/cm²
242.1 to 362.0…	20	25	32	51	76
	30	16	19	31	47
	40	11	14	22	33
	50	8	10	16	25
362.1 to 420.0…	20	12	15	25	37
	30	8	10	15	23
	40	5	7	11	16
	50	4	5	8	12
420.1 to 550.0…	20	23	29	47	70
	30	14	18	29	43
	40	10	13	20	30
	50	8	9	15	23
550.1 to 800.0	20	25	31	50	75
	30	15	19	31	46
	40	11	13	21	32
	50	8	10	16	24

Notes:

* This table is for open-air, phase-to-ground electric-arc exposures. It is not for phase-to-phase arcs or enclosed arcs (arc in a box).

† The table assumes the arc length to be the sparkover distance for the maximum phase-to-ground voltage of each voltage range (see Appendix B to this subpart). The table also assumes that the employee will be the minimum approach distance minus twice the arc length from the electric arc.

‡ The Occupational Safety and Health Administration calculated the values in this table using the ARCPRO method listed in Table 2.

\# For voltages of more than 72.6 kV, employers may use this table only when the minimum approach distance established under § 1926.960(c)(1) is greater than or equal to the following values:

72.6 to 121.0 kV 1.02 m
121.1 to 145.0 kV 1.16 m
145.1 to 169.0 kV 1.30 m
169.1 to 242.0 kV 1.72 m
242.1 to 362.0 kV 2.76 m
362.1 to 420.0 kV 2.50 m
420.1 to 550.0 kV 3.62 m
550.1 to 800.0 kV 4.83 m

** The voltage range is the phase-to-phase system voltage.

B. Selecting Protective Clothing and Other Protective Equipment

Paragraph (g)(5) of § 1926.960 requires employers, in certain situations, to select protective clothing and other protective equipment with an arc rating that is greater than or equal to the incident heat energy estimated under § 1926.960(g)(2). Based on laboratory testing required by ASTM F1506-10a, the expectation is that protective clothing with an arc rating equal to the estimated incident heat energy will be capable of preventing second-degree burn injury to an employee exposed to that incident heat energy from an electric arc. Note that actual electric-arc exposures may be more or less severe than the estimated value because of factors such as arc movement, arc length, arcing from reclosing of the system, secondary fires or explosions, and weather conditions. Additionally, for arc rating based on the fabric's arc thermal performance value [5] (ATPV), a worker exposed to incident energy at the arc rating has a 50-percent chance of just barely receiving a second-degree burn. Therefore, it is possible (although not likely) that an employee will sustain a second-degree (or worse) burn wearing clothing conforming to § 1926.960(g)(5) under certain circumstances. However, reasonable employer estimates and maintaining appropriate minimum approach distances for employees should limit burns to relatively small burns that just barely extend beyond the epidermis (that is, just barely a second-degree burn). Consequently, protective clothing and other protective equipment meeting § 1926.960(g)(5) will provide an appropriate degree of protection for an employee exposed to electric-arc hazards.

Paragraph (g)(5) of § 1926.960 does not require arc-rated protection for exposures of 2 cal/cm² or less. Untreated cotton clothing will reduce a 2-cal/cm² exposure below the 1.2- to 1.5-cal/cm² level necessary to cause burn injury, and this material should not ignite at such low heat energy levels. Although § 1926.960(g)(5) does not require clothing to have an arc rating when exposures are 2 cal/cm² or less, § 1926.960(g)(4) requires the outer layer of clothing to be flame resistant under certain conditions, even when the estimated incident heat energy is less than 2 cal/cm², as discussed later in this appendix. Additionally, it is especially important to ensure that employees do not wear undergarments made from fabrics listed in the note to § 1926.960(g)(3) even when the outer layer is flame resistant or arc rated. These fabrics can melt or ignite easily when an electric arc occurs. Logos and name tags made from non-flame-resistant material can adversely affect the arc rating or the flameresistant characteristics of arc-rated or flameresistant clothing. Such logos and name tags may violate § 1926.960(g)(3), (g)(4), or (g)(5).

Paragraph (g)(5) of § 1926.960 requires that arc-rated protection cover the employee's entire body, with limited exceptions for the employee's hands, feet, face, and head. Paragraph (g)(5)(i) of § 1926.960 provides that arc-rated protection is not necessary for the employee's hands under the following conditions:

---

[5] ASTM F1506-10a defines "arc thermal performance value" as "the incident energy on a material or a multilayer system of materials that results in a 50% probability that sufficient heat transfer through the tested specimen is predicted to cause the onset of a second-degree skin burn injury based on the Stoll [footnote] curve, cal/cm²." The footnote to this definition reads: "Derived from: Stoll, A.M., and Chianta, M.A., 'Method and Rating System for Evaluations of Thermal Protection,' Aerospace Medicine, Vol 40, 1969, pp. 1232-1238 and Stoll A.M., and Chianta, M.A., 'Heat Transfer through Fabrics as Related to Thermal Injury,' Transactions-New York Academy of Sciences, Vol 33(7), Nov. 1971, pp. 649-670."

For any estimated incident heat energy	When the employee is wearing rubber insulating gloves with protectors
If the estimated incident heat energy does not exceed 14 cal/cm².	When the employee is wearing heavy-duty leather work gloves with a weight of at least 407 gm/m² (12 oz/yd²)

Paragraph (g)(5)(ii) of § 1926.960 provides that arc-rated protection is not necessary for the employee's feet when the employee is wearing heavy-duty work shoes or boots. Finally, § 1926.960(g)(5)(iii), (g)(5)(iv), and (g)(5)(v) require arc-rated head and face protection as follows:

Exposure	Minimum head and face protection		
	None *	Arc-rated faceshield with a minimum rating of 8 cal/cm²*	Arc-rated hood or faceshield with bala-clava
Single-phase, open air	2-8 cal/cm²	9-12 cal/cm²	13 cal/² or higher.†
Three-phase	2-4 cal/cm²	5-8 cal/cm²	9 cal/cm² or higher.‡

* These ranges assume that employees are wearing hardhats meeting the specifications in § 1910.135 or § 1926.100(b)(2), as applicable.

† The arc rating must be a minimum of 4 cal/cm² less than the estimated incident energy. Note that § 1926.960(g)(5)(v) permits this type of head and face protection, with a minimum arc rating of 4 cal/cm² less than the estimated incident energy, at any incident energy level.

‡ Note that § 1926.960(g)(5) permits this type of head and face protection at any incident energy level.

IV. Protection Against Ignition

Paragraph (g)(3) of § 1926.960 prohibits clothing that could melt onto an employee's skin or that could ignite and continue to burn when exposed to flames or to the available heat energy estimated by the employer under § 1926.960(g)(2). Meltable fabrics, such as acetate, nylon, polyester, and polypropylene, even in blends, must be avoided. When these fibers melt, they can adhere to the skin, thereby transferring heat rapidly, exacerbating burns, and complicating treatment. These outcomes can result even if the meltable fabric is not directly next to the skin. The remainder of this section focuses on the prevention of ignition.

Paragraph (g)(5) of § 1926.960 generally requires protective clothing and other protective equipment with an arc rating greater than or equal to the employer's estimate of available heat energy. As explained earlier in this appendix, untreated cotton is usually acceptable for exposures of 2 cal/cm² or less.[6] If the exposure is greater than that, the employee generally must wear flame-resistant clothing

[6]See § 1926.960(g)(4)(i), (g)(4)(ii), and (g)(4)(iii) for conditions under which employees must wear flame-resistant clothing as the outer layer of clothing even when the incident heat energy does not exceed 2 cal/cm².

with a suitable arc rating in accordance with § 1926.960(g)(4) and (g)(5). However, even if an employee is wearing a layer of flame-resistant clothing, there are circumstances under which flammable layers of clothing would be uncovered, and an electric arc could ignite them. For example, clothing ignition is possible if the employee is wearing flammable clothing under the flame-resistant clothing and the underlayer is uncovered because of an opening in the flame-resistant clothing. Thus, for purposes of § 1926.960(g)(3), it is important for the employer to consider the possibility of clothing ignition even when an employee is wearing flame-resistant clothing with a suitable arc rating.

Under § 1926.960(g)(3), employees may not wear flammable clothing in conjunction with flame-resistant clothing if the flammable clothing poses an ignition hazard.[7] Although outer flame-resistant layers may not have openings that expose flammable inner layers, when an outer flame-resistant layer would be unable to resist breakopen,[8] the next (inner) layer must be flame-resistant if it could ignite.

Non-flame-resistant clothing can ignite even when the heat energy from an electric arc is insufficient to ignite the clothing. For example, nearby flames can ignite an employee's clothing; and, even in the absence of flames, electric arcs pose ignition hazards beyond the hazard of ignition from incident energy under certain conditions. In addition to requiring flame-resistant clothing when the estimated incident energy exceeds 2.0 cal/cm^2, § 1926.960(g)(4) requires flameresistant clothing when: The employee is exposed to contact with energized circuit parts operating at more than 600 volts (§ 1926.960(g)(4)(i)), an electric arc could ignite flammable material in the work area that, in turn, could ignite the employee's clothing (§ 1926.960(g)(4)(ii)), and molten metal or electric arcs from faulted conductors in the work area could ignite the employee's clothing (§ 1926.960(g)(4)(iii)). For example, grounding conductors can become a source of heat energy if they cannot carry fault current without failure. The employer must consider these possible sources of electric arcs [9] in determining whether the employee's clothing could ignite under § 1926.960(g)(4)(iii).

[79 FR 20736-20742, July 10, 2014]

Reprinted from www.osha.gov

Occupational Safety & Health Administration
200 Constitution Avenue, NW
Washington, DC 20210

---

[7] Paragraph (g)(3) of § 1926.960 prohibits clothing that could ignite and continue to burn when exposed to the heat energy estimated under paragraph (g)(2) of that section.

[8] Breakopen occurs when a hole, tear, or crack develops in the exposed fabric such that the fabric no longer effectively blocks incident heat energy.

[9] Static wires and pole grounds are examples of grounding conductors that might not be capable of carrying fault current without failure. Grounds that can carry the maximum available fault current are not a concern, and employers need not consider such grounds a possible electric arc source.

REGULATIONS (STANDARDS - 29 CFR)

- **Part Number:** 1926
- **Part Title:** Safety and Health Regulations for Construction
- **Subpart:** V
- **Subpart Title:** Electric Power Transmission and Distribution
- **Standard Number:** 1926 Subpart V App F
- **Title:** Appendix F to Subpart V of Part 1926 - Work-Positioning Equipment Inspection Guidelines
- **GPO Source:** e-CFR

Appendix F to Subpart V of Part 1926 - Work-Positioning Equipment Inspection Guidelines

I. Body Belts

Inspect body belts to ensure that:

A. The hardware has no cracks, nicks, distortion, or corrosion;

B. No loose or worn rivets are present;

C. The waist strap has no loose grommets;

D. The fastening straps are not 100-percent leather; and

E. No worn materials that could affect the safety of the user are present.

II. Positioning Straps

Inspect positioning straps to ensure that:

A. The warning center of the strap material is not exposed;

B. No cuts, burns, extra holes, or fraying of strap material is present;

C. Rivets are properly secured;

D. Straps are not 100-percent leather; and

E. Snaphooks do not have cracks, burns, or corrosion.

III. Climbers

Inspect pole and tree climbers to ensure that:

A. Gaffs are at least as long as the manufacturer's recommended minimums (generally 32 and 51 millimeters (1.25 and 2.0 inches) for pole and tree climbers, respectively, measured on the underside of the gaff);

Note: Gauges are available to assist in determining whether gaffs are long enough and shaped to easily penetrate poles or trees.

B. Gaffs and leg irons are not fractured or cracked;

C. Stirrups and leg irons are free of excessive wear;

D. Gaffs are not loose;

E. Gaffs are free of deformation that could adversely affect use;

F. Gaffs are properly sharpened; and

G. There are no broken straps or buckles.

[79 FR 20742, July 10, 2014]

Reprinted from www.osha.gov

Occupational Safety & Health Administration
200 Constitution Avenue, NW
Washington, DC 20210

REGULATIONS (STANDARDS - 29 CFR)

- **Part Number:** 1926
- **Part Title:** Safety and Health Regulations for Construction
- **Subpart:** V
- **Subpart Title:** Electric Power Transmission and Distribution
- **Standard Number:** 1926 Subpart V App G
- **Title:** Appendix G to Subpart V of Part 1926 - Reference Documents
- **GPO Source:** e-CFR

Appendix G to Subpart V of Part 1926 - Reference Documents

The references contained in this appendix provide information that can be helpful in understanding and complying with the requirements contained in Subpart V of this part. The national consensus standards referenced in this appendix contain detailed specifications that employers may follow in complying with the more performance-based requirements of Subpart V of this part. Except as specifically noted in Subpart V of this part, however, the Occupational Safety and Health Administration will not necessarily deem compliance with the national consensus standards to be compliance with the provisions of Subpart V of this part.

ANSI/SIA A92.2-2009, *American National Standard for Vehicle-Mounted Elevating and Rotating Aerial Devices.*

ANSI Z133-2012, *American National Standard Safety Requirements for Arboricultural Operations-Pruning, Trimming, Repairing, Maintaining, and Removing Trees, and Cutting Brush.*

ANSI/IEEE Std 935-1989, *IEEE Guide on Terminology for Tools and Equipment to Be Used in Live Line Working.*

ASME B20.1-2012, *Safety Standard for Conveyors and Related Equipment.*

ASTM D120-09, *Standard Specification for Rubber Insulating Gloves.*

ASTM D149-09 (2013), *Standard Test Method for Dielectric Breakdown Voltage and Dielectric Strength of Solid Electrical Insulating Materials at Commercial Power Frequencies.*

ASTM D178-01 (2010), *Standard Specification for Rubber Insulating Matting.*

ASTM D1048-12, *Standard Specification for Rubber Insulating Blankets.*

ASTM D1049-98 (2010), *Standard Specification for Rubber Insulating Covers.*

ASTM D1050-05 (2011), *Standard Specification for Rubber Insulating Line Hose.*

ASTM D1051-08, *Standard Specification for Rubber Insulating Sleeves.*

ASTM F478-09, *Standard Specification for In-Service Care of Insulating Line Hose and Covers.*

ASTM F479-06 (2011), *Standard Specification for In-Service Care of Insulating Blankets.*

ASTM F496-08, *Standard Specification for In-Service Care of Insulating Gloves and Sleeves.*

ASTM F711-02 (2007), *Standard Specification for Fiberglass-Reinforced Plastic (FRP) Rod and Tube Used in Live Line Tools.*

ASTM F712-06 (2011), *Standard Test Methods and Specifications for Electrically Insulating Plastic Guard Equipment for Protection of Workers.*

ASTM F819-10, *Standard Terminology Relating to Electrical Protective Equipment for Workers.*

ASTM F855-09, *Standard Specifications for Temporary Protective Grounds to Be Used on De-energized Electric Power Lines and Equipment.*

ASTM F887-12^{e1}, *Standard Specifications for Personal Climbing Equipment.*

ASTM F914/F914M-10, *Standard Test Method for Acoustic Emission for Aerial Personnel Devices Without Supplemental Load Handling Attachments.*

ASTM F1116-03 (2008), *Standard Test Method for Determining Dielectric Strength of Dielectric Footwear.*

ASTM F1117-03 (2008), *Standard Specification for Dielectric Footwear.*

ASTM F1236-96 (2012), *Standard Guide for Visual Inspection of Electrical Protective Rubber Products.*

ASTM F1430/F1430M-10, *Standard Test Method for Acoustic Emission Testing of Insulated and Non-Insulated Aerial Personnel Devices with Supplemental Load Handling Attachments.*

ASTM F1505-10, *Standard Specification for Insulated and Insulating Hand Tools.*

ASTM F1506-10a, *Standard Performance Specification for Flame Resistant and Arc Rated Textile Materials for Wearing Apparel for Use by Electrical Workers Exposed to Momentary Electric Arc and Related Thermal Hazards.*

ASTM F1564-13, *Standard Specification for Structure-Mounted Insulating Work Platforms for Electrical Workers.*

ASTM F1701-12, *Standard Specification for Unused Polypropylene Rope with Special Electrical Properties.*

ASTM F1742-03 (2011), *Standard Specification for PVC Insulating Sheeting.*

ASTM F1796-09, *Standard Specification for High Voltage Detectors-Part 1 Capacitive Type to be Used for Voltages Exceeding 600 Volts AC.*

ASTM F1797-09 ^{81}, *Standard Test Method for Acoustic Emission Testing of Insulated and Non-Insulated Digger Derricks.*

ASTM F1825-03 (2007), *Standard Specification for Clampstick Type Live Line Tools.*

ASTM F1826-00 (2011), *Standard Specification for Live Line and Measuring Telescoping Tools.*

ASTM F1891-12, *Standard Specification for Arc and Flame Resistant Rainwear.*

ASTM F1958/F1958M-12, *Standard Test Method for Determining the Ignitability of Non-flame-Resistant Materials for Clothing by Electric Arc Exposure Method Using Mannequins.*

ASTM F1959/F1959M-12, *Standard Test Method for Determining the Arc Rating of Materials for Clothing.*

IEEE Stds 4-1995, 4a-2001 (Amendment to *IEEE Standard Techniques for High-Voltage Testing*), *IEEE Standard Techniques for High-Voltage Testing.*

IEEE Std 62-1995, *IEEE Guide for Diagnostic Field Testing of Electric Power Apparatus-Part 1: Oil Filled Power Transformers, Regulators, and Reactors.*

IEEE Std 80-2000, *Guide for Safety in AC Substation Grounding.*

IEEE Std 100-2000, *The Authoritative Dictionary of IEEE Standards Terms Seventh Edition.*

IEEE Std 516-2009, *IEEE Guide for Maintenance Methods on Energized Power Lines.*

IEEE Std 524-2003, *IEEE Guide to the Installation of Overhead Transmission Line Conductors.*

IEEE Std 957-2005, *IEEE Guide for Cleaning Insulators.*

IEEE Std 1048-2003, *IEEE Guide for Protective Grounding of Power Lines.*

IEEE Std 1067-2005, *IEEE Guide for In-Service Use, Care, Maintenance, and Testing of Conductive Clothing for Use on Voltages up to 765 kV AC and ±750 kV DC.*

IEEE Std 1307-2004, *IEEE Standard for Fall Protection for Utility Work.*

IEEE Stds 1584-2002, 1584a-2004 (Amendment 1 to IEEE Std 1584-2002), and 1584b-2011 (Amendment 2: Changes to Clause 4 of IEEE Std 1584-2002), *IEEE Guide for Performing Arc-Flash Hazard Calculations.*

IEEE C2-2012, *National Electrical Safety* Code.

NFPA 70E-2012, *Standard for Electrical Safety in the Workplace.*

[79 FR 20742-20743, July 10, 2014]

Reprinted from www.osha.gov

Occupational Safety & Health Administration
200 Constitution Avenue, NW
Washington, DC 20210

Index